FLORA OF BHUTAN

INCLUDING A RECORD OF PLANTS
FROM SIKKIM AND DARJEELING

VOLUME 2 PART 3

A. J. C. GRIERSON & D. G. LONG
EDITED BY L. S. SPRINGATE

Illustrations by M. Bates, Chen Yo-Jiun, C. Oliver,
L. Olley, G. Rodrigues
& M. Tebbs

ROYAL BOTANIC GARDEN EDINBURGH
ROYAL GOVERNMENT OF BHUTAN
2001

This volume is dedicated to the memory of

Andrew J. C. Grierson (1929–1990) and Pauline Grierson (née Shillabeer, 1927–1996)

Published by the Royal Botanic Garden Edinburgh, 20A Inverleith Row, Edinburgh EH3 5LR, UK and the Royal Government of Bhutan

ISBN 1 872291 93 7

Typeset, printed and bound by
The Charlesworth Group, Huddersfield, UK, 01484 517077

CONTENTS

FIGURES

INTRODUCTION

by David G. Long, Henry J. Noltie & Lawrence S. Springate

This part of the *Flora of Bhutan* completes the treatment of the Dicotyledons and includes the important families Solanaceae, Scrophulariaceae, Acanthaceae, Gesneriaceae, Caprifoliaceae and Compositae. The first drafts of many of the accounts were completed between 1990 and 1994, although funding for publication has only recently become available through the generous support of Danida. Consequently the early drafts (particularly Compositae, Scrophulariaceae and Solanaceae) have undergone thorough review recently. Amongst these families the Compositae (also known as Asteraceae) are pre-eminent as the largest family of Dicotyledons, encompassing 126 genera and 372 species, outnumbered in the angiosperms only by two Monocotyledon families: Gramineae and Orchidaceae. Compositae was the special interest of the late Andrew Grierson, whose vision and enthusiasm had a major influence in the inauguration and running of the *Flora of Bhutan* from its beginning until his untimely death in 1990. This account of Compositae stands as a tribute to his enormous contribution. The Compositae is particularly well represented at high altitudes in the East Himalaya; it shows many strong links with the Tibetan flora, and for that reason does not show much endemism within the political boundaries of the *Flora*. Amongst other families, Scrophulariaceae is noteworthy as containing the largest single genus in the *Flora of Bhutan*: *Pedicularis*, with 76 species represented. Many of these are local East Himalayan endemics. Several of the family accounts were contributed by the late Rose A. Clement (née King) and these constitute her remaining posthumous publications.

ACKNOWLEDGEMENTS

The editors of the *Flora of Bhutan* wish to acknowledge our deep gratitude to Danida, the Overseas Aid Division of the Royal Danish Ministry of Foreign Affairs, for the Flora of Bhutan Completion Project. The present volume is published under their auspices; this support has covered editing and printing costs and the preparation of illustrations. We are also pleased to acknowledge the continued interest and help of the Royal Government of Bhutan, in particular the Minister of Agriculture, Lyonpo Dr Kinzang Wangdi and Yeshey Dorji of the National Biodiversity Programme.

We also wish to thank the authors of accounts contained in this part, most of whom have had to put up with a very lengthy delay in publication of their work; this delay was due to constraints of funding for publication and we are grateful for their forbearance: Eona M. Aitken, Olive M. Hilliard, Paul W. Meyer, Robert R. Mill, Sally J. Rae, John R.I. Wood and Jenny Wright. Thanks are also due to the artists for the drawings they have contributed: Mary Bates, Chen Yo-Jiun, Christina Oliver, Louise Olley, Glenn Rodrigues and Margaret Tebbs.

The Directors and Curators of the Herbaria at the Royal Botanic Gardens, Kew and the Natural History Museum, London, have continued to provide facilities for study visits and many very large loans of specimens. Some interesting nineteenth-century collections were supplied by Manchester Museum. The staff of these organisations are also thanked for taxonomic advice and assistance, as are the following specialists for advice on particular groups: Charles Jeffrey (Compositae), Jan Kirschner (*Taraxacum*), Valery Malecot (*Viburnum*) and Peter Taylor (*Utricularia*).

We also wish to thank Eona Aitken, Norma Gregory and Kim Howell for technical and editorial help with this part.

As this part concludes both the treatment of Dicotyledons, and is also the final outstanding part of the *Flora* to be written at the Royal Botanic Garden Edinburgh, it is appropriate now to acknowledge those colleagues and friends in Britain, Bhutan and elsewhere who have contributed to the *Flora of Bhutan* project over the years in many different ways, as collectors, contributors of revisions, referees, taxonomic and nomenclatural advisers, horticulturists and general enthusiasts for Himalayan plants and Bhutan. We are grateful to them for their continued support: Peter Baxter, Prof. Stephen Blackmore, Simon Bowes Lyon, B.L. Burtt, James Cullen, Lyonpo C. Dorji, Yeshey Dorji, Norma Gregory, Chris Grey-Wilson, D.M. Henderson, Prof. David Ingram, Roy Lancaster, Ron J.D. McBeath, Prof. David Mann, Robert R. Mill, Chris Parker, Tony Peers, Rebecca Pradhan, Sally J. Rae, Keith Rushforth, Tony Schilling, Ian W.J. Sinclair, Philippe de Spoelberch, W.T. Stearn, Sonam Tshering, Tandin Wangdi, Mark F. Watson and John R.I. Wood.

We wish also to correct an earlier omission of acknowledgement to Prof. David Mabberley, for refereeing Andrew Grierson's account of Meliaceae in Volume 2 Part 1 (1991), which contained several previously unpublished observations and conclusions contributed by Prof. Mabberley in his review of the manuscript.

ABBREVIATIONS

Abbreviations for languages and dialects of common names of plants used in this volume are:

Dz: Dzongkha language (W Bhutan and official language)
Eng: English
Med: Bhutanese medicinal name
Nep: Nepali
Sha: Shachop dialect (E Bhutan)

For other abbreviations, e.g. botanical authorities, users are referred to Volume 1 Part 1, p. 34.

Family 171. SOLANACEAE

by R.R. Mill

Herbs, shrubs, trees or lianas. Leaves usually simple and alternate (sometimes appearing falsely paired because of suppression of lateral branches), exstipulate, but often with smaller leaves (*minor leaves*) in their axils or intercalary. Flowers often showy, solitary and axillary or extra-axillary, or in racemes, spikes, panicles, corymbs or cymes, usually hermaphrodite, actinomorphic or somewhat zygomorphic. Calyx (3–)5(–10)-lobed, cupular, campanulate or tubular, usually persistent, often enlarged in fruit. Corolla rotate, campanulate, infundibular or salverform; lobes 5(–10), equal or somewhat unequal, often hairy, plicate or valvate in bud. Stamens (4–)5(–6), epipetalous and alternating with corolla lobes, sometimes didynamous; anthers sometimes connivent, dehiscing by terminal (and sometimes also basal) pores or by introrse longitudinal slits; fifth anther sometimes replaced by a staminode. Ovary superior, 2–4(–5)-locular; placentation axile. Fruit a berry or capsule; seeds many, small, compressed, reniform or discoid; testa smooth or ornamented.

Genera 1–16 belong to subfamily Solanoideae, 17–22 to subfamily Cestroideae.

1. Anthers connivent, forming a cone or tube around style 2
\+ Anthers not connivent, free from style 7

2. Anthers opening by pores 3
\+ Anthers opening by longitudinal slits 5

3. Anther connective thickened and evident throughout whole of back of anther; berry 50–70mm, dull red, pendent (cultivated shrub or small tree with pinkish-white flowers in scorpioid cymes) .... **12. Cyphomandra**
\+ Anther connective not thickened, not evident throughout whole of back of anther; berry usually 3–25mm, if larger than 30mm (up to 23cm) then dark purplish, not red (habit and flower colour various) 4

4. Inflorescence a pedunculate cyme or raceme, or flowers solitary; calyx with 5 ribs and 5 teeth .... **9. Solanum**
\+ Inflorescence a sessile fascicle of pedicellate flowers; calyx with 10 ribs and 5–10 teeth .... **10. Lycianthes**

5. Corolla pendulous, 20–30cm, infundibular .... **16. Brugmansia**
\+ Corolla not as above 6

6. Leaves imparipinnate; anthers yellow; fruit succulent, locules without air-spaces .... **11. Lycopersicon**
\+ Leaves entire; anthers bluish; fruit a coriaceous or fleshy berry, locules with large air-spaces .... **8. Capsicum**

7. Stamens didynamous, with 4(–5) fertile stamens and 1 or more staminodes ... 8
\+ Stamens not didynamous, always 5 fertile ... 10

8. Annual herb 20–35cm; corolla tube 12–16mm; fruit a dry capsule **20. Browallia**
\+ Shrub to 3m (often much less); corolla tube c 20–25mm; fruit a berry or capsule ... 9

9. Branches glabrous; corolla mauve (in ours); fruit a berry .... **22. Brunfelsia**
\+ Branches viscid-pubescent; corolla orange (in ours); fruit a capsule **21. Streptosolen**

10. Corolla 6–27cm ... 11
\+ Corolla less than 6cm ... 12

11. Flowers erect, nodding or borne at an angle; corolla limb 5.5–8cm diameter in our species; anthers free; pericarp of fruit spiny or tuberculate **15. Datura**
\+ Flowers pendulous, corolla limb 11–17cm diameter; anthers coherent at first, sometimes free later; pericarp of fruit smooth and unarmed **16. Brugmansia**

12. Fruit a capsule ... 13
\+ Fruit a berry, or fruits not formed in our area ... 16

13. Flowers solitary and axillary (in *Przewalskia* appearing clustered amongst leaves) ... 14
\+ Flowers in panicles, racemes or spikes ... 15

14. Aerial stems present (40cm–2m), leaves clearly alternate; corolla yellowish-green or brownish-white with pinkish or purplish lobes, campanulate, more than 30mm ... **3. Anisodus**
\+ Stems mainly underground, with rosette of leaves at top; corolla yellow, narrowly infundibular, less than 25mm ... **4. Przewalskia**

15. Capsule opening by an operculum; corolla broadly campanulate, lurid yellow with a dark purple centre and network of fine purplish veins **5. Hyoscyamus**
\+ Capsule opening by valves; corolla infundibular, not coloured as above **18. Nicotiana**

16. Shrubs ... 17
\+ Herbs ... 19

17. Stems spiny ... **2. Lycium**
\+ Stems unarmed ... 18

18. Pedicels 0.5–2mm, or flowers sessile .... **17. Cestrum**
+ Pedicels long (12–18mm in our species) .... **14. Iochroma**

19. Corolla salverform, 50–65mm, never nodding .... **19. Petunia**
+ Corolla much less than 50mm, campanulate, often nodding 20

20. Pedicels strictly erect at anthesis from a rosette of apparently sessile leaves, deflexed and touching ground in fruit; stem greatly elongating and over-topping fruit with petiolate leaves; calyx enlarging only slightly in fruit (high alpine plant not found below 3300m) .... **13. Mandragora**
+ Pedicels ± pendent at anthesis, solitary or paired in axils of cauline leaves; calyx greatly enlarging in fruit and becoming membranous or muricate (weeds of waste places and fields at low altitudes, not occurring above c 2600m) 21

21. Leaves cuneate-attenuate at base; flowers usually paired in leaf axils; calyx lobes becoming muricate outside in fruit .... **6. Physaliastrum**
+ Leaves cordate, truncate, obtuse or oblique at base; flowers in our species solitary in axils; calyx reticulate in fruit but not becoming muricate 22

22. Corolla with bluish base, white tube and pale blue or mauve lobes; calyx segments 15–20mm in flower; ovary 3–5-locular, style 3–5-lobed **1. Nicandra**
+ Corolla yellow, sometimes with purple basal spots; calyx segments 2–9mm in flower; ovary bilocular, style 2-lobed .... **7. Physalis**

1. NICANDRA Adanson

Robust annual or perennial herb. Leaves alternate, lower ones long-petiolate. Flowers solitary in axils of most leaves, showy, nodding. Calyx deeply divided into 5 cordate sepal-like lobes, becoming membranous in fruit and strongly accrescent, enveloping the fruit. Corolla actinomorphic, campanulate; limb 5-plicate. Stamens attached near base of corolla; anthers oblong, dehiscing longitudinally. Style linear, stigma indistinctly 3–5-lobed. Ovary 3–5-locular; ovules very numerous. Fruit a pendulous globose berry. Seeds discoid, pitted, very numerous. A monotypic genus in its own tribe (Nicandreae).

1. N. physalodes (L.) Scopoli; *N. physaloides* (L.) Gaertner in error. Nep: *Dong dongey bin.* Fig. 92c–d.

Stems (20–)100–200cm, minutely but densely white-glandular at petiole bases, otherwise glabrous. Petioles 20–80mm; lamina elliptic to broadly ovate, (25–)80–220 × (15–)65–210mm, sinuate-dentate or -crenate, rather bluntly acuminate, base obtuse to truncate, upper surface densely but minutely sessile-glandular, lower surface glabrous, margins minutely scabrid. Pedicels 15–25mm.

Calyx segments sepal-like, 15–20 × 10–12mm in flower, to 30mm in fruit, cordate at base, green becoming whitish and membranous in fruit, prominently reticulate. Corolla blue at base, with whitish tube and pale blue or mauve lobes, 25–35mm across. Anthers pale yellow, c 3.5mm. Berry yellow or brownish-yellow, pendulous. Seeds red-brown, 1.5–1.8 × 1.3–1.6mm, reticulately pitted.

Bhutan: S – Gaylegphug district (Betni Bridge (38), Rama Camp (38)), **C** – Punakha district (Norwaung, Thinleygang) and Tongsa district (Shamgong, Dakpai (38)); **Darjeeling:** Labdah, Lebong, St Mary's (97a), Rakti (97a), Kurseong–Tindaria (97a); **Sikkim:** unlocalised (Watt collection). Roadsides, field margins and secondary scrub, 1070–1950m. May–October; fruiting July–December but often persisting until the following spring.

Native of Peru, now a casual throughout much of the warmer parts of the world.

2. LYCIUM L.

Spiny shrubs. Leaves usually clustered, shortly petiolate or appearing subsessile, lamina entire. Flowers solitary or in small, sometimes pedunculate clusters on short and long shoots. Calyx cup-shaped, 2-lipped. Corolla infundibular to campanulate, lilac or purple; tube slender or broad, dilated upwards or same width throughout. Stamens inserted near top of corolla tube; filaments glabrous or hairy. Fruit a small berry, red or black.

Field notes on fruit colour are vitally important. The species are difficult to identify; frequent misidentifications mean that literature records cannot be relied upon, unless made by a specialist.

1. L. armatum Griffith

Very spiny low shrub. Young branches markedly zigzag, older ones much less so. Spines very numerous, 5–40mm, with prominent tubercles (like small cedar cones) at base. Leaves in fascicles of (3–)5(–6) from tubercles on short shoots, and alternating with spines on long shoots, subsessile; lamina c 12–14 × 4mm, linear-spathulate (those alternating with spines often obovate-orbicular), obtuse or subacute, base attenuate, subglabrous. Flowers solitary from leaf fascicles, shortly pedicellate; pedicel suberect in flower, recurved and nodding in fruit. Calyx cup-shaped to shortly campanulate, tube c 2–3 × lobes, lobes unequal, ovate-triangular. Corolla purple; tube c 1½–2 × calyx and subequalling corolla lobes; lobes patent, obtuse; throat hairy. Filaments with tuft of long hairs near base. Style longer than stamens. Berry globose, red.

Bhutan: C – Punakha district (Pho Chu River near Punakha). Riversides, probably c 1400m. April.

Endemic to C Bhutan.

Further collections of *L. armatum* are needed to complete its description and properly assess its taxonomic status, as both type specimens are in poor condition and now lack flowers and berries. The description of *L. armatum* given here is based upon an examination of these specimens together with Griffith's plate (65) and type description (66). Griffith's drawings, although excellent, lack any indication of scale and therefore measurements of corollas etc. cannot be given here.

This material has previously been determined as *L. ruthenicum* Murray. This cannot be so, as *L. ruthenicum* has black berries whilst Griffith described those of *L. armatum* as red. *L. ruthenicum* also has much larger (10–13mm) corollas with a long, narrowly infundibular tube. *L. armatum* seems much more closely allied to *L. barbarum* L. and *L. chinense* L., both of which resemble it in having filaments with a dense tuft of hairs at their base. The very prominent tubercles, resembling small cedar cones, at the base of the spines of *L. armatum* have not been seen in any specimens of *L. barbarum* or *L. chinense*, and *L. armatum* also has far more numerous, longer spines than either of those species. Therefore, for the present it is maintained here as a distinct, endemic species.

L. depressum Stocks (*L. turcomanicum* Miers) has also been recorded from Punakha district (Punakha and Bhotokha areas, riversides, 1400–1450m, early May (71)). It differs from *L. armatum* by its corolla with glabrous throat, stamens with glabrous filaments, and ovoid (not globose) berry. The specimens on which the records were based have not been seen and confirmation of their identity is required, especially because they were collected at more or less the same locality and habitat as the type of *L. armatum*.

3. ANISODUS Sprengel

Perennial herbs with branched aerial stems. Leaves petiolate, alternate, entire. Flowers solitary and axillary, pendent, pedicellate. Calyx lobes unequal, exserted in bud, much enlarged in fruit and conspicuously ribbed. Corolla actinomorphic, campanulate or campanulate-cylindrical, less than twice as long as calyx; lobes auriculate, imbricate in bud, free. Stamens inserted near base of corolla; filaments erect; anthers longitudinally dehiscent. Stigma 2-lobed. Fruit a spherical apically circumscissile capsule, enclosed within calyx. Seeds reticulate, with curved embryo.

Arguments for the generic distinctness of *Anisodus* [*Whitleya* Sweet] from *Scopolia* (326, 401) have been followed here.

1. A. luridus Sprengel; *Scopolia lurida* (Sprengel) Dunal, *Physalis stramonifolia* Wall., *S. stramonifolia* (Wall.) Shrestha, *S. anomala* (Link & Otto) Airy-Shaw. Fig. 91g.

Stems (40–)100–200cm, often appearing rather succulent, branched, angled, shortly pubescent with fasciculate hairs mainly along angles. Petiole 10–80mm; lamina ovate or elliptic, 45–240 × 30–130mm, subacute or obtuse, base obtuse or cuneate, ± glabrous above, densely creamy-tomentose beneath. Flowers very foetid. Pedicels pendent, 10–35mm, glandular-villous. Calyx bright green and 30–35mm in flower, broadly campanulate, greatly enlarging in fruit (to 50 × 30mm) and then conspicuously 10-ribbed; tube long-villous, lobes unequal and shortly pubescent. Corolla 35–50 × 35–40mm, yellowish-green to brownish-white below, with pinkish or purplish lobes only slightly longer than calyx; tube densely pilose within in lower half. Filaments and anthers yellowish-green; anthers c 5 × 1.5mm. Seeds irregularly ovoid, c 2.5 × 2mm, brown, obscurely reticulate.

Bhutan: C – Ha, Thimphu, Punakha, Tongsa and Bumthang districts, **N** – Upper Mo Chu, Upper Pho Chu, Upper Bumthang Chu and Upper Kuru Chu districts; **Sikkim:** Chamnago, Natu La; **Chumbi:** Do-ree-chu. Open situations, especially in and around villages and yak huts but also in cloud forest clearings etc., 2285–3960m. June–July; fruiting until October.

Administration of a tincture of the leaves to the eyes causes considerable pupil dilation after a short time, and sometimes temporary blindness. The stems and leaves contain the alkaloids hyoscyamine, himaline, atropine and scopolamine.

Plants, originally described from 'Sikkim–Tibet', with corollas not or scarcely exceeding the calyx and leaves dull above, have been called *A. luridus* var. *fischerianus* (Pascher) Wu & Chen [*A. fischerianus* Pascher]. An example is the specimen from Chumbi (*King's collector* 347, K), which might in fact be type material of *A. fischerianus*. Their status requires further consideration.

4. PRZEWALSKIA Maximowicz

Low perennial herbs of high-alpine habitats with long, stout rhizomes. Stem short, mainly underground, producing a rosulate cluster of abbreviate-alternate ovate or oblanceolate leaves at top. Leaves ovate to oblong, petiolate. Flowers numerous, each solitary and axillary but appearing to form a cluster in the centre of the tuft of leaves, pedicellate. Calyx at anthesis tubular, 5-toothed, reticulate, greatly accrescent in fruit and becoming inflated and cartilaginous. Corolla narrowly infundibular, yellow (in our species). Stamens inserted above middle of corolla tube; anthers oblong, dehiscing by longitudinal slits. Fruit a glabrous capsule enclosed within the persistent calyx. Seeds numerous, compressed-reniform. A small genus of two (possibly three) species, mainly Tibetan.

1. P. shebbarei (C.E.C. Fischer) Grubov; *Mandragora shebbarei* C.E.C. Fischer, *P. tangutica* sensu Sanjappa & Raju (402) and Zhang, Lu & D'Arcy (105) non Maximowicz

Rosulate perennial. Stem to 10cm above ground. Leaves subcoriaceous, narrowly oblanceolate to narrowly elliptic, 40–120 × 12–30mm (excluding petiole), subacute, base decurrent into thick petiole 40–90(–110)mm, margins undulate, surfaces almost glabrous. Pedicels 6–7mm in flower, to 20mm in fruit, sparsely pilose with gland-tipped hairs. Calyx at anthesis rather fleshy, 8–15mm, glandular-pubescent outside; teeth lingulate, unequal, longest 2.5–3mm, other 4 c 1mm shorter. Corolla yellow; tube 15–20mm, glandular-puberulent outside; lobes triangular or rounded-oblong, c 3 × 1.5mm, irregularly crenate, plicate and reticulate inside. Filaments glabrous, c 3.5mm; anthers 1.5–2mm, oblong. Fruiting calyx 25–50(–60) × 13–35(–40)mm, goblet-shaped, narrowed towards apex but with relatively wide, open mouth; teeth triangular, 3–8mm deep, not closing mouth. Capsule 10–15 × 10–12mm, globose or subglobose. Seeds c 3 × 2.5mm, deep brown, testa granulate.

Bhutan: N – Upper Mo Chu district (Chomo Lhari); **Sikkim:** Tista River–Dorji La, Tista River–Kerang (402); **Chumbi:** Chomo Lhari, Phari. Sandy micaceous soils in cold arid high-alpine areas, 4420–5400m. June–July; fruits July–September.

Sanjappa & Raju (402) provide the first records of this species from Sikkim, together with a good illustration, but misidentified their material as *P. tangutica*. *P. tangutica* Maximowicz is endemic to the Qinghai-Xizang Plateau (China); it is easily distinguished by its perfectly ovoid or pyriform (not goblet-shaped), usually larger fruiting calyx 6–11cm which has a very narrow mouth almost closed by very small teeth 1–2mm.

The habit, and the shape and colour of the flowers, show a remarkable convergence with *Scrophularia przewalskii* and the genus *Oreosolen* (Scrophulariaceae), both of which have similar habitats to *P. shebbarei*.

5. HYOSCYAMUS L.

Villous or glabrescent, annual, biennial or perennial herbs. Leaves entire to pinnatipartite. Inflorescence a bracteate spike or raceme. Flowers weakly zygomorphic. Calyx tubular-campanulate, 5-dentate, accrescent in fruit. Corolla infundibular or campanulate with oblique limb. Stamens inserted near base of corolla, included or exserted. Fruit an operculate capsule.

1. H. niger L.; *H. agrestis* Kitaibel ex Schultes

Large, erect, widely spreading, coarse annual or biennial. Stems 20–100cm, often somewhat zigzag above, shortly glandular and villous with longer hairs sometimes up to 7mm. Basal leaves to 55 × 25cm, shortly petiolate; cauline 5–18 × 2–9cm, sessile, amplexicaul; lamina ovate-oblong to oblong-lanceolate, entire or usually with 1–3 pairs of ± large, triangular-lanceolate acute lobes, shortly glandular-pilose on veins and margins, surfaces minutely glandular. Inflorescence a bracteate raceme, elongating to 30cm in fruit; bracts leaf-like. Flowers sessile, ± secund. Calyx 10–15mm in flower, 20–40mm in fruit, constricted in middle; lower part ventricose, densely villous, upper part membranous, prominently reticulate-veined, minutely glandular, with 5 short teeth. Corolla pale lurid yellow with dense network of purple veins and dark purple throat, broadly campanulate, 25–35mm, shortly pubescent outside on tube and lower part of lobes. Stamens included; anthers yellow. Capsule included in calyx; seeds pale brown, c 1.5 × 1.2mm, ± reniform, coarsely reticulate.

Sikkim: Tindi. Field margins, farmyards and wasteground, 1980–3353m. May–July.

Habitat data above based on Nepalese material.

6. PHYSALIASTRUM Makino

Perennial, shortly rhizomatous unarmed herbs. Stems with lax bifurcating branching pattern. Leaves alternate, entire, petiolate, alternately with and without an axillary lateral branch, those where the branch is suppressed appearing falsely paired. Flowers (1–)2(–3) in axils of the paired leaves or lateral branches, pedicellate. Calyx shortly campanulate or obconical-campanulate, shortly 5-toothed at apex, pilose outside, greatly enlarging in fruit but not inflated, becoming muricate. Corolla rotate-campanulate; tube short, with 5 fascicles of barbules within; lobes broadly triangular, spreading at anthesis. Stamens 5; anthers opening by introrse longitudinal slits. Ovary bilocular; style filiform. Fruit a globose or ellipsoid berry, completely enclosed within accrescent calyx. Seeds numerous, orbicular-reniform, rugose-faveolate.

Following (332), the Asiatic genus *Physaliastrum* is here treated as distinct from the American *Leucophysalis* Rydberg.

1. P. yunnanense Kuang & Lu subsp. **bhutanicum** (Grierson & Long) D'Arcy & Zhang; *Leucophysalis yunnanensis* (Kuang & Lu) Averett subsp. *bhutanica* Grierson & Long

Perennial herb 60cm–1m. Stems sulcate, pilose when young, later glabrescent. Petioles short (10–20mm); lamina elliptic or broadly elliptic, 40–160 × 20–100mm, shortly acuminate, base cuneate-attenuate and ± equal, surfaces sparsely pilose, becoming glabrous. Flowers usually in axillary pairs; pedicels c 12mm in flower, c 25–30mm in (young) fruit. Calyx lobes c 4mm in flower, densely pilose outside, accrescent in fruit and then globose-ellipsoid, 14–25 × 8–20mm, muricate externally. Corolla whitish, c 18mm diameter, tube and lobes densely and minutely pubescent outside; tube c 4.5mm, lobes 6–7mm. Anthers ovoid, c 2.5 × 1mm. Style c 13mm. Berry globose, to 17mm diameter.

Bhutan: C – Punakha district (Rinchu–Kancham, type) and Mongar district (Khosa–Tamji). 1525–2000m. May–July.

No other ecological information available for subsp. *bhutanicum*; subsp. *yunnanense* (endemic to Yunnan) occurs in thickets from 1800–2560m. No mature fruit of subsp. *bhutanicum* has been seen although small, immature fruits are present; details have been taken from material of subsp. *yunnanense*.

7. PHYSALIS L.

Annual or perennial herbs. Leaves usually alternate, petiolate, soft-textured, entire, sinuate or shortly lobed. Flowers solitary (our species) or paired in leaf axils, pedicellate, nodding. Calyx campanulate, 5-lobed, much enlarged in fruit and becoming membranous and sometimes brightly coloured. Corolla broadly campanulate, white or yellow (sometimes with purplish spots near base), 5-angled or 5-lobed. Stamens inserted near base of corolla; anthers dehiscing longitudinally, shorter than filaments. Ovary bilocular. Fruit a globose berry enclosed within persistent calyx. Seeds numerous, reniform, compressed, with curved embryo.

P. alkekengi L., with an orange or red fruiting calyx and white corolla, is widely cultivated but so far does not seem to have been recorded from our area.

1. Calyx at anthesis c 2.5mm, corolla 4–6mm, usually without purple spots; plant subglabrous or pubescent but not villous .... **1. P. divaricata**
\+ Calyx at anthesis 4–9mm; corolla with purple spots at base or a dark centre; plant villous or densely pubescent 2

2. Calyx 7–9mm at anthesis, pedicels 8–12mm (slightly longer than calyx) **2. P. peruviana**
\+ Calyx 4–6mm at anthesis, pedicels subequal to calyx .... **3. P. pubescens**

1. P. divaricata D. Don; *P. minima* sensu F.B.I. non L., *P. indica* Lamarck var. *indica*, *P. indica* var. *microcarpa* Nees. Fig. 91d–f.

Diffusely branched annual 15–45cm, pubescent or usually subglabrous. Petiole to 40mm; lamina ovate, 30–85(–110) × 15–40(–70)mm, acute or acuminate, base cordate or oblique, margin sinuate, repand or subentire. Pedicels 5–10(–12)mm at anthesis. Calyx campanulate, c 2–2.5mm at anthesis, 15–25mm in fruit and globose-ovoid, pubescent; lobes triangular, 0.5–1mm and c ⅕–⅓ × tube at anthesis. Corolla yellow, without basal spots, 4–5(–6)mm; lobes acute, pubescent. Filaments c 2mm; anthers c 1mm. Berry orange, globose, c 10mm diameter. Seeds c 2mm, brownish-yellow, minutely reticulate.

Bhutan: S – Phuntsholing district (near Phuntsholing), Gaylegphug district (Gaylegphug), **C** – Punakha district (Wangdu Phodrang), Mongar district (Autsho), Tashigang district (Kheri (38), Shalli); **Darjeeling:** Rakti (97a); **Darjeeling/Sikkim:** unlocalised (Hooker collection). Wasteground, edges of cultivation (e.g. of maize and foxtail millet) and river shingle, 200–1400m. April–July(–November in Kurseong (97a)).

P. angulata L., similar to *P. divaricata* in its nearly glabrous stems and leaves but differing in its larger calyx c 7mm at anthesis with narrowly linear-lanceolate lobes almost as long as the tube, occurs in Nepal but has not yet been recorded from Sikkim or Bhutan. It is native to the Americas (Manitoba to Argentina) but is widely naturalised in the Old World.

2. P. peruviana L.; *P. edulis* Sims. Eng: *Cape gooseberry.*

Erect, branched perennial 30–100mm, villous-pubescent, indumentum often dense. Leaves ovate or ovate-cordate, 45–140 × 35–105mm, shortly acuminate, base cordate to obtuse, margin sinuate to repand or dentate or almost entire, villous-pubescent on both surfaces; petiole 10–60mm. Pedicels 8–12mm at anthesis (slightly longer than calyx), slightly lengthening in fruit. Calyx 7–9mm at anthesis, 35–40 × c 25mm in fruit, divided to near middle; lobes 3–4mm at anthesis, ovate-triangular. Corolla yellow with conspicuous, pubescent chocolate-brown or purple blotches (c 3 × 4–5mm) at base, 10–12mm (12–20mm diameter), sparsely pubescent outside and with shallowly triangular, shortly acute, ciliate lobes. Filaments 3–4mm; anthers 2.3–3.5mm, bluish. Berry orange, globose, 12–14mm diameter. Seeds c 2.4mm, brown, minutely reticulate.

Darjeeling: Darjeeling, Kariabasti (97a); **Sikkim:** Great Rangit Valley and below Gangtok. River shingle and wasteground, and as an escape from cultivation, 1220–2135m. April–November; fruiting October–January.

Native to tropical America; widely cultivated on account of its edible fruits.

3. P. pubescens L.

Similar to *P. peruviana* in being villous-pubescent, its corollas having a dark, pubescent eye, but calyx at anthesis only 4–6mm long; pedicels mostly subequal to the calyx.

Bhutan: C – Mongar district (Autsho). Weed of maize, c 1000m. June.

Only one specimen seen from Bhutan; first recorded in 1992. It has also been collected in C Nepal.

A native of the Americas from Massachusetts to Argentina, *P. pubescens* is naturalised in many of the warmer parts of the Old World.

8. CAPSICUM L.

Short-lived perennial herbs or low shrubs, often cultivated as annuals. Stems glabrous or pubescent; hairs simple, sometimes glandular. Leaves petiolate, simple, entire, ovate or elliptic. Inflorescence of 1 or more axillary fasciculate flowers; pedicels usually recurved at anthesis, erect or recurved in fruit. Calyx cyathiform or shortly tubular, truncate or with 5 or 10 short teeth, not enclosing fruit. Corolla small, rotate, white or yellowish, deeply 5-lobed. Stamens 5; filaments inserted at base of corolla tube; anthers usually blue or purplish, opening by longitudinal slits. Ovary bilocular, glabrous, many-ovulate. Fruit a dry, coriaceous or fleshy berry; loculi with large air-spaces. Seeds lenticular, flattened.

Both species are native to C and S America, widely cultivated elsewhere for their hot, edible fruits (peppers or chilli).

1. Pedicels 1(–2) at each node after the first; calyx conical; corolla white without yellow tinge .... **1. C. annuum**
\+ Pedicels (1–)2–4(–6) at each node after the first (care! look for scars where pedicels have fallen); calyx cup-shaped; corolla yellowish-green or cream **2. C. frutescens**

1. C. annuum L. Eng: *Chilli pepper.*

Short-lived perennial usually cultivated as an annual, 30–75cm, herbaceous or slightly shrubby; branches glabrous or shortly pubescent. Petioles 5–40mm;

lamina ovate or elliptic, 15–80 × 5–45mm, acute, usually cuneate or attenuate, less commonly obtuse or truncate, glabrous or glabrate except sometimes for spreading hairs on veins beneath. Pedicels 1 (very rarely 2) per node, 5–30mm in fruit. Calyx conical, c 2.5mm, sometimes with rows of papillae outside. Corolla pure white, c 9mm; lobes 4.5–5mm, acute. Anthers blue or turquoise, 1–1.4mm. Berry ovoid-acuminate and sometimes falcate (var. *acuminatum* Fingerhuth) or globose (var. *aviculare* (Dierbach) D'Arcy & Eshbaugh [var. *baccatum* (L.) Kuntze]), green at first, turning yellow then red when mature, 10–50 × 8–35mm.

Bhutan: S – Phuntsholing district (Torsa River), Gaylegphug district (Taklai Khola E of Gaylegphug), Deothang district (Deothang); **Sikkim:** Tista. On dry shingle, and also cultivated, 200–1425m. May–June.

According to J.R.I. Wood (pers. comm.), this species is only grown in expatriates' gardens and is not eaten by the Bhutanese.

2. C. frutescens L.

Like *C. annuum* but a diffuse shrub 1–1.5(–2)m, the branches usually more zigzag and more densely pubescent; leaves acuminate at apex; pedicels 2–3(–6) per node, 10–25mm in fruit; calyx cup-shaped; corolla pale yellowish-green or cream; fruit usually narrowly cylindrical, turning red or rarely black when mature.

Bhutan: S – Deothang district (Deothang); **Darjeeling/Sikkim:** unlocalised (Treutler collection). Cultivated, c 1000m. June (but probably flowering and fruiting all year).

The fruits of this species take longer to mature than those of *C. annuum* and it can only successfully be grown in more tropical regions. They are used medicinally to cure headaches.

The 'national vegetable' of Bhutan, grown everywhere. The plants are at first grown in nurseries in warm valleys, and are transplanted as seedlings into frost-sensitive regions in May. The fruits are ripe in September–October and are dried on house roofs in the autumn. The usual varieties grown in Bhutan have large, elongate-ovoid fruits (J.R. I. Wood, pers. comm.).

9. SOLANUM L.

Herbs, lianas, shrubs or small trees, unarmed or armed with prickles or spines, often with stellate hairs. Leaves simple or lobed (rarely pinnate). Flowers usually in cymes or racemes, often extra-axillary but also sometimes axillary or leaf-opposed, rarely solitary. Calyx campanulate or cup-shaped, 5-lobed, enlarging in fruit or not. Corolla rotate or shortly campanulate, actinomorphic or slightly zygomorphic; lobes 5, spreading or recurved. Stamens attached near mouth of short corolla tube, exserted, the anthers often forming a cone, opening by 2 apical pores and occasionally also by longitudinal slits. Fruit a variously coloured berry. Seeds compressed, discoid.

S. wendlandii Hook.f., a native probably of C America, has been recorded from E Nepal (Sanguri Lekh) and might occur in our area as an ornamental which could become naturalised. It is a very distinctive climbing species with small (1–1.5mm) prickles on the stems, petioles and midribs, variable but mainly imparipinnate leaves (usually 2–4 pairs of leaflets), and showy cymose inflorescences of large deep mauve corollas c 45mm diameter which are completely glabrous and whose midribs are produced into a pointed tip c 5mm. It has been keyed out below.

1. Plants unarmed 2
\+ Plants armed with spines or prickles at least on stem 13

2. Leaves and stems woolly or stellate-pubescent 3
\+ Leaves and stems neither woolly nor stellate-pubescent, usually glabrous 7

3. Herb; berry 25–230mm, often elongate, globose or ellipsoid **17. S. melongena**
\+ Shrub or small tree; berry 6–15mm, globose 4

4. Leaves lobed, with ± truncate base .... **3. S. kurzii**
\+ Leaves entire, with cuneate or obtuse base 7

5. Berries green, 6–8mm diameter; leaf base ± truncate .... **3. S. kurzii**
\+ Berries yellow, 18–22mm diameter **S. mirikense** (see note under *S. mauritianum*)

6. Flowers creamy white; small auricle-like leaves absent .... **1. S. erianthum**
\+ Flowers bluish-violet; small auricle-like leaves present in leaf axils **2. S. mauritianum**

7. Shrub or liana 8
\+ Herb 11

8. Liana or scrambling shrub 9
\+ Erect non-scrambling shrub 10

9. Corolla 7–10mm, purple or violet, each lobe with 2 green basal spots **8. S. dulcamara**
\+ Corolla 20–50mm, white, mauve or pale purple .... **9. S. seaforthianum**

10. Leaves 40–80 × 10–20mm; calyx lobes triangular-lanceolate; inflorescence a 1–3-flowered cyme .... **7. S. pseudocapsicum**
\+ Leaves 80–220 × 30–70mm; calyx cupular with very shallow lobes; inflorescence a 3–12-flowered raceme .... **4. S. spirale**

11. Tuberous; corolla lobes 8–12mm .... **10. S. tuberosum**
\+ Not tuberous; corolla lobes c 5mm 12

12. Peduncle 8–35mm, longer than 3–10mm pedicels; mature berry black
5. S. americanum subsp. **nodiflorum**
\+ Peduncle 2–6mm, shorter than 4–9mm pedicels; mature berry red
6. S. villosum subsp. **puniceum**

13. Most or all stem prickles pubescent or villous towards base 14
\+ All stem prickles completely glabrous 15

14. Corolla white; stem glandular viscid with simple hairs only; upper surface of leaves with simple hairs, lower surface stellate-pubescent .. .. **14. S. viarum**
\+ Corolla mauve; young stem stellate-pubescent; both leaf surfaces with stellate hairs, lower surface tomentose .... **12. S. anguivi**

15. Corolla white 16
\+ Corolla blue, mauve or violet 17

16. Stems very densely patent-hirsute with yellowish simple hairs c 1.5mm; leaves with simple hairs only; anthers c 3.5mm .... **13. S. aculeatissimum**
\+ Stems densely stellate-pubescent; leaves stellate-pubescent; anthers 5.5–7.5mm .... **11. S. torvum**

17. Climber with many leaves imparipinnate; corolla 40–45mm diameter, glabrous; prickles 1–1.5mm
S. wendlandii (see note at end of genus description, above)
\+ Combination of characters not as above 18

18. Berry enclosed in calyx; undershrubs 19
\+ Berry not enclosed in the calyx; herbs 20

19. Inflorescences less than 5cm; peduncles less than 5mm; corolla white
16. S. griffithii
\+ Inflorescences more than 5cm; peduncles more than 5mm; corolla mauve or blue .... **15. S. barbisetum**

20. Very densely spiny prostrate herb; berry 15–20mm, globose, yellow
18. S. virginianum
\+ Sparsely spiny or unarmed erect cultivated herb; berry 25–230mm, usually dark purplish, globose or usually ellipsoid and elongate .. .. **17. S. melongena**

1. S. erianthum D. Don; *S. verbascifolium* sensu F.B.I. non L. Dz: *Namphai* (38); Nep: *Burbee* (34).

Unarmed shrub or small tree (60cm–)1–6m. Young branches terete, densely stellate-tomentose, older ones with greyish bark; hairs shortly stalked (0.3–0.5mm) with numerous (15–20) rays. Leaves lanceolate-elliptic, 100–300

× 50–130mm, long-acuminate, base obtuse or shortly cuneate, dull dark green or brownish-green and sparsely stellate above, greyish-tomentose beneath with stellate hairs; petiole 10–55mm; axils lacking small auricle-like leaves. Inflorescence a dense, axillary, long-pedunculate corymbose cyme near tip of branch; peduncle 3–14cm. Calyx 6–8mm, stellate-tomentose outside and less densely inside; tube 3.5–5mm, lobes ovate, 2.5–3mm. Corolla white (often creamy), 8–12mm; lobes 4–5mm, densely stellate outside and much more sparsely (but not glabrous) within. Anthers yellow, dehiscing by terminal pores. Berry globose, c 10mm diameter, yellow; seeds ovate-discoid, c 1.5 × 1.2mm, reddish-brown, reticulate.

Bhutan: **S** – Deothang district (Diwangiri), **C** –Thimphu, Punakha, Tongsa and Tashigang districts; **Darjeeling:** Kalimpong, Munsang, Pul Bazar; **Sikkim:** Great Rangit Valley etc.; **W Bengal Duars:** Jalpaiguri. Shrubberies, 250–1450m. Flowering and fruiting most of year.

The fruit is used as an edible vegetable in Sikkim (352).

2. S. mauritianum Scopoli; ?*S. mirikense* Mukhopadhyay

Stout unarmed shrub or small tree 2–3(–4)m. Stems terete, bifurcating, brittle, densely floccose-pubescent; hairs dendroid, pale yellowish-green or yellowish-brown, long-stalked (0.5–1mm) with relatively few (7–10) rays. Leaves ovate-elliptic, 140–290 × 70–110mm, gradually acuminate, base cuneate, entire, dark green above with numerous branched hairs, paler and densely felted beneath with stellate and dendroid hairs; petiole 15–45mm; each leaf axil subtended by a pair of small auricle-like cordate leaves 8–20mm. Inflorescence a long-pedunculate, erect corymbose cyme 4–7(–15)cm across, of numerous small flowers; peduncle 5–13cm. Calyx c 5mm, campanulate, densely stellate outside; tube c 3mm, lobes c 2mm. Corolla pale bluish-violet or lilac, 10–11mm; tube stellate above, glabrous below; lobes stellate outside, glabrous inside, becoming patent. Anthers c 1mm, dehiscing by an introrse apical pore. Berry 10–15mm diameter, orange yellow turning dull brown, densely stellate-pubescent. Seeds discoid, c 2 × 1.5mm, cinnamon-coloured, reticulate.

Darjeeling: Observatory Hill, Singtam, Lebong, Ging etc. (all 333). Cultivated for ornament and now sometimes naturalised, 1200–1700m. March–April.

Native of Uruguay and S Brazil.

S. mirikense, recently described from Mirik, Darjeeling (1700–1900m), and characterised by unarmed habit, sinuate-lobed acuminate leaves, blue flowers c 1cm, and yellow berries 1.8–2.2cm diameter, is possibly synonymous with this recent alien introduction; however, its berries were described as slightly larger than given here for *S. mauritianum*, its leaves were described as sinuate-lobed, not entire, and its flowering time is reputedly different (August–December, rather than March–April). Its lobed leaves are shared by *S. kurzii*, but the berries of the latter are much smaller and green, not yellow. Mukhopadhjay allied *S. mirikense* with the armed species *S. indicum* L. (probably meaning *S. anguivi* Lamarck); its taxonomic status needs further investigation.

3. S. kurzii Prain

Unarmed shrub or small tree 1–1.3m. Young branches densely rusty stellate-tomentose all round, older parts grey with 2 broad rows of stellate hairs. Leaves ovate, (30–)45–100 × (20–)30–85mm, deeply and obtusely lobed with 2–4 wide lobes per side, obtuse or subacute at apex of all lobes, base truncate, upper surface stellate-pubescent (rays 5–7, unequal and all orientated ± towards leaf apex), lower surface densely stellate-tomentose; petiole 15–40mm, stellate-tomentose. Inflorescence a pedunculate, extra-axillary racemose cyme; peduncle 15–30mm; pedicels 10–15mm, rusty-tomentose. Calyx small, appressed to fruit and not reflexed. Corolla blue, with white margins to lobes and green mid-veins, to c 9mm; lobes oblong-lanceolate, acute, one longer than other 4, mealy outside and stellate along mid-vein inside. Anthers c 5mm, oblong-linear. Fruit (not seen) a globose, glabrous, bitter, green berry 6–8mm diameter.

Bhutan: C – Tongsa district (Birti (38)); **Darjeeling:** Rishep; **Sikkim:** unlocalised (391). Roadsides etc., 650–1220m. April–May.

In Assam, the fruit is cooked and eaten, and is also used medicinally for liver complaints and malaria, and to cure eye diseases in poultry (352).

4. S. spirale Roxb.

Unarmed shrub 60cm–4m. Branches erect, greyish, glabrous, with 2 sharp longitudinal ridges. Leaves elliptic or elliptic-obovate, 80–220 × 30–70mm, acute or acuminate, base cuneate-attenuate and often slightly unequal, minutely puberulent above but glabrous beneath; petiole 10–40mm; each leaf subtended by a much smaller, caducous elliptic leaf. Flowers in numerous, spirally arranged, extra-axillary, 3–12-flowered pedunculate racemes; peduncle 5–20mm; pedicels 5–15mm in flower, 20–30mm in fruit. Calyx cupular with very shallow, broad, minutely ciliate lobes. Corolla white, 8–10mm, deeply divided into 5 elliptic lobes 6–8mm, nearly glabrous. Anthers c 3.5mm. Fruit a globose, yellow or orange, bitter berry (8.5–)11–15mm. Seeds discoid, somewhat concave with thickened margins, c 4 × 3.5mm, obscurely and finely reticulate.

Bhutan: C – Tongsa district (below Shamgong), Mongar district (Lingtsi, Shonga, Mongar), Tashigang district (Kulong Chu near Shali). Warm broad-leaved and mixed evergreen forest, 1675–1750m. June–August; fruits August–April.

The leaves and fruits are eaten as a vegetable in Assam.

5. S. americanum Miller subsp. **nodiflorum** (Jacquin) Henderson; *S. americanum* var. *nodiflorum* (Jacquin) Edmonds, *S. americanum* var. *patulum* (L.) Edmonds, *S. nigrum* auct. ind. non L. Fig. 90a–c.

Much-branched unarmed weedy annual with suffrutescent base, 20–70cm. Stems sparsely to moderately appressed-pubescent, hairs all eglandular. Leaves ovate to ovate-rhombic, 15–100 × 6–50mm, sometimes with a pair of basal leaflets, acute or acuminate, base cuneate, margin sinuate-dentate or entire, surfaces subglabrous to sparsely pubescent; petioles 8–40mm, a petiolate simple leaf arising opposite a compound leaf or branch ('maurelloid branching'). Inflorescence a few (3–8)-flowered, pedunculate, umbellate extra-axillary cyme, arising a short distance below, and on same side of stem as, each compound leaf or leafy branch. Peduncle 8–35mm. Pedicels 3–10mm, shorter than peduncle. Calyx 0.6–2mm, cupular; lobes 0.3–0.6mm, ciliate-margined. Corolla white (lobes sometimes with purple mid-vein), c 5mm; lobes triangular-acute, c 2.5mm. Stamens 1.5–2mm, with ventro-apical pore and finally opening also by a longitudinal slit. Berry green at first, finally black, 5–9(–10)mm diameter, globose. Seeds c 1.5mm, discoid, minutely reticulate-foveolate.

Bhutan: S – Samchi (117), Phuntsholing, Chukka, Gaylegphug (117) and Deothang (117) districts, **C** – Thimphu and Tongsa districts; **Darjeeling:** Darjeeling, Lebong, Sukna terai etc.; **Sikkim:** Tista, Yoksam–Bakhim. Roadsides, field margins, vegetable plots and waste places near habitation, 200–2800m. Almost throughout the year; flowering of young plants appears to begin in August–September.

All material seen from Darjeeling, Sikkim and Bhutan labelled '*S. nigrum* L.' belongs to *S. americanum*, which differs from *S. nigrum* by its umbellate cymes and smaller flowers.

Fruit edible; fruit and leaf used against coughs in our area, various parts used for a wide variety of medicinal purposes in much of India (352, as '*S. nigrum*').

6. S. villosum Miller subsp. **puniceum** (Kirschleger) Edmonds; *S. alatum* Moench, *S. rubrum* Miller, *S. miniatum* Bernhart

Weedy unarmed annual 10–60cm, drying yellowish-green or somewhat glaucous. Stems subglabrous to moderately densely appressed-pubescent; glandular hairs absent. Leaves ovate, 15–150 × 10–90mm, coarsely sinuate-dentate (especially in lower half) or subentire, acute or bluntly acuminate, base cuneate to truncate, sometimes with a pair of small leaflets; upper surface subglabrous or sparsely pilose, lower surface sparsely pilose; petiole 5–60mm. Inflorescence an extra-axillary, sub-umbellate or condensed-racemose cyme; peduncle 2–6mm, shorter than pedicels; pedicels 4–9mm, appressed-pubescent. Calyx 2–2.5mm, divided to near base; lobes elliptic, acute, ciliate-margined, reflexed in fruit. Corolla white, 4–5mm; lobes ovate-elliptic, acute, shortly pubescent outside. Stigma pubescent, 2-lobed. Berry globose, red, 6–8mm diameter. Seeds ovoid, compressed, c 2 × 1.2mm, reticulate-foveolate.

Bhutan: C – Thimphu district (Nimchling (71)). In similar habitats to *S. americanum*, c 2200m. Fruits June onwards.

Only one record of this species is known from our area; it has been much confused in the past with *S. nigrum*, which like the closely allied *S. americanum*

FIG. 90. **Solanaceae.** a–c, *Solanum americanum* subsp. *nodiflorum*: a, flowering shoot (× ½); b, flower (× 6); c, fruit (× ⅔). d–f, *Solanum torvum*: d, flowering shoot (× ½); e, flower (× 3); f, fruit (× ⅓). g–i, *Lycianthes crassipetalum*: g, leaf and inflorescence (× ½); h, flower (× 3); i, fruit (× 2). Drawn by Glenn Rodrigues.

a
b
c
d
e
f
g
h
i

belongs to sect. *Solanum* of which it is the type species. Subsp. *villosum*, with yellow berries, has not been recorded from our area so far but should be searched for.

7. S. pseudocapsicum L. Eng: *Jerusalem cherry.*

Unarmed shrub (sometimes herbaceous), 90–120cm. Branches ascending, striate, greyish, appearing glabrous although actually very minutely puberulent at least when young. Leaves lanceolate or elliptic-ovate, rather narrow, 40–80 × 10–20mm, entire or weakly repand, acute or obtuse, base cuneate or attenuate, decurrent, surfaces glabrous; petiole 5–10mm. Inflorescence a 1–3-flowered pedunculate axillary or extra-axillary cyme; peduncle short; pedicels 6–7mm at anthesis, 8–10mm in fruit. Calyx lobes 3–5mm in flower, triangular-lanceolate, accrescent in fruit. Corolla white, rotate, 10–14mm diameter; lobes c 6 × 2.5mm, glabrous. Anthers orange-yellow, c 2.5mm, oblong. Style c 5mm. Berry globose, 8–15mm diameter, scarlet.

Darjeeling: Kurseong, St Mary's (both 97a). Cultivated as an ornamental; flowering and fruiting almost throughout the year.

8. S. dulcamara L. Eng: *Woody nightshade.*

Scrambling unarmed shrub 60–200cm or more. Stems slender, woody at base, glabrous or minutely pubescent. Leaves ovate or oblong-lanceolate, 20–90 × 5–65mm, acuminate (sometimes rather bluntly), base cordate, truncate or obtuse (occasionally with 1–2 pairs of lateral leaflets), ± glabrous; petiole 8–30mm, slender. Inflorescence a lax, pedunculate, extra-axillary repeatedly dichotomous cyme, 6–25-flowered; peduncle 10–55mm. Pedicels 5–6mm in flower, to 15mm in fruit, pendent. Calyx cupular, 1.5–2mm, lobes very shallow and short (c 0.2mm) forming a remotely scalloped margin, minutely pubescent on margins. Corolla purple or violet, 7–10mm, lobes triangular-lanceolate, becoming recurved, each with 2 green basal spots, pubescent on margins and near tips. Anthers bright yellow, 2.5–4mm, connivent forming a cone. Style c 5mm, exceeding anther cone. Berry 8–10mm, ovoid-globose, green at first, turning yellow and finally scarlet (all colour stages often present in same cyme). Seeds 2–2.5mm, minutely reticulate.

Bhutan: C – Tongsa district (Rukubje); **Darjeeling:** Rumman, Rithu Valley; **Sikkim:** Chunthang. Wasteground, edges of cultivation etc., 1830–2590m. June–July; fruits September–November.

9. S. seaforthianum Andrews; *S. jasminoides* sensu Mukherjee (218) non Paxton

Unarmed, high-climbing liana, 1–5m; stems green at first, becoming woody, almost glabrous. Leaves ovate, 45–70 × 15–45mm, usually a mixture of simple, pinnatifid or pinnately lobed on same plant, acute to shortly acuminate, base cuneate, truncate or very shallowly cordate, sparsely and minutely puberulent above, subglabrous beneath; petiole 15–30mm. Inflorescence a ternately branched panicle, leaf-opposed or appearing terminal; pedicels slender, 4–10mm, thickened and clavate below calyx. Calyx 1–3mm, divided almost to base into broadly ovate lobes 1–3 × 1–2.5mm with extremely narrow whitish margins. Corolla white (Darjeeling (218)), bluish, pinkish or purplish, 2–5cm diameter; lobes narrowly elliptic, 12–15 × c 3mm. Berry (not known from Darjeeling) 10–20mm diameter, scarlet, juicy.

Darjeeling: Government College Campus (218), Darjeeling (71), both as *S. jasminoides*.

Although the specimens on which the Darjeeling records are based have not been seen, they are referred to *S. seaforthianum* because Mukherjee (218) described the Darjeeling plants as having some leaves irregularly pinnatifid or pinnate, and on leaf size. *S. jasminoides* is like *S. seaforthianum*, but in the material seen none of the leaves are compound and the leaves are all smaller (15–30 × 8–16mm) than described by Mukherjee (50–60 × 25–37.5mm). The correct identity of the Darjeeling plants is uncertain but they seem to fall more within the range of *S. seaforthianum* and that name has therefore been provisionally used here. *S. seaforthianum*, which is probably native to the Caribbean, is widely cultivated and naturalised in many countries of warm regions.

10. S. tuberosum L. Eng: *Potato*

Unarmed tuberous perennial 60–90cm. Stems erect, branched, pubescent. Leaves imparipinnate, 70–220 × 50–150mm, with 3–5 pairs of leaflets and much smaller, ± orbicular intercalary leaflets; principal leaflets increasing in size towards leaf apex (terminal one 30–100 × 12–60mm), abruptly short-acuminate, base shortly cuneate, petiolulate; petiole 10–100mm. Inflorescence a terminal, few-flowered paniculate cyme, c 4–5 × 6–7cm; peduncle 7–16cm; pedicels 8–25mm, articulated just below calyx. Calyx campanulate, 6–8mm, divided to c ⅔, pubescent; lobes ovate, c 4mm. Corolla white or pale mauve, pendent, 12–25mm diameter; lobes 8–12mm, ovate, pubescent. Anthers yellow, 4–5mm. Style minutely but densely pubescent as far as anther apices, uppermost part exceeding anthers and glabrous; stigma minutely puberulent. Fruit a rarely produced berry.

Bhutan: C – Tongsa district (Shamgong). Cultivated as a crop (potato) in fields and gardens, 1980–2590m. June.

A major commercial crop in Bhutan, the tubers being exported by truck to India. The principal areas of commercial cultivation are all at c 2500–2800m, around Chapcha, Phubjikah, Bumthang, Dremtse and Khaling. Potatoes are also grown for local consumption throughout Bhutan (J.R.I. Wood, pers. comm.).

11. S. torvum Swartz; *S. stramonifolium* sensu Roxb. non Jacquin. Nep: *Bin* (38). Fig. 90d–f.

Sparingly armed shrub (sometimes flowering as a tall herb), 60cm–2.5m. Branches densely stellate-pubescent when young, sometimes ferrugineous. Prickles few, usually confined to stems, occasionally also on petioles and near base of midribs, glabrous. Leaves ovate in outline, 70–200 × 40–120(–180)mm, subentire or usually deeply 2–3-sinuate, lobes (when present) acute, apex acute or shortly acuminate, base unequally cuneate, upper surface stellate-pubescent (hairs usually with 8 unequal rays), densely stellate-tomentose beneath; petiole 20–40mm. Inflorescence a dense, pedunculate, extra-axillary, branched, unarmed, many-flowered cyme; peduncle 5–30mm; pedicels 6–10mm in flower, 15–20mm in fruit, with stalked stellate hairs and stalked simple glandular hairs. Flowers c 10mm, corolla slightly zygomorphic. Calyx 3.5–4mm, lobes ± obovate, abruptly acuminate, not accrescent. Corolla white, one lobe slightly longer than other 4, all densely stellate-tomentose outside and with broad band of stellate pubescence along mid-vein inside. Anthers yellow, 5.5–7.5mm, linear, slightly unequal. Berry 10–15mm diameter, green turning bright yellow and finally orange, glabrous. Seeds c 2.2 × 2mm, discoid, pale olive, almost smooth.

Bhutan: S – Samchi district (Dorokha (38), Kalapani), Phuntsholing district (Phuntsholing area, frequent), Gaylegphug district (Gaylegphug), Deothang district (Deothang), **C** – Tongsa district (Birti (38)); **Darjeeling:** Lohagarhi, Rishep, Manjitar etc.; **Sikkim:** Tista, Sombari, Pakhyong, etc.; **W Bengal Duars:** Jalpaiguri. Weed of disturbed ground by roadsides, plantations, forest margins and secondary growth, 200–1250m. Flowering almost throughout the year.

Root used as an antidote to poison, leaves used against snake bite, enlargement of the spleen, etc.; fruits eaten (352).

12. S. anguivi Lamarck; *S. indicum* auct. non L., *S. incanum* auct. non L. Nep: *Bin* (38).

Spiny shrub 50cm–3m. Older branches with greyish, ± glabrous bark; younger stems densely greyish stellate-tomentose. Prickles ± numerous on stem, sparse or ± absent on leaves, pedicels and calyces; stem prickles short (2.5–5.5mm), sparsely stellate towards base, usually hooked and often stout (thickest ones 4–5.5mm broad at base, c 0.8–0.9 × as broad as long). Leaves ovate or oblong-ovate, 40–150 × 35–90mm, repand or lobed, acute, base unequally truncate; upper surface dark green, stellate-pubescent (rays 6–8, mostly equal but one usually up to twice as long as others); petiole 5–30mm. Inflorescence an extra-axillary cyme of 4–15 flowers all of which are bisexual. Pedicels to 14mm in flower, 10–20mm in fruit, unarmed or with a few small prickles. Calyx 3–4mm, cup-shaped, densely stellate-pubescent. Corolla 18–20mm across; lobes 11–14 × 4.5–5.5mm, densely stellate-tomentose outside, more sparsely stellate-pubescent inside. Anthers 5–7mm, equal. Ovary and style pilose. Berry globose, 8–10mm diameter, orange.

Bhutan: S – Samchi district (Dorokha (38)), Gaylegphug district (Gaylegphug forests (38)), **C** – Punakha district (Punakha, Wangdu Phodrang, Khuru), Tongsa district (Tama); **Darjeeling:** Darjeeling, Munsang etc.; **Sikkim:** Pemayangtse; **W Bengal Duars:** Jalpaiguri. In thickets and wasteground usually near habitation, and by roadsides, 900–2300m. April–July; fruits May–January.

Cultivated and often escaping. The fruits are used in curries in Sikkim.

Sometimes confused with *S. incanum* L.; the latter species, which has not been recorded from our area except as a misidentification, differs in having longer calyces (6–7mm), inflorescences with a mixture of solitary bisexual flowers and racemes of male flowers, and much larger yellow berries (25–30mm diameter).

13. S. aculeatissimum Jacquin; *S. khasianum* Clarke var. *khasianum*

Stout herb or undershrub 1–1.3m. Stems very densely patent-hirsute with yellowish simple hairs c 1.5mm, lacking stellate bases, and a much sparser indumentum of very short appressed hairs. Prickles fairly numerous, on stems, petioles and leaf veins only, straight or curved (especially below), glabrous. Leaves ovate, 80–220 × 60–170mm, lobed to c halfway with lobes often incised, acute, base slightly unequal, truncate or subcordate, veins very densely patent-hirsute, surfaces much less densely pilose with simple hairs; petiole 30–70mm. Inflorescence a 1–4-flowered lateral extra-axillary or almost axillary raceme; peduncle almost absent; pedicels 12–25mm. Calyx lobes c 6mm, densely hirsute but not prickly. Corolla white, lobes c 9mm, narrowly triangular-lanceolate, almost glabrous outside. Stamens all equal, anthers c 3.5mm. Style equalling tips of anthers. Berry globose, to c 20mm, green mottled white, turning yellow, glabrous. Seeds discoid, somewhat reniform, c 3 × 2.5mm, orange-brown, faintly ridged.

Darjeeling: Kurseong (69); **Sikkim:** Gangtok–Saramsa (69). Roadsides etc., 1500–1600m. April–June.

Records of this species from Sikkim and Darjeeling require confirmation; they may be referable to *S. viarum*. *S. aculeatissimum* seems to be restricted to Khasia, the Brahmaputra Valley and northern Myanmar (Burma) but might occur in easternmost Bhutan; a description, based on material from Khasia and Myanmar, has therefore been provided. The species is a native of C America which is widely cultivated in tropical countries as a source of solasodine.

14. S. viarum Dunal; *S. myriacanthum* auct. non Dunal, *S. khasianum* Clarke var. *chatterjeeanum* Sen Gupta. Nep: *Kachera kanra.*

Spiny herb or undershrub 50–150cm. Stems densely patent-villous with whitish-grey, thin, simple, viscid-glandular hairs (stellate hairs absent). Prickles present on stems, petioles, both leaf surfaces, pedicels and calyces (very few), pale yellow, 6–25mm, mostly acicular but with few shorter hooked prickles on stem only, the acicular ones villous towards base. Leaves broadly ovate to suborbicular, 45–140 × 45–125mm, shallowly or deeply lobed, lobes and apex subobtuse or acute, base almost equal, truncate or subcordate; veins densely patent-villous with simple hairs, upper surface less densely villous, lower surface with scattered 3–4-branched weak, silvery stellate hairs. Flowers extra-axillary, solitary or in groups of up to 3, andromonoecious (at least some apparently male-only) with only basal ones fertile. Pedicels 6–12mm in flower, 12–20mm in fruit, villous and with a few short slender prickles. Calyx c 4.5mm, lobes triangular-lanceolate, c 3mm, villous with very few small prickles. Corolla white, rotate, lobes c 7mm, sparsely villous outside especially towards apex, glabrous inside. Anthers narrowly flask-shaped, 5.5–6.5mm, equal, glabrous, pale yellow. Style c 9mm and slightly exceeding anthers. Berry globose, densely pubescent and green with white mottling when young, yellow when mature. Seeds c 2mm, discoid, strongly flattened, finely reticulate.

Bhutan: S – Phuntsholing district (Torsa River), Chukka district (E of Jumuag), Deothang district (near Deothang), **C** – Punakha district (Lobesa area), Mongar district (Rewan); **Darjeeling:** Labha, Algarah; **W Bengal Duars:** Jalpaiguri. Evergreen oak forest, fields and field margins, shingle and roadsides, 200–2100m. May–October; fruits June–February.

Native of S America (Brazil, Paraguay, Mexico); introduced in India, Nepal, Myanmar (Burma) etc. and grown for medicinal purposes (a source of solasodine). The fruit is used to treat a variety of ailments, including asthma, bronchitis, coughs and colds, toothache and tooth decay, and as an abortifacient (352). Seeds used as a contraceptive, for menstrual complaints and for 'worms in the teeth'; fruits used to treat toothache and wounds (352).

Frequently misidentified as *S. khasianum* (*S. aculeatissimum*), *S. myriacanthum* or *S. melongena*. The following literature records of *S. myriacanthum* may also all belong to *S. viarum*: from **Bhutan: S** – Samchi district (Chongkar and Dorokha (38)), Gaylegphug district (Sureylakha (38), Deothang district (38)), **C** – Thimphu district (Paro Valley (38)), Punakha district (Punakha (38)), Tongsa district (Birti, Rani Camp–Tama (38)); **Darjeeling:** Kurseong (69); **Sikkim:** Gangtok–Saramsa (69).

15. S. barbisetum Nees

Spiny undershrub 30–100cm. Stems sparsely stellate-pubescent (rays 6–7, somewhat unequal, often one prolonged into a setule, all weak and slender). Prickles numerous, present on stems, petioles, main leaf veins and calyces, 2–8mm, straight, tapering from broad base, glabrous. Leaves paired; petiole 45–60mm; blade rather broadly ovate, 75–280 × 40–210mm, sinuate with 4–5 coarse acute lobes per side, acute, base almost equally obtuse, shortly cuneate or subtruncate, surfaces hirsute to villous with long thin setules 1–2mm arising from stellate bases (rays 4–9). Inflorescence an extra-axillary shortly pedunculate scorpioid raceme 5–10cm, of up to 15 flowers; peduncles and pedicels prickly and stellate-setulose; pedicels 4–14mm in flower. Calyx c 5–6.5mm in flower, 10–16mm in fruit, lobes densely stellate-setulose with yellow hairs and with erect prickles which become reflexed in fruit. Corolla mauve, 10–12 × c 15mm; lobes 7–8 × 3mm, ovate-acuminate, setulose outside. Anthers 7–7.5mm, linear-ovoid (abruptly narrowing near middle), pubescent. Ovary glabrous. Style c 7mm, shorter than total length of stamens by 1.5–2mm. Berry globose, c 12mm diameter, glabrous, mostly enveloped by persistent fruiting calyx.

?Bhutan: unlocalised (Griffith collection, K.D. 5911); **Darjeeling:** Riyang, Kalimpong; **Sikkim:** Tista. In thickets of bamboo and secondary growth, 300–1220m. May–October.

Most specimens from our area labelled '*S. barbisetum*' belong, or appear to belong, to *S. griffithii* (see below). *Griffith* K.D. 5911/1, from Assam, is a syntype of *S. griffithii*; the other specimen of this number seen (5911) is *S. barbisetum* but its locality is not precisely known. *Clarke* 11772 from Riyang was noted as having mauve corollas and is therefore assigned to *S. barbisetum*.

16. S. griffithii (Prain) Wu & Huang; *S. barbisetum* Nees var. *griffithii* Prain

Like *S. barbisetum* but with shorter inflorescences (less than 5cm), shorter peduncles (4–5mm, not c 2cm), shorter calyx (c 7mm, not 10–12mm) and white (not mauve) corollas; filaments of stamens slightly longer (c 1.5mm, not less than 1mm); whole plant usually more densely stellate-hairy.

Darjeeling: Kanding; **Darjeeling/Sikkim:** unlocalised (Hooker collections). In thickets of bamboo and secondary growth, 1220–1830m.

The differences between this species and *S. barbisetum* seem weak, especially in our area; however, they were kept separate by Zhang, Lu & D'Arcy (433) and their treatment has been adopted here. Prain originally erected his var. *griffithii* purely on one character, 'all parts densely softly stellate woolly'; this is true for all specimens here assigned to *S. griffithii* but some specimens referred to *S. barbisetum* on the basis of the inflorescence and corolla characters used by Zhang *et al.* also approach *S. griffithii* in the density of their indumentum. Moreover, some specimens here assigned to *S. griffithii* on inflorescence characters appear to have had mauve corollas (e.g. *Hooker s.n.* 'Sikkim Himalaya', 4000–6000ft, K); unfortunately, flower colour notes are present on very few specimens seen of either taxon.

In Assam, the fruit of *S. barbisetum* (sensu lato) is considered edible (352). The information probably refers properly to *S. griffithii.*

17. S. melongena L.; *S. longum* L., *S. insanum* L., *S. ovigerum* L., *S. esculentum* Dunal. Eng: *Egg-plant.*

Armed or unarmed large suffruticose annual herb 60–250cm; stems with 2 rows of rather sparse stellate hairs. Prickles, when present, 4–9mm, slender but stout, straight or slightly curved, glabrous. Leaves ovate, 40–150 × 30–100cm, subentire to sinuate or obtusely lobed, acute, base ± unequal, obtuse to shallowly cordate, upper surface stellate-pubescent, more densely stellate-tomentose and greyish beneath; petiole 5–40mm. Inflorescences often paired, one peduncle bearing a single perfect flower and the other a short raceme of male-sterile flowers; peduncles mostly extra-axillary, 20–25mm at first, lengthening and to 50mm in fruit; pedicels 5–20mm. Calyx lobes 6–12mm in flower, 16–25mm in fruit, elliptic or oblong, acute. Corolla blue-violet, to c 15mm long and c 35mm diameter, lobes to c 10mm, stellate-pubescent outside and sparsely pubescent on mid-vein inside. Anthers to c 7mm. Style stellate-pubescent or glabrous. Berry globose to ellipsoid, often large and elongate especially in cultivation, usually some shade of dark purple, 25–230mm, edible.

Bhutan: S & C – cultivated occasionally on a small scale in the warmer parts, mainly in gardens (J.R.I. Wood, pers. comm.).

No specimens seen from our area; the description is based on Indian specimens. An edible vegetable. The fruit stalks are used for fistula and piles (352).

18. S. virginianum L.; *S. surattense* Burman f., *S. xanthocarpum* Schrader & Wendland, *S. jacquinii* Willdenow, *S. diffusum* Roxb.

Very spiny diffuse prostrate herb 30–120cm across. Stems sparsely stellate-pubescent to almost glabrous (excluding prickles). Prickles very numerous, present on all parts except corolla, to 18mm, straight, straw-yellow, glabrous. Leaves ovate or elliptic-oblong, 30–80 × 25–50mm, sinuate or deeply lobed (lobes 3–4 per side, often themselves toothed or lobed), sparsely stellate-pubescent above (rays 8–9, slightly unequal) and less so beneath, acute, base unequally truncate. Inflorescence a 2–4-flowered racemose pedunculate cyme; peduncles 10–20mm; pedicels 5–8mm in flower, 10–15mm in fruit. Calyx c 5mm, lobes acute. Corolla mauve or violet, 20–28mm across; lobes 10–12mm, ovate-triangular, stellate-pubescent in a broad band outside and a narrow line inside along mid-vein, with simple hairs on inner surface and margins. Anthers linear, 7–8mm, yellow. Berry globose, 15–20mm, green turning yellow. Seeds discoid, smooth or faintly reticulate.

Darjeeling/Sikkim: unlocalised (Treutler collection).

Known in our area at present only from a single unlocalised collection; *Gamble* 3401A, labelled 'Myanoung, B.B.', is more likely to be from 'British Burma' than Darjeeling ('British Bhutan'). Very common in NW Himalaya and extending to Myanmar (Burma) and Bangladesh. In these territories it is a weed of open, disturbed habitats up to c 1300m and flowers throughout much of the year; various parts are used medicinally. The roots and fruits used for asthma and other chest complaints and for toothache (352).

10. LYCIANTHES (Dunal) Hassler

Unarmed shrubs or herbs. Leaves simple, usually entire, upper ones often in false pairs and unequal. Inflorescences axillary; peduncle absent or obsolete; pedicels 1–8 together forming ± sessile umbels or fascicles. Calyx ± cup-shaped, 10-ribbed, with 5–10 usually minute teeth (when 10, usually in 2 series) which arise below the apex. Corolla stellate or rotate; lobes glabrous outside. Anthers opening by oblique, introrse terminal pores. Fruit a berry. Seeds numerous.

Although the American species of this large genus are very distinct from *Solanum*, the Asiatic members treated here much more closely resemble certain species of *Solanum* (especially *S. spirale*) and could possibly be transferred, although they were retained in *Lycianthes* by Zhang *et al.* (433). The most reliable distinction is the axillary (not extra-axillary) flowers.

1. Procumbent creeping herb with ascending branches ... **3. L. lysimachioides**
+ Small shrub 2

2. Calyx teeth 5–7, linear-triangular, c 2mm .... **1. L. crassipetalum**
+ Calyx teeth 10, subulate, 3.5–6mm (if teeth 10 and only 1.5–2mm, cf. *L. biflora*) .... **2. L. macrodon**

1. L. crassipetalum (Wall.) R.R. Mill; *Solanum crassipetalum* Wall., *S. pachypetalum* Sprengel, *L. pachypetala* (Sprengel) Bitter, *L. laevis* (Dunal) Bitter subsp. *crassipetala* (Wall.) Deb. Fig. 90g–i.

Shrub 60–300cm. Branches sparsely and shortly appressed-pubescent or ± glabrous, somewhat flexuous. Leaves elliptic-lanceolate, 70–220 × 35–80mm, abruptly narrow-acuminate into a drip-tip with some leaves usually obtuse, base unequally cuneate, upper surface usually nearly glabrous (hairs few, multicellular), lower surface appressed-pubescent on veins, with numerous minute whitish pustules but otherwise glabrous on lamina. Inflorescences axillary; pedicels (1–)3–8 together, 6–16mm in flower, thickened above and 10–22mm in fruit. Calyx 2–4mm; teeth 5–7, spreading, linear-triangular. Corolla pink, purple or whitish; lobes oblong-triangular, c 6mm, acute. Anthers c 3.5mm, oblong. Berry red, 6–11mm diameter. Seeds pale buff, triangular, c 2.3 × 1.9mm, very obscurely reticulate.

Bhutan: S – Samchi district (Changda Chungli Hills (38)), Gaylegphug district (Gaylegphug (38)), **C** – Mongar district (Kuru Chu Valley near Mongar); **Darjeeling:** Kurseong, Balasun, Dumsong, Rishap, Mongpu, Kodabari; **Sikkim:** Upper Rammam Bridge, Rishi etc. Dry or damp hillsides and forest gullies, and amongst shrubs in village terraces, (90–)900–1830m. July–October; fruits August–November or later.

The leaves are cooked and eaten in Sikkim.

L. laevis (Dunal) Bitter is a mainly S Indian taxon which does not occur in the Himalaya.

2. L. macrodon (Nees) Bitter; *Solanum macrodon* Nees, *L. biflora* (Loureiro) Bitter subsp. *macrodon* (Nees) Deb

Shrub (sometimes scrambling), 60cm–2m or more; young stems usually hispidulous with spreading, often fulvous hairs mostly exceeding 0.5mm, sometimes almost glabrous. Leaves elliptic-lanceolate to ovate-lanceolate, 30–130 × 20–70mm, acuminate, base cuneate, upper surface sparsely pilose with yellowish, moniliform, jointed hairs, lower surface with spreading hairs (usually exceeding 0.5mm) on veins. Flowers axillary; pedicels (1–)2–4(–6) together, 4–11mm at anthesis and 10–22mm in fruit, ± densely patent hispidulous with jointed hairs. Calyx tube 1.5–2mm, green with 10 pinkish ribs; teeth 10, subulate, 3.5–6mm, alternate ones often shorter. Corolla 7–12mm, purple, pink or whitish, each lobe with 2 green gland-spots near base; lobes acute, pubescent or almost glabrous outside. Anthers yellow, with long oblong terminal pores. Berry red, globose, 6–9.5mm diameter.

Bhutan: **S** – Deothang district (near Khaling), **C** – Tashigang district (Diri Chu, Tashiyangsi Chu); **Darjeeling:** Darjeeling, Kurseong, Mongpu etc.; **Sikkim:** Panding, Yoksam, Lachen etc. Amongst shrubs in warm broad-leaved forest and secondary growth, 1220–2440m. May–August.

Shoots eaten.

Deb (334) treated this as a subspecies of *L. biflora* (Loureiro) Bitter but gave no supporting reasons. The latter species is reputed to occur in our area (cf. 135) but no specimens or literature records have been seen to substantiate this; *Solanum decemdentatum* Roxb. has been treated as a synonym of *L. biflora* (433) but all material labelled '*S. decemdentatum*' from our area has turned out to be *L. macrodon. L. biflora* differs from *L. macrodon* principally by its shorter calyx teeth c 1.5–2mm, and shorter, usually less dense hairs on pedicels and calyx (but the indumentum is very variable). Subspecific rank for *L. macrodon* may indeed be appropriate but further study of the taxa is needed before a convincing argument can be made. It was treated as a full species in Zhang *et al.* (433) and that ranking is followed here.

3. L. lysimachioides (Wall.) Bitter; *Solanum lysimachioides* Wall., *S. macrodon* Nees var. *lysimachioides* (Wall.) Clarke, *L. macrodon* (Nees) Bitter subsp. *lysimachioides* (Wall.) Deb

Gregarious procumbent creeping perennial herb with ascending stems 15–30cm; stems shortly crispate-pilose. Leaves in false pairs, shortly petiolate (petiole 3–25mm); lamina ovate, 8–55 × 6–30mm, acute or bluntly acuminate, base cuneate, sparsely pilose above (hairs stoutish, multicellular) and with more slender hairs on veins beneath. Flowers solitary in axils; pedicels 3–10mm, straight, suberect at first but deflexed later. Calyx tube cup-shaped, 1–2.5mm; lobes 10, linear-triangular, spreading, in 2 slightly unequal series. Corolla white or pale yellow, 10–20mm across, stellate; lobes 8–11 × 2–4mm, narrowly lanceolate, minutely puberulent on margins. Anthers 3–4mm, narrowly oblong, with large oblique terminal pores. Style 6–7mm. Berry scarlet, soft and fleshy, with pointed apex. Seeds white, suborbicular with one side nearly straight, smooth.

Darjeeling: Darjeeling, Rishap; **Sikkim:** Talung Chhu, Tholung Valley. Wet forest, field margins etc., 900–2440m. May–August.

A variable species; however, material from our area all belongs to var. **lysimachioides**.

Has been confused with *Lysimachia evalvis* Wall. (Primulaceae) which has very similar procumbent stoloniferous habit and superficially similar flowers. *L. lysimachioides* can be distinguished from *L. evalvis* by the shorter, straight (not curved) pedicels which are suberect at first, the calyx with 10 short spreading teeth at top (not deeply 5-lobed) and by the anthers lacking sagittate bases. The flowers of *L. evalvis* are bright yellow, not white or pale yellow as in *L. lysimachioides*.

11. LYCOPERSICON Miller

Unarmed, aromatic, annual or perennial herbs with soft, sprawling stems. Leaves pinnately lobed with interstitial leaflets between main leaflets and on petiole; leaflets petiolulate; minor leaves present at base of petiole. Inflorescences lateral, pedunculate, racemose. Flowers 5–6-merous (in Himalaya). Calyx divided almost to base, lobes lanceolate, reflexed in fruit. Corolla rotate-stellate. Anthers coherent into a tube by a mesh of lateral hairs, densely pubescent within tube, opening by introrse lateral slits. Ovary bilocular; placentae axile, often branching in cultivated plants to divide fruit into several loculi. Fruit an edible, juicy berry.

1. L. esculentum Miller; *Solanum lycopersicum* L., *L. lycopersicum* (L.) Karsten. Eng: *Tomato.*

Sprawling annual herb to 1.5m. Stems weak, densely but very shortly viscid-pubescent with scattered longer eglandular simple hairs. Leaves ovate in outline, mostly 90–190 × 60–100mm (including petiole), imparipinnate with usually 3 principal pairs of leaflets and numerous much smaller interstitial leaflets, all of which are petiolulate; leaflets acute, base unequally obtuse to truncate, very shortly pubescent. Minor leaves often conspicuous. Cymes 2–5-flowered; pedicels densely viscid-pubescent. Calyx lobes 5, narrowly triangular-lanceolate, Corolla yellow, 9–15(–20)mm diameter; lobes 5(–6), triangular, 5–6.5mm, acute, pubescent outside. Anthers deeper yellow than corolla lobes, c 4.5mm. Berry to 8cm diameter, scarlet when mature. Seeds 2–3mm, discoid with broad wing.

Bhutan: C – Thimphu district (Motithang) and Punakha district (Wangdu Phodrang). Cultivated, c 2400m. April–September.

The tomato, a native of C and S America, is commercially grown in the Wangdu Valley for sale in June, and is also grown widely in Bhutan as a kitchen garden plant (J.R.I. Wood, pers. comm.). The commonest variety grown in Bhutan has ovoid, not globose, berries. The description applies to specimens seen from Nepal, Bhutan and other parts of the Himalaya, all of which have small flowers. Plants cultivated in other parts of the world can have flowers 6–9-merous, up to 9mm with anthers up to 14mm.

12. CYPHOMANDRA Sendtner

Unarmed, softly pubescent shrub or small tree with a single trunk which divides trichotomously (rarely dichotomously) in flowering region. Indumentum of unequal eglandular, acute and glandular, finger-like hairs, and shortly stipitate glands which release a pungent odour. Leaves petiolate (petiole sometimes amplexicaul), large, entire. Inflorescence a pedunculate scorpioid cyme or cincinnus, often arising at a stem trichotomy (or dichotomy). Pedicels secund, articulating at base, leaving scars when shed. Flowers 5-merous. Calyx campanulate. Corolla stellate or stellate-campanulate. Filaments free, or connate only at base; anthers attenuate towards apex, opening by 2 small terminal pores, and sometimes finally also longitudinally; connective gibbous at base, forming a hump. Fruit a large, ellipsoid, pendent, fleshy, fragrant, edible berry. Seeds reniform, flattened, testa reticulate often with hyaline margin.

A poorly delimited, variable genus of 50–60 species of the mountains of S America, whose distinction from *Solanum* (mainly based on stamen characters) requires clarification. The above generic description applies to the one cultivated species found in our area but does not cover the entire range of variation found in the genus.

1. C. betacea (Cavanilles) Sendtner; *C. crassicaulis* (Ortega) Kuntze. Eng: *Tree tomato*.

Large shrub or small tree c 2m. Branches densely subsessile-glandular and minutely pubescent. Leaves glossy, 75–140 × 25–75mm or larger, abruptly short-acuminate, base cordate, minutely puberulent beneath; petiole 35–40mm. Cymes up to 15-flowered; pedicels 15–20mm at anthesis. Calyx c 3mm, lobes broadly ovate, c 2 × 3mm, obtuse. Corolla pinkish-white, lobes c 11 × 3mm, pubescent along dorsal mid-line. Stamens c 5mm. Berry pendent, ellipsoid-pyriform, turning dull red or orange, 50–70mm.

Bhutan: S – Chukka district (Raidak Valley near Tabji Khola); **Darjeeling:** Birch Hill (218), Singtan (333); **Sikkim:** unlocalised (218, 333). Cultivated on terraced subtropical and warm temperate hillsides, 1250–1900m. August–February; fruits from September onwards.

Native of S America, cultivated as a food crop in Bhutan, Sikkim and Darjeeling for its edible fruits.

13. MANDRAGORA L.

Caulescent perennial with long, often forked or branched massive tap-root. Stem short at anthesis, rapidly elongating later; lower part with a few small scale-like leaves, terminating in a rosette of leaves, upper part not evident at anthesis but rapidly lengthening afterwards, becoming leafy and much over-topping fruiting pedicels. Leaves of flowering rosette ± sessile, those of post-anthesis stem petiolate. Pedicels extra-axillary, erect at anthesis, deflexed later. Flowers 5–6-merous, nodding. Calyx and corolla subequal (in plants from our area). Calyx campanulate, somewhat unequally lobed. Corolla yellow or purple, campanulate, lobed to 1/5–1/2, lobes obtuse. Filaments with lower 1/3–1/2 dilated and shortly pilose, upper part slender and glabrous. Anthers attached near their base. Ovary massive, hemispherical. Style short, with 2-lobed stigma. Fruit a berry loosely enclosed in persistent calyx. Seeds many, reniform with short beak.

The generic description applies only to the Bhutan species, which differs markedly from the European members of the genus. A higher rank for it may be appropriate although it was retained in *Mandragora* by Ungricht (418).

1. M. caulescens Clarke

Stems 1–23cm from top of root to leaf rosette and base of pedicels; scale leaves 7–20 × 5–8mm. Leaves obovate-elliptic, 15–180 × 5–60mm, obtuse or subacute, base attenuate and sessile (rosette leaves) or petiolate (later cauline leaves), softly pilose especially at first, usually glabrescent later. Pedicels 3–12cm at anthesis. Flowers strongly but pleasantly scented, nodding. Calyx 10–23mm, green or greenish-purple. Corolla 8–20mm, yellow or purplish but very variable in colour, with 5(–6) rounded lobes. Anthers 2.5–5mm, white or yellow. Berry rarely collected, usually 10–20mm diameter.

1. Corolla yellowish or greenish-yellow, c 10mm .... **a.** subsp. **flavida**
\+ Corolla usually purple or purplish-green (occasionally yellowish-black, if so then more than 15mm), 8–20mm 2

2. Calyx 14–23mm; corolla 16–20mm; anthers (3.5–)4–5mm **b.** subsp. **purpurascens**
\+ Calyx 10–12mm; corolla 8–10mm; anthers 2.5–3.5mm .. **c.** subsp. **caulescens**

a. subsp. **flavida** Grierson & Long. Fig. 91c.

Flowers always 5-merous, c 10mm; corolla pale yellow, divided for ¼–⅓ into rounded lobes. Filaments 3.5–5.5mm, noticeably expanded below; anthers 2–3mm.

Bhutan: N – Upper Bumthang Chu district (Kantanang, Pangotang); **Darjeeling/Nepal:** Phalut–Sandakphu; **Sikkim:** Zongri; **?Chumbi:** Jhangkhar Chu Valley. Open hillsides (sometimes among shrubs), above fir zone, often in loose soil and/or in wet situations, 3500–4100m. May–July.

b. subsp. **purpurascens** Grierson & Long

Flowers 5–6-merous, 16–20mm; corolla very deep purple, dark yellowish-black or dark livid green, divided for ⅕–½. Filaments 7–10mm, only slightly expanded below; anthers (3.5–)4–5mm.

Bhutan: C – Tongsa district (Dungshinggang, Pobjeka), **N** – Upper Kulong Chu district (Shingbe); **Darjeeling:** Phalut, Sandakphu; **Sikkim:** Lachung, Gnatong. Stony alpine meadows, only occasionally in boggy ground, 3353–4877m. April–July.

c. subsp. **caulescens**

Flowers 5–6-merous; corolla 8–10mm, blood purple, slate-green or purplish-black, very variable in colour, divided to c ½. Filaments 6.5–10mm, noticeably expanded below; anthers 2.5–3.5mm.

Bhutan: C – Ha district (Ha, Tare La), Thimphu district (Chelai La, Pajoding, Paro Chu), Punakha district (Byasu La), **N** – Upper Mo Chu district (Laum Thang); **Darjeeling:** Tanglu; **Darjeeling/Nepal:** Sandakphu; **Darjeeling/Sikkim:** Phalut; **Sikkim:** Kupup, Zongri, Meguthang, Gamothang etc.; **Chumbi:** Dothag,

Chulong. Among dwarf rhododendrons and juniper bushes above fir zone, and on open stony alpine hillsides, 3700–4570m. May–June.

Intermediate between subsp. *purpurascens* and subsp. *flavida*.

14. IOCHROMA Bentham

Unarmed shrubs. Leaves entire, petiolate. Flowers in pendent cymose clusters; pedicels long. Calyx cup-shaped, unequally 5-toothed. Corolla scarlet, yellow or shades of purple and blue, tubular, ventricose at middle and slightly constricted at mouth, tube long and somewhat curved, 2–6 × calyx, limb short, subcampanulate, lobes acute. Stamens 5, included or exserted; filaments adnate to base of tube but free from below middle; anthers bilocular, longitudinally dehiscent. Ovary obovoid, bilocular; style filiform, thickened at apex. Fruit an ovoid berry, included in inflated calyx.

A small S American genus, several of whose species are cultivated as ornamentals elsewhere.

1. I. coccinea Scheidweiler

Shrub 1.5–5m. Young branches shortly pubescent, older ones ± glabrous. Petiole 15–30mm, shortly pubescent. Leaves elliptic, 40–100 × 15–45mm, pale green, shortly acuminate, base cuneate to attenuate, margin entire, both surfaces sparsely and finely pubescent. Flowers in cymes of 4–12. Pedicels 12–18mm, slender, shortly pubescent. Calyx cup-shaped, small (c 3mm), teeth obtuse, less than 1mm. Corolla salmon-scarlet outside, paler inside except for a band of deeper red near top of tube, 25–35 × 5–6mm, slightly curved, long-tubular, broadening slightly above, pubescent outside. Stamens shortly exserted; filaments deep cream; anthers coffee-coloured, 2.5–3.5 × c 0.5mm. Style creamy-white; stigma green. Not fruiting in our area.

Darjeeling: Observatory Hill (333). Gardens and roadside plantations, c 2100m. May–June.

Native of Mexico; cultivated in our area but not escaping (333).

I. cyaneum (Lindley) M.L. Green (*I. tubulosum* Bentham), with deep purplish-blue flowers, is commonly cultivated in India but has not so far been recorded from Bhutan or Sikkim.

15. DATURA L.

Annual or perennial herbs or soft-wooded shrubs. Leaves alternate. Flowers solitary, erect. Calyx tubular, elongate, circumscissile at base leaving a disc under the fruit, with 5 teeth at apex. Corolla large (in Bhutan species), infundibular or tubular, ± conspicuously cuspidate. Stamens 5, attached near middle of corolla tube or towards base; anthers oblong, slightly curved, basifixed, free. Style filiform, stigma 2-lobed. Ovary bilocular above, often 4-locular below due to presence of false septum. Fruit a globose or ovoid capsule, dehiscing irregularly or by 4 valves; pericarp spiny or tuberculate. Seeds laterally compressed, numerous.

1. Soft-stemmed shrub 1–2.5m; leaves entire, repand or angulate with very asymmetrical base; capsule ± pendulous, dehiscing irregularly, largest spines or tubercles on pericarp not more than 4mm long and 1.5mm wide at base .... **1. D. metel**
\+ Erect annual herb 30–100cm; leaves always coarsely acuminately toothed, with nearly symmetrical base; capsule erect, dehiscing by 4 valves, largest spines on pericarp 6–10mm long and at least 2mm wide at base **2. D. stramonium**

1. D. metel L.; *D. fastuosa* L., *D. alba* Nees. Nep: *Kalo-daduna.*

Erect, soft-stemmed shrub 1–2.5m. Branches ± zigzag, green, black or tinged with red or purple, minutely pubescent to almost glabrous. Leaves with foetid, acrid smell when crushed; petiole 30–150mm, ⅓–½ × lamina; lamina elliptic to broadly ovate, 60–250 × 30–230mm, acute or shortly acuminate, base obtuse, very asymmetrical, margin entire, repand or angulate. Flowers solitary in axils, erect. Pedicels 5–20mm, pubescent. Calyx tubular, 60–80mm; tube 45–65mm, minutely pubescent; teeth unequal, 8–12mm, triangular-acuminate. Corolla white (sometimes with veins streaked red or purple) or purple; tube 95–130mm, infundibular, ± pubescent outside; limb 55–80mm diameter, 5-cuspidate. Fruit a ± pendulous spherical capsule, dehiscing irregularly; pericarp with blunt spines or tubercles to 4mm long and up to 1(–1.5)mm wide at base. Seeds brown, c 5mm.

Darjeeling: Siliguri, Kalimpong, Darjeeling; **Sikkim:** Great Rangit Valley, Jorethang (188); **W Bengal Duars:** Judgal, Jalpaiguri. Path-sides, wasteground and cultivated land, 100–1220m. April–October.

Narcotic. Many medicinal uses claimed, including for dropsy, epilepsy, leprosy, headache, hydrocephalus, mumps, smallpox, syphilis and madness in humans and animals (352).

Native of the W Indies; introduced to India (from where the type was described) and tropical Asia many centuries ago and now thoroughly naturalised. Very variable. Plants with red-tinged stems also have corollas streaked red or purple on the veins; the name *D. fastuosa* L. applies to these, which seem to be the commonest variant in Sikkim and Darjeeling. Experiments have shown that the four main variants segregate following Mendelian patterns and warrant at most varietal or cultivar rank.

2. D. stramonium L.; *D. tatula* L. Eng: *Thorn apple.* Fig. 91a–b.

Erect branched annual, sometimes becoming (soft) woody at base. Leaves with petiole 10–60(–100)mm, c ½ × lamina; lamina ovate, rhomboid or elliptic, 70–210 × 35–150mm, acuminate, base cuneate, almost symmetrical, margin deeply sinuate-dentate, teeth sharply acuminate; surfaces ± glabrous. Flowers solitary, axillary, erect. Pedicels 5–10mm. Calyx tubular, 40–45mm, equally 5-toothed (teeth 6–8mm), circumscissile. Corolla white or less commonly purplish, narrowly infundibular-cylindrical but dilated near mouth, 60–100mm; limb 60–70mm diameter, with 5 narrow cusps. Capsule ovoid, 50–70 × 30–50mm, erect; pericarp very sharply spiny, longer spines 6–10mm long and more than 2mm wide at base. Seeds black, c 3 × 2.7mm, slightly reniform, minutely pitted.

Bhutan: S – Sarbhang district (Sarbhang), **C** –Thimphu district (Mirichoma, Dotena, Thimphu), Tongsa district (Dakpai), Tashigang district (Pam Camp); **Darjeeling:** Kalimpong. Fields, stony roadsides, camps and villages, 1200–2440m. March–October (possibly almost throughout the year).

All parts of the plant are narcotic, especially the seeds, which have a stupefying effect and can be fatal.

Plants with purple flowers have been called *D. tatula* L.

16. BRUGMANSIA Persoon

Small trees or large shrubs with very soft wood, so appearing herbaceous, suckering to form large clones. Leaves alternate. Inflorescence predominantly monochasial. Flowers solitary, pendulous at anthesis (pedicels sometimes erect in bud), remaining open during the day. Calyx tubular, elongate, with 5 teeth (Bhutan species) or spathaceous and splitting down one side, persistent, not circumscissile. Corolla very large, infundibular or tubular, cuspidate. Stamens 5, alternating with corolla lobes; anthers basifixed, linear, coherent into a narrow cylindrical tube, sometimes finally becoming free. Style filiform; stigma 2-lobed. Ovary bilocular at apex, bicarpellate; false septa absent. Fruit a fusiform berry-like capsule; pericarp unarmed, smooth. Seeds lacking a caruncle, usually with corky seed coat.

Brugmansia is sometimes included in *Datura* (e.g. 381) but is separated by Symon & Haegi (414).

1. B. suaveolens (Willdenow) Berchtold & Presl; *Datura suaveolens* Willdenow. Dz: *Dhokrey phul.*

Shrub or small tree, 1–6m; branches brittle, young stems minutely pubescent. Leaves ovate-elliptic, 90–250 × 50–120mm, acute or shortly acuminate, base cuneate, subglabrous except for minutely pubescent veins; petiole densely and minutely puberulent, 30–50mm. Flowers pendulous from horizontal branches,

FIG. 91. **Solanaceae.** a–b, *Datura stramonium*: a, flowering shoot (× ½); b, fruit (× ⅔). c, *Mandragora caulescens* subsp. *flavida*: habit (× ⅔). d–f, *Physalis divaricata*: d, flowering shoot (× ½); e, flower (× 5); f, fruit (× 1). g, *Anisodus luridus*: flowering shoot (× ⅔). Drawn by Glenn Rodrigues.

a
b
c
d
e
f
g

fragrant or not. Calyx tube 65–115mm, subglabrous; lobes 5, 10–20 × 8–10mm. Corolla cream at first, turning white, 20–30cm; tube 14–27cm, with narrowly cylindrical lower part exserted from calyx, abruptly becoming infundibular above; limb 11–17cm diameter, pleated, the 5 main veins ending in projecting recurved cusps. Anthers c 35mm, coherent. Anthers 25–35mm, coherent into a narrow, cylindrical tube. Fruit fusiform, pendulous at an angle, c 90 × 30mm.

Bhutan: S – Samchi district (Chepuwa Khola), Gaylegphug district (Surey); **Darjeeling:** St Mary's, Kariabasti, Kurseong (all 97a); **Sikkim:** Gangtok (field note). Ravines in subtropical forest and by streams, near villages, always as an escape from cultivation, 500–1800m. February–September; fruiting September–April.

The species, which is narcotic, is used as an ornamental hedge plant in Bhutan, Nepal and adjacent mountainous regions of India; it is native to coastal rain forests of SE Brazil.

17. CESTRUM L.

Unarmed shrubs with slender twigs. Leaves simple, entire, alternate, petiolate; stipule-like leaves present or absent. Inflorescences axillary (often appearing terminal), racemose, paniculate, cymose or corymbose. Pedicels very short or absent; bracteoles present. Calyx cyathiform or campanulate (Himalayan species), divided less than halfway into 5 sometimes slightly unequal lobes. Corolla infundibular or salverform, tube obconical, sometimes constricted below limb; limb of 5 short, often recurved or spreading lobes which are variously pubescent or glabrous. Stamens 5, ± equal; filaments adnate to corolla tube, often pubescent and/or geniculate or appendaged. Fruit a fleshy or juicy berry.

A large S American genus of over 200 species, of which five are commonly cultivated in the Himalayas and elsewhere.

1. Corolla pale yellowish-green 2
+ Corolla orange, scarlet or purplish-pink 3

2. Leaves oblong-elliptic to lanceolate, c 2½ × as long as wide; flowers fragrant by night, corolla tube narrow, 0.5–1mm wide at top of calyx, 0.9–1.2mm wide at middle, 2.5–3mm wide below limb .... **4. C. nocturnum**
+ Leaves narrowly ovate-lanceolate or oblong, c 4 × as long as wide; flowers not scented but plant foetid, corolla tube broader, c 1–1.2mm wide at top of calyx tube, 2–2.5m wide at middle, 3.5–4mm wide below limb **5. C. parqui**

3. Adult leaves and young twigs glabrous; corolla orange .. **1. C. aurantiacum**
+ Adult leaves hairy; young twigs tomentose-pubescent; corolla scarlet or purplish-pink 4

4. Corolla deep rose pink or purplish-pink .... **2. C. elegans**
\+ Corolla scarlet .... **3. C. fasciculatum**

1. C. aurantiacum Lindley

Shrub 1.8–2.4m. Branches brownish-grey, sulcate, glabrous. Petioles 15–35mm, glabrous. Leaves elliptic, lanceolate or ovate, 50–130 × 28–70mm, shortly acuminate, base obtuse to cuneate, surfaces minutely puberulous when young but adult leaves glabrous. Stipule-like leaves lanceolate, 6–9 × c 2mm, brownish-tomentose. Inflorescences appearing terminal, racemose, (4–)7–13cm, 7–25-flowered, rather congested; individual panicles 7–8cm. Calyx tube 5.5–6.5mm, glabrous outside, puberulent within; lobes 1.5–3mm, subulate, tomentose at base. Corolla orange, (18–)22–25(–27)mm; tube 16–21mm, infundibular, lower half narrow, gradually dilated above and not constricted below limb, glabrous outside; lobes 2.7–3.5mm, acute, becoming reflexed, shortly pubescent outside and inside. Berry unknown.

Darjeeling: Bhotia Basti, St Mary's (97a), Kurseong (97a); **Sikkim:** Gangtok. 1500–2100m. August–February; fruiting September–April.

Native of C America; cultivated for ornament elsewhere, now naturalised.

2. C. elegans (Neumann) Schlechtendal; *Habrothamnus elegans* Neumann

Shrub 1.5–3.5m. Branches greyish-brown, striate; twigs densely glandular pubescent-tomentose to hirsute. Petioles 3–13mm. Leaves elliptic or lanceolate-elliptic, 35–105 × 13–45mm, shortly acuminate, base obtuse, sparingly pilose above, shortly hirsute beneath especially on veins. Stipule-like leaves absent. Inflorescences corymbose, appearing terminal, 35–85mm, 11–27-flowered, composed of groups of 6–7-flowered cymes. Bracteoles ovate-lanceolate, 3–6 × 1–2mm, hirsute. Pedicels 1–2mm. Calyx obconical, subcampanulate, 4.5–5.5mm; lobes ovate-triangular, c 1.5mm, pubescent within. Corolla deep rose-pink or pinkish-purple; tube clavate, ventricose, 15–16mm, constricted below limb; lobes c 2mm, spreading at anthesis, glabrous outside, papillose-tomentose within. Berry reddish-purple, obovoid, 10–13 × 6–8mm.

Darjeeling: Darjeeling, Kurseong–Balasan, St Mary's. September–February; fruiting December–March.

Native of Mexico; cultivated for ornament elsewhere.

3. C. fasciculatum Miers

Shrub 70–360cm. Branches brownish-grey, twigs blackish-tomentose when young. Petioles 5–22mm. Leaves ovate, elliptic to almost orbicular, 20–130 × 15–60mm, acute or shortly acuminate, base subtruncate, obtuse or shortly cuneate, shortly pilose above, lower surface puberulent and very densely spreading white-villous on veins. Stipule-like leaves absent. Inflorescences appearing terminal, condensed, 3.5–5cm, composed of 2–3 5–11-flowered cymes forming a subglobose head. Bracteoles elliptic, 6–8 × 2–3mm. Pedicels c 1mm. Calyx campanulate, 5–6.5mm, pubescent and ± heavily tinged purple or reddish.

Corolla scarlet, 18–21mm; tube clavate-infundibular, 15–17mm, constricted below limb; lobes c 2mm, white-tomentose on margin. Berry ellipsoid, c 9 × 6mm.

Darjeeling: Darjeeling (218). The Darjeeling record has not been verified; it may be a misidentification of *C. elegans*, as the corolla was described as purplish-red.

Native of S America.

4. C. nocturnum L. Eng: *Queen-of-the-night.*

Shrub (1.3–)2–3(–4)m. Branches arching, glabrous. Petioles 10–16mm, glabrous. Leaves oblong-elliptic to lanceolate, 50–120(–140) × 20–52mm, acute or shortly acuminate, base obtuse or shortly cuneate, young leaves shortly pilose, adult ones glabrous. Stipule-like leaves absent. Inflorescences clearly axillary, long-pedunculate, paniculate. Bracteoles linear, 2–4mm. Pedicels 0.5–1.5mm. Calyx campanulate, (2.2–)2.5–3.2(–3.5)mm, glabrous. Corolla pale yellow-green, 18–25mm, very strongly fragrant by night; tube narrowly infundibular, 0.5–1mm wide at top of calyx, 2.8–3.5mm wide at base of limb; lobes 2.5–3 × 0.8–1mm, puberulent outside and tomentose on margins. Berry white, 8–10 × 4.5–6.5mm. Seeds c 8, c 3mm, rectangular.

Darjeeling: Mongpu, Kurseong (97a), Karibasti (97a). Cultivated for ornament and for its night scent, 900–1525m, and naturalised along hedges, etc. August–February; fruiting December–March.

Native of W Indies and C America.

5. C. parqui L'Héritier

Very foetid shrub 1–3m. Branches pale greyish-brown, glabrous; twigs terete, virgate, flexuous, purplish near tips, glabrous. Leaves alternately approximate; petioles 5–10mm, glabrous; lamina narrowly ovate-lanceolate or oblong, 35–105(–130) × 8–24(–40)mm, acuminate, base cuneate-attenuate, glabrous, with prominent thickened margin. Inflorescences appearing terminal, racemose or paniculate, pyramidal, 35–90mm. Bracteoles linear, c 6mm. Pedicels absent or to 0.5mm. Calyx campanulate, 3.5–5mm, glabrous or sparsely tomentose. Corolla yellow-green, with broad dark stripe on back of lobes, 16–18mm; tube narrowly cylindrical-infundibular, scarcely constricted below limb, glabrous outside; lobes of limb 2.5–3.5mm, densely tomentose outside. Berry brownish-violet, ovoid, 8–10 × 6mm.

Native of temperate S America (Chile, Bolivia, Argentina, Uruguay, Paraguay and Brazil); cultivated in Nepal, c 1400m, where it flowers from July to October. It may also be cultivated in our area but so far there are no records.

18. NICOTIANA L.

Erect, viscid pubescent herbs (our species). Leaves entire or sinuate. Inflorescence a terminal panicle or of subterminal racemes. Calyx ovoid or tubular, 5-fid. Corolla infundibular; tube sometimes differentiated into 2–3 ± distinct sections, the uppermost termed the *throat*; lobes 5, ± spreading, often acuminate. Stamens 5, attached in lower part of corolla tube, usually unequal or ± didynamous with a shorter fifth stamen; filaments filiform, anthers ovate, longitudinally dehiscent. Ovary bilocular; style filiform, stigma shortly bifid. Fruit a 2- or partly 4-celled capsule, 2-valved to middle, valves often again splitting. Seeds small, very numerous.

All three species originated in America, but widely cultivated elsewhere.

1. Branches of inflorescence almost as long as its main axis; stems slender, wiry; corolla tube scarcely or not differentiated into 2 or more parts **3. N. plumbaginifolia**
+ Branches of inflorescence, when present, much shorter than its main axis; stems stout; corolla tube differentiated into 2–3 distinct parts 2

2. All leaves shortly petiolate, base cordate or obtuse; corolla 12–17mm excluding limb, greenish-yellow, tube differentiated into short cylindrical lower part and much broader campanulate-obconical throat .... **1. N. rustica**
+ At least some cauline leaves sessile, base attenuate or cuneate; corolla (30–)35–55mm excluding limb; limb usually pink or whitish, tube greenish-white, differentiated into lower narrowly cylindrical, middle more broadly cylindrical part and upper cup-shaped throat .... **2. N. tabacum**

1. N. rustica L.

Coarse annual 50–150cm. Stems and branches ± stout, densely white glandular-pubescent. Leaves fleshy, all petiolate; petioles not winged, basal ones to 80mm, cauline 10–20mm; lamina ovate-triangular, basal ones to 27 × 22cm, cauline 30–160 × 25–135mm, obtuse to acute, base cordate, obtuse-cordate or obtuse, minutely puberulent or viscid-puberulent. Inflorescence a panicle, branches much shorter than main axis. Pedicels 3–4mm in flower, 5–7mm in fruit. Calyx c 9mm; lobes broadly triangular, one much larger than others. Corolla greenish-yellow, 12–17mm excluding limb, puberulent outside; tube differentiated into 2 parts, lower part cylindrical, c 3 × 2mm, upper part (throat) broadly campanulate-obconical, 7–10 × 6–8mm, slightly contracted at mouth; limb 3–6mm wide, lobes very short, obtuse, entire or apiculate. Capsule ellipsoid-ovoid or subglobose, 7–16mm, ± included in calyx. Seeds dusky brown, ellipsoid or ovoid, 0.7–1.1mm.

Bhutan: S – Chukka district (W slope of Raidak Valley near Tabji Khola). Cultivated on terraced hillsides, 1300m. February.

A cultigen not now known in the wild state, probably originating in Mexico or the southern United States. Only one specimen has been seen from Bhutan; it was previously misidentified as *N. tabacum*.

2. N. tabacum L. Eng: *Tobacco*. Fig. 92a–b.

Erect annual or short-lived perennial 1–2m. Stems and branches stout, densely glandular-pubescent. Leaves pale green, lower with short, winged petiole, most cauline ones sessile and ± amplexicaul or decurrent; basal leaves 28–50 × 5–19cm, cauline 9–30 × 4–9cm, acute or acuminate, base cuneate or attenuate, entire, shortly pubescent. Inflorescence an often compound panicle, branches when present shorter than main axis. Pedicels 5–10(–15)mm in flower, 10–20(–25)mm in fruit. Calyx pale green, glandular-pubescent, tube 12–20(–25)mm, lobes triangular-acuminate. Corolla (30–)35–55mm excluding limb, with greenish-white tube and pink, red or whitish limb and throat, straight or gently curved, glandular-puberulent outside; tube differentiated into 3 parts, lowest part (7–)10–15 × c 2mm, middle part 11–16 × 2.5–5mm, upper part (throat) cup-shaped or obconical, 10–14 × 7–12mm; limb 10–15mm diameter, pentagonal or acuminately 5-lobed. 2 longer pairs of stamens extending to near corolla mouth, fifth stamen shorter. Capsule narrowly ellipsoid, ovoid or orbicular, 15–20mm, included in calyx. Seeds spherical or broadly ellipsoid, c 0.5–0.6 × 0.3–0.4mm, brown, ridges fluted.

Bhutan: S – Chukka, Manas and Deothang districts, **C** – Thimphu, Punakha and Tongsa districts; **Darjeeling:** Takdah, Darjeeling; **Darjeeling/Sikkim:** unlocalised (Hooker collection). Cultivated near houses; also on rubbish heaps, 305–2500m. April–November.

Cultivated for tobacco. A cultigen not known in the wild state, probably cultivated in Mexico or United States in pre-Columbian times, which does not persist long outside of cultivation.

3. N. plumbaginifolia Viviani

Erect annual 30–100cm; branches slender, wiry, shortly tuberculate-hispid. Lowest leaves in a rosette, with petioles 25–30mm, lamina broadly oblong-spathulate to obovate, 60–110 × 40–75mm, obtuse, base cuneate or attenuate, undulate, shortly hispid on veins; cauline 20–100 × 2–50mm, lower ones similar to basal, upper rapidly becoming smaller and uppermost narrowly lanceolate or linear-lanceolate, acuminate. Inflorescence of lax few-flowered false racemes, branches nearly as long as main axis. Pedicels 3–7mm in flower, 5–10mm in fruit. Flowers opening in the evening. Calyx 8–13mm; tube 4–6mm, elliptic-ovoid, hispid; lobes subulate-filiform, 6–8mm, subequalling tube. Corolla with greenish-ivory or purplish tube and ivory or lavender limb; tube not clearly differentiated into lower part and throat, 25–35mm, slightly swollen near apex but contracted near mouth; limb c 10mm diameter, deeply lobed, lobes ovate, acute, with 5 dark stripes outside. Anthers purple, 4 ± sessile near mouth and

FIG. 92. **Solanaceae and Buddlejaceae. Solanaceae.** a–b, *Nicotiana tabacum*: a, leaf (× ½); b, apex of inflorescence (× ½). c–d, *Nicandra physalodes*: c, flower (× 1); d, fruit (× 1). **Buddlejaceae.** e–h, *Buddleja asiatica*: e, inflorescence (× ½); f, flower (× 8); g, dissected corolla (× 8); h, fruit with corolla removed (× 8). Drawn by Glenn Rodrigues.

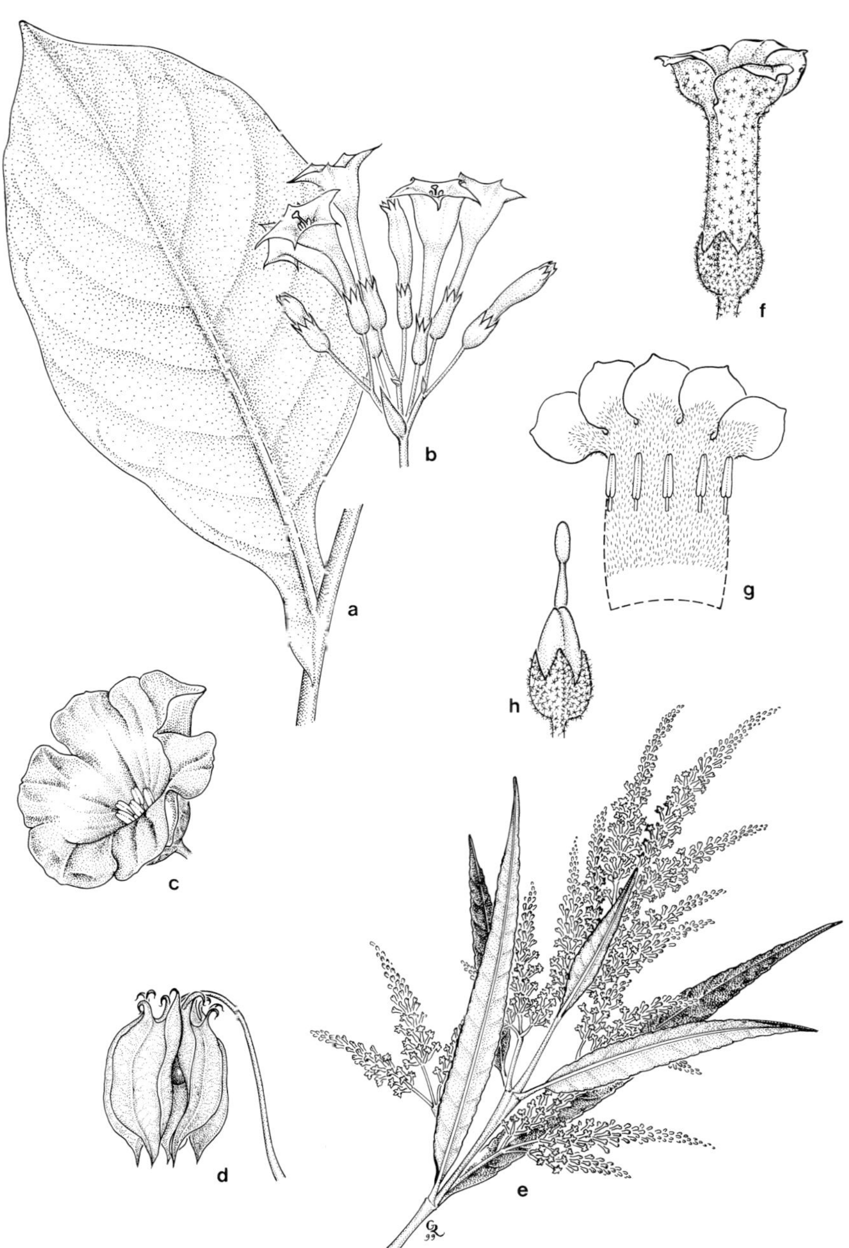
a
b
c
d
e
f
g
h

often ± didynamous, fifth inserted well below others. Capsule narrowly ovoid, 8–11mm, ± included. Seeds roundish-ellipsoid, c 0.5mm, reticulate.

Darjeeling/Sikkim: unlocalised (Hooker and Treutler collections). Weed of cultivated ground, 600–1500m. March–May.

Native to S America east of the Andes from NW Argentina to SW Brazil, and Peru, Ecuador, the Caribbean, Mexico and Florida. Tends to become established or naturalised, the only species of the genus to behave so in the Indian subcontinent.

19. PETUNIA L.

Viscid, erect or sprawling annual herb. Leaves alternate, sessile, simple, entire; stipule-like leaves absent. Inflorescence a short few-flowered raceme, the flowers appearing solitary, sometimes reduced to a single flower; bracts leaf-like, opposite; bracteoles 2, opposite, equal. Pedicels sturdy, articulating at base. Flowers showy, 5-merous, slightly zygomorphic. Calyx unequally 5-lobed to near base; lobes narrowly ovate. Corolla salverform, limb sinuate-lobed. Fruit a dry capsule, apically septicidally dehiscent. Seeds minute, globose, reniform or prismatic, foveate.

The description above applies to the widely cultivated species described below.

1. P. hybrida Vilmorin; *P. violacea* auct. non Lindley

Annual 30–50cm. Stems somewhat flexuous, shortly glandular-pilose with ascending spreading hairs. Leaves sessile or subsessile, ovate, 10–25 × 5–18mm, obtuse, base obtuse or shortly cuneate, glandular-pilose. Pedicels 20–35mm, glandular-pilose. Calyx lobes narrowly oblong, recurved, three c 11 × 1.5mm, two c 14 × 3.5mm. Corolla pinkish-purple or white suffused pink, sometimes with alternating white and coloured bands, 50–65mm; tube pilose, gradually dilated upwards, 30–35mm, lobes pubescent outside, c 30mm.

Bhutan: C – Thimphu district (Paro), Tashigang district (Tashigang). Cultivated in gardens as an ornamental, 1350–2370m. June–October.

A cultigen; origin reputedly *P. axillaris* (Lamarck) Britton, Sterns & Poggenburg × *P. inflata* Fries (cf. 410).

20. BROWALLIA L.

Unarmed annual herb. Leaves alternate, entire or nearly so, petiolate. Flowers solitary, axillary in axils of sometimes reduced leaves; pedicels elongating in fruit. Calyx tubular-campanulate, prominently longitudinally veined when dry, lobed to less than ⅓; lobes 4–5, triangular to lanceolate, enlarging in fruit and longer than but not enclosing capsule. Corolla salverform, zygomorphic; tube swollen at top, limb conspicuous or not, with small mouth. Stamens 4, didynamous, with a fifth staminode or rarely a fifth fertile anther; upper pair of stamens with flattened, curved, pilose filaments and 1 fertile theca (the other obsolete), lower pair with basally geniculate filaments and 2 thecae. Stigma compressed between anthers, expanded into a furrowed wafer-like structure. Fruit a 2-valved, septicidally dehiscent capsule. Seeds very numerous, minute, prismatic, foveate.

1. B. americana L.; *B. demissa* L., *B. elata* L.

Erect, often much-branched annual 20–35cm. Stems shortly glandular pubescent. Leaves alternate or subopposite; petioles 3–20mm; lamina 10–35 × 5–15mm, acute, acuminate or obtuse, base cuneate, obtuse or truncate, glabrous or with short simple hairs above, puberulent on veins beneath. Pedicels 3–6mm in flower, lengthening and up to 10(–20)mm in fruit, finely puberulent. Calyx 8–9mm, densely glandular pubescent on lobes, less so on 10-veined tube; lobes 2–2.5mm. Corolla bright violet, blue or white; tube 12–16mm; limb zygomorphic, shorter lobes 2–7mm, longer ones 6–12mm. Stamens 4; staminode absent. Capsule c 5mm.

Darjeeling: Observatory Hill (218). Cultivated; flowering August–December, fruiting September–January.

Native of Peru.

21. STREPTOSOLEN Miers

Evergreen unarmed shrub. Leaves alternate or subopposite, shortly petiolate, entire. Inflorescence terminal, composed of rather conferted, subcorymbose, ± dichotomously branched cymes. Calyx tubular-campanulate, shortly 5-lobed. Corolla zygomorphic, infundibular, somewhat twisted, limb with 3 smaller posterior lobes and 2 larger anterior lobes. Stamens 4, didynamous. Fruit a capsule. Seeds numerous.

1. S. jamesonii (Bentham) Miers; *Browallia jamesonii* Bentham

Straggling shrub 1.5–2m; branches rugose, densely viscid-pubescent with tuberculate-based spreading hairs. Leaf lamina elliptic, 20–40 × 12–18mm, acute or subobtuse, base shortly cuneate, upper surface sparsely and minutely puberulent with strongly impressed veins, lower surface shortly hispidulous especially on prominently raised veins; petiole 5–15mm, pubescent. Pedicels 5–7mm, subequalling calyx, densely glandular-pubescent. Calyx 7–13mm, bluish-tinged; lobes 2–5mm, triangular-lanceolate, glandular-ciliate. Corolla 26–35mm, orange-yellow at first, limb turning orange-red; tube slightly curved downwards, pilose outside, 20–25mm; limb 15–20mm diameter, shortly pubescent outside, 2 anterior lobes much larger than 3 posterior, one posterior lobe emarginate, others entire. Capsule not seen.

Darjeeling: Observatory Hill (333). Cultivated as an ornamental, c 2000m. March.

22. BRUNFELSIA L.

Unarmed shrub. Leaves coriaceous, entire, glabrate, often apparently clustered at ends of twigs, shortly petiolate. Flowers solitary or in few-flowered clusters, appearing terminal. Calyx tubular to cyathiform, 5-lobed, somewhat accrescent in fruit. Corolla salverform, very showy; tube narrow; lobes slightly zygomorphic. Stamens 5, with 1 or more sometimes reduced to staminodes or entirely absent. Ovary shortly stipitate, 2-locular but apically 1-locular. Fruit a berry.

1. B. pauciflora (Chamisso & Schlechtendal) Bentham; *B. calycina* (Hook.) Bentham

Shrub to 3m but often much smaller. Branches glabrous, glossy, brown. Lamina of upper leaves 25–40 × 15–25mm (lower ones can be 7–15cm), undulate or subentire, very minutely puberulent on veins otherwise glabrous. Pedicels c 5mm. Calyx c 4 × 2mm in flower, to c 10 × 6mm in fruit. Corolla mauve with yellow eye, c 30mm; tube c 25mm, glabrous; limb c 35mm diameter, lobes slightly undulate, c 15 × 15mm.

Bhutan: S – Samchi district (Samchi). Cultivated in gardens, c 500m. February–March.

Native of Brazil.

Family 172. BUDDLEJACEAE

by S.J. Rae

Erect shrubs or small trees, hairs often stellate. Leaves simple, opposite, pinnately veined; stipules leafy, often interpetiolar, sometimes reduced to a line or absent. Flowers in terminal and axillary, often spike-like panicles, bisexual, actinomorphic. Calyx campanulate, shortly 4(–5)-lobed. Corolla narrowly or broadly tubular, with 4(–5) spreading or erect lobes. Stamens 4, inserted within corolla tube. Ovary superior, 2-celled, ovules numerous, placentation axile, style simple, short, stigma clavate or capitate. Fruit a capsule, surrounded by persistent calyx, weakly laterally compressed. Seeds many, compressed.

1. BUDDLEJA L.

Description as for Buddlejaceae.

1. Plants glabrous; leaves connate at base in pairs; flowers 2–4mm **1. B. bhutanica**
+ Plants pubescent to tomentose, rarely almost glabrous; leaves not connate at base in pairs; flowers 5–30mm 2

2. Corolla tubular-campanulate, 17–30mm; panicle branches lax, 3-flowered; capsules 10–16 × 5–8mm .... **2. B. colvilei**
\+ Corolla cylindric, 3–13mm; panicle branches dense, usually more than 3-flowered; capsules 3–12 × 1.2–4mm 3

3. Leaves broadly rounded or cordate at base, thickly woolly-tomentose on both surfaces; flowers sometimes appearing before leaves ...... **4. B. crispa**
\+ Leaves cuneate or attenuate at base, pubescent or tomentose beneath, never thickly woolly above; flowers appearing after leaves 4

4. Corolla tube stellate-tomentose 5
\+ Corolla tube glabrous or sparsely pubescent 7

5. Corolla tube 3–4mm; leaves narrowly lanceolate .... **5. B. asiatica**
\+ Corolla tube 5–10mm; leaves lanceolate, elliptic or narrowly ovate 6

6. Mature leaves finely crenate-serrate, ± glabrous above, very closely tomentose beneath; corolla tube 5–11mm .... **6. B. macrostachya**
\+ Mature leaves subentire or denticulate, stellate-pubescent or thinly tomentose above, thickly woolly-tomentose beneath; corolla tube 5–7mm **7. B. paniculata**

7. Corolla broadly cylindric, tube 2–3.5mm wide; calyx 4–5mm **3. B. forrestii**
\+ Corolla narrowly cylindric, tube c 1mm wide; calyx 2.5–3mm .. **8. B. davidii**

1. B. bhutanica Yamazaki

Shrub 1–2m, glabrous. Leaves connate at base in pairs, amplexicaul, narrowly oblong-lanceolate, 6–12(–16) × 1–2.5cm, acuminate, margins minutely serrulate or subentire; sessile; stipules absent. Panicles terminal and axillary, 6–17cm, spike-like; flowers 3–6 in sessile clusters, lemon-scented. Calyx 1–2mm, glabrous, lobes triangular. Corolla white, tube cylindrical 2–4mm, glabrous outside, pilose within near mouth, often with some glandular hairs on the lobes; lobes broadly elliptic, 1.3–1.5mm, spreading. Stamens inserted above middle of tube. Capsules ellipsoid, 3–5 × 2–3mm.

Bhutan: C – Punakha district (Lobeysa, Tinleygang and Ngawang). In scrub on river banks in chir pine forest, and in scrubby, warm, broad-leaved forest, 1310–2000m. April–June.

Endemic to Bhutan.

2. B. colvilei Hook.f. & Thomson

Shrub or small tree 2–8m. Leaves elliptic-oblanceolate, 6–20 × 1.5–5.5cm, acuminate, base attenuate, margins shallowly serrate, glandular and sparsely to densely stellate hairy especially beneath; petioles 0–8mm; stipules reduced to a line between opposing petioles. Panicles 7–20cm, terminal and axillary, branches cymose, lax, 3-flowered, not fragrant. Calyx 5–7mm, glandular and subglabrous to densely stellate-tomentose, lobes short. Corolla wine-red, tubular-campanulate, tube 15–22mm, 6–10mm diameter at mouth, glandular and sparsely stellate hairy outside, sparsely pilose within; lobes erect, suborbicular 5–8mm. Stamens inserted ⅔ up tube. Ovary stellate-tomentose. Capsules broadly ellipsoid, 10–16 × 5–8mm.

Bhutan: C – Thimphu, Punakha, Tongsa, Bumthang, Mongar, Tashigang and Sakden districts, **N** – Upper Mo Chu and Upper Kuru Chu districts; **Darjeeling:** Sandakphu–Garibans; **Sikkim:** Bakkim–Chhokha, Fiyengong, Kalipokhri, Phusum; **Chumbi:** Langrang. Amongst shrubs on open hillsides and margins of hemlock and mixed forests, 2100–5000m. June–August.

Prized as an ornamental species on account of its showy inflorescences.

3. B. forrestii Diels; *B. cooperi* W.W. Smith

Similar to *B. colvilei* but leaf margins crenate-serrate or sometimes subentire, thinly stellate hairy beneath; panicles 6–25cm, often densely flowered, fragrant; calyx sparsely stellate tomentose; corolla wine-red, tube orange or bluish, broadly cylindrical, 8–10 × 3mm, glabrous or stellate-tomentose near mouth; lobes spreading, suborbicular 2–4mm; capsules ellipsoid, 6–10mm.

Bhutan: C – Bumthang district (Rudong La). Scrub near river banks, 2000–4000m. July.

4. B. crispa Bentham; *B. tibetica* W.W. Smith, *B. tibetica* var. *grandiflora* Marquand, *B. whitei* Kraenzlin

Deciduous shrub 1.5–3(–5)m, densely stellate-tomentose. Leaves ovate, 3–7(–12) × 1.5–3(–8)cm, acute, base cordate or truncate, sometimes narrowly decurrent on petiole, margins crenate-dentate, both surfaces rugose and densely whitish woolly tomentose (brownish when dry); petiole 1–1.5cm, stipules variable, sometimes large and united. Panicles appearing before or after leaves, terminal and lateral, dense or spike-like, branched, 2–30cm, sweet scented. Calyx 2–4mm, tomentose; lobes triangular c 1.5mm. Corolla pinkish-purple with orange throat, tube narrow 6–11 × 1–2mm, stellate-tomentose and with some glandular hairs outside, pilose within; lobes spreading, suborbicular 1–2mm, subentire. Stamens inserted near middle of tube. Ovary stellate-tomentose. Capsule ellipsoid, 5–6mm.

Bhutan: S – Chukka district (Chima Kothi), **C** –Ha district (Ha Valley), Thimphu district (Paro Chu and Thimphu Chu Valley from Kyapcha to Thimphu). Amongst scrub in blue pine forest, on open hillsides and on dry roadsides, 2000–2500m. April–June.

B. crispa is treated here in a broad sense on account of its variability in leaf and inflorescence characters. Plants from Bhutan have broader and more densely tomentose leaves than typical *B. crispa* specimens from Nepal and NW Himalaya; they may be subspecifically distinct. Most Bhutan plants are deciduous with flowers appearing before the leaves and have sometimes been treated as a distinct species, *B. tibetica*. Plants with flowers and leaves appearing together occur only in the Paro Chu Valley. Surprisingly absent from Sikkim and Darjeeling.

5. B. asiatica Loureiro. Dz: *Kang shing chuwa*. Fig. 92e–h.

Shrub to 5m. Leaves narrowly lanceolate, 6–15 × 1–4cm, acuminate, base cuneate; margins minutely dentate, closely white stellate-pubescent or tomentose beneath, glabrous to stellate-pubescent above; petioles 3–10mm with stipular line between opposing petioles. Panicles dense, spike-like, terminal and axillary, 5–20cm, not fragrant. Calyx 2–3mm, stellate-pubescent, lobes triangular, short. Corolla white, sometimes violet, tube 3–4 × 1–1.5mm, stellate-pubescent outside, woolly inside; lobes orbicular, 1.5mm. Stamens inserted near middle of tube. Capsules ovoid, compressed, 3–5 × 1.5–3mm.

Bhutan: S – Samchi, Phuntsholing, Chukka, Sarbhang, Gaylegphug and Deothang districts, **C** – Punakha, Tongsa, Mongar and Tashigang districts; **Darjeeling:** Darjeeling and Tista Valley. In scrub on river banks, open hillsides and margins of subtropical and warm broad-leaved forests, 200–1500m. January–March.

Sometimes cultivated in villages in foothills.

6. B. macrostachya Bentham; *B. griffithii* (Clarke) Marquand, *B. macrostachya* var. *griffithii* Clarke

Shrub or small tree 2–3(–7)m, stellate-pubescent. Leaves elliptic, 6–20 × 1.5–7cm, acuminate, base attenuate into petiole, margins finely crenate-serrate, white stellate-tomentose beneath (brownish when dry); sessile or with short winged petiole; stipules united, interpetiolar, rounded. Panicles dense, elongated, spike-like, terminal, 5–20cm, sweet-scented. Calyx 2–4mm, tomentose, lobes triangular, 1–2mm. Corolla cream-purple, with deep orange throat; tube 8–10 × 2–3mm, tomentose and with some glandular hairs outside, villous within; lobes spreading, suborbicular 2–4mm. Stamens inserted near the mouth of the tube. Ovary stellate-tomentose, with glandular hairs. Capsules ovoid, compressed, 7–8mm.

Bhutan: S – Chukka district (Jumudag–Tala), **C** – Tongsa district (Chendebi, Tashiling), Mongar district (?Latun La), Tashigang districts (Tashi Yangtsi); **Sikkim:** Chungthang. Amongst scrub on streamsides and margins of warm broad-leaved forest, 2000–2600m. August–October.

7. B. paniculata Wall.; *B. acutifolia* Wright

Similar to *B. macrostachya* but leaves ovate-lanceolate, margins denticulate or subentire, white or greyish stellate-pubescent or thinly tomentose above, thickly woolly tomentose beneath (brown when dry); corolla white-lavender; tube 4–7mm, lobes white, 1.5–2mm; capsules 4–5mm.

Bhutan: S – Chukka district (Takhti Chu), **C** –Punakha district (Wangdu Phodrang, Mendegang, Wache), Tongsa district (Kinga Rapden); **Darjeeling:** Takvar. Warm broad-leaved forests, 1200–2100m. April–May.

8. B. davidii Franchet

Shrub 1–4m, pubescent. Leaves narrowly elliptic-lanceolate, 4–20 × 1–5cm, acuminate, base cuneate, margins serrate or serrulate, dark green and glabrous above, closely white tomentose beneath (brown when dry); petioles 0–4mm; stipules interpetiolar, suborbicular. Panicles terminal, sometimes axillary, many-flowered, 10–30cm, fragrant. Calyx 2–3mm, glabrous or stellate pubescent, lobes triangular, c 1mm. Corolla pinkish-purple with orange throat; tube 6–11 × c 1mm, glabrous or stellate-pubescent, outside, pilose within; lobes spreading, orbicular, 1–3mm. Stamens inserted near middle of tube. Ovary glabrous or pubescent. Capsules ellpsoid, 5–6mm, glabrous or pubescent.

Bhutan: C – Thimphu district (Taba, Paro). Cultivated in gardens, 2400m. July– October.

Native of China, cultivated for its showy fragrant panicles.

Family 173. SCROPHULARIACEAE

by R.R. Mill

Herbs, more rarely shrubs or trees (occasionally epiphytic), autotrophic, hemiparasitic or parasitic. Leaves alternate, opposite or sometimes whorled, simple, lobed or pinnately dissected. Inflorescence a thyrse, raceme or spike or flowers solitary. Flowers hermaphrodite, usually distinctly zygomorphic, sometimes ± actinomorphic. Calyx ± deeply divided into (2–)4–5 segments. Corolla sympetalous, (3–)4–5-lobed, bilabiate or not, sometimes spurred or saccate. Stamens attached to corolla tube and alternating with lobes, 2, 4 or 5, often didynamous; fifth (adaxial) stamen sometimes replaced by a staminode (e.g. *Scrophularia*) or absent; thecae sometimes unequal, often divergent and/or confluent at tips. Ovary superior, bilocular; style terminal; stigma capitate, punctiform, or 2-lobed, usually wet. Ovules ± numerous. Fruit usually a septicidal, loculicidal or poricidal capsule, less commonly a berry, rarely (*Lagotis*) a schizocarp. Seeds ± numerous, angular or winged.

Recent molecular research has shown that Scrophulariaceae as traditionally defined (as in *Flora of Bhutan*) is paraphyletic. These studies restrict the family to *Scrophularia*, *Verbascum*, *Selago* and its allies, and the mainly African tribe

Manuleae (*Manulea* etc.). Most genera, such as *Calceolaria*(?), *Digitalis*, *Kickxia*, *Limnophila*, *Lindernia*, *Mimulus*, *Scoparia* and *Veronica*, are transferred to Plantaginaceae; the name Antirrhinaceae Persoon has recently been proposed for conservation over Plantaginaceae (399). The position of some genera is still unclear and it is possible that *Calceolaria* should be excluded from the expanded Plantaginaceae (354). The hemiparasitic genera such as *Euphrasia* and *Pedicularis* are all transferred under this new classification to Orobanchaceae. *Wightia* should presumably be transferred to Bignoniaceae under the new classification because all members of the three families as newly defined are herbs or shrubs only, not trees or lianas. The genera in our area would divide up as follows: Scrophulariaceae sensu stricto (genera 1 and 2); Plantaginaceae (3–28, but *Ellisiophyllum* and *Calceolaria* of uncertain position); Orobanchaceae (29–36); Bignoniaceae(?) (37 and 38). For further discussion see (354).

1. Trees or shrubs 2
\+ Herbs 4

2. Leaves, calyces and young twigs without stellate hairs ... **11. Lindenbergia**
\+ Leaves, calyces and young twigs stellate-tomentose 3

3. Tree (sometimes epiphytic), 3–15m; leaves 70–190mm broad; flowers in many-flowered inflorescences .... **38. Wightia**
\+ Shrub, 0.6–2.5m; leaves 8–30mm broad; flowers solitary or in pairs **37. Brandisia**

4. Procumbent annual of rice-fields and ponds, with filiform stems and minute leaves, forming moss-like mats; corolla pink, c 2.5mm .. **20. Microcarpaea**
\+ Not as above 5

5. Fertile stamens 2 6
\+ Stamens 4–5, all with at least 1 fertile loculus 12

6. Two staminodes or sterile stamens present as well as 2 fertile stamens ... 7
\+ Only 2 fertile stamens present 9

7. Leaves entire, all but the lowest 1–2 pairs reduced and not more than 2mm **16. Dopatrium**
\+ Leaves serrate or finely dentate, all well developed and usually more than

8mm 8

8. Calyx flattened, of 4 segments in 2 series, one outer segment much larger than other 3 .... **17. Picria**
\+ Calyx tubular or campanulate, with 5 teeth and 5 wings or ribs, all segments more or less equal .... **19. Lindernia**

9. Corolla yellow or greenish-yellow 10
\+ Corolla blue, pink, purple or white 11

10. Leaves lanceolate, simple; stout-stemmed perennial; corolla greenish-yellow, not slipper-like, with exserted stamens .... **26. Calorhabdos**
\+ Leaves broadly ovate, pinnatisect; slender-stemmed annual; corolla deep yellow, slipper-shaped with lower lip 3 × as long as upper; stamens included **10. Calceolaria**

11. Corolla rotate, not or scarcely zygomorphic (except that one lobe is sometimes smaller than other 3) .... **27. Veronica**
\+ Corolla bilabiate, strongly zygomorphic .... **28. Lagotis**

12. Fertile stamens 5; corolla rotate .... **1. Verbascum**
\+ Fertile stamens 4, usually didynamous, sometimes with a fifth staminode; corolla usually bilabiate, sometimes rotate 13

13. Corolla tube spurred or saccate at base 14
\+ Corolla tube not spurred or saccate at base 16

14. Corolla 30–45mm, tube saccate at base (cultivated) .... **6. Antirrhinum**
\+ Corolla 10mm or less (excluding spur), tube spurred (native) 15

15. Corolla violet or white, with yellow palate; leaves reniform or semicircular, palmately 5–9-lobed .... **5. Cymbalaria**
\+ Corolla entirely yellow (except for small red spots on palate); leaves triangular or ovate, never palmate but often hastate or sagittate .... **4. Kickxia**

16. Inflorescence a large terminal panicle of numerous cymes .. **2. Scrophularia**
\+ Inflorescence various but not a compound panicle of cymes 17

17. Prostrate, creeping herbs, stems rooting at nodes 18
\+ Erect or acaulescent herbs, stems not rooting at nodes 22

18. Stems completely glabrous 19
\+ Stems hairy (sometimes sparsely) 20

19. Corolla yellow; leaves serrate in upper ½–⅔ .... **15. Mecardonia**
\+ Corolla bluish, mauve or lilac; leaves entire .... **14. Bacopa**

20. Leaves dimorphic, suborbicular and petiolate (wet season) and fasciculate, needle-like (dry season); flowers sessile in leaf axils ...... **22. Hemiphragma**
\+ Leaves all similar, not as above; flowers usually pedicellate, axillary or not 21

21. Corolla subrotate, nodding; each node emitting 1 leaf and 1 long-pedicellate flower; leaves deeply pinnatifid, resembling *Geranium* ... **21. Ellisiophyllum**
\+ Corolla tubular, not nodding; growth pattern not as above; leaves simple **19. Lindernia**

22. Corolla subrotate 23
\+ Corolla tubular, campanulate or bilabiate 24

23. Corolla white or pale green, 4-lobed, posterior lobe emarginate; capsule 2-valved; stems glabrous except at nodes .... **23. Scoparia**
\+ Corolla yellow or purple, 5-lobed; capsule 4-valved; stems scabrid or pilose throughout .... **33. Sopubia**

24. All aerial leaves 3-sect, pinnatifid or pinnatisect 25
\+ All aerial leaves simple (sometimes finely pinnatisect submerged leaves present: *Limnophila*) 26

25. Upper lip of corolla with reflexed margins .... **35. Phtheirospermum**
\+ Upper lip of corolla often hood-like, never with reflexed margins **36. Pedicularis**

26. Anterior stamens exserted from corolla; flowers pale blue to dark violet **24. Neopicrorhiza**
\+ All stamens included in corolla; flower colour various 27

27. All stem leaves alternate; corolla 40–50mm, purple, pink or white outside with dark purple spots within .... **25. Digitalis**
\+ All or most stem leaves opposite; corolla usually less than 40mm (if 40–50mm, then yellow: *Centranthera grandiflora*) 28

28. More or less acaulescent plants (stems 0–12cm) of high-alpine mobile scree; flowers bright yellow; leaves all in a rosette, stems with scale-like leaves 29
\+ Habit and habitat not as above; flower colour various; leaves all cauline or if some basal then stem leaves not scale-like 30

29. Corolla tube relatively broad .... **2. Scrophularia** (*S. przewalskii*)
\+ Corolla tube narrow .... **3. Oreosolen**

30. Calyx subtended by 2 bracteoles (and sometimes also a bract) 31
\+ Calyx not subtended by a pair of bracteoles but sometimes bracteate .. 36

31. Each anther with 2 separate thecae; plant often strongly aromatic, not parasitic; inflorescence various, but not a spike 32
+ Anther thecae joined at tips, or each anther with a single theca; plant not aromatic, always wholly or hemiparasitic; inflorescence a spike 33

32. Aquatic or marsh plants; calyx regular or rarely with only 1 lobe (adaxial) larger than others; all stamens with both thecae fertile ..... **13. Limnophila**
+ Terrestrial roadside plant; calyx irregular, with 2 inner lobes much smaller than outer 3; 2 or all 4 stamens with only one of the 2 thecae fertile **12. Adenosma**

33. Anther loculi 2 34
+ Anther loculus 1, by abortion 35

34. Corolla 6–7mm .... **29. Alectra**
+ Corolla 12–50mm .... **32. Centranthera**

35. Corolla white or yellow, tube decurved near apex; inflorescence lax, only 1–2 flowers open at one time .... **31. Striga**
+ Corolla bluish-purple or violet, tube straight; inflorescence dense, several flowers open at one time .... **30. Buchnera**

36. Corolla mainly or wholly yellow, sometimes with brown or red spots .. 37
+ Corolla mainly or wholly white, mauve or bluish, sometimes with yellow throat 38

37. Calyx glabrous except along 5 ribs and margins of teeth; anther thecae confluent (straggling herbs, wet places).......................... **7. Mimulus**
+ Calyx pubescent or hirsute all over; anther thecae separate (erect herbs woody at the base, forests and dry slopes) .... **11. Lindenbergia**

38. Leaves mainly or all in a basal rosette 39
+ Leaves mainly or all cauline 40

39. Plant usually ± glabrous and acaulescent; raceme very condensed, flowers appearing subsessile amongst leaves; fruit an indehiscent berry... **9. Lancea**
+ Plant usually ± hairy and caulescent; raceme well developed, scapose; fruit a loculicidal capsule .... **8. Mazus**

40. Corolla white or lilac with fine purple longitudinal veins and yellow throat; fruit a loculicidal capsule; hemiparasitic .... **34. Euphrasia**
+ Corolla variously coloured but lacking fine purple veins; fruit a septicidal capsule; autotrophic 41

41. Ovary entirely glabrous; calyx unwinged or with narrow wings **19. Lindernia**
+ Ovary shortly scabrid-pilose in upper part; calyx with 5 ± broad wings or keels .... **18. Torenia**

1. VERBASCUM L.

Eng: *Mullein.*

Biennial herbs, often pubescent, floccose or tomentose. Leaves alternate, simple or divided; basal ones in a rosette, sessile or shortly petiolate; cauline numerous, smaller. Plants glabrous or with indumentum of eglandular or glandular, simple, branched or dendroid hairs. Inflorescence a branched or unbranched spike or raceme; flowers solitary or in clusters in bract axils. Calyx 5-lobed, regular or ± zygomorphic; lobes lanceolate or narrowly lanceolate. Corolla yellow, ± actinomorphic or slightly zygomorphic, flattened-rotate, tube very short or almost absent; lobes rounded or broadly obovate, posterior ones slightly smaller. Stamens 5. Filaments inserted at base of corolla, ± covered with whitish or purple woolly indumentum; 2 posterior longer than 3 anterior, ± glabrous above or throughout, 3 posterior woolly up to anthers. Anterior anthers decurrent and obliquely inserted on to filament, posterior ones medifixed. Style filiform or thickened towards apex. Ovary bilocular. Capsule globose or oblong-ovoid, septicidal. Seeds numerous, obconical-prismatic, foveolate.

The above description applies only to the two following species. Other species of this very large genus (some 400 species, mostly in SW Asia) vary in stamen number (4–5), and in the form of the anterior and posterior anthers and the distribution of wool on the filaments.

1. Leaves tomentose; flowers in clusters of 2–7 in bract axils, completely hiding inflorescence axis; filament wool whitish-yellow .... **1. V. thapsus**
+ Leaves glabrous; flowers solitary in bract axils, axis of inflorescence easily seen; filament wool purple-violet .... **2. V. blattaria**

1. V. thapsus L. Dz: *Kachum.* Fig. 93a–b.

Stout, erect biennial, 30–150cm, densely tomentose, eglandular. Stem narrowly winged at least above. Basal leaves ovate to oblong, 8–50 × 2.5–14cm, crenate to subentire, tomentose. Cauline leaves decurrent, oblanceolate to obovate, acute or shortly acuminate. Inflorescence dense, simple or very rarely branched, spike-like, 15–30(–90)cm, tomentose; axis scarcely visible. Flowers in clusters of 2–7, sweet-scented. Bracts ovate- to lanceolate-acuminate, lower ones decurrent. Bracteoles 2; pedicels partly adnate to stem, free part 1–5mm. Calyx 7–12mm. Corolla yellow, 12–20mm diameter, punctate with pellucid glands, stellate-pubescent outside, ciliate to base of upper lobes within. Filament wool whitish-yellow; 2 anterior filaments glabrous or sparingly hairy in middle. Anthers orange. Capsule broadly ovate-elliptic, 7–10 × 4–6mm, stellate-tomentose.

Bhutan: S – Chukka district, **C** – Thimphu to Bumthang districts; **Darjeeling/ Sikkim:** unlocalised Hooker collection; **Chumbi:** Chumbi, Lingmatang etc. Dry, often abandoned fields, clearings in *Pinus wallichiana* forest, often on sandy soil, 1220–3048m. June–October.

The seeds are used for stupefying fish; the plant is used for pulmonary disease, asthma, diarrhoea, and bleeding of the lungs and bowel.

2. V. blattaria L.

Erect biennial, 30–150cm, glabrous below, glandular in inflorescence. Stem angled above. Basal leaves broadly or narrowly oblong, 10–25cm, crenate or sinuately lobed, glabrous. Uppermost cauline leaves triangular with cordate semi-amplexicaul base. Inflorescence lax, simple or with a few branches, spike-like, 30–60cm; axis clearly visible. Flowers solitary in bract axils. Bracts lanceolate, acuminate. Bracteoles absent; pedicels free from stem, in fruit 5–25mm and much longer than calyx. Calyx 5–8mm; lobes glandular. Corolla yellow, 20–30mm diameter, ± glandular outside, villous with violet papillae to base of upper lobes within. Filament wool purple; 2 anterior filaments glabrous in upper ⅓. Capsule globose, 5–8mm, glandular.

Darjeeling: Singalila Ridge.

Known in our area only from a fruiting specimen collected as long ago as 1855 somewhere between 360m and 2743m. Probably introduced; rarely persisting long in the same locality.

2. SCROPHULARIA L.

Perennial herbs, in Bhutan usually tall (except *S. przewalskii*). Stems quadrangular, sometimes winged. Leaves petiolate or subsessile, usually ovate, toothed. Inflorescence usually a large terminal panicle of cymes, sometimes with axillary inflorescences below, in *S. przewalskii* reduced to a terminal cluster amongst upper leaves. Cymes pedunculate or sessile, sometimes dichotomously branched. Bracteoles present. Calyx lobes oblong, ovate or suborbicular, often with narrow scarious margin. Corolla usually inconspicuous and greenish, rarely larger and yellow (*S. przewalskii*). Stamens 4, didynamous, included or exserted. Staminode present or absent. Fruit a capsule.

Important identification characters include the presence or absence of glandular hairs on the bracteoles and parts of the inflorescence; the relative length of the *alar pedicel* (that arising in the fork between the 2 primary branches of a dichasium, or the first-formed flower of a monochasial cyme) to the first pair of bracteoles subtending the same cyme; staminode presence or absence and its shape.

S. przewalskii is very anomalous, and much more closely resembles *Oreosolen*

FIG. 93. **Scrophulariaceae.** a–b, *Verbascum thapsus*: a, flowering shoot (× ½); b, lower leaf (× ½). c, *Kickxia ramosissima*: habit (× ⅓). d–e, *Scrophularia urticifolia*: d, flowering shoot (× ⅓); e, dissected corolla (× 5). f–g, *Mimulus bhutanicus*: f, flowering shoot (× ½); g, dissected calyx, exterior (× 2). h, *Mimulus nepalensis*: dissected calyx, exterior (× 2). i–j, *Lancea tibetica*: i, habit (× ⅔); j, fruit (× 1). Drawn by Glenn Rodrigues.

a
b
c
d
e
f
g
h
i
j

in many characters (e.g. yellow corolla, adaptation to scree habitat with very long roots, presence of scale-like leaves on stem etc.). Corolla structure seems closer to *Scrophularia* and meantime it is retained in that genus.

1. Corollas yellow, at least 12mm, the upper lip deeply 2-lobed; stem 3–12cm, with 2–3 pairs of very small scale-like leaves and a group of much larger leaves at top .... **7. S. przewalskii**
\+ Corollas greenish, not more than 8mm, the upper lip not deeply 2-lobed; stem 50–150cm, with numerous pairs of similar large leaves throughout .. 2

2. Cymes dense, scarcely or not pedunculate 3
\+ Cymes lax, dichotomously branched, distinctly pedunculate 4

3. Leaves petiolate; corolla c 4mm .... **1. S. pauciflora**
\+ Leaves subsessile; corolla c 6mm .... **2. S. subsessilis**

4. Stamens included; cymes (1–)3–9-flowered; petioles not auriculate at base **3. S. urticifolia**
\+ Stamens finally exserted; cymes (5–)9–19-flowered; petioles auriculate at base 5

5. Leaf margin coarsely doubly dentate; bracteoles glabrous ..... **4. S. elatior**
\+ Leaf margin finely serrate; bracteoles glandular or glabrous 6

6. Bracteoles glandular; petioles winged and auriculate; peduncles glabrous; scarious margin of calyx lobes scarcely evident; capsule globose **5. S. cooperi**
\+ Bracteoles glabrous; petioles unwinged, exauriculate but sometimes dilated; peduncles pubescent and sometimes glandular; scarious margin of calyx lobes conspicuous; capsule ovoid .... **6. S. himalayensis**

1. S. pauciflora Bentham; *S. sikkimensis* Yamazaki

Stems 50–90cm or more, shortly glandular-pubescent (especially on angles) or almost glabrous. Petioles 20–50mm, glandular-pubescent or glabrous. Leaves ovate or oblong-lanceolate, 35–110 × 20–80mm, acute to acuminate, base abruptly truncate-cordate, margin coarsely dentate, sparsely puberulent above and on veins beneath or glabrous. Flowers in distant terminal leafless condensed cymes. Inflorescence axis, peduncle and pedicels ± densely glandular-pubescent or -puberulent; peduncle very short and not clearly visible; pedicels 2.5–4mm; bracteoles linear-lanceolate, 2.5–3mm. Calyx lobes ovate or ovate-lanceolate, c 4mm, sparsely glandular-puberulent outside, narrowly scarious-margined. Corolla green, c 4mm, all 4 upper lobes subequal. Stamens included; staminode usually absent, rarely present and subulate. Capsule 5.5–6.5 × c 4mm, dark olive; persistent style c 2mm.

Bhutan: C – Thimphu, Punakha, Tongsa, Bumthang and Mongar districts, **N** – Upper Mo Chu and Upper Bumthang districts; **Darjeeling:** Phalut; **Sikkim:** Lachen, Gopethang, Dzongri etc.; **Chumbi:** Natu La–Champitang. Moist ravines in *Abies/Rhododendron* and *Abies/Tsuga* forest, 3200–4270m. June–September; fruits till November.

Yamazaki (423) originally distinguished *S. sikkimensis* by the presence of a small subulate staminode, lanate petioles and longer, narrower sepals. Later (101) he reduced it to a synonym of *S. pauciflora* as his field observations had revealed that the presence or absence of the staminode was variable and did not correlate with other characters.

2. S. subsessilis R.R. Mill

Stem c 60cm, subglabrous below, very sparsely pilose above, sparsely glandular-pubescent above uppermost leaf-pair. Leaves subsessile (petioles less than 10mm), upper middle ones largest; lamina ovate, 45–90 × 25–60mm, acute, base shallowly cordate-cuneate, margin coarsely doubly dentate, each tooth with a stout mucro; surfaces very sparsely white-pilose and with numerous pale yellow-green dots; veins conspicuous, blackish. Cymes dense, shortly pedunculate. Pedicels short, densely sessile-glandular. Calyx lobes ovate, c 3mm, shortly acuminate, sessile-glandular on central part of lamina outside, without a distinct scarious margin. Corolla green, c 6mm, margins of lobes paler; 2 posterior lobes largest. Stamens included; ?staminode absent. Capsule not seen.

Bhutan: N – Upper Kulong Chu district (Shingbe, Me La). Clearing in *Abies* forest, c 3500m. June(?–July).

Endemic to Bhutan; known only from the type, which is in bud with a few very immature flowers. More material is needed to complete the description.

3. S. urticifolia Bentham. Fig. 93d–e.

Stems 60–90cm, sparsely or less commonly densely short-pilose, angled but not winged. Petioles usually short (2–10mm) but sometimes distinct and 20–50mm. Lamina ovate, 35–95 × 25–60mm, acute, base obtuse-truncate or very slightly cordate, margin coarsely doubly dentate, surfaces sparsely glandular-puberulent. Inflorescence a terminal panicle, usually leafless above, with several pairs of cymes arising from axils of floral leaves below; lower peduncles 10–40mm, slender; cymes usually divaricately branched, 3–9-flowered, sometimes reduced to 1–2 flowers; pedicels filiform, alar one 5–12mm; axis, peduncles and pedicels ± glandular-pubescent. Calyx lobes broadly ovate, c 2.5 × 1.5mm, very obtuse, narrowly scarious-margined. Corolla greenish, 6–7mm, 2 posterior lobes longer than others. Stamens included; staminode conspicuous, obovate to reniform. Capsule ovoid to subglobose, c 4mm, acuminate, dark brown; beak c 0.5mm, style finally falling.

Bhutan: C – Thimphu district (Dotena, Lamnakha, Zado La), Punakha district (Dochu La), Tongsa district (Tratang), Mongar district (Senghor); **Darjeeling:** Darjeeling, Labha, Tonglu etc.; **Sikkim:** Yoksam, Bakhim, Chia

Bhanjang, Phusum etc. *Tsuga* and other evergreen forest, grassy slopes, rocks and gravelly scree, 1830–3350m. May–August.

4. S. elatior Bentham

Stem 90–150cm, ± stout, winged throughout, glabrous or sparsely white-pilose below, glandular above. Petioles 20–50mm, with conspicuous basal auricles which almost unite to form a connecting ring around stem; small axillary leaves often present. Lamina ovate, acute, base shallowly cordate, margin sharply and coarsely doubly dentate, surfaces nearly glabrous. Inflorescence a large terminal many-flowered panicle with few axillary inflorescences below; lower peduncles 15–25mm; cymes divaricately dichotomous, (5–)9–15-flowered; pedicels rather stout, alar one 4–8mm; axis sparsely pilose, peduncle and pedicels densely patent-glandular-pubescent, glandular hairs c 0.3mm. Bracteoles linear-lanceolate or narrowly ovate, 4–7mm, glabrous. Calyx lobes ovate-lanceolate, 2.5–4mm, acute or subacute, with narrow scarious margin, glabrous. Corolla green, c 6mm, 2 posterior lobes larger. Stamens finally long-exserted; staminode spathulate or absent. Capsule ovoid to subglobose, c 6 × 5mm, with long persistent style 6–8mm.

Bhutan: C – Tashigang district (Gomchu (117)); **Darjeeling:** Senchal, Darjeeling; **Sikkim:** Lachen. Woods, shrubby places and near cliffs, 2400–3050m (Nepal).

The record from eastern C Bhutan needs confirmation and no recently collected material from Sikkim has been seen.

5. S. cooperi R.R. Mill

Stout perennial. Stems to 120cm, narrowly winged throughout, glabrous. Petioles 4–20mm, winged and basally auriculate. Lamina (only upper ones seen) oblong-ovate, 35–60 × 12–25mm, acute, base cuneate or obtuse, margin finely serrate, very sparsely short-pilose above, glabrous beneath. Inflorescence a large terminal panicle with few axillary inflorescences below; lower peduncles 20–30mm; cymes divaricately dichotomous, mostly 13–19-flowered; pedicels fairly stout, alar one 4–9mm; axis and peduncles glabrous, pedicels with sparse very short-stalked glands. Bracteoles narrowly oblong, 3.5–4.5mm, glandular. Calyx lobes ovate, c 2.5 × 1.5mm, glandular, with narrow, slightly paler but scarcely scarious margins. Corolla greenish-white, 5–6mm. Stamens exserted; staminode broadly spathulate. Capsule globose, c 5mm.

Bhutan: C – Thimphu district (Dochong La), Mongar district (Pimi). 2135–2440m. July–August.

Resembling *S. elatior* but the leaves are more finely toothed and the glandular bracteoles shorter. A Bhutan endemic.

6. S. himalayensis Royle

Stout perennial. Stems to 1m or more, obtusely angled but scarcely if at all winged, glabrous. Petioles (10–)20–70mm, ⅓–⅔ × lamina, slender, unwinged, exauriculate although sometimes with dilated base, glabrous or puberulent. Lamina ovate to oblong-ovate, 30–120 × 12–65mm, acute, base truncate to shallowly cordate, margin ± coarsely dentate or double-dentate, ± glabrous above and beneath. Inflorescence a terminal panicle; lower peduncles 10–20mm; cymes dichotomous, mostly 5–11-flowered; pedicels slender, alar one 2–9mm; axis glabrous or pubescent, peduncles pubescent (at least in lines) and sometimes glandular, pedicels densely glandular. Bracteoles narrowly lanceolate, 3–7mm, glabrous. Calyx lobes oblong-ovate, c 2.5 × 2mm, glabrous, with conspicuous scarious margins. Corolla greenish, c 4mm. Stamens very long-exserted. Capsule ovoid, acute, c 6 × 4–5mm, with long persistent style.

Bhutan: S – Chukka district (Chima Kothi–Bunakha). Moist, shaded rocks in damp broad-leaved forest, c 2000m. September.

A mainly NW Himalayan species which extends to Nepal; the single Bhutan record is exceptional and represents the easternmost limit of its range.

7. S. przewalskii Batalin

Low perennial with very long vertical roots to several metres. Stems 3–12cm, lower part almost naked except for 2–3 pairs of very reduced ± scale-like leaves. Leaves ovate to suborbicular, 15–30 × 10–25mm, obtuse, base obtuse, margin shallowly dentate, surfaces very sparsely pilose. Flowers in a rosette-like cluster among apical leaves. Pedicels 8–12mm, thickened distally. Calyx lobes oblong-lanceolate, 7–8mm, rather densely white-glandular-pilose outside, without scarious margin. Corolla yellow, 16–20mm, glandular-pubescent, with long broad tube; posterior lobes larger. Filaments unequal, c 7–8mm; stamens included; staminode absent. Style 14–17mm. Capsule globose, c 7 × 6mm.

Sikkim: Lachung, Goecha La. Alpine and glacial scree, c 4600m.

Much more closely resembling *Oreosolen* than the other species of *Scrophularia* in its habit and relatively large yellow flowers.

3. OREOSOLEN Hook.f.

Small short-stemmed or almost acaulescent perennial herbs. Leaves opposite or rosulate, lower pairs when present often very reduced. Flowers few to several in a cluster arising from uppermost leaf pair. Calyx divided to base into linear-lanceolate segments. Corolla yellow, with long ± narrow tube; limb bilabiate, lips ± unequal; upper lip emarginate or 2-lobed, lower 3-lobed. Stamens 4, fertile, didynamous; staminode 1 or 0. Fruit a septicidal capsule. Seeds ellipsoid with reticulate testa.

Scrophularia przewalskii Batalin is very similar to *Oreosolen* species in its general habit and yellow flowers; *Oreosolen* is often named as '*S.* cf. *przewalskii*' by field botanists. However, the form of the corolla in *S. przewalskii* is typical of other members of the genus *Scrophularia* apart from the deeply lobed upper lip.

1. Staminode absent; corolla 11–13mm .... **3. O. williamsii**
\+ Staminode present; corolla 14mm or more 2

2. Leaves minutely glandular-puberulent; calyx lobes c 7 × 0.6mm; corolla 14–16mm .... **1. O. wattii**
\+ Leaves ± densely covered with short, rather thick, white hairs; calyx lobes c 5 × 1mm; corolla 20–25mm .... **2. O. unguiculatus**

1. O. watti Hook.f.

Low short-stemmed or almost acaulescent herb from rather stout vertical root-stock; plant blackening when dry. Stem to 2cm, ± glabrous, sheathed by a pair of reduced scale-like leaves. Leaves ovate to obovate, 15–30 × 10–40mm, with rather broad dark veins arising ± palmately from leaf base and anastomosing near margin, base cuneate, margins rather coarsely dentate, surfaces minutely glandular-puberulent. Calyx lobes linear to linear-spathulate, c 7–8 × 0.6–1.2(–1.6)mm, acute, minutely glandular-ciliate. Corolla yellow, 14–17(–18)mm; upper lip obcordately 2-lobed, lower lip shorter with 3 subequal rounded lobes. Filaments attached in upper ⅔ of tube, c 6mm, gradually broadening towards base of anther. Staminode linear-spathulate, attached to middle of upper lip. Style c 11mm, broadened towards stigma. Capsule c 3mm, ovoid-globose. Seeds ovoid-elliptic, 0.9–1 × c 0.6mm.

Bhutan: N – Upper Mo Chu district (Kohina, Shingche La); **Sikkim:** Dzongri, Lachen, Lhonak, Bikbari Valley etc.; **Chumbi:** upper Chumbi Valley, Phari etc. Gravelly scree, 3000–5200(?–5800)m. May–August.

Used in Tibetan medicine. Spermicidal saponins have been recently extracted by Chinese phytochemists (432). A very broad view of this species is taken in *Flora of China* (351), regarding it as the only species of the genus, including *O. unguiculatus* in synonymy but seemingly overlooking *O. williamsii.*

2. O. unguiculatus Hemsley

Similar to *O. wattii* but usually more robust; leaves 15–50 × 15–50mm, rather rugose, veins much less conspicuous, margins shallowly crenate-dentate, surfaces ± densely short white-glandular-pubescent; calyx lobes relatively shorter and broader, c 5 × 1mm; corolla longer, 20–25mm; filaments attached near top of tube, c 6mm, less conspicuously dilated distally; style longer, c 23mm, scarcely or not broadened below stigma.

Bhutan: N – Upper Mo Chu district (Singche La (71), doubtful record). Alpine turf, c 3000m. May.

O. wattii has been collected at the same locality and Yamazaki's identification needs confirmation. The species is otherwise known from Tibet and W and C Nepal. *O. unguiculatus* has recently been considered a synonym of *O. wattii* (351) but the distinctions are normally clear.

3. O. williamsii Yamazaki

Similar to *O. wattii* but plant only 2–3cm tall; leaves 6–12 × 5–9.5mm; calyx lobes linear-oblong, c 3–5 × 0.7–1mm; corolla 11–13mm, with long tube 8–9mm and short upper and lower lips, upper lip 2-lobed, c 1.5–2mm, lower with 3 rounded lobes, 2 lateral lobes c 2–2.5mm, middle lobe slightly shorter; staminode absent; style c 8mm.

Sikkim: Dzongri. In Nepal this species occurs on high-altitude recent screes, c 4570–4720m. June–July (Sikkim and Nepal).

Little material has been seen. Yamazaki described the calyx lobes as 1–1.5mm wide, but this not only disagrees with the shape he stated they were, but also with the actual dissection on the type sheet.

4. KICKXIA Dumortier

Suffrutescent, glabrous or hairy perennial herbs. Stems many, branched mainly from base, slender. Leaves petiolate, mainly alternate, lamina ovate or oblong-ovate, all similar or lower and upper differing in shape, often hastate or sagittate. Flowers axillary, pedicels filiform with recurved tip. Calyx deeply divided into 5 subequal, narrow, entire or irregularly lobed segments. Corolla tube cylindrical, spurred; limb unequally bilabiate, adaxial lip shorter and sub-erect, lower lip larger, spreading and with low palate. Stamens 4, didynamous, included. Fruit an ovoid or subglobose valvate bilocular capsule with thin membranous or papery walls; loculi slightly unequal. Seeds small, ± ellipsoid or ovoid, tuberculate.

All the species in our area belong to *Kickxia* sect. *Valvatae* (Wettstein) Janchen. Betsche (321) divided this into two segregate genera: *Nanorrhinum* Betsche and *Pogonorrhinum* Betsche. The Bhutanese species would be referable to *Nanorrhinum*. Following Sutton (413), this view is not adopted here.

1. Plant densely greyish-villous or hirsute .... **1. K. incana**
\+ Plant glabrous or almost so 2

2. Calyx lobes 0.9–1.2mm broad, irregularly lobed near base; corolla 8–10mm **2. K. membranacea**
\+ Calyx lobes 0.4–0.8mm broad, entire; corolla 5–8mm 3

3. Seeds with large, blunt, whitish papillae; leaves papillose beneath **3. K. papillosa**
\+ Seeds with small barbate tubercles; leaves not papillose beneath **4. K. ramosissima**

1. K. incana (Wall.) Pennell; *Linaria incana* Wall. (but not sensu F.B.I.)

Prostrate, greyish-green, suffrutescent perennial, with many diffuse branches. Stems to 40cm, slender, villous with unequal, white, eglandular hairs mostly

0.5–1mm, intermixed with some glandular hairs. Leaves shortly petiolate (petiole 0.5–3mm); lamina ovate or oblong-ovate, upper sometimes deltoid, 3–16 × 2–12mm, obtuse or subacute, base hastate or obtuse (basal lobes if present very short and spreading), entire or with 2 lateral teeth near base, eglandular-villous on both surfaces. Pedicels 7–19mm, filiform, erecto-patent, ± straight except for abruptly reflexed tip. Calyx lobes linear-lanceolate, c 3.5 × 0.7mm, scarious-margined, shortly eglandular-villous. Corolla yellow, 5.5–7mm (excluding spur), densely pilose outside; spur 2–3mm. Capsule subglobose, 2.5–4mm; seeds 0.35–0.5mm, tuberculate.

Bhutan: C – Tashigang district (Tashigang–Yadi). Roadside rocks and cliff-slopes in very arid valley, 800–950m. October.

2. K. membranacea Sutton

Diffusely branched, suffrutescent, scandent or prostrate perennial. Stems many, slender, wiry, 20–50cm, glandular-villous near base, otherwise glabrous. Leaves all with short slender petiole; lamina thin and membranous, narrowly ovate-lanceolate, base sagittate with narrow sharply deflexed or subpatent lobes, glabrous, tips of main and basal lobes sharply acuminate, ± whitish-scarious. Pedicels 6–18mm, straight or slightly curved with sharply recurved distal end, glabrous. Calyx lobes linear-lanceolate, 3.5–5 × 0.9–1.2mm, abruptly broadened in lower part with broad scarious margin and often irregular lobes, upper part acuminate. Corolla bright yellow with small red spots at throat, 6–8(–9)mm excluding spur, hairy outside; spur 1.5–3mm, acute, making obtuse angle with corolla tube. Capsule oblong-ovoid to subglobose, 3–4mm, glabrous; seeds 0.4–0.6mm, reddish-brown, with numerous short acute or obtuse tubercles lacking barbulate papillae.

Bhutan: C – Punakha district (Wangdu Phodrang, Punakha and Samtengang areas, locally frequent). Roadside rock-faces and crevices in dry valleys and open semi-desert country, chir pine forest zone, 1200–2175m. June–October.

Endemic to dry valleys in western C Bhutan. Replaced in the east by *K. papillosa* and *K. ramosissima*.

3. K. papillosa R.R. Mill

Suffrutescent many-stemmed perennial. Stems 25–40cm, glabrous, ribbed. Leaves all similar, with short, slender petiole 2–4mm; blade ovate, 4–25 × 2–15mm, apex acute-mucronate, base obtuse or very shallowly cordate but always lacking lateral lobes, with scattered appressed hairs near midrib above, glabrous but densely papillose beneath. Pedicels filiform, twining, 11–12mm, recurved distally only. Calyx lobes linear-ovate, 3–3.5 × 0.6–0.8mm, with broadly scarious-margined basal half and abruptly tapered, narrow, long-acuminate proximal half. Corolla tube whitish tinged with pink, lips bright yellow, spotted dark red in throat; corolla 7mm excluding spur, sparsely pilose outside and inside; spur 1.5mm, at c 90° to corolla tube. Capsule broadly ellipsoid to globose, c 3 ×

3.5mm, valvate, glabrous except for a few hairs at apex. Seeds oblong, c 0.4 × 0.25mm, covered with large, whitish, blunt, non-barbate papillae.

Bhutan: C – Mongar district (Gorgon–Tangmachu). Crevices on steep rocky slopes in dry chir pine valley, c 1200m. October.

Endemic to Bhutan and known only from the type specimen. Similar to *K. membranacea* in its seeds with blunt whitish papillae, but leaves not long-sagittate, papillose beneath and with a few appressed hairs above; seeds smaller (c 0.4 × 0.25mm).

4. K. ramosissima (Wall.) Janchen; *Linaria ramosissima* Wall. Fig. 93c.

Prostrate suffrutescent perennial. Stems to 40cm, slender, glabrous except near base. Leaves ovate or narrowly ovate, occasionally almost orbicular in outline, 5–25 × 2–25mm, acute, base obtuse or hastate, sometimes with small lateral teeth just above base, usually glabrous. Pedicels 8–12mm, ± straight except for recurved distal end, filiform. Calyx linear-lanceolate, 3–3.5mm, gradually acuminate upwards from scarious-margined base. Corolla pale yellow with brown spots in throat, c 7mm excluding spur, hairy outside and inside; spur c 2mm. Capsule subglobose, 2.5–3mm; seeds dark brown, covered with barbate tubercles.

Bhutan: C – Tashigang district (Cha Zam near Tashigang and Tashigang Yadi). Dry rock-faces in arid valley and on rocks by roadsides, 900–1000m. June–October.

A very variable species, much confused in the literature, which needs further taxonomic study. The description applies to plants from the E Himalaya only. Yamazaki's record of this species from Punakha district (Wangdu Phodrang (71)) almost certainly refers to *K. membranacea*.

Used in Uttar Pradesh as a cure for diabetes (352).

5. CYMBALARIA Hill

Creeping perennial herbs. Leaves opposite below, alternate above, lamina cordate-orbicular to reniform, palmately veined, petiolate. Flowers solitary in leaf axils, long-pedicellate. Calyx deeply 5-lobed, lobes ± unequal. Corolla strongly 2-lipped, lower lip 3-lobed with projecting palate closing mouth of corolla tube; tube spurred. Fertile stamens 4, didynamous, included, adjacent pairs of anthers coherent at margins. Capsule opening by 2 lateral pores, each 3-valved. Seeds numerous, ovoid with strong wavy ridges and usually some tubercles, black.

1. C. muralis Gaertner, Meyer & Scherbius. Eng: *Ivy-leaved toadflax.*

Glabrous trailing or drooping perennial. Stems 10–80cm, often purplish, slender. Leaves mostly alternate; petiole longer than lamina; lamina usually reniform to semicircular, (6–)12–25(–55) × (10–)15–30(–65)mm, palmately 5–9-lobed, often purplish beneath; lobes rounded to triangular, often mucronate, middle one ± larger than others. Calyx 2–2.5mm. Corolla usually violet (rarely white) with yellow palate, 6–9mm excluding spur; spur 1.5–2.5mm, subequal to calyx. Capsule globose, c 4mm, ± longer than calyx, glabrous; seeds coarsely rugose.

Darjeeling: Darjeeling. Introduced from Europe and naturalised on walls, c 2100m. June.

6. ANTIRRHINUM L.

Penerennial herb, usually suffruticose but often scarcely so in cultivars. Leaves opposite below, usually alternate above, simple, entire, pinnately veined. Inflorescence a terminal bracteate raceme. Flowers zygomorphic. Calyx deeply and ± equally 5-lobed, lobes much shorter than corolla tube. Corolla glandular-pubescent outside; tube cylindrical, wide, abaxially gibbous at base; limb bilabiate, upper lip 2-lobed, lower 3-lobed and with a prominent palate at base which closes the mouth of the tube. Stamens 4, didynamous, included. Stigma capitate. Capsule beaked, with 2 unequal loculi; adaxial loculus longer, narrower above, opening by a single apical pore; abaxial loculus shorter, broader above, with 2 apical pores. Seeds numerous, reticulate-rugose.

1. A. majus L. Eng: *Snapdragon.*

Stems glandular-pubescent, ± erect. Leaves linear to ovate, 10–70 × 1–10mm, subacute to obtuse, base cuneate, shortly petiolate. Raceme dense to lax, densely glandular-pubescent or sometimes glabrous. Bracts ovate, 2–10mm. Pedicels 8mm in fruit. Calyx lobes ovate-oblong, 5–8mm. Corolla violet, purple, pink, yellow or white, 30–45mm; palate yellow. Capsule oblong, 8–10mm, glandular-pubescent. Seeds oblong, c 0.7–1 × 0.5–0.8mm, with fuscous, deeply foveolate testa.

Bhutan: C – Tashigang district (Tashigang). Cultivated in gardens, c 1350m. June.

Native to W Mediterranean area. Cultivars of subsp. *majus*, or of hybrid origin, are commonly grown as ornamentals. Only one specimen, with white flowers, has been seen from Bhutan.

7. MIMULUS Adanson

Perennial herbs of wet places. Stems erect or prostrate, quadrangular. Leaves opposite, shortly petiolate or sessile, toothed. Flowers solitary, axillary. Calyx cylindrical to campanulate or infundibular, 5-angled, 5-toothed, teeth equal or unequal, mouth of calyx straight and ± truncate, or oblique. Corolla showy, yellow, bilabiate; upper lip erect or reflexed, 2-lobed; lower lip spreading, 3-lobed; throat with a swollen 2-lobed palate and often spotted red or brown within. Stamens 4, didynamous, inserted at base of corolla; anther thecae divergent. Style with 2 expanded lips. Fruit a compressed loculicidal capsule, valves separating from placentiferous axis. Seeds numerous, oblong, minute.

1. Calyx cylindrical-campanulate or campanulate, 6–10mm in flower; corolla 10–20(–23)mm; pedicels longer than leaves .... **1. M. nepalensis**
\+ Calyx broadly infundibular, 11–17mm in flower; corolla (18–)23–35mm; pedicels shorter than or subequalling leaves .... **2. M. bhutanicus**

1. M. nepalensis Bentham. Fig. 93h.

Diffusely branched perennial, often forming wide carpets. Stems 2–25cm, usually almost completely glabrous, green-winged on angles throughout their length. Leaves shortly petiolate (petioles 1–7mm, sparsely pubescent or glabrous), lamina ovate or ovate-oblong or in smallest leaves ovate-orbicular, 4–20 × 3–12mm, obtuse or subacute, base cuneate or shortly attenuate, margin shallowly and remotely serrate in upper ⅔, upper surface glabrous or sparsely pubescent, lower surface subglabrous. Pedicels 10–30mm, longer than leaves. Calyx cylindrical-campanulate, 6–10mm in flower, 10–14mm in fruit and appearing truncate; teeth subequal, short and broad with abruptly pointed tips. Corolla yellow, sometimes with reddish or brownish spots in throat, 10–20(–23)mm. Filament pairs 8–9mm and 5.5–6.5mm, all flattened ± throughout their length. Style 5–7mm. Capsule narrowly ellipsoid, 8–10 × 2.5–3.5mm.

Bhutan: S – Chukka and Gaylegphug districts, **C** – Ha, Thimphu, Punakha, Tongsa and Bumthang districts; **Darjeeling:** Ghoom, Darjeeling, Kurseong etc. By streams or on moist banks in clearings of subtropical and coniferous (*Abies*) forest, 900–2800m. March–September.

A specimen from Ha district (*Ludlow & Sherriff* 116), which was identified by Tsoong as a small form of *M. tenellus* var. *procerus* (= *M. bhutanicus*, below), falls entirely within the range of variation of *M. nepalensis* but also approaches the Chinese species *M. bodinieri* Vaniot in some characters such as its subentire, very shortly petiolate leaves.

2. M. bhutanicus Yamazaki; *M. nepalensis* var. *procerus* Grant, *M. tenellus* Bunge var. *procerus* (Grant) Handel-Mazzetti. Fig. 93f–g.

More robust than *M. nepalensis*. Stems 14–50cm or more, sparsely pubescent or glabrous, angles narrowly winged for only a short distance below each node or not at all. Leaves usually subsessile, sometimes shortly petiolate (petiole 3–15mm); lamina ovate to oblong-ovate, 15–45 × 6–20mm, acute, base cuneate, margin ± coarsely serrate, shortly pubescent above and on veins beneath or subglabrous, margins often minutely scabrid. Pedicels 12–25(–30)mm, shorter than or subequalling leaves, erecto-patent. Calyx broadly infundibular, 11–17mm in flower and fruit, slightly oblique, shortly eglandular-pubescent and with a few glandular hairs especially near base; teeth short- to long-acuminate, one often longer than others. Corolla bright yellow usually with brownish or orange spots inside throat, (18–)25–35mm. Filament pairs 12–14mm and 9–10mm, flattened in lower ⅔ and c 0.5mm wide, abruptly becoming filiform and c 0.1mm wide in upper ⅓. Style 10–12mm. Ovary 4–5.5 × 1.5–2mm. Capsule ellipsoid but tapered apically, 5–8 × 2–3.5mm, c ½ × calyx.

Bhutan: C – Tongsa, Bumthang, Mongar and Sakden districts, **N** – Upper Pho Chu and Upper Kuru Chu districts; **Darjeeling:** Tiger Hill; **Sikkim:** Lachen, Lagyap–Phusum, Yumthang etc. Damp alpine meadows and streamsides in coniferous (*Abies*) and mixed forest, 2800–3800m. Late June–September.

More northern in its distribution in Bhutan and Sikkim than *M. nepalensis*, and tending to occur at higher altitudes and so flowering later. The two species overlap in C Bhutan (e.g. Tongsa) and occasional intermediates occur (e.g. Chukka, 1830m, *Cooper* 1306, which is like *M. nepalensis* but has a somewhat infundibular calyx).

8. MAZUS Loureiro

Annual or perennial herbs, sometimes stoloniferous. Leaves mainly crowded in a basal rosette; cauline when present opposite below, ± alternate above. Inflorescence an often secund, scapose, terminal, bracteate raceme. Calyx infundibular to campanulate, 5-fid, lobes ± equal. Corolla personate, tube very short; limb bilabiate; upper lip erect, entire or shortly bifid, much smaller than lower lip; lower lip spreading, 3-lobed, its middle lobe with 2 deep channels on underside and 2 ridges clothed with clavate glandular hairs on upper side. Stamens 4, didynamous, inserted in corolla tube; anthers 2-celled, cells divergent. Fruit a loculicidal 2-valved capsule. Seeds very numerous, minute.

1. Flowering stems few, decumbent to erect; calyx tube longer than lobes, lobes remaining suberect in fruit 2
\+ Flowering stems several to many, ± prostrate; calyx tube shorter than or subequalling lobes, lobes becoming ± stellately spreading in fruit 3

2. Stoloniferous; inflorescence glandular; corolla 7–11mm ... **1. M. surculosus**
\+ Non-stoloniferous; inflorescence eglandular; corolla (11–)15–22mm **2. M. dentatus**

3. Stem and leaves ± densely eglandular- and glandular-pubescent, longest eglandular hairs at least 0.4mm and up to 0.9mm; fruiting pedicels usually shorter than or subequal to calyx; bracts green, not hyaline .. **3. M. delavayi**
\+ Stem and leaves subglabrous or sparingly eglandular- and glandular-puberulent, longest eglandular hairs not more than 0.3(–0.4)mm; fruiting pedicels usually longer than calyx; bracts green but hyaline at least in part **4. M. pumilus**

1. M. surculosus D. Don. Nep: *Dhabre*.

Perennial with arching aerial stolons, hardly blackening when dry. Flowering stems decumbent to suberect, 4–14cm, sparingly eglandular-pubescent below, more shortly glandular-puberulent above. Leaves ± all basal; petioles 5–20mm; lamina ± obovate (smallest ones suborbicular), 4–40 × 4–25mm, with 5–7 crenations per side, larger ones also usually with 1–2 pairs of basal lobes; both surfaces sparingly eglandular-pilose and with sessile glands, veins sparsely whitish-hyaline eglandular-pilose beneath. Inflorescence up to 14-flowered, subsecund. Bracts 2–4mm, linear. Pedicels 2.5–8mm in flower, 7–22mm in fruit, sparsely hairy. Calyx infundibular-campanulate, 4–6.5mm in flower, c 8mm in fruit, with a few sessile glands on lobes and scattered hairs on veins; tube c twice lobes; lobes suberect, scarcely accrescent in fruit, remaining ± erect. Corolla 7–11(–13)mm; upper lip mauve, lower lip white with 2 yellow or orange ridges, throat with red or brownish-yellow spots. Capsule ovoid to subglobose, 4–4.5 × 3mm. Seeds lenticular, 0.3–0.4 × 0.15–0.2mm, light brown, each end with a tiny projection.

Bhutan: S – Phuntsholing and Chukka districts, **C** – Thimphu, Punakha and Tongsa districts, **N** – Upper Kulong Chu district; **Darjeeling:** Ghoom, Jallapahar, Kurseong, Mahaldiram etc.; **Sikkim:** Gangtok, Phusum, Sangachoiling etc.; **Chumbi:** Chumbi. Warm broad-leaved forest, steep grassy hill slopes, field margins, short grassland, road- and ditch-sides, 1260–3050m. April–August.

A specimen from C Bhutan (Shamgong Dzong, 600m, *Ludlow, Sherriff & Hicks* 18563, BM) was determined as *M. henryi* Tsoong by Tsoong himself in 1958. In his original description (416) he described the flowers as being 2–3, axillary, and distinguished it from *M. surculosus* on indumentum and leaf shape. The Bhutan specimen examined by him has a terminal inflorescence of up to 9 flowers with corollas up to 13mm. These are slightly larger than in most material of *M. surculosus*, but in other respects the specimen cannot be separated from *M. surculosus.* Another specimen which is unequivocally *M. surculosus* (*Bartholomew* 1587, E) has also been seen from the Shamgong area. *M. henryi* would therefore appear to be restricted to China (Yunnan).

2. M. dentatus Bentham

Non-stoloniferous perennial, turning ± black when dry. Stems ascending to erect, 7–20cm, leafy only towards base, rather densely eglandular-pubescent below, much less so and sometimes subglabrous above; glandular hairs absent. Basal leaves with pubescent or pilose petiole 10–90mm; lamina oblong-elliptic or oblong-ovate, 25–100 × 7–50mm, subacute, base obtuse to truncate, margin undulate or with a few broad shallow crenations, upper surface with scattered short hairs, lower surface glabrous except for densely pubescent veins. Inflorescence 2–8-flowered, subsecund. Bracts 1.5–4mm, linear. Pedicels 4–6mm in flower, 6–8mm in fruit, ± pubescent. Calyx infundibular-campanulate, 5–8mm in flower, scarcely accrescent in fruit and remaining erect, tube with short hairs and sessile glands, lobes triangular, almost glabrous. Corolla mauve or blue-violet (sometimes with darker spots), (11–)15–22mm; upper lip erect, lower much larger, with paler or whitish throat and a 2-lobed bright orange densely pubescent swelling within; lobes of lower lip often fimbriate at margin. Seeds oblong, c 0.6 × 0.3mm, blackish, with finely reticulate surface.

Bhutan: S – Gaylegphug district (Gale Chu), **C** –Punakha district (Mishichen–Khosa (Upper Mo Chu district) (71)), Mongar district (Saling, Sengor), **N** – Upper Mo Chu district (Tamji–Gasa (71)); **Darjeeling:** Bhotia Basti, Ghumpahar, Rimbick etc.; **Sikkim:** Karponang, Lagyap. On and under damp mossy rocks and boulders in *Quercus* and *Tsuga* forest, 1670–3050m. April–June.

3. M. delavayi Bonati; *M. japonicus* (Thunberg) Kuntze var. *delavayi* (Bonati) Tsoong, *M. pumilus* (Burman f.) van Steenis var. *delavayi* (Bonati) Wu. Fig. 94g.

Annual, remaining ± green when dry. Stems numerous, procumbent to suberect, 6–25cm, ± densely patent-pubescent with unequal glandular and eglandular hairs; eglandular hairs to 0.7mm; glandular hairs shorter, most numerous in inflorescence, ± absent elsewhere. Basal leaves cuneate into indistinct petiole 8–30mm, cauline subsessile; lamina of basal leaves 15–40 × 9–16mm, oblong-ovate, obtuse, margin shallowly bicrenate or crenate, teeth (3–)5–9 per side; upper surface with sessile glands and usually some eglandular hairs; lower surface glabrous except for shortly hairy veins. Main inflorescences ± secund, 8–15-flowered, later developed ones shorter, fewer-flowered, not secund. Bracts 1–2.5mm in flower, 4–5.5mm in fruit, linear-triangular, green. Pedicels shorter than or subequalling calyx. Calyx 2–4.5mm in flower, 6–9.5mm in fruit; tube shorter than lobes; lobes erect and 2–3mm in flower, stellate-spreading and 4–6mm in fruit, ovate-triangular, acute; veins prominent, eglandular hairy. Corolla 5.5–7.5mm, white or pale mauve, palate with yellow to brown markings. Capsule obovoid to subglobose, apiculate, c 3.5 × 3mm including apiculus. Seeds 0.3 × 0.15–0.2mm, oblong-cuneate, with a tiny projection from one basal corner.

Bhutan: S – Chukka district (71), **C** –Ha, Thimphu (71), Punakha (71) and Tongsa (71) districts; **Darjeeling:** Kalimpong (69); **Sikkim:** Lachen. Streamsides, roadsides and as a wheat- and rice-field weed, 250–2800m. March–June.

4. M. pumilus (Burman f.) van Steenis; *M. rugosus* Loureiro, *M. japonicus* (Thunberg) Kuntze, *M. goodeniifolius* (Hornemann) Pennell

Similar to *M. delavayi* but usually drying rather dark brownish-green; stems very sparingly eglandular-puberulent throughout and with sparse sessile glands in inflorescence, eglandular hairs very short, unequal, 0.05mm (or less)–0.4mm; petiole of basal leaves 0–11(–14)mm, lamina unlobed at base, 7–30 × 2.5–15mm, margin shallowly dentate or crenate with (1–)2–3 teeth per side (sometimes subentire), surfaces subglabrous or with scattered hairs; inflorescence usually 2–11-flowered; bracts 0.5–1.5mm in flower, to 3mm in fruit, at least partly hyaline; lowest pedicels usually longer than fruiting calyx; corolla white or whitish-blue with upper lip purplish or blue and with orange spots on palate of lower lip; seeds with 2 tiny projections diagonally opposite each other.

Bhutan: S – Samchi district (Samchi), Chukka district (Chasilakha (117)), **C** – Tongsa district (Birti (117)); **Darjeeling:** Char Churabhandar; **Sikkim:** Lachen, Reinak. Weed of gardens and in moist places, 500–1220(?–2000)m. March–May.

M. pumilus is treated here in a broad sense, yet excluding the much more hairy plant which is readily separated as *M. delavayi.* The latter is by far the commoner species of this group in Bhutan. The literature records cited above may apply to it and consequently require confirmation.

9. **LANCEA** Hook.f. & Thomson

Low, perennial herb with slender root-stock and sometimes underground stolons. Stem usually absent, occasionally developed. Leaves rosulate, and opposite when stem present. Inflorescence a usually very condensed raceme; flowers shortly pedicellate, ± sunk among leaves. Bracteoles present. Calyx broadly campanulate, 5-fid, lobes equal. Corolla bilabiate; upper lip suberect, oblong, bifid; lower lip much larger, spreading, 3-lobed, longitudinally bigibbous at throat. Stamens 4, didynamous, all fertile; anther thecae contiguous. Ovary broadly oblong; style slender with 2-lamellate stigma. Fruit a fleshy bilocular berry. Seeds very numerous, brown.

Hooker & Thomson described the root as being annual but this seems unlikely as underground stolons frequently occur. *Flora of China* (351) regards it as perennial.

1. L. tibetica Hook.f. & Thomson. Med: *Payak*. Fig. 93i–j.

Usually glabrous and acaulescent herb with slender root-stock and thin fibrous roots. Stem, when present, 3–10(–15)cm. Leaves shortly petiolate (petiole 5–20mm); lamina obovate to elliptic, 20–50(–70) × 8–35mm, subacute, base cuneate to attenuate, glabrous or sparsely pubescent above, usually glabrous beneath, margin ± ciliate, entire or obscurely serrate. Pedicels 2–10mm. Bracteoles linear-triangular, acute. Calyx 6–8mm, lobes acute, ⅓–½ × tube. Corolla purple or violet-blue, paler near base of tube and lower lip paler than upper; tube 10–12mm; upper lip c 3mm; lower lip 5–8 × 7–12mm, with white hairs near mouth of tube. Filaments white; anthers pale yellow. Style and stigma white. Berry crimson at first, brown or black at maturity, 5–7mm diameter, shortly apiculate.

Bhutan: C – Thimphu district (Kumathang), **N** –Upper Mo Chu district (Lingshi, Chebesa), Upper Bumthang Chu district (Damakura, Pangotang), Upper Kulong Chu district (Me La); **Sikkim:** Lachen–Thanggu, Llonakh, Yumchho; **Chumbi:** Phari, Chumbi. In bogs and swamps, and by sandy and gravelly stream banks and irrigation channels, 3800–5270m. May–September.

10. CALCEOLARIA L.

Annual herb with somewhat weedy habit. Stems ± flaccid or succulent, much branched from near base and above, glandular-pilose at least above. Leaves opposite, decussate, petiolate, pinnatifid. Inflorescence a terminal thyrse often with ± numerous axillary flowers below. Calyx 4-partite. Corolla slipper-shaped, bilabiate; upper lip hooded, smaller than saccate, inflated lower lip. Stamens 2; anthers relatively large, glabrous, with elongated connective; anterior theca smaller than posterior one or reduced and sterile. Fruit a capsule which splits longitudinally along 4 sutures.

Calceolaria is an extremely large genus, many species of which are horticulturally valuable. The description above applies only to *Calceolaria* sect. *Calceolaria*, to which our species belongs. The north-west S American species of this group, including ours, have been carefully revised by Molau (379), who has determined much of the material seen for *Flora of Bhutan*.

1. C. tripartita Ruiz & Pavón; *C. mexicana* sensu F.B.I. non Bentham, *C. gracilis* Humboldt, Bonpland & Kunth

Stems erect to decumbent, 8–50cm or more, sparsely short glandular-pilose. Leaves petiolate (petioles 1–4cm), lamina ± broadly ovate in outline, c 20–100 × 8–85mm, pinnatisect with 1–2 pairs of leaflets and a larger terminal lobe; lateral leaflets ovate-elliptic, 8–25mm; terminal leaflet lanceolate to ovate or slightly obovate and sometimes 3-lobed, 15–35mm; all leaflets serrate, glandular-pilose, paler beneath. Pedicels 1.5–5cm, glandular-hirsute. Sepals ovate, 4–8 × 1.5–6.5mm, acuminate. Corolla deep yellow (including when dried); upper lip ± globose, 3–5 × 3–6mm; lower lip almost orbicular, 9–15mm, saccate for at least ⅘ of its length. Anthers with white posterior theca 1–2mm and vestigial, sterile anterior theca. Capsule broadly conical or subglobose, 5–8mm, glandular-hirsute, tips of valves reflexed after dehiscence.

Darjeeling: Darjeeling–Tonglu, Ghoom, Senchal etc. Usually in damp, shady places, 1980–2440m. April–November.

Native of southern Mexico, C and tropical S America (especially the Andes), south to Bolivia; introduced and cultivated as an ornamental in parts of tropical Asia including Nepal and Sikkim, where it has escaped from gardens. Naturalised since at least 1860 in Sikkim. Watt (in sched.) observed that 'it forms a striking feature of spring vegetation' around Darjeeling.

11. LINDENBERGIA Lehmann

Annual or perennial herbs, or ± scandent small shrubs. Stem erect or ascending (sometimes pendent), branched or less commonly unbranched, pubescent or glabrous. Leaves simple, opposite or uppermost alternate, petiolate, serrate. Flowers in terminal spike-like racemes or axillary. Bracts leaf-like but usually smaller. Calyx campanulate or crateriform, divided to middle or less into 5 acutely triangular lobes. Corolla bilabiate, yellow (sometimes tinged purple or red at least on upper lip), corolla throat or tube with orange or brown markings; tube cylindrical; upper lip erect, pubescent in middle; lower lip longer than upper, hooded, with 3-lobed apex. Stamens 4, didynamous. Style filiform, often dilated towards apex, hairy at least near base (in ours). Fruit a loculicidally dehiscent, ellipsoid to ovoid capsule; seeds numerous, elongate-ovoid or ellipsoid, with reticulate testa.

1. Leaves glabrous on both sides .... **2. L. hookeri**
\+ Leaves hairy on both sides or at least beneath 2

2. Calyx 6–8mm; corolla 23–45mm .... **3. L. grandiflora**
\+ Calyx 3–5mm; corolla not more than 15mm 3

3. Style hairy throughout; stem and leaves with silky hairs ..... **1. L. griffithii**
\+ Style hairy only at base; stem and leaves with non-silky hairs 4

4. Perennial; flowers in compact terminal racemes; style 8–10mm **4. L. titensis**
\+ Annual; flowers axillary; style 3.3–6.5mm .... **5. L. muraria**

1. L. griffithii Hook.f. Fig. 94c–e.

Small weak shrub or tall herb to 3m. Stems unbranched, erect (soon becoming decumbent) or somewhat scandent, densely appressed-sericeous with soft silky yellowish-white hairs, glabrescent in inflorescence after flowering. Leaves very shortly petiolate; lamina elliptic, 50–110 × 25–50mm, acuminate or sharply acute, base cuneate, margin serrate, surfaces silky-villous. Racemes solitary or usually in fascicles in leaf axils, often branched. Peduncle 1–3cm; axis 5–10cm, yellowish-pilose. Bracts sessile, ovate, 2.5–4mm. Calyx 3–5mm, densely short-hirsute; lobes shortly triangular, acute. Corolla bright yellow (drying blackish) with small purplish-brown speckles on lower lip, 6.5–8.5mm; tube 3–4mm, sparsely pilose at first, glabrescent; upper lip 2–2.5mm, triangular, abruptly tapering from broad base to emarginate apex; lower lip 2–2.8mm, 3-lobed with middle lobe largest. Ovary glabrous; style hairy throughout, 5–5.5mm. Capsule broadly ellipsoid, 4–4.5 × 2.6–2.9mm, pilose along line of dehiscence when immature, finally completely glabrous.

Bhutan: C – Tashigang district (two collections near Tashigang). Among shrubs on arid hillsides, 1220–1350m. February.

Apparently local in eastern Bhutan; otherwise known only from Assam.

2. L. hookeri Hook.f.; *L. bhutanica* Yamazaki. Nep: *Hik-shut-up.*

Suffrutescent perennial herb or small shrub, 60–180cm. Stems much-branched, glabrous, inflorescence branches pubescent with minute ascending-erect hairs. Leaves shortly petiolate or subsessile; lamina elliptic, 14–60 × 7–18mm, acuminate, base cuneate, margin ± remotely but usually coarsely and sharply serrate, both surfaces glabrous. Flowers fragrant, opposite but secund in relatively short terminal racemes 4–9cm; hairs of racemes glandular, clavate. Bracts sessile, ovate, acute, shorter than calyx. Calyx 7–10mm, bowl-shaped, divided almost to middle, sparsely pubescent outside and inside; lobes triangular, acute. Corolla golden yellow with purplish or reddish upper lip, 14–26mm; tube 7–14mm, broadened upwards, glabrous outside, pilose inside but glabrescent; upper lip c 4mm, ± glabrous; lower lip c 6mm, middle part deeply hooded and densely pubescent, apex 3-lobed. Style 12–16mm, hairy only at base. Capsule ovoid, 6–8 × 3.5–4.5mm.

Bhutan: S – Samchi district (Tamangdhara Forest), Chukka district (Gedu–Kharbandi), Phuntsholing district (Phuntsholing, Rinchending), Gaylegphug district (Sher Camp), Deothang district (N of Deothang), **C** – Tashigang district (Pintsogong); **Darjeeling:** Punkabari–Kurseong; **Darjeeling/Sikkim:** Sura and unlocalised (Hooker and Clarke collections). Subtropical forest, wet montane broad-leaved forest and secondary jungle, on shaded slopes and damp roadside banks, 650–1800m. February–May.

3. L. grandiflora (D. Don) Bentham

Perennial herb or small shrub, often scandent or pendent, (12–)30–250cm or more. Stems branched (occasionally simple if depauperate), rather flexuous, with ± dense, short, spreading glandular hairs. Leaves petiolate (petiole 5–25mm); lamina broadly to narrowly elliptic-ovate, 45–130 × 25–80mm, acute to acuminate, base usually obtuse, margin serrate, both surfaces densely short-glandular-pubescent. Flowers axillary, racemose or solitary. Bracts subsessile, ovate or oblong-ovate, acute, smaller than leaves. Calyx 6–9mm, densely spreading glandular-hirsute inside and outside. Corolla golden yellow, with a brown band on upper side of tube and sometimes with small orange spots within, 23–45mm; tube 12–20mm, gradually broadening from narrow base, glabrous outside in lower half but pilose above, sparsely hairy within; upper lip 3–6mm, obcordate, hairy; lower lip spreading, 3-lobed, lobes 3–7mm diameter. Anthers c 0.8mm, whitish. Ovary densely hairy; style 19–20mm, hairy towards base. Capsule ovoid, 3–5mm, hairy.

Bhutan: S – Samchi district (W bank of Torsa River), Deothang district (Chunkar); **Darjeeling:** Bhotia Batasi, Darjeeling, Rambi etc. Evergreen jungle, shady banks and cliffs in subtropical zone, and in open *Quercus*/*Rhododendron* forest, 200–2135m. Often hanging from other vegetation or from steep banks (including railway embankments). August–April.

4. L. titensis Sikdar & Maiti; *L. philippensis* auct. non (Chamisso) Bentham

Suffrutescent perennial. Stems (15–)30–60cm, slender, strongly branched from base (occasionally unbranched when depauperate), densely eglandular- and glandular-pubescent, glabrescent. Leaves shortly petiolate (petiole 5–10mm); lamina elliptic or ovate-elliptic, 25–60 × 3–25mm, acute, base cuneate, margin acuminately dentate. Flowers numerous, in dense, compact terminal and axillary racemes. Bracts ovate-elliptic, 5–9 × 3–5mm, acute, longer than calyx. Calyx scarcely zygomorphic, 3.5–4mm, densely pilose outside and inside; lobes ovate-lanceolate, acute. Corolla yellow, with red dots in throat, 10–15mm; tube 5–8mm, glabrous or sparsely pubescent; upper lip 4–6mm, abruptly curved, with 2-lobed apex; lower lip 5–8mm, obovate, 3-lobed. Ovary densely pilose; style hairy near base, 8–10mm. Capsule ovoid, 4–5mm, densely pilose, later glabrescent.

Bhutan: S – Samchi district (Khagra Valley near Gokti; Daina Khola); **?Darjeeling** (see below); **W Bengal Duars:** Jalpaiguri district, Maderihat Forest Range, Titi (type). Roadsides and river beds in secondary subtropical forest, on limestone, c 400m. February–March.

The Darjeeling record is based on a Cowan specimen (E), determined as *L. philippensis* (Chamisso) Bentham by M. Hjertson (Uppsala); it bears an annotation suggesting that the plant was actually collected near Chittagong (Bangladesh). This record should therefore be treated with caution, as should all records of *L. philippensis* from our area. *L. titensis* is similar to *L. macrostachya* Bentham and to *L. philippensis*; from the latter it can be distinguished by its ovate leaf-like bracts longer than the calyx, ovate, acute calyx lobes and densely pilose ovary. The Bhutan specimens were recently determined as *L. titensis* by Hjertson.

5. L. muraria (Roxb.) Brühl. *L. urticifolia* Lehmann, *L. ruderalis* sensu Pennell non (Retzius) Voigt, *L. indica* sensu 69 and 101 p.p. non (L.) Vatke. Nep: *Beduwar jhar.* Fig. 94b.

Annual, 10–40cm. Stems usually branched from base, 6–25cm, ± densely eglandular- and glandular-hairy. Leaves shortly petiolate; lamina ovate to elliptic, 7–38 × 4–25mm, crenate-serrate in upper ⅔ (teeth 3–13 per side, rather acute), obtuse or subacute, base cuneate, both surfaces with long glandular and eglandular hairs. Flowers solitary in axils, subsessile in flower but with pedicels to c 3mm in fruit. Calyx 3–5mm; lobes ovate, acute, middle lower lobe shorter and narrower than others, all lobes densely hirsute. Corolla yellow with red markings on tube and orange palate; tube 4–8mm; upper lip 2–4mm, with long simple eglandular hairs inside along middle, tip obcordate or 2-lobed; lower lip 2.5–7.5mm, densely hairy along middle. Ovary densely hairy; style with dense hairs at base, 3.3–6.5mm. Capsule ovoid, 2.5–4 × 2.5–3.5mm, densely pilose.

Bhutan: S – Samchi district (Torsa River), **C** –Tongsa district (Tongsa Dzong, above Dakpai, near Pertimi), Mongar district (above Mongar), Tashigang district (above Khaling), **N** – Upper Kuru Chu district (Denchung, Khoma Chu); **Darjeeling:** Darjeeling, Kodabarry, Mongpu, Tista; **Sikkim:** Chuentong, Tumlang. Dry roadside rock-faces and walls, often limestone, 200–2550m. February–October.

Closely allied to *L. indica* (L.) Vatke and by some authors included in the latter. However it is easily distinguished by the hairy ovary, style and capsule and I have followed Prijanto (394) in keeping them as separate species. *L. indica* sensu stricto is primarily a species of Pakistan and India (Punjab to Calcutta). There are several literature records of *L. indica* from Bhutan (Thimphu, Tongsa and Tashigang districts (117)) as well as records from Sikkim and Darjeeling (69). These are not accepted here and no specimens of true *L. indica* from Bhutan or Sikkim (or indeed Nepal) have been seen. It is likely that all such records apply to *L. muraria. L. indica* is used to treat skin eruptions, sore throats and toothache (352).

12. ADENOSMA R. Brown

Aromatic annual herb. Leaves opposite, simple, serrate. Inflorescences terminal and axillary, densely capitate, many-flowered. Bracteoles 2, filiform, at base of calyx. Calyx with 3 large outer lobes and 2 much smaller linear-lanceolate inner lobes. Corolla cylindrical, bilabiate; upper lip exterior in bud, 2-lobed; lower lip equalling upper, deeply 3-lobed, lobes linear-oblong. Stamens 4, didynamous, included; thecae separate, one loculus of all 4 anthers, or of 2 anterior ones, sterile. Style filiform; stigma 2-lobed. Fruit a capsule, dehiscing both septicidally and loculicidally. Seeds numerous, small, with reticulate testa.

Superficially resembling *Lindenbergia* in habit and in its serrate leaves, but differing by the violet (not yellowish) corollas, the outer lobes of which (not the inner) are exterior in bud, and by the bracteolate calyx.

1. A. indianum (Loureiro) Merrill; *A. capitatum* (Bentham) Hance

Annual, robust especially in fruit, smelling of eucalyptus. Stems simple or branched above, 10–70cm, villous-hirsute especially above with white erecto-patent eglandular hairs. Leaves subsessile or with short hairy petiole 0.5–9mm; lamina ovate-oblong or ovate, 1–7 × 0.5–2cm, subacute, base cuneate, finely serrate, hirsute above and on veins beneath, lower surface with numerous large ± circular sessile glands. Inflorescences globose or shortly ovoid-cylindrical, largest heads 0.5–3 × c 1cm, axillary ones usually smaller. Upper bracts lanceolate to linear-lanceolate. Bracteoles filiform, c 4mm, hairy. Calyx c 4mm, densely long-villous; upper lobe broadly lanceolate, c 0.5mm broad, others half as broad, all acute. Corolla violet, 5–6mm, glabrous outside. Capsule ovoid, c 3.5 × 2.5mm, acuminate.

Darjeeling: Balasun, Darjeeling, Siliguri etc. Sal forest, roadsides and banks, 150–300m (but to c 1000m in Nepal). September–January.

Hooker incorrectly described the indumentum as being glandular-villous.

13. LIMNOPHILA R. Brown

Aquatic or marsh plants, annual or perennial, aromatic and provided with pellucid gland-dots. Stems erect, diffuse or floating, terete. Submerged leaves, when present, whorled and pinnatifid to very finely divided into capillary segments. Aerial leaves opposite or whorled, sessile or more rarely petiolate; lamina dentate, serrate or pinnate. Inflorescence a raceme, cyme or panicle, or flowers solitary in axils. Bracteoles 2 (rarely 0) at base of calyx. Calyx tubular or tubular-campanulate, divided into 5 usually lanceolate lobes. Corolla tubular or infundibular, bilabiate; upper lip orbicular, emarginate or 2-lobed; lower lip 3-lobed. Stamens 4, didynamous, included; thecae stipitate, separate. Ovary glabrous. Style filiform with bilamellate stigma. Fruit an ellipsoid or globose capsule, dehiscing both septicidally and loculicidally with valves separating from placentiferous axis by 2 wings. Seeds cylindrical or ellipsoid, small, punctate.

All species occur in wet or muddy places, such as rice-fields. The genus is under-collected in our area. Efforts should be made by local botanists to establish the current frequency and distribution of all species. Nearly all records are based on material collected prior to 1915.

1. Finely divided submerged leaves present 2
+ Finely divided submerged leaves absent 3

2. Flowers sessile (rarely with very short pedicels not more than 1.5mm); bracteoles absent; aerial stems ± white-hirsute .... **1. L. sessiliflora**
+ Flowers distinctly pedicellate (pedicels 3–16mm); bracteoles present; aerial stems glabrous or glandular but not hirsute .... **2. L. indica**

3. Calyx ± zygomorphic with enlarged adaxial lobe .... **3. L. rugosa**
+ Calyx regular, all lobes ± equal 4

4. Corolla 2.5–7mm; pedicels absent or very short (0–4mm) 5
+ Corolla 10–17mm; pedicels usually distinct, (2–)5–18mm 6

5. Stems with sessile or stalked glands but with no eglandular hairs; leaves glabrous or minutely glandular; pedicels glabrous or with stalked glands **6. L. aromatica**
+ Stems ± lanate with flexuous eglandular hairs; leaves ± lanate; pedicels patent-hirsute .... **7. L. chinensis**

6. Corolla 6–7mm .... **8. L. repens**
+ Corolla 2.5–5mm 7

7. Stems densely glandular-pubescent; corolla 4–5mm; flowers numerous, in a panicle of cymes .... **4. L. polyantha**
+ Stems glabrous; corolla 2.5–4mm; flowers rather few, solitary or in small heads .... **5. L. micrantha**

1. L. sessiliflora (Vahl) Blume

Aquatic or amphibious perennial. Submerged stem to 50cm, glabrous or almost so; submerged leaves 5–30mm, divided into many flattened or capillary segments, glabrous. Aerial stem to 20cm, ± branched, usually white-hirsute, sometimes nearly glabrous; aerial leaves all in whorls of 3–6, lanceolate-elliptic, 4–12(–20) × c 3mm, crenate-serrate or dissected, glabrous, densely punctate. Flowers axillary, usually sessile (rarely pedicels to 1.5mm). Bracteoles absent. Calyx 4–7mm, shortly hirsute. Corolla blue, mauve or purple, 7–12mm. All filaments glabrous; posterior c 1mm, anterior c 3mm. Capsule 3.5–5.5mm.

Bhutan: S – Gaylephug district (Gaylephug River), **C** – Tashigang district (Gomchu–Rongthong (117)); **Darjeeling:** Darjeeling, Sukna; Siliguri (386), Mongpu (338); **W Bengal Duars:** Guzledubar, Rajabhat Khawa (338). Rice-fields and marshes, 60–750m. November–May.

Separable from *L. indica* by the ± hirsute aerial stems, sessile flowers and absence of bracteoles. The flowers are not infrequently cleistogamous. Used in Sikkim and Bengal to treat gastric complaints (352).

2. L. indica (L.) Druce

Aquatic or amphibious perennial. Submerged stem to 1m, much-branched, glabrous; submerged leaves in whorls of 6–12, pinnatisect with flattened or capillary segments, to 30mm. Aerial stems 2.5–14cm, ± branched, with sessile or stalked glands above or subglabrous, never hirsute; aerial leaves in whorls of usually 4–6, sometimes with 2–3 pairs of opposite leaves near apex; most aerial leaves usually ± dissected, opposite pairs (if present) undissected, crenate-serrate, with sessile glands or ± glabrous. Flowers axillary, distinctly pedicellate; pedicels usually 3.5–10mm, accrescent and to c 16mm in fruit, glandular. Bracteoles present, linear or linear-lanceolate, 2–4mm. Calyx 3.5–6mm, tube sessile-glandular, sometimes with a few stalked glands, teeth usually glandular. Corolla white or pale yellow, or yellow at base of tube and red or purplish above, usually 8–12mm, glabrous outside, villous within. All filaments glabrous; posterior c 2mm, anterior c 4mm. Capsule ellipsoid to subglobose, compressed, c 3.5mm.

Darjeeling: Phansidown; **Darjeeling/Sikkim:** unlocalised (Anderson collection (338)). Swamps and rice-fields, c 500m. October–March.

Cleistogamy is unknown in this species (cf. *L. sessiliflora* above). Antiseptic. Used in parts of S and W India to treat dysentery, skin eruptions and elephantiasis, but no medicinal uses have been recorded from our area (338, 352).

3. L. rugosa (Roth) Merrill; *L. roxburghii* sensu F.B.I. non G. Don

Terrestrial annual. Stems to 50cm, ± robust, glabrous or sparsely to densely hispid. Leaves opposite, petiolate; petiole 5–25mm, ± densely pubescent (often only towards base); lamina ovate-lanceolate to ovate-elliptic, 15–90 × 7–50mm, pinnately veined, obtuse, base cuneate-attenuate into petiole, margin crenate-serrate, upper surface scabrid or glabrous, lower surface shortly hirsute or scabrid on midrib and veins and densely punctate. Flowers solitary, sessile and axillary, or in sessile or pedunculate heads of 2–7 flowers subtended by 2 amplexicaul, glabrous, leaf-like bracts; peduncles 0–60mm. Bracteoles absent.

Calyx irregular; adaxial lobe ovate-lanceolate, 8–11 × c 4mm, other 4 lobes 6–9mm and narrower. Corolla blue, to 16mm, pilose outside and inside (especially on lower lip, hairs yellow); upper lip deeply emarginate. Posterior filaments hairy, c 3mm; anterior c 6mm, glabrous. Capsule ovoid, 5.5–6.5mm.

Bhutan: S – Samchi district (Samchi (117)); **Darjeeling:** Labha, Rishap, Rangit, Dalka Jhar, Darjeeling. Shady swamps and bogs, 180–1000m. August–December.

4. L. polyantha Hook.f.

Terrestrial annual. Stems erect or suberect, to 35cm, with ± long branches to 20cm, densely glandular-pubescent. Leaves opposite or in whorls of 3–4, sessile, linear or linear-lanceolate, 5–25 × 1–3.5mm, pinnately veined, base semi-amplexicaul or ± auriculate, margin serrulate, densely glandular-pubescent, glabrescent below. Flowers numerous, sessile or shortly pedicellate (pedicels 0.5–2mm), in a terminal panicle of ± flexuous, monochasial, 7–12-flowered cymes. Bracteoles 2, linear-subulate, glandular. Calyx 2.5–3mm, divided to just beyond middle; lobes lanceolate, shortly acuminate, striate when mature. Corolla pinkish-purple, 4–5mm, glabrous. All filaments glabrous; posterior c 0.2mm, anterior c 0.4mm. Capsule ovoid, c 2mm.

Darjeeling: Darjeeling, Siliguri, Singhi Jhora (338). Marshes in tropical zone, c 500m. November–January.

5. L. micrantha (Bentham) Bentham

Low terrestrial annual with creeping underground stems and many erect or ascending glabrous aerial stems 3–20cm. Leaves opposite or in whorls of 3, sessile, linear-oblong to oblong, 4–14 × 1–5mm, pinnately veined, obtuse, base attenuate, with 3–4 pairs of remote teeth in upper half, glabrous, punctate beneath. Flowers axillary, solitary or in heads of 2–4. Pedicels 0.5–1mm (very slightly longer in fruit). Bracteoles 2, linear, 1.5–3mm. Calyx 2.5–3mm, deeply divided into 5 equal linear-lanceolate lobes 1.5–2mm, glabrous, striate when mature. Corolla pale blue or mauve, or white, 2.5–4mm, glabrous outside and within except for sparsely villous throat. All filaments glabrous; posterior c 0.5mm, anterior c 1mm. Capsule broadly ellipsoid or compressed-globose, 2–2.5mm.

Darjeeling: Siliguri. Wet places including rice paddy-fields, c 150m. November–December.

No recent material seen.

The plant has a strong smell of balsam.

6. L. aromatica (Lamarck) Merrill; *L. gratissima* Blume, *L. chinensis* (Osbeck) Merrill subsp. *aromatica* (Lamarck) Yamazaki. Nep: *Haneya jhan*. Fig. 94e.

Terrestrial annual or perennial. Stems 15–60cm, erect, unbranched or much-branched from base, glabrous or minutely glandular, never with eglandular hairs. Leaves usually in whorls of 3, sometimes opposite, sessile but with narrowed semi-amplexicaul base, lanceolate-elliptic to ovate-lanceolate, 10–55 × 3–10mm, pinnately veined, acute or sometimes obtuse, shallowly serrate throughout, glabrous or minutely glandular, densely punctate beneath. Flowers in terminal racemes often with some solitary axillary flowers below. Pedicels 5–15(–20)mm, stalked-glandular or ± glabrous. Bracteoles 2, linear. Calyx 4–7mm, divided to c ¾ into linear-lanceolate, glabrous, punctate segments, striate when mature. Corolla purple, 10–13.5mm, minutely glandular outside, long-villous within especially adaxially. Posterior filaments villous, 2–2.5mm; anterior glabrous, c 4mm. Capsule ellipsoid, 5–6mm, compressed, brown.

Bhutan: C – Punakha district (Samtengang); **Darjeeling:** Sukna, Tashiding Forest. Paddy-field margins and muddy places, 600–1980m. October–December.

The plant has a strong smell of turpentine. The juice is used as an antiseptic and febrifuge.

7. L. chinensis (Osbeck) Merrill; *L. hirsuta* (Bentham) Bentham

Terrestrial or less commonly aquatic annual or perennial. Stems creeping below and rooting at nodes, then erect, 5–40cm, pinkish, densely lanate with eglandular hairs to subglabrous. Leaves opposite or in whorls of 3–4, sessile but distinctly narrowed into semi-amplexicaul sometimes petiole-like base, ovate-lanceolate to linear-oblong, usually 10–40 × 5–14mm, indistinctly pinnately veined, distinctly serrate in upper ⅔, subglabrous and indistinctly punctate above, glabrous or hairy on veins and rather densely punctate beneath. Inflorescence a terminal leafy panicle, or flowers solitary and axillary. Pedicels (2–)6–18mm, patent-hirsute. Bracteoles filiform or subulate, 1–2mm. Calyx (4–)6–9mm, divided to ½–¾ into long-acuminate lanceolate segments which become spreading in fruit, hirsute and with short-stalked glands, distinctly striate when mature. Corolla blue or purple with whitish tube, usually 10–17mm, thinly pilose outside, villous inside. All filaments glabrous; posterior 1.5–2.5mm, anterior 3.5–4.5mm. Capsule broadly ellipsoid, compressed, c 5mm.

Darjeeling: Siliguri (386), Darjeeling–Mongpu (338) etc.; **Darjeeling/Sikkim:** Rongpoghora. Terai, wet lowland swamps, rice-fields and meadows and occasionally in stagnant water, to 1000m. October–December.

8. L. repens (Bentham) Bentham; *L. sessilis* (Bentham) Fischer, *L. conferta* Bentham *nom. illeg.*

Terrestrial or amphibious annual. Stems usually procumbent or floating below and rooting at lower nodes, becoming erect, to 40cm, glabrous or shortly hairy at least above. Leaves smelling of camphor, opposite, sessile or subsessile, obovate or ovate-elliptic or linear-lanceolate, 7–30(–50) × 4–8(–15)mm, pinnately veined, obtuse or acute, base attenuate or amplexicaul, margin serrate above middle, glabrous or hairy, densely punctate. Flowers in short bracteate axillary cymes or racemes, or solitary-axillary. Pedicels 0.5–3mm. Bracteoles 2, linear-oblong, 1.5–3mm. Calyx usually 4–5.5mm, divided to c ⅔; lobes linear-acuminate, recurving at maturity, shortly hispid. Corolla white or mauve,

6–7mm, glabrous outside, villous within. All filaments glabrous; posterior 1–1.5mm, anterior 2.5–3.5mm. Capsule ovoid, 3–4mm.

Bhutan: S – Samchi district (Samchi (117)); **Darjeeling:** Sukna, Rangit, Singhi Jhora, Sivok–Siliguri (338); **W Bengal Duars:** Jalpaiguri. Rice-fields, at low altitudes (to c 500m). October–February.

The description applies to the typical variety, which occurs from E Nepal to Myanmar (Burma). Var. *brevipilosa* (Yamazaki) Yamazaki occurs in Vietnam and Thailand.

14. BACOPA Aublet

Glabrous, ± fleshy perennial herbs of wet places. Leaves opposite. Flowers solitary in leaf axils. Calyx deeply divided into 5 lobes unequal in width (posterior one broader). Corolla campanulate, subequally or ± unequally 5-lobed. Stamens 4, equal; anther loculi parallel, contiguous. Stigma subentire. Fruit a 4-valved capsule opening loculicidally and septicidally. Seeds very small, numerous.

1. B. monnieri (L.) Pennell; *Herpestis monnieri* (L.) Humboldt, Bonpland & Kunth, *Bramia monnieri* (L.) Pennell. Fig. 94h.

Stems rooting at nodes, procumbent but finally ascending, to 60cm, glabrous. Leaves sessile, decussate, spathulate to obcordate-cuneate, 6–25 × 2–10mm, obtuse or rounded at apex, entire; veins indistinct except for midrib beneath; lower surface gland-dotted. Pedicels to 15mm; bracteoles 2, just below flowers, linear-lanceolate, slightly unequal, 2–4.5mm. Calyx lobes 5–7mm, posterior one ovate, others narrower. Corolla bluish to mauve or lilac with darker lines, 8–11mm. Capsule ovoid, acute, 5–8mm.

Bhutan: S – Phuntsholing district (Toribar). Frequent weed of flooded rice paddies, c 500m.

Apparently locally frequent, but first recorded in August 1991 in the non-flowering state.

A very common weed of paddy-fields and damp places throughout most of India including parts of the Himalaya. It differs from *Mecardonia procumbens* in its completely entire, more obovate or spathulate leaves, and pale purple (not yellow) flowers. It is used medicinally as a tonic and as a cure for epilepsy.

15. MECARDONIA Ruiz & Pavón

Creeping annual glabrous herb. Leaves opposite, shortly petiolate. Flowers axillary, on pedicels exceeding leaves. Flowers zygomorphic. Calyx divided to base; lobes narrowly ovate, imbricate, arranged in 2 series: outer sepals (dorsal and 2 ventral) ovate or narrowly ovate, dorsal slightly longer than 2 ventral ones; 2 inner lateral sepals linear, subequalling ventral ones in length but much narrower. Corolla bilabiate; tube pubescent near base; upper lip emarginate, lower 3-lobed. Stamens 4, didynamous, and with a small staminode replacing the fifth stamen. Fruit a septicidal 2-valved capsule, valves shortly bifid at apex. Seeds numerous, scrobiculate.

1. M. procumbens (Miller) Small; *Bacopa chamaedryoides* (Kunth) Wettstein, *B. procumbens* (Miller) Greenman, *M. dianthera* (Swartz) Pennell

Stems 10–50cm, rather flaccid, prostrate or decumbent with tips ascending or suberect. Leaves ovate, 8–25 × 4–15mm, serrate in upper ½–⅔. Pedicels erecto-patent, 10–15mm. Calyx 6–8mm, all lobes obtuse; dorsal one 3–6mm broad, ventral ones 2–4mm broad, inner ones 0.5–1.2mm broad. Corolla yellow with brown veins; tube pubescent, 4–6mm; upper lip obovate, 1.2–2.5 × 2–4.5mm, emarginate; lower lip with obovate or suborbicular middle lobe. Capsule ovoid or ellipsoid, usually 5.5–8 × 3–4mm. Seeds ellipsoid, 0.3–0.5mm.

Bhutan: S – Samchi, Phuntsholing, Gaylegphug and Deothang districts, **C** – Tongsa district. Wet meadows, damp grassy roadsides and streamsides and as a weed of cultivated fields and parks, 200–1400m. March–May.

Native to southern USA, Mexico, the Caribbean and northern, eastern and central S America south to Argentina; presumably introduced in Bhutan but surprisingly common.

16. DOPATRIUM Bentham

Slender, glabrous, erect annual aquatic herbs. Leaves opposite, lower ones small, upper ones remote and minute. Flowers solitary in leaf axils. Bracteoles absent. Calyx small, membranous, divided to beyond middle into 5 imbricate lobes. Corolla infundibular, somewhat bilabiate; tube slender, exserted from calyx and obliquely dilated into a broad throat; upper lip short, bifid, lower lip larger, 3-lobed. Stamens 4, didynamous, very small, lower pair sterile and staminodial, upper pair fertile with anthers coherent. Fruit a loculicidal 4-valved capsule; valves carrying away the placentas. Seeds minute, very numerous, tuberculate.

1. D. junceum (Roxb.) Bentham. Fig. 94a.

Erect aquatic annual. Stems 10–30cm, broadened and rather flaccid below, otherwise slender, unbranched or with a few erecto-patent branches. Lower 2–4 pairs of leaves obovate to narrowly obovate, 8–25 × 2–5mm, approximate, others remote, narrowly ovate, much reduced and scarcely more than 2mm. Flowers borne in upper leaf axils, lower ones ± sessile, uppermost on filiform

---

FIG. 94. **Scrophulariaceae.** a, *Dopatrium junceum*: habit (× ⅔). b, *Lindenbergia muraria*: pistil (× 7). c–d, *Lindenbergia griffithii*: c, flowering shoot (× ¼); d, pistil (× 7). e, *Limnophila aromatica*: flowering shoot (× ⅔). f, *Picria fel-terrae*: flowering shoot (× ⅔). g, *Mazus delavayi*: habit (× ½). h, *Bacopa monnieri*: habit (× 1). Drawn by Glenn Rodrigues.

a
b
c
d
e
f
g
h

pedicels to 10mm. Calyx 1.5–2mm, segments obtuse. Corolla pale pinkish-violet, 4–7mm; tube 3–4mm; lower lip to 8mm wide, middle lobe with white blotch at base. Capsule broadly ellipsoid to subglobose, green becoming purplish when mature, 3–4mm, glabrous. Seeds c 0.5mm, blackish-brown, longitudinally ribbed and tranversely rugose between ribs.

Bhutan: C – Thimphu district (Simtokha); Punakha district (near Punakha). Abundant weed of flooded rice, 1100–2300m. August–September.

First recorded in 1988 for Bhutan, but apparently quite widespread. Not yet recorded from Darjeeling or Sikkim.

17. PICRIA Loureiro

Diffuse annual herb, often rooting at nodes. Stems quadrangular, branched. Leaves opposite, petiolate, lamina ovate, crenate. Inflorescences terminal or pseudo-axillary, racemose with opposite pedicels or often reduced to 1 flower or a dichotomous pair. Bracts lanceolate, small. Bracteoles absent. Calyx flat, consisting of 4 segments in 2 series: upper outer sepal largest, cordate; lower outer sepal ovate, shallowly emarginate, slightly smaller than upper; inner lateral pair smallest, filiform. Corolla bilabiate, tubular; tube short; upper lip entire, ligulate, lower lip 3-lobed, spreading. Posterior stamens 2, anthers cohering with divaricate thecae; anterior stamens replaced by 2 clavate staminodes each bearing a sterile anther rudiment. Style filiform. Ovary triangular-ovoid, glabrous. Fruit a septicidal 2-valved capsule. Seeds globose, rugose with 8 oblong hollows.

1. P. fel-terrae Loureiro; *Caranga* ('*Curanga*') *amara* Vahl. Fig. 94f.

Stems to 50cm, densely and minutely eglandular-pubescent. Petioles 2–15mm, often c 10mm, minutely pubescent; lamina ovate, 20–60 × 15–35mm, subacute, base cuneate to truncate or in largest leaves very shallowly cordate, finely crenate-serrate with usually 20–30 teeth per side, sparsely puberulent above, puberulent also beneath especially along veins and near margin. Inflorescences (1–)2–16-flowered, 2–9cm including peduncles. Pedicels 4–7mm increasing to c 8–12mm in fruit, broadening distally. Calyx pale green, minutely pubescent; upper outer sepal 7–9 × 6–8mm in flower, 12–17 × 8–13mm in fruit, lower outer sepal 5–6 × 4–5mm, accrescent to 7–9 × 6–7mm, inner lateral pair 2–3 × c 0.5mm, hardly accrescent. Corolla white with purple throat (appearing reddish-brown when dry), c 8mm; tube glabrous outside, 2 densely glandular-pubescent ridges inside; upper lip 2–3 × 2–3mm, lower 2.5–3.5 × 4–5mm with undulate margin. Fertile stamens c 1mm. Capsule obovate, compressed, c 4 × 3mm. Seeds c 0.6mm.

Darjeeling: Mongpu (several collections). Evergreen forest, 600(?–900)m. September; fruiting October–February.

Little material has been seen, much of it in fruit, and certain details, especially of floral morphology, are therefore based on examination of specimens from China, Assam, Myanmar (Burma), Indo-China and adjacent regions together with information in 427. The flower colour is taken from field notes on a *R.L. Keenan* specimen from Kachar; the flowers are often described as being reddish-brown but this would appear to refer to the dried state. The leaves are very bitter (hence the epithet, which means 'bile of the earth') and the plant is used medicinally as a febrifuge. The epithet is sometimes spelled 'felterrae' (e.g. 351) but ICBN (Tokyo) Art. 60.9 states that the hyphen is to be used.

18. TORENIA L.

Eng: *Wishbone flower*.

Terrestrial annual or perennial herbs. Stems prostrate to ± erect, sometimes rooting at nodes. Leaves opposite, petiolate or sessile; lamina ovate or lanceolate, usually crenate or serrate. Flowers axillary or in terminal sub-umbellate or racemose inflorescences. Pedicels subtended by inconspicuous, usually linear bracts; bracteoles absent. Calyx bilabiate, often deeply cleft, with 2–5 lobes or teeth, winged, keeled or plicate. Corolla bilabiate, upper lip external in bud, emarginate or bifid; lower lip larger, spreading, 3-lobed; tube somewhat curved, cylindrical but dilated above. Stamens 4, didynamous, all perfect; anterior (upper) 2 inserted near top of tube, with longer, sometimes appendaged filaments; posterior (lower) 2 inserted at throat, with short, never appendaged filaments and anthers ± larger than those of anterior stamens; anther cells divergent, anthers touching or coherent in pairs. Ovary bilocular; style slender. Fruit an oblong septicidal capsule enclosed within calyx. Seeds rugose or cancellate.

T. cordata (Griff.) Dutta is doubtfully recorded from our area, based on a collection from Kurseong (Darjeeling district) cited by Dutta (337) and on several records from Gangtok (Sikkim) and Darjeeling Town (69). *T. cordata* was described from S Myanmar (Burma). It has large royal blue and white corollas c 35–40mm; the anterior filaments are appendaged. Material from Nepal, determined as *T. cordata* by Yamazaki, in no way matches the type and other material from Myanmar, and the Sikkim and Darjeeling identifications must therefore be considered suspect.

1. Calyx 5–6mm in flower; corolla 10–12mm; flowers in axillary groups of (2–)3–6 .... **3. T. thouarsii**
+ Calyx 10–25mm in flower; corolla usually at least 14mm; flowers solitary and axillary, sometimes with a terminal sub-umbel of 2–4 but other axillary flowers never more than 2 per node 2

2. Corolla (30–)35–50mm .... **4. T. burttiana**
+ Corolla (12–)14–30mm 3

3. Calyx ovoid with subcordate, truncate or obtuse base and 3 broad wings; corolla (12–)14–20mm .... **1. T. cordifolia**
\+ Calyx ± ellipsoid with ± cuneate or obtuse base and 5 keels or wings; corolla 17–30mm 4

4. Calyx 2-lobed and 5-ribbed; anterior filaments with conspicuous sterile appendage 0.5–1mm; all corolla lobes deep violet, without darker blotches **2. T. diffusa**
\+ Calyx 5-toothed and 5-winged; anterior filaments with no sterile appendage (except sometimes a minute tooth); corolla lobes violet, lower 3 with deep mauve or purple tips .... **5. T. violacea**

SECT. **Torenia.** Anterior filaments with a spur-like appendage at base.

1. T. cordifolia Roxb.

Annual herb, much branched from just above base; main stem usually erect, lower branches decumbent; stem sometimes prostrate below and rooting at lower nodes. Branches quadrangular or narrowly winged, sparsely hirsute on angles or almost glabrous. Petioles 3–15mm, sparsely hirsute; at least upper leaves usually subsessile. Lamina ovate, 18–38 × 8–20mm, acute or subacute, base obtuse or shortly cuneate, margin sharply serrate, upper surface sparsely puberulent or glabrous, lower surface puberulent on veins. Flowers in a terminal, 1–6- (often 3-)flowered, umbel-like inflorescence and in opposite axillary pairs (or solitary) below. Pedicels 10–30mm in flower, 20–35mm in fruit, usually 2–2½ × calyx, erecto-patent. Calyx 9–14 × 4.5–7mm in flower, ovoid and broadest towards truncate or subcordate base, reticulate-veined in fruit, divided ± to base into 2 broad lobes, one 2-toothed, the other one 3-toothed, and with 3 rather broad wings, 2 decurrent on to pedicel. Corolla pale bluish-purple or white with darker blotches at tips of lower lobes, (12–)14–20mm, glabrous. Anterior filaments with short spur c 0.5mm. Capsule oblong, c 8mm, acute, enclosed by persistent calyx. Seeds yellow, c 0.7mm, truncate at both ends.

Bhutan: C – Tongsa district (Tongsa–Shemgang); **Darjeeling:** Kalimpong, Darjeeling, Rayeng, Pankhabari. Wet cliff-faces and wet places in forest, 300–2175m. June–September.

2. T. diffusa D. Don. Fig. 95d–f.

Weak sprawling or creeping branched annual. Stems 20–60cm, sharply quadrangular, sparsely short-pilose or almost glabrous. Petioles 5–10mm, usually shortly pilose; lamina ovate or ovate-lanceolate, 10–35 × 7–21mm, acute or subacute, base obtuse or cuneate, margin serrate (teeth sometimes tinged red, with minute, slightly down-curved mucros), almost glabrous above, sparsely pilose on veins and punctate on lamina beneath. Flowers axillary, usually solitary; pedicels 10–16mm, sparsely pilose; bracts c 6mm, linear. Calyx 12–15 × 4–4.5mm, broadest near middle, divided almost to middle into 2 acuminate lobes, pale green with 5 darker, purple-tinged ribs, glabrous, narrowly winged especially in fruit, base cuneate. Corolla 17–25mm; tube pale purplish (usually drying cream-coloured), lobes deep violet-purple. Anterior filaments 4.5–7mm, with conspicuous, often purple-tipped sterile branch 0.5–1mm; posterior filaments c 2mm, clavate. Capsule c 10mm, ellipsoid, acute.

Bhutan: S – Phuntsholing district (Phuntsholing), Gaylegphug district (Tatapani), Deothang district (Samdrup Jongkhar, Deothang); **Darjeeling:** Mongpu, Kalimpong, Siliguri etc.; **Sikkim:** Gongchungchu and Gangtok. Subtropical jungle, among grass and on open banks, (?270–)550–2185m. (?April–)May–September.

T. vagans Roxb. is probably synonymous with this species; a specimen of *T. diffusa* from Deothang was named *T. vagans* on its field label.

3. T. thouarsii (Chamisso & Schlechtendal) Kuntze; *T. parviflora* Bentham

Annual herb. Stems 10–40cm, usually branched from just above base, branches decumbent or ascending, quadrangular, very sparsely short-hispidulous or almost glabrous; lower part of stem occasionally prostrate and rooting at nodes. Petioles 3–6mm. Lamina broadly lanceolate to ovate-lanceolate, 10–22 × 6–15mm, subacute to acute, base obtuse to shortly cuneate, margin crenate-serrate, upper surface sparsely hirsute, lower surface glabrous except on veins. Flowers in axillary groups of 2–3(–6). Pedicels 3–10mm in flower, usually shorter than calyx but occasionally to 1½ × as long, ascending in flower, reflexed in fruit, glabrous or very sparsely puberulent. Bracts c 1mm, linear. Calyx 5–6mm in flower, 8–10mm in fruit, divided nearly to base, sparsely hirsute, unwinged but with 5 narrow keels shortly decurrent on to pedicel; teeth c 2mm. Corolla pink, reddish-violet or bluish, without dark blotches, 10–12mm; tube puberulent outside. Anterior filaments with spur c 0.3mm. Capsule 7–9.5 × 2.5–3.2mm. Seeds pale yellow.

Darjeeling: Pankhabari, Rayeng, Sukna. On banks, etc., 150–250m. April–December.

Widespread in tropical Africa. Despite the disjunction, there seem to be no significant differences between the tropical Himalayan plants and those from Africa and the Mascarenes, although the little Himalayan material seen has slightly shorter pedicels. The species is treated in *Flora of China* (351) under the name *T. parviflora.*

A specimen from Assam (Digboi, *Barnard* G.D. 3) is similar except in its much longer pedicels to 18mm, and larger calyces.

4. T. burttiana R.R. Mill

Prostrate (?)annual, occasionally rooting at some nodes. Stems 15–c 40cm, unbranched, shortly hispidulous-pubescent along angles. Petioles 2–7mm, pubescent. Lamina ovate to lanceolate, 16–27 × 9–15mm, acute, base obtuse or shortly cuneate, margin finely serrate with c 9–12 serrations per side, upper surface sparsely pubescent, lower surface heavily tinged purple and glabrous except for short, spreading hairs on veins. Flowers (1–)2 or 4 in axils of terminal pair of leaves. Pedicels (25–)35–50mm, much longer than leaves, 4-angled below, weakly 10-ribbed above, minutely hirsute. Bracts linear-lanceolate, 5–6.5mm. Calyx 16–18mm in flower, split nearly to middle, sparsely pubescent all over; lobes 7–8mm, lanceolate; wings c 0.5mm broad, purplish-tinged. Corolla very pale violet with deep violet throat, (30–)35–41mm, without dark blotches; tube 25–30mm, very sparsely puberulent outside. Anterior filaments with linear-clavate appendage c 2.5mm, anthers c 2.5 × 0.8mm; posterior filaments c 2mm, anthers c 4.5 × 1.6mm (including well-developed connective). Style c 24mm; ovary c 5.5 × 1.5mm, hairy at apex. Capsule and seeds unknown.

Bhutan: S – Deothang district (Keri Gompa). Clearings in jungle, c 1070m. June.

SECT. **Anodous** Fournier. Anterior filaments without a basal spur-like appendage.

5. T. violacea (Blanco) Pennell; *T. peduncularis* Hook.f., *T. edentula* sensu Bentham non Griff. Fig. 95g–h.

Erect or prostrate annual. Stems 10–40cm, branched, prominently angled, sparsely hirsute on angles. Petioles 5–15mm, narrowly winged, often purplish. Lamina ovate, 12–30 × 7–20mm, subacute, base obtuse or obtuse-truncate, margin serrate or crenate-serrate, upper surface sparsely short-pubescent and often tinged purplish, lower surface paler, glabrous except for pubescent veins. Flowers solitary and axillary in upper axils, or 2–4 in a terminal sub-umbel. Pedicels 10–25mm, with 2 rows of very short spreading hairs. Bracts 1–4mm, linear. Calyx oblong-ovoid or ovoid, 5-winged, 10–15 × 5–6mm in flower, to 18 × 8mm in fruit; wings often crimson- or purplish-tinged, 1–1.5(–1.8)mm broad, decurrent on to pedicel, pubescent on ridges; teeth 5, 4 of them lanceolate but unequal, the fifth much smaller and bifid. Corolla 20–30mm; tube violet, purplish-blue or white, lobes violet, lower 3 with deep mauve or purple tips. Anterior filaments c 7mm, not spurred but bent at c 90° in lower ¼ and with a minute tooth less than 1mm; posterior filaments c 2mm. Capsule narrowly lanceolate-ellipsoid, 8–10 × 2–3mm.

Bhutan: S – Gaylegphug district (Gaylegphug (117)), Deothang district (Diu Ri Valley, Demri Chu), **C** – Tongsa district (Kinga Rapden), Mongar district (Shersing Thang); **Darjeeling:** Darjeeling, Mongpu, Bamanpokhri etc.; **Darjeeling/Sikkim:** unlocalised (Hooker collections); **Sikkim:** Gangtok, Lachen etc. Streamsides and wet rocks, wet broad-leaved forest, damp roadside banks, 200–2185m. April–November.

Ludlow, Sherriff & Taylor 7245 from Demri Chu is described on the field label as having 'corolla tube white, segments pale lilac, 2 laterals with purple spot, petals pale yellow'; this disagrees somewhat with most descriptions, which are of flowers having a purplish-blue or violet tube.

Some Hooker collections from Sikkim ?and Darjeeling (e.g. Lachen, 6000ft, 5 viii 1849) approach *T. cordifolia* in having subsessile leaves. *T. cordifolia*, however, has appendaged anterior filaments.

19. LINDERNIA Allioni

Annual or perennial herbs. Stems erect to prostrate, quadrangular. Leaves opposite, simple, pinnately or palmately veined. Flowers in terminal or axillary racemes, or solitary and axillary or in umbel-like clusters. Calyx 5-veined, each vein with a ± distinct rib or wing. Corolla tubular, bilabiate; upper lip emarginate or shallowly 2-lobed, lower lip 3-lobed, spreading. Stamens 2 or 4; posterior pair usually fertile, anterior pair fertile, sterile or reduced to staminodes. Fruit a septicidal 2-valved capsule. Seeds small, ellipsoid, scrobiculate or reticulate.

This genus seems to have been under-collected in our area. At least one other species may be expected to occur: *L. parviflora* (Roxb.) Haines. This differs from *L. procumbens* by its smaller calyx 1.5–2mm, and white (not pink) corolla; it occurs in paddy-fields and on river mud in E Nepal, etc. and should be searched for in Sikkim and western Bhutan.

1. Leaves palmately 3–5-veined but lamina not suborbicular **1. L. procumbens**
\+ Leaves pinnately veined (if *appearing* palmately veined, then lamina suborbicular) 2

2. Stamens 2, posterior; anterior stamens replaced by a pair of staminodes .. 3
\+ Stamens 4, either all fertile or posterior pair fertile, anterior pair sterile with reduced anthers; staminodes absent 5

3. Corolla 12–15mm .... **10. L. ruellioides**
\+ Corolla 5–10mm 4

4. Teeth of leaves long-aristate; corolla 5.5–8mm .... **9. L. ciliata**
\+ Teeth of leaves not aristate; corolla 8–10mm .... **8. L. antipoda**

5. Calyx shortly 5-lobed, lobes shorter than calyx tube 6
\+ Calyx deeply 5-lobed, lobes longer than calyx tube 7

6. Capsule 5–6mm, longer than fruiting calyx; leaves subsessile, suborbicular **6. L. nummularifolia**
\+ Capsule c 3.5mm, shorter than fruiting calyx; leaves with 1–7mm petiole, broadly elliptic or lanceolate .... **7. L. crustacea**

7. Calyx 2–3mm; stems densely hispid with spreading white hairs; leaves hispid **2. L. viscosa**
\+ Calyx 4–7mm; stems and leaves glabrous or hairy but not hispid 8

8. Leaves greyish, densely hispid; stems densely patent-villous ... **3. L. mollis**
\+ Leaves glabrous or ± sparsely pubescent, not greyish; stems glabrous, scabrid or sparsely hirsute 9

9. Only 2 stamens (posterior) fertile, anterior pair sterile and much smaller; stems sparsely spreading-hirsute .... **4. L. hookeri**
\+ All 4 stamens fertile, anterior filaments with short spur; stems glabrous or pubescent only at nodes .... **5. L. anagallis**

1. L. procumbens (Krocker) Borbás; *Vandellia erecta* Bentham, *L. erecta* (Bentham) Bonati

Annual, 5–25cm, branched from base. Stems erect or ascending, indistinctly 4-angled, glabrous. Leaves sessile, elliptic or ovate-lanceolate, 7–25 × 3–10mm, palmately 3–5-veined, entire, glabrous. Flowering shoots many-flowered; flowers solitary, axillary. Pedicels erecto-patent, 5–20mm, glabrous. Calyx deeply 5-lobed ± to base, 2.5–3.5mm, lobes linear-lanceolate, glabrous below, sparsely short-hirsute near apex. Flowers white, 3–6mm, glabrous except for pubescent throat; upper lip broadly ovate to suborbicular, emarginate, lower lip with 3 rounded lobes. Fertile stamens 4; anterior filaments spurred. Capsule ellipsoid, 3–4 × 2.5–3mm, subequalling or slightly longer than persistent calyx. Seeds yellow, oblong, c 0.3 × 0.15mm.

Bhutan: C – Punakha district (Wangdu Phodrang); **Darjeeling:** Mahanadi. A frequent weed of flooded rice paddy, 300–1250m. May–July.

First recorded from Bhutan in 1991 but known from Darjeeling since 1850.

2, L. viscosa (Hornemann) Boldingh; *Vandellia hirsuta* Bentham, *L. hirsuta* (Bentham) Wettstein

Diffusely branched erect, ascending or ± prostrate annual 5–25cm. Stems white-hispid with long spreading hairs. Lower leaves shortly petiolate (petiole 3–8mm, hairy), others ± sessile; lamina ovate, oblong-ovate or elliptic, 5–45 × 3–20mm, pinnately veined, obtuse, base cuneate or truncate, margin undulate or shallowly crenate at least below, shortly hirsute above and beneath. Flowers in terminal lax racemes. Pedicels 5–10mm, sparsely glandular, stiff and finally slightly deflexed in fruit. Calyx 5-lobed ± to base, 2–3mm; lobes linear-lanceolate, hirsute. Corolla yellow, white or pink, c 5mm; upper lip broadly ovate, emarginate, lower lip 3-lobed. Fertile stamens 4; filaments of anterior pair with geniculum near base. Capsule globose, 2–2.5mm, equalling or slightly shorter than calyx. Seeds c 0.3 × 0.2mm, scrobiculate.

Darjeeling: Barnesbeg; Sukna; **Darjeeling/Sikkim:** Tista and unlocalised (Hooker collection). Terai, in sal forest, tea gardens and on roadsides and banks, 150–915m. August–October (June–December in Nepal), in the rainy season.

Little material has been seen from our area.

3. L. mollis (Bentham) Wettstein; *Vandellia mollis* Bentham. Fig. 95c.

Greyish, densely patent-villous perennial, creeping and rooting at nodes, 20–50cm. Leaves shortly petiolate, ovate to ovate-lanceolate, 15–40 × 8–30mm, pinnately veined, obtuse or subacute, base truncate or cuneate, crenate-serrate, hirsute above and beneath with long white hairs. Inflorescences terminal at ends of sympodial branches, sub-umbellate, 1–6-flowered. Pedicels 10–20mm, densely hirsute, slender. Bracts linear, 1–3mm. Calyx 4–7mm, densely hirsute; lobes linear-lanceolate, acuminate. Corolla white or pale violet, 8–11mm; upper lip triangular-ovate, shortly 2-lobed, lower lip 3-lobed. Stamens 4, all fertile; anterior filaments with papillose basal geniculum. Capsule ellipsoid, 4–5 × 2.5–3mm, shorter than calyx. Seeds c 0.3 × 0.2mm, scrobiculate.

Darjeeling: Rambi, Labha, Rishap, Kalimpong etc. 915–1220m. February–December.

4. L. hookeri (Hook.f.) Wettstein; *Vandellia hookeri* Hook.f.

Perennial, diffusely branched from base. Stems not rooting at nodes, 7–13(–30)cm, sparsely patent-hirsute. Leaves sessile or very shortly petiolate; lamina ovate or lanceolate, 12–23(–28) × 5–8(–12)mm, subacute, base cuneate, margin shallowly serrate, sparsely hirsute above and on veins beneath. Inflorescences terminal and sub-umbellate and also axillary; pedicels shorter or longer than leaves. Calyx bilabiate, 4.5–6mm and divided to middle in flower but often splitting ± to base in fruit and then 6.5–8mm; lobes hispid. Corolla blue-purple, 6.5–8mm. Stamens 4; posterior pair fertile, each with a basal geniculum almost as large as anther; anterior pair sterile with reduced anthers. Capsule ellipsoid, c 4mm, shorter than calyx.

Darjeeling: Siliguri, Kurseong. Sal forest, steep slopes, among shrubs and in rice-fields, c 500m. August–October; fruiting October–December.

5. L. anagallis (Burman f.) Pennell; *L. cordifolia* (Colsmann) Merrill

Annual, 20–60cm. Stems creeping and rooting at nodes, then ascending, branched, glabrous or occasionally with pubescent nodes. Leaves subsessile or very shortly petiolate, linear, linear-lanceolate, ovate or broadly deltoid-ovate, 5–25(–30) × 1–15(–18)mm, pinnately veined, entire or crenate, glabrous. Flowering shoots laxly several-flowered; flowers solitary, axillary. Pedicels 10–25(–50)mm, broadest just below calyx, ascending but finally often deflexed, glabrous. Calyx 4–5mm, deeply 5-lobed, glabrous. Corolla white to pale purple, 7.5–10(–16)mm; tube glandular outside; posterior lip 2.5–3mm, glabrous, entire; anterior lip with 3 rounded lobes. Stamens 4, all fertile; connective of posterior anthers produced into a long tail. Capsule cylindrical, 8–10(–13)mm, acuminate, glabrous.

Bhutan: **S** – Gaylegphug district (Gaylegphug), **C** – Tashigang district (Tashi Yangtsi Dzong); **Darjeeling:** unlocalised (Cowan collection); **Darjeeling/Sikkim:** unlocalised (Hooker collection); **Sikkim:** Rangpo–Tista. 300–1600m. May–September.

Plants with very narrow linear leaves have been called *L. angustifolia* (Bentham) Tsoong (*Vandellia angustifolia* Bentham). Such plants are common in Assam but the only example from the *Flora of Bhutan* area seen is the one from Tashi Yangtsi Dzong (*Ludlow, Sherriff & Hicks* 21024). Further research is needed to assess whether these plants deserve separation as a distinct species.

A leaf paste from this species is used in Assam to treat headaches (352).

6. L. nummularifolia (D. Don) Wettstein; *Vandellia nummularifolia* D. Don (as '*numularifolia*')

Erect annual herb, 4–10cm. Stem simple or laxly branched, with short retrorse scabrid hairs on angles. Leaves subsessile, semi-amplexicaul, ovate-orbicular or suborbicular, 3–25 × 3–25mm, pinnately veined (appearing palmately veined), acutely crenate-serrate, glabrous above, sparsely hirsute on veins beneath, margins shortly ciliate. Flowers solitary and axillary or in lax racemes. Pedicels usually 3–15mm, glabrous below, very sparsely pubescent below calyx. Bracts setaceous, 0.7–1mm. Calyx campanulate, 1.5mm in flower, accrescent in fruit to 2–3mm, lobed to near middle; lobes subacute, sparsely pubescent on veins. Corolla c 7mm, violet or pink, anterior lip with white lateral lobes; tube sparsely glandular outside, densely woolly inside on ventral surface; posterior lip ovate, c 3 × 2.5mm, anterior lip c 4mm, 3-lobed. Stamens 4, all fertile; anterior filaments with densely papillose basal geniculum. Capsule 5–6 × 2mm, linear-cylindrical, glabrous, translucent at maturity.

Darjeeling: Rongsong; **Sikkim:** Gangtok (69), unlocalised (Hooker collection). Damp places, fields, 600–2135m. March–August.

Some authors, including Philcox (385) follow Don's spelling of the epithet but I am treating it as an orthographic error to be corrected to *nummularifolia.*

7. L. crustacea (L.) Mueller; *Vandellia crustacea* (L.) Bentham. Eng: *False pimpernel.*

Diffusely branched annual, 5–20cm. Stems erect or prostrate and rooting at nodes, sparingly pubescent on angles and nodes or entirely glabrous. Leaves shortly petiolate; lamina broadly elliptic, broadly ovate or lanceolate, 5–17(–30) × 3–15mm, pinnately veined, bluntly serrate or subentire, glabrous except for remotely ciliate margins; petioles 1–7mm, glabrous or sparsely pubescent. Flowers solitary, axillary. Pedicels 6–20mm, glabrous or minutely pubescent. Calyx tubular, 3–4mm in flower, 4–6mm in fruit, shortly 5-lobed; lobes ± connivent in fruit, finely pubescent on veins or subglabrous. Corolla blue, purple or lilac, 6.5–11mm, glabrous; posterior lip ovate, subtruncate to ± emarginate; anterior lip with 3 subtruncate lobes. Stamens 4, all fertile. Filaments glabrous, anterior c 2½ × posterior with a blunt spur at or above base. Capsule oblong-ovoid to subglobose, c 3.5mm, obtuse, glabrous, pale brown. Seeds pale orange, c 0.5mm, scrobiculate.

Darjeeling: Dalkajhar, Manjitar, Mongpu, Darjeeling. Terai, on damp ground on river banks, 150–900m. August–October.

8. L. antipoda (L.) Alston; *Bonnaya veronicifolia* (Retzius) Sprengel, *B. grandiflora* (Retzius) Sprengel, *B. verbenifolia* (Colsmann) Sprengel. Fig. 95a–b.

Diffuse, much-branched annual, sometimes rooting at nodes. Stems 5–20cm, glabrous. Leaves sessile, obovate-oblong or oblanceolate, 10–40 × 3–10mm, acute, base attenuate, margin serrate with acute non-aristate teeth, glabrous above and beneath. Flowers in bracteate, terminal, lax racemes or solitary and axillary. Pedicels glabrous, stout, erecto-patent and 2–6mm in flower, spreading and 4–15mm in fruit. Calyx deeply 5-lobed, glabrous, c 4mm in flower, 5–6mm in fruit; lobes linear-lanceolate, acuminate, scabrid near apex. Corolla pale purple and white, 8–10mm. Fertile stamens 2, posterior; staminodes 2, anterior, filiform with hooked apex. Capsule cylindrical, 10–16 × 1–1.5mm, acuminate.

Bhutan: C – Punakha district (Changhyo); **Darjeeling**. Rice paddy-fields, c 1250m. July.

Used in Sikkim and Bengal for abdominal pain, cholera, convulsions, cough, delirium, night fever, ringworm, smallpox, snake bite and syphilis (352).

9. L. ciliata (Colsmann) Pennell; *Gratiola ciliata* Colsmann, *Bonnaya brachiata* Link & Otto

Erect annual herb, 5–20cm. Stems ascending to erect, glabrous except for sparsely hairy nodes. Leaves sessile, oblong or elliptic, 10–30 × 3–12mm, pinnately veined, rounded at apex except for a terminal tooth, base cuneate, semi-amplexicaul, margin sharply serrate with many long-aristate teeth, glabrous above, sometimes sparsely hispid beneath. Inflorescence a terminal, bracteate, leafless raceme. Bracts linear-lanceolate to subulate, 2–4mm, aristate. Pedicels slender, 1–5mm in flower, slightly longer in fruit. Calyx linear-cylindrical, 3–6mm, deeply 5-lobed; lobes purplish at least in upper half, aristate, glabrous, scarious-margined. Corolla white, pink, lilac or pale bluish, sometimes with rose centre or brownish blotches, 5.5–8mm; tube minutely glandular outside; posterior lip emarginate, anterior 3-lobed. Fertile stamens 2, posterior; 2 staminodes anterior, filiform, with hooked apex. Capsule cylindrical, 2–3 × persistent calyx, deep brown. Seeds yellow, ellipsoid, alveolate.

Darjeeling: Kurseong, Manjitar, Sukna. Damp ground on river banks, 150–450m. August–October.

10. L. ruellioides (Colsmann) Pennell; *Bonnaya reptans* (Roxb.) Sprengel

Prostrate perennial, creeping and rooting at nodes. Stems 10–50cm, sparsely pubescent. Leaf pairs distant; petioles 4–20mm, pubescent; lamina elliptic or orbicular-elliptic, 10–50 × 8–25mm, pinnately veined, obtuse, base attenuate into petiole, finely dentate, teeth incurved and acute; surfaces scabrid. Flowers in lax racemes; pedicels 3–5mm at anthesis, lengthening in fruit to 5–12mm. Bracts linear-lanceolate, 4–7mm, acuminate. Calyx cylindrical, deeply 5-lobed, 7–9mm, glabrous. Corolla pale purple, 12–15mm. Fertile stamens 2, posterior; staminodes 2, anterior, 2–3mm, hooked at apex. Capsule cylindrical, 10–20 × 1–2mm, acute, c twice calyx.

Bhutan: S – Chukka, Sarbhang, Gaylegphug and Deothang districts, **C** – Tongsa district; **Darjeeling:** Lebong. Pashok, Mongpu etc.; **Sikkim:** Yoksam. Woods and roadsides, hot jungle, 270–2135m. May–October.

Used in Assam to treat urinary complaints (352).

20. MICROCARPAEA R. Brown

Very small annual herbs with much-branched, ± glabrous, creeping stems rooting at nodes. Leaves opposite, sessile, linear-oblong or linear-oblanceolate, glabrous. Flowers solitary in axil of usually only 1 leaf of each opposite pair, sessile, ebracteolate. Calyx tubular-campanulate, persistent in fruit, 5-ribbed, margin ciliate. Corolla very small; tube shorter than calyx, relatively broad; limb bilabiate. Stamens 2, anterior, inserted near top of corolla tube, included; anthers 1-locular. Fruit a loculicidal 2-valved capsule. Seeds ellipsoid, with thin, indistinctly reticulate testa.

A monotypic genus sometimes included in *Peplidium* Delile but here kept separate following the recent treatments in *Flora of Thailand* (427) and *Flora of Ceylon* (330), as well as F.B.I.

1. M. minima (Retzius) Merrill; *M. muscosa* R. Brown, *Ammannia dentelloides* Kurz

Stem 3–15cm, procumbent and filiform, much-branched and forming pale green moss-like mats, glabrous or sometimes ciliate on angles. Leaves 2–5 × 1–2mm, entire, obtuse, base attenuate, both surfaces glabrous, upper surface punctate. Calyx 2–2.5mm; teeth ciliate. Corolla pink, ovoid, c 2.5mm, equalling or slightly longer than calyx; upper lip slightly 2-lobed, lower lip longer than upper, 3-lobed, middle lobe largest, all with ciliate margins. Capsule c ½ calyx. Seeds brownish.

Darjeeling: Siliguri (80). Rice-fields and dried-up ponds in terai, flowering shortly after the rains.

The plant is similar in habit to *Dentella repens* J. & G. Forster (Rubiaceae).

21. ELLISIOPHYLLUM Maximowicz

Creeping perennial herb, rooting at nodes. Roots short, very slender and fibrous. Stems hairy, prostrate. Leaves simple, alternate, deeply pinnatifid. Flowers solitary, axillary. Pedicels slender, coiling in fruit. Calyx campanulate,

---

Fig. 95. **Scrophulariaceae.** a–b, *Lindernia antipoda*: a, flowering shoot (× ⅔); b, dissected corolla (× 4). c, *Lindernia mollis*: dissected corolla (× 4). d–f, *Torrenia diffusa*: d, flowering shoot (× ⅔); e, dissected calyx (× 2); f, anterior stamen (× 6). g–h, *Torrenia violacea*: g, dissected calyx (× 2); h, anterior stamen (× 6). i, *Hemiphragma heterophyllum* var. *heterophyllum*: habit (× 1). j, *Ellisophyllum pinnatum*: habit (× ⅔). Drawn by Glenn Rodrigues.

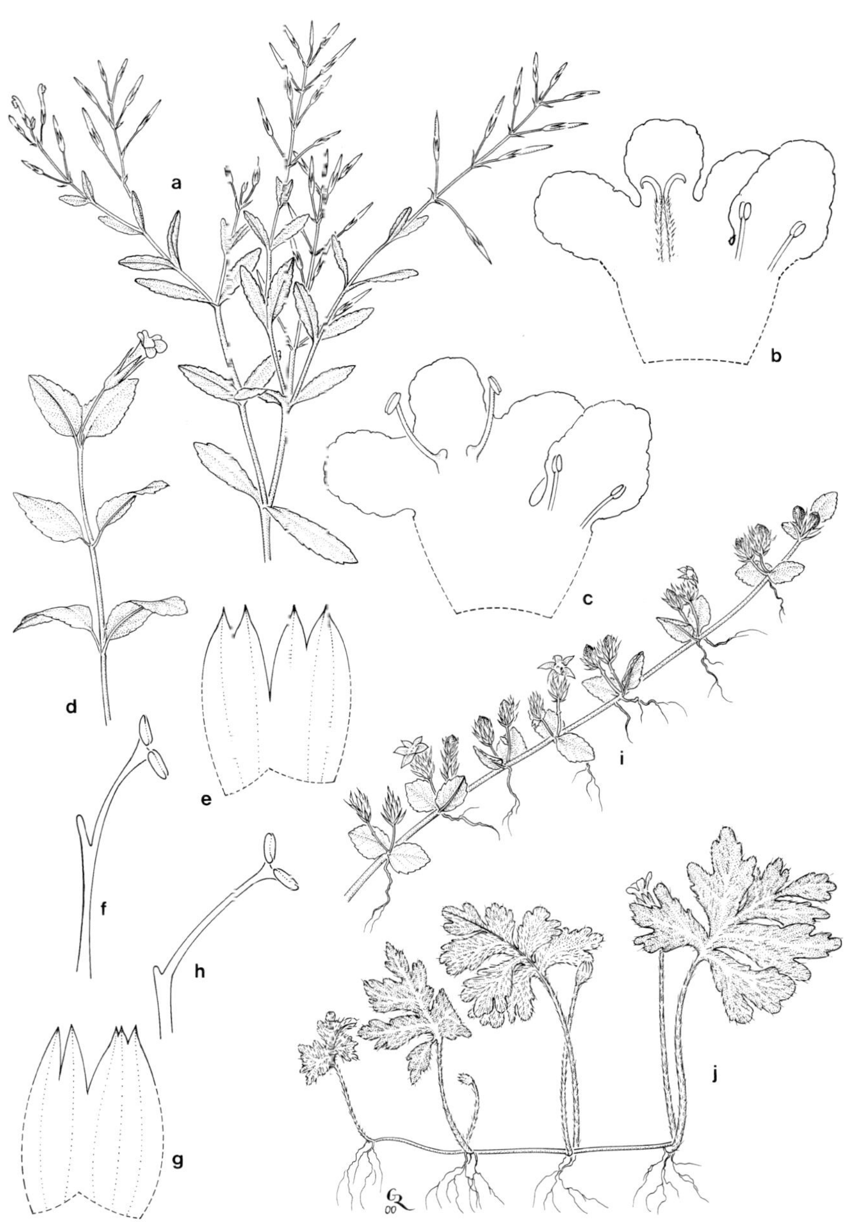
a
b
c
d
e
f
g
h
i
j

membranous; lobes 5, ovate, acute. Corolla subrotate; lobes 5, pubescent inside but glabrous outside. Stamens 4, didynamous, epipetalous; thecae divergent, united above. Ovary bilocular, glabrous; septum conical. Fruit a subglobose capsule enclosed by persistent calyx. Seeds 1–2 per loculus, rugulose.

On the basis of fruit and embryological characters this genus has sometimes been placed in its own family, Ellisiophyllaceae Honda, or as a monotypic subfamily of Scrophulariaceae (Ellisiophylloideae), or in the Hydrophyllaceae. The leaves bear a striking superficial similarity to some species of *Geranium* (Geraniaceae) or *Anemonopsis* (Ranunculaceae); this has led to fairly frequent misidentifications in the field.

1. E. pinnatum (Bentham) Makino; *Sibthorpia pinnata* (Bentham) Hook.f., *E. reptans* auct., non Maximowicz. Fig. 95j.

Stems 3–25cm or longer, prostrate, often forming wide carpets, shortly hirsute to subglabrous. Leaves borne erect on long ± hairy petioles 10–100mm; lamina broadly oblong to ovate in outline, 10–50 × 10–45mm, membranous, with 2–3 pairs of broadly adnate, obovate lateral segments crenulate or lobulate towards apex and a smaller, obovate, crenulate terminal segment; surfaces shortly and softly hirsute with jointed hyaline hairs, the cells of which progressively but abruptly become more slender towards apex and bear a very fine longitudinal white line; veins on lower surface more densely and shortly pubescent. Pedicels erect, 15–95mm, suddenly coiling spirally from base of ovary downwards when capsule matures. Flowers nodding, white with orange centre, c 10mm; lobes c 4.5–5mm, ± spreading. Filaments white when fresh; anthers 1.2–2.5mm. Style c 5mm, slightly exserted from corolla tube. Capsule c 6mm. Seeds c 4mm, glabrous.

subsp. **pinnatum**

Stems and petioles densely spreading-villous with hairs 0.5–1.1mm. Pedicel subequalling or slightly longer than petiole. Anthers 2.1–2.5mm.

Bhutan: S – Chukka district (Chimakothi (71, 117)), **C** – Punakha district (Punakha, Rinchu–Mishichen (71)), **N** – Upper Kuru Chu district (Shabling), ?Upper Mo Chu district (Mishichen (Punakha district)–Khosa (71)); **Darjeeling:** Tonglu. Moist earth banks etc. in warm broad-leaved forest and *Rhododendron* forest, 1676–3050m. April–August; fruiting September–October.

subsp. **bhutanense** R.R. Mill

Stems and petioles sparsely adpressed-pubescent to nearly glabrous, any stem hairs only 0.35–0.5mm. Pedicel shorter than petiole. Anthers 1.3–1.4mm.

Bhutan: C – Tashigang district (Chorten Gora–Dechen Phodung). Moist soil in wet broad-leaved forest, c 2440m. June (only collected once).

Endemic subspecies.

22. HEMIPHRAGMA Wall.

Slender creeping perennial. Leaves dimorphic in varying proportions according to season, either opposite, very shortly petiolate and ± orbicular (*wet season leaves*), or fasciculate and needle-like (*dry season leaves*). Flowers sessile, axillary. Calyx lobes 5, narrow. Corolla shortly campanulate with 5 small, spreading, subequal lobes. Stamens 4, inserted at base of corolla; anthers reaching nearly to mouth of tube, shortly sagittate with tips of thecae confluent. Style shorter than corolla tube; stigma minute. Fruit a fleshy, berry-like septicidal capsule. Seeds numerous, smooth, minute.

A monospecific genus extending across much of the Himalaya. The seasonal nature of the leaf dimorphism was first discussed by Maury in 1887 – see (384). According to Maury, the orbicular leaves are formed during the hot wet season and are most numerous in September, while the small needle-like ones develop in the cool dry season and persist over winter until May. Correlating the date of collection with the frequency of the two leaf types in the specimens examined seems to support this conclusion.

1. H. heterophyllum Wall. var. **heterophyllum**. Fig. 95i.

Slender creeping perennial. Stems to 60cm, sparsely pubescent, with numerous short shoots, both main stem and shoots bearing either orbicular or needle-like leaves or a mixture of both according to season. Wet season leaves orbicular or ovate-orbicular, 5–20 × 6–15mm, subacute, base truncate, margins 5–7-serrate, sparsely pubescent above, less so beneath, lower surface pale green. Dry season leaves in small, dense ovoid-globose fascicles of up to 30, rigid and needle-like, convex on back, deeply channelled ventrally, greyish-green, ciliate. Flowers most commonly arising from axils of needle-like leaves but sometimes from orbicular ones, sessile. Calyx lobes narrowly elliptic, c 2.5mm. Corolla pink (white towards base), c 4mm; lobes ovate, c 1.5mm. Capsule green at first, turning scarlet (drying blackish), shining, c 8mm diameter.

Bhutan: S – Chukka and Deothang districts, **C** – Ha, Thimphu, Tongsa, Mongar and Tashigang districts, **N** – Upper Mo Chu district; **Darjeeling:** Darjeeling, Sandakphu, Phalut etc.; **Sikkim:** Dzongri, Jakeyupyak, Phusum etc. Mixed broad-leaved (oak, birch etc.), fir and pine/oak forests, *Rhododendron*, juniper and bamboo scrub, open grassy hillsides and mountain rocks, 1830–4270m. February–June (buds forming July); fruits February–October.

Plants from our area belong to var. *heterophyllum*, which is the most widespread of 3 varieties recognised. Hooker's statement (80) that the fruit is black is clearly based on examination of dried material. Field notes on a large number of specimens consistently describe the fresh colour as scarlet or red (one collector described it as resembling sealing-wax). The fruit is frequently, but incorrectly, described as a berry.

23. SCOPARIA L.

Erect, glabrous or pubescent herbs or undershrubs. Stems angled. Branches slender, alternate below, opposite above. Leaves opposite or whorled, gland-dotted. Flowers axillary, small. Pedicels ebracteolate. Calyx divided into 4–5 imbricate, almost free segments. Corolla subrotate, 4-lobed, lobes equal, posterior one emarginate; throat densely bearded. Stamens 4, equal; anthers ovoid, subsagittate at base, versatile, exserted; thecae parallel or divergent, distinct. Style pubescent, clavate; stigma punctiform, emarginate, exserted. Capsule subglobose or broadly ovoid, 2-valved, membranous, dehiscing septicidally and loculicidally from apex. Seeds many, obovoid, angular, scrobiculate.

1. S. dulcis L. Eng: *Sweet broom.*

Bushy annual to perennial, 15–150cm. Stems ± woody at base, occasionally minutely pubescent at nodes. Leaves linear-oblanceolate, narrowly elliptic or narrowly obovate, 5–40 × 1–18mm, at least larger ones shallowly serrate distally, subacute, base cuneate into indistinct short petiole, bases of adjacent leaves connected by a membrane. Inflorescence of many-flowered leafy racemes; flowers (1–)2–4 per axil. Pedicels filiform, c 5mm. Calyx divided ± to base, lobes oblong, 1.5–2mm, glandular-punctate. Corolla white or pale greenish, 3–5mm diameter, lobes ± reflexed, c 2–3.5mm, glabrous outside, with dense tuft of long silky white hairs surrounding stamens. Filaments c 1mm; anthers c 0.7–1mm, yellow. Style c 1.5mm. Capsule 2–3.5 × 1.5–3mm, yellowish-brown, glabrous. Seeds c 0.5 × 0.15mm, brown, prismatic, rather irregular in shape, coarsely reticulate.

Bhutan: S – Samchi, Phuntsholing and Gaylegphug districts, **C** – Thimphu, Tongsa and Mongar districts; **Darjeeling:** Bhotia Basti, Rayeng; **W Bengal Duars:** Jalpaiguri. Damp grassland, shingle and as a weed of roadsides, paddy-fields, apple orchards and other cultivated land, 90–2300m. November–August.

A native of tropical America, now a pantropical weed. In parts of India it is used to sweeten well water. It has many medicinal uses; in Bengal and Sikkim, used to treat gout, urinary complaints and impotence, rheumatism, tongue ulcers and as an aid to childbirth; tied on the arm, it is said to act as a galactagogue (352).

24. NEOPICRORHIZA Hong

Perennial rhizomatous herbs. Leaves basal. Inflorescence a terminal spike; peduncle scape-like, naked. Calyx subequally 5-partite, middle (posterior) lobe narrowly elliptic, acute, other 2 upper lobes oblanceolate, obtuse, 2 lower lobes oblanceolate with acute hooked tip. Corolla much longer than calyx, zygomorphic, bilabiate, glandular outside but not long-ciliate; tube curved, slightly exceeding calyx; lobes 4, posterior one (formed by fusion of 2 lobes) much longer than other 3. Stamens didynamous, posterior 2 slightly shorter than upper lip, anterior 2 exserted. Capsule turgid, tapered to apex, first dehiscence loculicidal. Seeds ellipsoid, turgid, with transparent alveolate testa.

Recently separated from the W Himalayan monotypic genus *Picrorhiza* Bentham; the two are very close in their vegetative morphology and in seed characters but differ markedly in floral morphology. *Picrorhiza* has an inflorescence subtended by a few small leaves (peduncle not completely naked); an actinomorphic, 5-lobed, ± glabrous, eglandular corolla shorter than the much thinner calyx whose lobes are all ± equal in width; and all stamens long-exserted on filaments c 3–4 × corolla; the capsule dehisces septicidally distally at first, then somewhat loculicidally. For discussion see Hong (348).

1. Leaves 20–60 × 10–20mm (excluding petiole or petiole-like base) with 20–40 teeth; anthers longer than broad, thecae parallel to filament **1. N. scrophulariiflora**
\+ Leaves 5–15 × 2–8mm (excluding petiole) with 7–9 teeth; anthers broader than long with thecae tips spreading outwards from filament .. **2. N. minima**

1. N. scrophulariiflora (Pennell) Hong; *Picrorhiza kurrooa* sensu Hook.f. (80) p.p. ('short-stamened form') non Royle, *P. scrophulariiflora* Pennell. Dz: *Puti sing*; Nep: *Kutki*, *Kutaki.* Fig. 95h–i.

Leaves usually 10–20 per rosette; petioles often indistinct (lamina with long-attenuate but relatively broad base) or if distinct from lamina, 15–40mm; lamina narrowly obovate-oblong, oblanceolate or narrowly spathulate, 20–60 × 10–20mm, serrate in upper half (total number of teeth (20–)23–35(–40)), teeth mostly 1–3mm deep with longest side 2.5–6mm and with apical mucro; surfaces glabrous or sparingly short-glandular-hairy. Scape in flower 2–6cm, in fruit 4–9cm, densely short-pilose with brownish-white thin hairs to 0.5mm. Bracts ovate, acuminate with blunt tip Inflorescence oblong-ovoid, with usually 12–20 flowers. Calyx lobes 4–6 × c 1–2.5mm in flower, with marginal hairs c 0.5mm. Corolla pale blue to dark violet-purple, 8–12mm; upper lip very shallowly emarginate. Anthers oblong-reniform with tips of thecae parallel to filament, 0.9–1.2 × 0.7–0.8mm; pollen yellow. Capsule 10–14mm, ovoid-acuminate. Seeds c 1.7mm.

Bhutan: C – Thimphu and Punakha districts, **N** –Upper Mo Chu, Upper Bumthang Chu and Upper Kulong districts; **Sikkim:** Dzongri, Chho La, Tsomgo etc. Usually among rocks and boulders on open grassy hillsides, 3500–4880m. May–July according to altitude.

The root is bitter and is chewed to cure headaches; an extract is used as a febrifuge and remedy for stomach illness, and to expel intestinal worms. It is also used for asthma, jaundice and coughs (352). In Nepal, the dried plant is used as incense by monks. All records of *Picrorhiza kurrooa* from Sikkim refer to this species.

2. N. minima R.R. Mill; *N. scrophulariiflora* auct. non (Pennell) Hong. Fig. 96j.

Leaves usually 5–6 per rosette; petioles 3–15mm; lamina 5–15 × 2.5–8mm, obovate to shortly spathulate or suborbicular, total number of teeth 7–9, teeth 2–5mm deep with longest side 1–4mm; surfaces glabrous. Scape 1–5.5cm in flower, 4.5cm or more in fruit, densely puberulent especially when young with very short glandular hairs which appear bluish-white under low-power dissecting microscope because of purple (not brown) joints. Inflorescence shortly capitate, few-flowered (4–9). Calyx lobes 3.5–4mm with marginal hairs c 0.1mm. Corolla 8–10.5mm, violet-purple. Upper lip deeply emarginate, minutely glandular on back. Anthers shallowly reniform, 0.5 × 1mm, pollen whitish-buff. Capsule 6mm, ovoid-acuminate.

Bhutan: C – Tongsa district (Padima Tso near Thampe La, Wartang), **N** – Upper Bumthang Chu district (Waitang). Steep rocky and grassy hillsides, 4420–4725m. June–August.

Endemic to Bhutan.

25. DIGITALIS L.

Tall biennial to perennial herb. Leaves simple, alternate, lowest in a basal rosette. Inflorescence a long, terminal, bracteate, secund raceme. Calyx equally deeply 5-fid, shorter than corolla tube. Corolla much longer than calyx, with long cylindrical-campanulate tube constricted at base; limb ± 2-lipped, lips scarcely spreading, erecto-patent, upper shorter than lower. Stamens 4, didynamous, included. Style filiform, unequally bifid at apex. Capsule ovoid to conical, septicidal. Seeds oblong, numerous.

1. D. purpurea L. Eng: *Foxglove.*

Biennial or rarely perennial, occasionally flowering in first year. Stem (25–)60–180cm, ± densely short-tomentose above, often subglabrous towards base. Basal leaves with long winged petiole, lamina ovate to lanceolate, 4–30 × 2–13cm, ± rugose, crenate to serrate, ± hairy above and beneath, with usually a mixture of glandular and eglandular hairs. Raceme simple or slightly branched, many-flowered, secund; bracts variable, often minute. Pedicels tomentose, longer than calyx. Lower calyx lobes ovate, 10–12 × 5.5–7mm, upper lanceolate, 10–12 × c 4mm, all acute. Corolla purple, mauve, pink or white outside, usually whitish with darker purple spots inside, 40–50mm, glabrous or pubescent to villous outside, shortly ciliate and with a few long hairs inside. Style 15–25mm. Capsule ovoid, obtuse, 8–12mm, at least as long as calyx.

Darjeeling: Darjeeling, Singalila Ridge. 2100m. May.

Native of W, SW and western C Europe; cultivated in Pakistan, India, China, Myanmar (Burma) and elsewhere as a medicinal herb and sometimes occurring as an escape in forest clearings; usually calcifuge. The drug digitalis, obtained from the leaves, is used in the treatment of cardiac illness. For a full account of its cultivation and medicinal properties see (403).

26. CALORHABDOS Bentham

Tall, stout-stemmed, erect perennial herb. Leaves alternate, serrate. Inflorescence of dense, bracteate terminal and axillary spikes. Bracteoles absent. Sepals lanceolate, acuminate. Corolla 4-lobed, somewhat bilabiate, lobes erecto-patent, 3 lower ones smaller than upper; tube incurved. Stamens 2, exserted; anther thecae parallel, finally divergent with confluent tips. Style filiform. Capsule acute, opening both septicidally and loculicidally; valves 4, with inflexed margins. Seeds numerous, minute.

Hong (348) reduced *Calorhabdos* to one of four sections of *Veronicastrum* Fabricius. This treatment was also used in *Flora of China* (351). If this generic concept is adopted, the correct name for our species is *Veronicastrum brunonianum* (Bentham) Hong.

1. C. brunoniana Bentham

Stem 30–100cm, ± glabrous except in inflorescence, ± purplish. Leaves lanceolate, 50–140 × 10–40mm, long-acuminate, base obtuse to shortly attenuate, ± sessile, serrulate with small forward-pointing teeth, glabrous. Spike 1(–2), terminal, very long, narrow and dense, 8–25cm × 6–12mm, usually ± densely and shortly spreading-pilose; flowers imbricate. Bracts lanceolate, c 3mm. Calyx lobes all unequal, narrowly lanceolate, shortest c 1–1.5mm, longest c 2–3.5mm. Corolla greenish-yellow, 3.5–5mm; lobes sparsely pubescent inside. Anthers relatively large (c 2mm), greenish-black. Capsule 5–6 × 3–3.7mm, dark brown. Seeds pale brown, c 0.7mm.

Bhutan: C – Thimphu district (Thimphu). Grassy places near streams, 2500–3050m. July–August; fruiting November.

A second subspecies endemic to Sichuan is recognised by Hong (351).

27. VERONICA L.

Erect or prostrate perennial or annual herbs. Leaves opposite or upper ones alternate, petiolate or sessile, simple or sometimes with 3–5 lobes, entire or usually crenate or serrate. Flowers in terminal and/or axillary racemes or spikes (sometimes reduced and umbel-like) or apparently solitary in axils of leaf-like bracts. Calyx divided ± to base, lobes 4–5, 2 lower ones usually larger than upper, uppermost lobe much smaller and often absent. Corolla rotate, slightly zygomorphic, usually blue or pinkish, often with one lower lobe smaller and paler than others; tube broader than long; lobes 4, longer than tube, ± unequal. Stamens 2. Fruit a bilocular capsule dehiscing loculicidally and often also septicidally, often transversely flattened, apex usually notched or 2-lobed. Style persistent, not lengthening in fruit. Seeds few or numerous, flat or boat- or cup-shaped, often with an elevation (podium) on ventral surface.

1. At least some leaves alternate 2
+ All leaves opposite 3

2. Leaves triangular-ovate, crenate-serrate with 3–4 teeth per side; flowering pedicels usually shorter than subtending leaf; capsule ciliate, c twice as long as broad .... **11. V. persica**
+ Leaves reniform or orbicular in outline, 3–5-lobed; flowering pedicels ± longer than subtending leaf; capsule glabrous, less than twice as long as broad .... **12. V. hederifolia**

3. Flowers sessile or subsessile 4
+ Flowers with obvious although sometimes very short pedicels 5

4. Leaves woolly, middle and upper ones densely imbricate and ± hiding stem; flowers 7–8mm .... **8. V. lanuginosa**
+ Leaves pilose, not imbricate, stem clearly visible; flowers 3–4mm **10. V. cephaloides**

5. Leaves petiolate 6
+ Leaves sessile 9

6. Inflorescence umbel-like, consisting of a pair of sessile racemes **5. V. umbelliformis**
+ Inflorescence not umbel-like, racemes clearly pedunculate 7

7. Capsule ± globose, c 3 × 3mm; corolla c 3mm, scarcely longer than calyx; pedicels 0.7–1mm .... **6. V. javanica**
+ Capsule broadly deltoid, 3.5–5.5 × 7–10.5mm; corolla 6–7mm, c twice as long as calyx; pedicels 1.5–3mm 8

8. Calyx lobes minutely retuse (use good lens); lower part of stem crisply white-pilose; leaves with 3–7(–10) teeth per side .... **3. V. cana**
+ Calyx lobes never retuse; lower part of stem reddish-brown-hirsute; leaves with 9–15 teeth per side .... **4. V. robusta**

9. Stems succulent; bracts 1–2mm, shorter than or subequalling flowering pedicels .... **1. V. anagallis-aquatica**
+ Stems not succulent; bracts 3–13mm, longer than flowering pedicels ... 10

10. Capsule longer than broad, or subglobose 11
\+ Capsule distinctly broader than long 12

11. Middle leaves 40–90mm; corolla lobes hairy outside, 8–11mm, longer than 4–8mm calyx; capsule 7–8 × 4mm .... **2. V. himalensis**
\+ Middle leaves 5–40mm; corolla lobes glabrous, hardly longer than c 3mm calyx; capsule c 3 × 3mm .... **6. V. javanica**

12. Leaves 10–45mm, serrate; corolla 4–5mm; capsule broadly deltoid, c 4 × 8.5mm, only hairy near apical notch .... **7. V. deltigera**
\+ Leaves 6–8mm, ± entire; corolla c 3mm; capsule obcordate, c 2.5–3 × 3.5–4mm, ciliate .... **9. V. serpyllifolia** subsp. **humifosa**

1. V. anagallis-aquatica L.

Erect, usually almost glabrous perennial with creeping root-stock, usually ± blackening when dry. Stems to 1m, rather flaccid and succulent, unbranched (except inflorescence). Leaves opposite, sessile, semi-amplexicaul, oblong to oblong-lanceolate, 20–70 × 5–22mm, acute, remotely and shallowly serrate, glabrous. Inflorescence of several pairs of opposite racemes in axils of leaves in upper half of stem; racemes 3–11cm. Bracts linear-lanceolate, c 1–2mm, shorter than or subequalling flowering pedicels. Pedicels 1–2mm in flower, to 5mm in fruit, with very few short jointed hairs. Calyx lobes ovate-lanceolate, acute, c 3mm in flower, 4mm in fruit. Corolla white or pale mauve with darker purple or magenta veins, 4–5mm diameter. Capsule orbicular, c 3mm, with shallow apical notch c 0.5mm, glabrous.

Bhutan: C – Thimphu district (Thimphu, Paga, Paro Chu), **N** – Upper Mo Chu district (Gasa Dzong); **Chumbi:** Chumbi. Sides of ditches and in moist sand, 2180–2450m (but to 4725m in Nepal). April–August.

Used in Uttar Pradesh to cure skin disease and purify the blood (352).

The variation of this very widespread species, and its relationships with other species of sect. *Beccabunga* that are absent from our area, are not fully understood.

2. V. himalensis D. Don

Erect perennial. Stems rather stout, 20–70cm, unbranched except in inflorescence, shortly pilose to almost glabrous. Leaves opposite, sessile, 7–12 pairs, ± distant, lowest 3–6 pairs much smaller than other ones and soon falling; lamina triangular-ovate, middle and upper ones 40–90 × 10–30mm, acute, base obtuse, margin ± coarsely serrate (teeth 8–14 per side, shortly mucronate), shortly pubescent above and on veins beneath. Racemes usually 2 or 4, in axils of upper 1–2 pairs of leaves; peduncles arcuate-ascending, 7–12cm, appressed-pilose. Flowers 8–15, in upper ⅓ of each inflorescence; pedicels 4–8mm; bracts linear-oblong, 5–13mm. Calyx lobes ovate-oblong, 4–8mm, united at base, pubescent. Corolla blue or blue-violet, occasionally white; lobes 8–11mm, pilose. Filaments and anthers blue (white when corolla white). Style 6–9mm with prominent subcapitate or slightly 2-lobed stigma. Capsule ovoid, acute or shortly acuminate, 7–8 × c 4mm, sparsely hairy mainly near valve margins, sometimes glabrous. Seeds c 0.6 × 0.2mm, ovoid-elliptic, plano-convex, brown.

Bhutan: C – Ha district (Ha, Chelai La), Thimphu district (Pajoding, Pumo La–Pyemitangka), Bumthang district (Rudo La), **N** – Upper Mangde Chu district (Saga La, Namdating); **Sikkim:** Ningbil, Lachen, Phedup etc.; **Chumbi:** Dotag, Natu La–Chubitang, Yatung etc. Steep grassy open hillsides and mountain tracks in and above *Abies* forest, 2750–4270m. July–September(–November).

3. V. cana Bentham. Fig. 96f.

Rhizomatous perennial. Stems 5–17cm, erect, unbranched below, shortly white-crispate pilose below in 2 rows or all round. Leaves petiolate, all opposite, 3–4 pairs mostly in upper half of vegetative part with lowest pair usually more remote and smaller; adjacent petioles nearly free, (2–)5–15mm; lamina ovate (lowest shortly triangular), 7–35 × 5–25mm, subacute, base obtuse to truncate, shallowly dentate (teeth 3–7(–10) per side), sparsely hirsute above, usually glabrous beneath except on veins but occasionally with a few hairs on lamina. Inflorescence of 1–2 terminal bracteate racemes, axis and pedicels shortly white crispate eglandular-pubescent or -pilose or sometimes shortly patent glandular-pilose, hairs usually 0.2–0.5mm (rarely to 0.7mm). Bracts usually shorter than or subequalling 2–3mm pedicels, rarely exceeding them. Calyx lobes oblong-linear, (2.5–)3–3.5mm in flower, almost always minutely retuse (use × 20 hand lens or low-power microscope!), glandular-pilose. Corolla pale blue or mauve, c 9mm diameter; lobes c 6mm. Anthers cordate, c 0.5mm or less. Style 2–3mm in fruit. Capsule broadly deltoid, 3.5–5.5 × 7–10.5mm, shortly ciliate; apical notch usually 0.5–2mm deep.

Bhutan: C – Tongsa district (Yuto La), Mongar district (Sengor), **N** – Upper Mo Chu district (Kohina), Upper Kulong district (Lao); **Darjeeling:** Singalila; **Sikkim:** Lachen, Lagyap, Dzongri etc. Path-sides and marshy ground in and at edges of *Abies/Rhododendron* and *Tsuga/Rhododendron* forest, and on wet cliffs, 2743–4115m. May–August.

Both the specimens on which the C Bhutan records are based have long-glandular pilose racemes, as in *V. robusta*, but are otherwise similar to more typical *V. cana.*

4. V. robusta (Prain) Yamazaki; *V. cana* Bentham var. *robusta* Prain. Fig. 96e.

Very similar to *V. cana* but stems usually taller (12–40cm), reddish-hirsute all round below; leaves 3–5 pairs, petioles 10–30mm, adjacent ones ± distinctly united by a membrane densely ciliate on upward-facing surface, lamina 25–65 × 20–35mm, acute, base truncate or usually shortly cuneate, shallowly crenate-dentate to coarsely dentate (teeth 9–15 per side), hirsute on both surfaces (lower surface rarely almost glabrous except for veins); inflorescence patent-glandular-pilose with hairs usually 0.5–1.2mm; bracts longer than 1.5–2.5mm pedicels and sometimes reaching beyond middle of calyx; calyx lobes elliptic or lanceolate-elliptic, 2–2.5mm, broadest in middle or below, obtuse or subacute (never retuse); style (3–)4–5mm in fruit; capsule 3.8–4.5 × 8.5–9mm, apical notch usually absent or very shallow (not more than 0.5mm deep) but occasionally to 1mm.

Bhutan: C – Thimphu district (Sinchu La, Dochong La), Punakha district (Mara Chu Valley); **Darjeeling:** Sandakphu, Tonglu, Garibans–Tonglu etc. Wet banks in dense mixed forest, 2700–3658m. May–August.

Intermediates between this and *V. cana* have been seen from eastern Nepal, having the robust habit and glandular inflorescence of *V. robusta* but stems with white, not reddish, hairs in vegetative part and leaves which are glabrous beneath (except on veins), as in *V. cana*. They may represent hybrids; two of the specimens seen bear fruit but a third is totally sterile. It is possible that further study of a larger suite of specimens may result in the re-uniting of these two very closely allied species under a broader concept of *V. cana*.

5. V. umbelliformis Pennell; *V. capitata* var. *capitata* sensu F.B.I. p.p. non Bentham, *V. capitata* var. *sikkimensis* Hook.f., *V. szechuanica* Batalin var. *sikkimensis* (Hook.f.) Tsoong. Fig. 96a–b.

Perennial. Stems erect or slightly decumbent, 3–18(–23)cm, shortly pubescent or villous; hairs in 2 rows, longest and densest at and just below upper nodes. Leaves opposite, 3–4 pairs, mainly towards stem apex with lowest pair remote; petioles 1–6mm, much shorter than lamina; lamina ovate, 6–22(–30) × 4–22mm (upper pairs largest), obtuse, base obtusely truncate, margin shallowly crenate-dentate, surfaces thinly eglandular-pilose-hirsute. Inflorescence umbel-like, consisting of a single pair of sessile racemes in axils of leaf-like bracts. Pedicels 4–10mm, ascending. Calyx lobes oblong-lanceolate, 2.5–3mm in flower, to 4mm in fruit, hirsute. Corolla pale blue, lavender or mauve (sometimes white), lobes 3.5–5mm, broadly rounded. Filaments white; anthers yellow. Style 2–2.5mm. Capsule obcordate, c 5 × 6.5mm, with shallow apical notch, shortly ciliate round apical margin, otherwise glabrous.

Bhutan: C – Bumthang district, **N** – Upper Mo Chu, Upper Pho Chu, Upper Bumthang Chu and Upper Kulong Chu districts; **Sikkim:** Bikbari, Tsomgo, Tari, Yume Samdong etc. Open grassy alpine meadows and shady places in *Abies* forest, 2985–4270m. Late May–mid August but mainly June–July.

6. V. javanica Blume. Fig. 96c.

Much-branched annual with bushy habit. Stems decumbent or ascending, 15–45cm, pubescent with 2 longitudinal rows of very dense short white hairs and very few hairs elsewhere. Leaves opposite; petioles very short, (0–)1–7mm; lamina ovate, 5–40 × 4–30mm, obtuse or subacute, base truncate or shallowly cordate, margin crenate-or bicrenate-serrate with (6–)9–13 small subacute teeth per side, shortly and sparingly pubescent above and beneath. Racemes axillary, 5–9-flowered; axis and pedicels densely but very shortly white-pubescent all round. Bracts linear to linear-spathulate, 3–5mm, much longer than 0.7–1mm pedicels. Calyx lobes linear-oblong, c 3mm in flower and c 4mm in fruit, obtuse, with rather long setiform marginal hairs c 0.5mm. Corolla whitish or pale blue, lobes scarcely longer than calyx. Capsule obcordate, c 3 × 3mm, shorter than calyx, with shallow ciliate apical notch, otherwise glabrous. Seeds c 0.5mm, elliptic, convex on both surfaces, with very finely reticulate testa.

Bhutan: S – Chukka district (Chasilakha (117)), Gaylegphug district (Sher Camp), **C** – Thimphu district (Taba, Namselling), Punakha district (Tinlegang), Mongar district (Lingmethang); **Darjeeling:** Happy Valley (69), Kurseong (69); **Sikkim:** Thinglen–Yoksam (69), Gangtok (69), Pedong etc. Damp roadside banks and ditches in subtropical forest and also open habitats, 800–2370m. March–May.

7. V. deltigera Bentham. Fig. 96d.

Perennial. Stems erect from ascending base, 9–40cm, slender, with 2 rows of dense white spreading to ± deflexed hairs longest just below prominent nodes. Leaves opposite, sessile; lamina ovate, ovate-oblong or lanceolate-oblong, 10–45 × 5–25mm, subacute, base obtuse or truncate, margin slightly thickened beneath, shallowly serrate with 6–10 teeth per side, glabrous above, sparsely pubescent (sometimes only towards margin) or glabrous beneath; veins prominent beneath. Racemes terminal (sometimes also 1–2 axillary), (3–)4–9(–10)-flowered, densely glandular-pubescent. Bracts linear, 3–6mm, much longer than pedicels. Pedicels 1–2.5mm, glandular. Calyx lobes linear-oblong, 2–4mm (usually c 3mm), subacute. Corolla pale blue or mauve with darker blue or purple stripes; lobes 4–5 × 5–6mm, very obtuse, occasionally 5. Filaments and anthers white. Style 3.5–4mm, filiform, often thickened towards capitate stigma. Capsule broadly deltoid (resembling a moth with smaller hind wings apical), c 4 × 8.5mm, pilose near apical notch otherwise glabrous; apical notch shallow, not more than 0.5mm deep. Seeds ovoid, convex on both surfaces, 0.8–1mm, golden-yellow.

Sikkim: Lagyap. Grassy and rocky hillsides, clearings in and margins of oak and rhododendron forest, 2440–4150m. June–August.

A mainly Nepalese species of which only one Sikkim specimen has been seen; the description and ecological information is largely based on specimens from Nepal. Confused by both Hooker (80) and Dutta (336) with *V. lanosa* Bentham,

---

FIG. 96. **Scrophulariaceae.** a–b, *Veronica umbelliformis*: a, flowering stems (× ½); b, fruit (× 5). c–f, fruit of *Veronica* species: c, *V. javanica* (× 6); d, *V. deltigera* (× 5); e, *V. robusta* (× 5); f, *V. cana* (× 5). g, *Lagotis kunawurensis* var. *sikkimensis*: habit (× ½). h–i, *Neopicrorhiza scrophulariifolia*: h, habit (× ½); i, stamen (× 10). j, *Neopicrorhiza minima*: stamen (× 10). Drawn by Glenn Rodrigues.

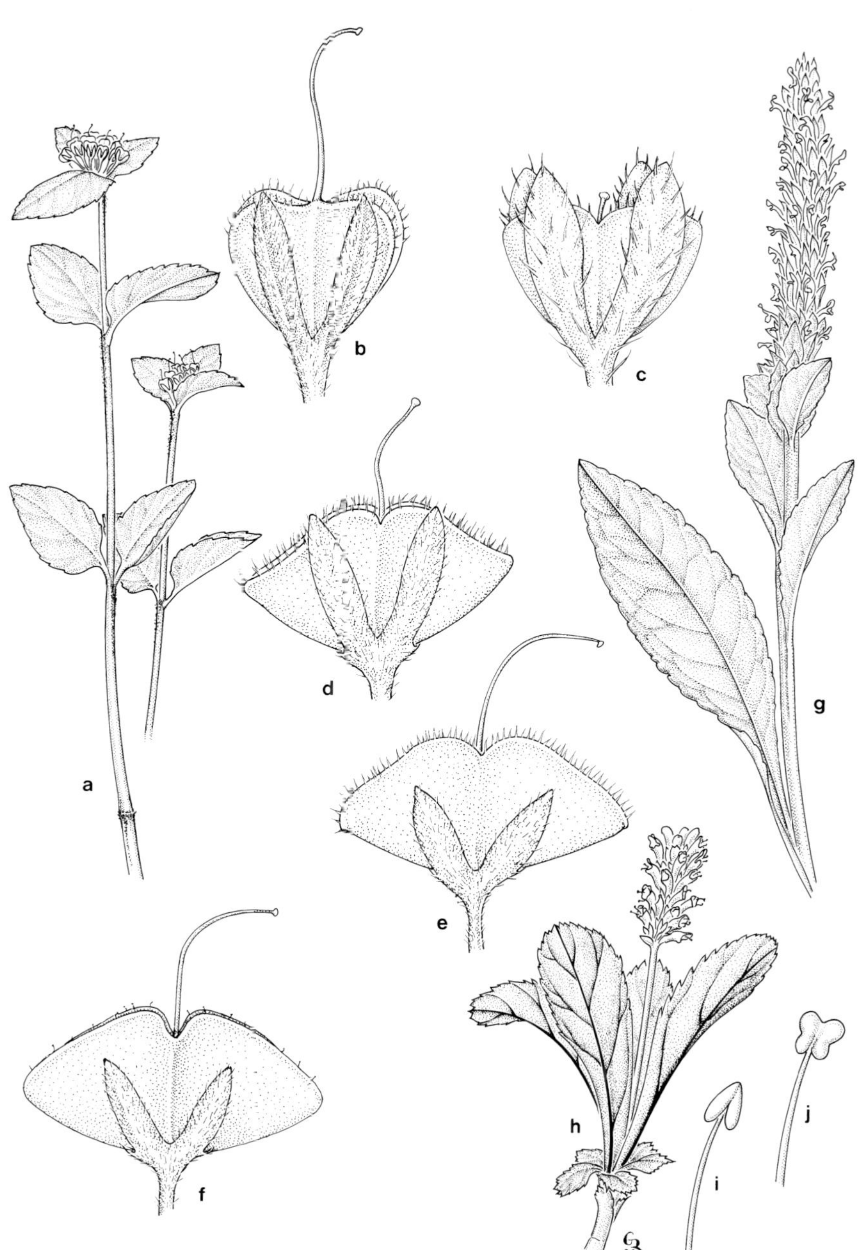
a
b
c
d
e
f
g
h
i
j

a W Himalayan species which has an ovoid, not broadly deltoid, capsule. The true affinities of *V. deltigera* are with *V. cana* and *V. robusta*, which differ in their petiolate leaves. Elenevsky (340) inexplicably treated *V. deltigera* as synonymous with *V. javanica* Blume; these two species differ markedly in capsule shape, leaf, stem and petiole indumentum, and flower size and colour.

8. V. lanuginosa Hook.f.

Perennial. Stems 3–10cm, erect, blackish, unbranched or with a single much shorter branch from base, nearly glabrous in lower half, becoming progressively more densely white-woolly upwards and in inflorescence. Leaves all opposite, lowest 2–3 pairs minute (c 2 × 1mm) and subglabrous, middle and upper ones imbricate, sessile, orbicular or uppermost broadly ovate, very obtuse, densely white-woolly above and beneath, progressively becoming larger upwards, middle 6–10 × 6–13mm, upper c 12 × 14–15mm. Flowers ± sessile in axils of upper leaves, forming a short subcapitate inflorescence. Calyx lobes oblong-obovate, c 4.5mm, woolly. Corolla dark blue or gentian blue, 7–8mm, upper lobe orbicular-spathulate, c 5mm wide, c twice as broad as other 3 obovate-spathulate lobes. Style 8–9mm. Capsule elliptic, emarginate, equalling calyx lobes, pubescent. Seeds few, large, oblong, plano-convex.

Bhutan: N – Upper Mo Chu district (Chumolari), Upper Pho Chu district (Kangla Karcha La), Upper Mangde Chu district (Gyophu La); **Sikkim:** Chemathang, Samdong, Goecha La, Kangpupchuthang etc.; **Chumbi:** Ghora La. Scree, sandy banks and glacial grit, often on peaty soil, 4530–5500m. June–September.

9. V. serpyllifolia L. subsp. **humifusa** (Dickson) Syme; *V. humifusa* Dickson

Perennial; vegetative stems creeping and rooting at nodes, sparsely pubescent, flowering stems ascending, 10–30cm, glandular-puberulent especially above, hairs c 0.2mm. Leaves sessile or very shortly petiolate, ovate, oblong or suborbicular, 6–8 × 3–8mm, apex and base obtuse, margin entire or very indistinctly crenate, glabrous. Flowers in bracteate, ± dense, terminal racemes with lowest 1–2 flowers sometimes remote. Lower bracts ± leaf-like but smaller, upper narrowly oblong, all slightly longer than their pedicel. Pedicels glandular-puberulent, 2–4mm in flower, 4–6mm in fruit. Calyx lobes ovate-oblong, c 2mm. Corolla pale blue, often with a slate or lavender tinge and darker veins, c 3mm. Filaments white; anthers bluish. Style 2–2.5mm. Capsule obcordate, c 2.5–3 × 3.5–4mm, ciliate.

Bhutan: N – Upper Kulong Chu district (Lao); **Darjeeling:** Happy Valley, Senchal, Tiger Hill. Moist stony places in mixed forest, c 1900–2600m. April–June.

Subsp. *humifusa*, to which all material seen belongs, is widespread throughout the Himalaya and China. Subsp. *serpyllifolia*, native to Europe and occasionally naturalised in parts of the Himalaya, differs by the hairs of the stem and pedicels being shorter and curved upwards, and by the smaller corolla c 2mm.

10. V. cephaloides Pennell; *V. ciliata* Fischer subsp. *cephaloides* (Pennell) Hong, *V. ciliata* sensu Dutta (336) non Fischer, *V. nana* Pennell

Erect perennial, sometimes branched below. Stems 3–26cm, often purplish below, slender, white-pubescent or -villous with hairs in 2 rows. Leaves opposite, sessile or at least lower ones very shortly petiolate (petioles to 1.5mm), lanceolate-ovate, 9–15 × 4–9mm, acute or subacute, base cuneate to obtuse, margin serrate (usually only above broadest part of leaf), surfaces pilose or almost glabrous. Racemes terminal (sometimes suppressed) and in axils of 1–2 uppermost pairs of leaves, dense and head-like at first but elongating slightly in fruit and then shortly cylindrical. Flowers subsessile (pedicels 0–0.5mm). Bracts oblong-linear, to 4mm, subequalling calyx. Calyx lobes narrowly oblong, 2–3.5mm, obtuse, ± densely white-hirsute. Corolla deep blue or slate-blue, 3–5mm; 3 lobes c 1.5 × 1mm, largest lobe c 2.5 × 2.2mm. Filaments and anthers white. Style c 1mm with prominent subcapitate stigma. Capsule ovoid-oblong, c 5.5–6 × 3mm, shallowly notched, hairy.

Bhutan: N – Upper Mo Chu district (above Laya); **Sikkim:** Samiti Lake, Yampung, unlocalised (Hooker and Gammie collections); **Chumbi:** Phari, Chumbi, Tang La. Grassy banks and exposed earthy or sandy slopes, 4000–4600m (2100–4950m in Nepal). July–September.

V. ciliata sensu stricto occurs mainly in C Asia and Mongolia extending south to China (as far as Xinjiang, Qinghai, Sichuan and Yunnan); all records from E Himalaya apply to *V. cephaloides*, which has been treated as a subspecies of *V. ciliata* in *Flora of China* (351).

11. V. persica Poiret

Prostrate or decumbent creeping annual, branched from base. Stems 10–40cm, hairy, with many very short hairs c 0.3mm mostly in 2 rows and fewer much longer stouter curved hairs. Lowest leaves opposite and without axillary flowers, all others alternate, each bearing a long-pedicellate flower in its axil. Leaves shortly petiolate, lamina triangular-ovate, 5–30mm, coarsely crenate-serrate with 3–4 teeth per side, with sparse rather stout hairs on veins beneath, margin ciliate. Pedicels 15–22mm, up to 2½ × subtending leaf, erecto-patent, tip decurved in fruit. Calyx lobes 4–6.5mm, ovate, ciliate, strongly accrescent and divaricate in fruit. Corolla 8–12mm diameter, bright Prussian blue with lower lobe paler or white. Capsule 2.5–4 × 9mm, broadly deltoid with 2 sharply keeled, strongly divergent, ciliate lobes c 4.5–6mm. Seeds 1.4–2.3 × 0.9–1.6mm, slightly boat-shaped, testa dorsally rugose.

Bhutan: C – Tashigang district (Khangma Research Station). Field weed, 1220–3600m. March–October.

Possibly introduced; first recorded in 1991. Ecological notes partly based on Nepalese material.

12. V. hederifolia L.

Prostrate or decumbent annual, with numerous branches from base. Stems 10–60cm, ± deflexed-pilose, hairs mainly in 2 rows. Leaves mostly alternate but lowest opposite, shortly petiolate (petiole 3–10mm), lamina reniform or suborbicular in outline, usually with 3–5 shallow lobes (terminal largest), sometimes subentire, 5–18 × 8–25mm, rather thick, pale green, palmately 3-veined, ciliate and with a few scattered hairs beneath. Flowers solitary in axils; pedicels 5–25mm, usually shorter than leaves in flower, lengthening in fruit and then sometimes exceeding leaves. Calyx lobes 4–5, persistent, ovate, 2.5–5mm in flower but accrescent in fruit, broadly and shortly acuminate, base ± deeply cordate, margin long-ciliate. Corolla blue or pale lilac, 4–9mm diameter, lobes 4–5, usually shorter than calyx. Style c 0.3–1.1mm. Capsule globose, scarcely compressed, c 3.5 × 4.5mm, glabrous, with shallow apical notch. Seeds subglobose, c 2.5 × 2.5mm, rugose, with deep pit on inner face, black.

Sikkim: unlocalised (336).

The record is based on a specimen (*R.S. Rao* 326) in the Calcutta herbarium which has not been examined. The description is based on European material and Dutta (336).

Four subspecies have been recognised in Europe, on the basis of length of pedicel relative to the calyx, shape of the middle lobe of the leaf, pedicel indumentum and style length. Dutta's description combines characters of subsp. *hederifolia* (leaves 3–5-lobed, flowers pale blue) and subsp. *lucorum* (Klett & Richter) Hartl (leaves 5–7-lobed, flowers pale lilac); both are common European weeds of cultivation.

28. LAGOTIS Gaertner

Perennial fleshy herbs, sometimes stoloniferous. Leaves mainly basal, entire to pinnatifid. Flowering stems scape-like, leafy above (leaves sometimes reduced and bract-like) or entirely naked; inflorescences sometimes produced directly from stolons. Inflorescence a dense spike-like raceme or ± globose head; bracts present, bracteoles absent. Bracts usually broad, imbricate and equalling or longer than flowers, sometimes shorter. Calyx spathaceous or of 2 sepals, imbricate in bud. Corolla blue, purple or white; tube curved; limb bilabiate, upper lip entire or bifid, lower lip 2–4-lobed. Stamens 2, inserted at corolla throat; anthers reniform. Ovary bilocular; style jointed at base; stigma capitate or bifid. Fruit a small, bilocular, 1–2-seeded drupe.

A difficult genus centred in the mountains of Tibet and China. There are several complex groups of species which show puzzling variation in flower colour and other characters, one of these being the *L. glauca* Ruprecht aggregate to which *L. kunawurensis* belongs. Field notes should include details of colour and texture of leaves (including the veins, which may or may not be red or purple, contrasting with the lamina) and bracts, flower colour, and notes on flower scent (if any).

The genus was included in the Selaginaceae in F.B.I., but, although anomalous and isolated, it seems better placed in Scrophulariaceae, within which it has affinities with *Neopicrorhiza*, *Picrorhiza* and *Wulfeniopsis* of the tribe Veroniceae.

1. Stoloniferous; acaulescent, scapes shorter than leaves ...... **2. L. chumbica**
+ Not stoloniferous; caulescent, scapes ± longer than leaves 2

2. Leaves pinnately lobed; calyx of 2 oblong ± free sepals ...... **1. L. pharica**
+ Leaves entire or dentate; calyx gamophyllous 3

3. Calyx spathaceous, hooded, 10–12mm, longer than corolla .. **3. L. clarkei**
+ Calyx not spathaceous or hooded, shorter than exserted corolla 4

4. Crown of root-stock naked (not clothed with sheaths); lower lip of corolla 3-cleft .. .. **4. L. crassifolia**
+ Crown of root-stock clothed with sheaths; lower lip of corolla 2-cleft **5. L. kunawurensis** var. **sikkimensis**

1. L. pharica Prain; *L. ramalana* sensu Yamazaki p.p. non Batalin

Perennial, flowering before the leaves expand, with many long stout pale greyish roots often longer than plant height. Petioles 3–5cm, uniformly narrow throughout; lamina of leaves ovate-oblong in outline, pinnately lobed, 25–40 × 15–30mm, base truncate. Scape 4–8cm, naked except for 3 small unequal leaves just below inflorescence which are 5–10mm and finely toothed. Spike ovoid, 13–15 × 10–13mm. Bracts 6–8mm, with 7 teeth in upper half. Calyx c 4mm; sepals oblong-linear. Corolla dark blue, 5.5–6.5mm; upper lip 2–3-fid or -cleft, lower lip 3-cleft; all segments of both lips linear, inner segment of lower lip shorter than outer ones. Filaments longer than upper lip. Anthers c 1mm, pale blue. Style included, 2.5–4mm; stigma notched.

Bhutan: C – Thimphu district (Taglung La), **N** –Upper Mo Chu district (Yale La); **Chumbi:** Tern La (390), Chumegata. On loose scree, c 4570m. May; fruiting September.

An eastern counterpart of *L. globosa* (Kurz) Hook.f. of W Tibet and Chitral, differing in its shorter bracts and ovoid spikes.

2. L. chumbica R.R. Mill

Very small, stoloniferous almost acaulescent perennial, drying blackish. Leaves all in a basal rosette, c 7–10 per plant, linear-elliptic, 10–25 × 1–2.5(–4)mm, obtuse, base attenuate, outer ones becoming reflexed, almost glabrous; margins entire, involute beneath; petiole indistinct, 3–5mm. Scapes naked, 3–8(–10)mm, shorter than leaves, soon becoming strongly reflexed, puberulent. Inflorescences 6–8 × 6.5–8mm, ovoid-globose, spicate, dense. Bracts oblong to ovate-lanceolate, equalling the flowers. Calyx split almost to base into 2 ovate-oblong, obtuse lobes and with hyaline keels on each side. Corolla pinkish-white, c 6.5mm, less than twice length of calyx; lips equal, upper one ovate, c 2 × 1mm, lower reflexed and divided into 2 lobes c 2 × 0.5mm. Fruit not seen.

Chumbi: Phari Plain. Marshy alluvial soil in river beds, 4420–4877m. May–July.

Differs from *L. brachystachya* Maximowicz (China: Gansu) in its more linear, smaller, narrower leaves which are canaliculate beneath with ± involute (not flat) margins, shorter scapes and inflorescences, and anthers subexserted (not included). Further specimens of *L. chumbica* are known only from sites in Tibet near Chumbi; data from these are incorporated in the description and ecological details above.

3. L. clarkei Hook.f.

Crown of root-stock naked, lacking fibrous sheaths. Petioles of basal leaves red at flowering time, 2–7cm, lengthening to c 12cm in fruit; lamina ovate-elliptic or ovate-oblong, 3.5–12 × 1.5–5cm, glaucous, obtuse or subacute, base obtuse to shortly attenuate, margin indistinctly crenate or nearly entire; veins red like petiole. Scape 7–12cm at anthesis, 20cm or more in fruit; cauline leaves 2–3(–5), in upper part of scape below inflorescence, subsessile, 2–4.5cm. Inflorescence oblong, 1.5–4 × 1–1.5cm in flower but longer in fruit. Bracts c 7mm, shorter than calyx. Calyx gamosepalous, spathaceous, greenish-white or yellowish-green, ovate-lanceolate, 10–12 × c 6mm, hooded, longer than corolla. Corolla mauve or purple, 8–8.5mm; both lips short and entire, lower broader than upper. Filaments adnate to margins of upper lip; anthers bluish.

Bhutan: C – Ha district (Tare La), Thimphu district (Pajoding); **Sikkim:** Yakla, Chho La, Tosa etc.; **Chumbi:** Kupup (390), Lam Pokri. Alpine turf and scree, 3800–4870m. July–September.

Outside our area, only occurring in easternmost Nepal (Arun/Tamur watershed area).

4. L. crassifolia Prain

Non-stoloniferous, scapose rosulate perennial; crown of rhizome naked (lacking sheaths). Rosette leaves 6–8 per plant; petiole 2–7cm, glabrous; lamina oblong-elliptic, 3–7 × 0.8–2cm, thick-textured, rugulose, obtuse, base cuneate or shortly attenuate, margin crenate or crenate-serrate (7–12 crenations per side). Scapes red, 8–13cm (excluding inflorescence), glabrous; scape leaves 1–2 subopposite pairs, situated just below inflorescence, sessile, ovate, shallowly crenate or subentire. Inflorescence a narrowly oblong spike (thimble-shaped when young), 50–110 × c 12mm at anthesis (to 18mm wide in fruit). Bracts 8–10 × 5–6mm, broadly obliquely ovate with indistinct shoulder or lateral tooth on one margin. Calyx 8.5–9mm, divided in upper ⅓ into 2 obtuse, shortly mucronate lobes, one lobe emarginate with mid-vein excurrent from apex of one emargination. Corolla blue or reddish, 11–13mm; tube c 6mm, included in calyx; upper lip 4.5–6 × 2–3mm, oblong, acute or indistinctly emarginate; lower lip 3-sect, segments linear, 4.5–7 × 0.4–0.7mm, middle one slightly narrower than laterals. Style 4–5mm, stigma globose or slightly bifid. Drupe ellipsoid, c 6 × 2.5mm, sparsely pilosulous on sides.

Bhutan: **N** – Upper Mo Chu district (Yale La); **Sikkim:** Tankra (390); **Chumbi:** Kalaeree, Phari and neighbouring area. Stony or swampy alpine turf, 4575–4880m. June–September.

5. L. kunawurensis (Bentham) Ruprecht var. **sikkimensis** (Hook.f.) Yamazaki; *L. glauca* Gaertner var. *sikkimensis* Hook.f., ?*L. spectabilis* Hook.f. Fig. 96g.

Usually an erect, ± glabrous perennial. Crown of root-stock with non-fibrous sheaths. Rosette leaves several borne ± erect; petioles 6–14cm, distinct; lamina elliptic to narrowly oblong, 6–19(–23) × 3–8.5cm, obtuse or subacute, base shortly attenuate or cuneate, margin ± crenate (occasionally ± entire). Scapes normally 15–30cm (but occasionally flowering when only c 2cm tall), typically with few to several leaves gradually becoming smaller towards base of inflorescence (sometimes almost naked except for a whorl below inflorescence); scape leaves sessile, ovate, shallowly crenate to subentire. Inflorescence an oblong-ovoid, finally narrowly oblong dense spike, (4.5–)8–12 × 1.5–2cm. Bracts rhombic to elliptic, 6–12 × 2.2–6.5mm, variably dentate to almost entire. Calyx 5.5–10 × (when opened out) 3.5–5(–9.5)mm, divided in upper ⅓ into 2 obtuse, shortly mucronate lobes. Corolla blue or pale blue-violet, turning white; tube ± straight; upper lip 4–6 × 1.5–3.5mm, obtuse, entire or rarely 3-notched; lower lip composed of 2 linear, strongly recurved segments 3–5 × 0.4–0.8mm. Anthers oblong, thecae longer than broad but becoming divergent; filaments 0.5–0.7mm. Style (1.5–)3–6mm; stigma 0.6–1.1mm across, of 2 ± unequal branches, both usually horizontal, rarely erecto-patent. Ovary c 1.5mm.

Bhutan: **C** – Thimphu district (Tremo La–Sharna), Bumthang district (Bumthang, Rudo La, Tripte La etc.), **N** – Upper Mo Chu district (Yale La, Nelli La, Shingche La), Upper Bumthang Chu district (Marlung); **Sikkim:** Sebu La, Sherabthang, Lhonak, Tsomgo etc. Wet semi-stable scree and marshland, 2700–4900m. June–August.

Very variable in relative length of bract and calyx, width of calyx, toothing of bract, and form of stigma. The variation in the characters appears to be uncorrelated and consequently the subspecies is here not further divided, but further studies are needed to understand the nature of the variation. As circumscribed here, var. *sikkimensis* occurs mainly in Bhutan and N Sikkim but extends into easternmost Nepal. *L. spectabilis* Hook.f. (described from the Nepal/Darjeeling/Sikkim border) may be a local variant of it; Prain (390) noted that it differed from all other *Lagotis* species in the area by drying straw-yellow, not black.

29. ALECTRA Thunberg

Hemiparasitic annual herb. Stem simple or branched above, quadrangular. Leaves opposite, sometimes alternate above or almost throughout, shortly petiolate, serrate. Flowers in loose terminal spikes or solitary and axillary. Bracteoles 2, at base of calyx. Calyx campanulate, membranous, 5-lobed. Corolla tubular or campanulate with indistinctly bilabiate subequally 5-lobed limb; lobes spreading. Stamens 4, included, connivent in pairs, lower 2 larger than upper ones; filaments adnate to base of corolla tube, filiform; anther thecae parallel, confluent towards apex, mucronate at base. Ovary subglobose. Fruit a globose, loculicidal and septicidal 4-valved capsule. Seeds linear-cylindrical, numerous; testa hyaline, reticulate.

1. A. avensis (Bentham) Merrill; *A. indica* Bentham, *Melasma avense* (Bentham) Handel-Mazzetti ('arvensis'). Nep: *Gulit gheim*. Fig. 97b.

Erect annual. Stems (10–)15–90cm, shortly subappressed- or deflexed-patent pubescent, hairs mainly in 2 rows, white. Petioles 1–4mm; lamina oblong-ovate, ovate-oblong or ovate-rhombic, 7–35(–50) × 3–15(–20)mm, acute, base cuneate-attenuate, coarsely and sharply serrate with 3–7 teeth per side, sparsely hirsute above, glabrous beneath except for sparsely hirsute veins. Flowers in terminal and axillary bracteate spikes becoming lax below but condensed above, or solitary and axillary. Pedicels not more than 1mm. Bracteoles 4–6mm, with long marginal hairs. Calyx 4–5mm, 10-veined; lobes triangular-ovate, acuminate, c 2 × 2.5mm. Corolla yellow (rarely pink), 6–7mm. Posterior stamens c 3mm, with almost glabrous filaments; anterior pair c 4mm, with woolly filaments; anthers yellow. Capsule subglobose, 5–6mm.

Bhutan: S – Chukka district (near Chimakothi); **Darjeeling:** Mongpu, Darjeeling, Takdah etc.; **Sikkim:** Tista, Lachung, Lusing. Steep stony and dry grassy slopes and by roadsides, 300–1900m (possibly to 2100 or 2400m). August–November.

Host plant not stated on any material seen. There is an unlocalised Sikkim specimen collected by Hooker with a vague altitudinal range of '6,000–8,000ft', and the Lachung specimen is labelled '6,000–7,000ft', but no accurately localised specimen has been seen from above 1900m. The specific epithet has frequently been mis-spelt 'arvensis', but is derived from Ava (an old name for northern Myanmar (Burma)), where Wallich collected the type specimen.

30. BUCHNERA L.

Erect annual herb, blackening when dry. Leaves rosulate below (rosette withering before fruiting), opposite or subopposite on stem. Inflorescence a lax terminal spike; flowers alternate or lower subopposite, subtended by a bract and 2 bracteoles. Calyx tubular, 10-veined, indistinctly ribbed, with 5 apical teeth. Corolla tubular; limb with 5 subequal spreading lobes. Stamens 4, included, slightly didynamous; anthers dorsifixed, each with a single theca. Style thickened or clavate above. Fruit a loculicidal capsule; seeds numerous.

1. B. hispida D. Don (non Bentham)

Rigid annual. Stems slender, 15–45cm, hispid below with long white hairs, more shortly spreading-hispid above. Rosette leaves soon withering, obovate to elliptic, 15–20 × 8–12mm, obtuse, base cuneate into short petiole, entire, hispid on veins. Lower cauline oblong-elliptic to elliptic, c 3 × as long as broad, upper becoming linear, to 50 × 3mm, usually with a few shallow teeth, hispid. Spike lax especially in fruit, lower flowers becoming remote. Bracts 3.5–5 × 0.8–1mm, ovate to lanceolate. Bracteoles linear-lanceolate to subulate, c 3–4 × 0.15–0.4mm. Calyx 4–8mm, narrowly cylindrical, rather suddenly expanded into lanceolate apical teeth c 1mm, white-hispid. Corolla deep mauve, bluish-purple or violet, 8–12mm; tube slender, 6–7.5mm, exserted from calyx, ± pilose outside or glabrous; limb 3–5mm diameter, lobes obovate or obovate-oblong, 2–3mm; throat villous. Capsule 5–6 × c 3mm, equalling calyx, shortly beaked.

Bhutan: C – Punakha district (several collections from Wangdu Phodrang/ Chusom area). Dry rocky banks, 1350–2400m. October.

B. hispida Bentham is a synonym of *Striga densiflora* (Bentham) Bentham. *B. cruciata* D. Don (*B. densiflora* Hooker & Arnott, non *Striga densiflora* (Bentham) Bentham) is common in Assam, Nepal and N India and might occur in our area. It differs from *B. hispida* D. Don in stems not hispid but pubescent with short hairs directed backwards; calyx teeth twice as long (c 2mm), linear; inflorescences much shorter and more condensed even in fruit, 15–30 × 7–15mm.

31. STRIGA Loureiro

Eng: *Witchweed.*

Annual root-parasitic herbs. Stems stiff, erect, greenish or greyish, 4-angled and often ridged. Leaves opposite, small, relatively few and at least lower ones reduced and scale-like. Inflorescence a spike; flowers in axils of long, alternate sessile leaf-like bracts much longer than true leaves. Bracteoles 2. Calyx tubular, 5-lobed, with either 5 or at least 10 veins. Corolla bilabiate; tube long and narrow, with very narrow hairy mouth; limb expanded at right angles to tube, upper lobes fused, lower lobes 3, spreading. Stamens 4, didynamous, inserted in upper part of tube; anthers basifixed, 1-locular; pollen sticky and transferred to style in a mass. Nectary present, at base of ovary. Style terete, stigma bifid. Fruit a loculicidal capsule. Seeds minute, prominently ridged.

Parasites of grain crops and other plants, often causing serious damage and consequently of economic importance. Very few collections are known from our area and efforts should be made to assess the distribution and frequency of each species.

1. Calyx ribs 5, or as many as calyx lobes .... **1. S. densiflora**
\+ Calyx ribs 10–15, or at least twice as many as calyx lobes 2

2. Calyx ribs usually 10, only 1 per lobe extending to tip of lobe, others ending at sinuses .... **2. S. asiatica**
\+ Calyx ribs usually 15, each lobe with 1 central and 2 marginal ribs all extending to tip .... **3. S. angustifolia**

1. S. densiflora (Bentham) Bentham; *Buchnera densiflora* Bentham, non Hooker & Arnott. Fig. 97g.

Stem 15–45cm, simple or with erect-ascending sometimes fastigiate branches, scabrid or shortly strigose. Vegetative leaves linear, c 35–50 (or more) × 2–4mm, with rather dense white calcified pustules especially on upper surface and on midrib beneath, otherwise ± glabrous. Bracts linear, 10–25 × 1.5–3mm, with white scabrid pustules especially along margins. Calyx c 7mm, 5-ribbed, teeth linear-triangular, shortly hispidulous, finally recurved. Corolla white, 10–16mm; tube exserted from calyx, 8–10 × c 0.5mm, only slightly broadening below limb; upper lip of limb c 3mm; lower lip 5–7 × c 6mm, lobes obovate, middle one c 5 × 2.5mm. Capsule ovoid-globose, c 4mm, dark brown.

Darjeeling/Sikkim: Tista River. Parasitic on grasses including *Sorghum* and other grain crops, c 450–600m. March–April.

Known in our area only from a single collection, made by Gamble in 1875. The ecological notes refer to N India.

2. S. asiatica (L.) Kuntze; *Buchnera asiatica* L., *S. lutea* Loureiro. Fig. 97f.

Hispid annual. Stems branched above middle, 7–30cm, densely short-hispid and scabrid. Vegetative leaves linear, c 5mm, often withered by flowering time. Bracts numerous, alternate, erecto-patent to spreading, linear-lanceolate, 9–35mm, hispid and scabrid. Calyx 6–7mm including teeth, usually 10-ribbed (but sometimes with more), translucent between ribs; ribs stout, dark green, one of each group of 3 extending to tip of tooth, other 2 ending at sinus between teeth; teeth c 2mm, narrow, erect, tips becoming slightly deflexed. Corolla yellow (in Bhutan material seen; white forms also known from India), 11–15mm; tube 6–9 × c 0.5mm, bent near limb, shortly glandular-pubescent; upper lip of limb emarginate, 1–1.5mm, lower lip 3-lobed, lobes 2–4mm.

Bhutan: C – Punakha district (Sankosh Valley S of Wangde Phodrang, and near Chusomsa). Around rock outcrops in dry open bush-land, c 1000–1200m. August–October.

Usually only 2 flowers are open at one time on each inflorescence branch.

3. S. angustifolia (D. Don) Saldanha; *S. euphrasioides* sensu F.B.I. non Vahl. Fig. 97d–e.

Similar to *S. asiatica* but calyx ribs 15, all extending to tips of teeth, which each have 1 central and 2 marginal ribs; stems much less hispid; bracts ascending, usually held close to stem (not patent or subpatent).

Bhutan (unlocalised). Dry grassland and slopes, 150–1675m. August–October.

Known in our area from a single unlocalised specimen collected by Griffith (K.D. 3934, K). The description and ecological information are based mainly on Nepalese material. In Nepal, the flowers are most commonly white, only rarely yellow, whereas *S. asiatica* in our area typically has yellow, rarely white, flowers.

32. CENTRANTHERA R. Brown

Hemiparasitic annual herbs. Stems ascending or erect, terete, simple or branched. Leaves opposite or upper ones alternate, sessile. Flowers solitary and axillary or in terminal spikes. Bracts leaf-like. Bracteoles 2, small, below calyx. Calyx spathe-like, split along lower margin, obliquely acute with shortly 3-lobed adaxial apex. Corolla bilabiate; tube curved, dilated upwards; upper lip deeply 2-lobed, lower lip ± equal to upper, deeply 3-lobed. Stamens 4, didynamous, included; filaments adnate to base of corolla tube, ciliate; anther thecae 2, joined at tips, with basal spur or mucro. Fruit a loculicidal 2-valved capsule. Seeds oblong, spirally reticulate.

1. Inflorescence a terminal spike; corolla yellow, 30–50mm; stem scabrid, hairs directed downwards, very short .... **1. C. grandiflora**
\+ Inflorescence of solitary axillary flowers; corolla pink, 12–20mm; stem hispid, hairs directed upwards, at least some of them c 1mm **2. C. nepalensis**

1. C. grandiflora Bentham

Robust annual. Stems rather stout, 30–70cm, scabrid with short downward-directed hairs. Leaves shortly petiolate or sessile, narrowly elliptic or linear-oblong, 30–65 × 3–12mm, subacute, base attenuate, upper surface with dense many-celled cystolith tubercles, lower surface shortly pubescent mainly near midrib and ± inrolled margin. Inflorescence a terminal bracteate spike. Flowers shortly pedicellate. Calyx 12–15mm, inflated, segments cohering, scabrid on veins otherwise ± glabrous. Corolla yellow, 30–50mm, pubescent. Capsule globose, c 8mm.

Bhutan: C – Punakha district (Rinchu) and Mongar district (Kuru Chu and Ngasamp), **N** – Upper Mo Chu district (Gasa); **Darjeeling:** Badamtam, Rangit; **Darjeeling/Sikkim:** Tista, Monglong. Damp slopes, *Pinus longifolia* forest, 300–2135m. May–August, fruiting January–March.

Used in Assam as a snake-bite antidote and for insect stings (352).

2. C. nepalensis D. Don; *C. hispida* sensu F.B.I. non R. Brown, *C. cochinensis* (Loureiro) Merrill var. *nepalensis* (D. Don) Merrill

Viscid annual. Stems slender, 8–60cm, rather densely hispid with upward-directed hairs. Leaves sessile, narrowly elliptic to linear, 10–30 × 1–5mm, obtuse or subacute, base attenuate, hispid above and beneath with hairs arising from cystolith bases, viscid. Flowers axillary, sessile, opposite or usually alternate. Calyx c 5mm, densely hispid all over. Corolla deep pink to dull crimson, 12–20mm, sparsely pubescent. Capsule globose, 4–5mm.

Darjeeling: Kuprail terai. Marshy places and damp grassland, up to c 1600m. September.

33. **SOPUBIA** D. Don

Hemiparasitic erect annual herbs. Stems quadrangular, simple or often branched at least above. Leaves opposite or upper sometimes alternate, very narrow, often laciniate. Flowers in bracteate terminal racemes or spikes. Pedicels with 2 filiform bracteoles at base of calyx. Calyx campanulate or tubular-campanulate, with 5 teeth or lobes. Corolla rotate or subcampanulate; limb indistinctly bilabiate, with 5 subequal lobes. Stamens 4, didynamous, included, inserted in lower or middle part of short corolla tube; filaments glabrous, filiform; anther thecae equal, or unequal with one much smaller and sterile. Ovary subglobose or ovoid; style filiform with clavate stigma. Fruit a 4-valved capsule, dehiscing loculicidally and septicidally. Seeds cylindrical, numerous, testa striate-reticulate.

1. Pedicels 5–15mm; corolla yellow, rotate, 8–10mm long and 10–13mm diameter; calyx lobes triangular .... **1. S. trifida**
\+ Pedicels 0.5–2mm; corolla purplish, tubular-campanulate, 7–8mm long and 4.5–5.5mm across at mouth; calyx lobes lanceolate .... **2. S. stricta**

1. S. trifida D. Don. Fig. 97h.

Stems 20–90cm, minutely scabrid-pubescent. Leaves opposite, often fasciculate; lower ones 3-fid into linear-filiform segments 5–40 × 0.2–1mm, upper entire, all ± convex, midrib broad and prominent beneath, glabrous; margins narrowly revolute and scabrid with small white plate-like protuberances. Inflorescence a lax almost glabrous raceme of usually 7–12 nodes, flowers in false whorls of often 3 pedicels or some alternate. Pedicels 5–15mm, arcuately erecto-patent, narrowly clavate. Bracteoles 1–2mm, scabrid on margins like leaves. Calyx 3–4mm, densely scabrid, protuberances largest on lobes; lobes c 1mm, triangular, fringed with dense creamy-white hairs. Corolla yellow with red centre, rotate, 8–10mm long and 10–13mm diameter; lobes orbicular-ovate, 5–6mm. Filaments and anthers red or reddish-brown; anther thecae unequal, one in each anther being much smaller, sterile and clavate. Capsule globose-ovoid, 3–4mm. Seeds 0.6 × 0.4mm.

Bhutan: C – Punakha district (Sena Thang–Samtengang, Wangdu Phodrang etc.), Thimphu district (Simtokha–Talukha Gompa), Tongsa district (Tongsa Bridge), Tashigang district (Tashi Yangtse), **N** – Upper Kuru Chu district (Denchung); **Darjeeling:** Darjeeling; **Darjeeling/Sikkim:** unlocalised (80). Open dry grassy or gravelly hillsides, and in *Pinus–Quercus* scrub, 1830–2600m; common. July–early September. Occurring in similar habitats and flowering till early October in Nepal.

2. S. stricta (Bentham) G. Don

Similar to *S. trifida* but stems often taller, to 120cm, densely appressed-pilose with long thin retrorse hairs; leaves undivided, narrowly linear, 10–55 ×

0.5–2mm, midrib pubescent beneath; margins more strongly revolute and even more densely white-scabrid; inflorescence with usually more than 15 nodes, all flowers ± alternate; pedicels 0.5–2mm; bracteoles c 4mm, subequalling or exceeding calyx; calyx c 4mm, lobes ovate-lanceolate, not fringed with hairs; corolla pale mauve to deep purple, tubular-campanulate, 7–8mm long and 4.5–5.5mm across at mouth; all anther thecae fertile and subequal.

Darjeeling/Sikkim: unlocalised (Hooker collection). 600m. October–November.

Apparently rare, or at least rarely collected; the description is based on scanty Burmese material and the descriptions by G. Don (335) and Hooker (80). A mainly Indo-Chinese species which is at its north-west limit in Darjeeling/Sikkim.

34. EUPHRASIA L.

Eng: *Eyebright.*

Hemiparasitic, annual herbs, usually less (often much less) than 30cm. Stem pubescent, ± branched, with numerous internodes. Indumentum of eglandular and/or glandular hairs. Leaves subsessile, opposite or upper ones subopposite, crenate or dentate. Flowers axillary, solitary, subsessile or sessile. Calyx 4-toothed. Corolla white (sometimes tinged pink or lilac), usually with dark violet stripes along lobes and 3 yellow spots in throat, bilabiate; upper lip hooded, shortly 2-lobed, lower lip out-spread, 3-lobed, the lobes emarginate. Stamens 4; filaments inserted in corolla throat, curved; anthers dark red, partly hidden under hood of upper lip, hairy, their thecae spurred, the posterior spurs of the posterior anthers being longest. Ovary 2-locular, laterally compressed; stigma capitate. Capsule opening in upper half; seeds fusiform, striate.

The genus description applies to sect. *Semicalcaratae* Bentham, a section containing annual species of the north temperate zone and to which the Bhutanese species belong.

Species delimitation in *Euprasia* sect. *Semicalcaratae* is extremely difficult, because of apomixis which has led to the recognition throughout the northern hemisphere (but especially in Europe) of a very large number of 'microspecies', based on very small differences. Of some 200 of these, three have so far been recorded from our area, with a considerably larger number known from the more western parts of the Himalaya. The genus is under-collected in Bhutan and it is possible that more microspecies may be discovered, either new to science or as new records for the area. (*E. himalayica* Wettstein and *E. platyphylla* Pennell both extend to eastern Nepal but have not yet been correctly identified from Sikkim or Bhutan.)

The key and descriptions below should be used with caution. Misidentifications are not uncommon. The possibility that two (or more) microspecies are growing together should not be overlooked when making identifications. Determination as far as the aggregate level will suffice for many purposes, in which case the name '*Euphrasia officinalis* L., agg.' may be used. The generic description applies equally to the aggregate as represented in our area. The species descriptions follow, in an abbreviated form, the model used by Yeo (430).

When collecting specimens of *Euphrasia*, a minimum of six examples and preferably 12 should be collected from a given locality. Nodes are counted omitting the cotyledonary node. Corollas are measured along the upper (dorsal) side; if not dissected, the length from the base of the calyx to the tip of the upper corolla lip may be taken and 0.5–1mm deducted from the value obtained.

1. Some hairs on stem, floral leaves and calyx arising from ± conspicuous black bases .... **2. E. melanosticta**
\+ No hairs on stem, floral leaves or calyx arising from black bases 2

2. Flowering beginning at node 4–7(–10); stem with a few glandular hairs as well as eglandular hairs; cauline internodes 2–5 × subtending leaf; calyx c 5mm in flower, divided to c ⅘; capsule borne at an acute angle to stem **1. E. bhutanica**
\+ Flowering beginning at about node 3; stem lacking glandular hairs; cauline internodes not more than 1½ × subtending floral leaf, mostly equalling or shorter than ripe capsule; calyx 3.5–4mm in flower, divided to c ½; capsule borne closely appressed to stem .... **3. E. chumbica**

1. E. bhutanica Pugsley; *E. jaeschkei* auct. non Wettstein. Fig. 97a.

Stems weak, 5–20cm, 0.4–0.8(–1)mm diameter, unbranched or with 1–2(–4) branches from lower nodes, with short retrorse-appressed eglandular hairs and a few usually 3-celled spreading glandular hairs. Vegetative leaves soon falling. Cauline internodes 2–5 × subtending leaf. Flowering beginning at node 4–7(–10); lower flowering nodes 1.2–3 × subtending floral leaf. Lower floral leaves broadly trullate, 4–6 × 4–6mm, with 3–4 pairs of acute teeth 1–1.5mm deep, glandular-hairy especially on veins and margin. Calyx c 5mm in flower, divided to c ⅘, each lobe with 2 narrowly triangular-acuminate teeth c 2.5mm; teeth glandular-hairy, hairs lacking black bases. Corolla white or mauve, 6–8mm; tube and back of upper lip shortly pubescent, lower lip glabrous; upper lip 3–3.5mm, with purple lines; lower lip 3-lobed, each lobe conspicuously emarginate; throat with yellow blotch within. Anthers dark chocolate, white-pilose. Capsule borne at an acute angle to stem, oblong, 4–5.5(–6) × 1–1.5mm,

FIG. 97. **Scrophulariaceae.** a, *Euphrasia bhutanica*: habit (× ⅔). b, *Alectra avensis*: flowering stem (× ⅖). c, *Phtheirospermum tenuisectum*: habit (× ⅔). d–e, *Striga angustifolia*: d, habit (× ½); e, dissected calyx, exterior (× 3½). f, *Striga asiatica*: dissected calyx, exterior (× 5). g, *Striga densiflora*: dissected calyx, exterior (× 4). h, *Sopubia trifida*: habit (× ⅔). Drawn by Glenn Rodrigues.

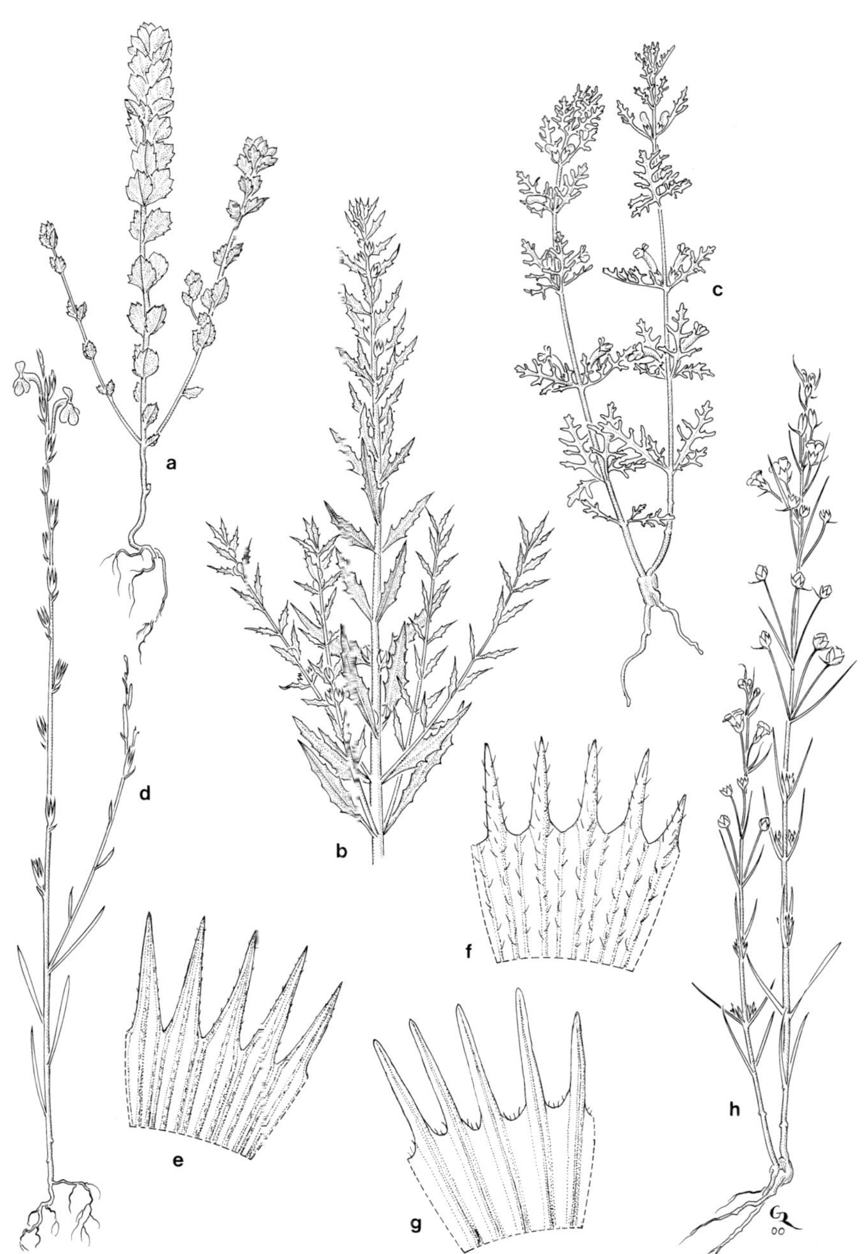
a
b
c
d
e
f
g
h

subequalling or scarcely exceeding calyx, eglandular-pilose; apex shallowly retuse.

Bhutan: C – Ha district (Ha), Bumthang district (Kitiphu, Tang, Tashiling etc.), Mongar district (Rip La), **N** –Upper Mo Chu district (Sinchu La, Gasa (71)); **Chumbi:** Dotag. Open, usually damp, grassy hillsides or marshes, 2750–4000m. May–September.

Pugsley's original description of *E. bhutanica* disagrees not only with that given above but also with the characters of the holotype and paratype. In particular, he described the leaves as numerous, not easily falling, and exceeding the internodes, while the flowers were said to be only 5mm. This is not the case; the type agrees with all the other material seen in having characters as given above.

A very dwarf specimen from Khamba Dzong (Tibet, immediately north of Sikkim, *Younghusband* 154, K) has been determined as *E. jaeschkei* Wettstein by Siddiqi, but has the thickened leaf margins of *E. bhutanica* which are apparently lacking in *E. jaeschkei.* The latter is a NW Himalayan species.

2. E. melanosticta R.R. Mill

Stems weak, 2–18cm, 0.5–1mm diameter, purplish at least above and with black blotches at bases of hairs, usually unbranched, sometimes with 1–3 branches towards base, shortly eglandular crispate-puberulent, with few spreading 2–3-celled glandular hairs above. Vegetative and lower floral leaves very easily falling; vegetative leaves c 2 pairs, broadly ovate-orbicular, c 3 × 3mm, the lower pair entire, the upper pair with 2 pairs of lateral lobes, their surfaces sparsely hairy. Flowering beginning at node 2–5; lower floral internodes 1½–2½ × subtending leaf. Lower floral leaves broadly ovate, c 6 × 6mm, with 3–4 pairs of acutely triangular teeth c 1mm deep and a broader, more obtuse terminal lobe, sparsely setulose and glandular-pilose, hairs arising from black bases. Calyx c 4mm in flower, divided to c ¾, each lip with 2 narrowly triangular teeth c 1mm; teeth glandular-pilose, hairs arising from ± conspicuous black bases. Corolla 5–6.5mm, white, upper lip with purple streaks, lower lip with central yellow blotch and 2 purple streaks on ventral side; tube shortly pubescent, upper and ventral surface of lower lips pilose, some hairs arising from small dark blotches; upper lip 2.5–3mm, emarginate, lower lip unequally 3-lobed, each lobe broadly and shallowly emarginate, the notch ± acute at base. Capsule borne at acute angle to stem, oblong, 3.5–5.5(–6) × 1.8–2.5mm, subequalling fruiting calyx, apex broadly obtuse or almost truncate, shortly mucronate, with rather long setules.

Bhutan: N – Upper Mo Chu district (Gasa Dzong, above Kohina), Upper Bumthang Chu district (Lhabja); **Sikkim:** Yumthang, Kupup, Meguthang; **Chumbi:** Chumegati. Damp grassy banks, yak pasture and open, often peaty hillsides, 2800–4270(–4600)m. Mid July–September.

Noltie 169 is exceptionally young and is the only specimen seen with vegetative leaves. A specimen of this species from Bumthang has been misidentified as *E.*

himalayica Wettstein. The black glands, absent in that species, are very distinctive. A specimen from Chumegati, collected at 4600m, is extremely dwarf and forms very dense, small cushions but is otherwise similar to taller plants from lower altitudes.

3. E. chumbica R.R. Mill; *E. platyphylla* auct. non Pennell, *E. kurramensis* auct. non Pennell

Stems 17–27cm, (0.6–)1–1.5mm diameter, usually with several strong erect branches except in weakest plants, shortly crispate-pubescent, lacking glandular hairs. Vegetative leaves caducous. Flowering beginning at about node 3; inflorescence dense, lower internodes not more than 1½ × subtending floral leaf and mostly equalling or even shorter than ripe capsule. Lower floral leaves ovate or broadly ovate, 5.5–6 × c 6mm, with 3–4 pairs of subacute teeth c 1mm deep, eglandular-setulose on margin and veins. Calyx 3.5–4mm in flower, divided to c ½, each lobe with 2 narrowly triangular-acuminate teeth c 1.5mm; teeth eglandular-setulose, hairs without black bases. Corolla 7–7.5mm, white, upper lip with purple streaks; tube and upper lip appressed-pubescent; lower lip 3-lobed, lobes broadly and shallowly emarginate. Capsule appressed close to and ± parallel to stem, 4.5–5.5 × c 1.8mm, sparsely pilose; apex obtusely truncate and mucronate.

?Bhutan (see below); **Sikkim:** Lachen, Nathang, Takti; **Chumbi:** Chumbi. Gravelly hill slopes and meadows, 3353–3660m. July–September; fruits July–December.

A very late-collected specimen from C Bhutan (Thimphu, *Cooper* 3551, E) resembles *E. melanosticta* but has subsessile glands on the calyx. More material is needed to assess its status. A Hooker specimen from Lachen was misidentified by Siddiqi as the NW Himalayan *E. platyphylla* Pennell. The same worker identified another specimen of *E. chumbica* (Chumbi, x 1904, *Bell* s.n.) as another NW Himalayan species, *E. kurramensis* Pennell, which is sometimes treated as a synonym of *E. himalayica* Wettstein. The latter species extends to Nepal and is very variable.

35. PHTHEIROSPERMUM Bunge

Viscid annual, biennial or perennial herbs. Leaves opposite, small, pinnatisect or 3-sect. Flowers axillary, solitary, sessile or shortly pedicellate, ebracteolate. Calyx campanulate, lobes 5, subequal or unequal, narrow. Corolla 2-lipped; tube dilated above; upper lip short, with 2 lobes folded back; lower lip longer, 3-lobed; throat open with bigibbous palate. Stamens 4, didynamous; anthers positioned below upper lip, glabrous, thecae parallel. Stigma spathulate, shortly 2-lobed. Fruit a compressed, beaked, loculicidal capsule. Seeds numerous, ovoid, reticulate.

1. Leaves 2–3-pinnatisect; corolla tube 10–13mm, c twice length of calyx **1. P. tenuisectum**
+ Leaves 3-sect; corolla tube 2.5–3mm, hardly longer than calyx **2. P. glandulosum**

1. P. tenuisectum Bureau & Franchet. Fig. 97c.

Perennial, somewhat suffruticose at base. Stems numerous, slender, 10–40cm, simple or branched mainly from above middle, sparsely glandular-pubescent especially above, hairs in 2 rows, glands not conspicuous. Leaves dark green or purplish-green drying blackish, ovate in outline, 15–20 × 10–15mm, 2–3-pinnatisect with usually 4–5 pairs of linear primary segments, glabrous. Pedicels c 1mm. Calyx lobes linear, 4.5–6mm, subequal, ± purplish-green or brownish-purple. Corolla tube orange-red or yellowish-purple outside, yellow inside with upper lip blotched reddish or brownish at throat; tube 10–13mm, almost twice length of calyx; lower lip 3-lobed, c 3–6 × 7–10mm, upper shorter. Capsule blackish, 5–6mm, rather densely pilose.

Bhutan: C – Thimphu district (Bela La, Tsalimaphe, Taba etc.), Punakha district (Sena Thang), **N** – Upper Mo Chu district (near Tatsi Markha); **Chumbi:** Tassi-chen-doom. Shaded and open grassy banks and hillsides, 2135–3353m. June–August.

2. P. glandulosum Bentham

Similar to *P. tenuisectum* but plant probably annual or biennial; stems only 7–10cm, reddish-tinged, much more densely viscid-glandular-pilose all round, the glands black, very conspicuous; leaves less than half as large, 3-sect; calyx lobes unequal, from 2.5–3.5mm; corolla c 6mm, tube c 2.5–3mm, hardly longer than calyx; lower lip c 2 × 3.5mm; capsule c 3.5mm.

Bhutan: C – Punakha district (Sena Thang–Samtengang (71)). Dry open grassy slopes, c 1830m. October.

The record needs confirmation although the two species are very distinct from one another. Ecological notes are based on Nepalese material.

36. PEDICULARIS L.

Pedicularis or annual, hemiparasitic herbs, associated with grasses. Stems single or forming tufts, erect, decumbent or prostrate. Radical leaves usually present (often clustered), occasionally absent; cauline leaves alternate, opposite or whorled; lamina usually variously pinnatifid, pinnatipartite (divided halfway to midrib) or pinnatisect, sometimes crenate only. Inflorescence a lax or condensed sometimes spike-like raceme, or flowers axillary or arising directly from crown. Bracts present, similar or dissimilar to leaves. Flowers pedicellate or rarely sessile. Calyx cylindrical, tubular, campanulate or urceolate, tube often membranous, its apex with 2–5(–6) teeth, which are variously divided or toothed or entire. Corolla usually pink to purple, sometimes white or yellow, very variable in shape, comprising a tube equal to or ± longer than calyx; a 3-lobed lower lip; and a hooded upper lip (*galea*) which may be erect or more commonly strongly bent at middle and is divided into a lower *erect part*, a ± inflated middle part containing the anthers (*anther-bearing part*) and usually a ± well-developed *beak* (sometimes absent). Stamens 4, didynamous, on long, glabrous or hairy filaments, anthers concealed within galea which is split on its ventral margin. Style very long, equalling or just exserted from tip of galea. Fruit a capsule, often with acute or acuminate apex. Seeds numerous, small, with striate or reticulate testa.

The 76 Bhutanese species fall into two major groups, 21 sections, and 46 series. Many of the series are represented by only a single species in our area, as the centre of diversity of the genus is SW China. The classification of the genus is still very artificial. A conspectus of the taxa in our area is presented below. Brief descriptions of each section and series are provided; these apply only to those species occurring in our area.

CONSPECTUS OF SECTIONS, SERIES AND SPECIES OF *PEDICULARIS* IN THE *FLORA OF BHUTAN* AREA

GROUP I. *Cyclophyllum Li*

SECT. 1. *Orthosiphonia* Li
- Ser. 1. *Molles* Tsoong
  - 1. *P. mollis* Bentham
- Ser. 2. *Gibberae* Prain
  - 2. *P. gibbera* Prain
  - 3. *P. polygaloides* Hook.f.
- Ser. 3. *Denudatae* Prain
  - 4. *P. denudata* Hook.f.
- Ser. 4. *Porriginosae* Tsoong
  - 5. *P. porriginosa* Tsoong
- Ser. 5. *Graciles* Maximowicz
  - 6. *P. gracilis* Bentham
  - 7. *P. instar* Maximowicz
- Ser. 6. *Myriophyllae* Maximowicz
  - 8. *P. alaschanica* Maximowicz
- Ser. 7. *Semitortae* Prain
  - 9. *P. oliveriana* Prain

SECT. 2. *Sigmantha* Li
- Ser. 8. *Verticillatae* Maximowicz
  - 10. *P. roylei* Maximowicz
  - 11. *P. diffusa* Prain
  - 12. *P. nana* C.E.C. Fischer

SECT. 3. *Brevispica* R.R. Mill
- Ser. 9. *Collatae* Prain
  - 13. *P. collata* Prain

SECT. 4. *Nothosigmantha* R.R. Mill
- Ser. 10. *Cheilanthifoliae* Maximowicz
  - 14. *P. globifera* Hook.f.

SECT. 5. *Orthorrhynchus* Prain
- Ser. 11. *Pseudasplenifoliae* Prain
  - 15. *P. schizorrhyncha* Prain
  - 16. *P. gammieana* Prain
- Ser. 12. *Debiles* Prain
  - 17. *P. confertiflora* Prain
  - 18. *P. paradoxa* (Prain) Yamazaki
  - 19. *P. heydei* Prain
  - 20. *P. inconspicua* Tsoong
- Ser. 13 *Sikkimenses* Yang
  - 21. *P. sikkimensis* Bonati

SECT. 6. *Asthenocaulus* R.R. Mill
- Ser. 14. *Flexuosae* Prain
  - 22. *P. chumbica* Prain
  - 23. *P. flexuosa* Hook.f.
  - 24. *P. tenuicaulis* Prain

SECT. 7. *Elephanticeps* Yamazaki
- Ser. 15. *Integrifoliae* Prain
  - 25. *P. integrifolia* Hook.f.

SECT. 8. *Cryptorhynchus* Yamazaki
Ser. 16. *Ludlowianae* Tsoong
26. *P. ludlowiana* Tsoong

SECT. 9. *Brachyphyllum* Li
Ser. 17. *Lyratae* Maximowicz
27. *P. lyrata* Maximowicz
Ser. 18. *Urceolatae* Yang
28. *P. xylopoda* Tsoong

SECT. 10. *Dolichophyllum* Li
Ser. 19. *Tantalorhynchae* Yang
29. *P. tantalorhyncha* Bonati

GROUP II. *Allophyllum* Li

SECT. 11. *Lasioglossa* Li
Ser. 20. *Imbricatae* Yang
30. *P. clarkei* Hook.f.
31. *P. imbricata* Tsoong
Ser. 21. *Craspedotrichae* Li
32. *P. mucronulata* Tsoong
33. *P. melalimne* R.R. Mill
34. *P. sanguilimbata* R.R. Mill
Ser. 22. *Trichoglossae* Li
35. *P. trichoglossa* Hook.f.
Ser. 23. *Lachnoglossae* Prain
36. *P. lachnoglossa* Hook.f.
Ser. 24. *Rudes* Prain
37. *P. prainiana* Maximowicz

SECT. 12. *Brachychila* Li
Ser. 25. *Aloenses* Li
38. *P. dhurensis* R.R. Mill
39. *P. kingii* Maximowicz

SECT. 13. *Dolichomischus* Li
Ser. 26. *Corydaloides* Li
40. *P. cryptantha* Marquand & Airy-Shaw
Ser. 27. *Regelianae* Yamazaki
41. *P. regeliana* Prain

SECT. 14 *Rhizophyllum* Li
Ser. 28. *Flammeae* Prain
42. *P. oederi* Vahl
Ser. 29. *Pseudomacranthae* Yang
43. *P. fletcheri* Tsoong
Ser. 30. *Pumiliones* Prain
44. *P. bella* Hook.f.
45. *P. przewalskii* Maximowicz
Ser. 31. *Merrillianae* Yang
46. *P. yarilaica* R.R. Mill
Ser. 32. *Pseudo-oederianae* H. Limpricht
47. *P. pseudoversicolor* Handel-Mazzetti

SECT. 15. *Pedicularis*
Ser. 33. *Furfuraceae* Prain
48. *P. pantlingii*
49. *P. furfuracea* Bentham
50. *P. microcalyx* Hook.f.
Ser. 34. *Carnosae* Prain
Ser. 35. *Curvipes* Prain
51. *P. bifida* (D. Don) Pennell
52. *P. amplicollis* Yamazaki
53. *P. curvipes* Hook.f.

SECT. 16. *Rhizophyllastrum* R.R. Mill
Ser. 36. *Asplenifoliae* Prain
54. *P. albiflora* (Hook.f.) Prain
55. *P. cooperi* Tsoong
56. *P. microloba* R.R. Mill
57. *P. longipedicellata* Tsoong
Ser. 37. *Filiculae* Li
58. *P. trichodonta* Yamazaki
59. *P. filiculiformis* Tsoong
Ser. 38. *Perpusillae* Yamazaki
60. *P. hicksii* Tsoong
61. *P. perpusilla* Tsoong
Ser. 39. *Odontophorae* Prain
62. *P. odontophora* Prain

SECT. 17. *Leptochilus* Yamazaki
Ser. 40. *Excelsae* Maximowicz
63. *P. excelsa* Hook.f.

SECT. 18. *Phaneranta* Li
Ser. 41. *Macranthae* Prain
64. *P. scullyana* Maximowicz
Ser. 42. *Robustae* Prain
65. *P. nepalensis* Prain
66. *P. daltonii* Prain
67. *P. robusta* Hook.f.
68. *P. elwesii* Hook.f.
Ser. 43. *Garckeanae* R.R. Mill
69. *P. garckeana* Maximowicz

SECT. 19. *Schizocalyx* Li
Ser. 44. *Longiflorae* Prain
70. *P. longiflora* Rudolph
71. *P. siphonantha* D. Don

SECT. 20. *Botryantha* Li
Ser. 45. *Rhinanthoides* Prain
72. *P. rhinanthoides* Schrenk

SECT. 21. *Saccochilus* Yamazaki
Ser. 46. *Megalanthae* Prain
73. *P. megalantha* D. Don
74. *P. megalochila* Li
75. *P. pauciflora* Maximowicz
76. *P. woodii* R.R. Mill

The following key first leads to the groups, separated by Li (367, 368) on leaf and flower arrangement. Each group has its own key (Keys A and B) leading to the sections within that group. Many sections are keyed out more than once, to allow for variation and aid identification. Once a specimen has been keyed out to section, one should proceed to the relevant section key; these are arranged alphabetically by section name at the end of each group key and lead to the series and species within each section. Species of monotypic sections are keyed out in Keys A or B as appropriate.

1. Plant stemless or appearing so, or if caulescent then without even small cauline leaves 2
\+ Plant with a distinct though often short stem bearing at least 1 leaf or pair of leaves (sometimes much smaller than radical leaves) 3

2. Flowers obviously opposite or in whorls of 3–6 **Key A (Group *Cyclophyllum*)**
\+ Flowers not opposite or whorled: inflorescence a raceme of alternate flowers, or flowers axillary or solitary or arising directly from crown **Key B (Group *Allophyllum*)** (p. 1164)

3. All cauline leaves opposite or in whorls of 3–6 **Key A (Group *Cyclophyllum*)**
\+ At least some cauline leaves alternate **Key B (Group *Allophyllum*)** (p. 1164)

KEY A (GROUP *CYCLOPHYLLUM*)

1. Leaves crenate but not pinnately divided (if appearing crenate to more than halfway, treat as pinnatifid and proceed to couplet 3) 2
\+ Leaves pinnatifid, pinnatipartite or pinnatisect 3

2. Leaves linear-lanceolate with attenuate base; corolla wine-red or magenta (Sect. *Elephanticeps*) **25. P. integrifolia**
\+ Leaves ovate or oblong-ovate with cordate base; corolla yellowish or white **Sect. *Brachyphyllum*** p.p. (p. 1162)

3. At least some cauline leaves in whorls of 3–6 4
\+ All cauline leaves opposite (very rarely a few alternate) or cauline leaves absent 11

4. Galea erect or arched, not decurved at right angles to corolla tube 5
\+ Galea strongly decurved at middle, anther-bearing part and beak (when present) ± at right angles to corolla tube 6

5. Corolla tube bent, 8–13mm .... **Sect. *Sigmantha*** (p. 1163)
\+ Corolla tube straight, 4–8mm .... **Sect. *Orthosiphonia*** p.p. (p. 1163)

6. Flowers yellow .... **Sect. *Orthosiphonia*** p.p. (p. 1163)
\+ Flowers not yellow 7

7. Flowers in whorls of 3, or few and axillary; leaves in whorls of 3 8
\+ Flowers in whorls of 2 or 4; leaves usually in whorls of 4 (sometimes 3), or at least some of them opposite 9

8. Leaves with 11–14 pairs of segments (Sect. *Dolichophyllum*) **29. P. tantalorhyncha**
\+ Leaves with 3–5 pairs of segments ..... **Sect. *Orthorrhynchus*** p.p. (p. 1162)

9. Beak of galea very short (c 0.5mm); leaves usually with 9–15 pairs of segments .... (Sect. *Nothosigmantha*) **14. P. globifera**
\+ Beak of galea distinct (2.5–8mm); leaves with 4–8 pairs of segments ... 10

10. Annual, usually 10–100cm in flower; calyx sparsely hairy or glabrous **Sect. *Orthosiphonia*** p.p. (p. 1163)
\+ Perennial, usually not more than 10cm in flower; calyx hirsute or villous **Sect. *Orthorrhynchus*** p.p. (p. 1162)

11. Whole of galea erect or inclined upwards, not making a right angle with corolla tube 12
\+ Anther-bearing part of galea, or its beak, more or less at right angles to corolla tube 14

12. Plant 30–100cm, annual .... **Sect. *Orthosiphonia*** p.p. (p. 1163)
\+ Plant 1–20cm, perennial 13

13. Calyx with oblong apical lobes, the posterior one 1.5–3mm (Sect. *Brevispica*) **13. P. collata**
\+ Calyx with ovate or triangular apical lobes, the posterior one very small or absent .... **Sect. *Orthosiphonia*** p.p. (p. 1163)

14. Calyx 10–13mm 15
\+ Calyx 3–9mm 16

15. Corolla tube 8.5–11 × 2–2.5mm, hardly longer than calyx (Sect. *Cryptorhynchus*) **26. P. ludlowiana**
\+ Corolla tube (12–)17–45 × c 1.5mm, usually considerably longer than calyx .... **Sect. *Asthenocaulus*** p.p. (below)

16. Stems villous-pubescent with brownish-white hairs often arising in small groups .... **Sect. *Orthorrhynchus*** p.p. (p. 1162)
\+ Stems glabrous or, if hairy, not villous with hairs as above 17

17. Corolla tube usually 16mm or longer when full-grown **Sect. *Asthenocaulus*** p.p. (below)
\+ Corolla tube less than 15mm when full-grown 18

18. Annual, 10–90cm (usually more than 25cm) **Sect. *Orthosiphonia*** p.p. (p. 1163)
\+ Perennial, 2–30cm (usually less than 25cm) 19

19. Stems glabrous in vegetative part .... **Sect. *Asthenocaulus*** p.p. (below)
\+ Stems hairy (sometimes sparsely) in vegetative part 20

20. Four calyx teeth divided clover-leaf fashion into 3, 3-lobulate lobes; beak of galea 4–5mm .... **Sect. *Brachyphyllum*** p.p. (p. 1162)
\+ Calyx teeth not 3-lobed as described above; beak of galea c 3mm **Sect. *Orthorrhynchus*** p.p. (p. 1162)

Sect. *Asthenocaulus*

1. Flowers 2, terminal; lamina of radical leaves 6–10 × c 3mm; calyx 6–6.5mm **22. P. chumbica**
\+ Flowers more than 2, axillary or in a terminal raceme; lamina of radical leaves 20–90 × 10–25mm; calyx usually 7.5–10mm 2

2. Calyx lobes all entire; corolla tube 8–9mm, hardly longer than calyx; anterior filaments glabrous .... **24. P. tenuicaulis**
\+ Calyx lobes (except posterior) acute-serrate; corolla tube (12–)17–25(–45)mm, usually at least twice calyx; anterior filaments barbate **23. P. flexuosa**

SECT. ***Brachyphyllum***

1. Corolla deep red or crimson; calyx teeth (except posterior) divided into 3 lobes (clover-leaf-like), which are themselves 3-lobulate; leaves pinnatifid; plant apparently perennial .... (Ser. *Urceolatae*) **28. P. xylopoda**
\+ Corolla yellowish or white; calyx teeth (except posterior) elliptic and serrate, not 3-lobed; leaves undivided, crenate; plant apparently annual (Ser. *Lyratae*) **27. P. lyrata**

SECT. ***Orthorrhynchus***

1. Corolla tube 20–25mm and finally c 4 × calyx (Ser. *Sikkimenses*) **21. P. sikkimensis**
\+ Corolla tube 6–18mm and 1–2½ × calyx 2

2. Beak of galea 3–4mm, 3-toothed or emarginate at apex (Ser. *Pseudasplenifoliae*) 3
\+ Beak of galea 4–8mm, truncate and not toothed at apex ..(Ser. *Debiles*) 4

3. Corolla tube 12–14mm, not more than c 1½ × calyx; beak of galea c 3mm, 3-dentate at apex; all filaments completely glabrous... **15. P. schizorrhyncha**
\+ Corolla tube 14–18mm, more than twice calyx; beak of galea c 4mm, emarginate at apex; anterior filaments barbate above ... **16. P. gammieana**

4. Corolla tube 6–7mm 5
\+ Corolla tube 8–14mm 6

5. Lamina of radical leaves 7–16mm with 6–15 pairs of segments **19. P. heydei**
\+ Lamina of radical leaves 3–8mm with 4–7 pairs of segments **18. P. paradoxa**

6. Lamina of radical leaves 5–20 × 2–6mm; beak of galea (6.5–)7–8mm **17. P. confertiflora**
\+ Lamina of radical leaves 20–40 × 10–13mm; beak of galea 5–6mm **20. P. inconspicua**

Sect. *Orthosiphonia*

1. Corolla yellow .... (Ser. *Myriophyllae*) **8. P. alaschanica**
+ Corolla some shade of red pink or purple, or bicoloured (sometimes white, at least in part, but never yellow) 2

2. Galea beakless 3
+ Galea with at least a very short beak 4

3. Corolla tube 10–11mm; corolla with darker spots and streaks on a lavender or reddish ground; perennial 7–20cm .. (Ser. *Porriginosae*) **5. P. porriginosa**
+ Corolla tube 4–5mm; corolla dark wine-red, dark purple, or maroon, lacking spots or streaks; annual 30–100cm .... (Ser. *Molles*) **1. P. mollis**

4. Beak of galea long, slender and shallowly S-shaped (Ser. *Semitortae*) **9. P. oliveriana**
+ Beak of galea straight or falcate, never S-shaped 5

5. Annual 6
+ Perennial .... (Ser. *Gibberae*) 8

6. Lower lip of corolla shorter than galea; filaments pilose (Ser. *Denudatae*) **4. P. denudata**
+ Lower lip of corolla subequalling or longer than galea; all filaments glabrous (Ser. *Graciles*) 7

7. Beak of galea 2.5–3.5mm, very slightly declinate but up-curved at extreme tip which is truncate and acute .... **7. P. instar**
+ Beak of galea 4–7mm, porrect and not curved upwards at tip which is sometimes 2-lobed or emarginate .... **6. P. gracilis**

8. Corolla white; all pedicels 1–2.5mm, shorter than calyx; beak of galea erect with obtuse apex .... **2. P. gibbera**
+ Corolla purple; at least lower pedicels 6–10mm and longer than calyx; beak of galea horizontal with 2-lobed apex .... **3. P. polygaloides**

Sect. *Sigmantha*

1. Inflorescence a condensed head-like raceme of only 2 or 4 flowers **12. P. nana**
+ Inflorescence of 2–8 whorls each of 4 flowers 2

2. Stem pubescent all round (if rarely in 4 rows, with a few hairs between rows); lower lip of corolla 7–10 × 8–20mm; all filaments glabrous **10. P. roylei**
\+ Stem pubescent in (2 or) 4 rows, with no hairs between rows; lower lip of corolla c 4–6 × 7–10mm; 2 filaments pilose .... **11. P. diffusa**

KEY B (GROUP *ALLOPHYLLUM*)

1. Anther thecae with long subulate basal tails 0.5–1mm, which protrude from the galea .... **Sect. *Brachychila*** (p. 1165)
\+ Anther thecae obtuse to acuminate at base but lacking tails, whole anther concealed within galea 2

2. Galea of corolla sparsely to densely hairy on ventral margins and usually also on dorsal surface .... **Sect. *Lasioglossa*** (p. 1166)
\+ Galea of corolla without long hairs on ventral margins and usually also glabrous dorsally 3

3. Beak of galea absent, or straight to declinate 4
\+ Beak of galea strongly coiled or falcately curved 12

4. Stems prostrate or diffuse, sometimes almost absent; flowers arising directly from crown (if axillary go to second lead of couplet) **Sect. *Dolichomischus*** (p. 1166)
\+ Stems not prostrate, usually erect or ascending and ± stout but sometimes decumbent and very short or ± absent; flowers in spike-like racemes, or axillary, or in small groups at top of a short stem 5

5. Galea of corolla without a beak, or with a very short beak not more than 1mm (leaves always mainly radical) ..... **Sect. *Rhizophyllum*** p.p. (p. 1169)
\+ Galea of corolla with a usually distinct beak at least 2mm (if beak indistinct and apparently less than 2mm, leaves all cauline) 6

6. Leaves mainly cauline, scattered along stem; flowers axillary or in lax racemes; plants caulescent, stem always at least 6cm and usually 10cm or more in flower 7
\+ Leaves mainly radical, cauline few or several, or absent; inflorescence subcapitate, or a short, dense raceme, or flowers axillary; plants often acaulescent or with short stems rarely more than 16cm 8

7. Corolla tube 20mm or longer .... **Sect. *Rhizophyllum*** p.p. (p. 1169)
\+ Corolla tube 7–18mm .... **Sect. *Pedicularis*** (p. 1167)

8. Beak of galea 2–4mm .... **Sect. *Rhizophyllastrum*** p.p. (p. 1168)
\+ Beak of galea (4–)5–8mm 9

9. At least posterior filaments glabrous or nearly glabrous
Sect. *Rhizophyllastrum* p.p. (p. 1168)
\+ All filaments hairy (barbate, lanate, villous or pubescent) at least towards apex, though posterior sometimes less so than anterior 10

10. Calyx glabrous; corolla tube always glabrous
Sect. *Rhizophyllastrum* p.p. (p. 1168)
\+ Calyx hairy, at least on veins and/or margins; corolla tube hairy at least above, or glabrous 11

11. Calyx hoary-pubescent; corolla tube at least twice calyx
Sect. *Rhizophyllum* p.p. (p. 1169)
\+ Calyx hirsute or pubescent but not hoary; corolla tube 1–1.5(–2) × calyx
Sect. *Phanerantha* p.p. (p. 1167)

12. Middle lobe of corolla lower lip abruptly narrowed into a long, linear projection with spoon-like tip in which galea rests when young, before springing apart later; lateral lobes of corolla lower lip obsolete; plant 1m or more .... (Sect. *Leptochilus*) **63. P. excelsa**
\+ Lower lip of corolla large and broad, distinctly 3-lobed with large lateral lobes; plant less than 1m 13

13. Inflorescence opening from top downwards 14
\+ Inflorescence opening from bottom upwards or flowers opening almost simultaneously 15

14. Corolla yellow .... **Sect. *Phanerantha*** p.p. (p. 1167)
\+ Corolla crimson, red, purple, or partly (rarely entirely) white
Sect. *Saccochilus* (p. 1169)

15. Corolla tube relatively stout, 1–2 × calyx
Sect. *Phanerantha* p.p. (p. 1167)
\+ Corolla tube slender, 2–10 × calyx 16

16. Calyx with 2 apical lobes; corolla tube usually 3–10 × calyx
Sect. *Schizocalyx* (p. 1170)
\+ Calyx with 5 apical lobes or teeth; corolla tube usually 2–3 × calyx ... 17

17. Corolla tube greenish, turning yellow; beak of galea c 8mm with obtuse, entire or shallowly emarginate apex ..... **Sect. *Rhizophyllum*** p.p. (p. 1169)
\+ Corolla tube pink; beak of galea c 15mm, with acute, 2-lobed apex
(Sect. *Botryantha*) **72. P. rhinanthoides**

Sect. *Brachychila*

1. Corolla tube 2.5–3mm, equalling calyx tube; anther-thecae tails 0.7–1mm **38. P. dhurensis**
+ Corolla tube 6.5–8.5mm, c twice calyx; anther-thecae tails c 0.5mm **39. P. kingii**

Sect. *Dolichomischus*

1. Corolla sulphur yellow, 16–19mm; diffuse perennial with numerous long sterile stems and a central fertile stem (Ser. *Corydaloides*) **40. P. cryptantha**
+ Corolla bright pink, 23–36mm; low, almost acaulescent plant forming small tufts .... (Ser. *Regelianae*) **41. P. regeliana**

Sect. *Lasioglossa*

1. Corolla, including galea, yellow 2
+ At least galea of corolla some shade of red, violet, purple or lilac 5

2. Bracts linear-oblong; anther-bearing part of galea arcuate forwards (not horizontal), not ending in a beak .... (Ser. *Rudes*) **37. P. prainiana**
+ Bracts with ovate lower part which ± abruptly terminates in a ± long acuminate or caudate apex; anther-bearing part of galea horizontally decurved, ending in a short beak .... (Ser. *Craspedotrichae*) 3

3. Total length of lower bracts c 30mm, caudate apical part c 20mm **33. P. melalimne**
+ Total length of lower bracts 9–18mm, caudate apical part 2–8mm 4

4. Lobes of corolla lower lip glabrous; bracts c 9 × 2.5mm with short apical part c 2mm; leaves with 10–14 pairs of segments ..... **34. P. sanguilimbata**
+ Lobes of corolla lower lip long-ciliate; bracts c 18 × 6mm with apical part c 8mm; leaves with 15–22 pairs of segments .... **32. P. mucronulata**

5. Corolla white or pale yellow with wine-red galea .... (Ser. *Imbricatae*) 6
+ Corolla entirely lilac, violet, wine-red or purple 7

6. All bracts linear or narrowly linear, with base not or scarcely broadened; posterior calyx tooth 1.5–2mm, others c 3mm .... **30. P. clarkei**
+ Middle and upper bracts ovate with broadened lower part and caudate-acuminate upper part; calyx teeth subequal, 4–5mm ...... **31. P. imbricata**

7. Leaves mainly clustered near base; corolla lilac, violet or purple; galea, including beak, lanate with pale pinkish hairs (Ser. *Lachnoglossae*) **36. P. lachnoglossa**
+ Leaves all cauline; corolla deep wine-red; galea, except for glabrous beak, lanate with reddish-purple hairs .... (Ser. *Trichoglossae*) **35. P. trichoglossa**

Sect. *Pedicularis*

1. Calyx either not split at all on anterior side (*P. pantlingii*) or split to at least ⅔ on anterior side 2
+ Calyx split less deeply (to ⅓) on anterior side .... (Ser. *Curvipes*) 5

2. Radical leaves persistent, ± slightly larger than cauline; leaves coarsely crenate and elliptic; calyx with 2 lobes at top of tube (Ser. *Carnosae*) **51. P. bifida**
+ Radical leaves if present soon withering, smaller than cauline; leaves pinnatifid or pinnatisect, or if crenate not elliptic; calyx with 4–5 lobes at top of tube .... (Ser. *Furfuraceae*) 3

3. Leaves lanceolate or oblong, about half as wide as long; corolla bright pink, without pale whitish mark on middle part of galea ...... **50. P. microcalyx**
+ Leaves ovate, orbicular-ovate or broadly triangular, about as wide as long; corolla pale pinkish-purple, with a whitish or greenish mark on middle part of galea 4

4. Calyx split almost to base on anterior side, 4.5–5mm, with 4 small teeth at apex; pedicels with short deflexed eglandular hairs; middle part of galea greenish-white .... **49. P. furfuracea**
+ Calyx not split on anterior side, 6–8mm, with 5 ± recurved lobes at apex 1.5–2mm; pedicels densely glandular-lanate; middle part of galea pure white **48. P. pantlingii**

5. Lower lip of corolla c 7 × 12mm; stems puberulous with 2 narrow rows of hairs; fruiting pedicels soon becoming decurved .... **53. P. curvipes**
+ Lower lip of corolla c 12 × 18mm; stems lanate; fruiting pedicels remaining erecto-patent .... **52. P. amplicollis**

Sect. *Phanerantha*

1. Corolla lemon- or primrose-yellow, with galea tipped purple (Ser. *Macranthae*) **64. P. scullyana**
+ Corolla purple, pink, crimson, magenta or rarely pure white 2

2. Lower lip of corolla erecto-patent and ± enclosing galea 3
+ Lower lip of corolla spreading, not enclosing galea ... (Ser. *Robustae* p.p.) 4

3. Corolla tube 22–30mm, puberulent, c 1½ × calyx; beak of galea c 7.5mm, strongly curved to form an almost complete circle (Ser. *Garckeanae*) **69. P. garckeana**
\+ Corolla tube c 10mm, glabrous, subequalling calyx; beak of galea 4–5mm, hooked but not forming a circle .... (Ser. *Robustae* p.p.) **68. P. elwesii**

4. Lower lip of corolla 15–16mm wide; all 5 calyx teeth subequal **67. P. robusta**
\+ Lower lip of corolla 18–23mm wide; posterior calyx tooth smaller than other 4 5

5. All calyx teeth elliptic, crenate, not reflexed; beak of galea with bifid apex **66. P. daltonii**
\+ Four larger calyx teeth obovate-orbicular or obovate, incised-serrate, reflexed; beak of galea with deeply 2-lobed apex .... **65. P. nepalensis**

Sect. *Rhizophyllastrum*

1. Flowers solitary and axillary, or arising ± directly from crown on long pedicels; plant always almost acaulescent 2
\+ Flowers in small terminal heads or a short terminal raceme; plant caulescent or acaulescent 5

2. Flowers on pedicels arising ± directly from crown or at the top of very short stem rarely more than 1cm .... (Ser. *Asplenifoliae* p.p.) 3
\+ Flowers solitary and axillary .... (Ser. *Perpusillae*) 4

3. Petioles 10–30mm; lamina linear-oblanceolate, 10–30mm; pedicels 20–70mm; corolla 33–43mm; roots ± fusiform ..... **57. P. longipedicellata**
\+ Petioles 5–7mm; lamina ovate or oblong-ovate, 4–10mm; pedicels 8–14mm; corolla 25–28mm; roots fibrous .... **56. P. microloba**

4. Petioles sparsely pilose; flowers 1–3 per plant; calyx teeth 3 **61. P. perpusilla**
\+ Petioles glabrous; flowers 7–20 or more per plant; calyx teeth 5 **60. P. hicksii**

5. Calyx 10–13mm, tube split shallowly on anterior side .... (Ser. *Filiculae*) 6
\+ Calyx 5–10mm, tube not split on anterior side 7

6. Stems glabrous, 10–20cm; corolla purple, c 28mm, lobes of lower lip entire, glabrous .... **59. P. filiculiformis** var. **dolichorhyncha**
\+ Stems pubescent, (1–)3–9(–11)cm; corolla pink, mauve or wine-red, 20–25mm, lobes of lower lip denticulate, ciliate .... **58. P. trichodonta**

7. Corolla tube 10–12mm, c 1½ × calyx (Ser. *Odontophorae*) **62. P. odontophora**
\+ Corolla tube 15–25mm, at least twice calyx ..... (Ser. *Asplenifoliae* p.p.) 8

8. Leaves with 2–3 pairs of segments; erect part of galea with 2 blunt teeth on ventral margin just below base of anther-bearing part ... **55. P. cooperi**
\+ Leaves with 5–10 pairs of segments; erect part of galea without teeth on ventral margin .... **54. P. albiflora**

Sect. *Rhizophyllum*

1. Corolla tube 15–45mm; galea with distinct beak 3–8mm 2
\+ Corolla tube 6–12mm; galea without beak, or with very short beak not more than 1mm 4

2. Plant with distinct stems (4–)7–36cm and few radical leaves; corolla tube scarcely longer than calyx; leaves glabrous, except on main veins beneath (Ser. *Pseudomacranthae*) **43. P. fletcheri**
\+ Plant almost stemless (stems 0–3cm), leaves all basal or sub-basal; corolla tube 2–3 × calyx; leaves hoary-tomentose on both surfaces (Ser. *Pumiliones*) 3

3. Corolla wholly claret, wine or cyclamen-coloured, the galea darker red with a bluish bloom when dry; calyx 9–11mm .... **45. P. przewalskii**
\+ Corolla tube greenish, turning yellow, galea maroon with no bluish bloom, lower lip purple, or pink with purple margin; calyx 12–16mm **44. P. bella** subsp. **bella**

4. Corolla purple or purplish-pink; galea with 2 subulate teeth on anterior side of beak .... (Ser. *Merrillianae*) **46. P. yarilaica**
\+ Corolla lemon-yellow, the galea with crimson blotches halfway up and a purplish, red or blackish tip; galea without teeth on beak but sometimes with projections on ventral margin near base 5

5. Ventral margins of galea each with a small hump-like projection about ⅓ way up; calyx with some short hairs persistent in fruit (Ser. *Flammeae*) **42. P. oederi** subsp. **branchiophylla**
\+ Ventral margins of galea without projections; calyx glabrate in fruit (Ser. *Pseudo-oederianae*) **47. P. pseudoversicolor**

Sect. *Saccochilus*

1. Corolla 27–31mm with tube 14–20mm; beak of galea with entire, truncate apex .... **74. P. megalochila** var. **megalochila**
\+ Corolla 35–90mm with tube 25–75mm; beak of galea with emarginate apex 2

2. Galea and throat white, rest of corolla reddish-purple (rarely white: var. *alba*); leaf petioles usually completely glabrous (sometimes sparsely pilose only at base); pedicels glabrous .... **73. P. megalantha**
\+ Galea and throat purple, like rest of corolla; leaf petioles and pedicels sparsely pubescent or pilose 3

3. Corolla tube 35–75mm, 4½–6 × calyx; lobes of lower corolla lip pilose **75. P. pauciflora**
\+ Corolla tube 25–30mm, c twice calyx; lobes of lower corolla lip glabrous or at most very sparsely pilosulous .... **76. P. woodii**

SECT. ***Schizocalyx***

1. Corolla yellow, with 2 brown, chocolate or purple comma-like marks in throat; calyx lobes 3-fid .... **70. P. longiflora** subsp. **tubiformis**
\+ Corolla rose-purple or pink with white throat; calyx lobes orbicular **71. P. siphonantha**

GROUP I. **Cyclophyllum** Li

All cauline leaves, when present, in whorls of 3–6. Flowers opposite, or in whorls of 3–6.

SECT. 1. **Orthosiphonia** Li

Annual or perennial. Leaves opposite or whorled. Corolla tube straight in the calyx but sometimes bent above it. Calyx teeth 5. Galea beaked or beakless.

Ser. 1. **Molles** Tsoong

Annual. Leaves normally in whorls of 3–4, pinnatisect, cauline ones numerous. Inflorescence a long, interrupted spike-like raceme. Corolla dark reddish; tube bent above calyx; galea beakless.

1. P. mollis Bentham. Fig. 98k–l.

Tall annual, 30–100cm. Stem usually rather stout, erect, unbranched or branched from near base, with 4 rows of hairs below (sometimes nearly glabrous), rather densely greyish short-lanate above. Radical leaves withered at anthesis; cauline leaves in several whorls of 3–4 or sometimes opposite pairs; petioles of lower ones 5–20mm, opposite pairs scarcely or not connate at base, upper leaves sessile; lamina ovate-triangular or ovate-oblong, 15–70 × 8–25mm, pinnatisect with 7–12(–15) pairs of ovate-oblong or oblong, pinnatifid or pinnately lobed segments, ultimate divisions acutely serrate; upper surface sparsely pilose, lower surface reticulate-veined, nearly glabrous except on midrib and veins, sometimes with some scurfy scales. Inflorescence a terminal raceme of numerous 2–4-flowered whorls, upper ones approximate, lower ± distant. Bracts leaf-like but much smaller, sessile, lanate, deciduous in fruit. Pedicels 0.5–1mm. Calyx campanulate, 4.5–5.5mm, pubescent, deeply 5-lobed; lobes unequal, posterior linear-oblanceolate, 2 × 1mm, entire, others oblanceolate,

2.5–3 × c 1mm, crenate-dentate towards apex. Corolla dark wine-red, dark purple or maroon, 9–11mm; tube 4–5mm, slightly bent just above calyx but ± straight, glabrous; galea erect, straight, 3–4mm, obtuse at apex, ventral margin with minute tooth-like hairs; lower lip shorter than galea, 2–2.5mm, 3-lobed; lobes pilose on margins, suborbicular, middle lobe only slightly smaller than laterals. Stamens inserted opposite middle of ovary; all filaments glabrous; anthers broadly ellipsoid or almost orbicular, 0.9–1.2 × 0.7–0.9mm, thecae apiculate at base. Capsule ovoid, 6–8.5 × 3–4.5mm, obliquely acuminate. Seeds brown, ovoid, 1.4–1.6 × 0.8mm, coarsely reticulate with c 16 longitudinal ridges and c 17 lacunae per row.

Bhutan: C – Thimphu, Tongsa and Bumthang districts, **N** – Upper Mo Chu, Upper Pho Chu, Upper Bumthang Chu and Upper Kulong Chu districts; **Sikkim:** Dzongri, Lachen, Yumthang etc.; **Chumbi:** Bakcham, Champitang–Yatung, Phari etc. Meadows and steep rocky and grassy slopes, under *Abies* and *Berberis* etc., open *Rhododendron* scrub, 2750–4270m. June–August.

Ser. 2. **Gibberae** Prain

Low perennials. Basal leaves soon withering; cauline opposite (very rarely whorled in *P. polygaloides*), petiolate; lamina pinnatilobed or pinnatisect. Inflorescence congested above, flowers axillary. Calyx lobes 5, the posterior minute. Corolla white or purple; tube finally slightly longer than calyx, straight, ± glabrous; galea straight, sometimes ventrally toothed, shortly beaked.

2. P. gibbera Prain

Delicate perennial 3–13cm. Rhizome slender, with long filiform roots, crown bearing numerous small scales and buds. Stems prostrate below, ascending above, eglandular-pubescent with 2 rows of rather long hairs. Leaves basal and cauline, all similar, basal ones soon withering, cauline opposite; petioles 3–15mm, glabrous or sparsely hairy; lamina ovate, 5–13 × 3–8mm, pinnately lobed with 2–4 pairs of small, acutely incised-serrate segments; surfaces sparsely white-pilose. Flowers solitary, axillary, upper ones ± congested; pedicels 1–2.5mm, pilose. Bracts leaf-like. Lower flowers opening first. Calyx campanulate, membranous, 3–4 × 1–1.5mm, sparsely hairy, 5-lobed; posterior lobe very small (c 0.5mm), setaceous, other 4 ovate-oblong, c 1.2 × 0.5mm, acute-serrate. Corolla white, c 12mm; tube c 5 × 1mm, mainly straight but bent at junction with galea, glabrous; galea straight, c 5mm, glandular-punctate on hood, with very small tooth-like hairs on ventral (forward) side of beak; beak erect, obtuse at apex; lower lip equalling galea, c 5 × 5mm, 3-lobed, lateral lobes reniform, c 3 × 3mm, middle lobe broadly orbicular, c 2.5 × 3mm, emarginate, margins glabrous. Stamens inserted in middle of corolla tube; all filaments glabrous; anthers elliptic, 1–1.2mm, thecae subacute at base. Capsule lanceolate, 6–7 × 2mm, attenuate-acuminate. Seeds few and large, ovoid, c 3mm, testa blackish and very minutely punctulate.

Bhutan: N – Upper Kulong Chu district (Me La); **Sikkim:** Dzongri,

Esan-Nangi, Yampung. Streamsides, in *Rhododendron* forest, c 4270m. August–September.

3. P. polygaloides Hook.f.

Low diffusely branched perennial with short, slender, woody root-stock. Stems 1.5–12cm, slender, shortly pubescent (sometimes glabrate) with white eglandular hairs mainly in 2 rows. All leaves similar, very small; radical forming a tuft, soon withering; cauline opposite or very rarely in whorls of 3–4, shortly petiolate (2.5–5mm) or upper ± sessile, lamina ovate or ovate-oblong, 4–8(–10) × 2.5–3.5mm, pinnatisect with 3–5 pairs of small, oblong-ovate, acutely incised-serrate segments; surfaces pale green (tips of segments darker), sparsely white-pilose. Flowers axillary, lower ones remote, upper conferted, all pedicellate; pedicels shortly pilose, lower ones 6–8(–10)mm, longer than calyx, upper shorter. Lower flowers opening first. Calyx oblong-campanulate, 4–6mm, sparsely white-villous, 5-lobed; posterior lobe minute, entire, other 4 ovate, cristate, 2 middle slightly longer and broader than 2 anterior. Corolla grape-purple, 12–17mm; tube c 4mm and subequalling calyx at first but soon lengthening to c 6–8mm and then 1½ × calyx, straight, erect, dilated at top, almost glabrous outside; galea straight, inclined forwards and upwards, c 6mm, c 1½ × as broad as tube, ending in very short, horizontal, conical, subacute, 2-lobed beak; lower lip slightly longer than galea, c 6 × 9mm, 3-lobed, lateral lobes suborbicular, c 4.5 × 3.7–4mm, middle lobe ovate, c 2.5–3 × 2–2.5mm. Stamens inserted almost at top of corolla tube; all filaments glabrous; anthers very small (less than 0.5mm). Capsule narrowly ovate, c 8 × 3.5–4mm, obliquely acute. Seeds broadly ovoid, c 2mm, greyish, testa loose, distinctly reticulate.

Bhutan: C – Thimphu district (Pajoding), **N** – Upper Kuru Chu district (Singhi, Narim Thang); **Sikkim:** Yumthang, Kupup, Yakla etc.; **Chumbi:** Natu–Champitang, Tangkar La, Ta-chey-kung etc. *Rhododendron* scrub and exposed grassy hill slopes, 3800–4573m. July–August.

Classified in the next series by Yang *et al.* (429).

Ser. 3. **Denudatae** Prain

Annual. Leaves in whorls of 4, pinnatifid, cauline ones few. Inflorescence an interrupted spike; flowers whorled. Calyx lobes 5. Corolla red-purple; tube straight, glabrous; galea with very short beak.

4. P. denudata Hook.f.

Annual, 9–40cm. Stem slender, erect, unbranched, sparsely short-pubescent on 4 sides. Radical leaves present but soon withering; petiole c 1cm, pilose especially near base; lamina ovate in outline, c 6–8 × 4.5–5mm, deeply crenate. Cauline leaves in 1(–2) whorl(s) of 4, usually above middle of stem; petioles 2.5–3mm, opposite pairs connate at base; lamina oblong-ovate, 10–20 × 6–12mm, obtuse, pinnatifid with 3–5 pairs of obtuse lobes, sparsely pilose above with denser pubescence on midrib, sparsely short-pilose beneath. Inflorescence an interrupted spike of 2–6 4-flowered whorls. Bracts leaf-like, sessile. Flowers sessile. Calyx broadly campanulate, 4–6mm, membranous, densely pilose, 5-lobed; lobes ovate-oblong, 1–1.5mm, posterior one much smaller than other 4, narrowly triangular, entire, other 4 oblong-spathulate, 2 laterals c 1.5 × 0.8–1mm, subentire, 2 anterior c 2.5 × 1.5mm, entire, dentate, all becoming reflexed. Corolla pale purple-red; tube paler than lips (almost white), straight, 4.5–5.5 × 1.5mm, glabrous; galea c 5 × 1.5mm, abruptly deflexed at base, arching towards shortly beaked apex, glabrous; lower lip c 4 × 5–6mm, shorter than galea, shortly stipitate, deeply 3-lobed with deep sinuses between lobes, lateral lobes c 2 × 3mm, reniform, middle lobe c 1.5 × 1.8mm, suborbicular, all glabrous on margins but with subsessile glands near throat. All filaments pilose; anthers c 1 × 0.6mm, thecae acute at base. Capsule broadly ellipsoid to subglobose, 6–7 × 4.5–5mm, shortly acuminate, warm brown. Seeds ± ellipsoid, c 2mm, very irregularly and sharply angled (sometimes appearing whelk-shaped), coarsely hollow-reticulate all over with 6–7 longitudinal striae per face and c 20 lacunae per longitudinal row, the angles conspicuously thickened.

Sikkim: Patang La, Natu La, Dik Chhu, Changu, Lachen Valley. Open, often peaty turf slopes among boulders, 3660–4573m. July–August.

Ser. 4. **Porriginosae** Tsoong

Perennial. Leaves radical and cauline; cauline opposite, lamina pinnatipartite. Inflorescence ± contiguous; flowers opposite, with lowest pair remote. Calyx lobes 5, the uppermost very small and sometimes absent. Corolla lavender or pinkish; tube bent forwards, glabrous; galea suberect, beakless.

5. P. porriginosa Tsoong

Perennial. Stem 1, erect, 7–20cm, simple or rarely sparsely branched, with few linear scales at base, with 4 rows of appressed hairs. Leaves radical and cauline; radical leaves petiolate (petiole to 20mm), smaller than cauline and often withered, lamina ovate, 5–7 × 3–4mm; cauline leaves 1–3 pairs, opposite, very shortly petiolate or upper pair subsessile, lamina oblong-lanceolate or lanceolate, 12–35 × 7–12mm, pinnatipartite with 6–12 pairs of oblong, incised-crenate or -serrate segments; both surfaces sparsely covered with white scurfy scales, especially along midrib. Flowers in a terminal ± contiguous inflorescence with a remote pair of flowers in axils of upper leaf-pair. Calyx 6.5–8mm, subcylindrical; teeth 5 (upper smallest, sometimes absent), triangular-lanceolate, entire. Corolla lavender pink or red with darker throat and darker spots and striations; tube 10–12mm, bent forwards and inflated in upper ⅓, glabrous; galea almost erect, 3–4mm, without beak; lower lip shortly stipitate, c 8 × 13mm, deeply 3-lobed, lateral lobes suborbicular, c 6–6.5 × 6–6.5mm, middle lobe broadly obovate, c 6–7 × 5.5–6mm, all entire. Filaments all densely pilose. Capsule not known.

Bhutan: C – Thimphu district (Lunama Tso), Tongsa district (Rinchen Chu, Omta Tso). Grassy and open stony hill slopes and grass-covered cliff-ledges, 4300–4725m. July–August.

Endemic to Bhutan.

Ser. 5. **Graciles** Maximowicz

Annual, but stems rigid, often woody and tall. Leaves cauline, opposite or in whorls of 4; lamina pinnatifid to pinnatipartite. Inflorescence an interrupted raceme, lax. Corolla pink to purple; tube straight, glabrous; galea strongly decurved at middle, anther-bearing part porrect, ending in short, slightly falcate beak.

6. P. gracilis Bentham

Rather coarse annual herb. Stem erect, 10–90cm or more, usually ± much-branched, shortly pubescent to shortly villous with hairs in 2 or 4 rows. Leaves opposite or in whorls of 4; petioles of lower ones 5–15mm, upper sometimes sessile; lamina lanceolate-oblong, oblong, oblong or ovate, 10–40 × 5–20mm, pinnatifid to pinnatipartite with 4–8 pairs of oblong-linear, acutely incised-serrate segments, sparsely white-pilose on both surfaces or glabrous. Inflorescence a lax terminal raceme of 3–10, 2- or 4-flowered whorls. Bracts similar to leaves but shorter and broader. Pedicels 0.5–2mm. Calyx tubular, 3–6 × 2–2.5mm; tube membranous, sparsely white-lanate or glabrous, with 5 subequal lobes at apex; lobes either broadly triangular to semi-orbicular and obtuse, or oblong-ovate, reflexed and acute-dentate at least near tip, ciliate on margin and in upper part within. Corolla pink or rose, often with white throat, 10–18mm; tube 4–9mm, straight, 1.2–1.5 × calyx, glabrous; galea strongly decurved at middle, erect part 2–3mm, anther-bearing part c 2mm broad, gradually narrowed into slender porrect beak 4–7mm with obtuse, truncate, sometimes shallowly 2-lobed or emarginate apex; lower lip ± longer than galea, 7–10 × 7–10mm, 3-lobed, middle lobe orbicular, 2.5–3 × 2.5–3mm, lateral lobes reniform or semi-orbicular, 5–7mm wide. Stamens inserted near middle of tube; filaments all glabrous; anthers elliptic or oblong, 1.2–1.6 × c 0.6mm, thecae acute at base. Capsule ovate or narrowly ovate-oblong, 6–10 × 3–4mm, acuminate at apex. Seeds ellipsoid-reniform, rather short and broad, yellowish-grey, reticulate with c 10 curved rows of rather square lacunae.

1. Calyx teeth shallowly triangular or semi-orbicular, usually broader than long; beak of galea 4–5mm .... **a.** subsp. **stricta**
+ Calyx teeth oblong-ovate or ovate-triangular, usually longer than broad; beak of galea (5–)6–7mm .... **b.** subsp. **macrocarpa**

a. subsp. **stricta** (Prain) Tsoong; *P. gracilis* Bentham var. *gracilis* fm. *stricta* Prain. Fig. 98c–d.

Bhutan: C – Thimphu, Punakha, Tongsa and Bumthang districts, **N** – Upper Mo Chu district; **Darjeeling:** Char Churabhandar, Darjeeling, Jalapahar etc.; **Sikkim:** Lachung; **Chumbi:** Galing, Chumbi, Phari. Open rocky hillsides and rough grassland; 1800–4200m. June–September.

Subsp. *gracilis* appears to be confined to the W Himalayas; it has bracts longer than the flowers which they subtend but otherwise closely resembles subsp. *stricta*. Subsp. *stricta* has been subsumed within subsp. *gracilis* by some authors, however.

b. subsp. **macrocarpa** (Prain) Tsoong; *P. gracilis* var. *macrocarpa* Prain, *P. brunoniana* Pennell, *P. pennelliana* Tsoong *nom. superfl. et inval.*

Bhutan: C – Thimphu, Punakha, Tongsa, and Mongar districts, **N** – Upper Bumthang Chu district (Pangothang); **Darjeeling:** Phalut, Tonglu etc.; **Sikkim:** Changu; **Chumbi:** Chumbi, Taong-shong. Rough grassland and clearings in *Abies* and *Picea* forest; 2900–3600m. July–August.

Here treated as a subspecies of *P. gracilis*, following Yang *et al.* (429), although most recent Floras regard it as a distinct species (usually as *P. pennelliana*).

7. P. instar Maximowicz

Annual with very short straight roots. Stems erect, 20–50cm, simple, very sparsely and shortly white-crispate-pubescent, sometimes nearly glabrous. Radical leaves rapidly withering and usually absent at flowering time; petiole to 10mm, lamina ovate or ovate-lanceolate, to 20 × 16mm, obtuse, pinnate with 4–6 pairs of petiolulate segments; segments oblong-linear, to 10 × 4mm, pinnatifid and runcinately acute-dentate, white-furfuraceous beneath. Cauline leaves remote, in whorls of 4; petiole 1–4mm; lamina oblong-linear or oblong-lanceolate, obtuse, 10–20 × 3–8mm, sparsely white-pilose above, white-furfuraceous beneath. Inflorescence a terminal raceme of 3–6, 4-flowered whorls, lower whorls ± remote. Bracts leaf-like, subsessile, exceeding calyces. Calyx urceolate-tubular, 4–5.5mm, sparsely pilose or almost glabrous, 5-lobed at apex; posterior lobe setaceous, other 4 oblong, c 1–2mm, crenate-dentate towards apex, becoming reflexed. Corolla reddish-purple, 12–15mm, microscopically glandular-punctate; tube straight, 5–6mm (scarcely exceeding calyx), glabrous; galea strongly decurved at middle, erect part c 2mm, anther-bearing part porrect, c 2.5 × 1.8–2.2mm, gradually narrowing into slender beak c 2.5–3.5mm which is very slightly declinate but curved upwards at extreme tip; apex of galea truncate, acute; lower lip c 5 × 3mm, subequalling galea, deeply 3-lobed, middle lobe obovate, lateral lobes obovate-orbicular, all glabrous on margins. Stamens inserted near middle of corolla tube; all filaments glabrous; anthers ellipsoid, c 1.2 × 0.8mm, thecae acute at base. Capsule oblong-ovoid, 8–12 × 4–5m, acuminate. Seeds ovoid, yellowish-brown, reticulate.

Sikkim: Onglakthang, Peyk ong La. *Rhododendron* scrub, 3660–4200m. July–August.

Few collections known.

Ser. 6. **Myriophyllae** Maximowicz

Perennial. Leaves basal and cauline, the basal withering very early; proximal cauline leaves opposite, distal ones in whorls of 3–4; lamina pinnatisect. Inflorescence an interrupted spike; flowers whorled. Calyx lobes 5. Corolla yellow; tube erect, slightly bent near apex, glabrous; galea bent apically, beaked.

8. P. alaschanica Maximowicz

Perennial from slender woody root-stock. Stems several, branching from base, erect or outer ones widely decumbent, 5–35cm, densely and finely appressed white-hirsute. Radical leaves withering early. Cauline leaves petiolate; lamina linear-oblong in outline, 20–45 × 5–13mm, pinnatisect with 5–8(–12) pairs of very narrow serrate segments 2–10 × 0.4–0.7mm which often have ± callous, whitish margins, glabrous. Inflorescence an interrupted spike of usually 5–8 whorls, lower ones becoming distant in fruit. Lower bracts leaf-like, upper ones with membranous entire base and lanceolate, pinnatilobed and serrate upper half. Flowers subsessile. Calyx ovoid, membranous, c 11mm, 10-veined, hirsute or villous, with 5 teeth; posterior tooth deltoid and entire, others lanceolate and shallowly serrate. Corolla yellow, c 25mm; tube c 11–12mm, equalling calyx, glabrous; galea with erect part c 4mm, slightly bent backwards with respect to tube, anther-bearing part inflated and arcuate-incurved, narrowed to short, straight, porrect, narrowly conical beak 1–3mm with truncate entire apex; lower lip 9–11 × 10–12.5mm, lateral lobes c 9mm wide, bent upwards, much larger than the flat, forwards-directed, very small, suborbicular central lobe c 2–2.5 × 2.3–2.7mm. Stamens inserted opposite top of ovary; anterior filaments hirsute (rarely all glabrous); anthers c 2 × 0.6mm, thecae obtuse at base. Capsule broadly ovoid, c 11mm, acuminate. Seeds ovoid, c 3 × 1.8mm, pale, testa rugose and deeply reticulate.

1. Leaf segments 2–4mm, with conspicuous, whitish thickened margin .... **a.** subsp. **tibetica**

\+ Leaf segments (4–)5–10mm, with no or very indistinct pale, unthickened margin .... **b.** subsp. **alaschanica**

a. subsp. **tibetica** (Maximowicz) Tsoong

Plant 5–20cm, at least some stems strongly decumbent. Most leaves opposite, 20–25 × 5–7.5mm. Leaf segments 2–4mm, straight, with conspicuous whitish, thickened margins. Calyx villous. Beak of galea c 1mm.

Chumbi: Do-tho. Open hillsides and alpine plateaux, 4265–4573m. July–August.

A collection from Tibet north of Lhasa (*Ludlow & Sherriff* 11138) differs from other material of this subspecies (and subsp. *alaschanica*) in having all filaments glabrous (not the anterior pair hirsute). The name *P. alaschanica* subsp. *tibetica* was misapplied by Yamazaki (135) to *P. anas* Maximowicz subsp. *nepalensis* (Yamazaki) Yamazaki, which belongs to ser. *Cheilanthifoliae* and has pink corollas with darker beak.

b. subsp. **alaschanica**

Plant 25–35cm; stems erect or ascending. Most leaves in whorls of 3–4, 30–45 × 7–13mm. Leaf segments 4–10mm, at least the longer ones typically slightly curved near apex; margin not or hardly thickened, not whitish. Calyx hirsute. Beak of galea 1.5–3mm.

Sikkim: Naku Chhu (113), Lhonak (113). Alpine meadows and rocky slopes, 2590–3261m. June–September.

(The above ecological information refers to material from Gansu and E Tibet.)

The records of this subspecies from Sikkim require confirmation; they are more likely to belong to subsp. *tibetica*. Subsp. *alaschanica* seems to be a taller, more erect plant of considerably lower altitudes than subsp. *tibetica*. It occurs in Mongolia, on the Tibetan plateau, and in the Chinese provinces of Gansu, Nei Mongol, Qinghai and Ningxia; with such a distribution it would seem unlikely to be present in Sikkim.

Ser. 7. **Semitortae** Prain

Perennial (our sp.). Radical leaves withering early; cauline in whorls of 4–6, lamina pinnatipartite. Inflorescence a long spike; flowers whorled. Calyx lobes 5, teeth subequal (our sp.). Corolla bicoloured (white/purple); tube erect, glabrous; galea bent apically, produced into a long, slender, coiled beak (in our sp. S-shaped).

9. P. oliveriana Prain; *P. oliveriana* subsp. *lasiantha* Tsoong

Tall perennial 30–75cm. Stem rather stout, minutely pubescent on 2 sides. Radical leaves withering early. Cauline leaves in whorls of 4–6; petioles 5–15mm, erecto-patent; lamina lanceolate, oblong-lanceolate or ovate-lanceolate in outline, 25–60 × 6–20mm, pinnatipartite with winged rachis and 7–15 pairs of segments; segments oblong-lanceolate, pinnatifid, each pinnule with short fine mucro. Inflorescence opening upwards from below, of 5–7 distant 4-flowered whorls and a contiguous upper portion of several whorls. Bracts of lowermost 1–2 whorls leaf-like and pinnatifid, those of upper whorls linear, dentate. Flowers sessile. Calyx c 5mm, campanulate, 5-toothed, teeth subequal, each with 2–3 minute sharp teeth towards apex. Corolla with white tube and rich magenta-purple lobes and galea, 14–16mm; tube 5–7mm, only slightly exceeding calyx, glabrous; galea with short erect lower part, inflated at top of hood and produced into a long shallowly S-shaped beak, deflexed upwards, with acute, entire tip.

Chumbi: Natu La–Champitang, Phari. Lush grassland and scrub, 2900–3400m. June–July.

Habitat details refer to Tibetan plants, mostly previously referred to subsp. *lasiantha*; this subspecies was reduced to synonymy within *P. oliveriana* by Yang *et al.* (429).

Sect. 2. **Sigmantha** Li

Description as for the only series in our area.

Ser. 8. **Verticillatae** Maximowicz

Perennials (ours). Radical leaves clustered, sometimes withering early (*P.*

diffusa); cauline in whorls of 3–5 with the lowest leaves sometimes opposite; lamina pinnatifid or pinnatipartite. Inflorescence a raceme or spike; flowers whorled. Calyx lobes 5 (our spp.), posterior one small and often setaceous. Corolla mainly pink; tube decurved basally and often sigmoid, glabrous or glandular-punctate; galea without ventral teeth or beak.

10. P. roylei Maximowicz; *P. verticillata* sensu F.B.I. non L. Fig. 98g–h.

Perennial with tapering, fibrous roots. Stems 1–numerous from base, erect or ascending, sometimes from prostrate or decumbent base, 2–22cm, sparsely to densely pubescent with hairs present all round or, if mainly in 4 rows, with few hairs between rows. Leaves clustered and opposite near base, in few whorls of 3–4 on stem; petioles of lowest leaves 10–40mm, glabrous or white-pubescent; lamina linear-lanceolate or lanceolate-oblong, 8–30 × 2–7mm, pinnatifid or pinnatipartite with 6–10 pairs of segments; segments shortly oblong or oblong-lanceolate, acute, doubly incised-serrate with acute teeth. Inflorescence a short rather condensed terminal raceme of 2–5 4-flowered whorls. Bracts leaf-like, at least as long as calyx, with broadened, marginally white-pilose petioles. Calyx ± tubular-campanulate, c 6mm; tube membranous, prominently 10-veined, 5-lobed; posterior lobe setaceous, lateral lobes obovate-oblong, 1.5–3 × 0.7–1.5mm, acute-serrate towards apex. Corolla reddish-pink or reddish-purple, 15–25mm; tube 8–13mm, decurved at mouth of calyx, almost same diameter throughout, microscopically glandular-punctate, otherwise ± glabrous; galea subequalling or shorter than lower lip, slightly arched towards apex, 4–10 × 1.5–2.5mm, with truncate apex; lower lip 7–10 × 8–20mm, ± longer than galea, 3-lobed, lateral lobes reniform, middle lobe orbicular, all glabrous. Stamens inserted at top of ovary; filaments all glabrous; anthers ellipsoid, c 2 × 1.2mm, thecae acute at base. Capsule oblong-ellipsoid, 10–11 × 4–5mm, shortly and obliquely acuminate. Seeds ovoid, c 1.25mm, testa pale, distinctly reticulate.

1. Corolla 20–25mm, galea 8–10mm, lower lip 15–20mm wide ..... **c.** var. **speciosa**
\+ Corolla 15–20mm, galea 4–7(–9)mm, lower lip 8–12mm wide 2

2. Galea only about half as long as lower lip .... **b.** var. **brevigaleata**
\+ Galea about ¾ length of lower lip or subequal to it .... **a.** var. **roylei**

a. var. **roylei**

Bhutan: N – Upper Mo Chu, Upper Pho Chu, Upper Bumthang Chu, Upper Kuru Chu and Upper Kulong Chu districts; **Sikkim:** Bikbari–Chaunrikhiang, Goechar La, Samiti etc. **Chumbi:** unlocalised (collected for King). Alpine peat meadows and grassy mountain slopes, 3960–4880m (ascending to 5500m in Nepal). June–September.

Occasional variants occur with stem hairs mainly in 4 rows (e.g. *AGSES* 228 from Sikkim). These can be difficult to distinguish from *P. diffusa* except by the glabrous anterior filaments.

b. var. **brevigaleata** Tsoong

Bhutan: N – Upper Kulong Chu district (Me La). Alpine streamsides and *Rhododendron* moorland, 3960–4270m. July–August.

Also in Tibet.

c. var. **speciosa** (Prain) Yamazaki; *P. roylei* subvar. *speciosa* Prain, *P. roylei* subsp. *megalantha* Tsoong, *P. rupicoloides* Yamazaki, *P. saipalensis* Yamazaki

Sikkim: Peykiong La. Here reaching the limit of its range; more frequent in C and E Nepal.

11. P. diffusa Prain subsp. **diffusa**

Perennial; root-stock unbranched, vertical, with fibrous rootlets. Stems 6–45cm, erect or less commonly decumbent, shortly pubescent with 4 rows of hairs. Radical leaves soon withering, cauline in whorls of 4; petioles 5–10mm; lamina ovate or ovate-oblong in outline, 8–40 × 4–16mm, pinnatipartite with 5–8 pairs of segments; rachis narrowly winged between segments especially on largest leaves; segments oblong or ovate-oblong, acutely incised-serrate, sparsely pubescent above and beneath. Inflorescence subcapitate and condensed in flower (lowest whorl sometimes ± remote), of up to 8 whorls each of 4 flowers, axis pilose with 4 rows of rather long weak hairs; internodes considerably elongating in fruit. Bracts leaf-like, shortly petiolate, ± longer than calyx, reticulate-veined beneath. Flowers subsessile, pedicels to 2mm in fruit; lowest flowers opening first. Calyx campanulate, 5.5–6mm, membranous, inflated especially in fruit, 10-ribbed, very finely reticulate between ribs; teeth 5, posterior one smallest, triangular, entire, other 4 minutely crested. Corolla 15–20mm, pink with whitish tube, throat whitish with purple streaks; tube c 10mm, sigmoid, gradually broadening upwards from a narrow base, glabrous; galea shorter than lower lip, erect, 3.5–5mm, truncate at apex; lower lip c 4–6 × 7–10mm, 3-lobed; lobes widely spreading, laterals broadly orbicular, 3–4 × 3–4mm, middle lobe orbicular, 2–3 × 2–3mm, all entire and glabrous. Stamens inserted at top of ovary; 2 anterior filaments pilose, other 2 glabrous; anthers elliptic, c 1mm, thecae acute at base. Capsule narrowly lanceolate, 12 × 5mm, acute, reticulate.

Bhutan: C – Thimphu district (near Darkey Pang Tso), Mongar district (Thrumse La), **N** – Upper Kuru Chu district (Singhi), Upper Kulong Chu district (Shingbe); **Sikkim:** Mt Tankra, Ningbil, Tong etc. Usually by streams, in *Abies*/*Rhododendron* or *Picea* forest and on damp, sometimes boggy ground on open hillsides, 3050–4573m. May–August.

Subsp. *elatior* Tsoong, which occurs in SE Tibet (Kongbo province), is weakly delimited by its larger leaves, calyces and corollas; the specimen from Thrumse La approaches this in leaf size but the corollas are no larger than in other more typical specimens of subsp. *diffusa*.

Sometimes difficult to distinguish from *P. roylei*, especially from forms of that species with stem hairs mainly in 4 rows. The galea is usually shorter and the lower lip smaller; the most diagnostic character is the pilose anterior filaments.

12. P. nana C.E.C. Fischer

Dwarf, densely caespitose perennial. Stems 1.5–4cm, glabrous below, spreading-pubescent above. Leaves mostly towards base, each stem with a lower opposite pair and a whorl of (3–)4(–5) further up; petioles 5–15mm, white-pilose; lamina oblong or linear-oblong in outline, 4–10 × 1.5–3mm, pinnatifid with 3–6 pairs of oblong to subrotund, lobulate segments, sparsely white-pilose above and beneath. Inflorescence a terminal, subcapitate raceme of 2 or 4 flowers. Bracts leaf-like with broad membranous petioles. Pedicels 1–3mm. Calyx tubular-campanulate, 5–6mm, tube membranous with 10 dark veins; teeth 5, posterior one small and setaceous, others narrowly oblong, obtuse, entire, rather thick. Corolla crimson with whitish throat; tube c 10mm, abruptly bent and narrowed below, slightly broadened below throat, glabrous below, minutely glandular-punctate above; galea erect, c 5mm, straight, with obtuse truncate apex; lower lip c 6 × 8–9mm, lateral lobes c 3.5mm wide, middle lobe suborbicular, c 2.5–3 × 3–3.5mm, emarginate, all lobes with denticulate, glabrous margins. Stamens inserted near base of corolla; all filaments glabrous; thecae tapered to obtuse base.

Bhutan: N – Upper Bumthang Chu district (Dole La, Marlung). Grassy swards on peaty open hillsides, 3963–4573m. Late June–late August.

Surprisingly rare (or under-collected) given its wide distribution outside Bhutan (from Nepal to Myanmar (Burma)).

SECT. 3. **Brevispica** R.R. Mill (378); sect. *Brachystachys* (Tsoong) Yamazaki *nom. inval.*

Description as for only series in our area.

Ser. 9. **Collatae** Prain

Low perennials, caespitose. Leaves mostly basal; cauline 1 opposite pair; lamina pinnatisect. Flowers few. Calyx lobes 5, posterior one linear. Corolla pink; tube straight (our sp.), hairy; galea with 2 marginal teeth near apex on ventral side, beakless.

13. P.collata Prain. Fig. 98a–b.

Delicate low perennial (1–)2–12cm; roots slender, fleshy; crown covered with scales. Stems erect, very slender, 0.5–1mm diameter, broadest just below flowers,

FIG. 98. **Scrophulariaceae.** a–p, flower and inner face of dissected calyx of *Pedicularis* species: a–b, *P. collata* (a × 2, b × 4); c–d, *P. gracilis* subsp. *stricta* (c × 3½, d × 4½); e–f, *P. ludlowiana* (e × 3, f × 4); g–h, *P. roylei* (g × 3, h × 5); i–j, *P. xylopoda* (i × 3, j × 4); k–l, *P. mollis* (k × 3, l × 4); m–n, *P. integrifolia* (m × 2, n × 3); o–p, *P. schizorrhyncha* (× 3). Drawn by Glenn Rodrigues.

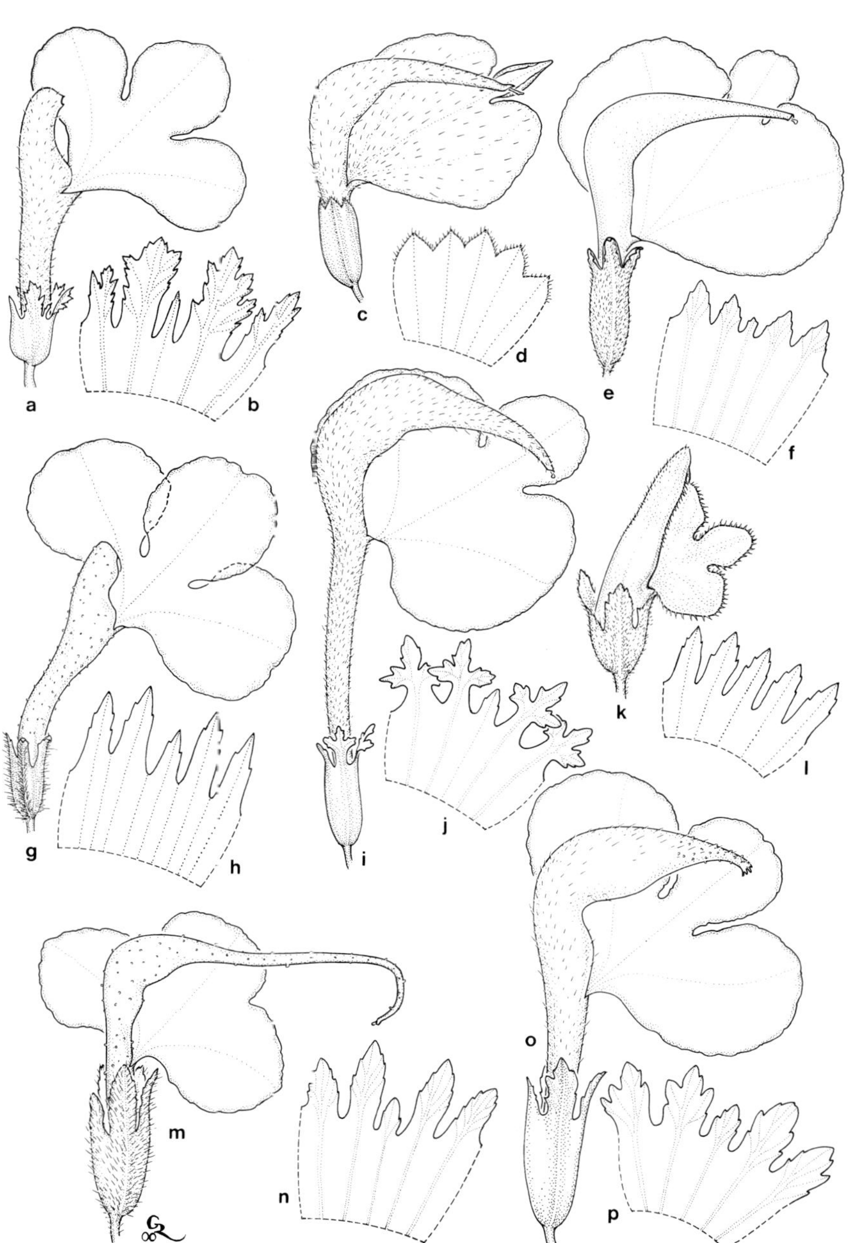
a
b
c
d
e
f
g
h
i
j
k
l
m
n
o
p

glabrous. Radical leaves 2 or 4, petioles filiform, 20–90mm, glabrous; cauline 1 pair, opposite, petioles slightly thicker, c 10mm; laminas of all leaves oblong-lanceolate or ovate-oblong, 6–30 × 4.5–16mm, pinnatisect with 4–7 pairs of ovate, pinnatifid segments, glabrous above and beneath. Flowers 1–3; pedicels 2–8mm; bracts leaf-like. Calyx rather broadly tubular-campanulate, 7–12mm, membranous, glabrous, 5-lobed; posterior lobe linear, 1.5–3mm, entire, acute; other 4 lobes oblong, entire below and dentate above, 2 laterals c 4mm, twice as long as 2 anterior. Corolla pink, c 20–25mm; tube 11–15 × 3mm, straight, sparsely pilose; galea straight with slightly arched apex, suberect (directed slightly backwards), with 2 small apical teeth on ventral side, beakless; lower lip 10–12 × 14–15mm, glabrous on margins, 3-lobed, lateral lobes reniform, 8mm, middle lobe obovate-orbicular, c 6 × 6mm, entire. Stamens inserted opposite middle of ovary; 2 anterior filaments hirsute, 2 posterior glabrous; anthers ellipsoid, c 1.5 × 0.8mm, thecae acute at base. Capsule broadly ellipsoid, c 6 × 5mm (excluding apex), with long, attenuate-acuminate, curved, very finely pointed whiplash-like apex to 5mm.

Bhutan: C – Punakha/Tongsa district (Dungshinggang), **N** – Upper Pho Chu district (Chojo Dzong), Upper Mangde Chu district (Saga La), Upper Bumthang Chu district (Marlung); **Sikkim:** Goecha La, Lam Pokhri, Meguthang; **Chumbi:** Chho La. Among moss on cliffs and ledges, 3960–4725m. June–August.

Sect. 4. **Nothosigmantha** R.R. Mill (378)

Description as for the only series in our area.

Ser. 10. **Cheilanthifoliae** Maximowicz

Perennial, caespitose. Radical leaves whorled, clustered, persistent; cauline in whorls of 4; lamina pinnatisect. Inflorescence subcapitate becoming spicate at least proximally. Calyx teeth 5. Corolla bicoloured (pink/white) or white; tube decurved in calyx, glabrous; galea falcate or suberect, toothless, shortly beaked.

14. P. globifera Hook.f.

Bushy perennial from ± stout, straight, simple or occasionally 2-branched root-stock. Stems numerous, (3–)5–40cm, erect often with markedly decumbent base, or decumbent, or sometimes ± prostrate, shortly pubescent with 4 lines of white hairs. Radical and cauline leaves in whorls of 4; petioles of lower leaves 5–30mm, those of cauline shorter, upper whorls ± sessile; lamina linear-lanceolate in outline, 8–32 × 2–11mm, pinnatisect with (7–)9–15 pairs of pinnatilobed segments, the lobes with whitish, callous margins, usually at least sparsely pilose on both surfaces; rachis not winged. Inflorescence of several 4-flowered whorls, very condensed and ± globose at first, becoming lax at least below. Lower bracts leaf-like, upper ones with very broad lower petiolar portion tapering upwards and passing into pinnatisect lamina. Pedicels 0.5–2mm, pilose. Calyx tubular-campanulate, 7–8mm, tube split to c ⅓ on anterior side, 5-veined, villous; teeth 5, oblong or ovate, with 2–4 acute teeth at apex. Corolla either pink with paler, whitish throat or entirely creamy white, 16–20mm; tube bent forwards near mouth of calyx, 7–11mm, glabrous; lower part of galea c 8mm, slightly inclined backwards then erect, anther-bearing part decurved, c 2mm, abruptly narrowed to very short beak c 0.5mm with acute apex; lower lip c 6 × 9–10mm, equalling or shorter than galea, 3-lobed, lateral lobes reniform, 4.5–6mm wide, middle lobe oblong-orbicular to orbicular, c 3 × 3mm, all glabrous on margins. Stamens inserted near base of tube; all filaments glabrous; anthers c 1.8 × 1.2mm, thecae acute at base. Capsule obliquely ovate, c 10 × 5mm, acuminate. Seeds ovoid, with basal appendage, c 2mm, testa pale and distinctly reticulate.

Sikkim: Donkung (80), Lhonak; **Chumbi:** Phari, Chelung, Tang La. Most frequently on open, usually ± dry, grassy hillsides; also on stabilised scree, and in scrub jungle, 3660–4570m (–5500m in Tibet). July–September (flowering also in May and June in Lhasa area, Tibet).

SECT. 5. **Orthorrhynchus** Prain (‘*Orthorrhynchae*’); sect. *Orthorrhynchus* (Prain) Yamazaki *comb. superfl.*

Low perennials or annuals Cauline leaves, when present, opposite or in whorld of 3. Inflorescence racemose, few-flowered. Calyx teeth 5. Corolla tube straight; galea shortly beaked.

Ser. 11. **Pseudasplenifoliae** Prain (‘*Pseudasplenifolia*’); ser. *Pseudo-asplenifoliae* in error (426), ser. *Pseudorostratae* Limpricht

Low perennials, caespitose. Leaves mainly radical, long-petiolate; cauline 0 or 1 opposite pair; lamina pinnatipartite. Inflorescence short, few-flowered, flowers in 1–2 whorls. Corolla red or red-purple; tube less than 1½ × calyx, hairy; galea strongly arcuate, with ventral marginal auricles and short, straight beak erose-denticulate at its tip.

15. P. schizorrhyncha Prain. Fig. 98o–p.

Low, loosely caespitose perennial. Stems 3–7cm, slender, suberect, unbranched, reddish, shortly glandular- and eglandular-pubescent below, pilose to villous with longer hairs below inflorescence. Leaves mainly in a basal tuft, cauline absent or 1 opposite pair. Radical leaves with long slender petioles 5–18mm; lamina linear-oblong, 4–20 × 2–5mm, pinnatipartite with 3–8 pairs of ovate or ovate-oblong, acutely incised-serrate segments and narrowly winged rachis; surfaces uniformly rather dark green, sparsely white-pilose. Cauline leaves, if present, similar but with much shorter petiole slightly broadened at base. Flowers 1–4 in 1–2 whorls at top of scape, opening almost simultaneously. Bracts leaf-like. Pedicels 0.5–2mm. Calyx tubular-campanulate, membranous, 5–7 × 2–3mm, split ventrally to about ¼, 5-lobed; posterior lobe very small, others rather broadly oblong, 1–2 × c 1mm, with dark tips, crenate-dentate. Corolla crimson, 20–25mm; tube 7–12 × c 1.5mm, straight, sparsely pilose outside; galea strongly arched, erect part c 3–4mm, arched part 8–9mm, with sparse appressed hairs on dorsal surface, directed forwards ± horizontally, gradually tapered to short, truncate beak c 3mm with 3-dentate apex, minutely glandular-punctate, ventral margins auriculate; lower lip conspicuous, longer than galea, c 8–10 × 9–13mm, 3-lobed, lateral lobes reniform, c 6 × 7mm, indistinctly and remotely dentate, middle lobe orbicular, c 3 × 4mm, all sparsely ciliate on distal but not lateral margins. Stamens inserted in middle of corolla tube; all filaments glabrous; anthers large (c 1.8–2 × 1–1.1mm), pale yellow, thecae acute at base. Capsule ovoid, c 8 × 4mm, pale whitish-brown, obliquely truncate at apex. Seeds subtriquetrous, black, c 1.25mm, very minutely reticulate.

Bhutan: S – Sankosh district (Lawgu–Paga La), **C** – Thimphu district (Tremo La–Kang La, Paro area, around Darkey Pang Tso), **N** – Upper Kulong Chu district (Me La); **Sikkim:** Natu La, Kan-ker-teng etc.; **Chumbi:** Tangkar La. Moist cliffs and gullies and in boggy moorland on open hillsides, frequently associated with mosses, 3960–4450m. June–September.

16. P. gammieana Prain; *P. flagellaris* sensu F.B.I. p.p. and sensu Prain (392) p.p. non Bentham

Perennial with slender rhizome. Stem 6–22cm, slender, arcuate or flexuous, unbranched below but with short lateral branches in inflorescence, villous-pubescent with brownish-white hairs 0.5–1.5mm tending to arise in small groups. Radical leaves with petiole to 30mm, cauline leaves opposite or in whorls of 3 with petiole 3–8mm; lamina of all leaves ovate in outline, 6–13 × 4–6mm, pinnatipartite with 3–5 pairs of segments, lower segments petiolulate; segments ovate-oblong, pinnatifid or ± deeply incised-serrate with acute teeth; upper surface glabrous, lower surface brownish-furfuraceous. Inflorescence terminal, of few axillary flowers; bracts leaf-like. Pedicels 2–3mm. Calyx campanulate, 7mm, split to c ⅓ on anterior side, hirsute, with 5 teeth; posterior tooth lanceolate, entire, others with attenuate base and expanded, elliptic, deeply incised-dentate lamina. Corolla reddish-purple with darker galea, c 25mm; tube 14–18 × 1–2mm, more than twice calyx, not dilated upwards, shortly pilose; galea inflated, arcuate-incurved, erect part 5–6mm, anther-bearing part c 6 × 3mm, attenuate into narrow, straight, porrect beak c 4mm with deeply emarginate apex. Stamens inserted in middle of tube; anterior filaments barbate above. Style exserted. Capsule unknown.

Sikkim: Lang-mang nang-zo, Lachung. 3050–3350m. May–August.

Endemic to Sikkim, whence it is still known only from the type and one other gathering. Prain's later reduction of this species to synonymy with *P. flagellaris* Bentham of Assam was incorrect; the two are distinct species, distinguished by several characters of flowers and indumentum. It is possible that *P. flagellaris* (which belongs to ser. *Microphyllae* Prain, a series that is otherwise not represented in our area) might occur in SE Bhutan but for the present it is not included in the *Flora* treatment. It can be distinguished from *P. gammieana* by the following key:

1. All leaf segments sessile; corolla tube 10–13mm, less than 1½ × calyx; beak of galea indistinct, scarcely 2mm, with bifid apex .... *P. flagellaris*
\+ Lowest leaf segments petiolulate; corolla tube 14–18mm, at least twice calyx; beak of galea c 4mm, with deeply emarginate apex .. *P. gammieana*

Ser. 12. **Debiles** Prain

Low perennials (sometimes caespitose) or annuals. Leaves basal and cauline, the latter opposite or rarely in whorls of 3; lamina pinnatifid to pinnatisect. Inflorescence subcapitate or racemose; flowers opposite. Corolla shades of red; tube less than twice length of calyx, glabrous; galea decurved at middle, margin toothless, beak slender, usually inclined downwards.

As circumscribed here, this series excludes *P. chumbica* and *P. tenuicaulis* which were included in it by Yang *et al.* (429). Both are here placed in ser. *Flexuosae*. Their taxonomic position is unclear.

17. P. confertiflora Prain subsp. **confertiflora**; *P. brevifolia* sensu F.B.I. p.p., non D. Don

Low annual with fibrous roots. Stems 1 or several, 2–20cm (rarely more than 10cm when in flower), erect or decumbent, white-villous. Radical leaves with petioles 4–15mm; lamina lanceolate or oblong-lanceolate, 5–20 × 2–6mm, deeply pinnatifid with 4–11 pairs of short, ovate to obovate, incised-serrate or -lobed segments; rachis unwinged; upper surface sparsely villous, lower surface sparsely white scurfy. Cauline leaves opposite, (1–)2–3 pairs, mainly on lower part of stem with 1 pair at middle or above. Bracts broadly obovate, membranous below, 3–7-crested above. Flowers conferted in short subcapitate terminal inflorescences, sometimes with ± remote pair or whorl below. Pedicels 2–5mm. Calyx tubular-campanulate, 6–8mm, 10-veined, hirsute or villous; teeth 5, posterior deltoid, entire, other 4 lanceolate and irregularly dentate above. Corolla rich magenta or rose, scented; tube 9–12mm, straight, glabrous; galea with ± erect part c 2mm, anther-bearing part c 3–4 × 2–2.5mm, minutely glandular, gradually tapering into slender, straight or ± curved, 6.5–8mm beak which is directed downwards and truncately acute at apex; lower lip 8–10 × 10–12mm, 3-lobed, lateral lobes c 6 × 4.5mm, larger than middle lobe, all entire, sparsely ciliate on margin. Stamens inserted near middle of tube; all filaments glabrous; anthers c 1.5mm, thecae acute at base. Capsule obliquely ovate, acuminate, c 10 × 5mm. Seeds ovoid, c 1.3 × 0.8mm, reticulate.

Bhutan: C – Thimphu district (between Pajoding and the Lakes); **Bhutan/Chumbi** Chomo Lhari; **Sikkim:** Lhonak, Kupup, Chho La, Tangkar La, N of Zelep La; **Chumbi:** Phari, Chaerlung. Open grassy slopes and bushy banks, 3400–4877m. Mid June–September.

Subsp. *confertiflora* is distributed from Nepal to China; subsp. *parvifolia* (Handel-Mazzetti) Tsoong is endemic to Yunnan.

18. P. paradoxa (Prain) Yamazaki; *P. instar* Maximowicz var. *paradoxa* Prain, ?*P. porrecta* var. *sikkimensis* Tsoong *nom. inval.*

Low perennial with several short fusiform roots. Stems erect, 5–10cm, simple, densely short white-crispate-pubescent all round. Leaves radical and cauline; radical leaves several, with petioles 5–10mm, cauline leaves few (1–3 pairs), opposite or in whorls of 3, with short petioles 0.5–2mm; lamina of all leaves linear or linear-oblong, 3–8 × 1–3mm, obtuse, white-villous on both surfaces, pinnatifid with 4–7 pairs of segments; segments short, oblong-ovate, shallowly acute-serrate, reflexed. Inflorescence a short terminal raceme of 1–3 whorls of 2 flowers. Bracts leaf-like, shorter than calyces. Pedicels 2–6mm. Calyx tubular-campanulate, 4–5mm, rather densely white-villous, 5–6-lobed at apex, posterior lobe c 0.8mm, setaceous, 2(–3) lateral lobes oblong-ovate, c 1–1.5mm, obtuse, subentire except for denticulate tip, 2 anterior lobes smaller, c 0.6–0.7mm, obtuse, subentire. Corolla reddish-purple, c 15mm, microscopically glandular; tube straight, 6–7mm, 1.2–1.5 × calyx, glabrous; galea strongly decurved at middle, erect part c 2.5mm, anther-bearing part porrect, c 2mm wide, gradually narrowing into slender straight beak 4–5mm with slightly broadened, truncate, emarginate apex; lower lip c 8 × 10mm, deeply 3-lobed, middle lobe orbicular-ovate, lateral lobes reniform, all glabrous along slightly undulate margins. Stamens inserted near middle of tube; all filaments glabrous; anthers ellipsoid, c 1.4–1.5 × 0.9mm, thecae acute at base. Capsule narrowly oblong-ovoid, c 9 × 3mm, acuminate with reflexed tip. Seeds with ivory testa and dark interior, lozenge-shaped, c 1.2 × 0.5mm, with c 6 longitudinal striae per face and c 12 rectangular lacunae per row; 2 dorsal longitudinal ridges thickened and wing-like.

Sikkim: Bijan, Meguthang, Lachung Valley, ?Dzongri. Alpine ridges etc., 3600–4400m. August.

P. porrecta var. *sikkimensis* Tsoong, a manuscript name for part of material on a very mixed sheet of specimens collected by Hooker in Sikkim (K), seems to be unpublished; the specimen bearing the determinavit slip with this name closely matches the type of *P. paradoxa*.

19. P. heydei Prain

Low perennial with several slender, fleshy, elongate-fusiform roots. Stems branched from base, 3–14cm, erect (sometimes decumbent at base), rather densely pubescent or pilose with thin greyish-white often spreading hairs. Radical leaves with slender petiole 4–30mm; lamina linear or linear-oblong, 7–16 × 2–3mm, pinnatifid or pinnatilobed with 6–15 pairs of segments; segments oblong or ovate, those in middle part of lamina ± equal, larger than basal ones and those towards apex, all shortly pubescent. Cauline leaves opposite, usually 3 remote pairs, lower shortly petiolate, upper ± sessile, all held erect and close to stem; lamina similar to radical leaves. Inflorescence a condensed terminal head of 2–3 nodes each with 2 flowers. Bracts ovate or broadly ovate, shorter than calyx, entire and membranous below, green and deeply cristate-lobed above. Calyx tubular-campanulate, 4–6mm, villous; teeth 5, posterior narrowly triangular, 1–1.5mm, subacute, others ovate-oblong, 1.5–2mm, obtuse. Corolla dark red or crimson; tube 6–7mm, straight, glabrous; galea strongly decurved at middle, erect part c 2mm, anther-bearing part c 2mm wide, gradually narrowed into slender beak 4–5mm which is directed slightly downwards and slightly broadened at truncate apex; lower lip 3–5 × 3.5–5mm, margin erose-undulate, glabrous. Stamens inserted above middle of tube; all filaments glabrous; anthers ellipsoid, c 1.2mm, thecae apiculate at base. Capsule ovoid, acuminate, 11 × 5mm.

Sikkim: Lhonak. In turf on open hillsides, 4725–5325m. July–August.

Here reaching its easternmost limit. The Sikkim specimen is slightly less hairy than plants from the western Himalaya from where the type was collected.

20. P. inconspicua Tsoong

Slender caespitose perennial with filiform roots. Stems several, 11–15cm, ascending, shortly white-pubescent with hairs in 2 rows. Radical leaves with long filiform petioles 15–55mm; lamina oblong or elliptic in outline, 20–40 × 10–13mm, pinnatisect with 5–8 pairs of oblong, pinnatifid segments; rachis unwinged; upper surface of segments blackish when dry, glabrous; lower surface mottled light and dark brown and with ± dense white granular indumentum. Cauline leaves opposite, 2–3 pairs, remote, similar to radical but lower more shortly petiolate, uppermost pair almost sessile. Flowers few, opposite, ± sessile, lowest pair remote, opening first. Bracts c 7mm, membranous below with 2 stipitate, ovate, pinnatifid lateral lobes and larger terminal lobe with 2 pairs of acute lateral teeth and larger obtuse terminal tooth. Calyx tubular-campanulate, c 7mm, 10-veined, loosely white-hirsute; teeth 5, posterior subulate, half as long as others, other 4 linear, obtuse and ± dilated at tip, c 1.5mm. Corolla deep wine-red; tube c 11mm, straight, glabrous; galea with ± erect part c 4mm, anther-bearing part curved forwards, 3–4.5mm, attenuate into gradually tapered straight beak 5–6mm which is slightly inclined downwards; lower lip large, c 12 × 13mm, 3-lobed, lateral lobes c 8 × 6mm, ± twice as large as middle lobe, all rounded and entire, ciliate on margin. Stamens inserted near middle of tube; anterior filaments thinly pilose, posterior glabrous; anthers 1.8–2mm, thecae rounded at base.

Bhutan: N – Upper Bumthang district (Kantanang). Beside stream on steep slope, 3963m. June.

Endemic to Bhutan.

Ser. 13. **Sikkimenses** Yang

Perennials. Leaves mostly radical and opposite-verticillate, cauline opposite or rarely alternate; lamina pinnatisect. Inflorescence racemose; flowers opposite. Calyx lobes 5, shorter than the hairy tube. Corolla pink to purple; tube very narrowly cylindrical, much longer than calyx; galea without marginal teeth on ventral side, with straight beak.

21. P. sikkimensis Bonati (non C.B. Clarke in sched.)

Caulescent loosely caespitose perennial with creeping rhizome and fibrous roots. Stems 4–16(–20)cm, decumbent to ascending, slender, rather flexuous, glabrous below, sparsely pilose above with brownish hairs. Radical leaves numerous, whorled but fasciculate, forming a tuft; petioles 15–90mm, slender, sparsely pilose, unwinged; lamina linear or oblong, 10–50 × 5–10mm, pinnatisect with 7–10 pairs of oblong-ovate, pinnatifid or crenate segments, lower ones shortly petiolulate; surfaces subglabrous. Cauline leaves 1–2 pairs, opposite or rarely alternate, shortly petiolate, lamina triangular, pinnatifid with sessile segments. Flowers in a short terminal inflorescence of usually 3 opposite pairs, upper 2 conferted, lowest remote. Bracts similar to upper cauline leaves, petiolate, 2–3 × calyx. Pedicels 2–6mm. Calyx cylindrical, 5–7mm; tube 3.5–4mm, sparsely pilose; lobes 5, subequal, linear-oblong, 1.5–2mm, middle lobe acute, others dilated and acutely dentate. Corolla pink or purplish-red, 26–30mm; tube very narrowly infundibular-cylindrical, considerably elongating during anthesis, finally 20–25 × 1–1.5mm, rather densely pubescent; galea with erect part 4–6mm, anther-bearing part porrect, falcate, c 6mm, gradually tapering into short slightly down-curved beak 2–3mm, beak noticeably broader at its proximal end, with non-parallel sides and blunt, fimbriate apex; lower lip 8–9 × 9–10mm, margins crenate, glabrous. Stamens inserted in upper ⅓ of tube; 2 anterior filaments sparsely villous; anthers c 1.5 × 0.8mm, thecae obtuse at base. Capsule not seen.

Sikkim: Tsomgo, Chomnagu. Moist, peaty meadows, 3960–4115m. July–early August.

Otherwise known only from SE Tibet (Deyong La, Kongbo province).

SECT. 6. **Asthenocaulus** R.R. Mill (378)

Description as for the only series in our area.

Ser. 14. **Flexuosae** Prain

Caespitose or ascending perennials. Radical leaves tufted, cauline opposite or absent; lamina pinnatifid to pinnatisect. Inflorescence a lax raceme or very reduced; flowers opposite. Calyx cleft to ⅓ anteriorly, lobes 5, posterior smallest. Corolla pink to red-purple; tube more than twice as long as calyx (except *P. tenuicaulis*), straight, usually pilose; galea bent at right angles, toothless, with slender porrect beak.

22. P. chumbica Prain

Dwarf caespitose perennial. Stems 2–6cm, slender, erect, glabrous. Radical leaves with petioles to 15mm; lamina ovate-oblong, 6–10 × c 3mm, pinnatisect with 5–10 pairs of contiguous sharply serrate segments, teeth tapering to sharp, curved point; upper surface sparsely white-pilose, lower surface white-scurfy. Cauline leaves 0 or 1 opposite pair, subsessile or with petiole to 5mm, lamina oblong, pinnatisect with 4–5 pairs of segments similar to radical. Bracts leaf-like. Flowers 2, opposite and terminal; pedicels 3–6mm. Calyx cylindrical-campanulate, cleft in upper ⅓, 6–6.5mm, 5-toothed, upper tooth linear and entire, others narrowly lanceolate, cristate at apex, ± glabrous. Corolla reddish-purple; tube narrowly cylindrical, 16–20mm × c 1.2mm, straight, sparsely pilose outside; galea with erect part 2.5–3mm, anther-bearing part minutely glandular, arched, somewhat inflated, incurved, c 4mm, tapering to a slender beak c 6.5mm which is entire and acute at apex; lower lip 3-lobed, c 7 × 12–14mm, lateral lobes at c 90° to middle lobe, all subequal with sparsely ciliate margins. Anterior stamens inserted near middle of tube, posterior near top of ovary, all filaments glabrous. Capsule lanceolate, 10.5–14 × 4.5–6mm, obliquely acute or shortly acuminate, c twice calyx. Seeds with ivory testa and blackish interior, oblong-ellipsoid, obliquely tapered to obtuse apex, with short basal peg-like appendage, testa striate and hollow-reticulate with c 9–10 conspicuous longitudinal striae per face and c 25 rather square lacunae per row.

Sikkim: Sherabthang, Tankra La; **Chumbi:** Za-ne-gang, Put-lo, Natu La–Champitang. 3353–4573m. July–August.

The description of seeds is based on material from SE Tibet. A very rarely collected species, very similar to *P. schizorrhyncha* of sect. *Orthorrhynchus* ser. *Pseudasplenifoliae* in its leaves, but our sp. can be distinguished by its glabrous stems and entire, not 3-toothed, beak of the galea. Yang *et al.* (429) place it in series *Debiles*, which belongs to sect. *Orthorrhynchus*.

23. P. flexuosa Hook.f.; *P. sikkimensis* C.B. Clarke in sched. non Bonati

Perennial. Stems numerous, 6–30(–40)cm, ascending, unbranched, slender, thinly pilose throughout. Radical leaves forming a loose tuft, with very slender, pilose petioles 30–140mm; lamina ovate-oblong or linear-oblong, 20–90 × 10–25mm, pinnatipartite with 5–10 pairs of lanceolate or oblong-linear, pinnatilobed, acutely serrate segments, sparsely pilose. Cauline leaves opposite, with shorter petioles and shorter, more ovate lamina than radical ones. Flowers in lax, terminal, leafy, spike-like racemes sometimes with few solitary axillary flowers below. Bracts petiolate, similar to cauline leaves. Pedicels 2.5–6mm. Calyx tubular, 8.5–10mm, posterior lobe c 1.5mm, setaceous, others c 3.5mm, acutely serrate with mucronate teeth. Corolla pink to reddish-purple; tube (12–)17–25(–45) × c 1.5mm, straight, shortly pilose outside; galea strongly decurved at middle, erect part 3–4mm, anther-bearing part c 2.5mm wide, minutely glandular-punctate, tapering to 4–6mm horizontal beak truncate and emarginate at apex; lower lip 3-lobed, 10–12 × 12–14mm, lateral lobes reniform, c 6–7 × 8–10mm, middle lobe orbicular, 4–5 × 4–5mm. Stamens inserted near top of tube; anterior filaments sparsely hairy, posterior glabrous; anthers oblong, c 1.8mm, thecae obtuse at base. Capsule oblong-lanceolate, 9–14 × 3–4mm, acuminate. Seeds blackish, irregularly ellipsoid, 1.2–1.5 × c 0.7mm, minutely verruculose, with very faint longitudinal ridges.

Bhutan: C – Thimphu district (Pajoding, Bela La, Motithang etc.), ?Mongar district (Sawang–Dzulu (Upper Kuru Chu district)), **N** – Upper Mo Chu district (Gasa–Pari La); **Sikkim:** Dzongri, Changu, Lhonak etc.; **Chumbi:** Yatung, unlocalised (Cooper collection). *Tsuga*, bamboo, *Rhododendron* and mossy *Abies*/*Juniperus* or mixed conifer forest, open hillsides, peaty meadows, 2135–4270m. Late June–September.

Plants from Nepal with corolla tube 40–45mm have been distinguished as var. *longituba* Tsoong, but do not appear to occur in Bhutan or Sikkim, where the corolla tube is generally less than 25mm long. Clarke applied the name *P. sikkimensis* to a specimen of *P. flexuosa* that he collected at Dzongri (*Clarke* 25850, K). This name was never validly published and the epithet was later used by Bonati for a different species (species 21 here).

24. P. tenuicaulis Prain

Perennial. Stems 5–30cm, decumbent or ascending, glabrous. Radical leaves in a tuft, long-petiolate (petiole to 8cm), lamina ovate-oblong, 20–30 × 10mm, pinnatifid with 6–8 pairs of serrate-dentate segments, glabrous. Cauline leaves opposite, petiolate (petiole 20–40mm), lamina similar to radical leaves. Bracts leaf-like but often slightly larger, petiolate, glabrous but white-furfuraceous. Flowers axillary, lower remote, upper conferted; pedicels 4–6mm. Calyx campanulate, c 7.5mm, glabrous, cleft in upper ⅓; teeth 5, posterior one subulate, others ovate-lanceolate, all entire. Corolla crimson; tube 8–9mm, hardly longer than calyx; galea with erect part c 4 × 2mm, anther-bearing part c 5 × 2.5mm, somewhat inflated at right angles to erect part, produced into deflexed-porrect beak c 5mm with entire apex; lower lip 3-lobed, c 11mm broad, lateral lobes rounded, middle lobe smaller. Stamens inserted in middle of tube, anterior filaments barbate. Capsule lanceolate, 10–13 × 4mm, straight, acuminate. Seeds ovoid, blackish, c 1.5 × 0.75mm, minutely reticulate.

Sikkim/Nepal border: Bijan; **Chumbi:** Tangkar La, Pan-ka-be-see-mo. Open hill slopes, 3353–3660m. July–August.

A Pantling collection from Peykiong La, included here by Prain (388), was excluded in his later account (389).

Sect. 7. **Elephanticeps** Yamazaki

Description as for the only series in our area.

Ser. 15. **Integrifoliae** Prain

Caespitose perennials drying black. Leaves mainly radical; cauline opposite, sessile; lamina undivided, crenate-margined. Inflorescence of 1–2 pairs of opposite flowers. Calyx lobes 5, posterior smallest. Corolla dark red; tube about 1½ × calyx, with sessile glands; galea erect then strongly decurved, toothless, suddenly narrowed into long beak which is down-curved distally and resembles an elephant's trunk.

25. P. integrifolia Hook.f. Med: *Langna*. Fig. 98m–n.

Caespitose perennial from stout fleshy root-stock. Stems several, 3–16cm, decumbent below, finally ascending or suberect, densely and shortly white-hirsute. Leaves mainly radical, cauline opposite, 2–3 pairs. Radical leaves clustered, petiole 5–18mm; lamina linear-lanceolate, 10–45 × 3–6(–8)mm, undivided, obtuse or subacute, base attenuate, margin reflexed, crenate (crenations sometimes purple-tinged); upper surface shortly hirsute, lower surface with white scurfy scales mainly on crenations. Cauline leaves narrowly oblong, sessile, 10–14 × 4–7mm, crenate-dentate. Flowers in 1–2 opposite subsessile pairs, lower pair opening first. Bracts with broad, very short petiole; lamina broadly ovate or orbicular-ovate, 9–10.5mm, crenate, convex, cradling calyx, shortly hirsute outside, glabrous inside. Calyx tubular, 12–15 × 3–4mm, densely hirsute, 5-lobed; posterior lobe ovate, c 2 × 1.5mm, other 4 oblong or oblong-lanceolate, c 4 × 1.5–2mm, obtuse, shallowly crenate to subentire, hirsute. Corolla rich wine-red or magenta, with whitish tube; tube cylindrical, 17–22 × 0.9–1.5mm, straight, sparsely sessile-glandular; galea erect for c 3mm, then strongly decurved, anther-bearing part c 6–7 × 4.5–5mm, ventrally slightly inflated, apex suddenly narrowed into long beak; beak c 16mm, directed forwards at slight upwards angle, suddenly curved downwards at c 90° just beyond middle, with truncate tip; lower lip shorter than galea (including its beak), 10–11 × 15–18mm, 3-lobed, lateral lobes rhomboid, c 10 × 7–8mm, middle lobe suborbicular, 5.5 × 6mm, all entire and glabrous. Stamens inserted at top of corolla tube; all filaments villous (mainly near middle); anthers oblong, c 3 × 1.5mm, thecae obtuse at base. Capsule ovoid-ellipsoid, c 11 × 5mm, paper-white, acute at apex, valves strongly diverging after dehiscence. Seeds not seen.

Bhutan: C – ?Thimphu district (Barshong–Lingshi), **N** – Upper Mo Chu district (Lingshi, Zambuthang); **Sikkim:** Thanggu; **Chumbi:** Phari, Chelung, Kalaeree. Alpine meadows and open grassy hillsides, 3660–4877m. July–September.

SECT. 8. **Cryptorhynchus** Yamazaki

Description as for the only series in our area.

Ser. 16. **Ludlowianae** Tsoong

Stout caespitose perennial. Leaves mainly radical; cauline opposite, short-stalked; lamina bipinnatisect or bipinnatipartite. Inflorescence a dense short spike; flowers opposite. Calyx lobes 5, posterior tiny, subulate. Corolla reddish; tube subequalling calyx, thinly pilose within near throat, otherwise glabrous; galea bent at right angles, toothless, scarcely beaked.

26. P. ludlowiana Tsoong. Fig. 98e–f.

Sturdy caespitose perennial from stout fleshy root-stock. Stems 1–4, 8–20cm, arcuate-ascending or suberect, densely pilose with long pale brown hairs. Leaves mainly radical, cauline opposite, usually 2 pairs. Radical leaves with 20–35mm petiole; lamina ovate-elliptic in outline, 35–70 × 15–25mm, bipinnatisect or bipinnatipartite with 15–20 pairs of primary segments; rachis winged, with small interlobes between segments; primary segments oblong-lanceolate in outline, each with 3–5 pairs of secondary segments; surfaces with long brownish setae and sparse, small white scales; tips of segments tinged purple when young.

Cauline leaves much smaller, shortly petiolate; petioles of upper pair more strongly dilated at base than lower pair; lamina 11–20 × 6–10mm, pinnatisect with 6–9 pairs of deeply crenate to pinnatifid segments, hairy along rachis and centres of segments. Flowers in a dense, few-flowered spike opening from below. Bracts leaf-like, dilated at base. Pedicels very short (less than 1mm). Calyx tubular-campanulate, 10–13mm; tube c 9mm, split ventrally to ¼–⅓, brownish-pilose; teeth 5, posterior very small and subulate, others oblong-elliptic, 2.5–4mm, incised-lobed, brownish-ciliate. Corolla rose- or wine-red, c 30mm; tube 8.5–11mm, scarcely exceeding calyx, abruptly narrowed at extreme base (c 0.5mm broad) but otherwise straight and 2–2.5mm broad, slightly broadened at throat, glabrous outside, thinly pilose at throat inside; galea erect for c 5mm, then bent at right angles with anther-bearing part directed horizontally forwards, 8–10mm, gradually tapered to bifid apex; lower lip very large, 12–18 × 16–20mm, 3-lobed, lateral lobes 8–11 × 9–12mm, middle lobe reniform, emarginate, 3.5–6 × 6–8mm, all lobes with repand or erose margins. Stamens inserted above middle of tube; anterior filaments pilose at base and on upper part, posterior completely glabrous; posterior anthers c 3.3mm, anterior c 2.5mm, thecae acute at base.

Bhutan: C – Punakha/Tongsa district (Dungshinggang), **N** – Upper Mo Chu district (Kangla Karchu La). Wet scree and amongst grass in dwarf *Rhododendron* scrub, c 4570m. Late June.

Endemic to Bhutan.

Known with certainty from only two gatherings. A third (Thimphu district: above Tataka, rocky grassland, 3800m, *Wood* 7306, E), flowering in August, is similar but taller, with much larger cauline leaves which are less hairy above. It may represent a later stage of development, or may be a new, closely allied species.

Sect. 9. **Brachyphyllum** Li

Annual or perennial. Leaves opposite (our spp.), sessile or shortly petiolate. Calyx campanulate, teeth 5. Corolla tube distinctly longer than calyx; galea beaked or beakless, if beakless then with small apical teeth.

Ser. 17. **Lyratae** Maximowicz

Annual (our sp.). Leaves radical and cauline, the latter opposite, short-stalked or sessile; lamina deeply crenate-dentate. Inflorescence racemose, few-flowered. Corolla yellowish or whitish (in ours), tube c 1½ × calyx, glandular inside; galea falcate distally, with 3 small subapical teeth, beakless.

Only one species occurs in our area, all others being endemic to China. This series has sometimes been placed in sect. *Hyporrhyncholophae* Hurusawa; if sect. *Brachyphyllum* Li and sect. *Hyporrhyncholophae* Hurusawa are united, the latter, being earlier, is the correct name for the combined section. However, sect. *Brachyphyllum* is here considered distinct; see Mill (378).

27. P. lyrata Maximowicz

Dwarf annual, unbranched or with several branches from crown, outermost ones often procumbent; root-stock slender, somewhat woody. Stems 3–6cm, sometimes appearing acaulescent, pubescent. Leaves radical and opposite; petioles of radical leaves 1–1.5cm, subequalling lamina, lamina ovate to oblong-ovate, 9–13(?–20) × 3–7(–10)mm, obtuse, base cordate, margins crenate almost to middle, crenations denticulate; surfaces hirsute-pubescent. Bracts leaf-like. Calyx c 8mm, tubular, 5-toothed to c ¼, upper tooth triangular-subulate and shortest, 2 adjacent longest with short entire basal portion and elliptic, serrate lamina, 2 lower teeth intermediate in length, all hirsute. Corolla yellowish or white; tube straight, c 12mm (c 1½ × calyx tube), slightly broadened upwards, glandular-ciliate inside; lower part of galea c 5mm, erect, upper part falcate, c 4mm, apex cucullate with 3 small teeth on each side just below tip; lower lip 3-lobed, lobes subequal, erose-crenulate. Filaments inserted near base of corolla tube just above ovary, all glabrous. Capsule oblong-lanceolate, almost twice calyx, acute. Seeds ovoid, c 1.3mm, reddish-brown, reticulate.

Sikkim: Samdong; **Chumbi**: Phari, Chugya, Ting. 4270–4877m. August–September.

Ser. 18. **Urceolatae** Yang

Dwarf perennials, base ± woody. Basal leaves few and withering early; cauline opposite, petiolate; lamina pinnatifid or pinnatisect. Inflorescence a short raceme of opposite flowers. Corolla deep red; tube 3–4 × calyx, hairy; galea beaked.

28. P. xylopoda Tsoong. Fig. 98i–j.

Dwarf perennial with short cylindrical root-stock. Stems 3–7cm, branched from somewhat woody base, outside branches often procumbent, pubescent with hairs in 2 rows. Cauline leaves opposite, with petioles to c 13mm; lamina to c 18 × 7mm, oblong-lanceolate in outline, pinnatifid or pinnatisect with 5–7 coarsely lobulate or incised-dentate segments usually 3-toothed at apex; both surfaces, but especially lower, covered with whitish scurfy scales. Pedicels c 5mm. Calyx tube 4.5–5.5mm; teeth 5, c 2.5mm, posterior tooth subulate, other 4 each divided in clover-leaf fashion into 3 lobes which are themselves 3-lobulate. Corolla deep velvety wine-red or crimson; tube 15–22mm, narrowly cylindrical, slightly bent forwards at top, pilose outside; vertical part of galea c 4mm, tip attenuate into 4–5mm beak; lower lip 10–11 × 12–13mm, deeply cordate at base, 3-lobed, middle lobe hooded and half size of laterals. All filaments glabrous. Capsule c 12 × 5mm, elliptic, slightly asymmetrical.

Bhutan: N – Upper Bumthang Chu district (Pangotang). Open grassy slopes, 4260–4573m. September.

Endemic to Bhutan and known only from the type gathering.

Sect. 10. **Dolichophyllum** Li

Description as for the only series in our area.

Ser. 19. **Tantalorhynchae** Yang

Caespitose perennial. Leaves mainly basal; cauline few, in whorls of 3; lamina pinnatipartite. Inflorescence spicate; flowers in whorls of 3. Calyx lobes 5, very unequal. Corolla red-purple; tube slightly longer than calyx, gently inclined forwards, glabrous; galea beaked, the beak deeply bifid or notched.

29. P. tantalorhyncha Bonati

Loosely caespitose perennial with very long tapering root to 25cm. Stems numerous, erect, suberect or ascending, 13–20cm, sulcate, pilose to lanate in the grooves with 4 rows of unequal greyish-white hairs. Leaves mainly basal, cauline in 1–2 whorls of 3. Petiole of radical leaves 20–70mm, flattened, narrowly winged, hirsute; lamina narrowly oblong or oblong-lanceolate in outline, 25–50 × 9–17mm, pinnatipartite with 11–14 pairs of segments, lowermost 1–2 pairs often very reduced, middle ones c 4–4.5 × 2–2.5mm, ovate or ovate-oblong, pinnatifid and incised-serrate, thinly pilose above, pilose and white-furfuraceous beneath at least at first. Inflorescence a terminal spike of 3–5 rather distant 3-flowered whorls; lowest whorl opening first. Lower bracts ± leaf-like, upper with membranous lower part and 3-sect apical half, with foliaceous, pinnately lobed segments. Flowers ± sessile. Calyx ovoid, c 10mm, stiffly pilose, divided to c ⅓ into 5 very unequal lobes, each lobe with narrowly linear stipe-like base and pinnately lobed foliaceous apex. Corolla rich purple at first, fading to rose, 22–26mm; tube 12–15mm, very gently inclined forwards and slightly broadening towards top, glabrous; galea darker than rest of corolla, abruptly bent at right angles, erect part 5–6mm, anther-bearing part porrect, c 6 × 2.5mm, microscopically glandular on sides and along ventral margins, gradually tapering into slender straight beak 4.5–5mm with deeply emarginate to bifid apex. Stamens inserted near top of corolla tube; 2 anterior filaments sparsely lanate towards apex, 2 posterior glabrous; anthers ellipsoid, thecae obtuse at base. Capsule narrowly ovoid, c 11 × 3mm, shortly acuminate.

Sikkim: Yume Samdong. Alpine meadows, banks and among bushes, 3050–4880m. July.

The specimen was collected in September; it bears leaves only. Though a good match, collection of flowering material to confirm that this species actually occurs in Sikkim would be very desirable. The description is based mainly on material from Yunnan and SE Tibet.

GROUP 2. **Allophyllum** Li

At least some cauline leaves alternate (sometimes appearing subopposite). Flowers neither opposite nor whorled, usually in spicate or racemose inflorescences of alternate flowers, sometimes axillary or solitary or arising directly from the crown.

Sect. 11. **Lasioglossa** Li

Perennials, usually tall and 1-stemmed. Cauline leaves sessile or subsessile. Calyx sometimes cleft anteriorly, teeth 5. Corolla with long, dense hairs on lower margin of galea and also often hairy elsewhere on galea (especially dorsally).

Ser. 20. **Imbricatae** Yang

Moderately tall perennials, usually 1-stemmed, drying black. Radical leaves absent at anthesis; cauline numerous, alternate, ± amplexicaul; lamina pinnatifid. Inflorescence racemose. Corolla white or pink with dark red galea; tube slightly shorter than calyx, straight, glabrous; galea crescent-shaped, pubescent, distinctly beaked.

30. P. clarkei Hook.f.

More or less tall perennial with straight, branched, fleshy roots. Stems 1 or few, erect, stout, 20–85cm, sparsely white-pilose with lines of hairs along angles, sometimes almost glabrous except near base. Radical leaves withering early. Cauline leaves numerous, alternate, linear or linear-lanceolate, 40–100 × 10–20mm, acute, base truncate and semi-amplexicaul, pinnatifid with 7–17 pairs of segments, veins sometimes red; segments 1–3mm apart, ovate, sharply doubly incised-serrate or sometimes merely shallowly lobed, 4–8 × 2–3.5mm, with few scattered hairs above, glabrous beneath. Inflorescence a rather densely lanate, terminal spike 8–25cm. Bracts leaf-like, all linear or narrowly linear, lower pinnatifid, upper caudate-acuminate from only slightly broadened base, c 20 × 3mm. Calyx green or reddish-green, 10–14mm in flower, with 5 strong veins running to tips of teeth, with no distinct intercalary veins running to sinuses, pilose or lanate; teeth 5, triangular or lanceolate, posterior c 1.5–2mm, others c 3 × 1.5mm, minutely denticulate. Corolla white or very pale pink with wine-red galea, c 27mm; tube 13–15 × c 2.5mm, bent backwards near middle, glabrous; galea with erect part c 5mm, anther-bearing part 4–6mm, creamy-pilose or -lanate dorsally and with shorter silvery-white hairs ventrally, gradually narrowed to short, straight, subacutely or obtusely truncate beak; lower lip stipitate, 10–11 × 8–10mm, stipe c 2mm, lip 3-lobed, the lobes all ± parallel and directed forwards, c 5 × 2.5mm, with subentire margins. Stamens inserted near top of tube; all filaments glabrous; anthers c 2 × 1.2mm. Capsule obliquely ovoid, 13–15 × 7–8mm, acuminate. Seeds yellowish-white, irregularly ellipsoid-angled, c 2.5 × 1.2mm, coarsely hollow-reticulate all over with no longitudinal striae.

Bhutan: C – Bumthang district (Kitiphu); Bumthang/Mongar district (Rudo La), **N** – Upper Pho Chu district (Wachey, Lhedi), Upper Bumthang Chu district (Pangotang, Kantanang); **Sikkim:** Changu, Onglakthang, Ningbil, Dzongri, Yumthang etc.; **Chumbi:** Pun-ka-bee-see-mo, Sham-dhu, Yatung. Moss-covered boulders, steep banks and cliffs, less commonly in marshes, 3660–4300m. May–August; fruiting September–October.

The degree of development of the leaf pinnae and their toothing is very variable.

31. P. imbricata Tsoong

Tall perennial with thickened root-stock and ± fusiform roots. Stems 1–2, erect, 65–70cm, moderately pilose with silvery-white hairs. Radical leaves withering early. Cauline leaves numerous, alternate, oblong-lanceolate in outline, 65–90 × 11–14(–22)mm, acute, base subamplexicaul and truncate-cordate, pinnatifid with 13–19 pairs of ovate, denticulate segments 3.5–4mm and 3–4.5mm apart, with a few scattered hairs above, glabrous beneath. Lower bracts leaf-like, middle and upper ovate with broadened entire lower part and caudate-acuminate denticulate upper part. Calyx wine-red (drying greenish), 15–16mm, membranous-reticulate with 5 strong veins running to tips of teeth and 5 weaker ones running to sinuses, pilose; teeth 5, ovate-deltoid, subequal, 4–5mm, acute, minutely denticulate. Corolla white with wine-red galea, c 35mm; tube 13–14 × c 3mm, straight, glabrous; galea with erect part c 5mm, anther-bearing part 6–8mm, loosely pilose dorsally and densely so along involute lower margin, gradually narrowed to short beak c 4mm with obliquely truncate apex; lower lip stipitate, c 14 × 10mm, stipe c 4mm, lip 3-lobed, the lobes all directed forwards, ovate-elliptic, c 6 × 3mm, with undulate margins. Stamens inserted at 2 levels near top of tube; all filaments glabrous; anthers ovoid, c 2 × 1.2mm. Capsule obovate, 16 × 9mm, scarcely apiculate; seeds ovate-elliptic, compressed, 3–3.5 × 2.5mm, reticulate.

Bhutan: N – Upper Bumthang Chu district (Ju La). Amidst grass in shrubs, 4270m. July.

Endemic; known only from the type gathering.

Very close to *P. clarkei*, from which it differs principally by the shape of the middle and upper bracts and larger floral parts. It may only be worthy of varietal rank. It is, however, to be regarded as the type species of ser. *Imbricatae*.

Ser. 21. **Craspedotrichae** Li

Tall perennials, 1-stemmed, drying black. Leaves apparently all cauline, alternate, ± amplexicaul; lamina pinnatisect or pinnatifid. Inflorescence a long raceme. Corolla yellow; tube subequalling calyx, ± straight; galea boat-shaped, shortly beaked.

32. P. mucronulata Tsoong

Perennial with group of short slightly fleshy fibrous roots at top of slender vertical root-stock. Stem 1, 40–60cm, unbranched, sparsely white-pilose with lines of white hairs along angles. Radical leaves absent. Cauline leaves numerous, alternate, linear-lanceolate in outline, 30–55 × 2–9(–11)mm, obtuse, base cordate-amplexicaul, deeply pinnatisect with 15–22 pairs of segments, midrib prominently raised beneath; segments 0.5–1(–1.5)mm apart, oblong or ovate-triangular, 1–4 × 1–2mm, obtuse or rounded with tip sometimes revolute, shallowly crenate with 1–2 pairs of crenations, with scattered short white hairs on lamina and denser, shorter buff hairs on midrib above, glabrous beneath. Inflorescence a fairly lax spike up to 13cm. Bracts c 18 × 6mm, with broad, ovate base abruptly narrowed into long, acuminate apex c 8mm with crenate margin. Calyx membranous, 12–14mm, with 5 strong veins running to apices of teeth and no intercalary veins, long-pilose; teeth 5, subequal, broadly triangular, acute, remotely and shallowly denticulate. Corolla pale yellow, 30–32mm; tube 12–15mm, slightly longer than calyx, almost straight; galea with erect part c 6mm, anther-bearing part horizontally decurved, densely cream-villous dorsally, gradually narrowed to short, broadly conical beak with truncate apex; lower lip stipitate-cuneate, obovate in outline, 13–14 × 10–11mm, 3-lobed, lobes all directed forwards, with long-ciliate, undulate margins, lateral lobes c 3 × 3mm, smaller than broadly ovate, shortly cuspidate middle lobe c 6 × 5mm. Stamens inserted near middle of tube; all filaments glabrous; anthers c 2 × 1.5mm, thecae obtuse at base. Capsule unknown.

Bhutan: N – Upper Kuru Chu district (Singhi Dzong). Open hill slopes among bushes, 2440m. July.

Endemic to Bhutan. Known only from the type gathering. Other specimens tentatively referred to this species do not belong to it.

33. P. melalimne R.R. Mill (378)

Stout perennial with thick, branched root-stock. Stem 1, 60–80cm, unbranched, very shortly pubescent below, pilose with c 5 lines of longer hairs above. Radical leaves absent. Cauline leaves alternate, linear-oblong in outline, 70–120 × 15–22mm, subacute-mucronate, base amplexicaul, coarsely pinnatifid with 8–14 pairs of segments; segments 2–5mm apart, ovate-triangular, 2.5–8 × 3–5mm, subacute, doubly acute-serrate with 3–4 pairs of mucronate teeth, midrib densely and minutely pubescent, with sparse punctate hairs on either side above (confined to central area, not on pinnae), glabrous beneath. Inflorescence a long, dense spike c 16cm. Lower bracts c 30mm, with broad asymmetrical base c 10mm and long caudate apex c 20mm, pinnatifid to c ¼ and with teeth along apical edge of basal portion; upper bracts shorter, ovate without long-caudate apex, middle and upper bracts strongly tinged red. Calyx membranous, c 16mm, with 5 strong veins running to apices of teeth and very weak intercalary veins to sinuses, reticulate, villous with greyish hairs; teeth 5, posterior tooth lanceolate, c 2 × 1.5mm, others deltoid, c 3 × 3mm, acute, remotely and very shallowly denticulate. Corolla yellow, c 27mm; tube c 14 × 2mm, slightly shorter than calyx, straight, abruptly becoming slightly broader (c 3mm) above middle, sparsely pilose; galea with erect part c 6mm, gently curved, anther-bearing part smoothly curved horizontally forwards, c 7 × 2.5mm, densely appressed cream-villous dorsally, gradually tapered to a short pale beak c 3mm with obliquely truncate apex; lower lip stipitate, c 12 × 8mm, lateral lobes elliptic, c 4 × 1.5mm, middle lobe ovate-elliptic, c 6 × 3.5mm, margins undulate, densely ciliate. Anterior filaments hairy. Capsule unknown.

Bhutan: N – Upper Kulong Chu district (Shingbe). Marsh, c 3960m. August.

Endemic to Bhutan; known only from the type gathering.

34. P. sanguilimbata R.R. Mill (378)

Robust but rather slender perennial. Stem 1, c 60cm, unbranched, subglabrous below, with 4–5 lines of short deflexed hairs above. Radical leaves absent. Cauline leaves alternate, narrowly oblong in outline, 30–35 × 6–8mm, acute, base cordate-amplexicaul, deeply pinnatisect with 10–14 pairs of segments, midrib not very prominent beneath; segments of middle leaves 1–2.5mm apart, oblong-ovate in outline, 2.5–4.5 × 1.5–2.5mm, acute, doubly acute-serrate with 2–3 pairs of teeth, with scattered short white hairs on lamina (including pinnae) above, glabrous beneath. Inflorescence a lax spike, rather short and few-flowered. Bracts c 9 × 2.5mm, rather narrowly ovate, villous, blood-red on margins, abruptly narrowed to short, caudate glabrous apex c 2mm. Calyx membranous, barrel-shaped, 11–15mm, with 5 very strong ribs running to tips of teeth and no intercalary veins, long-pilose; teeth 5, triangular, subequal in length but 2 posterior broader than 3 lower, 2.5–3 × 1–2mm, finely denticulate. Corolla yellow, 22–32mm; tube 12–15mm, subequalling calyx, almost straight, slightly constricted at top of ovary, minutely pilose above constriction but glabrous below; galea with erect part 4.5–6mm, anther-bearing part horizontal and directed forwards, buff-villous dorsally, gradually narrowed to indistinct, short, broadly conical beak c 2mm with broad, obtuse apex; lower lip stipitate-cuneate, obovate in outline, 10–11 × c 6.5mm, 3-lobed, the lobes all directed forwards, lateral lobes c 6 × 2mm, larger than narrowly elliptic, acute middle lobe c 4.5 × 1.5mm, margins glabrous, those of laterals undulate or almost denticulate. Stamens inserted near middle of tube; all filaments glabrous; anthers c 2 × 1.6mm, thecae obtuse at base. Capsule unknown.

Bhutan: N – Upper Kuru Chu district (Singhi). 3800m. July–August.

Endemic to Bhutan; known only from the type gathering.

Ser. 22. **Trichoglossae** Li

Erect usually 1-stemmed perennial. Stem lanate. Leaves all cauline, alternate, amplexicaul; lamina pinnatifid. Inflorescence racemose, condensed at first. Corolla dark red, laterally twisted; tube subequalling calyx, glabrous; galea densely lanate, with long pendent beak.

35. P. trichoglossa Hook.f.

Perennial with ball of short fibrous roots near soil surface and longer tap-root. Stem usually solitary, erect, (10–)20–80cm, unbranched, reddish or purplish especially above, ± densely lanate with spreading, crispate, usually fuscous (sometimes whitish) hairs. Leaves numerous, all cauline, alternate, sessile or subsessile; lamina linear-oblong to linear, 20–85 × 3–20mm, acute, pinnatifid with 12–18 pairs of ovate, obtuse, irregularly incised-serrate segments, sparsely white pilose above and beneath; midrib and/or margins often tinged red or purple. Inflorescence a terminal spicate raceme to 25cm, condensed at first but flowers soon becoming remote; axis lanate like stem. Bracts leaf-like, linear, crenate-serrate. Calyx campanulate, 10–12 × 6–8mm, lanate, 5-lobed to almost halfway, often darker purple than corolla; lobes subequal, ovate-oblong, obtuse, crenate-serrate. Corolla deep wine-red, c 25mm, twisted 90° laterally; tube c 10 × 3.5mm, glabrous, slightly arcuate near middle; galea with erect part c 4mm, glabrous, anther-bearing part arched and decurved horizontally to close throat, turgid, c 6 × 4mm, very densely lanate with long reddish-purple jointed hairs 2.5–3.5mm, attenuate into slender, glabrous beak c 5mm with apex hanging vertically downwards within one side of lower lip; lower lip cowl-shaped with throat closed by arch of galea, c 7 × 15mm, lateral (upper and lower) lobes reniform, 8–9mm wide, twice as large as broadly orbicular middle (side) lobe c 6 × 7mm, glabrous on margins. Stamens inserted near top of tube; all filaments glabrous; anthers oblong-lanceolate, c 3 × 1.2mm, thecae acute at base. Capsule obliquely ovoid, 10–15 × 6–7mm, attenuate into acute apex. Seeds ivory-white, irregularly ovoid-tetrahedral, 2–2.2(–2.5) × 1.5–1.6mm, with no longitudinal ridges but very coarsely reticulate all over, lacunae ± hexagonal.

Bhutan: C – Ha, Thimphu and Bumthang districts, **N** – Upper Mangde Chu, Upper Bumthang Chu and Upper Kulong Chu districts; **Sikkim:** Lachen, Yak La, Chulong, Thanggu etc.; **Chumbi:** Chumbi, Chumegati, Dotag etc. Steep, grassy rocky and stony slopes and scree, in and above dwarf *Rhododendron* zone, 3660–4970m. July–October.

Ser. 23. **Lachnoglossae** Prain

Erect perennial. Stem pilose. Leaves mainly basal, long-petiolate; cauline few and small; lamina pinnatisect, feather-like. Inflorescence racemose. Corolla lilac to violet, tube slightly longer than calyx, straight, glabrous; galea densely lanate, shortly beaked.

36. P. lachnoglossa Hook.f. Fig. 99e–f.

Perennial with an often long, stout or slender, vertical root-stock clothed with scales near crown. Stems 1 or few, 11–35cm, unbranched, sparsely white-pilose and -puberulent (hairs mainly in 2 rows) in flower, glabrescent in fruit. Leaves several to numerous, mainly clustered at base; petioles of radical leaves 12–40mm, channelled, pubescent on the concave upper surface; lamina linear-lanceolate, 25–110 × 4–13mm, acute, pinnatisect with 30–50 pairs of narrowly lanceolate, obtuse, serrate segments, callose on margins beneath, larger ones with mucronate tip, sparsely pubescent or subglabrous above and beneath; cauline leaves very few and small. Inflorescence a terminal raceme 3–12cm, pyramidal at first then oblong with flowers remote ± from first; axis sparsely pilose to glabrous. Bracts dissimilar to leaves, linear-lanceolate, ± longer than calyx, attenuate-acuminate, with recurved teeth, lanate on margin with cream-coloured hairs. Calyx tubular, 8.5–10 × 4mm, with 5 prominent ridges, sparsely pilose on tube; lobes 5, linear-lanceolate, unequal, posterior c 2mm, acuminate, laterals c 2.5mm, obtuse, anterior ones c 3mm, acuminate, all lanate on margins. Corolla lilac, violet or purple with paler tube, c 23mm; tube 12–15 × c 2mm, glabrous, straight below but slightly deflexed above; galea abruptly curved near middle, with erect part c 2.5mm, anther-bearing part arched, c 4 × 2.5mm, turgid, densely lanate with pale pinkish hairs c 1mm, ending in beak c 3mm lanate on dorsal side and at apex; lower lip 5–6 × 4.5–5mm, deeply 3-lobed, lobes subequal, ± parallel, elliptic, lanate on margins, c 4 × 2mm. Stamens inserted near middle of tube; filaments glabrous or sparsely hairy; anthers oblong, c 2 × 0.8mm, thecae acute at base. Capsule oblong-lanceolate, 8–16 × 3–6mm, with recurved, sharply acuminate apex; surface reticulate-rugose. Seeds narrowly ovoid, c 2.5 × 1mm, greyish-ochre, testa loosely perforate-reticulate.

Bhutan: C – ?Ha district (Ha–Tremo La), Thimphu district (Tremo La, Thimphu, Chekha, Kang La, Pumo La, Shodug), **N** – Upper Kulong Chu district (Me La); **Sikkim:** Lachen Valley, Lonakh etc.; **Chumbi:** Phari, Lingmatang, Cho-le-la etc. Open, dry grassy meadows and hillsides, often on well-drained, gravelly soil, 3195–4270m. June–August.

Ser. 24. **Rudes** Prain

Tall perennial, drying black. Stem 1, sparsely hairy. Leaves cauline, auriculate; lamina pinnatisect. Inflorescence a dense raceme, lanate. Calyx lobes 5, subequal. Corolla yellow; tube slightly longer than calyx, arcuate, glabrous; galea boat-shaped, pilose dorsally, beakless.

37. P. prainiana Maximowicz; *P. rudis* Prain in sched. non Maximowicz

Tall perennial, with dense tuft of short fibrous roots near crown. Stem 1, erect, rather stout, 40–60cm, striate, sparsely white-lanate in the shallow channels between the ridges, densely leafy. Radical leaves absent or soon withering; lowest cauline leaves also soon withering. Cauline leaves alternate, linear-oblong, 70–100 × 10–20mm, ± obtuse at apex, pinnatisect with 10–12(–15) pairs of segments and winged rachis; segments oblong, 4.5–6 × 3–4mm, acute-dentate with mucronate tips, sparsely pilose above in flower, glabrous beneath except for puberulent midrib and main veins of segments, ± glabrous on both surfaces in fruit. Inflorescence a creamy-lanate terminal spike-like raceme c 10cm in flower, c 30cm in fruit. Bracts linear-oblong, acute, serrate, lower ones longer than flowers, uppermost subequalling calyx. Pedicels 2–3mm in flower, 6–11mm (uppermost shorter) in fruit, lanate. Calyx membranous, campanulate, 10–14(–16)mm, 5-veined, lanate; teeth 5, subequal, posterior deltoid and entire, others lanceolate and doubly serrate. Corolla yellow, c 28mm; tube 17–18 × c 2.5mm, shallowly arcuate-incurved, glabrous; galea with erect part c 5mm, anther-bearing part arcuate forwards, boat-shaped, c 8 × 5mm, sparsely creamy-pilose dorsally, glabrous ventrally, without beak; lower lip cuneate-obovate, c 10 × 6mm, with 3 obovate-elliptic ciliate lobes and 2 intercalary umbonate teeth. Stamens inserted in middle of corolla tube; all filaments glabrous; anthers c 4 × 1mm. Capsule ovoid, 12–14 × 6–7mm, with recurved acuminate tips. Seeds pale buff, irregularly ellipsoid-tetrahedral, 2–2.5 × 1.2–1.5mm, coarsely hollow-reticulate all over with no longitudinal striae.

Bhutan: N – Upper Mo Chu district (Chhew La); **Chumbi:** Lu-ma-poo. Among shrubs by streams in alpine zone, c 4100m. July; fruiting September–October.

P. rudis Maximowicz is endemic to northern China and the Tibetan plateau.

SECT. 12. **Brachychila** Li

Description as for the only series in our area.

Ser. 25. **Aloenses** Li

Perennials, not drying black. Stems numerous. Branches and leaves opposite or sometimes alternate. Leaves petiolate; lamina pinnatisect. Flowers paired in all leaf axils. Calyx lobes 5, entire. Corolla whitish- or pale yellow; tube subequalling calyx; galea straight or ± falcate, toothless, beakless.

38. P. dhurensis R.R. Mill (378). Fig. 99i–j.

Perennial. Stems 14–30cm, numerous, branched from base, decumbent to suberect, shortly pilose in 2 rows, hairs c 0.3mm. Leaves opposite; petioles of lower ones 30–60mm, of upper ones 8–16mm, narrowly green-winged; lamina of lower leaves ovate-triangular, 35–50 × 20–25mm, pinnatisect with 5–6 pairs of opposite or slightly alternate segments, the lowest segments subequal to or not more than 1½ × the length of the next highest pair; lower 2–3 pairs of segments shortly petiolulate and ± remote, others sessile, contiguous and decurrent, laminas elliptic or elliptic-ovate, deeply pinnatifid with 4–5 pairs of acute-dentate lobules and 3-dentate terminal lobule; upper surface subglabrous except for puberulent veins and midrib, lower surface glabrous, reticulate-veined. Inflorescence of pairs of flowers in axils of each leaf (even the lowest ones); pedicels 1.5–2mm in flower, c 2.5mm in fruit, suberect. All bracts leaf-like but uppermost reduced to 3 lobes or to 1–2 pairs of leaflets. Calyx campanulate, 3.3–3.7mm, pale green with dark longitudinal streaks extending ½–¾ down tube; teeth narrowly subulate-triangular, glabrous, separated by broad, very shallow sinuses with ciliate margins; tube glabrous. Corolla white or very pale yellow, 14–17.5mm, erecto-patent with respect to axis (mostly 20–45°); tube 2.5–3mm, equalling calyx tube; galea 10–13mm, erect with shallowly incurved, beakless upper part which is sparsely short-pubescent; lower lip 8–10mm, lower ⅔ adnate to galea, upper ⅓ free, slightly out-curved, shallowly 3-lobed; middle lobe oblong-triangular, subacute, lateral lobes broadly deltoid, acute, all sparsely hairy. Stamens inserted at base of tube; all filaments sparsely hairy especially near top; anthers c 2mm, thecae with long subulate tails 0.7–1mm. Capsule (immature) c 8 × 3.5mm, obliquely lanceolate, gradually narrowed to long acuminate tip.

Bhutan: C – Bumthang Chu district (Dhur Chu Valley near Bumthang). On wet banks and track-sides in *Abies* forest, 3660–3810m. July.

Endemic to Bhutan. The gathering was originally named as *P. aloensis* Handel-Mazzetti, a Chinese endemic.

39. P. kingii Maximowicz

Perennial. Stems 15–30cm, slender and weak, flexuous, pilose with 2 rows of hairs 0.3–0.5mm. Leaves opposite or subopposite; petioles 10–60mm; lamina ovate-triangular in outline, 30–40 × 20–30mm, pinnatisect with 5–6 pairs of subalternate segments, the lowest ones subequal to the next highest pair; lower 2–3 pairs of segments shortly petiolulate and ± remote, upper ones contiguous and slightly decurrent on to narrowly winged rachis; laminas of segments ovate to elliptic, 8–12 × 3.5–6.5mm, lower ones deeply pinnatifid to pinnatisect, upper ones pinnatifid or coarsely dentate, teeth or lobules usually 6 pairs; upper surface puberulent on midrib and veins and sparsely pilose with some scattered longer hairs, lower surface reticulate-veined, glabrous. Flowers in axils of upper leaves; pedicels 1–3mm, suberect. Calyx ovate-campanulate, 4–4.5mm, deeply split on anterior side, pale green and membranous, with 5 triangular teeth c 0.5 × 0.5mm, apparently lacking subulate points. Corolla apparently white, 14.5–16mm, erecto-patent; tube broadly cylindrical, 6.5–8.5mm, c twice calyx, with lines of short hairs on sides; galea 7.5 × 3.5mm, ± straight (only very slightly inclined at top), sparsely pubescent; lower lip 4–4.5mm, subquadrate with 2 folds, 3-lobed towards apex, lateral lobes acutely ovate, twice as large as the rounded, concave middle lobe. Stamens inserted near base of corolla tube; anterior filaments barbate near top; anthers c 2mm, thecae with subulate tails c 0.5mm. Capsule unknown.

Sikkim: Gangtok. August.

Endemic and known only from the type.

Sect. 13. **Dolichomischus** Li

Caespitose perennials (our spp.). Leaves mostly clustered, long-petiolate. Inflorescence racemose or flowers axillary. Calyx not cleft anteriorly, teeth 5. Galea of corolla beaked or beakless.

Ser. 26. **Corydaloides** Li

Diffusely caespitose perennial with one fertile and several sterile stems. Leaves mainly radical and long-petiolate; cauline absent or 1 pair (subopposite); lamina pinnatifid or pinnatisect. Inflorescence racemose, secund. Corolla yellow; tube slightly longer than calyx, bent forwards above, glabrous; galea erect, toothless, beakless.

40. P. cryptantha Marquand & Airy-Shaw subsp. **cryptantha**

Diffuse, loosely caespitose perennial with numerous sterile stems from the middle of which arises a single inflorescence. Radical leaves with long slender petioles 20–70mm; lamina ovate or oblong in outline, 20–60(–90) × 7–25mm, pinnate with 5–9 pairs of ovate-elliptic, pinnatifid or pinnatisect segments which are shallowly incised-lobate usually with curved, mucronate tips; upper surface very sparsely white-hirsute or subglabrous, lower surface white-furfuraceous. Sterile stems slender, narrowly green-winged, glabrescent below, pubescent above. Flowering stems with a single pair of subopposite leaves similar to radical but smaller. Inflorescence a subsessile, rather dense ± secund raceme; pedicels 5–25mm, slender, sparsely hirsute. Calyx cylindrical, 5–7mm; tube 4–5mm, sparsely villous; lobes 5, narrowly lanceolate, shallowly incised-lobate. Corolla sulphur-yellow with tip of galea brownish, 16–19mm; tube c 9mm, glabrous, erect and c 1.5mm broad below, bent forwards and dilated above; galea erect, c 8–10mm, with slightly incurved, darker, acute tip but lacking a distinct beak; lower lip 3-lobed, middle lobe orbicular, c 3mm, lateral lobes reniform, c 5mm, all with sparse minute marginal cilia. Stamens inserted in middle of tube; all filaments glabrous or 2 anterior hirsute; anthers c 1.5 × 0.7mm, thecae subacute at base. Capsule unknown.

Bhutan: N – Upper Pho Chu district (Lhedi). Usually on damp, moss-covered rocks, in alpine pasture and under *Rhododendron* scrub, 2740–3960m. May–July.

Only one specimen seen from our area, but this subspecies is frequently collected in SE Tibet; the description is thus mainly based on Tibetan specimens. Although at least two dozen specimens are in British herbaria, none has been collected in fruit. Subsp. *erecta* Tsoong, with shorter, erect stems to 7cm tall, smaller leaves barely 45mm long and densely pubescent calyx, is endemic to SE Tibet.

Ser. 27. **Regelianae** Yamazaki

Nearly acaulescent perennial, caespitose. Leaves clustered, alternate, long-petiolate; lamina pinnatisect. Flowers axillary, solitary, appearing as if arising from crown. Corolla large, red-purple; tube straight, hairy; galea decurved, beaked.

41. P. regeliana Prain. Fig. 99c–d.

Low, almost acaulescent perennial forming small, loose tufts; roots ± fleshy. Stems extremely short (0.3–2(–4)cm) with abbreviated internodes, prostrate then ascending or erect. Leaves alternate (but appearing as a radical tuft), 4–9 per plant; petioles 10–50(–70)mm, sparsely pubescent; lamina oblong in outline, 8–30(–60) × 5–15mm, pinnatisect with 4–6 pairs of ovate or oblong-ovate segments which are sharply acute-serrate in their upper half; upper surface dark green, pubescent along midrib, otherwise glabrous, lower surface pale greyish, glabrous. Flowers solitary and axillary (appearing to arise ± from crown); pedicels 10–60mm, ± erect from decumbent or ascending base, slender, sparsely pilose. Calyx narrowly tubular-campanulate, 7–12mm; tube 5.5–7mm, sparsely pilose; lobes 5, posterior lanceolate-oblong, 1.5–2mm, acute-serrate, laterals oblong-ovate, 2–4 × 1–2mm, acutely incised-serrate. Corolla bright pink, carmine or purple, 23–36mm; tube 15–23mm, 2–3 × calyx, straight, sparsely long-pilose, not dilated upwards; galea decurved, erect part c 3mm, anther-bearing part falcate, c 6mm, cucullate at apex ending in short, truncate, 2-lobed beak; lower lip c 11 × 16mm, middle lobe broadly orbicular, c 5 × 6mm, lateral lobes reniform, c 10 × 16mm, all sparsely ciliate. Stamens inserted near middle of tube; 2 anterior filaments pilose, 2 posterior glabrous; anthers oblong, 2.2–2.5

× 1–1.5mm, thecae obtuse at base. Capsule oblong, 10–18 × 4–5mm, acuminate. Seeds narrowly ovoid, c 2mm, greyish, striate but scarcely reticulate.

Bhutan: C – Thimphu district (lakes region beyond Pajoding), **N** – Upper Kulong Chu district (Shingbe); **Sikkim:** Chamnago, Yampung etc.; **Chumbi:** Natu La–Champitang, Zelep La. On and under mossy rocks and in bogs on open hillsides, (3350–)3800–4570m. June–August.

Sect. 14. **Rhizophyllum** Li

Perennials or rarely annuals. Roots often fleshy and enlarged. Cauline leaves when present alternate. Inflorescences racemose or spicate, or flowers axillary and few. Calyx split or unsplit anteriorly, teeth 2, 4, or 5. Corolla tube straight or bent; galea beaked or beakless; middle lobe of lower lip smaller, more distal than the laterals.

As circumscribed here, following Yamazaki (426), only series 28–32 below belong to this section. *P. elwesii* is referred to sect. *Phanerantha* ser. *Robustae* despite its placement in sect. *Rhizophyllum* ser. *Pseudomacranthae* together with *P. fletcheri* by Yang *et al.* (429).

Ser. 28. **Flammeae** Prain

Low perennial. Leaves mainly sub-basal, petiolate; lamina pinnatifid. Inflorescence a centrifugal spike. Corolla yellow; tube straight, glabrous; galea arched forwards, toothless, beakless.

42. P. oederi Vahl subsp. **branchiophylla** (Pennell) Tsoong; *P. branchiophylla* Pennell. Fig. 99k–l.

Perennial with rather long, fleshy fusiform roots 2.5–14cm, sometimes widely divergent. Stems 1 or few, 1–5cm in flower, densely lanate-villous when young, soon glabrate except for 4 lines of very short pubescence. Leaves mainly radical or sub-basal; petioles 10–25(–40)mm, slender, with narrow crenate wing, channelled and pubescent above; lamina narrowly oblong or oblong-lanceolate in outline, 20–40 × 3–8mm, pinnatifid with 20–30 pairs of segments, purplish-green; segments at first flat and at least some usually overlapping, soon becoming transversely raised and arranged ± vertically (resembling gills), all sessile, crenate or occasionally pinnately lobed, ± thickened, with inrolled margins; upper surface glabrous except for minutely puberulent midrib, lower surface brownish-furfuraceous. Cauline leaves few, alternate, similar to basal. Inflorescence a terminal spike of 4–10 flowers, lanate-villous at anthesis (hairs greyish-white). Pedicels very short. Calyx tubular, c 10mm, densely greyish-lanate in flower, at least some ± dense, short hairs persistent in fruit; lobes 5, posterior one small and setaceous, laterals entire below, dilated and toothed above. Corolla lemon-yellow with deeper yellow throat and deep yellow streak down middle of ventral side, tip of galea streaked dull red or purple and with crimson spot on each side of galea on ventral side near middle; tube straight, c 12 × 2mm, glabrous; galea erect, slightly arched forwards, 8–11mm, beakless, ventral margins each with a row of glands but lacking any tooth or projection, dorsal tip rounded, ventral tip subacute and making an angle of c 90° with dorsal side; lower lip shorter than galea, lateral lobes spreading, c 8 × 11mm, middle lobe porrect, c 3 × 2mm. Stamens inserted at or just above middle of tube; anterior filaments pubescent; anthers oblong, c 2 × 1mm, thecae obtuse at base. Style included or very shortly exserted (by up to 1.5mm). Capsule c 12mm, lanceolate-ellipsoid with obliquely truncate apex. Seeds ellipsoid, pale greyish, distinctly reticulate.

Bhutan: C – Ha district (Kang La), Thimphu district (Paro Chu), **N** – Upper Bumthang Chu district (Taasiegem, Tolegang); **Sikkim:** Lhonak, Theu La, Naku La etc.; **Chumbi:** Phari–Tremo La. Open alpine meadows and short turf, also on cliffs, 4115–4870(–5200)m. May–July.

No fruiting material has been seen; the capsules and their indumentum were described by Pennell (384). This is apparently the only taxon within the *P. oederi* complex that occurs in our area. Subsp. *heteroglossa* (Prain) Pennell occurs in the C Himalayas (Theri, Nepal and Tibet) while subsp. *multipinna* (Li) Tsoong is found in Tibet and W China. Subsp. *oederi* is widespread from N and C Europe and Asia eastwards to Kashmir and Mongolia. Some material previously identified as subsp. *branchiophylla* differs in having a small hump-like projection ⅓ way up the ventral margin of the galea, and calyx indumentum not persistent in fruit. These specimens are here referred to *P. pseudoversicolor* Handel-Mazzetti (species 47).

Ser. 29. **Pseudomacranthae** Yang

Robust perennial. Radical leaves few; cauline alternate, petiolate; lamina pinnatisect. Inflorescence racemose, condensed. Corolla mainly white; tube subequalling calyx, straight, glabrous; galea distally arched, beaked.

43. P. fletcheri Tsoong; *P. fletcheriana* Tsoong *nom. illeg.*

Robust, caespitose perennial with ± fleshy, branched root-stock. Stems few to numerous (up to 10 per plant), middle ones erect, lateral ones prostrate at base then ascending or erect, (4–)7–36cm, stout, glabrous. Radical leaves few; petioles 25–90mm, glabrous or hirsute; lamina oblong, 50–120 × 15–28mm, pinnatisect with winged rachis and 7–11 ± distant pairs of segments; segments oblong-ovate, 6–12 × 2–8mm, pinnatipartite or pinnatifid with coarsely acute-serrate margins; upper surface glabrous, lower surface reticulate-veined and sparsely white short-pilose along main veins, otherwise glabrous. Cauline leaves alternate, (2–)4–6, all petiolate, upper scarcely smaller than lower; petioles, especially of upper leaves, ciliated at base. Inflorescence a ± condensed terminal raceme, lower flowers ± remote but others ± contiguous, shorter than or scarcely exceeding long, leaf-like bracts. Pedicels 6–14mm, glabrous. Calyx cylindrical, 20–26mm, split to ¼ on anterior side, white-hirsute mainly on posterior side; lobes 2 or 4, foliaceous, ovate, 5–12 × 5–8mm, deeply pinnatifid with 2–4 pairs of coarsely serrate segments. Corolla white with purple throat and galea, 30–42mm; tube c 22 × 1.5mm, scarcely longer than calyx, glabrous, straight, dilated near top; galea with erect part 6–7mm, anther-bearing part c 6mm, arched, attenuate into short, slightly declinate, conical, deeply 2-lobed beak c 3mm; lower lip large, enclosing much shorter galea, c 16–22 × 20mm, middle lobe oblong, slightly longer than lateral lobes, each lobe with shallowly retuse apex. Stamens inserted near middle of tube; anterior filaments slightly pilose, posterior glabrous; anthers c 2 × 0.8mm, thecae shortly acuminate at base. Capsule lanceolate-ellipsoid, 18–23 × 8–10mm, shortly acuminate. Seeds blackish, obovoid-ellipsoid, c 2.5 × 1.2mm (excluding basal peg) with truncate base and short peg-like basal projection c 0.2mm, indistinctly solid-reticulate with c 15 longitudinal striae and c 27 very indistinct transverse connections per row.

Bhutan: N – Upper Kuru Chu district (Singhi area), Upper Kulong Chu district (Tashi Yangtsi Chu at Lao). Meadows and open places in forest, usually close to streams, 2440–3800m (ascending to 4570m in SE Tibet). June–July.

Ser. 30. **Pumiliones** Prain

Dwarf perennials or annuals, almost stemless. Leaves basal or sub-basal, petiolate; lamina crenate or shallowly pinnatifid. Flowers axillary. Calyx tube very narrow. Corolla tube very long, at least twice calyx, densely pilose; galea porrect, beaked.

44. P. bella Hook.f. subsp. **bella**. Fig. 99a–b.

Dwarf, apparently annual, almost stemless (0–3cm) from slender vertical rootstock 3–5cm. Leaves nearly all basal or sub-basal; petiole 4–18mm, somewhat yellowish, broadened and vaginate towards base, with pubescent narrowly winged margin; lamina oblong-lanceolate, obovate-oblong or almost spathulate, 5–21 × 3–6mm, obtuse, base cuneate, margin crenate or pinnatilobed to less than halfway, lobes themselves crenate when well developed, densely hoary-tomentose on both surfaces. Flowers axillary, peidellate, very large and conspicuous for size of plant, 1–10(–14) per plant; pedicels to 6mm. True bracts absent. Calyx cylindrical-campanulate, 12–16mm, hoary-pubescent, split in upper ⅓, with 5 teeth; anterior teeth suborbicular, posterior lanceolate, all crenate. Corolla tube greenish turning yellow, very slender, 25–42 × 0.7–1.5mm, densely pilose; galea maroon, erect part c 3mm, anther-bearing part c 10mm, sigmoid, somewhat abruptly tapered into slender falcate beak c 8mm with obtuse, entire or shallowly emarginate but not bifid apex; lower lip 3-lobed, pink with broad, deep claret margins (sometimes wholly deep purple), erect, infundibular, forming wide cup in which galea rests; lateral lobes 10–14mm broad, c 4 × as broad as small ovate middle lobe c 3 × 3.5mm; margins entire, glabrous. Stamens inserted at top of tube; all filaments

---

FIG. 99. **Scrophulariaceae.** a–n, flower and inner face of dissected calyx of *Pedicularis* species: a–b, *P. bella* subsp. *bella* (× 2½); c–d, *P. regeliana* (c × 3, d × 3½); e–f, *P. lachnoglossa* (e × 3½, f × 5); g–h, *P. hicksii* (g × 3, h × 4); i–j, *P. dhurensis* (i × 4, j × 6); k–l, *P. oederi* subsp. *branchiophylla* (× 2½); m–n, *P. furfuracea* (m × 3, n × 5). Drawn by Glenn Rodrigues.

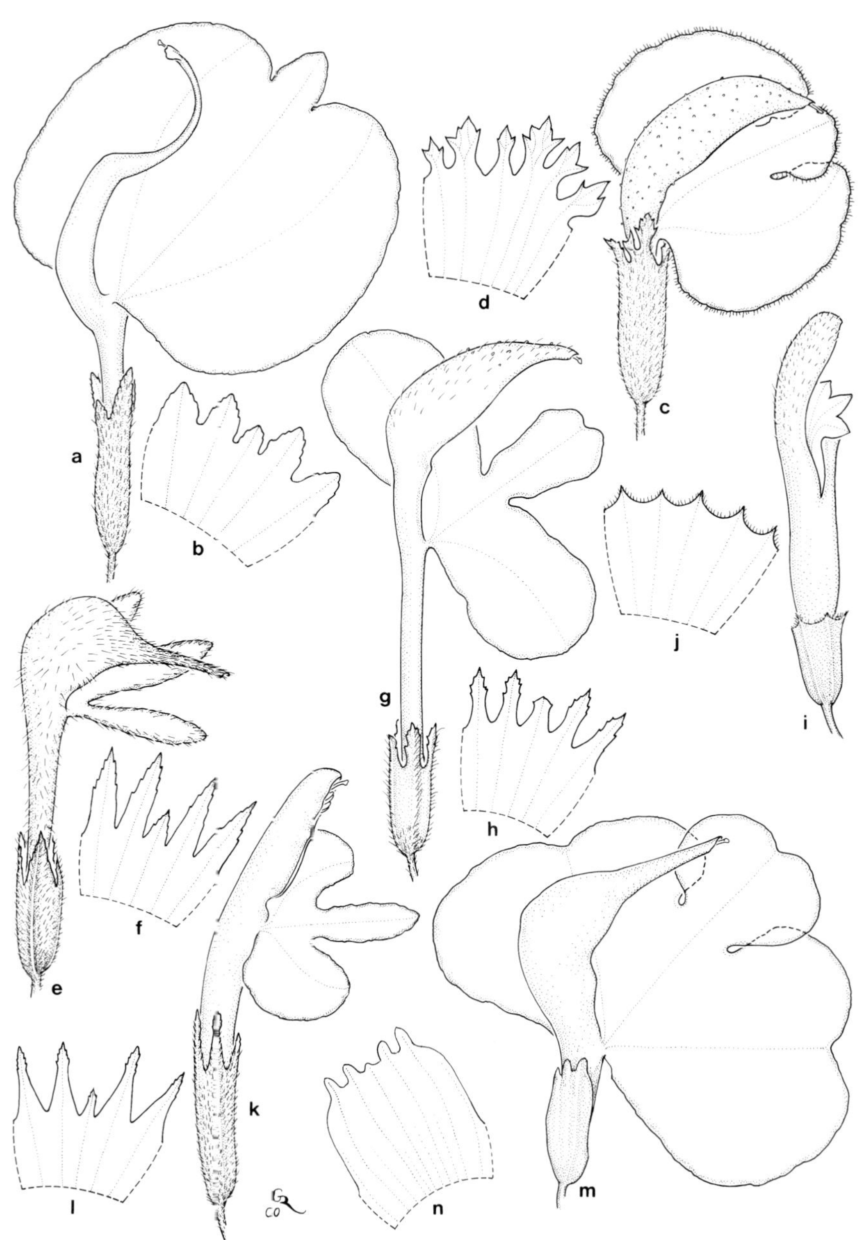
a
b
c
d
e
f
g
h
i
j
k
l
m
n

barbate; anthers narrowly ellipsoid, c 2.5 × 0.5mm, thecae acute at base. Capsule narrowly ellipsoid, obliquely acute at apex; seeds blackish, striate.

Bhutan: **C** – ?Thimphu district (Tremo La–Yale La (Upper Mo Chu district)); **Sikkim:** Dokung, Dongkya La, Temu La, Yume Samdong; **Chumbi:** Chumolari, E of Phari on Bhutan/Tibet border. Grassy hillsides and on rocks, near snows, 4420–5180m. June–early September.

Subsp. *holophylla* (Marquand & Airy-Shaw) Tsoong is apparently endemic to SE Tibet; it differs from subsp. *bella* in having completely entire leaves and seems to prefer slightly lower altitudes (3660–4635m).

45. P. przewalskii Maximowicz

Dwarf, apparently perennial, almost stemless from slender vertical root-stock 1–5cm. Leaves all basal or sub-basal; petiole 4–16mm, dark, scarcely or not broadened at base, pubescent, unwinged; lamina narrowly oblong or obovate, 6.5–16 × 1.5–3.5mm, obtuse, base cuneate, margin crenate or pinnatilobed to less than halfway, lobes themselves sometimes crenate, densely hoary-pubescent on both surfaces, ciliate on margins. Flowers axillary, 1–5 per plant, pedicels (4–)6–10mm. True bracts absent. Calyx cylindrical-campanulate, 9–11mm, hoary-pubescent, with 5 teeth; posterior tooth very small, others 1.5–2mm, crenate. Corolla claret, wine or cyclamen-coloured, ± concolorous but sometimes paler in throat and beak darker; tube cylindrical, not dilated upwards, 20–30 × 1.2–2mm, densely pilose; galea very dark red with a strong bluish bloom when dry; erect part c 7mm, anther-bearing part horizontal, c 5mm, tapered into slender, deeply bifid, straight beak 5–6mm; lower lip spreading, 3-lobed; lateral lobes 8–9 × 10–11mm, middle lobe scarcely smaller, 5.5–7 × 7–9mm; margins entire, glabrous. Stamens inserted at top of tube; all filaments woolly, especially above; anthers ovoid, c 2 × 1.2mm, thecae obtuse at base. Capsule obliquely ellipsoid, shortly cuspidate. Seeds pale, distinctly reticulate.

Chumbi: Phari. Moist alpine meadows and grassy hillsides, 3963–4725m. Late June–late July.

The account given above applies to plants from our area and Tibet; these appear to belong to subsp. **australis** (Li) Tsoong of SE Tibet and Yunnan. Subsp. *przewalskii*, which also occurs in S Tibet, has longer leaves (30–35mm) which are only sparsely pubescent or glabrous. A variable taxon, especially in leaf indumentum and flower colour; several other subspecies, some of which have been recognised as species, occur in China: cf. Li (368) and Tsoong (417).

Ser. 31. **Merrillianae** Yang (ser. *Rhynchodontae* auct. p.p. non Prain)

Caespitose perennial. Leaves mainly basal, petiolate; lamina pinnatisect. Flowers few, terminal. Calyx tube shallowly split anteriorly, nearly glabrous. Corolla tube subequalling calyx tube, glabrous; galea erect then inclined forwards, with 2 teeth on anterior side and ending in a very short beak.

46. P. yarilaica R.R. Mill (378)

Low, rather densely caespitose perennial with stout, fusiform, scarcely tapered fleshy roots. Stems several, 1–6cm, minutely pubescent with 2 rows of hairs c

0.1mm. Leaves mainly in basal tuft, alternate; petioles 4–18mm, subglabrous; lamina ovate-oblong in outline, 6–13 × 2.5–6mm, ± runcinate-pinnatisect with 5–10 pairs of triangular-oblong, subentire lateral segments with broad, rounded, decurved tips and inrolled margins and a relatively large, suborbicular, ± entire terminal segment; upper surface glabrous, lower surface brownish-furfuraceous near strongly serpentine-cristate midrib. Inflorescence of 2(–3) terminal flowers. Pedicels c 4mm. Bracts c 12mm, with short broad petiole and deeply 3-fid lamina, each lobe petiolulate and pinnatifid. Calyx cylindrical-campanulate, 8–10.5mm, tube 10-veined, glabrate or very sparsely pubescent, split to c ¼ on anterior side, with narrow, triangular-acute posterior tooth and obtuse recurved lateral teeth, pilose especially in sinuses. Corolla purplish-pink, 22–25mm; tube 6–8mm, equalling calyx tube glabrous; galea with erect part c 6mm, anther-bearing part suberect and slightly inclined forwards, suddenly narrowed into very short, slightly downwards-directed, truncate beak c 1mm with 2 horizontally projecting subulate teeth c 0.5–0.6mm on anterior side; lower lip c 10 × 12.5mm, middle lobe oblong, c 3 × 2.5mm, smaller than laterals, laterals spreading at angle of c 45°, c 3 × 4.5mm, indistinctly undulate, ± glabrous on margins. Stamens inserted near base of corolla tube; 2 anterior filaments exceedingly sparsely short-pilose in upper ⅔, posterior glabrous; anthers c 2.5 × 1.2mm, thecae triangular-obtuse at base. Style finally exserted by 1.5–2.5mm. Capsule unknown.

Bhutan: N – Upper Mo Chu district (Jari La, Chumiten–Tharizaj Thang). Damp hill slopes, c 4300m. August.

Endemic to Bhutan.

Ser. 32. **Pseudo-oederianae** Limpricht (ser. *Rhynchodontae* auct. p.p. non Prain)

Perennial. Leaves mainly radical; cauline alternate; lamina pinnatifid. Inflorescence spicate. Corolla yellow; tube straight, glabrous; galea suberect, inclined forwards, with 2 anterior humps, beakless.

47. P. pseudoversicolor Handel-Mazzetti; *P. oederi* auct. non Vahl

Perennial with long, stout, fleshy, fusiform roots 5–13cm, sometimes widely divergent. Stems 1 or few, (3–)5–10cm in flower and to 15cm in fruit, at first fuscous-lanate, finally pubescent with 4 lines of short fuscous hairs. Leaves mainly radical, cauline few, alternate; petioles of radical leaves (10–)20–60mm, with narrow crenate wing, channelled and fuscous-pubescent above; lamina narrowly oblong or oblong-lanceolate in outline, 25–60 × 5–11mm, pinnatifid with (12–)15–25(–30) pairs of segments; segments often overlapping, usually remaining flat, sometimes becoming vertically orientated, oblong-triangular, crenate-lobed towards apex, ± thickened and with inrolled margins; upper surface with midrib minutely pubescent, otherwise glabrous, lower surface fuscous-furfuraceous. Inflorescence a terminal spike of usually 8–20 flowers, lanate-villous at anthesis (hairs brownish-grey). Pedicels c 3mm at anthesis. Calyx tubular,

11–12mm, densely greyish-lanate in flower, glabrate in fruit; lobes 5, narrowly oblong-triangular, two c 3 × 0.7mm with obscurely lobed, recurved tip, other three c 2 × 0.6mm, subentire. Corolla pale lemon-yellow with darker yellow throat, with tip of galea blackish, steel-grey or purple and purple lines down rest of galea, sometimes with crimson blotches at margin of each side of galea; tube straight, c 12 × 2mm, glabrous; galea suberect, inclined forwards, 11–14(–16)mm, beakless, ventral margins each with shallow hump-like projection about ⅓ way up and ± dark glands between this projection and lower lip, glabrous above, dorsal tip rounded, ventral tip obtuse and making an angle of c 90° with dorsal side; lower lip c 6.5–8 × 10–13mm, lateral lobes spreading, c 6 × 6mm, middle lobe porrect, c 2 × 2mm. Stamens inserted in lower ⅓ of tube; filaments glabrous; anthers c 2.5 × 1mm, thecae obtuse at base. Style exserted. Capsule 16–19 × 5–6mm, lanceolate-ellipsoid with shortly acuminate apex.

Bhutan: C – Thimphu district (Kumathang), Tongsa district (Rinchen Chu), **N** – Upper Mo Chu district (Yale La, Jari La); **Chumbi:** Phari–Tremo La.

Previously confused with *P. oederi* subsp. *branchiophylla*, from which it is distinguished by the small hump-like projection on each side of its galea at the base of the ventral margin.

SECT. 15. **Pedicularis**; sect. *Cladomania* Li *nom. inval.*, sect. *Phyllomania* Li p.p. (Bhutan spp.)

Perennials. Cauline leaves alternate. Calyx often but not always split anteriorly; teeth 2, 4 or 5. Corolla tube equalling or longer than calyx; galea shortly beaked.

Ser. *Carnosae* is here included in this section, following Yamazaki (426).

Ser. 33. **Furfuraceae** Prain

Moderately tall perennials. Radical leaves small, soon withering; cauline alternate, petiolate; lamina pinnatilobed or pinnatifid. Inflorescence a lax raceme of solitary, axillary, remote flowers. Calyx either deeply split anteriorly with 4 teeth, or not split with 5 teeth. Corolla pink or purple; tube subequalling or longer than calyx, glabrous or with dorsal line of hairs; galea decurved near middle, shortly beaked.

48. P. pantlingii Prain; *P. furfuracea* Bentham var. *integrifolia* Hook.f.

Perennial with short root-stock and fibrous roots. Stems 1 or few from crown, simple, 12–60cm, subglabrous or sparsely pubescent below, rather more densely pilose above, hairs often in 3–5 lines. Radical leaves smaller than cauline but similar, soon withering. Cauline leaves alternate; petiole 5–70mm, glabrous or sparsely hairy, distinctly winged and, especially lower, with base dilated and membranous; lamina broadly ovate or orbicular-ovate in outline, 15–42 × 15–50mm, uppermost sometimes more narrowly ovate-lanceolate, obtuse or subacute, base truncate or cordate, margin with 3–7 pairs of rounded lobes or pinnatilobed to ¼–½ leaf width, segments or lobes acutely serrate or obtusely crenate; upper surface glabrous or sparsely pilose, lower surface reticulate-veined and white-furfuraceous. Bracts leaf-like, small. Flowers solitary, alternate and axillary, forming a lax but nevertheless somewhat conferted terminal raceme; axis densely lanate with long ± appressed grey-brown glandular hairs. Pedicels 1–4mm, densely glandular-lanate. Calyx tubular-campanulate, 6–8mm, not cleft on anterior side, with 5 lobes which become recurved, spreading-hirsute or villous on tube, lobes glabrous; posterior lobe c 1.5mm, lanceolate, ± entire, other 4 ovate, incised-dentate. Corolla pale pinkish-purple, white on middle part of galea, 20–25mm; tube c 8mm, slightly dilated towards top, glabrous; galea ± inflated, basal part erect, c 4mm, abruptly bent at top in right angle, anther-bearing part horizontal, c 5.5 × 3mm, tapering into slender porrect but slightly incurved beak c 2.5–5mm with 2-lobed apex, each lobe minutely emarginate; lower lip 7–10 × 14–17mm, lateral lobes broadly orbicular, c 8mm wide, middle lobe small, orbicular, c 3 × 3mm, all with glabrous or slightly pilose margins. Stamens inserted in middle of tube; anterior filaments pilose near middle or all filaments glabrous; anthers narrowly ovoid-oblong, c 2 × 1mm, thecae acute at base. Capsule narrowly ovoid, c 20 × 6.5mm, acute. Seeds broadly ovoid, c 2 × 1mm, testa blackish, very finely reticulate.

Bhutan: C – Thimphu, Punakha, Tongsa, Bumthang and Mongar districts, **N** – Upper Mo Chu, Upper Mangde Chu and Upper Bumthang Chu districts; **Darjeeling:** Sandakphu; **Sikkim:** Lagyap, Lungthung, Tong etc. Open, moist *Abies* forest and woods, 2135–4270m. May–September.

As in *P. furfuracea*, the degree of incision of the leaf margin is very variable. Plants from Sikkim with very pale flowers whose lower lips are almost glabrous have been named, at least in manuscript, as 'var. *minor* Prain' but it is not known if this name has been validly published.

49. P. furfuracea Bentham. Fig. 99m–n.

Perennial with many elongate-fusiform roots. Stems numerous, branched at base and simple or sparsely branched above, 10–60cm, glabrous below, weakly long-pilose above with hairs in 2 rows. Radical leaves small, soon withering. Cauline leaves alternate; petiole 10–60mm, pilose on upper surface especially near base, narrowly winged; lamina ovate or ovate-triangular in outline, 10–70 × 10–60mm, subacute, base cordate or truncate, variably pinnatilobed or pinnatifid for ¼–⅘ of leaf width or occasionally only deeply crenate, segments 3–6 pairs, oblong-ovate or ovate-lanceolate and subacute with doubly acute-serrate margins when well developed; upper surface usually glabrous except for sparsely hirsute midrib, sometimes scabrid-puberulent with short brownish hairs, lower surface reticulate-veined and white-furfuraceous. Bracts leaf-like. Flowers solitary, alternate, axillary, rather remote, forming lax leafy raceme. Pedicels 2–3mm, with short deflexed hairs. Calyx pale green, with purplish tinge towards apex, obliquely ovate, split almost to base on anterior side, 4.5–5mm, with 4 small acute glabrous teeth at apex. Corolla c 20mm, pale purple or pink, galea paler whitish-green on middle part and apex often tinged bluish; tube c 6 × 1mm, glabrous; galea decurved near middle, erect part 3.5–5mm, anther-bearing part c 8mm, gradually tapering into slender beak c 4mm with truncately acute, shallowly 2-lobed apex, the lobes themselves shallowly emarginate, upper pair much larger than lower; lower lip 7–8(–10) × 10–12(–15)mm, lateral lobes reniform, 6–8mm wide, larger than orbicular, emarginate middle lobe 3–5 × 4–6mm, all with glabrous margins. Stamens inserted in middle of tube opposite top of ovary; all filaments glabrous; anthers c 2 × 0.8mm, thecae obtuse at base. Capsule lanceolate, 10–15 × 3–4mm, acuminate. Seeds acutely ovoid, c 2 × 1mm, blackish, very finely reticulate.

Bhutan: C – Ha, Thimphu, Punakha, Tongsa, Bumthang and Mongar districts, **N** – Upper Mo Chu and Upper Kulong Chu districts; **Darjeeling:** Sandakphu, Tonglu; **Sikkim:** Lagyap, Chho La, Bakhim etc.; **Chumbi:** Yatung. *Abies* forest and bamboo and *Rhododendron* scrub, often on humus-rich soil, 2600–3960m. Mid June–mid August.

The lower lip has previously been described (by Prain and Yamazaki) as being much larger, 10 × 17–18mm, but there is no evidence for this from the substantial number of specimens examined by the author. The specimen from Mongar (*Grierson & Long* 1937) is somewhat aberrant, having large broadly elliptic leaves with 4–6 pairs of lobes obovate-spathulate in outline and twice 3-lobed at their apex.

50. P. microcalyx Hook.f.

Perennial. Roots spreading, narrowly elongate-fusiform. Stems 12–30cm, erect, simple or branched from base, glabrous below, shortly pubescent above with 2 lines of hairs. Radical leaves absent at anthesis, cauline also soon falling, usually 2–3 in upper half of stem, alternate, long-petiolate (petioles of middle leaves 15–60mm); lamina lanceolate or oblong in outline, 10–45 × 6–25mm, deeply pinnatifid with 5–10 pairs of ovate-oblong, acute, doubly acute-serrate segments, dark and glabrous above, paler beneath with conspicuous reticulate veins and sparsely white-furfuraceous. Bracts leaf-like but ± sessile, exceeding calyx. Inflorescence a loose head of several alternate axillary flowers. Pedicels 1–6mm, puberulent. Calyx campanulate, 5–6mm, with 5 subequal lobes; posterior one linear, others oblanceolate, entire below, sparsely acute-serrate above. Corolla bright pink or purple, c 20mm; tube straight, c 10 × 1.2–1.5mm, c twice calyx, with line of hairs down back; galea decurved near middle, erect part c 3mm, anther-bearing part c 5–6mm, deep purple, gradually narrowed into tapering ± straight beak c 4mm with acute shallowly 2-lobed apex; lower lip c 10 × 10–12mm, lateral lobes reniform, c 6mm wide, only slightly larger than orbicular middle lobe c 4 × 4.5mm; margins entire, sparsely pilose. Stamens inserted in middle of tube opposite top of ovary; all filaments glabrous; anthers c 1.8 × 0.8mm, thecae obtuse at base. Capsule oblong, 7–10 × 3–4mm, apiculate; inner margin of valve lips conspicuously thickened. Seeds oblong-ovoid, c 1.8 × 1mm, very finely reticulate.

Bhutan: S – Sankosh district (Lawgu), **C** – Thimphu district (Semi La), **N** – Upper Pho Chu district (Chojo Dzong); **Sikkim:** Lachen, Samdong, Changu,

Dzongri, Bikbari etc. On moss-covered rocks and among grass in *Rhododendron* scrub, also occasionally in river grit, (2590–)3353–4300(–4725)m. June–August.

Ser. 34. **Carnosae** Prain

Perennial. Leaves alternate (sometimes subopposite), petiolate; lamina crenate-dentate. Inflorescence a lax raceme (in ours). Calyx deeply cleft anteriorly; teeth 2. Corolla pink; tube scarcely longer than calyx, glabrous; galea decurved at middle, recurved apically, indistinctly beaked.

51. P. bifida (D. Don) Pennell; *P. carnosa* Wall.

Somewhat canescent perennial (appearing annual on specimens lacking complete root system), with several short fleshy fusiform roots. Stems 8–40(–60 or more)cm, erect or with decumbent base, ± stout, branched or simple, white- or grey-pilose or -pubescent. Radical leaves with short, lanate petioles 4–20(–30)mm; lamina oblong to oblanceolate, 10–45 × 5–17mm, obtuse, base attenuate, margin crenate-dentate, surfaces pubescent. Cauline leaves alternate or occasionally subopposite, 7–9(–11) (excluding bracts), all petiolate (uppermost often very shortly), lamina linear to oblanceolate, (8–)12–40 × 3.5–12mm. Inflorescence a lax terminal raceme of 5–20 flowers. Bracts leaf-like, as long as or longer than calyx. Pedicels 1–4mm, pubescent. Calyx tubular-campanulate, deeply cleft on anterior side and appearing truncate in some lateral views, (6–) 7–10.5mm, loosely hirsute or pubescent mainly in lower part and along midrib; tube narrowing apically to narrow 2-lobed mouth, lobes orbicular, irregularly crenate-dentate, reflexed. Corolla pink with galea tipped deep purple or crimson, 20–23mm; tube whitish, 8–11mm, scarcely longer than calyx, glabrous; galea with erect part 5.5–6mm, anther-bearing part arched, decurved at middle, directed forwards, c 7 × 4mm, abruptly narrowed into slender glabrous beak 4–5mm with truncate denticulate apex; lower lip c 13 × 16mm, lateral lobes orbicular-ovate, c 7mm wide, middle lobe c 3 × 6mm, all minutely ciliate on margins. Stamens inserted at or just below middle of tube; anterior filaments very sparsely pilose, posterior glabrous; anthers 2–3 × 0.9–1.2mm, ellipsoid, thecae acute at base. Capsule oblong-ellipsoid, 10–13 × 3–4mm, acute. Seeds not formed in only fruiting specimen examined.

Bhutan: C – Tongsa district (Chendebi), Tashigang district (Tashi Yangtsi Dzong, Gumpa); **Sikkim:** Lachung (several collections). Grassy slopes and among shrubs in mixed 'rain' forest, 1400–2750m (ascending to 3500m in C and E Nepal). August–September.

Incompletely gathered specimens are deceptive, appearing annual. Leaf shape and stem and leaf indumentum are very variable. Several specimens from Sikkim and those from Bhutan differ from the majority in having dense, short, closely appressed pubescence on stems, leaves and calyx. Plants from the eastern part of the range (including Sikkim and Bhutan) tend to have longer calyces (8–10.5mm) than those from the western Himalaya (6–8mm). The species requires further study and could possibly be divided.

Ser. 35. **Curvipes** Prain

Probably perennials. Leaves all cauline, alternate or lower ones opposite, petiolate; lamina usually pinnatisect. Inflorescence racemose, or flowers axillary. Calyx split anteriorly (?or posteriorly); teeth 2 or 4. Corolla pink or purple; tube equalling or longer than calyx, straight, glabrous; galea arcuate, with declinate beak.

52. P. amplicollis Yamazaki

?Perennial. Stem 10–20cm in flower, slender, suberect at first then decumbent, lanate with curved hairs. Lower leaves opposite, upper alternate; petiole 8–30mm, lanate, slender; lamina ovate in outline, 10–30 × 8–20mm, pinnatisect with 3–5 pairs of lateral segments; segments oblong, coarsely acute-serrate; upper surface puberulous, lower surface pilose on midrib. Flowers in axils of alternate, leaf-like bracts. Pedicels 3–5mm in flower, to 10mm in fruit, erect, with decurved hairs. Calyx tubular-campanulate, 9.5–10 × 3.5mm; tube split on posterior side (fide Yamazaki), densely lanate, with usually 4 ovate incised-serrate teeth 1.5–2mm. Corolla purple, c 20mm; tube c 14 × 2–2.5mm, longer than calyx, glabrous, straight; galea glabrous, with vertical part bent slightly backwards with respect to tube, anther-bearing part arcuate-incurved, c 3mm, attenuate into very slender straight but ± declinate beak c 6mm with entire apex; lower lip c 12 × 18mm, lateral lobes obliquely rounded, much larger than small, reniform, emarginate middle lobe c 3 × 8mm, all lobes with minutely pilose margins. Stamens inserted below middle of tube; filaments sparsely pilose; anthers oblong, c 1.8mm, thecae subacute at base. Immature capsule obliquely oblong, c 9mm, acuminate.

Bhutan: C – Thimphu district (near Chapcha Dzong). 2800m. July.

Endemic to Bhutan and apparently known only from the type (not seen). The calyx was described as posteriorly split by Yamazaki but the usual condition in *Pedicularis* is for any cleft to be on the anterior side so his observation needs confirmation.

53. P. curvipes Hook.f.

?Perennial. Stem 30cm or more, suberect at first then decumbent, unbranched or branched from base, shortly puberulous with hairs mainly in 2 narrow rows. Leaves all cauline, alternate and petiolate; petiole puberulous; lamina ovate in outline, 10–35 × 8–25mm, pinnatisect (lowest leaves sometimes pinnatifid) with 3–5 pairs of lateral segments; segments 3–10mm, incised-dentate, -crenate or -lobate at least above; upper surface glabrous, lower surface sparsely white-furfuraceous. Inflorescence a lax terminal raceme, pyramidal at anthesis; flowers in axils of leaf-like bracts. Pedicels c 2mm in flower, 6–10mm in fruit, at first erecto-patent but very soon becoming decurved in fruit. Calyx ovoid, 6–8mm, tube split to c ⅓ on anterior side, membranous, glabrous except for cilia along margins of anterior fissure, with 2 small obovate-oblong, crenate apical lobes. Corolla pale rose with whitish throat and darker pink galea, c 15mm; tube c

7mm, not exceeding calyx, glabrous, straight; galea puberulous, with vertical part c 5mm bent slightly backwards with respect to tube, anther-bearing part arched and inflated, c 4mm, attenuate into slender, ± decurved beak c 5mm with emarginate apex; lower lip c 7 × 12mm, 3-lobed; lateral lobes c 5 × 5mm, obliquely rounded and very shallowly emarginate, middle lobe c 3.5 × 3.5mm, emarginate. Stamens inserted near middle of corolla tube; all filaments glabrous; anthers c 1.5 × 0.8mm, thecae shortly acuminate at base. Style scarcely exserted. Capsule 9–10 × 3–4mm, oblong-ovoid with obliquely acuminate apex.

Sikkim: Tumbok. 2750–3050m. May (in cultivation, Britain).

The plants from Naga Hills (Assam) that were formerly included under this species have been transferred to *P. nagaensis* Li; *P. curvipes* is endemic to Sikkim.

SECT. 16. **Rhizophyllastrum** R.R. Mill (378)

Perennials, often caespitose Leaves mostly radical; cauline, when present, alternate. Inflorescence racemose, fasciculate, or flowers axillary. Calyx usually not split anteriorly, with 5 teeth, but rarely cleft, then sometimes with 3 teeth. Galea of corolla beaked.

Ser. 36. **Asplenifoliae** Prain

Perennials. Leaves almost all radical, petiolate; lamina pinnatisect or pinnatifid. Inflorescence fasciculate, flowers often arising directly from crown. Corolla pink to dark red; tube 1.3–4 × calyx, often very narrow, straight, glabrous or hairy; galea shortly beaked, the beak straight or down-curved.

54. P. albiflora (Hook.f.) Prain; *P. asplenifolia* Floerke var. *albiflora* Hook.f.

Low caespitose perennial. Stems 0–4cm, decumbent or ascending, glabrous or pubescent. Radical leaves numerous, tufted; petioles 20–40mm, very slender; lamina oblong-lanceolate, 15–20 × c 5mm, pinnatisect with 5–8 pairs of ovate, rather distant, obtuse, mucronate-dentate segments; surfaces glabrous. Flowers in small terminal groups of usually 4. Pedicels 1.5–5mm. Bracts c 15 × 10mm, c 1½ × calyx, dilated and sheathing below, 3-sect above. Calyx narrowly cylindrical-campanulate, 7–10 × c 1.5mm, glabrous; teeth 5, unequal, posterior one very small, acute, others narrowly oblong, obtuse. Corolla crimson, or sometimes possibly white (galea darker, at least when dry); tube 20–24 × c 1mm, at least twice calyx, cylindrical, not dilated upwards, glabrous; galea arcuate, lower part 5mm, anther-bearing part c 6mm, attenuate into conical, deflexed porrect beak c 3.5mm with crenate apex; lower lip 3-lobed, c 16mm wide, lobes rounded, laterals 1½ × middle lobe, all with glabrous margins. Stamens inserted in middle of tube; all filaments very sparsely pilose in lower part and with a few more hairs below anthers; anthers 1.5 × 0.7mm, thecae bluntly apiculate at base.

Bhutan: N – Upper Kulong Chu district (Me La – doubtful); **Sikkim:** Kankola (type). Among short turf and grasses on open slopes and on cliffs, above treeline, 4265–4570m. Mid June–August.

Probably endemic to Sikkim and ?Bhutan. Although named '*albiflora*' by

Hooker, what appears to be type of his variety at Kew has corollas that have dried pale reddish-purple, especially on the lower lip (the galea is darker purplish), and thus little different to the dried appearance of the two Bhutan specimens which were described in field notes as having crimson corollas. If Hooker's type was really white-flowered when freshly gathered, the possibility that it was an albino form of a normally crimson-flowered species (whose name must unfortunately remain *albiflora*) should not be discounted. Prain described the filaments of *P. albiflora* as glabrous, which, for the specimens seen, is wrong. Yamazaki used the name *P. albiflora* for a taxon in Nepal with lemon-yellow lower lip and chocolate galea, which probably represents an undescribed species.

55. P. cooperi Tsoong

Densely caespitose, almost acaulescent, low perennial; roots very slender, fibrous, not fleshy. Stems several but very short, 0.5–2cm, nearly glabrous. Radical leaves numerous, forming dense tufts; petioles 6–13mm, glabrous, narrowly winged; lamina ovate or ovate-orbicular, 4–5 × 2.5–3mm, pinnatifid with only 2–3 pairs of rather broadly obovate or suborbicular, crenate segments; upper surface glabrous, lower surface reticulate-veined and with few white scurfy scales when young, later glabrous. Flowers few, in small fascicles of 2–3. Bracts leaf-like, opposite, petiolate, lamina ovate, incised-lobed with reflexed margins. Pedicels to 4mm, glabrous. Calyx narrowly cylindrical, 6–7mm; tube 3–3.5mm, glabrous; lobes 5, subequal, posterior lobe filiform-triangular, acicular, c 1.5mm, others narrowly oblong or oblong-triangular, c 2mm, 2 laterals crenate with subobtuse apex, 2 anterior entire with acute apex, all shortly pilose. Corolla pink, 32–35mm; tube very narrowly infundibular-cylindrical, 15–20 × c 1.2mm, sparsely pubescent in line down ventral vein, otherwise glabrous; erect part of galea 5mm, with 2 prominent small blunt teeth on ventral margin just below top, anther-bearing part horizontal, porrect, 3.5–4 × 2mm, abruptly narrowed into rather long straight beak 5–5.5mm, not noticeably broader proximally, with parallel sides and slightly dilated, 2-lobed apex, lobes oblique, rounded, one larger than other; lower lip c 11 × 13mm, lateral lobes c 6.5 × 8mm, middle lobe ovate, c 6 × 5mm, subacute; lobes shallowly undulate, pilose. Stamens inserted in lower part of corolla tube; filaments ± glabrous; anthers 1.5 × 0.7mm, thecae minutely apiculate at base. Capsule unknown.

Sikkim: below Chho La pass. On moss-covered scree, c 4270m. September.

Endemic to Sikkim.

Very similar to *P. sikkimensis* Bonati which differs principally in its much less abruptly narrowed beak with non-parallel sides; larger opposite (normally not alternate) leaves with more pairs of segments; lower lip with glabrous (not pilose) margins etc.

56. P. microloba R.R. Mill (378)

Almost acaulescent, low perennial; roots rather slender, fibrous, slightly fleshy near crown. Stems ± absent. Leaves all radical, numerous, forming loose tuft;

petioles 5–7mm, glabrous; lamina ovate or oblong-ovate, 4–10 × 2–5mm, pinnatisect with 3–5 pairs of ovate to suborbicular, incised-lobate mucronate segments; both surfaces glabrous. Flowers few; pedicels arising ± from crown, 8–14mm, glabrous. Bracts absent. Calyx cylindrical-campanulate, 6.5–8mm; tube 5–6mm, sparsely pilose on veins in basal half, upper half glabrous; lobes 5, posterior narrowly triangular with 1 lateral tooth, lateral lobes broadly obovate from broad stipitate base, with 2 lateral teeth and broad acute apex, anterior like lateral ones but smaller. Corolla pink, 25–28mm; tube 9–11 × 1.3mm, c 1.3 × calyx, sparsely long-pilose; erect part of galea c 4mm, anther-bearing part slightly inflated, c 5mm, horizontal, glabrous, rather abruptly tapered into short conical beak c 3mm with obliquely truncate, bluntly crenulate apex; lower lip 8 × 10mm, lateral lobes reniform, middle lobe small and indistinct, 2 × 2.5mm, ovate-suborbicular, margins ± entire, glabrous. Stamens inserted near top of corolla tube; 2 anterior filaments sparsely villous, 2 posterior glabrous; anthers 1.9 × 1mm, thecae obtuse at base. Capsule unknown.

Bhutan: C – Thimphu district (Pajoding). 3960m. August.

57. P. longipedicellata Tsoong

Low, acaulescent caespitose perennial; roots longer than aerial parts, ± fusiform. Leaves all radical, numerous; petioles 10–30mm, glabrous, winged; lamina linear-oblanceolate, 10–30 × 3.5–8.5mm, pinnatisect or deeply pinnatifid with 7–11 rather distant pairs of ovate-oblong or oblong-linear, obtuse, deeply crenate to pinnatifid (sometimes subentire) segments. Flowers several per tuft, each solitary from crown on long erect glabrous pedicel 20–70mm. Bracts absent. Calyx cylindrical, 12–14 × 3.5–4mm; tube 7–9(–10)mm, lanate, sparsely pilose or subglabrous; lobes 5, lower half stipe-like, ciliate on margins, foliar apical half 3-sect, segments coarsely crenate, reflexed at tip. Corolla pink or wine-red with paler throat and darker hood, 33–43mm; tube 18–25mm, cylindrical but abruptly widening slightly near middle, pubescent along lateral veins, otherwise glabrous; galea with long cylindrical vertical part c 8mm, anther-bearing part c 6–8 × 4.5mm, minutely papillose or with stalked glands, rather abruptly attenuate into short beak 1–4mm with crenulate, truncate apex; lower lip 12–15 × 18–22mm, 3-lobed, lateral lobes spreading, broadly elliptic, c 10 × 8mm, middle lobe orbicular, c 6 × 8mm, all entire and ciliate on margin. Stamens inserted in middle of upper (broader) section of corolla tube; anterior filaments sparsely pilose, posterior glabrous; anthers c 2.5 × 1.2mm, thecae with wide, very shallowly emarginate base. Capsule lanceolate-ellipsoid, 12–14 × 4.5–5.5mm, shortly obliquely acuminate. Seeds c 2.5 × 1.1mm, ellipsoid-reniform with concave ventral face, dorsal face with c 12 faint ± straight longitudinal ridges, ventral face with c 12 more distinct curved longitudinal ridges; reticulation absent.

1. Leaf segments glabrous at tips; calyx not lanate; back of galea with sessile papillae; beak of galea 2.5–4mm .... **a.** var. **longipedicellata**
\+ Leaf segments with tufts of fine greyish hairs at tips; calyx lanate; galea with short stalked glands; beak of galea only 1–1.5mm .. **b.** var. **lanocalyx**

a. var. **longipedicellata**

Bhutan: N – Upper Mangde Chu district (Saga La), Upper Bumthang Chu district (Kantanang, Marlung, Hoopkye La), 4270–4725m. June–July.

Endemic to Bhutan.

b. var. **lanocalyx** R.R. Mill (378)

Bhutan: C – Bumthang district (Penge La). Dry hilltop among dwarf rhododendrons, 4390m. June–July.

Ser. 37. **Filiculae** Li

Perennials. Leaves mostly radical with long filiform petioles; cauline absent or 1–3, alternate; lamina pinnatilobed or pinnatipartite. Inflorescence a short raceme. Calyx split anteriorly, lobes 5. Corolla pink to purple; tube slightly longer than calyx, glabrous or partly hairy; galea beaked, the beak with denticulate apex.

58. P. trichodonta Yamazaki; *P. asplenifolia* var. *asplenifolia* sensu F.B.I. p.p. (Sikkim plants) non Floerke, *P. wallichii* sensu Prain (389) p.p. (E Himalayan plants) and Yamazaki (135) non Bunge

Low perennial; roots rather long, fleshy. Stems (1–)3–9(–11)cm, erect, shortly but rather densely pubescent. Radical leaves alternate, clustered; petiole 25–60mm, often almost filiform, sparsely pubescent with purplish hairs; lamina linear or linear-oblong, 10–35 × 1.5–8mm, dark green above, pale green beneath, more or less runcinate-pinnatilobed with 7–16 pairs of ovate-triangular, obtuse to truncate, subentire to lobulate or dentate segments with ± revolute margins; upper surface shortly grey-pubescent along midrib, otherwise glabrous; lower surface glabrous; lobes ciliate on margin. Cauline leaves (0–)1–2, alternate. Inflorescence a short, lax terminal raceme of 1–5 flowers. Bracts with broad petiole, pilose to lanate on margin, and short leaf-like lamina, subequal to calyx. Pedicels 2–8mm, sparsely pubescent or glabrous. Calyx tubular-campanulate, 10–13mm; tube c 7mm, shallowly split on anterior side, glabrous; lobes 5, triangular-lanceolate, posterior c 1.2mm, acute, entire, others lobate, laterals c 3mm, anterior c 2mm, all lanate on margins. Corolla pink, mauve or wine-red with darker veins and galea, 20–25mm; tube c 12–14 × 1.5mm, glabrous, not dilated above; galea with erect part c 8mm, anther-bearing part falcate, directed forwards, c 5.5 × 3.5mm, ± glabrous (sometimes sparsely papillose), gradually attenuate into porrect beak 2.5–4mm with truncate, denticulate, ciliate apex; lower lip 12–13 × 13–15mm, longer than galea, middle lobe c 4 × 5mm, much smaller than reniform lateral lobes and apically truncate, all lobes denticulate and at least sparsely (and very shortly) ciliate. Stamens inserted near middle of corolla tube; all filaments lanate above but glabrous in lower ⅔; anthers c 3 × 1.2mm, thecae acute at base. Capsule oblong-lanceolate, 15–20 × 4–5mm, acuminate. Seeds not seen.

Bhutan: C – Tongsa district (Rinchen Chu); **Sikkim:** Dzongri, Peykiong La,

Bijan, Phullalong etc. Open grassy alpine hillsides and on moss-covered boulders, 3660–4573m. June–August.

59. P. filiculiformis Tsoong var. **dolichorhyncha** Tsoong

Bushy perennial. Stems numerous, 10–20cm, ascending or suberect, clothed with lanceolate scales at base, completely glabrous. Radical leaves long-petiolate (petiole 40–50mm), lamina oblong, 15–45 × 5–8mm, pinnatipartite with 7–15 pairs of oblong-triangular, obtuse, lobulate segments, lowest pair conspicuously shorter than those above; upper surface glabrous, lower surface reticulate-veined, white-furfuraceous at first, later glabrous. Cauline leaves 2–3, alternate, with rather long petioles. Inflorescence a short, rather conferted terminal raceme of (3–)6–9 flowers. Bracts leaf-like, with purplish petioles 8–10mm ± dilated at base. Pedicels silvery-pubescent, 1.5–10mm, lowest much longer than others. Calyx cylindrical-campanulate 10mm; tube 6–6.5mm, shallowly split on anterior side, sparsely pilose; lobes 5, unequal, spathulate-stipitate, obtuse, with ciliate margins, tips becoming ± reflexed. Corolla purple, faintly scented, c 28mm; tube c 14mm, pilose along 2 lateral veins, otherwise glabrous; galea with erect part c 5.5mm, anther-bearing part c 6 × 3.5mm, dark purple, papillose along ventral margin, gradually attenuate into forward-directed, straight, slender, sparsely papillose beak c 5mm with slightly dilated, denticulate apex, the teeth themselves shallowly emarginate; lower lip c 8 × 14mm, lateral and middle lobes subequal, entire, glabrous. Stamens inserted just below top of corolla tube; anterior filaments pilose, anthers c 1.8 × 0.8mm, thecae obtuse at base. Capsule rather broadly ellipsoid, c 10 × 6mm, straw-coloured, acute.

Bhutan: N – Upper Kulong Chu district (Me La, two collections). Moist cliffs and overgrown boulders, 4420m. June.

Var. *dolichorhyncha* is apparently endemic to Me La; var. *filiculiformis*, with corollas having shorter beak and ciliate lower lip, is restricted to SE Tibet.

Ser. 38. **Perpusillae** Yamazaki

Dwarf caespitose perennials, sometimes nearly acaulescent. Leaves radical (and alternate if stems present), petiolate; lamina pinnatifid or pinnatisect. Flowers solitary, axillary. Calyx split anteriorly, lobes 3 or 5. Corolla mainly crimson; tube very slender, c 3 × calyx, straight, ± hairy; galea beaked, beak bifid or erose.

60. P. hicksii Tsoong Fig. 99g–h.

Dwarf, caespitose, many-stemmed perennial 2–7cm; roots slender, fibrous. Stems very short, almost absent. Leaves all radical; petioles 10–20mm, glabrous; lamina lanceolate or oblong, 8–15 × 3–5.5mm, pinnatifid to near middle, with 4–5(–6) pairs of ovate-triangular, serrate segments, upper surface very loosely long-pilose (hairs cream), lower surface white-furfuraceous at first and with hairs on midrib. Flowers solitary, axillary, 7–20 or more per plant; pedicels 6–15mm. Calyx narrowly campanulate; tube 5.5–7mm, 5-ribbed, membranous, shallowly split on anterior side; teeth 5, all stipitate, posterior suborbicular, indistinctly toothed, other 4 flabellate, incised-serrate. Corolla crimson, 28–33mm; tube very slender, 16–19mm, c 3 × calyx tube, straight, scarcely dilated near top, with very few hairs, with few sessile glands near top; galea with erect part c 4mm, anther-bearing part inflexed, 5–6 × 3–3.5mm, attenuate into short beak c 3mm with erose apex; lower lip 9–10.5 × 13–15mm, 3-lobed ± to middle; lateral lobes c 6.5 × 6.5mm, orbicular, middle lobe c 5 × 4.5mm, obovate with retuse apex, all lobes sparsely short-ciliate. Stamens inserted near top of tube; 2 anterior filaments pubescent; anthers c 1.5 × 1mm, thecae obtuse at base. Capsule 9 × 6mm, ovoid, straight, obliquely acute.

Bhutan: N – Upper Kulong Chu district (Shingbe). Scree, c 4420m. August.

Endemic to Bhutan; known only from type gathering.

61. P. perpusilla Tsoong

Very dwarf, caespitose, few-stemmed perennial 1–3cm (excluding flowers); roots fusiform. Stems 0–15mm, when evident shortly pubescent with hairs in 2 rows. Leaves radical (and alternate on stems); petioles 6–13mm, ± dilated at base, sparsely pilose; lamina ovate-oblong, 6–12 × 2.5–4.5mm, pinnatisect with 5–7 rather distant pairs of ovate-oblong segments. Flowers solitary, axillary, 1–3 per plant; pedicels 3–5mm, glandular. Bracts leaf-like with broad membranous bases sheathing calyx. Calyx 6–7mm, membranous, split to ⅔ on anterior side, nearly glabrous; teeth 3, central one (posterior) very reduced, other 2 (laterals) scarcely 2mm, lanceolate, slightly sigmoid and directed forwards, finally patent with recurved tips. Corolla crimson with white markings; tube 16–18mm, c 2½ × calyx tube, straight, hardly dilated at top, pilose; galea with erect part c 3mm, anther-bearing part 6mm, inflexed, glandular dorsally, attenuate into short beak c 4mm with obtuse, shortly bifid apex; lower lip c 6 × 11mm, 3-lobed; lateral lobes elliptic, 10 × 6mm, middle lobe transversely elliptic, deeply retuse, c 4 × 5.5mm; all lobes densely short-ciliate. Stamens inserted near top of tube; anterior filaments densely pubescent near apex. Capsule unknown.

Bhutan: C – Mongar district (Pung La). Peaty soil on rocks, c 3660m. July.

Endemic to Bhutan; known only from type specimen.

Ser. 39. **Odontophorae** Prain

Low caespitose perennial. Leaves radical and cauline, petiolate; cauline subopposite or alternate; lamina pinnatisect. Inflorescence conferted, usually 3-flowered. Corolla bicoloured (pink/white); tube nearly twice calyx, partly hairy; galea shortly beaked.

62. P. odontophora Prain

Low, rather delicate caespitose perennial with creeping rhizome. Stems 2.5–6cm, prostrate or procumbent then suberect, pilosulous to puberulous with hairs mainly in 2 lines. Radical leaves few, long-petiolate; petioles 15–30mm, sparsely pubescent; lamina ovate to oblong-ovate in outline, 8–22 × 8–13mm, pinnatisect with 3–13 alternate, deeply pinnatifid, sharply dentate segments, the lower ones shortly petiolulate; upper surface dark green, glabrous, lower surface paler with dark reticulate veins, glabrous. Cauline leaves alternate or more often subopposite, similar to radical but smaller and more shortly petiolate. Flowers usually 3, in a small conferted terminal inflorescence, each subtended by a leaf-like bract. Pedicels 1–4mm. Calyx cylindrical-campanulate, 6–7mm, ± glabrous, 5-toothed, the posterior and 2 anterior teeth triangular, 2 lateral teeth subulate, all entire. Corolla 24–26mm; tube white, c 10–12 × 1mm, with dorsal line of short hairs continuing up erect part of galea, otherwise glabrous; erect part of galea white, 3–4mm; anther-bearing part of galea pink, c 6 × 2.5mm, subinflated, arcuate-incurved with a small, broad-based, obtusely triangular tooth at base, minutely puberulent dorsally, attenuate into slender, pink, horizontal, porrect beak 3.5–4mm with slightly dilated, truncate apex; lower lip white with reddish-pink margins, 7–8 × 9.5–11mm, middle lobe c 3 × 4mm, laterals c 4 × 6mm, all glabrous on margins. Stamens inserted in middle of tube; 2 anterior filaments densely hirsute throughout, posterior glabrous above, sparsely barbate below. Style shortly exserted. Capsule c 9 × 3mm, obliquely lanceolate-ellipsoid with broad shoulder near middle, puberulent on margins towards apex, obtuse.

Sikkim: Na-tong; **Chumbi:** Phari, Do-tho. c 3350m. July–August.

Endemic to Chumbi and Sikkim; known only from three gatherings.

SECT. 17. **Leptochilus** Yamazaki

Description as for the only series in our area.

Ser. 40. **Excelsae** Maximowicz

Tall perennial. Leaves cauline, alternate, large with long petioles; lamina pinnatisect. Inflorescence a long raceme; flowers alternate. Calyx spathaceous, with 3 very small apical teeth. Corolla pink or purple; tube equalling calyx, distally twisted, glabrous; galea villous, with strongly curved beak whose tip at first rests on spoon-shaped projection from lower lip, then springs apart.

63. P. excelsa Hook.f. Fig. 100c–d.

Tall erect perennial 100–150cm. Stem ± stout, purplish-brown, unbranched below, laxly branched in inflorescence, glabrate below, sparsely short-pubescent in 4 rows above, pubescent ± all round in inflorescence. Radical leaves apparently absent. Cauline leaves alternate; petioles 40–80mm, narrowly winged, rachis winged and toothed; lamina ovate-triangular, ovate-lanceolate or lanceolate-oblong, 50–180 × 30–110mm, pinnatisect with 12–15 pairs of segments; largest segments c 3 pairs up from base of lamina, 30–60mm, pinnatifid, segments then rapidly diminishing in size upwards, sparsely short-pilose above, glabrous beneath, margins doubly and acutely incised-serrate. Inflorescence a long, slender terminal raceme 7–20cm with many alternate flowers separated by short internodes; axis pubescent ± all round. Bracts small, equalling or shorter than calyx, all shortly petiolate, with ovate, incised-serrate lamina c 3–8 × 2–3mm. Pedicels 2–4mm, shortly pubescent. Calyx spathaceous, 5–6 × c 2.5mm in flower, very shortly 3-toothed at apex, posterior tooth c 0.4mm, narrowly oblong-triangular, lateral ones broadly triangular, c 0.7mm, each with rudimentary tooth on outer side and pilose margins. Corolla pale pink or grape-purple with darker galea, 16–24mm; tube c 5.5–7mm, glabrous, twisted above; galea with anther-bearing part 4–5 × c 2.5mm, villous dorsally and along ventral margins, produced into slender beak, young beak c 6–7mm, curved downwards and inwards, sometimes almost forming a circle, with shortly 2-lobed apex resting in dilated tip of middle lobe of lower lip, later springing apart, lengthening to 10–11mm, curving upwards and outwards, finally forming a shallow S; lower lip lanceolate-oblong, 14–18 × c 5mm, obsoletely 3-lobed, lateral lobes pilose on margin, middle lobe produced c 8mm beyond tips of lateral lobes, linear, with spoon-like dilated tip. Stamens inserted in lower part of tube; all filaments glabrous; anthers 1.8–2 × 1.1–1.5mm, ovoid-elliptic, thecae obtuse at base. Capsule c 16–18 × 6–7mm, lanceolate-ellipsoid, obliquely mucronate at apex. Seeds blackish, ovoid, 1.9–2 × c 1mm, with c 16 (8 per side) very indistinct, slender, rather irregular longitudinal ridges connected by numerous faint transverse slightly raised lines.

Bhutan: C – Ha, Thimphu, Tongsa, Bumthang and Tashigang districts, **N** – Upper Mangde Chu, Upper Kuru Chu and Upper Kulong Chu districts; **Sikkim:** Dzongri, Lachen etc.; **Chumbi:** near Rinchingong, Phari, near Galing. *Abies* and mixed *Betula*/conifer forest, *Rhododendron* scrub, 2600–3800m. July–August.

SECT. 18. **Phanerantha** Li

Perennials. Roots thickened. Stems several. Cauline leaves alternate. Calyx cylindrical to ellipsoid, split anteriorly; teeth 5. Corolla tube equalling or up to twice as long as calyx; galea strongly bent, beaked; middle lobe of lower lip smaller than others.

Ser. 41. **Macranthae** Prain

Robust perennials. Leaves basal and alternate on stem, petiolate; lamina pinnatipartite. Inflorescence a dense, conspicuous raceme of many, large flowers. Calyx shortly split anteriorly. Corolla yellow; tube equalling calyx, straight, glabrous; galea falcate, beaked.

64. P. scullyana Maximowicz. Fig. 100a–b.

Robust perennial. Stems erect, 15–100cm, stout, ± lanate with brownish-white hairs, sometimes nearly glabrous. Leaves alternate, clustered near base with several on the stem. Petioles of lower leaves 20–160mm, ± pilose on upper surface; lamina linear to linear-lanceolate, 50–210 × 12–50mm, with winged rachis and pinnatipartite with 10–20(–25) pairs of segments; segments oblong to oblong-lanceolate, coarsely and deeply lobed, the lobes mucronate-dentate;

FIG. 100. **Scrophulariaceae.** a–n, flower and inner face of dissected calyx of *Pedicularis* species: a–b, *P. scullyana* (× 2); c–d, *P. excelsa* (c × 3, d × 4); e–f, *P. elwesii* (× 2); g–h, *P. longiflora* subsp. *tubiformis* (g × 2½, h × 3); i–j, *P. megalantha* (i × 3, j × 4); k–l, *P. siphonantha* (× 3); m–n, *P. rhinanthoides* subsp. *labellata* (m × 3, n × 4). Drawn by Glenn Rodrigues.

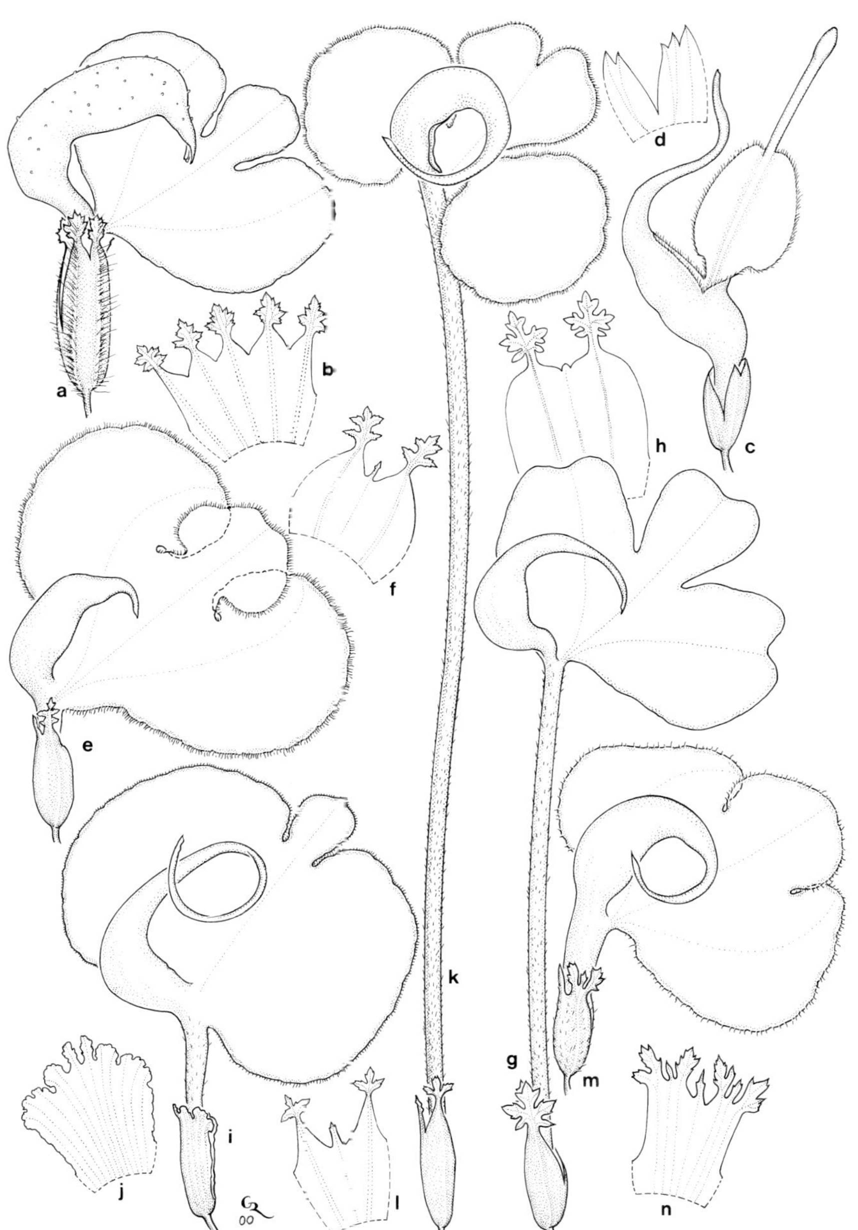
a
b
c
d
e
f
g
h
i
j
k
l
m
n

upper surface ± pilose, lower surface white-furfuraceous. Inflorescence a very conspicuous, usually long and dense, terminal raceme of many flowers. Bracts leaf-like, equalling or exceeding calyx, with broad petioles. Pedicels very short. Calyx urceolate-tubular, 15–22 × 7–8mm, sparsely long-pilose especially on 5 main longitudinal veins, shortly split on ventral side, with 5 ovate, very acutely dentate lobes 2–4 × 1–2mm, each tooth of the lobes long-aristate. Corolla pale lemon- or primrose-yellow with galea tipped purple, with faint lemon scent, 30–35mm; tube straight, 15–20mm, equalling calyx, glabrous; galea strongly falcate, erect part 6–8mm, anther-bearing part c 10mm, microscopically glandular-punctate, tapering to an abruptly incurved beak 4–5mm with 2-lobed apex; lower lip 12–18 × 25–30mm, middle lobe orbicular to obovate-orbicular, entire, c 8 × 10mm, lateral ones c 10 × 15mm, enclosing galea. Stamens inserted just above middle of tube; anterior filaments pilose right up to anthers, posterior sparsely pilose but glabrous in upper half; anthers c 3 × 1.2mm, thecae subacute at base. Capsule obliquely lanceolate, 20–25 × 6mm. Seeds ivory-white, irregularly ellipsoid, 3–3.5 × c 1.8mm, coarsely reticulate.

Sikkim: Dzongri, Meguthang.

Frequent in Nepal, reaching its eastern limit in Sikkim where it seems to have been rarely collected. The description applies to typical plants with pale yellow flowers and lemon scent, which constitute the only variant that has been collected in Sikkim. Other plants from Nepal, determined as *P. scullyana* but with pure white flowers, also differ in some other characters, including a 'strong rose scent'; they may merit recognition as a subspecies or separate species but further study is needed.

Ser. 42. **Robustae** Prain

Perennials, usually caespitose and often dwarf. Leaves mainly radical, petiolate, cauline alternate; lamina pinnatifid, pinnatipartite or pinnatisect. Inflorescence a short dense raceme of axillary flowers. Calyx split to ⅓ or more anteriorly, lobes usually 5, but 3 in most specimens of *P. elwesii*. Corolla pink to purple (rarely white in albinos); tube 1–2 × calyx, usually glabrous; galea porrect, beaked, the beak bifid or 2-lobed at apex.

65. P. nepalensis Prain

Somewhat caespitose perennial with short rhizome and fleshy fusiform roots. Plant to 24cm tall but stems usually much less (0–5cm); stems when developed erect, sparsely pilose. Leaves mainly radical, cauline alternate; petioles 10–50mm, slender, sparsely long-pilose; lamina linear or broadly linear, 10–100 × 4–20mm, obtuse or subacute, pinnatifid or pinnatisect with 5–15 pairs of segments; segments ovate, 1.5–8mm, acute-serrate or largest pinnatifid, mucronate; glabrous above except for long multicellular hairs at tips of segments, white-furfuraceous beneath at first. Flowers few to numerous, axillary or in dense terminal racemes to 8cm. Pedicels 5–25(–45)mm, sparsely pubescent. Calyx tubular-campanulate, 10–14mm (–16mm: fm. *alba*), cleft to c ⅓ on anterior side, strongly ribbed, sparsely white-hirsute on ribs; teeth 5, posterior oblanceolate, small, others obovate-orbicular or obovate, incised-serrate, reflexed, glabrous. Corolla purple (rarely pure white: fm. *alba*); tube 15–22mm, 1½–2 × calyx, sparsely short-pilose only in upper part; galea with erect basal part 6–7mm, anther-bearing part inflated, 6–8mm, hooked-decurved, glabrous, tapering to deeply 2-lobed short beak c 5mm; lower lip 12–13 × 18–20mm, spreading, lateral lobes reniform, 12–14mm wide, middle lobe broadly reniform-orbicular, 6–10 × 9–11mm, all lobes undulate, glabrous. Stamens inserted near middle of tube; 2 anterior filaments densely hairy, 2 posterior more sparsely hairy; anthers c 3 × 1.5mm, thecae acute at base. Capsule obliquely oblong, c 20mm, shortly cuspidate. Seeds broadly ovoid, c 2 × 1mm, pale, distinctly reticulate.

Bhutan: C – Tongsa district (Rinchen Chu at Lakse La); **Sikkim:** Choon-goo (389). Open alpine grassland, hillsides and scree, 3350–4725m. June–August(–early September).

Fm. **alba** Tsoong, based on the collection from Rinchen Chu, is endemic to Bhutan.

66. P. daltonii Prain; *P. asplenifolia* Floerke var. *pubescens* Hook.f.

Dwarf, shortly caulescent caespitose perennial with fleshy fusiform roots to 20cm. Stems erect, numerous, (0.5–)1–7cm, pilose. Leaves mainly in basal tuft, cauline leaves alternate; petiole of radical leaves 10–40mm, slender, becoming twisted, densely but minutely puberulent, with narrow glabrous wing; lamina linear-oblong, 10–30(–40) × 2.5–6mm, subacute, base cuneate or shortly attenuate, pinnatifid with 8–15(–20) pairs of segments; segments ovate, 1–2mm, crenate; upper surface sparsely long-pilose, lower surface sparsely white-furfuraceous. Flowers few, axillary, in dense raceme; pedicels 20–30(–60)mm, sparsely short-pilose. Calyx cylindrical-campanulate, 10–12mm, split to c ⅓ on anterior side, hirsute, 5-toothed; teeth elliptic, crenate, posterior one smallest. Corolla purple; tube 16–20mm long, 1.5–2mm diameter near middle but slightly dilated towards top, sparsely pilose, glabrescent, with few sessile glands near top; galea with erect basal part c 5mm, anther-bearing part inflated, 6–6.5 × 3–3.5mm, horizontal, porrect, abruptly tapered into straight declinate beak 4–5mm with acute bifid tip. lower lip spreading, 3-lobed, 14–18 × 19–23mm, lateral lobes reniform, 10–11 × 12–14mm, middle lobe 8–9 × 8–9mm, suborbicular with cuneate base, all lobes shallowly crenulate, glabrous. Stamens inserted just below apex of tube; all filaments hirsute; anthers 2.5 × 1mm, constricted towards apex, thecae obtuse at base. Capsule obliquely oblong, 15–20mm, shortly cuspidate. Seeds ovoid, c 1.8mm, pale, distinctly reticulate.

Bhutan: N – Upper Mo Chu district (Lingshi); **Sikkim:** Yakla, Tankra La, Chho La; **Chumbi:** Chakalung La and Kalaeree. Stony ground and alpine turf, 4420–4877m. July–September.

67. P. robusta Hook.f.

Dwarf perennial, with horizontal root-stock and longer roots. Stems ascending, 2–6cm, pubescent. Leaves mainly radical, cauline alternate; petioles 10–20mm, slender, subglabrous; lamina narrowly oblong, 13–30 × 4–6mm, subacute, base shortly attenuate, pinnatifid with 8–10 pairs of segments; segments ovate, 0.5–2mm, obtuse, at least larger ones crenate; both surfaces glandular-puberulent. Flowers axillary, in short dense raceme; pedicels 4–9mm. Calyx cylindrical-campanulate, c 10mm, split to ⅓ on anterior side, pubescent; teeth 5, subequal, oblong, serrate, spreading at anthesis, densely pubescent on margins. Corolla purple; tube c 9mm, glabrous; galea with basal part erect, c 4mm, anther-bearing part c 8mm, falcate, tapering into a short beak c 4mm with bifid apex, directed downwards; lower lip spreading, 15–16mm wide, middle lobe entire. Stamens inserted in middle of tube; all filaments barbate. Capsule obliquely oblong-ovoid, c 18 × 5mm, shortly cuspidate. Seeds ovoid, c 1.5mm, reticulate.

Sikkim: Lachung Valley, ?Samdong; **Chumbi:** Phari. 4270–5180m. (?July–)August; fruiting September.

Apparently very local and known from few collections, none of them recent. Records from Samdong probably apply to *P. daltonii* and/or *P. elwesii*; the sheets bearing the plants allegedly from Samdong are mixed gatherings, and it is not clear which labels go with which plants. The fruiting specimen from Singalelah (Darjeeling: Singalila, 3350m) mentioned in F.B.I. under *P. robusta* is, according to its annotation, *Clarke* 12571 (K). However, the leaves look more like those of *P. megalantha* and the capsules are 25–30mm long, thus matching those of *P. megalantha*. It bears the apparently unpublished name *P. fenestrata* Clarke.

68. P. elwesii Hook.f. Fig. 100e–f.

Perennial from long, stout, fleshy, straight root-stock. Stems several, decumbent or ascending, 5–20cm, densely puberulent all round. Leaves mostly radical and clustered, cauline alternate; petiole 10–70mm, densely puberulent; lamina broadly linear or oblong, 40–100 × 5–20mm, pinnatisect or pinnatipartite with narrowly winged rachis; segments (8–)12–22 pairs, oblong or ovate-oblong, obtuse, crenate-serrate, larger ones pinnatifid with crenate-serrate lobules, sparsely puberulent above and beneath, with white scurfy scales beneath at first. Inflorescence a short terminal conferted raceme borne on a long scape. Bracts leaf-like but with broad petioles expanded at base, lamina 6–14mm, pinnatifid, densely puberulent. Pedicels 2–15(–20)mm, densely puberulent. Calyx tubular-ventricose, deeply split on anterior side, 10–12mm, with narrow mouth and ± scarious margin; teeth 3(–5), posterior very small (c 1.5mm), linear, other 2 (or 4) ovate with short stalk, 3–4mm, pinnatifid. Corolla deep pink, crimson or magenta; tube c 10mm, subequalling calyx, mostly straight but slightly curved near top, glabrous; galea with erect basal part 3–5mm, anther-bearing part hooked and decurved near middle, turgid, c 8mm, enclosed by lower lip, glabrous, suddenly narrowed into downwards-directed, straight, twisted beak c 4mm with deeply 2-lobed apex; lower lip large, enclosing galea, 18–21 × 22–28mm, 3-lobed, with densely ciliate margin; lateral lobes reniform, 9–14mm wide, middle lobe orbicular or spade-shaped with shallowly emarginate apex. Stamens inserted in middle of tube; all filaments villous; anthers oblong, c 3.5mm, slightly curved, thecae acute at base. Capsule obliquely ovoid-oblong, 15–17 × 5–6mm, acute. Seeds buff, oblong-ellipsoid to ovoid with small basal peg-like appendage, 2.2–2.5 × 1.2–1.5mm, obtuse at apex, finely striate-reticulate with c 30 longitudinal striae on dorsal face and c 20 on ventral, with prominent ventral longitudinal groove (so that in ventral view seed resembles a bivalve).

Bhutan: C – Ha district (Tare La), **N** – Upper Mo Chu district (Yale La), Upper Bumthang Chu district (Marlung), Upper Kulong Chu district (Me La); **Sikkim:** Lachen, Dzongri, Yume Samdong, Gora La etc.; **Chumbi:** below Ghiva, Phari. Moist alpine grassland (often by streams), usually in open situations but sometimes among shrubs or under *Abies* forest, 3050–4877m. June–September.

The sectional and series placement of this species is disputed. Yamazaki (426) groups it in sect. *Phanerantha* ser. *Robustae* (the treatment followed here), whereas Yang *et al.* (429) consider it a member of ser. *Pseudomacranthae* together with *P. fletcheri*; that series is a member of sect. *Rhizophyllum.*

Ser. 43. **Garckeanae** R.R. Mill (378)

Perennial. Leaves radical and alternate, petiolate; lamina pinnatifid. Inflorescence a dense raceme. Calyx split to ⅓ anteriorly. Corolla pink or purple; tube 1½–2 × calyx, puberulent; galea circinnate, beak curved, almost forming a circle; lower lip erect, standard-like.

69. P. garckeana Maximowicz

Perennial, with fleshy fusiform roots. Stems erect, 6–16cm, pubescent, leafy. Leaves radical and alternate; petioles 20–35mm, slender, dilated at base; lamina linear-oblong, 20–60 × 3–6.5mm, obtuse, base shortly attenuate, pinnatifid with 12–20 pairs of segments; segments ovate, contiguous, 1–2.5mm, bluntly mucronate-dentate (mucro not always obvious); upper surface almost glabrous except for grey-puberulent midrib, lower surface minutely puberulent and with white scurfy scales. Flowers axillary, in dense 7–15-flowered raceme; pedicels 7–18mm. Calyx narrowly cylindrical-campanulate, 14–16mm, split to ⅓ on anterior side, hirsute mainly on ribs, 5-toothed; teeth elliptic, cristate, posterior smaller than other 4. Corolla deep pink or purple; tube 22–30 × c 2mm, not dilated upwards, puberulent; galea with erect basal part c 4 × 2mm, anther-bearing part 5–7 × 2.5–3.5mm, inflated, circinnate, tapering into a slender beak c 7.5mm with bifid apex, curved, nearly completing a circle; lower lip erecto-patent, standard-like, 3-lobed, 12–14 × 18–20mm, lateral lobes c 10 × 13mm, broadly ovate, middle lobe c 8 × 9mm, oblong-obovate or obovate, truncate, all with glabrous, entire margins. Stamens inserted in middle of tube; all filaments hairy but anterior more densely so than posterior; anthers ovoid, thecae acute at base. Capsule obliquely oblong, c 22mm, shortly cuspidate.

Sikkim: Dzongri (388); **Chumbi:** Phari, Chumbi, Na-toot, Tangkar La, Beongchin, Bootang, Lu-ma-poo. July.

Little material of this apparently local species has been seen.

SECT. 19. **Schizocalyx** Li; sect. *Phanerantha* Li p.p.

Description as for the only series in our area.

Ser. 44. **Longiflorae** Prain

Perennials (perhaps sometimes behaving as annuals). Leaves radical and cauline, cauline alternate or upper ones sometimes clustered, petiolate; lamina pinnatifid. Inflorescence racemose, the lowest flowers often remote. Calyx split to at least ⅓ anteriorly, lobes 2–3 (and sometimes rarely a posterior tooth in *P. siphonantha*). Corolla yellow or pink; tube 3–8 × calyx, hairy; galea twisted to form a ring or U, with or without ventral teeth, the beak slender and tapered, 2-lobed at apex; lower lip suberect.

70. P. longiflora Rudolph subsp. **tubiformis** (Klotzsch & Garcke) Pennell; *P. tubiflora* sensu F.B.I. non Fischer, *P. longiflora* var. *tubiformis* (Klotzsch & Garcke) Tsoong, *P. tubiformis* Klotzsch & Garcke. Med: *Lukro serpo*. Fig. 100g–h.

Perennial with several fleshy, elongate-fusiform roots to 8cm. Stems 1–50cm, usually several, simple, decumbent to erect, glabrous. Leaves alternate or upper sometimes subopposite and clustered; petioles 5–75mm, narrowly winged with broadened, flattened base, glabrous or sparsely pilose; lamina linear to oblong, 10–40 × 3–10mm, pinnatifid with 6–10 pairs of segments, segments ovate or ovate-oblong, acute, doubly callose-serrate, glabrous. Inflorescence a ± abbreviated, several-flowered terminal raceme often with some axillary flowers below. Bracts leaf-like with petioles broadened at base. Pedicels 2–15mm; flowers strictly erect at anthesis, strongly recurved to touch ground after pollination. Calyx urceolate-tubular, 8–11 × 4–5mm, split ± halfway on ventral side; mouth narrowed, 2-lobed, lobes shortly stipitate, ovate, 3-fid, lobules acutely dentate. Corolla bright golden yellow, with 2 reddish-brown, chocolate or purple comma-like marks in throat and pale yellow tube, slightly scented; tube 25–50 × 1.5mm, pubescent; galea with erect part c 4mm, abruptly bent at top, anther-bearing part 5–6 × 2.5–3mm, directed forwards but becoming strongly twisted, narrowed into slender beak c 5mm with 2-lobed apex, turning orange-red after pollination; lower lip 6–10 × 15–18mm, 3-lobed, middle lobe obovate-orbicular, 5–6.5 × 5.5–6mm, lateral lobes reniform, c 10 × 8mm, emarginate, all pilose on margin. Stamens inserted at apex of tube; filaments all pilose; anthers oblong, c 2 × 1mm, thecae acute at base. Capsule lanceolate-ellipsoid, 13–17 × 4–5mm, acuminate. Seeds ovoid, blackish, striate, somewhat crested at apex.

Bhutan: C – Ha and Thimphu districts, **N** – Upper Mo Chu, Upper Pho Chu, Upper Bumthang Chu and Upper Kulong Chu districts; **Sikkim:** Dzongri,

Lhonak, Lagyap, Yumthang, Dongkung; **Chumbi:** Lingmatang, Chho La etc. Swamps and marshy meadows, often in water, 3050–5560m. June–October.

Particularly long-stemmed plants (25–50cm) with long petioles to lower leaves (up to 75mm) have been called fm. *maxima* Bonati; they probably represent ecotypic variants adapted to a semi-aquatic habitat. They have been collected more than once at Lingmatang.

71. P. siphonantha D. Don; *P. hookeriana* sensu Tsoong p.p. (Bhutan and Sikkim plants) non Bentham. Med: *Lukro marpo*. Fig. 100k–l.

Perennial (?or sometimes annual) with ± stout vertical root-stock and fibrous roots. Stems erect, decumbent or sometimes prostrate (var. *prostrata* Bonati), 5–25(–35)cm, glabrous or pubescent. Leaves alternate, mostly radical except on plants with well-developed stems; petioles 10–40mm, glabrous or with long hairs especially near base; lamina linear-oblong, 10–40 × 2–15mm, with narrowly winged rachis, pinnatisect or pinnatifid with 5–8 pairs of segments; segments ovate or ovate-oblong, very variable in size and depth, doubly acute-serrate; upper surface glabrous or sparsely pilose, lower surface callose-reticulate and sparsely pilose. Inflorescence of several axillary flowers in upper part of stem (remote and axillary in var. *prostrata*). Bracts leaf-like with petioles broadened at base. Pedicels 2–5mm. Calyx urceolate-tubular, 10–12 × c 3mm, split to c ⅓ on ventral side; mouth strongly constricted, 2–3-lobed sometimes with a very small acuminate posterior tooth, the lobes shortly to distinctly stipitate, orbicular and doubly incised-serrate. Corolla bright rose-purple or pink with white throat and often whitish base of galea, 40–100mm; tube 30–90 × 0.5–2mm, straight, erect, pilose; galea twisted to form a U or incomplete ring with its tip pointing ± upwards, the anther-bearing part c 2mm wide, with tooth-like projection on each ventral margin, glandular-punctate, beak c 10mm, shallowly 2-lobed; lower lip orientated almost vertically, 8–10 × 15–18mm, pilose on margins, deeply 3-lobed, middle (lower) lobe obcordate, laterals reniform, spreading like wings. Stamens inserted at top of tube; 2 anterior filaments pilose, posterior glabrous; anthers narrowly ellipsoid, c 2 × 1mm, thecae subacute at base. Capsule lanceolate-ellipsoid, 10–15 × 4–5mm, shortly acuminate. Seeds broadly ellipsoid, greyish-black, 2.5–2.7 × 1–1.2mm, distinctly reticulate and with c 17 striae, abruptly narrowed basally into an obtuse appendage and sometimes also with a smaller, truncate apical projection.

Bhutan: C – Ha, Thimphu, Punakha, Tongsa, Bumthang and Mongar districts, **N** – all districts except Upper Kuru Chu district; **Sikkim:** Dotag, Dyalip La, Dzongri, Yampung; **Chumbi:** Bootang, unlocalised (collections for King). *Abies* forest, *Rhododendron* scrub, damp alpine meadows and sedge moorland, 3200–4270m. May–September.

Prostrate forms with long stems up to 35cm, which have been called var. *prostrata* Bonati or subsp. *prostrata* (Bonati) Tsoong), have frequently been confused with the W Himalayan species *P. hookeriana* Bentham which, however, has broader, more ovate leaves and corollas with shorter tube. They have been collected several times in Sikkim and C Bhutan and require further study, as do plants from Dzongri with exceptionally large corollas. Plants of 'var. *prostrata*' also differ in their seed morphology: their seeds are paler brown, narrowly ellipsoid, with a prominent basal appendage but no apical projection. *Cooper* 2157 from Sikkim is conspicuously reticulate with c 13 longitudinal rows per side, *Dhwoj* 18 from Nepal is much less so with c 17 longitudinal rows of striae.

SECT. 20. **Botryantha** Li

Description as for the only series in our area.

Ser. 45. **Rhinanthoides** Prain

Perennial. Leaves mostly radical, alternate, petiolate; cauline remote, small, alternate; lamina pinnatisect. Inflorescence racemose. Calyx split to at least halfway ventrally, lobes 5. Corolla reddish-purple; tube up to 2½ × calyx, straight, hairy; galea erect in centre of lower lip, twisted and S-shaped, ventrally toothed, beaked, the beak long and slender, 2-lobed at apex.

72. P. rhinanthoides Schrenk.

Perennial; roots elongate-fusiform. Stems 5–30cm, branched from base, ascending or somewhat decumbent, glabrous or very sparsely short-pilose. Leaves alternate, mostly radical, cauline remote, small; petioles of lower leaves 10–40mm, sparsely pubescent or glabrous; lamina linear or linear-lanceolate, 15–70 × 5–15mm, pinnatisect with 6–15 pairs of segments; segments ovate or oblong-ovate, acute, doubly incised-serrate, teeth usually acute or cuspidate, rarely obtuse; surfaces glabrous or sparsely pubescent. Inflorescence a short terminal raceme of 3–10 flowers. Bracts leaf-like but smaller (equalling or shorter than calyx), lamina ovate or ovate-triangular, sparsely white-pilose, with petiole broadened at base. Pedicels 4–15mm. Calyx urceolate-tubular, 10–13 × 4–5mm, tube membranous, sparsely pilose, narrowed towards top and split ventrally to ½–¾, mouth 5-lobed, posterior lobe usually subulate and entire, rarely ± resembling others and denticulate. Corolla pale reddish-purple with whitish throat and much darker purple galea, 25–35mm; tube 20–30 × c 1.5mm, straight, gradually lengthening up to 2½ × calyx, loosely pilose; galea borne erect in centre of lower lip, extreme base pale whitish and bearing small tooth on each ventral margin, remainder dark purple, twisted to form an incomplete ring or S-shape, anther-bearing part c 4mm, produced into long slender beak c 15mm with acute shallowly 2-lobed apex; lower lip 12–15 × 20–25mm, lateral lobes reniform, 10–15mm wide, middle lobe broadly orbicular, 5–6 × 5–10mm, all usually at least sparsely ciliate on margins, sometimes glabrous. Stamens inserted near top of corolla tube; 2 anterior filaments villous, 2 posterior glabrous; anthers oblong, c 2.5 × 1.3mm, thecae acute at base. Capsule oblong-ovoid, c 12 × 4mm, with oblique, shortly acuminate apex. Seeds ovoid, pale, distinctly reticulate, apically crested.

1. Leaf segments flat, not overlapping, with acute or cuspidate teeth; lower pedicels usually 8–15mm at anthesis; posterior calyx lobe subulate **a.** subsp. **labellata**
+ Leaf segments oblique, overlapping, with obtuse teeth; lower pedicels not more than 5mm at anthesis; all calyx lobes denticulate and revolute **b.** subsp. **revoluta**

a. subsp. **labellata** (Jacquemont) Pennell; *P. labellata* Jacquemont, *P. rhinanthoides* var. *labellata* (Jacquemont) Prain, *P. rhinanthoides* subsp. *tibetica* (Bonati) Tsoong. Fig. 100m–n.

Leaf segments flat, becoming ± separated; teeth acute or cuspidate. Lower pedicels usually 8–15mm at anthesis. Posterior calyx lobe subulate or linear-lanceolate, entire.

Bhutan: C – Thimphu district (near Paga La), **N** – Upper Pho Chu (Gyophu La), Upper Mangde Chu (Mangde Chu); **Chumbi:** (all 389) Phari, Kangma, Pe-namong Chu. Boggy moorland etc., 3960–4270m. June–August.

Rather variable in our area. Li (368) distinguished subspp. *tibetica* and *labellata* by their lower lips respectively ciliate on the margin and glabrous, but plants from the distributional range of subsp. *labellata* in the narrow sense (W Himalaya) can sometimes have ciliate lower lip, and some specimens identified as subsp. *tibetica* have glabrous lower lips. Yamazaki (426) treated the two as synonymous and his view is followed here.

b. subsp. **revoluta** Pennell

Leaf segments oblique and suberect, overlapping; teeth obtuse. Lower pedicels c 5mm at anthesis. Posterior calyx lobe ± similar to other 4, denticulate.

Sikkim: Naku Chho. 5000–5500m. August.

Known with certainty only from the type gathering, which has not been seen; the information given here is based on Pennell's description and key. The status of this subspecies requires investigation. The peculiar shape of the leaves recalls those of *P. oederi* Vahl subsp. *branchiophylla* (Pennell) Tsoong, described from the same area.

Sect. 21. **Saccochilus** Yamazaki

Description as for the only series in our area.

Ser. 46. **Megalanthae** Prain

Perennials. Leaves mainly clustered near base, alternate, petiolate; lamina pinnatifid to pinnatipartite. Inflorescence racemose. Calyx split to least ⅓, lobes 5. Corolla pink to purple; tube 1½–6 × calyx, straight, hairy; galea erect in centre of cup-like lower lip, its erect part very reduced or absent, anther-bearing part ± sessile on lower lip, strongly twisted into a ring or arc, with long coiled beak emarginate at apex.

73. P. megalantha D. Don. Fig. 100i–j.

Perennial; roots straight, simple or with few branches. Stems 1–several, erect or ascending, 10–80cm, unbranched or sparsely branched from base, glabrous or with a few thin hairs. Leaves alternate (often with at least some upper ones subopposite); petioles 20–80mm, narrowly winged, usually completely glabrous, sometimes sparsely pilose only on the broadened, flattened base; lamina linear-lanceolate to ovate-lanceolate, 40–120 × 10–40mm, pinnatifid to pinnatipartite with 7–12(–14) pairs of segments; segments ovate or oblong-ovate, pinnately lobed, the lobes crenate-serrate; both surfaces sparsely pilose, lower surface also ± white-furfuraceous at least when young. Inflorescence an elongated or ± condensed terminal raceme of 5–20 flowers, sometimes with a few axillary flowers immediately below, opening from top downwards. Bracts leaf-like but usually more ovate, with petioles broadened at base and ciliate on margin. Pedicels 2–12mm, glabrous. Calyx urceolate-tubular, 15–20 × 5–8mm, split to c ⅓; tube membranous, 10-veined, sparsely spreading-villous especially on veins; mouth with 5 orbicular-ovate, doubly incised-dentate apical lobes. Corolla deep reddish-purple or cerise with white galea and throat, or rarely entirely white, 45–75mm; tube 28–60mm, very slender (c 1–1.5mm diameter), straight, 2½–4 × calyx, sparsely pilose; galea erect in the centre of the lower lip which forms a deep cup around it, strongly twisted and forming a nearly complete ring, c 18mm, anther-bearing part ± sessile on top of corolla tube, c 4mm, minutely glandular-punctate, gradually attenuate into long coiled beak 10–14mm with obliquely acute, ± deeply emarginate apex; lower lip c 20 × 30mm, lateral lobes broadly orbicular-reniform, 15–20mm wide, middle lobe much smaller, ovate-triangular, c 7 × 4–6mm, all pilose on margin. Stamens inserted at top of tube; 2 anterior filaments pilose, posterior glabrous; anthers 2.5–4 × 1.5–2mm, elliptic, thecae acute at base. Capsule lanceolate, 25–35 × 8–9mm, apiculate at apex. Seeds straw-coloured with conspicuous blackish appendage, 3.3–4.1 × 1–1.5mm, reticulate with c 30 longitudinal rows (14–15 per face) of lacunae, c 40 per row; lacunae ± circular, with whitish inner margins.

Bhutan: C – Thimphu, Punakha and Tongsa districts, **N** – Upper Mo Chu, Upper Pho Chu, Upper Mangde Chu and Upper Bumthang Chu districts; **Darjeeling:** Richi La; **Sikkim:** Changu, Dzongri, Bikbari etc.; **Chumbi:** Natu La – Champitang, Yatung, Yuo-so etc. Clearings in *Abies* and bamboo forest and scrub, grassy alpine hillsides (often by streams), 2440–4420m. June–September.

A white-flowered plant has been collected from Bhutan (Ceukata, 3383m, *Bedi* 354, K); it was named var. *alba* (without authority) on the label but probably deserves no more than forma rank.

74. P. megalochila Li var. **megalochila** fm. **rhodantha** Tsoong

Perennial; roots slender, tending to become fusiform. Stems erect or ascending, 7–30cm, ± densely villous all round or nearly glabrous. Leaves alternate, clustered near base with several on the stem. Petioles of lower leaves 10–65mm, ± hairy; lamina narrowly oblong in outline, 40–90 × 8–17mm, pinnatipartite with 7–16 pairs of segments; segments oblong-triangular to ovate-triangular, incised-crenate; upper surface ± pilose, lower surface white-furfuraceous especially at first. Inflorescence a showy terminal raceme (½–¾ of plant height), opening from top downwards Bracts leaf-like but uppermost much shorter and more ovate, with flattened petioles. Pedicels 4–10mm. Calyx urceolate-ovoid, broadest at base, 9–13 × 2.5–4mm, split to well beyond middle, tube sparsely to densely white-pubescent or -villous (especially below) and frequently with 2 long dark streaks on anterior side, hairs in material from Bhutan usually dense and at least 1(–2)mm but shorter and sparser in Tibetan material, anterior half with 2 divergent apical lobes 1–1.5mm, posterior with 2 divergent apical lobes 1.8–2.5mm and much smaller linear tooth; lobes stipitate, spathulate-orbicular, subentire or shallowly crenate. Corolla crimson, cherry- or plum-red with darker galea, 27–31mm; tube straight, 14–20mm, slender but with distinctly broadened base, appressed-villous especially above or ± glabrous; galea very strongly falcate (almost forming a ring), enclosed by lower lip, c 14mm, anther-bearing part c 5mm, rather abruptly attenuate into long, curved, slender beak c 9mm with entire, truncate apex; lower lip very large and flag-like, 15–20 × 15–20mm, lateral lobes reniform, c 18mm wide, c twice as wide as spathulate orbicular middle lobe, all lobes ± densely ciliate on margins. Stamens inserted near top of tube; filaments densely hirsute or villous; anthers c 3 × 1.5mm, thecae produced at base into long, slender, curved tails c 1.5mm. Capsule ovoid, c 14 × 6mm, shortly acuminate. Seeds oblong-ellipsoid, straight, c 1.8 × 0.9mm, testa with c 30 longitudinal ridges connected by very fine, indistinct cross-links but not clearly reticulate.

Bhutan: C – Thimphu district (between Pajoding and the Lakes), Tongsa district (Rinchen Chu), **N** – Upper Kuru Chu district (Cang La, Narim Thang), Upper Kulong Chu district (Me La, Shingbe). Grassy alpine pasture and hillsides, often by streams or drippings from cliffs, 3900–4725m. June–August.

Var. *megalochila* fm. *megalochila* and var. *ligulata* Tsoong, both with yellow flowers, are restricted to SE Tibet. Plants of fm. *rhodantha* from Bhutan tend to have much hairier stems and slightly longer, hairier corolla tubes than those from SE Tibet. Plants with pale pink flowers from Tibet, included in fm. *rhodantha* by Tsoong, also differ in their ovate, sharply dentate apical calyx lobes and may represent an additional taxon.

75. P. pauciflora (Maximowicz) Pennell; *P. megalantha* D. Don var. *pauciflora* Maximowicz

Perennial (or possibly annual) with slender vertical root. Stems few, decumbent or ascending (central ones suberect), 4–8cm, sparsely puberulent. Leaves alternate, mainly clustered near base of plant; petioles 10–60mm, sparsely short-pilose throughout; lamina linear-lanceolate to ovate-lanceolate, 10–65 × 5–20mm, deeply pinnatifid to pinnatipartite with 4–8(–13) pairs of segments; segments oblong-ovate, doubly crenate-serrate; both surfaces white-furfuraceous at least when young, upper surface also sparsely pilose. Inflorescence of 3–10 axillary flowers in lax raceme. Bracts ovate, pinnatifid, on long pilose petioles, longer than calyx. Pedicels 4–10mm, shortly pubescent. Calyx obovoid-tubular with distinctly narrowed base especially in flower, 13–16 × 4–6mm, 10-veined, sparsely white-pilose especially on veins; mouth with 5 apical lobes, posterior one oblanceolate, others ovate, rather sharply crenate-serrate. Corolla purple throughout, 50–90mm; tube 35–75mm, 1–2mm diameter, straight, 4½–6 × calyx, sparsely pilose; galea erect in centre of lower lip, strongly falcate, forming a wide arc but not a nearly complete ring, c 20mm, anther-bearing part ± sessile on top of corolla tube, c 4mm, produced into slender beak c 15mm obliquely acute and emarginate at apex; lower lip 10–16 × 15–25mm, lateral lobes reniform, c 14mm wide, middle one ovate-oblong, c 8 × 5mm, all pilose on margin. Stamens inserted at top of corolla tube; all filaments villous; anthers oblong, c 2.5 × 1.2mm, thecae acute at base. Capsule lanceolate, 20–25 × 5–6mm, shortly apiculate at apex.

Sikkim: Dzongri, Kankergang; **Chumbi:** Phari. Grassy alpine slopes, 3350–4270m.

76. P. woodii R.R. Mill (378)

Perennial. Stems several, erect or ascending, 3–8cm, unbranched, shortly adpressed-pubescent all round, Leaves mainly radical with few alternate cauline; radical leaves numerous (12–25 per tuft), petioles 20–40mm, narrowly winged, shortly pubescent, lamina linear-oblong in outline, (15–)25–40 × 3–7mm, pinnatipartite with usually 7–15 pairs of segments; segments ovate-triangular, lowest pairs smaller and indistinct, more distant than upper pairs, each segment crenate-serrate; segments with long creamy caducous hairs when young, finally glabrous, also white-furfuraceous beneath when young; midrib with persistent dense short hairs on upper surface, glabrous and pale beneath. Inflorescence a few-flowered terminal raceme. Bracts leaf-like. Pedicels indistinct in flower, to c 10mm in fruit and then glabrous. Calyx narrowly urceolate-tubular, 15–17mm, split to c ⅓; tube membranous, 10-veined, sparsely spreading-villous on veins; mouth with 5 small oblong-ovate incised-dentate apical lobes c 1–1.5mm which tend to blacken and roll back when dry (especially the 3 anterior ones). Corolla concolorous but galea slightly darker, pale or deep pink, 35–50mm; tube 25–30mm, slender (c 2mm diameter near middle, widening slightly upwards), straight, c twice calyx, pilose; galea erect in centre of lower lip, strongly twisted, forming a wide arc but not an almost complete ring, c 14mm, anther-bearing part almost sessile on corolla tube, c 6mm, minutely glandular-punctate, gradually attenuate into coiled beak 6–7mm with obliquely acute, emarginate apex; lower lip 20–25 × 25–30mm, lateral lobes broadly orbicular-reniform, 8–10 × 14–18mm, middle lobe smaller, broadly obovate, 10–12 × 8–10mm, margins undulate and glabrous or with extremely sparse minute hairs. Capsule narrowly ovoid, c 20 × 5–6mm, shortly acuminate. Seeds not seen.

Bhutan: S – Sankosh district (Lawgu summit), **C** – Thimphu district (Dungtsho La). Rock-crevices, moist mountain cliffs, and open rocky moorland, 4400–4500m. August.

Endemic to Bhutan.

37. BRANDISIA Hook.f. & Thomson

Erect shrub with stellate tomentose indumentum. Leaves opposite or subopposite, shortly petiolate Flowers solitary or rarely in pairs, axillary, pedicellate; pedicels with 2 bracteoles. Calyx campanulate, 2-lipped, without prominent veins. Corolla bilabiate, infundibular, pink or rarely yellow. Stamens 4, didynamous; staminodes absent; anthers with long hairs at tip and margins. Ovary bilocular, hairy, with numerous ovules. Fruit a loculicidally dehiscent 2-valved capsule. Seeds many, linear, with thin membranous winged reticulate testa; endosperm absent.

The above description applies to the single species occurring in Bhutan, which is anomalous and has been placed in its own monotypic subgenus, subgen. *Rhodobrandisia* Li. For an account of the genus, see Li (366), who postulated a relationship with *Wightia* and *Paulownia*.

1. B. rosea W.W. Smith

Shrub, 60cm–2.5m. Young twigs rather densely white stellate-tomentose, older branches much more sparsely so. Leaves shortly petiolate (petioles 5–10mm), lamina lanceolate-elliptic, 30–100 × 8–30mm, acuminate, base attenuate or cuneate, margin remotely and finely denticulate or subentire, upper surface dark green (drying blackish), glabrous and rather glossy, lower surface densely grey stellate-tomentose. Pedicels 6–12mm, stellate-tomentose. Calyx 2-lipped, c 9mm, stellate-tomentose especially near base, margins and apex. Corolla deep rose-pink, greyish-brown, orange-yellow or pale yellow, 22–26mm, outside of tube shortly stellate-tomentose, lips hairy. Anthers densely bearded with long stout yellowish hairs to c 1.5mm. Capsule 10–18 × 7–11mm, dark blackish-brown, shortly acuminate; valves sulcate, stellate-hairy along sulcus.

Bhutan: C – Tashigang district (Balfi), **N** – Upper Mo Chu district (Mo Chu). Open, ± shady thickets and scrub, 2135–3353m. July–October.

The description and ecological notes are mainly based on material from Yunnan (China) and eastern Tibet. Two varieties occur in Bhutan. Var. **rosea** with rose-pink to yellowish-red flowers has been recorded from Tashigang. The yellow-flowered variants which occasionally occur (e.g. *Bowes-Lyon* 5180 from Mo Chu) have been distinguished as var. **flava** C.E.C. Fischer, described from Assam.

38. WIGHTIA Wall.

Evergreen, hemi-epiphytic tree or pseudo-liana. Branches scandent or ± pendulous, lenticellate. Leaves opposite, decussate, exstipulate, petiolate, entire, very coriaceous. Flowers in axillary thyrses; peduncles, pedicels and calyx all densely but minutely stellate-tomentose. Peduncles trichotomous. Pedicels with 2 small bracteoles near middle. Calyx campanulate, irregularly 3–5-lobed. Corolla large; tube incurved near base, dilated above; upper lip 2-lobed, erect, lower 3-lobed, reflexed or spreading. Stamens 4, didynamous, long-exserted, inserted near base of corolla; anthers sagittate, thecae parallel, dehiscence ± introrse. Style long, with incurved tip. Ovary glabrous. Fruit a 2-valved oblong or oblong-lanceolate septicidal capsule; valves separating from placentiferous axis. Seeds numerous, linear, with membranous broadly winged testa; endosperm absent.

For a detailed account of this genus and its disputed systematic position, see Maheshwari (376). Some authors (including both its discoverer, Wallich, and Maheshwari) assign it to Bignoniaceae, along with the closely allied genera *Paulownia* Sieber & Zuccarini and *Brandisia* Hook.f. & Thomson which together form Scrophulariaceae (or Bignoniaceae) tribe Paulownieae.

1. W. speciosissima (D. Don) Merrill; *Gmelina speciosissima* D. Don, *W. gigantea* Wall. Nep: *Lakori*, *Bauni-kath*, *Bop-kung*, *Zeru-kung*. Fig. 101a–b.

Small or usually large tree (3–)6–15m, often hemi-epiphytic and climbing up trunks of other trees by means of aerial roots arising from trunk; trunk often stout, sometimes exceeding 30cm diameter. Bark black; wood soft. Leaves with short stout petiole 1.5–3.5cm; lamina elliptic or ovate-elliptic, 10–28 × 7–19cm, acute or subacute, base obtuse or rounded, glabrous above, stellate-tomentose beneath. Thyrses borne on leafless shoots, 6–16cm, many-flowered; peduncles 6–8cm or more. Calyx 6–9mm, dull green tinged purple near base, stellate-tomentose outside and less densely towards apex of lobes within. Corolla white or very pale pink at first changing to bright, deep purplish-pink with paler tube and green throat, 25–35mm; tube very densely stellate-tomentose outside except near base, almost glabrous inside; lobes stellate-tomentose both outside and inside. Stamens much exserted; filaments whitish, anthers pinkish-brown, c 3mm. Style pale yellow or pinkish, c 40mm. Capsule oblong or linear-oblong, 25–40 × c 10mm.

Bhutan: C – Punakha district (near Khelekha) and Tashigang district (Jiri Chu Valley and Yonpu La); **Sikkim:** Ramom Valley etc.; **Darjeeling**. Warm broad-leaved forest and hot open hillsides, (?900–)1220–2100m; often epiphytic on trees or rocks. October–November.

The wood is soft and is used by priests to make temple idols.

Family 174. BIGNONIACEAE

by E. Aitken

Trees, shrubs and woody climbers, rarely herbs. Leaves opposite, decussate or alternate, rarely in whorls or rosettes, pinnately compound, 3-foliate with

FIG. 101. **Scrophulariaceae and Bignoniaceae. Scrophulariaceae.** a–b, *Wightia speciosissima*: a, leaf (× ⅓); b, flower (× 1½). **Bignoniaceae.** c–g, *Stereospermum colais*: c, leaf and inflorescence (×⅓); d, flower (× 3); e, flower with part of calyx and corolla cut away (× 3); f, fruit (× ¼); g, seed (× 7). h, *Incarvillea himalayensis*: habit (× ½). Drawn by Glenn Rodrigues.

c
f
g
e
d
h
b
a

branched or simple tendril often replacing terminal leaflet or undivided; glands often present at the base of the petiole. Flowers bisexual, zygomorphic, 5-merous, in panicles, cymes or solitary, terminal or on short lateral branches; bracts inconspicuous. Calyx campanulate, gamosepalous, spathaceous or 2–5-lobed. Corolla campanulate or tubular, with 2-lipped mouth or 5 slightly unequal lobes. Fertile stamens 4, occasionally reduced to 2 (5 in *Oroxylum*), missing stamens usually replaced by staminodes. Anthers each with 2 divergent locules, dehiscing by longitudinal slits. Ovary superior, 2-celled, subsessile, with nectariferous disc; style filiform; stigma 2-lobed, elliptic. Fruit an elongate capsule; seeds flat or ridged, prominently winged.

1. Herbaceous plants .... **1. Incarvillea**
+ Trees, shrubs or woody climbers 2

2. Leaf margins coarsely serrate .... **5. Tecoma**
+ Leaf margins entire 3

3. Climbers .... **7. Pyrostegia**
+ Trees or shrubs 4

4. Leaflets small, up to 2.5 × 0.5cm, capsule oblong or ovoid .. **6. Jacaranda**
+ Leaflets large, 2.5–20 × 1–9cm, capsule linear, elongate 5

5. Corolla 6.5–12.5cm; fertile stamens 5; capsule flattened, septum not thickened .... **4. Oroxylum**
+ Corolla 2–6.5cm; fertile stamens 4; capsule not flattened; septum thickened 6

6. Leaves 1-pinnate, leaflets 5.5–13.5(–20)cm; septum 0.8–1.5cm diameter; seeds ridged .... **2. Stereospermum**
+ Leaves 2-pinnate, leaflets 2.5–8cm; septum 3–4mm diameter; seeds not ridged .... **3. Radermachera**

1. INCARVILLEA Jussieu

Annual or perennial herbs. Roots simple or branched, woody or tuberous. Stems sometimes absent. Basal leaves in rosette, or stem leaves alternate, rarely opposite, without stipules, pinnate or pinnatisect, rarely undivided, puberulous. Flowers 5-merous, terminal, solitary or racemose, occasionally paniculate, zygomorphic. Pedicels subtended by a short bract. Calyx tube campanulate, lobes ovate, lanceolate, subulate or reduced to minute points. Corolla tube cylindrical at base, campanulate above, lobes subequal, rounded or emarginate. Stamens 4, in pairs, 2 long, 2 short, anthers 2-lobed, each with a single bristle; staminode absent or variable. Stigma horizontally elliptic. Capsule cylindrical, quadrangular or 6-winged. Seeds subdiscoid or ovate, surrounded by a wing or with fine hairs at each end.

In addition to the species with deep pink flowers described below, *I. longiracemosa* Sprague [*I. lutea* Bureau & Franchet subsp. *longiracemosa* (Sprague) Grierson] with yellow flowers may occur in Chumbi. It was described from a collection made between Phari and Shigatse, but all specimens seen from more precise localities were collected north of our area.

1. Lateral leaflets in 0–5 pairs; terminal leaflet 15–70mm; corolla tube 40–65mm .... **1. I. himalayensis**
\+ Lateral leaflets in 3–7 pairs; terminal leaflet 4–5mm; corolla tube 25–30mm **2. I. younghusbandii**

1. I. himalayensis Grey-Wilson; *I. mairei* (Léveillé) Grierson var. *grandiflora* (Wehrhahn) Grierson p.p. Fig. 101h.

Perennial, stems absent, leaves arising in rosette from fleshy root-stock. Leaves pinnate; lateral leaflets in 0–4 pairs, ovate, 7–40 × 6–20mm, acute, serrulate, ± sessile or petiolules c 1mm; terminal leaflet orbicular, (15–)25–40(–70) × 30–45(–65)mm; petiole (15–)40–100mm. Peduncles 22–28(–45)mm, unbranched or branching near base. Calyx tube 10–15mm; lobes ovate to lanceolate, 3–6 × 2–6mm, acuminate. Corolla deep pink, base of tube yellow or orange, throat with white streaks; tube 40–65mm; lobes orbicular, 10–15(–20)mm. Capsule woody, 4-angled in section, curved, 45–70 × 5–10mm. Seeds subdiscoid or ovate, densely covered in thick glandular hairs, 5–7 × 5mm including narrow wing.

Bhutan: N – Upper Pho Chu district (Chojo Dzong, Thanza). Mountain pastures, 3960–4100m. June–July.

2. I. younghusbandii Sprague

Similar in habit to *I. himalayensis* but smaller; lateral leaflets in 3–7 pairs, lanceolate, 3–5 × 1–2mm, terminal leaflet 4–5 × 6–7mm; petiole 5–10mm; calyx tube 4–5mm, lobes lanceolate, 2–3 × 1–1.5mm; corolla dark rose-pink, tube 25–30mm, lobes 5–7 × 6–9mm; capsule (not seen) curved, 25–30 × 8–9mm, narrowing towards apex.

Chumbi: Phari–Tuna (325). Open hillsides, 3000–4500m. June–July.

Collected several times immediately north of Sikkim and Chumbi.

2. STEREOSPERMUM Chamisso

Trees 8–20m. Leaves 1-pinnate (in Himalaya), entire or dentate. Flowers in large, lax terminal or axillary panicles. Calyx ovoid in bud, campanulate with short, rounded, unequal lobes when open. Corolla tubular-ventricose or subcampanulate, narrowing abruptly in the lower part, white with purple streaks, or crimson, lobes ovate, spreading, nearly equal, toothed or laciniate. Stamens 4, in pairs; staminode usually rudimentary. Capsule elongate, terete, slightly compressed or quadrangular, with small tubercules on outer surface; septum thickened, subterete, corky, transversely notched. Seeds with flat, oblong, membranous wing on each side; embryo with ridge embedded in notch in septum; wings of adjacent seeds not overlapping.

1. Young leaves and panicles ± glabrous; apex of leaves caudate-acuminate; capsule twisted, 0.8–1.2cm diameter .... **1. S. colais**
\+ Young leaves and panicles glandular or pubescent; apex of leaves shortly acuminate; capsule straight, 1.5–2cm diameter .... **2. S. chelonoides**

1. S. colais (Dillwyn) Mabberley; *S. personatum* (Hasskarl) Chatterjee, *S. chelonoides* sensu Clarke non (L.f.) DC., *S. tetragonum* A. DC. Nep: *Parari.* Fig. 101c–g.

Young leaves and panicles ± glabrous or sparsely, finely puberulous. Mature leaflets elliptic, 5.5–13.5(–20) × 2–6(–7)cm, caudate-acuminate, both surfaces glabrous or rarely finely puberulous beneath; petiolule conspicuous, 0.3–1.2(–2)cm. Calyx glabrous, 0.5–1cm. Corolla white with purplish or crimson markings; tube 1.2–1.8cm, glabrous outside, densely tomentose within; lobes 0.8–1 × 0.5–0.7cm. Capsule 30–80 × 0.8–1.2cm, prominently ridged, twisted when mature; septum c 8mm diameter. Seeds 5–6 × 13mm including wing; embryo c 5 × 2mm.

Bhutan: S – Phuntsholing and Sankosh districts, **C** – Tongsa, Mongar and Tashigang districts; **Darjeeling:** Rongtong, Sivok etc. On hillsides or river banks in subtropical forests, 150–1500m. May–July.

2. S. chelonoides (L.f.) DC.; *S. suaveolens* (Roxburgh) A. DC. Nep: *Parari.*

Similar to *S. colais* but young leaves and panicles densely glandular-tomentose. Mature leaflets broadly elliptic or broadly ovate, 10–15 × 6 × 9.5cm, shortly acuminate, ± glabrous above, pubescent beneath; petiolule stout or subsessile, 2–3mm. Calyx 0.6–0.8cm, tomentose. Corolla red; tube 1.8–2.2cm; lobes 0.5–1.2 × 0.4–0.9cm, pubescent outside, densely tomentose within. Capsule 35 × 1.5–2cm, not twisted, not prominently ridged; septum 1.2–1.5cm diameter. Seeds 7–9 × 20–25mm including wing; embryo 7–9 × 1–2mm.

Darjeeling: Simalbhutia, Katambari etc. Terai. Fruiting November.

The bark and roots are used for coughs and renal complaints.

3. RADERMACHERA Zollinger & Moritz

Similar to *Stereospermum* but leaves 2-pinnate (in Himalaya) and seeds in several overlapping rows, embryo flat, not embedded in septum of capsule.

1. R. sinica (Hance) Hemsley; *R. borii* C.E.C. Fischer, *Stereospermum sinicum* Hance

Leaflets glabrous, elliptic to lanceolate, 2.5–8 × 1–3cm, caudate-acuminate; petiolule 3–10mm. Calyx c 2.5cm, swollen and thickened at base, 5-ribbed and 3-lobed above; tip of ribs excurrent; lobes 3, unequal, ovate, obtuse, 4–8mm. Corolla white, funnel-shaped; tube 4–7cm; lobes 5, 1.5 × 2cm, crenate. Capsule 45–60 × 0.4–0.5cm, not ridged; septum 0.3–0.4cm diameter. Seeds 2–3 × 7–12mm including wing; embryo 2–3 × 3–4mm.

Bhutan: S – Chukka district (Marichong), **C** – Punakha district (Rinchu), Tongsa district (Mangde Chu, Shemgang), Tashigang district (Shali); **W Bengal Duars:** Buxa. On river banks in forests, 750–1675m. May.

Flowers not seen, details taken from materials from NE India and Myanmar (Burma), including the original description of *R. borii. Griffith* itin. no. 87 from Deothang (Bhutan) is likely to be this species, based on his notes (66).

4. OROXYLUM Ventenat

Glabrous trees. Leaves opposite, 2–3-pinnate; leaflets ovate, entire. Flowers in long terminal racemes. Calyx campanulate, leathery, truncate or obscurely toothed. Corolla campanulate-ventricose, creamy white or purplish, lobes subequal. Stamens 5, fertile, unequal. Capsule linear, compressed, 2-valved. Seeds flat, orbicular, with broad, oblong, papery wing.

1. Oroxylum indicum (L.) Ventenat; *Bignonia indica* L., *Calosanthes indica* (L.) Blume. Nep: *Totilla, Totola* (66).

Trees 4–6m. Leaflets 6.5–13 × 3.5–6.5cm, acute or apiculate; petiolules 5–20mm. Peduncle c 1m, erect. Flowers opening after sunset. Pedicels up to 7cm. Calyx 1.5–2.5cm. Corolla tube reddish-purple outside, yellow within, 4.5–8cm; lobes creamy, suborbicular, 2–4.5 × 2–4.5cm, reflexed, narrowing at base, margins crenate and irregularly lacinulate. Stamens 3–4.5cm, tomentose at base. Capsule 50–62 × 4–9cm. Seeds 3.5–4 × 6.5–9cm, including papery wing; embryo 1.5–2cm diameter.

Bhutan: S – Phuntsholing district (Phuntsholing), Gaylegphug district (Tori Bari), Deothang district (Samdrup Jongkar), **C** – Punakha district (Rinchu); **Darjeeling:** unlocalised (66); **Darjeeling/Sikkim:** Tista. In moist forest, 200–1525m. May–June.

5. TECOMA Jussieu

Shrubs or small trees. Leaves opposite, imparipinnate or digitate. Leaflets serrate. Flowers 5-merous, in panicles or terminal racemes. Calyx tubular-campanulate; lobes subequal. Corolla tubular or campanulate to narrowly infundibuliform, glabrous, base narrowing very abruptly, yellow, orange or red; lobes subequal. Fertile stamens 4, one long and one short pair; staminode 1, short. Capsule linear, woody, compressed, 2-valved, dehiscing perpendicular to septum. Seeds ovate or discoid, with broad papery wing.

1. Tecoma stans (L.) Humboldt, Bonpland & Kunth; *Stenolobium stans* (L.) Seeman, *Bignonia stans* L.

Shrub or small tree, 2–4m. Leaves 1–2-pinnate; leaflets lanceolate, 3–9 × 0.8–2.5cm, apex acuminate, margins deeply serrate, sparsely pilose beneath on midrib; petiolule 0–1cm. Pedicels 4–8mm. Calyx tube 4–5mm; lobes triangular, c 1mm, acuminate, margins finely glandular-pubescent. Corolla yellow, campanulate, sweet-scented; tube 4–4.5cm; lobes suborbicular, 1 × 1–1.5cm, overlapping, margin entire. Stamens 1.5–2cm. Capsule 16–19 × 0.5–1cm, tapering at ends; septum flat; seeds ovate, 22–25 × 4–6mm, including oblong wing; embryo 6–8 × 4–5mm.

Bhutan: S – Phuntsholing district. On roadside bank, 200m. May.

Cultivated, native of S America.

6. JACARANDA Jussieu

Trees. Leaves opposite, 2-pinnate. Flowers in large terminal panicles. Calyx short, campanulate. Corolla pubescent to glabrous, campanulate, narrowly tubular at base, mauve or blue, lobes ovate to suborbicular. Fertile stamens 4, one pair shorter than the other, glandular-pubescent at base; staminode 1, much longer, variously glandular-pubescent especially at apex. Capsule oblong or broadly ovate, woody, compressed, dehiscing perpendicular to septum; seeds obovate, flat with broad, irregularly discoid, papery wing.

1. Jacaranda mimosifolia D. Don; *J. ovalifolia* R. Brown

Trees 6–12m. Leaf raches finely winged; leaflets elliptic to lanceolate, sessile, 0.3–2.5 × 0.2–0.5cm, cuspidate, sparsely glandular-pubescent on upper surface, glabrous and with prominent vascular pattern beneath, margins revolute. Pedicels 3–6mm, glandular. Calyx 2–3mm, including triangular lobes, c 0.5–1mm, densely glandular-pubescent. Corolla densely glandular-pubescent, mauve, whitish within spreading, 2-lobed upper lip; tube 3.5–4.5cm; lobes 4–9 × 5–10mm. Fertile stamens 10–11mm, anthers with one fully formed cell, the other reduced to a minute appendage; staminode 25mm, with large tufts of glandular hairs at middle and apex. Capsule (not seen, measurements from non-Bhutan specimen) 5–8 × 4.5–6cm; seeds 1–1.5 × 2–2.5cm, including wing; embryo 7–9 × 5–7mm.

Bhutan: S – Phuntsholing district (Phuntsholing). In gardens, 200–250m. April.

Cultivated, native of tropical S America.

7. PYROSTEGIA Presl

Woody climber. Leaves opposite, 3-foliate, leaflets ovate, acuminate, undulate, branched or simple tendril often replacing terminal leaflet. Flowers in terminal and axillary corymbose fascicles. Calyx campanulate, lobes reduced to minute points. Corolla tubular, narrowing towards base, slightly curved upwards, 2-lipped, upper lip 2-lobed, lower lip 3-lobed. Stamens 4, shorter pair reaching to base of corolla lobes, longer pair to tip of lobes; staminode short, linear, included. Fruit (not seen) a 2-valved capsule. Fruit not usually produced outside native country.

1. P. venusta (Ker Gawler) Miers; *Bignonia venusta* Ker Gawler. Eng: *Golden shower*.

Stem ridged. Leaflets glabrous, 3–5 × 2–4cm; petiole 2–3cm; petiolules 0.5–2.5cm. Pedicels 0.7–2cm. Calyx tube 0.5–0.6cm; lobes less than 1mm, margin puberulous. Corolla orange, tube 4–5cm; upper lobes 0.7–0.8cm, lower lobes 1.7–2cm, oblong, margin and apex villous. Filaments 2.5–4cm, attached at middle of corolla tube; anthers 4–5mm. Style 5–6cm.

Bhutan: S – Phuntsholing district (Phuntsholing). On roof of hut in town, 200m. January–February.

A very common cultivated plant in all tropical countries. Native of Brazil.

Family 175. ACANTHACEAE

by J.R.I. Wood

Herbs or shrubs, rarely climbers (*Thunbergia*). Leaves opposite, decussate, simple, without stipules; cystoliths commonly present on leaves and other vegetative parts, appearing as white streaks. Flowers usually in cymes, racemes or spikes but sometimes solitary or in axillary whorls, bisexual, often zygomorphic. Bracts and usually bracteoles present, bracts often prominent and concealing calyx, sometimes arranged in 3–4 levels especially in those genera with axillary inflorescences. Calyx 4–5-lobed, lobes usually similar in genera with 5 lobes but often dissimilar in those with 4 lobes such as *Barleria*; much reduced in *Thunbergia*. Corolla usually 5-lobed but often 2-lipped with lower lip 3-lobed and upper lip notched or (in *Acanthus*) absent. Stamens attached to corolla, usually 4 in two dissimilar pairs, sometimes 2 with second pair reduced to staminodes; anthers 1–2-celled, sometimes spurred at base, cells at same level or one below the other; pollen very varied in appearance and often distinctive for a particular genus. Ovary superior, 2-celled; ovules 2–numerous. Style 1; stigma 2-lobed, posterior lobe often reduced and apparently absent. Fruit a loculicidal capsule, usually cylindrical or clavate in shape (globose and beaked in *Thunbergia*); seeds usually borne on short hook-like retinacula which eject the seed when ripe capsule opens. Seeds usually flattened, lens-shaped, glabrous or hairy, sometimes with elastic mucilagenous hairs.

A tropical family well represented in the warm subtropical valleys and foothills. Most species flower in the dry winter season.

1. Climbing plants; leaves palmately veined; calyx reduced to a ring; capsule with long beak .... **1. Thunbergia**
+ Herbs or shrubs, never climbing; leaves pinnately veined; calyx 4–5-partite; capsule oblong or clavate, never beaked 2

2. Calyx 4-lobed; plants often spiny 3
+ Calyx 5-lobed; plants rarely spiny 6

3. Leaves with spiny margins; corolla with a tube and a single (lower) lip; upper lip absent .... **12. Acanthus**
+ Leaves not spiny; corolla 2-lipped or 5-lobed 4

4. Calyx with 2 large and 2 small lobes .... **13. Barleria**
+ Calyx with 3 similar lobes and 1 larger one 5

5. Plant with axillary spines .... **4. Hygrophila**
+ Plant unarmed .... **2. Nelsonia**

6. Fertile stamens 4 7
+ Fertile stamens 2 16

7. Corolla distinctly 2-lipped 8
+ Corolla subequally 5-lobed 9

8. Capsule 4-seeded; flowers in dense 1-sided heads .... **19. Lepidagathis**
+ Capsule with 20 or more seeds; flowers in the leaf axils or in terminal bracteate spikes .... **4. Hygrophila**

9. Corolla less than 1cm 10
+ Corolla more than 1.2cm 11

10. Bracts orbicular; capsule 4-seeded, the seeds borne on hooked retinacula; flowers in short, dense, 1-lipped, bracteate heads .... **10. Phaulopsis**
+ Bracts linear-spathulate; capsule with 15 or more seeds, the seeds borne on short papillae; flowers in elongate terminal spikes .... **3. Staurogyne**

11. Anther cells spurred at the base; leaves entire 12
+ Anther cells not spurred at the base; leaves toothed or (rarely) entire .. 13

12. Capsule 4-seeded, with a long woody base; flowers in a terminal raceme **14. Asystasia**
\+ Capsule 8–16-seeded from the base; flowers in small cymes **5. Echinacanthus**

13. Capsule 4-seeded or capsule not available .... see **9. Strobilanthes**
\+ Capsule 8–16-seeded 14

14. Herb; corolla less than 1.5cm .... **8. Hemigraphis**
\+ Undershrub; corolla more than 2cm 15

15. Flowers in 1-sided axillary cymes; capsule 12–16-seeded ..... **7. Clarkeasia**
\+ Flowers in clusters either in leaf axils, or in racemes or on the branches of an open panicle; capsule 8-seeded .... **6. Aechmanthera**

16. Corolla subequally 5-lobed 17
\+ Corolla distinctly 2-lipped 21

17. Bracts conspicuous, ovate-elliptic; anthers usually exserted 18
\+ Bracts inconspicuous, linear-lanceolate; anthers included 19

18. Bracts conspicuously veined, at least 1cm; corolla blue or red, tube at least 1.5cm .... **11. Eranthemum**
\+ Bracts inconspicuously veined, less than 1cm; corolla white, tube less than 1cm .... **9. Strobilanthes**

19. Corolla tube narrowly cylindrical, c 3.5cm .... **15. Pseuderanthemum**
\+ Corolla tube strongly ventricose, less than 3cm 20

20. Shrub. Corolla elongate, 2.5–3cm .... **16. Mackaya**
\+ Herb. Corolla campanulate, c 1cm .... **17. Codonacanthus**

21. Inflorescence of small axillary cymes, flowers enclosed by pair of bracts; corolla purple (very rarely white) 22
\+ Inflorescence not as above, flowers not enclosed by pair of bracts; corolla variously coloured 24

22. Corolla 3–5cm .... **26. Peristrophe**
\+ Corolla less than 2.5cm 23

23. Anthers 2-celled; placenta rising up from base of ripe capsule **25. Dicliptera**
\+ Anthers 1-celled; placenta not rising up from base of ripe capsule **27. Hypoestes**

24. Corolla white; tube long, narrowly cylindrical, 3–4cm; upper lip entire, acuminate, c 1cm .... **23. Rhinacanthus**
\+ Corolla variously coloured and shaped but tube less than 3cm and upper lip notched or 2-lobed, rounded to acute, less than 1cm 25

25. Capsule 10 or more-seeded; stamens usually exserted .... **18. Phlogacanthus**
\+ Capsule 4-seeded; stamens included 26

26. Corolla large, curved, upper lip formed of 2 separate, nearly equal lobes, similar to the 3 lobes of lower lip .... **16. Mackaya**
\+ Corolla large or small, strongly 2-lipped, upper lip notched at the tip .. 27

27. Inflorescence an open lax panicle with flowers scattered on the branches **22. Isoglossa**
\+ Inflorescence of spikes, sometimes aggregated into a panicle 28

28. Bracts prominent, ovate or elliptic, at least 1cm .... **20. Justicia**
\+ Bracts small, linear, less than 1cm 29

29. Placenta rising up from base of ripe capsule .... **24. Rungia**
\+ Placenta not rising up from base of ripe capsule 30

30. Corolla white, less than 1cm, upper lip hooded .... **21. Leptostachya**
\+ Corolla variously coloured but if less than 1cm, pink and upper lip not hooded .... **20. Justicia**

1. THUNBERGIA L.f.

Climbing herbs and shrubs. Leaves petiolate, ovate or elliptic, usually cordate or hastate, entire or sinuate lobed, very variable in shape within individual species. Flowers pedicellate, solitary or paired in leaf axils or in racemes with small leaf-like bract at base of each pedicel. Bracteoles paired, prominent, enclosing calyx and corolla tube. Calyx a small ring. Corolla large, funnel-shaped, tube with short cylindrical base then widened, limb with 5 nearly equal lobes. Stamens 4; anthers 2-celled, cells at same height, spurred or not at base, sometimes bearded. Capsule globose with conspicuous, woody beak, 4-seeded; seeds compressed to spherical, glabrous.

1. Calyx distinctly toothed; most leaves less than 5cm; flowers axillary 2
\+ Calyx entire or crenulate; leaves at least 5cm, usually much more; flowers axillary or in racemes 3

2. Corolla white; petiole not winged .... **1. T. fragrans**
\+ Corolla yellow or white with a dark centre; petiole winged ..... **2. T. alata**

3. Corolla red; flowers in long, many-flowered, pendent racemes **5. T. coccinea**
+ Corolla white, bluish or yellow; flowers axillary or in few-flowered racemes which are not pendent 4

4. Corolla yellow, less than 4cm; leaves cuneate, decurrent onto petiole **3. T. lutea**
+ Corolla white or bluish, 5–9cm; leaves hastate, not decurrent at base **4. T. grandiflora**

1. T. fragrans Roxb.

Slender twining herb to 1.5m, usually roughly pilose but occasionally almost glabrous. Leaves narrowly ovate, 3–5(–11) × 1–3cm, usually hastate (rarely cordate or rounded at base), entire to unevenly sinuate; petiole 5–33mm. Flowers solitary or paired from leaf axils; pedicels 3–4.8cm. Bracteoles ovate to triangular, pale green, 20–28 × 5–15mm. Calyx a 12–16-toothed pubescent ring, teeth c 3mm. Corolla pure white, often fragrant, thinly pilose, tube 2.5–3cm, cylindrical for most of its length but widened just below mouth, lobes 1.5–2cm. Anther cells not spurred. Capsule 2.5–3cm, pubescent (rarely glabrous); seeds spherical, rugose.

Bhutan: C – Punakha district (20km S of Wangdu Phodrang), Mongar district (Saling). **Darjeeling:** without exact locality (34). Bush-land in deep, dry valleys, 1000–1300m. July–October.

2. T. alata Bojer

Very similar to *T. fragrans* but distinguished by its winged petiole and yellow (rarely white) corolla with dark centre.

W Bengal Duars: Buxa. A garden plant escaped and sometimes naturalised, 600m.

3. T. lutea Anderson

Vigorous climbing plant with stems up to 4m, glabrous or thinly pubescent except for prominent tuft of hairs at nodes. Leaves very thinly pilose on both surfaces, ovate to broadly elliptic, cuneate at base and usually shortly decurrent onto petiole, 9–19 × 3–15cm, small leaves entire, larger leaves repand or dentate; petiole 1–5cm. Flowers solitary in leaf axils, occasionally subracemose near tips of branches; pedicels 5–8cm. Calyx a glabrous, crenate ring, c 1mm deep. Corolla yellow, glabrous, tube ventricose from near base, 25–30mm, lobes very short, rounded, c 8mm. Anther cells with green, papillose basal appendage. Capsule c 3.5cm, glabrous; seeds spherical.

Darjeeling: Darjeeling, Manibhanjan, Chimli, Pankhashari, Pasting etc. Locally frequent in hill forest and in forest relics, 1800–2500m. July–October.

Endemic to Darjeeling district and eastern Nepal.

4. T. grandiflora Roxb.; *T. clarkei* Yamazaki. Nep: *Kanesi* (34).

Vigorous climbing or scrambling plant with stems several metres long; very variable in indumentum from glabrous to tomentose. Leaves ovate-triangular, rarely suborbicular, 5–19 × 4–14cm, acute or shortly acuminate, usually hastate, occasionally cordate and shortly cuneate at base, sinuate with few large teeth; petiole 2–6cm. Flowers borne on solitary pedicels 4–10cm from leaf axils on mature stems or in short bracteate racemes at tips of new branches; racemes 3–15cm with linear bracts c 5mm and pedicels 1–5cm. Bracteoles oblong-elliptic, 3–5cm, acute to apiculate. Calyx an entire glabrous ring, c 1mm. Corolla white or bluish (yellowish inside tube), 6–9cm, glabrous, tube broad, ventricose, about half as long as spreading limb. Anther cells with thin basal spur. Capsule c 3.5cm, glabrescent; seeds flattened.

Bhutan: S – Samchi district (Charmurchi River), Phuntsholing district (Phuntsholing, Phuntsholing–Rinchending), Gaylegphug district (Mao Valley), Deothang district (Samdrup Jongkhar–Deothang); **Darjeeling:** Great Rangit and Tista Valleys; **Sikkim:** Rangpo; **W Bengal Duars**. In dense subtropical scrub in gullies and on steep slopes, 300–800m. May–August.

5. T. coccinea D. Don. Nep: *Kanesi* (34). Fig. 102a–f.

Vigorous twining shrub with stems climbing to 12m and then pendulous. Stem variable in indumentum; glabrous, pubescent or setose especially near nodes. Leaves ovate-triangular, 4–18 × 3–10cm, acute, cordate to hastate at base, weakly sinuate with prominent teeth, glabrescent to thinly pilose; petiole 2–9cm. Flowers in long, pendulous, terminal racemes 10–50cm; bracts sessile, ovate, 0.5–3(–8)cm; pedicels (1–)2(–5)cm. Bracteoles ovate-elliptic, sometimes falcate, dark red, 1.8–3cm, acute or apiculate, densely puberulent, especially on margins. Calyx a minute rim. Corolla orange-red or crimson, glabrous or finely puberulent, tube widened from just above base, 22–28mm, lobes 8–11mm. Anthers with some cells spurred and some not spurred. Capsule c 4cm, glabrous; seeds flattened, rugose.

Bhutan: S – Samchi, Phuntsholing, Chukka, Gaylegphug and Deothang districts, **C** – Punakha, Tongsa, Mongar and Tashigang districts; **Darjeeling:** Darjeeling, Labha, Rangirun, Kurseong etc.; **Sikkim:** Karponang–Gangtok; **W Bengal Duars:** Sarugata (175). Common in moist hill forest, occurring wherever suitable habitats occur, 700–2200m. July–November.

FIG. 102. **Acanthaceae.** a–f, *Thunbergia coccinea*: a, leaf-pair (× ½); b, inflorescence (× ⅓); c, bract (× 1⅓); d, bracteole (× 1⅓); e, calyx (× 2); f, dehiscent capsule (× ½). g–k, *Nelsonia canescens*: g, habit (× 1); h, bract (× 4); i, calyx (× 7); j, dissected corolla (× 4); k, capsule valves (× 4). l–n, *Echinacanthus attenuatus*: l, habit (× ½); m, dissected corolla (× ⅔); n, anther (× 4). Drawn by Louise Olley.

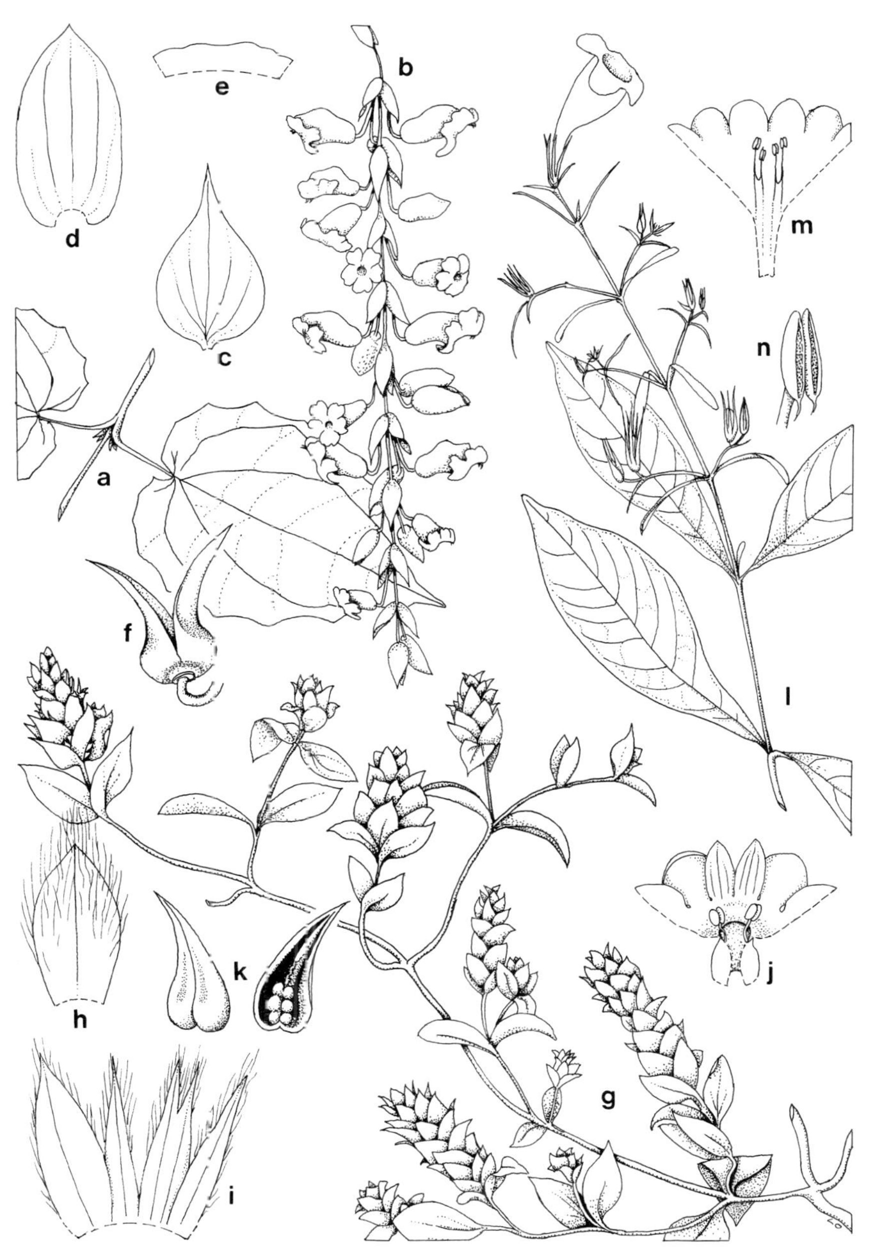
a
b
c
d
e
f
g
h
i
j
k
l
m
n

2. NELSONIA R. Brown

Prostrate herbs. Leaves entire. Flowers in short bracteate spikes. Bracts imbricate, alternate; bracteoles absent. Calyx 4-lobed, one lobe bigger than others, bifid. Corolla very small, tubular, 2-lipped; stamens 2; anthers 2-celled, not exserted, cells separate, muticous or minutely spurred. Capsule small, oblong, bearing seeds from base, many-seeded; seeds very small, glabrous, borne on minute points, retinacula absent.

1. N. canescens (Lamarck) Sprengel; *N. campestris* R. Brown. Fig. 102g–k.

Perennial herb. Stems radiating from central tap-root, trailing, densely and softly pilose, up to 50cm long. Leaves dimorphic, those on main stem petiolate (petiole 0–2.5cm), elliptic, 3–11 × 1.5–4.5cm, acute at both ends, sparsely pilose, those on flowering branches sessile, 1–3 × 0.3–1cm, densely pilose and much smaller. Inflorescence of dense, terminal, pilose, bracteate spikes, 1–3.5cm. Bracts ovate-elliptic, 5–9mm, acute. Calyx 4.5–7mm, pilose, lobes lanceolate. Corolla 4–6mm, lips purplish. Capsule 4–7mm, sparsely glandular near apex.

Bhutan: S – Samchi district (Buduni), Sankosh district (Balu Khola), Sarbhang district (Singi Khola), Gaylegphug district (Gaylegphug); **Darjeeling:** Siliguri; **W Bengal Duars:** Buxa, Jaldapara etc. Fields, river banks and disturbed ground in secondary scrub, 200–400m. February–April.

3. STAUROGYNE Nees

Herbs. Leaves entire. Flowers in terminal or axillary spikes. Bracts and smaller bracteoles present. Calyx 5-lobed, one lobe often larger than others. Corolla small with elongate, cylindrical tube and 5 short spreading lobes. Stamens 4, similar; anthers 2-celled, cells separate, muticous. Capsule small, oblong, bearing seeds from base, many-seeded; seeds very small, glabrous, borne on papillae, retinacula absent.

1. Leaves sessile, elliptic, all opposite; bracts shorter than flowers **1. S. polybotrya**
+ Leaves petiolate, obovate, alternate above; bracts longer than flowers giving flowering spike a leafy appearance .... **2. S. spatulata**

1. S. polybotrya (Nees) Kuntze; *Ebermaiera polybotrya* Nees

Slender herb; stem erect, 5–25cm, sparingly branched mostly from near base, glabrescent below, pilose above. Leaves all opposite, subsessile, elliptic, 2–4.5 × 1–2.5cm, obtuse, cuneate at base, glabrous except for few hairs on veins below. Inflorescence a dense terminal spike 1–4cm. Bracts oblong-linear, 5–7mm, glandular-pubescent; bracteoles linear, c 5mm. Calyx 5-lobed, 6–9mm, glandular-pubescent, lobes linear-oblong, one much longer and broader than others. Corolla purplish, 6–8mm, almost glabrous. Capsule 5–6mm, glabrous.

Darjeeling: Dalkajhar, Phansidewa Hat–Darjeeling, Siliguri. In terai swamps, 100–200m. October–January.

2. S. spatulata (Blume) Koorders; *S. glauca* (Nees) Kuntze, *Ebermaiera glauca* Nees

Similar to *S. polybotrya* but stem densely glandular-pilose above; leaves petiolate, obovate-spathulate, uppermost alternate; flowering spikes up to 15cm, usually interrupted below; bracts longer than flowers, lowermost leaf-like, oblong-spathulate, up to 5cm, so giving inflorescence a leafy appearance.

Darjeeling: terai (unlocalised Cowan collection).

4. HYGROPHILA R. Brown

Herbs growing in lowland marsh, rice-fields and similar wetland habitats. Leaves sessile, entire. Flowers in terminal spikes or axillary whorls. Bracts and bracteoles present, similar but bracteoles smaller. Calyx 5-lobed (4-lobed in *H. auriculata*), lobes linear to lanceolate. Corolla tubular, 2-lipped, upper lip shallowly notched, lower lip 3-lobed. Stamens 4 (2 in *H. polysperma*), similar or posterior pair smaller; anthers 2-celled, cells at same height. Capsule linear to narrowly oblong, bearing seeds from base, many-seeded (4–8-seeded in *H. auriculata*); seeds hairy; retinacula curved (short and straight in *H. helodes*).

1. Plant with axillary spines; calyx 4-lobed .... **5. H. auriculata**
\+ Plant unarmed; calyx 5-lobed 2

2. Flowers in terminal bracteate spikes 3
\+ Flowers in axillary whorls 4

3. Retinacula distinctly hooked; fertile stamens 2; corolla pubescent **2. H. polysperma**
\+ Retinacula very short, straight; fertile stamens 4; corolla glabrous or obscurely puberulent .... **1. H. helodes**

4. Leaves narrowly lanceolate, glabrous; calyx sparsely hispid-ciliate **3. H. salicifolia**
\+ Leaves elliptic, hirsute; calyx densely hispid-ciliate .... **4. H. phlomoides**

1. H. helodes Heine; *Cardanthera uliginosa* sensu F.B.I. non L.f.

Slender annual herb. Stems decumbent or ascending, 30–40cm, glabrous below, thinly pubescent above. Leaves oblong to subovate, 1–3.5 × 0.4–1cm, acute, usually glabrous. Flowers in terminal, bracteate spikes, somewhat interrupted below, 1–13cm. Bracts broadly elliptic, c 8 × 7mm, becoming imbricate in fruit. Calyx pubescent, lobes linear, 5–6mm. Corolla blue with paler tube, 11–13mm, glabrous or nearly so. Stamens 4, posterior pair smaller than anterior pair. Capsule glabrous except near apex, many-seeded, 6–8mm; retinacula very short, not hooked.

Darjeeling: terai (unlocalised Griffith collection).

2. H. polysperma (Roxb.) Anderson

Similar to *H. helodes* with which it is often confused but usually pubescent, sometimes densely so; bracts narrowly elliptic, not exceeding 5mm in width; corolla c 1cm, pubescent; fertile stamens 2; capsule with prominent, distinctly hooked retinacula.

Bhutan: S – Samchi district (Samchi (117)), Phuntsholing district (Torsa River); **Darjeeling:** terai (unlocalised Hooker collection); **W Bengal Duars**. 600m. January.

3. H. salicifolia (Vahl) Nees

Erect, short-lived perennial herb to 60cm, entirely glabrous except on bracts and calyx. Leaves narrowly lanceolate-elliptic, 2–8 × 0.3–1.2cm, attenuate at both ends. Flowers in dense axillary whorls. Bracts ovate, 4–7mm, acuminate, glabrous or sparsely hispid-ciliate. Calyx 5-lobed, lobes joined to halfway but splitting to base in fruit, linear-lanceolate, 7–8mm, sparsely hispid-ciliate. Corolla white, 12–15mm, glabrous. Fertile stamens 4, both pairs similar in size. Capsule many-seeded, longer than calyx, 9–11mm, glabrous; retinacula prominent, distinctly hooked.

Bhutan: S – Samchi district (Charmurchi–Daina River); Gaylegphug district (Gaylegphug); **Darjeeling:** Dalkajhar, Siliguri; **W Bengal Duars**. Swampy ditches, 300–500m. November–April.

4. H. phlomoides Nees

Similar to *H. salicifolia* but more robust; leaves pubescent to coarsely hirsute, oblong or elliptic, up to 2.5cm wide; bracts and calyx densely hispid-ciliate; corolla pale blue, 14–18mm; capsule 12–15mm.

Darjeeling: Dalkajhar, Katambari; **W Bengal Duars:** Jalpaiguri. October–December.

5. H. auriculata (Schumacher) Heine; *H. spinosa* Anderson, *Asteracantha longifolia* (L.) Nees

Vigorous perennial, 0.5–1.5m, with erect, woody, sparsely hispid stems and prominent axillary spines. Leaves in whorls of 6, two large, others reduced in size, lanceolate or oblong-lanceolate, 3–15 × 0.5–4cm, acute at both ends, sparsely hispid. Flowers in dense axillary whorls, each whorl usually with 6 spines. Spines 1–3cm. Bracts linear-lanceolate, 2–3cm, hispid. Calyx 4-lobed, lobes linear-lanceolate with one lobe broader than others, 15mm, coarsely ciliate. Corolla pale bluish-purple, 3–3.5cm, glabrous or finely puberulent. Fertile stamens 4, both pairs similar in size. Capsule 4–8-seeded, shorter than calyx, 9–11mm, glabrous; retinacula prominent, distinctly hooked.

Darjeeling: unlocalised (Treutler collection); **W Bengal Duars:** Jalpaiguri (174). 500m. October–February.

5. ECHINACANTHUS Nees

Perennial herbs or undershrubs. Leaves entire. Inflorescence of small cymes in axils of bracts forming an open panicle. Bracteoles linear, inconspicuous. Calyx 5-lobed, lobes equal. Corolla funnel-shaped, with ventricose tube, subequally 5-lobed. Stamens 4, anthers conspicuously spurred at base. Capsule oblong, 8–16-seeded.

1. E. attenuatus Nees. Fig. 102l–n.
Erect perennial herb to 60cm. Leaves petiolate, elliptic-oblanceolate, 5–18 × 2–4cm, acute, base attentuate, glabrous or thinly pubescent especially on veins; petioles 0.3–2cm. Inflorescence a narrow terminal panicle 6–20cm, branches formed from short few-flowered cymes arising in axils of bracts; cymes 3–6cm, usually 3-flowered; rhachis glandular-pubescent. Bracts narrowly oblong, sessile, 10–40 × 2–6mm; bracteoles linear, 8–15mm. Calyx 15–18mm, glandular-pilose, lobes linear, attenuate. Corolla purple, straight, 32–42mm, pubescent. Capsule c 18mm, glabrous.
Darjeeling/Sikkim: unlocalised (Treutler collection); **Sikkim:** Sosing. Subtropical forest, especially sal forest, 150–1300m. October–February.

6. AECHMANTHERA Nees

A small genus of hairy undershrubs similar to *Strobilanthes*. It has the same blue, funnel-shaped, 5-lobed corolla and toothed leaves but differs in its 6–8-seeded capsule. The genus is characterised by anther connectives usually but not always excurrent as short awns. Flowers are in clusters either in leaf axils or more commonly on branches of an open panicle.

1. A. gossypina (Wall.) Nees; *A. tomentosa* Nees
An isophyllous undershrub to 1m, usually grey tomentose but sometimes thinly pilose and greenish. Leaves (lanceolate-) ovate (-broadly elliptic), 2.5–13 × 1–6.5cm, acute, rounded, cuneate or subcordate at base, sometimes shortly decurrent, serrate; petioles 1–3cm. Inflorescence of dense, subsessile, 3–8-flowered clusters, sometimes in leaf axils of main stem, more commonly on short axillary branches, 3–8cm, where clusters are borne in axils of reduced leaves and may be aggregated together to form axillary thyrses or be more distant, so forming an open panicle. Bracts linear-ligulate, very variable in size, 6–12mm; bracteoles similar but only 5–7mm. Calyx deeply 5-lobed, 8–10mm,

pilose, lobes slightly unequal, linear, acute. Corolla blue (–purple), straight, funnel-shaped, 25–35mm, pubescent. Capsule 8-seeded, 7–9mm, pubescent; seeds hairy.

A. gossypina is very variable throughout its range. In Bhutan plants from Punakha district usually have narrowly ovate leaves and a lax open paniculate inflorescence with flower clusters distinctly shorter than subtending leaves. Plants from eastern Bhutan have broadly ovate leaves and more obviously axillary inflorescence with flowers in short dense thyrses about as long as subtending leaves.

Bhutan: S – Deothang district (Kheri Gompa), **C** – dry valleys of Punakha, Mongar and Tashigang districts; **Darjeeling:** Rangit. Locally abundant in dry grassland under chir pine, often with *Barleria cristata*, *Cymbopogon flexuosus* etc., 650–2200m. August–December.

7. CLARKEASIA J.R.I. Wood

Undershrubs similar in general appearance to *Strobilanthes*. Leaves slightly unequal in size, often asymmetrical at the base, toothed. Inflorescence of 1-sided axillary cymes. Bracts and bracteoles small, linear, similar, not deciduous. Calyx deeply 5-lobed with narrow equal segments. Corolla funnel-shaped, blue, 5-lobed as in *Strobilanthes*. Stamens 4, in 2 pairs, anther cells muticous. Capsule oblong, 16-seeded.

A characteristic feature of the genus is the production of curious non-developing leafy shoots at the main branches of the zigzag inflorescence.

1. C. parviflora (Anderson) J.R.I. Wood; *Echinacanthus parviflorus* Anderson, *E. andersonii* Clarke, *E. longistylus* Clarke, *Strobilanthes violifolia* Anderson, *Pteracanthus violifolius* (Anderson) Bremekamp

Small undershrub, 0.5–2m. Stem branched, wiry, glabrous or thinly pubescent, especially above. Leaves ovate (-elliptic), 5–12 × 1.5–6.5cm, shortly acuminate, serrate, base cuneate, rounded or cordate, usually asymmetric, glabrous or pubescent on veins, decreasing in size upwards, uppermost linear-lanceolate, bract-like, 12–25 × 2–5mm, sinuate-margined; petiole 1–2cm. Inflorescence of axillary, 1-sided cymes, becoming aggregated to form a panicle as plant matures; cymes often slightly zigzag, 2–9cm, flowers solitary or paired, somewhat remote, c 5–15mm apart. Bracts unequal in size, linear, 2.5–10mm; bracteoles similar, 2mm. Calyx 9–15mm, glabrous, strigose or glandular-pilose, united for c 3mm, lobes linear, acuminate. Corolla blue, 2.5–3cm, finely pubescent when young, glabrescent, lobes ovate, spreading, c 5mm. Capsule 1–1.2cm, glabrous or with few glandular hairs near tip.

var. **parviflora**

Stem, leaves and inflorescence glabrous or very sparsely pubescent.

Bhutan: S – Samchi district (Charmurchi and Torsa Rivers), Deothang district

(15km N of Deothang); **Darjeeling:** Kurseong, Pankhabari, Pomong, Selim; **Sikkim:** Linchyum, Rathang Chhu. On rocks and cliffs beside rivers in subtropical forest, 200–1300m, August–January.

var. **albescens** J.R.I. Wood

Short, dense, white, glandular indumentum on stem, inflorescence and undersurfaces of leaves.

W Bengal Duars: Buxa, 300m. February.

8. HEMIGRAPHIS Nees

Very similar to *Strobilanthes* but always isophyllous herbs; corolla small, straight, blue, less than 15mm; capsule 6–16-seeded.

1. H. hirta (Vahl) Anderson

Small grey herb with prostrate or ascending stems, densely covered in white hairs. Leaves broadly elliptic or obovate, 7–25 × 4–18mm, obtuse, base cuneate, margin crenate, densely pubescent on both surfaces, subsessile or shortly petiolate; petiole 0–5mm. Flowers in dense heads, terminal on main stems and on short branchlets from axils of upper leaves. Bracts elliptic, 11–15mm, pilose; bracteoles absent. Calyx 8–10mm, lobes linear-subspathulate, ciliate, subequal. Corolla blue, funnel-shaped, 12–15mm, pubescent, lobes spreading, c 3mm. Capsule 9–10mm, minutely glandular-puberulent, 12-seeded.

W Bengal Duars: Jalpaiguri district (174). Weed of grassy places in the plains. February–May.

9. STROBILANTHES Blume

Shrubs, undershrubs or (rarely) herbs. Leaves toothed (sometimes entire in species 5–7), sometimes oblique at base, usually unequal in each pair, occasionally with smaller leaf so reduced that leaves may appear alternate (especially in species 28 and 29). Inflorescence very varied, flowers in axillary and (less commonly) terminal spikes, heads and cymes or scattered in opposite pairs on a branched inflorescence sometimes forming an open panicle. Bracts usually distinct from leaves, persistent or deciduous as flowers open, varied in shape and texture, sometimes of two types in species with a capitate inflorescence: barren outer bracts differing from inner flower-bearing ones; bracteoles present except in *S. auriculata* and species 14–17. Calyx in our species 5-lobed to near base, lobes subequal or one distinctly longer than others. Corolla 5-lobed, blue (rarely white from cylindrical base), funnel-shaped, usually straight and gradually widened but sometimes bent and/or abruptly widened and strongly ventricose. Stamens 4, included (2 only, exserted in *S. khasyana*); anthers 2-celled, cells oblong, muticous, at same height. Capsule 2–4-seeded, oblong, seeds flattened, usually hairy, often with elastic hairs, sometimes with distinct glabrous areole.

Many species go in for periodic mass-flowering and cannot be found in flower every year. Some, such as *S. accrescens* and *S. echinata*, are definately monocarpic; they take about 12 years to mature, flower once and then die.

1. Fertile stamens 2, strongly exserted; corolla tube narrowly cylindrical for most of its length .... **1. S. khasyana**
\+ Fertile stamens 4, included; corolla tube distinctly widened just above base 2

2. Flowers imbricate or clustered in dense heads, spikes or reduced cymes, occasionally with 1–2 flowers below main head 3
\+ Flowers solitary or in opposite pairs, clearly separate and mostly at least 5mm apart, arranged laxly in spikes or cymes, but sometimes becoming confluent towards branch tips 19

3. Corolla less than 15mm; small herb with 12-seeded capsule **Hemigraphis hirta** (p. 1255)
\+ Corolla more than 15mm; herbs or undershrubs; capsule 4(–8)-seeded .. 4

4. Open corolla bent 90° just above base and then strongly inflated; leaves sessile, entire or obscurely serrate 5
\+ Open corolla gradually widened from base, straight or gently curved; leaves usually distinctly petiolate and serrate 6

5. Bracteoles present; capsule glandular-pubescent; leaves ovate, elliptic or subrhomboid .... **5. S. sabiniana**
\+ Bracteoles absent; capsule glabrous; leaves oblong-elliptic .. **4. S. auriculata**

6. One calyx lobe distinctly longer than others; bracteoles absent except in *S. lamiifolia*; bracts soon scarious and deciduous 7
\+ Calyx lobes all nearly equal in length; bracteoles present; bracts usually herbaceous and persistent 11

7. Corolla glabrous, even in bud and on lobes; leaves unequal in each pair 8
\+ Corolla hairy at least in bud and/or on lobes; leaves equal or unequal in each pair 9

8. Leaves strictly glabrous; bracts suborbicular, glabrous; flowers in dense elongate head .... **14. S. pentstemonoides**
\+ Leaves usually hairy at least on veins below; bracts ovate-suborbicular, sticky glandular-hairy; plant often with scattered flower pairs below main head .... **16. S. multidens**

9. Flower heads clearly axillary, always strictly capitate; leaves mostly more than 10cm .... **13. S. pubiflora**
+ Flower heads often appearing terminal or arising in branchlets from uppermost leaf axils only; heads usually elongate; leaves mostly less than 10cm 10

10. Bracts up to 1cm, outer ones ovate-suborbicular, inner ones often ± oblong; bracteoles absent; corolla with glandular-hairs, sometimes on lobes only; leaves usually unequal .... **15. S. oligocephala**
+ Bracts ovate, more than 1cm; bracteoles present, oblong-elliptic; corolla finely pubescent, not glandular-hairy; leaves equal .... **17. S. lamiifolia**

11. Decumbent herb densely covered in brown hairs; corolla white; inflorescence an apparently terminal spike .... **8. S. jennyae**
+ Erect undershrubs, glabrous or hairy but never covered in brownish hairs; corolla usually blue; inflorescence axillary 12

12. Leaves subsessile, obscurely serrulate, very unequal in each pair, one very small and often deciduous; bracts early deciduous .... **28. S. anisophylla**
+ Leaves petiolate, serrate, equal or somewhat unequal; bracts persistent (except sometimes in *S. capitata*) 13

13. Flowers in heads or clusters; bracts at least 1cm 14
+ Flowers in spikes; bracts less than 0.8cm 18

14. Bracts linear-ligulate; flowers sessile in clusters in leaf axils, sometimes aggregated to form a panicle; capsule 8-seeded **Aechmanthera gossypina** (p. 1253)
+ Bracts ovate; flowers in pedunculate heads; capsule 4-seeded 15

15. Flower heads ellipsoid, about twice as long as broad, mostly 2–2.5cm; corolla glabrous, white or pale blue 16
+ Flower heads ovoid, nearly as broad as long, usually 1–2cm; corolla thinly pubescent or pilose, deep blue (rarely white) 17

16. Outermost bract ovate, with long toothed appendage, much exceeding inner bracts; calyx lobes and bracteoles acuminate to fine point; corolla white **10. S. simonsii**
+ Outermost bracts ovate, acute without appendage, shorter than inner bracts; calyx lobes fimbriate, toothed or obtuse at tip; corolla pale blue **9. S. echinata**

17. Calyx lobes mucronate, mucro accrescent into long awn in fruit; leaf base decurrent onto petiole .... **12. S. accrescens**
+ Calyx lobes acute, not accrescent; leaf base rounded to attenuate but never decurrent .... **11. S. capitata**

18. Corolla short, less than 2cm; spikes subcapitate, 1–2cm .... **3. S. frondosa**
+ Corolla elongate, c 3cm; spikes elongate, 2–7cm .... **2. S. himalayana**

19. Flowers in an open leafless terminal panicle with capillary branches **29. S. hamiltoniana**
+ Flowers variously branched but if paniculate then branches spicate and/or leafy-bracteate 20

20. Flowers in leaf axils or in axils of petiolate, leaf-like bracts 21
+ Bracts sessile or absent at flowering time, not leaf-like 24

21. Calyx with one lobe much longer and broader than others; corolla straight, very large, up to 7cm .... **30. S cusia**
+ Calyx subequally 5-lobed or with one lobe slightly longer than others; corolla curved or bent, usually less than 4cm 22

22. Corolla glabrous; calyx 1–2.8cm .... **24. S wallichii**
+ Corolla hairy; calyx up to 1.4cm 23

23. Stem, leaves and bracts grey-pubescent; corolla abruptly inflated and bent just above base .... **26. S. inflata**
+ Stem, leaves and bracts glabrous; corolla gradually widened and bent **23. S. urophylla**

24. Flowers in 1-sided, axillary, zigzag spikes; bracts linear, usually unequal in each pair 25
+ Flowers in terminal and/or axillary spikes or cymes; bracts ovate, elliptic, oblong or absent 26

25. Corolla blue, straight, gradually widened from base; capsule 12–16-seeded **Clarkeasia parviflora** (p. 1254)
+ Corolla white or pale purplish, bent 90° and abruptly widened just above base; capsule 4-seeded .... **25. S. helicta**

26. Plant entirely glabrous except for few hairs on calyx lobes; leaves very unequal, often appearing alternate, green on both surfaces, larger in each pair lanceolate; calyx lobes finely acuminate .... **27. S. divaricata**
+ Plant variously hairy (check leaf veins, bracts, calyx and corolla lobes); if glabrous then leaves ovate-suborbicular, whitish below; calyx lobes never finely acuminate 27

27. Calyx lobes equal; corolla sharply bent above shortly tubular base; leaves subentire 28
\+ Calyx with one lobe distinctly longer than others; corolla gradually expanded from base, gently curved; leaves distinctly toothed 29

28. Corolla pilose; bracts persistent .... **6. S. tamburensis**
\+ Corolla glabrous; bracts early deciduous .... **7. S. rubescens**

29. Bracts all of same type, deciduous; leaves very unequal 30
\+ Bracts relatively persistent, lower ones ovate-lanceolate, leaf-like, those of upper flowers much smaller leaves equal or slightly unequal 31

30. Corolla glabrous; capsule 15–17mm .... **16. S. multidens**
\+ Corolla pubescent; capsule less than 10mm .... **18. S. thomsonii**

31. Inflorescence lax, not clearly spicate, with distinct internodes between uppermost flower pairs; bracts of upper flowers oblong-oblanceolate or -obovate; calyx lobes oblanceolate-spathulate, especially in fruit 32
\+ Inflorescence relatively dense, clearly spicate upwards, flower pairs confluent towards tips; bracts of upper flowers linear-oblong; calyx lobes linear, sometimes slightly spathulate in fruit 33

32. Leaves pilose on both surfaces, ovate, 1–5cm wide .... **19. S. extensa**
\+ Leaves almost glabrous, lanceolate or narrowly elliptic, less than 2cm wide **20. S. claviculata**

33. Leaves elliptic, 0.8–2cm wide, glabrous or sparsely pilose, often unequal; plant less than 20cm .... **22. S. subnudata**
\+ Leaves ovate, obovate or broadly elliptic, more than 3cm wide, pilose on both surfaces, equal; plant 0.3–1m .... **21. S. lachenensis**

1. S. khasyana (Nees) Anderson; *Endopogon khasyanus* Nees, *Listrobanthes khasyana* (Nees) Bremekamp

Undershrub 0.5–1.2m, densely covered in spreading brown hairs. Leaves unequal, narrowly ovate, 4–10 × 2.5–5cm, acuminate, base cuneate, densely brown pilose on both surfaces, crenate; petioles 1–6cm. Flowers in opposite pairs, c 5mm apart on axillary, leafless spikes; spikes in axillary tufts, unequal in length, 2–8cm. Bracts persistent, 4–11mm; lower bracts elliptic; upper bracts oblanceolate-spathulate, deciduous as flowers open; bracteoles linear, 5–7mm. Calyx 0.6–1cm, pilose, lobes linear, acute, subequal. Corolla white, c 1.5cm, pubescent, tube very slender, linear, widened only near mouth, c 1cm; lobes spreading; fertile stamens 2, exserted; staminodes also present. Capsule 7mm, glabrous except for few hairs at tip; seeds with abroad areole.

Darjeeling: unlocalised (Griffith collection). August–November.

2. S. himalayana J.R.I. Wood; *S. petiolaris* sensu F.B.I. non Nees

Undershrub 0.5–1m. Stems erect, glabrous except at the nodes. Leaves unequal, broadly elliptic, 1.5–12 × 1–5cm, acuminate, base cuneate and often decurrent, serrate, green on both surfaces, often black-spotted above, glabrous except for few hairs on veins below; petioles 0.3–3cm. Flowers in dense pedunculate spikes, terminal on main stem and on small, sometimes leafy axillary branchlets, often 3-forked near tip; spikes 2–7 × 0.7–1cm; peduncles 0–2cm, glandular-pubescent. Bracts obovate, persistent, 4–6mm, rounded, glandular-pubescent; bracteoles linear, c 4mm, obtuse. Calyx 7–8mm, glandular-pubescent, lobes linear, acute, subequal. Corolla blue, c 3cm, pilose, tube nearly straight, gradually widened to c 8mm. Capsule 7–8mm, pilose; seeds with large areole.

Bhutan: unlocalised (Griffith collection); **Darjeeling:** Kalimpong, Rambi, Kurseong, Mongpu, Rishap; **Sikkim:** Yakla, Rimbi. 900–1250(–3000)m. June–October, probably every year.

3. S. frondosa J.R.I. Wood

Perennial herb to 30cm. Stems woody below, decumbent, rooting at nodes and then ascending, often drooping at tips, pilose, sulcate. Leaves unequal in size, broadly elliptic, shortly acuminate, 0.8–12.5 × 1–6cm, base cuneate, green and pilose on both surfaces, serrate; petioles 0–4cm. Flowers in elongate, subcapitate, often nodding spikes terminal on simple or branched, leafy, axillary branchlets; spikes 1–2 × 0.8–1cm; peduncles 1.5–2cm, pilose. Bracts green, broadly elliptic-obovate with 1 tooth on either side, 7–8 × 4mm, subacute, pilose and gland-dotted, persistent; bracteoles linear, 5–8mm, subacute, pilose. Calyx 6–7mm, pale green, pilose, lobes linear, acute, equal. Corolla pale blue, 15–18mm, pilose on lobes, tube very short, cylindrical for 5mm and then abruptly widened to mouth, lobes spreading; anthers very shortly exserted. Capsule 5.5mm, pilose near tip; seeds with large areole.

Bhutan: S – Chukka district (Chukka–Chimakothi, Awakha–Takti Chu). By streams in shade in very moist broad-leaved forest relics, 1600–1700m. August–October, every year.

There is a specimen of *S. brunoniana* Nees at Kew which is said to have been collected at Darjeeling. This must be an error. *S. brunoniana* is similar to *S. frondosa* but flowers are in long spikes and leaves are narrowly oblong-lanceolate with subentire margins.

4. S. auriculata Nees; *S. edgeworthiana* Nees, *Perilepta auriculata* (Nees) Bremekamp, *P. edgworthiana* (Nees) Bremekamp

Much-branched undershrub 0.6–2m. Stems glabrous, quadrangular. Leaves subequal to very unequal, sessile, narrowly oblong-elliptic, 2–20 × 1.3–6cm, acuminate at both ends, or uppermost rounded at base, weakly serrate, green above, paler beneath, sparsely pilose on both surfaces, veins beneath highlighted. Flowers in dense, pilose, leafless, pedunculate, axillary and terminal spikes; spikes usually simple, sometimes sparingly branched, 3–10 × 1–2cm; peduncles

2–4cm. Bracts imbricate, obovate, 6–10mm, rounded, densely ciliate, persistent; bracteoles 0. Calyx glandular-pilose, 6–12mm, lobes linear, subacute, often ciliate, 2 lobes slightly shorter than others. Corolla pale purple, 20–26mm, glandular-pilose on lobes, tube cylindrical for c 4mm and then bent 90° and abruptly inflated. Capsule 9–10mm, glabrous.

Darjeeling: Darogadara, Peshok, unlocalised (Gamble collections). In terai, 450–950m. October–February, every year.

Some forms with long ciliate hairs on calyx and bracts have been called var. *edgeworthiana* (Nees) Clarke.

5. S. sabiniana (Lindley) Nees. Fig. 104a–f.

Undershrub 0.5–2m. Stems glabrous, reddish-brown, much branched, ascending. Leaves very unequal in size, ovate, elliptic or subrhomboid, 1.5–14 × 0.8–6cm, acute, base rounded or cuneate and decurrent, dark green above, white beneath, glabrous, entire or obscurely crenate. Flowers in dense, sticky, glandular, pedunculate, axillary spikes; spikes 3–8cm, usually solitary, simple, sometimes 1–2 branched; peduncles 1–1.5cm, glabrous; rhachis glandular-pubescent. Bracts obovate-suborbicular, 2.5–6mm, rounded, glandular-pubescent, persistent; bracteoles linear, c 4mm. Calyx 4–8mm, glandular-pubescent, pale green, lobes linear, obtuse, subequal. Corolla pale blue, 2.5–3cm, pubescent on lobes, tube cylindrical for c 5mm, then bent 90° and abruptly widened to 1cm at mouth. Capsule 6–7mm, glandular-pubescent.

Bhutan: S – Samchi district (Tamangdhanra Forest), Phuntsholing district (Rinchending–Sorchen), Sarbhang district (Sarbhang School, Kami Khola), Gaylegphug district (Tatopani), **C** – Tongsa district (Dakpai, Tintibi); **Darjeeling:** Sitong, Pasting; **W Bengal Duars:** Buxa. On steep scrubby banks and in well-drained clearings in subtropical forest, 400–1850m. January–April, flowering every year.

6. S. tamburensis Clarke

Perennial herb. Stems decumbent and rooting below, then ascending to 30cm, bifariously pilose. Leaves slightly unequal, merging upwards into bracts, ovate to suborbicular, 8–45mm long and wide, acute or obtuse, base rounded to subcordate, green above, whitish beneath, entire or obscurely crenate, glabrous or pilose on both surfaces, usually sessile but occasionally with petioles up to 3mm. Flowers in opposite, bracteate pairs, 25mm apart below, confluent above, forming lax, sticky-glandular terminal and, less commonly, axillary spikes 3–6cm. Bracts orbicular or obovate, 5–8mm long and wide, pilose, persistent till flowers fall; bracteoles obovate, c 3mm, pilose. Calyx 6–10mm, glandular-pilose, lobes linear-spathulate, subequal. Corolla blue, 2.5–3cm, pilose, tube cylindrical for c 3mm, then bent 90° and abruptly widened to 8–10mm at mouth. Capsule 10mm, glandular-pilose.

Bhutan: S – Chukka district (Chimakothi), **C** –Tashigang district (Tashi

Yangtsi–Bumdeling). In wet places in scrub or broad-leaved forest, 1800–2100m. August–November, probably not every year.

7. S. rubescens Anderson; *S. boerhaavioides* Anderson, *Pteracanthus rubescens* (Anderson) Bremekamp, *P. boerhaavioides* (Anderson) Bremekamp

Undershrub 0.5–2m. Stems, erect or ascending from woody root-stock, glabrous or sparsely pilose. Leaves unequal, ovate to suborbicular, 1–15 × 0.8–7.5cm, acute, cordate, rounded or cuneate at base, entire or crenate, glabrous or thinly pilose, dark green above, paler beneath, sessile or with petioles up to 3cm. Flowers in opposite pairs, 6–15mm apart in lax axillary and terminal spikes; spikes often curved, 4–12cm, usually with 2–4 side branches; peduncles 1–7cm. Bracts ovate, concave, 2.5–4mm, deciduous as flowers open; bracteoles oblong-elliptic, 1.5–2mm. Calyx 6–10mm, pale green, glabrous or thinly glandular-pilose, lobes oblong-elliptic, acute, subequal. Corolla blue, 32–40mm, glabrous, tube shortly cylindrical for 5mm then bent and widened to c 18mm at mouth. Capsule 1–1.2cm, glabrous except for tuft of hairs at tip.

Bhutan: S – Phuntsholing district (above Rinchending), Gaylegphug district (Chabley Khola), Deothang district (Deothang–Wamrong); **Darjeeling:** Sureil, Rambi Chhu, Jorbangala; **W Bengal Duars:** Buxa (231). On steep scrubby hill slopes, 750–2000m. October–May, probably not every year.

8. S. jennyae J.R.I. Wood

Small decumbent herb. Stem pilose with brown, spreading hairs. Leaves equal, ovate-elliptic, 3.5–6 × 1.5–5cm, acute, cuneate at base, roughly brown-pilose on both surfaces, sessile above, petiolate below; petioles up to 2.5cm. Flowers in short, dense apparently terminal spikes 2–3cm; peduncles c 1cm. Bracts oblong-elliptic, herbaceous, 9–15 × 4mm, acute, pilose, persistent; bracteoles linear, 6–9mm, acute, ciliate. Calyx 5–10mm, ciliate, very pale green, lobes linear, acute, equal. Corolla cream, straight, 28mm, gradually widened from base to 12mm at mouth, densely glandular-pilose. Capsule not known, presumably pubescent.

Bhutan: C – Tashigang district (Tashi Yangtsi–Bumdeling, Lumphi, E of Tashi Yangtsi). Streamsides and marsh in open broad-leaved forest, 2100–2350m. August–November.

Endemic to Bhutan.

9. S. echinata Nees; *S. pectinata* Anderson. Nep: *Kibu.*

Monocarpic shrub 0.5–3m. Stems erect, woody, pubescent above, glabrescent below. Leaves unequal, elliptic, slightly falcate, 3–14 × 1.5–6cm, acuminate, base cuneate, serrate, pilose above, softly green-tomentose beneath; petioles 0.5–4cm. Flowers in single, dense, ellipsoid, 2–3cm heads on short axillary branchlets with 1(–3) pairs of reduced leaves; peduncles pilose. Outer barren bracts 2, slightly unequal in size, ovate, 10–17mm, acuminate, serrate, persistent; inner bracts ovate-broadly elliptic, concave, 21–29mm, serrate in upper half, thinly pilose; bracteoles deeply concave, oblong, 20–25mm, fimbriate-tipped, sparsely pilose. Calyx 12–15mm, lobes linear with obtuse, toothed, or fimbriate tip, subequal, ciliate in upper half. Corolla pale lilac, glabrous, 4–6cm, tube straight, narrowly cylindrical for c 1cm then gradually widened to c 3cm at mouth. Capsule 13–18cm, glabrous.

Bhutan: S – Phuntsholing district (Sintalakha), Chukka district (Marichong, Gedu), **C** – Tongsa district (Tongsa–Changkha); **Darjeeling:** Darjeeling, Rungtong, Gumpahar etc.; **Sikkim:** Rimbi Chhu. On rocky slopes in moist broad-leaved hill forest, 1200–2300m. May–September, apparently flowering in a 12-year cycle, a flowering predicted for 1998, but a few flowering specimens are found occasionally in other years.

10. S. simonsii Anderson

Similar in general habit to *S. echinata*. Stems pilose. Leaves elliptic, falcate, not more than 9 × 3cm, pubescent on both surfaces but with veins beneath highlighted with longer hairs. Inflorescence similar to that of *S. echinata* but branchlets with 1–3 heads. Outer bracts 2, pilose, dissimilar, larger one 3–4cm, ovate, entire with elongate, lanceolate, falcate, dentate appendage, smaller one 1.3–2cm, oblong-lanceolate with dentate subfimbriate tip; inner bracts lanceolate, 2.5–3cm, acuminate, tip crenate; bracteoles lanceolate, 16–19mm, acuminate, pilose. Calyx 14mm, pubescent, lobes linear, acuminate into long fine point. Corolla pure white, 35–42mm, gradually widened from shortly cylindrical base to c 15mm, glabrous. Capsule c 15mm, glabrous.

Bhutan: S – Deothang district (Deothang). March–August.

11. S. capitata (Nees) Anderson; *Goldfussia capitata* Nees. Nep: *Kibu*. Fig. 103f–l.

Much-branched undershrub, 0.5–1.5m, often leafless when in flower. Stems decumbent or ascending, glabrescent below, thinly pilose above. Leaves unequal, (lanceolate-) broadly elliptic (-ovate), shortly acuminate at both ends, usually oblique at base, 4–14 × 1.5–9cm, serrate, thinly pilose above, pubescent on veins beneath, often glabrescent, green on both surfaces or whitish beneath; petioles 1–3cm. Flowers in dense, pedunculate, axillary heads which may be solitary, borne in 3s or borne on small axillary branchlets with reduced leaves forming axillary panicles; peduncles 1–2 in each axil, 1–6cm, glabrous or pubescent; heads ovoid or ellipsoid, 1.5–2 × 1–3cm. Outer barren bracts concave, ovate, 16–20(–23)mm, acuminate, crenate, pubescent (or glabrous), deciduous as flowers open; inner bracts oblong-elliptic, 13–16mm, entire or with 1–2 teeth, pilose, pale green, deciduous; bracteoles oblong-elliptic, 7–11mm, pilose. Calyx 11–13mm (accrescent to 2cm in fruit) pale green, lobes linear-oblanceolate, densely ciliate, equal. Corolla blue, 43–53mm, thinly pilose, tube straight, narrowly cylindrical for c 15mm, then gradually widened to c 15mm at mouth. Capsule c 14mm, glandular-pilose in upper half.

Bhutan: S – Samchi, Phuntsholing, Manas and Deothang districts, **C** – Tongsa and Mongar districts; **Darjeeling:** Peshok, Kurseong, Darjeeling etc.; **W Bengal Duars:** Gajalduba (175). Subtropical scrub and forest margins, particularly in moist gullies, locally frequent but mainly restricted to SW Bhutan, Kheng country and the Mongar region, 400–1700m. September–February, flowering every year.

12. S. accrescens J.R.I. Wood

Monocarpic shrub 0.5–2m, similar in appearance to *S. capitata*. Leaves equal, ovate or broadly elliptic, 5–25 × 2–11cm, not oblique at base, always decurrent into petiole, glabrous to tomentose; petioles 1–7cm. Inflorescence like *S. capitata*; outer bracts ovate or elliptic with distinct, dark green, crenate, terminal appendage, 13–18mm, persistent; inner bracts broadly elliptic, 12–14mm, acute, ciliate-margined, gland-dotted on dorsal surface; bracteoles gland-dotted, c 12mm. Calyx 11mm in flower, gland-dotted, lobes oblanceolate, ciliate, mucronate, subequal, in fruit mucro growing to 25mm, becoming curved and glandular-pubescent. Corolla blue-purple (rarely white), 2.5–3.7cm, sparsely pilose on middle of tube, tube slightly curved, gradually expanded to c 1cm at mouth. Capsule 12–13mm.

subsp. **accrescens**. Fig. 103q.

Stem, leaves and outer bracts glabrous. Leaves broadly elliptic, gradually narrowed to base, very large, up to 25 × 11cm, green on both surfaces with veins below brownish. Inflorescence relatively lax with peduncles up to 2cm, often compound; heads 1–2cm wide; outer bracts broadly elliptic, glabrous; corolla large, usually more than 3.3cm. Capsule glabrous except for 1–2 glands at apex.

Bhutan: S – Chukha district (Gedu, Asinabari, Takti Chu, Chukka, Chimakothi, Bunakha). Locally abundant in secondary scrub regenerating from moist broad-leaved forest, 1500–2200m. September–October, every 10–15 years.

subsp. **teraoi** J.R.I. Wood

Stem pubescent, gland-dotted, rarely exceeding 1m. Leaves ovate, abruptly rounded at base and then decurrent, up to 12 × 9cm, pubescent above, grey-tomentose below. Inflorescence very compact, peduncles up to 1cm; heads 8–12mm wide; outer bracts ovate, acuminate, grey-pubescent; bracteoles grey-pubescent; corolla 2.5–3cm; capsule pilose.

Bhutan: C – Punakha district (Mo Chu); Tongsa district (Tashiling, Tongsa). Locally abundant in rough marshy scrub and in clearings in dry broad-leaved forest, 1800–2650m. September–October, every 10–15 years.

S. accrescens is a remarkable species. It flowers every 10–15 years (the exact period is not known, but it flowered in 1992) and then dies. In flower it is almost identical to *S. capitata* except for the decurrent leaves but in fruit the mucros at the end of the calyx lobes elongate and become sticky, perhaps helping seed dispersal. It is extraordinarily abundant in the areas where it occurs. Both subspecies are endemic to Bhutan.

13. S. pubiflora J.R.I Wood; *S. pentstemonoides* auct. non (Nees) Anderson, *S. discolor* sensu F.B.I. non (Nees) Anderson

Undershrub 0.5–1m. Stems decumbent or ascending, glabrous. Leaves unequal, broadly elliptic, 6–20 × 2–7cm, shortly acuminate, base attenuate, serrate, glabrous, green above, paler beneath; petioles 0.3–2cm. Flowers in dense, suborbicular, pedunculate heads borne on simple or 2–3 forked axillary branchlets which are usually shorter than subtending leaves and leafless as leaves at branching point fall early; peduncles glabrous. Bracts suborbicular, concave, 2–9mm, rounded, glabrous, greenish when young, soon becoming scarious, soon deciduous; bracteoles absent. Calyx 5–7mm in flower, accrescent to 12mm in fruit, sticky glandular-pilose, lobes linear, acute, one longer than others. Corolla blue, 32–50mm, pubescent, tube whitish, slightly curved, gradually widened to 12–16mm at mouth. Capsule 14–15mm, glandular-pilose nearly to base.

Bhutan: S – Samchi district (Dorokha–Sengdhen), Phuntsholing district (Rinchending, Sorchen, Kamji), Sarbhang district (Loring Falls, Chirang road), Gaylegphug district (Batase, Rani Camp (117)), **C** – Tongsa district (Dakpai–Tintibi); **Darjeeling:** Kalimpong, Sureil, Takdah, Nangklas, Kurseong etc. Open subtropical forest and roadside scrub, 800–1700m. September–February, every year.

14. S. pentstemonoides (Nees) Anderson; *Goldfussia pentstemonoides* Nees

Similar to *S. pubiflora* with which it has often been confused. Leaves variable in shape, lanceolate, ovate or elliptic, often decurrent onto petiole, usually more coarsely serrate than in *S. pubiflora*. Flowers in distinctly elongate heads (up to 3cm) borne on very long, persistently leafy, axillary branchlets, usually much exceeding subtending leaves; peduncles glandular-pilose, rarely glabrous. Corolla blue, glabrous, tube curved. Capsule glandular-pilose at tip only.

Bhutan: C – Punakha district (Nobding), **N** – Upper Mo Chu district (Gasa). In moist oak forest, 2190–2450m. August–November.

Other records from lower altitudes in Bhutan and Darjeeling district refer to *S. pubiflora.*

15. S. oligocephala Clarke; *S. paupera* Clarke, *S. pentstemonoides* sensu Gamble (47) non (Nees) Anderson, *Goldfussia oligocephala* (Clarke) Bremekamp, *Diflugossa paupera* (Clarke) Bremekamp, *G. thomsonii* Hooker. Fig. 103a–e.

Much-branched undershrub 30–50cm. Stems decumbent and rooting below, eventually erect, pubescent or glabrous. Leaves unequal, elliptic or narrowly obovate, falcate, 4–13 × 1.5–4cm, abruptly acuminate, base attenuate, serrate, green above, whitish beneath, both surfaces glabrous, pilose or pubescent on veins beneath and margins only, sessile above, petiolate below; petioles 0–2cm. Flowers in pedunculate, shortly elongate heads, terminal on main stem or on simple or 2–3 branched, short, usually leafless branchlets arising from axils of upper leaves; sometimes a few flowers in opposite pairs below main head; peduncles glabrous or pubescent. Outer bracts ovate-suborbicular, 7–10mm, acute, mostly glabrous, greenish when young, soon scarious, soon deciduous; inner bracts similar but reduced, ± oblong, with more hair; bracteoles absent. Calyx 11–19mm, glandular-pilose, lobes linear (-oblong), acuminate, one slightly longer than others. Corolla blue or white, 3.5–5cm, sparsely or moderately glandular-pilose, tube gradually widened to c 1cm, bent at mouth. Capsule 13–14mm, glandular-pilose in upper part.

Bhutan: C – Thimphu district (Dodena, Hinglai La), Punakha district (Tinlegang); **Darjeeling:** Tanglu, Sandakphu, Neebay; **Sikkim:** Yoksam. On rocky slopes in moist oak forest, 2000–3000m. July–November, more frequently in some years than others.

16. S. multidens Clarke; *S. agrestis* Clarke, *S. laevigata* sensu Yamazaki (71) non Clarke, *Goldfussia multidens* (Clarke) Bremekamp, *Pteracanthus agrestis* (Clarke) Bremekamp

Undershrub, 1–2m. Stems erect, glabrous or pubescent. Leaves unequal or very unequal, often asymmetric and falcate, broadly elliptic (-ovate), 2–20 × 1–9.5cm, shortly acuminate, base attenuate and oblique, serrate, above glabrous or pilose, beneath pubescent on the veins or pilose (rarely glabrous), sessile above, petiolate below; petioles 0–5cm. Flowers usually in dense clusters at ends of trichotomously forked axillary branchlets, sometimes also with scattered flowers in opposite pairs on branches below; rarely only in scattered opposite pairs; peduncles (pubescent-) glandular-pilose when young, bearded when old. Bracts ovate-suborbicular, concave, 3–6(–10)mm, acute, sticky-glandular, purplish-green, soon deciduous; bracteoles absent. Calyx in flower 6–10mm, glabrous or glandular-pubescent, in fruit up to 16mm, sticky glandular-pilose, lobes linear, acute, one 2–3mm longer than others. Corolla blue, glabrous, 35–48mm, tube brownish, slightly curved, gradually widened to 15mm at mouth. Capsule 15–18mm, glandular-pilose at tip.

Bhutan: S – Phuntsholing, Chukka, Gaylegphug, Manas and Deothang districts, **C** – Punakha, Tongsa and Tashigang districts; **Darjeeling:** Darjeeling, Dumsong, Kurseong etc. Locally abundant by streams in moist broad-leaved hill forest, (1200–)1500–2000m. September–April every year.

17. S. lamiifolia (Nees) Anderson; *Goldfussia lamiifolia* Nees, *Ruellia rotundifolia* D. Don, *Pteracanthus rotundifolius* (D. Don) Bremekamp

Much-branched perennial herb up to 40cm. Stems decumbent, often rooting at nodes, glabrous or bifariously pubescent. Leaves equal, ovate-elliptic

---

FIG. 103. **Acanthaceae.** a–e, *Strobilanthes oligocephala*: a, habit (× ½); b, outer bract (× 2); c, inner bract (× 2); d, calyx (× 2); e, corolla (× ⅔). f–l, *S. capitata*: f, habit (× ½); g, outermost bract (inner face) (× 2); h, outer bract (× 2); i, inner bract (× 2); j, bracteole (× 2); k, calyx (× 2); l, dissected corolla (× ⅔). m–p, *S. helicta*: m, habit (× ½); n, bracts (× 2); o, calyx (× 1⅓); p, corolla (× ⅔). q, *S. accrescens* subsp. *accrescens*: calyx at fruiting stage (× 2). Drawn by Louise Olley.

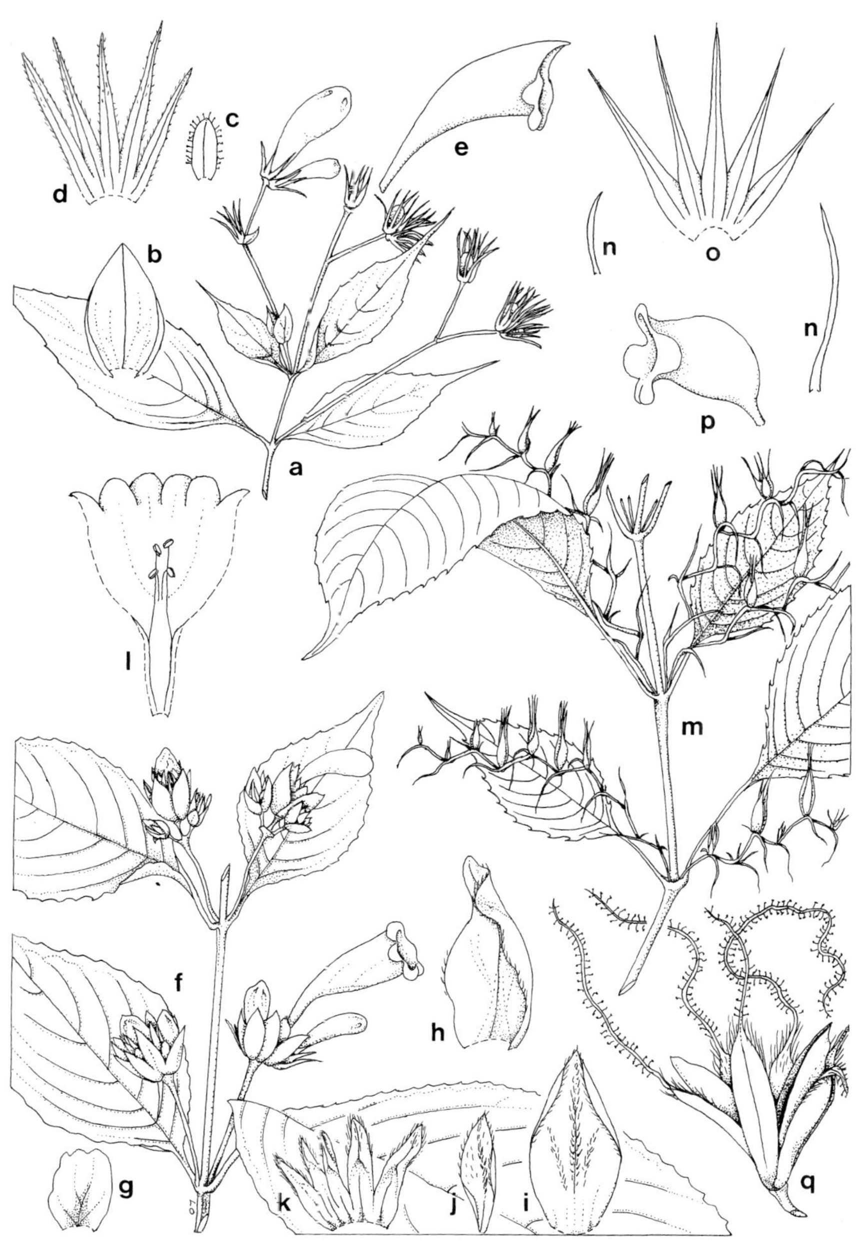
a
b
c
d
e
f
g
h
i
j
k
l
m
n
n
o
p
q

(-obovate), 1–7(–11) × 1–4(–6)cm, acute, narrowed at base and often decurrent, crenate or serrate, thinly pilose on both surfaces and pubescent on veins beneath, petiolate below, subsessile above; petioles 0–3cm. Flowers in short dense spikes, terminal on main stem and on short, simple branchlets from upper leaf axils; spikes 2–5cm; peduncles 0–6cm, pubescent. Bracts concave, ovate-elliptic, imbricate, 11–17mm, usually crenate, pubescent, dull green, slightly membranous, deciduous; bracteoles pale green, oblong-elliptic, 9–14mm, sparsely ciliate. Calyx 9–14mm, glandular-ciliate, lobes oblong-elliptic, one distinctly longer than others. Corolla blue, 38–45mm, pubescent, tube gently curved, gradually widened from base to c 12mm at mouth. Capsule 12mm, glabrous except for few hairs at tip.

Bhutan: S – Sankosh district (Daga Dzong), Manas district (Khomser), **C** – Punakha, Bumthang, Mongar and Tashigang districts; **Darjeeling/Sikkim:** unlocalised (Griffith Collection (316)). Common in grassland in open oak and chir pine forest in dry valleys of C Bhutan, 1000–2300m. August–November, every year.

18. S. thomsonii Anderson

Much-branched undershrub of untidy appearance, 0.6–2m. Stems erect, glabrous (occasionally hairs on upper nodes). Leaves unequal, ovate or elliptic, 2–20 × 0.6–8cm, acuminate, base cuneate or attenuate, serrate, glabrous or thinly pilose above, glabrous beneath, usually petiolate below, becoming sessile above; petioles 0–2.5cm. Flowers in opposite pairs, mostly 3–15mm apart, in lax (rarely dense) axillary spikes, these becoming branched and compound, sometimes developing into large leafless panicles in older plants; rhachis hairy. Bracts obovate-elliptic, 3–7mm, usually deciduous early; bracteoles linear, 2–3mm, deciduous before bracts. Calyx 4–6mm in flower, accrescent to 13mm in fruit, (glabrous-) glandular-pilose (-villous), lobes linear, long acuminate, one longer than others. Corolla blue, 25–33mm, pubescent, tube straight, gradually widened to c 10mm. Capsule 7–8mm, glandular-pilose.

Bhutan: S – Samchi, Sarbhang and Deothang districts, **C** – Punakha, Tongsa and Tashigang districts; **Darjeeling:** Dumsong, Kalimpong, Mintopang; **Sikkim:** Linchyum, Lusing, Singhik etc. In open broad-leaved forest, especially in rocky gullies in the dry valleys of C Bhutan, locally abundant, 800–1900m. September–November(–April), flowering gregariously at unknown intervals.

FIG. 104. **Acanthaceae.** a–f, *Strobilanthes sabiniana*: a, habit (× ½); b, bract (× 3); c, bracteole (×); d, calyx (× 3); e, corolla (× ½); f, dehiscent capsule (× 3). g–k, *S. extensa*: g, habit (× 1); h, lower bract (× 2); i, bract from top of inflorescence (× 2); j, calyx (× 2); k, corolla (× ½). l–p, *S. divaricata*: l, habit (× ½); m, bract (× 2); n, bracteole (× 2); o, calyx (× 2); p, corolla. Drawn by Louise Olley.

b
f
d
e
c
a
p
l
k
o
g
h
m
i
j
n

19. S. extensa (Nees) Nees; *Pteracanthus extensus* (Nees) Bremekamp. Fig. 104g–k.

Much-branched shrub, 0.5–2m. Stems erect, pilose. Leaves slightly unequal, ovate, cordate above, elliptic below, 1.5–8 × 1–5cm, shortly acuminate, base cuneate, serrate, pilose on both surfaces, whitish beneath, sessile, lower leaves decurrent onto pseudo-petiole. Flowers in opposite pairs, 0.4–3cm apart, in lax spikes terminal on main stems and on axillary branchlets; spikes 2–10cm; rhachis glandular-pilose. Bracts 6–20mm, diminishing in size upwards, herbaceous, deciduous as flowers open; all lower bracts ovate; upper bracts elliptic-obovate; bracteoles oblanceolate, 2–3mm. Calyx 10–16mm, glandular-pilose, lobes oblong-oblanceolate or spathulate, darker at tips, one 2–3mm longer than others. Corolla blue, glabrous except for few hairs on lobes, 32–37mm, tube gradually widened to 10–15mm, bent at mouth. Capsule 13–16mm, glandular-pilose in upper half.

Bhutan: C – Thimphu district (Chapcha, Ginekha, Gidakom, Dorji Drah). In mixed oak-pine forest, particularly on rocky slopes; very locally abundant, 2300–3000m. July–October, flowering gregariously at unknown intervals, last in 1992.

20. S. claviculata W.W. Smith; *S. duclouxii* Benoist

Very similar to *S. extensa*, differing by its glabrous stem; falcate, lanceolate or narrowly elliptic, nearly glabrous leaves, always less than 2cm wide; lowest bracts lanceolate-elliptic.

Bhutan: C – Thimphu district (Dodena), Punakha district (Tinlegang, Ranchu–Nobding), Tongsa district (Chendebi, Dorji Gompa). On rocky slopes and near streams in disturbed broad-leaved forest, 2200–2600m. July–October.

21. S. lachenensis Clarke; *Pteracanthus lachenensis* (Clarke) Bremekamp, *Sympagis petiolaris* sensu Hara (135) non (Nees) Bremekamp, *Strobilanthes glutinosa* sensu Yamazaki (69) non Nees

Perennial herb 0.3–1m, gregarious and probably rhizomatous. Stems erect from woody root-stock, glabrescent below, pilose above. Leaves equal, ovate, obovate or broadly elliptic, 3–13 × 2–6.5cm, acute, base rounded or tapered and often decurrent onto petiole, serrate to crenate, pilose on both surfaces, sessile or shortly petiolate; petioles up to 3.5cm. Flowers in opposite pairs on spikes terminal on main stem and on axillary branchlets, forming a diffuse panicle; spikes lax below with flowers 1cm apart, dense above with flowers imbricate, 3–15cm; rhachis glandular-pilose. Lowest bracts leaf-like, ovate, sessile, 3–7cm, persistent; middle and upper bracts oblong or elliptic, often toothed, 8–17 × 2–6mm, persistent till flowers fall; bracteoles linear, 4–5mm. Calyx 8–16mm, glandular-pilose, lobes linear with one longer and becoming spathulate in fruit. Corolla blue, glabrous except for few hairs on lobes, 32–36mm, tube slightly bent at mouth, gradually widened to 12mm. Capsule 11–14mm, glandular-pilose at tip.

The inflorescence is very variable sometimes with solitary terminal spike, sometimes with 2–3 spikes from uppermost leaf axils giving a strongly paniculate appearance and sometimes with spicate branches from all leaf axils.

Bhutan: C – Punakha district (Pele La), **N** – Upper Kuru Chu district (Denchung–Dzulu); **Sikkim**: Lachen. On grassy slopes amongst rhododendron and juniper, 2200–3200m. June–October, probably every year.

22. S. subnudata Clarke; *Pteracanthus subnudatus* (Clarke) Bremekamp

Perennial herb, 10–20(–30)cm. Stems glabrous (-thinly pilose), erect from creeping root-stock. Leaves unequal, narrowly elliptic, 2–6 × 0.8–2cm, shortly acuminate, base attenuate and often decurrent, both surfaces glabrous or thinly pilose, serrate, often deciduous at flowering time; petioles 0–2cm. Flowers in opposite pairs, 5–15mm apart in short lax spikes which terminate main stem and axillary branchlets; spikes 2–6cm, simple; rhachis glandular-pilose. Lower bracts ovate, sessile, 8–30 × 5–15mm, deciduous as flowers fall; upper bracts broadly elliptic, c 5mm; bracteoles linear-oblanceolate, 3–4mm. Calyx 12–16mm, pale green, glandular-pilose, lobes linear-oblanceolate, one nearly twice as long as others. Corolla deep blue, thinly pilose on lips, 3–4.4cm, tube gradually widened to c 1.2cm at mouth, bent in upper half. Capsule 11–14mm, glandular-pilose at tip.

Darjeeling: Bhotia Basti; **Sikkim:** Lachung. Grassland in scrub and open forest, 1800–2750(–3500)m. August–December(–April).

23. S. urophylla (Nees) Nees; *Pteracanthus urophyllus* (Nees) Bremekamp

Small branched undershrub. Stems glabrous. Leaves slightly unequal, narrowly ovate-elliptic, 5–14 × 1.5–5.5cm, shortly acuminate, base cuneate, serrate, paler beneath, with prominent veins, glabrous; petioles 0.5–4cm. Flowers in opposite pairs, 1–1.5cm apart in axils of leaf-like bracts on short axillary branchlets. Bracts ovate-elliptic, petiolate, 8–30mm, acute, glabrous, very persistent; bracteoles obovate-spathulate, petiolate, c 4mm, persistent, equalling their petioles. Calyx 7–8mm, pubescent, lobes linear, acuminate, one longer than others. Corolla blue (drying yellow), pilose, fragrant, 2–2.5cm, tube strongly but regularly widened and bent from base to 1.2cm at mouth. Capsule not known.

Darjeeling: Nagri Spur, Bhotia Basti, Rungyong etc. 1800m. May.

24. S. wallichii Nees; *S. atropurpurea* Nees, *Ruellia alata* Wall., *Pteracanthus alatus* (Wall.) Bremekamp

Perennial herb 0.3–0.5(–1)m. Stems glabrous (rarely pilose), erect from creeping woody root-stock. Leaves equal or slightly unequal, ovate or elliptic, 1.2–5(–16) × 1–3.5(–6.5)cm, acute, base rounded to attenuate and shortly decurrent, serrate to coarsely crenate, glabrous or thinly pilose, sessile above, petiolate below; petioles 0–2cm. Flowers in opposite pairs in axils of leaf-like bracts forming small axillary 1-sided spikes, but spikes often much reduced and flowers solitary in axils of leaves on main stem; bracts like small leaves, persistent; bracteoles linear, oblong, c 6mm. Calyx 1–1.5cm in flower, accrescent in fruit to 2(–2.8)cm, glabrous, pilose or glandular, lobes linear, obtuse (rarely subacute), equal. Corolla blue (rarely white), glabrous, 28–35(–50)mm, tube slightly inflated above base, widened to c 15mm at mouth, bent sharply below mouth. Capsule 13–19mm, glabrous.

Bhutan: C – Ha to Tashigang districts, **N** – Upper Mangde Chu, Upper Bumthang Chu and Upper Kuru Chu districts; **Darjeeling:** Darjeeling, Kurseong, Tanglu etc.; **Sikkim:** Lachen, Phusum, Zongri etc.; **Chumbi:** Chumbi, Yak La. Abundant and locally subdominant on the forest floor in moist fir and hemlock forest, extending upwards into open *Rhododendron* scrub (2200–)2600–3960m. June–September, every year in the E Himalayas.

25. S. helicta Anderson; *Echinacanthus calycinus* (Nees) Nees, *Pteracanthus calycinus* (Nees) Bremekamp. Nep: *Ankla*; Sha: *Khamtagmutsee*. Fig. 103m–p.

Monocarpic undershrub, 0.5–1m. Stems erect, usually glabrous, sometimes pubescent above. Leaves nearly equal, elliptic, 5–12 × 1–6cm, acuminate at both ends, serrate, dark green above, pale beneath, veins very prominent, glabrous except on margins near base; petioles 0.2–5cm, bifariously pubescent. Flowers usually solitary, 0.6–1cm apart, in 1-sided, axillary spikes; rhachis 2–8cm, often zigzag, glabrous or glandular pubescent. Bracts linear, 5–13mm, one in each pair shorter than other, glabrous or glandular-pubescent, persistent; bracteoles linear, 2–3mm. Calyx 13–25mm, glabrous, lobes lanceolate, acuminate, often keeled, subequal. Corolla white or flushed pale purple, glabrous, 35–40mm, tube cylindrical for c 5mm, then abruptly widened to c 18mm, narrowed slightly and bent 90° below mouth. Capsule 1.8–2cm, glabrous.

Bhutan: S – Deothang district (Yongla near Keri Gompa), **C** – Mongar district (Namning); **Darjeeling:** Jorpokhri, Baghora, Sureil, Debrepani; **Sikkim:** Rathang, Rangpo. Moist broad-leaved hill forest, 1600–2300m. September–October, flowering in 12-year cycles, a flowering expected in 1998.

26. S. inflata Anderson; *Pteracanthus inflatus* (Anderson) Bremekamp

Pubescent undershrub, 0.5–1m. Stem pubescent. Leaves as for *S. helicta* but pubescent on both surfaces, oblique and broadly cuneate or rounded at base. Flowers sometimes in 1-sided spikes like *S. helicta* but always paired and spikes often reduced with 1–2 flower pairs only. Bracts villous, petiolate, narrowly ovate below, ovate-spathulate above, 6–30m, resembling small leaves, persistent, with petioles 3–10mm; bracteoles linear, c 3mm, persistent. Calyx 11–14mm, grey-pubescent, lobes whitish inside. Corolla as for *S. helicta* but white in bud, purple in flower, pubescent. Capsule 14–17mm, glabrous.

Darjeeling: Jorpokhri, Ghumpahar, Mongpu; **Sikkim:** Rangpo. 1800–2350m. September–October, perhaps like *S. helicta* at long intervals.

27. S. divaricata (Nees) Anderson; *Diflugossa divaricata* (Nees) Bremekamp. Fig. 1041–p.

Branched undershrub 10–50cm. Stems dark green, glabrous, often zigzag above, erect from a creeping root-stock. Leaves very unequal, 1–15 × 0.5–4.5cm, obscurely serrulate, glabrous; smaller leaves ovate, acute, rounded at the base, sessile, often deciduous; larger leaves broadly lanceolate (-elliptic), acuminate, cuneate at base, subsessile or with petioles up to 3mm. Flowers in opposite pairs (sometimes solitary) on small axillary branchlets forming a lax spicate inflorescence. Bracts elliptic, 2–5mm, fugacious; bracteoles obovate, 1–2mm, fugacious. Calyx 8–15mm, lobes linear-lanceolate, acuminate, subequal, glabrous when young, pilose at tips later. Corolla deep purple (rarely white), glabrous, 31–37mm, tube straight or curved, gradually widened to c 10mm at mouth. Capsule 12–15mm, glabrous.

Bhutan: S – Samchi district (Sangura (117), Sengdhen), Chukka district (Gedu–Bunakha), **C** – Thimphu district (Chapcha); **Darjeeling:** Tanglu, Sureil, Darjeeling, Senchal etc.; **Sikkim:** Rathang, Yoksum etc. Moist broad-leaved hill forest, locally abundant, 1700–2700m. July–November, flowering gregariously, last in 1992.

28. S. anisophylla (Loddiges) Anderson; *Goldfussia anisophylla* (Loddiges) Nees

Small undershrub 30–60cm high. Stems erect, glabrous, much branched, sometimes zigzag, often with knobbly remains of old leaf nodes below. Leaves very unequal, the smaller very small and soon deciduous (rarely, in cultivated fm. *isophylla* (Nees) J.R.I. Wood, equal), lanceolate or oblong-lanceolate, 1.5–9 × 0.4–1.3cm, acuminate, base cuneate, glabrous, obscurely serrulate, sessile or with petioles up to 9mm. Flowers in small, shortly pedunculate, axillary clusters on main stem and on axillary branchlets, clusters sometimes developing into short 1-sided cymes; peduncles 2–12mm. Bracts elliptic, herbaceous, 6–7mm, obtuse, deciduous; bracteoles oblong-lanceolate, 3–4mm. Calyx 6–9mm, glandular-pubescent with reddish glands, lobes linear, acute, subequal. Corolla pale blue or white, 2.5–2.8cm, glandular-pubescent, tube slightly curved, gradually widened to 1cm. Capsule 9–10mm, glandular-pubescent.

Bhutan: S – Samchi district (Charmurchi River), Sarbhang district (Phipsoo), Deothang district (Deothang (276)); **W Bengal Duars:** Buxa (229). On rocks and cliffs beside rivers in subtropical forest, 300–1000m. October–December.

29. S. hamiltoniana (Steudel) Bosser & Heine; *S. colorata* sensu F.B.I. non Nees, *S. crinita* (Nees) Anderson, *S. laevigata* Clarke, *Goldfussia colorata* Nees, *G. crinita* Nees, *Diflugossa colorata* (Nees) Bremekamp, *D. crinita* (Nees) Bremekamp

Branched herb 0.3–1.5m. Stems glabrous. Leaves equal or slightly unequal, broadly elliptic, 5–19 × 2–cm, acute or shortly acuminate, base cuneate (subcordate on very large leaves), serrate, glabrous; petioles 0–8cm. Flowers in a large, lax, open, leafless terminal panicle; branches capillary, glabrous or finely glandular-pilose. Bracts oblong-elliptic, 3–5mm, herbaceous, soon deciduous; bracteoles similar but narrower. Calyx 0.8–1cm, greyish, glabrous, lobes linear-oblong, obtuse, subequal. Corolla white often faintly flushed with mauve, 3.2–4cm, glabrous, tube straight, gradually widened to c 1.2cm at mouth. Capsule 1–1.4cm, glabrous.

Bhutan: S – Samchi, Phuntsholing, Chukka, Gaylegphug, Manas and Deothang districts, **C** – Mongar district (Shongar); **Darjeeling:** Dumsong, Git Jhora, Pakhihaga etc.; **W Bengal Duars**. Shaded streamsides in very moist, subtropical forest, 270–1400m. August–November, every year.

30. S. cusia (Nees) Kuntze; *S. flaccidifolia* Nees, *Baphiacanthus cusia* (Nees) Bremekamp. Dz: *Ran*; Khengkha: *Sangja*; Sha: *Yangsawa*; Eng: *Assam indigo*.

Erect, branched, dark green undershrub, 0.5–1.5m. Stems glabrous or minutely brown-puberulent. Leaves equal, broadly elliptic, 4–18 × 1.7–8.5cm, acute, base attenuate, serrate, dark above, paler beneath, glabrous or minutely puberulent on veins beneath; petioles 0.5–3cm. Flowers in opposite pairs on short leafy axillary branchlets 3–10cm, often forming a leafy terminal panicle. Bracts oblanceolate-obovate, petiolate, 12–23mm including petiole, leaf-like, deciduous with flowers; bracteoles linear-oblanceolate, 2–3mm, deciduous before bracts. Calyx 4–15mm in flower, accrescent in fruit to 25mm, minutely puberulent, 4 lobes linear, acute, one spathulate and much longer. Corolla blue, glabrous, 3.5–7cm, tube straight, cylindrical for c 1.5cm, then gradually widened to 1–2cm at mouth. Capsule 12–24mm, glabrous.

Bhutan: S – Manas district (Panbang (165)), Deothang district (Mikuri, N of Decheling), **C** – Punakha district (Mendegang), Mongar district (Mongar). Native of the Indo-Burmese frontier but widely cultivated in eastern Bhutan, 500–1800m. December–February, rarely and irregularly in Bhutan.

Leaves are fermented to produce blue dye (125, 165) used in traditional cloth making.

10. PHAULOPSIS Willdenow

Small herbs. Leaves petiolate, entire or indistinctly toothed. Flowers in short, dense, 1-sided, bracteate, leafy terminal spikes. Bracts orbicular, imbricate, usually bearing 3 flowers each with its own bracteole. Calyx 5-lobed, one lobe ovate, much larger than the other 4 linear lobes. Corolla small, tubular with 5 subequal spreading lobes; stamens 4, 2-celled, cells at same height. Capsule clavate, 4-seeded from base, placenta rising from base when fruit is ripe; seeds lens-shaped with elastic hairs. *Phaulopsis*, *Rungia* and *Dicliptera* are the only genera in which the placenta rises up in this way.

1. P. imbricata (Forsskål) Sweet: *P. parviflora* Willdenow, *P. dorsiflora* (Retzius) Santapau

Low, much-branched, decumbent or ascending herb. Stems pubescent, rooting at nodes, rarely longer than 30cm. Leaves diminishing in size upwards, elliptic, asymmetric, 15–80 × 3–35mm, acuminate at both ends, very sparsely pubescent especially on veins and margins, leaves in each pair often unequal in size; petiole 5–35mm. Flowering spikes 1–8cm. Bracts pale green, becoming scarious with age, orbicular, 7–11 × 10–15mm, ciliate and glandular-pilose; bracteoles elliptic, c 9mm, ciliate. Calyx 7mm, glandular-pilose, lobes linear, acuminate. Corolla white, 9mm, sparsely pilose. Capsule shiny, c 6mm, glandular-pilose near tip.

Bhutan: S – Samchi district (Chengmari, Samchi), Phuntsholing district (Phuntsholing), Gaylegphug district (Gaylegphug, Aie Bridge); **Darjeeling:** terai (unlocalised Ribu & Rohmoo and Cowan collections); **W Bengal Duars**. Disturbed ground by roads and around buildings, 250–500m. October–February.

11. ERANTHEMUM L.

Undershrubs. Leaves petiolate, entire or obscurely toothed; cystoliths very prominent. Inflorescence of terminal spikes, sometimes branched and paniculate. Bracts large, prominently veined and often variegated green, yellow and white; bracteoles inconspicuous, usually slightly shorter than calyx. Calyx 5-lobed to about halfway, lobes linear-lanceolate. Corolla with long narrow tube and 5 subequal spreading lobes, usually blue or red. Stamens 2, anthers 2-celled, muticous, exserted; style long. Capsule clavate with solid base, 4-seeded; seeds compressed, lens-shaped, elastically hairy.

1. Corolla pink or red, tube ventricose from about middle 2
+ Corolla blue, tube narrowly cylindrical for almost all its length 5

2. Corolla pink; bracts and leaves almost glabrous .... **1. E. griffithii**
+ Corolla crimson to orange-red; bracts and leaves distinctly pubescent at least on veins 3

3. Corolla tube less than 3cm; bracts obovate, abruptly contracted to point **4. E. splendens**
+ Corolla tube 3–4.5cm; bracts elliptic to obovate, gradually narrowed to point 4

4. Corolla tube more than 3.9cm; stamens exserted 1–2cm; leaves pilose above **3. E. erythrochilum**
+ Corolla tube less than 3.4cm; stamens exserted only slightly; leaves glabrous above .... **2. E. tubiflorum**

5. Bracts ciliate, acuminate but not mucronate; inflorescence of solitary terminal spikes .... **6. E. ciliatum**
\+ Bracts glabrous to pubescent but never ciliate, mucronate; inflorescence of panicled spikes .... **5. E. pulchellum**

1. E. griffithii (Anderson) Bremekamp & Nannenga-Bremekamp; *Daedalacanthus griffithii* Anderson. Fig. 105a–g.

Erect undershrub to 2.5m; stems glabrous except for few hairs near nodes. Leaves lanceolate or narrowly elliptic, slightly falcate, 5–26 × 2–8cm, attenuate at both ends, entire, glabrous; petiole 5–15mm, nearly glabrous. Inflorescence of terminal and axillary spikes, forming a small panicle; spikes 4–15cm; peduncles 0–3cm. Bracts elliptic or nearly so, 20–30 × 11–15mm, tapering to a point, glabrous, variegated green and yellow; bracteoles linear, c 6mm. Calyx split to halfway only, c 8mm, sparsely glandular-hairy, lobes linear-acuminate. Corolla curved, pink, glabrous except for few glandular hairs; tube 20–30mm, cylindrical in the lower half, gradually widened to c 8mm at mouth; lobes ovate, 4–6 × 3–4mm. Anthers exserted c 5mm. Capsule c 2cm, glabrous.

Bhutan: S – Samchi district (Torsa River), Phuntsholing district (Rinchending–Sorchen), Sankosh district (Pinkua River), Sarbhang district (Sarbhang). **W Bengal Duars:** Chokerboo. Locally abundant in secondary scrub on well-drained slopes, 200–700m. November–February (for a few weeks only).

2. E. tubiflorum (Anderson) Lindau; *Daedalacanthus tubiflorus* Anderson. Fig. 105j.

Similar to *E. griffithii* but stems bifariously pubescent, eventually glabrescent; leaves up to 23 × 8cm, distinctly toothed, glabrous above but sparsely pubescent below, especially on veins; petiole pubescent; bracts elliptic, 14–22(–28) × 9–14mm, tapering to a point, densely pubescent on veins; calyx 9–10mm; corolla crimson, sparsely glandular–pubescent, tube 30–34mm, lobes oblong, 6–9 × 4–6mm. Capsule not seen.

Bhutan: S – Gaylegphug (117). This record is improbable and requires confirmation.

3. E. erythrochilum J.R.I. Wood; *E. splendens* sensu Sikdar (231) non (Anderson) Siebert & Voss. Fig. 105k.

Similar to *E. griffithii* and more particularly to *E. tubiflorum* but differing in its simple inflorescence, larger flowers and strongly exserted anthers. Stem

---

FIG. 105. **Acanthaceae.** a–g, *Eranthemum griffithii*: a, habit (× ½); b, bract (× 1); c, calyx (× 2⅔); d, dissected corolla (× ⅔); e, upper part of stamen (× 2); f, capsule (× 1⅓); g, seed (× 1⅓). h–l, bracts of *Eranthemum* species (×1): h, *E. ciliatum*; i, *E. splendens*; j, *E. tubiflorum*; k, *E. erythrochilum*; l, *E. pulchellum*. m–u, *Acanthus carduaceus*: m, habit (× ½); n, leaf (× ⅓); o, bract (× ⅔); p, bracteole (× ⅔); q, upper calyx segment (× ⅔) r, lower calyx segment (× ⅔); s, dissected corolla, partly spread out (× ⅔); t, dehisced capsule (× ⅔); u, seed (× ⅔). Drawn by Louise Olley.

c
b
a
e
d
f
g
h
i
j
k
l
p
q
r
s
t
u
m
n
o

pubescent. Leaves subentire, thinly pilose on both surfaces and pubescent on veins below; petiole pilose. Inflorescence a solitary terminal spike 5–14cm. Bracts elliptic-obovate, 17–22 × 9–13mm, gradually narrowed to point, glandular-pubescent all over. Calyx 12–13mm. Corolla brilliant crimson, glandular-pubescent outside, tube 39–45mm, widened to c 10mm at mouth, lobes 10–12 × 7–8mm. Anthers and style exserted 1.5–2.5cm. Capsule not seen.

W Bengal Duars: Buxa; between Larrong and Sarbhanga Rivers. 300–1200m. January–February.

4. E. splendens (Anderson) Siebert & Voss; *Daedalacanthus splendens* Anderson. Nep: *Arklejhar* (34). Fig. 105i.

Similar to the three previous species but readily recognised by its soft yellow-green pubescence covering vegetative parts. Leaves entire, sparsely pilose on both surfaces and densely pubescent on veins. Inflorescence of panicled spikes; spikes pedunculate, 5–35cm, sometimes interrupted below; peduncle 2–5cm. Bracts obovate, 15–25(–45) × 8–17(–21)mm, abruptly narrowed to point, glandular-pubescent, strongly variegated-reticulate with green veins. Calyx c 9mm. Corolla orange-red, pubescent, tube 23–28mm, widened to 7mm at mouth, lobes elliptic, 5–6 × 0–4mm. Anthers and style exserted 1–2cm. Capsule c 1.8cm, glandular-pilose.

Bhutan: S – Samchi (117); **Darjeeling:** Panchkilla, Pankhabari, Sivok hills etc.; **W Bengal Duars:** Sealduba (175). Common in terai, 300–900m. December–February.

5. E. pulchellum Andrews; *E. nervosum* (Vahl) Roemer & Schultes, *Daedalacanthus nervosus* (Vahl) Anderson, *D. scaber* (Nees) Anderson. Nep: *Arklejhar* (34), *Keboo*. Fig. 105l.

Erect undershrub to 2m; stems glabrous. Leaves entire, elliptic (-ovate), 5–25 × 1.5–9cm, acute, attenuate at base, glabrous; petiole 0–3cm, glabrous. Inflorescence a trichotomously branched panicle formed of short subsessile spikes terminal on branches; spikes 3–8cm; peduncles 0–5mm. Bracts oblong to oblanceolate, 15–21 × 6–11mm, narrowed into distinct mucro, glabrous, variegated white and green; bracteoles linear, c 6mm. Calyx scarious, split to a little more than halfway, c 7mm, puberulent, lobes linear, acuminate. Corolla blue, glabrous, tube narrowly cylindrical for whole length, 15–20mm, lobes ovate, 6–8 × 5–6mm. Anthers shortly exserted. Capsule 1–1.1cm, glabrous.

Bhutan: S – Samchi district (Daina Khola), Sankosh district (Sankosh), Deothang district (Deothang), **C** – Punakha district (Sankosh Valley); **Darjeeling:** Rangit, Sivok, Tista, Mahanadi, Bagrakot etc.; **Sikkim:** Kerang. In moist gullies and by streams in dry subtropical forest, 300–1100m. February–April.

6. E. ciliatum (Craib) Benoist; *Daedalacanthus ciliatus* Craib. Fig. 105h.

Very similar to *E. pulchellum* but slenderer. Leaves narrowly elliptic, not exceeding 5cm wide, acuminate, attenuate at base and decurrent on petiole. Inflorescence of simple, terminal spikes (never paniculate); bracts elliptic, 24–33mm, gradually tapered to long acute point, margins ciliate. Corolla as for *E. pulchellum* but tube up to 30mm.

Bhutan: C – Mongar district (Lingmethang). Weed in a citrus orchard, 850m. March.

E. purpurascens Nees [*E. strictum* sensu Yamazaki (71) non Roxb., *Daedalacanthus purpurascens* (Nees) Anderson] is said to occur in Bhutan and Sikkim (71). Although *E. purpurascens* occurs in Nepal this is almost certainly an error. *E. purpurascens* is very similar in its inflorescence to *E. ciliatum* but can be distinguished by its ovate, acuminate, obtuse tipped bracts which are strongly divergent, those near base of spike often spreading at about 90° from axis. Spikes are often, but not always, long-pedunculate.

12. ACANTHUS L.

Herbs or shrubs. Leaves simple or pinnatifid, toothed, usually spiny. Inflorescence of dense terminal or, less commonly, axillary spikes. Bracts large, ovate, spiny-margined, growing in size as plant matures; bracteoles lanceolate, unarmed. Calyx 4-lobed, with 2 outer lobes larger than 2 inner lobes, one of the larger lobes ending in 2 points, the other in 1. Corolla with short ovoid tube and 3-lobed lower lip, upper lip absent. Stamens 4, included; anthers 1-celled, densely hairy. Capsule ellipsoid, shiny, 4-seeded; seeds glabrous, orbicular.

1. Shrub; stem glabrous; leaves deeply pinnatifid, lobes narrowly triangular, viciously spine-tipped, mostly 3–10cm; bracts usually 1.5–3cm wide **1. A. carduaceus**

\+ Herb; stem rufous-villous; leaves spiny-serrate, sometimes sinuate but lobed to less than 1cm; bracts less than 1cm wide .... **2. A. leucostachyus**

1. A. carduaceus Griff. Fig. 105m–n.

Shrub 2–3m, usually gregarious and probably rhizomatous. Stems erect, glabrous. Leaves sessile, elliptic, 30–60 × 10–20cm, pinnatifid with narrowly triangular lobes 3–12cm, glabrous, main veins terminating in spines. Inflorescence of dense terminal and axillary spikes 10–30cm, terminal spikes often sessile, axillary spikes borne on peduncles up to 20cm. Bracts ovate, 2–4 × 1.5–3cm, pilose to lanate, hairier on upper surface, spiny-margined; bracteoles linear-lanceolate, c 3 × 0.8cm, acuminate, villous or ciliate-margined. Calyx 4-lobed, ciliate, 2 larger lobes ovate-elliptic, 22–30 × 7–10mm, 2 smaller lobes linear-lanceolate, 15–18 × 5mm. Corolla white, 2.5–3cm, glabrous outside, pilose inside, tube c 1cm. Capsule brown, 2.5–3cm, glabrous.

Bhutan: S – Chukka district (Sinchu La), Sankosh district (above Goshi, above Daga Dzong), **C** – Punakha district (Tinlegang), Tongsa district (Dakpai, Pertimi), Mongar district (Unjar, Namning–Lingmethang); **Darjeeling:** Birch Hill, near Lloyd Botanical Gardens; **W Bengal Duars:** Ramiti (231), Buxa. A spectacular but very local species of scrubby cliff slopes and rocky gullies, 900–1800m. November–March.

Endemic to Bhutan and Buxa; records from Darjeeling are all recent and presumably represent introductions.

2. A. leucostachyus Nees

Herb. Stem rufous-villous, slightly succulent, decumbent and rooting at nodes, up to 50cm. Leaves shortly petiolate, elliptic, 5–25 × 3–8cm, glabrous above, white-scabrid and reticulate below, margin spiny-dentate, sometimes sinuate. Inflorescence of subsessile, terminal spikes 8–18cm. Bracts oblong-obovate, 15–20 × 5–8mm, spiny-margined, pubescent; bracteoles linear-lanceolate, c 15mm, acuminate, pilose. Calyx 4-lobed, scarious, pilose, 2 larger lobes oblong, 15–20mm, obtuse, mucronate, 2 smaller (inner) lobes linear-lanceolate, c 15mm, long acuminate. Corolla white, 20–25mm, pilose inside and outside, tube c 7–8mm. Capsule brown, 1–1.5cm, glabrous.

Bhutan: locality unknown (Griffith collection, ?Deothang district). Shaded banks on forest margins, 900–1300m. January–March.

13. BARLERIA L.

Undershrubs, rarely herbs, often armed. Leaves entire, shortly petiolate, usually with prominent cystoliths. Inflorescence of axillary cymes, sometimes secund or reduced to dense clusters or flowers solitary. Bracts present or absent; bracteoles 2. Calyx deeply 4-lobed, 2 outer lobes larger (one usually bifid), becoming membranous, 2 inner lobes much smaller. Corolla usually large, funnel-shaped, 5-lobed, lobes subequal. Fertile stamens 2, included; anthers 2-celled with cells oblong, at same height, acute or muticous at base; staminodes present, 2. Capsule 2–4-seeded; seeds flattened, lens-shaped, hairy.

1. Corolla yellow; interpetiolar spines present; capsule 2-seeded **1. B. prionites**
\+ Corolla blue or purple; interpetiolar spines absent; capsule 4-seeded 2

2. Flowers in dense axillary clusters; bracteoles linear-lanceolate, erect; bracteoles and calyx spiny-margined and spinescent .... **2. B. cristata**
\+ Flowers in 1-sided axillary cymes; bracteoles ovate-lanceolate, deflexed; bracteoles and calyx unarmed .... **3. B. strigosa**

1. B. prionites L.

Usually strigose undershrub to 1.5m, bearing prominent, greyish, 3-forked, interpetiolar spines. Stem pale grey, much branched. Leaves elliptic-obovate,

6–12 × 2–5cm, acute, base attenuate, glabrous above, pubescent or glabrous beneath; petiole 0–2cm. Flowers solitary in axils of upper leaves, becoming denser upwards and forming short terminal spikes. Bracts oblong (-ovate), 10–25mm, acute, spine-tipped; bracteoles linear, 8–15mm, spinescent. Outer calyx lobes ovate, 15–17mm, spine-tipped, glabrous; inner calyx lobes ovate-lanceolate, 10–12mm, spine-tipped. Corolla, pale yellow, 2.5–4cm, pubescent, 2-lipped, upper lip formed of 4 equal elliptic lobes 1–1.5cm, lower lip formed of 1 lobe 2–2.5cm; tube 1.5–2cm, slightly widened only. Capsule ovoid, 1.7–2cm, long-beaked, glabrous, 2-seeded.

W Bengal Duars: Apalchand, Baikunthapur Forest Division (175). Planted in hedges. September–February.

2. B. cristata L.

Pilose undershrub to 50cm. Stem wiry, much branched. Leaves (ovate-) elliptic (-oblong), 3–14 × 1–3.5cm, acute at both ends, pilose on both surfaces; petiole 0–1cm. Flowers subsessile in dense axillary clusters becoming capitate at branch ends. Bracts absent; bracteoles very variable, linear (-lanceolate), 13–22 × 1–3mm, acuminate, usually spiny-margined but sometimes bristly pilose-margined, spinescent with age. Outer calyx lobes lanceolate or narrowly elliptic, 20–25 × 5–13mm, pilose, reticulate, scarious, spiny-margined, one mucronate at tip, the other bimucronate; inner calyx lobes (linear-) lanceolate (-ovate), scarious-margined, 6–12mm. Corolla pinkish-purple, 55–65mm, tube pilose, narrowly cylindrical in the lower half and then gradually widened, lobes oblong-elliptic, c 15mm, glabrous. Capsule oblong, 15–17mm, glabrous, 4-seeded.

Bhutan: S – Deothang district, **C** – Punakha, Tongsa, Mongar and Tashigang districts; **Darjeeling:** Cok, Bamanpokhri; **W Bengal Duars:** Jalpaiguri. Common and characteristic plant of chir pine formations in all deep dry valleys of C Bhutan, usually growing as scattered individuals in dry grassland with *Cymbopogon*, *Aechmanthera* etc.; less common in foothills, 100–2000m. August–November.

3. B. strigosa Willdenow

Strigose undershrub to 50cm. Leaves ovate or broadly elliptic, 4–19 × 2–10.5cm, acute, attenuate at base and decurrent on petiole, strigose especially on margin and veins below; petiole 0–3cm, winged. Flowers subsessile in dense, shortly pedunculate, 1-sided, axillary cymes 3–6cm. Bracts absent; bracteoles green, lanceolate-ovate, 8–20 × 4–9mm, acute or with strap-shaped tip, spreading or reflexed, strigose, margin undulate, unarmed. Outer calyx lobes ovate, greenish-purple, 18–28 × 14–18mm, strigose on veins and margins, reticulate, becoming scarious, one obtuse, the other 2-lobed; inner calyx lobes linear, c 10mm, acuminate. Corolla blue (-purple), 5–6.5cm, tube glandular-pubescent, cylindrical in lower half, then gradually widened; lobes obovate-oblong, c 2cm, thinly pilose. Capsule oblong, 15–17cm, glabrous, 4-seeded.

Bhutan: S – Samchi district (Samchi), Phuntsholing district (Torsa River);

Darjeeling: Rilli Forest; **W Bengal Duars:** Jalpaiguri, Bhutanghat (231), Apalchand (175). In subtropical scrub and forest, 200–400m. September–December.

14. ASYSTASIA Blume

Herbs or undershrubs. Leaves entire, petiolate. Flowers in simple or compound spikes or racemes. Bracts and bracteoles linear, shorter than calyx. Calyx 5-lobed, lobes linear, equal. Corolla with ventricose or cylindrical tube, 5-lobed, lobes subequal. Stamens 4, not exserted, 2-celled, cells oblong, at same height, base muticous or spurred. Capsule 4-seeded, clavate with a long seedless base; seeds flattened, lens-shaped, glabrous.

1. A. macrocarpa Nees; *Mackaya macrocarpa* (Nees) Das

Decumbent or ascending undershrub to 2m, but usually less. Stem glabrous below, sparsely pilose above. Leaves lanceolate (ovate or oblong), 2–13 × 1–4.5cm, acuminate at both ends, dark green, sparsely pilose on both surfaces and especially on veins below; petiole 0–1.5cm, pilose. Inflorescence of short, solitary, terminal racemes 2–6cm; rhachis pilose; flowers in opposite pairs, internodes very short above but up to 2cm below; pedicels 5–6mm. Bracts linear, 2–3mm. Calyx 5–7mm, glandular-pubescent, lobes lanceolate, acuminate. Corolla pale purple, drying white, 28–38mm, thinly glandular-pubescent, tube ventricose c 10mm above base, lobes elliptic, c 10mm. Anther cells shortly spurred. Style persistent. Capsule 28–36mm, glandular-pubescent.

Bhutan: S – Samchi district (Torsa River), Phuntsholing district (Phuntsholing), Sankosh district (W of Pinkua River), Sarbhang district (Singi Khola); **Darjeeling:** Rangit; **W Bengal Duars:** Jaldapara. Open, subtropical forest, often in seasonally moist places, 200–600m. February–April.

15. PSEUDERANTHEMUM Radlkofer

Shrubs or, less commonly, herbs. Leaves entire, petiolate. Flowers in simple or compound racemes. Bracts and bracteoles linear, much shorter than calyx. Calyx 5-lobed, lobes linear, equal. Corolla with long narrowly cylindrical tube and 5 subequal, spreading lobes. Fertile stamens 2; staminodes 2; anthers not exserted, 2-celled, cells at same height, oblong, muticous or acute at base. Capsule woody, 4-seeded, clavate with long, seedless base; seeds flattened, lens-shaped, glabrous.

1. P. palatiferum (Nees) Radlkofer; *Eranthemum palatiferum* Nees

Undershrub, 0.3–1m. Stem usually glabrous, sometimes pubescent above. Leaves elliptic(-obovate), 5–20 × 1.5–8cm, acuminate at both ends, often abruptly so at apex, dark green above, paler below, glabrous except on veins;

petiole 0.5–2cm. Inflorescence of long terminal racemes, sometimes solitary, often 2–5 arising from tip of stem, rarely compound and paniculate; racemes 3–30cm, very dense above but with internodes up to 3.5cm below, flowers arranged in small clusters; rhachis glandular-pubescent; pedicels 0–5mm. Bracts linear-triangular, c 2.5mm. Calyx 5–7mm, glandular-pubescent, lobes linear. Corolla tube pubescent, c 35mm, 2mm wide at mouth; corolla lobes elliptic, c 10 × 7mm, 2 slightly smaller than other 3 giving a slightly 2-lipped appearance, mauve with a yellow spot, glabrous. Anthers acute at base. Capsule c 2cm, pubescent.

Bhutan: S – Gaylegphug district (Gaylegphug, Surey), **C** – Mongar district (Shongar); **Darjeeling:** Rayeng, Dalkajhar. In moist, subtropical forest, 300–1600m. August–April.

16. MACKAYA Harvey

Small shrubs similar to *Pseuderanthemum* in having a racemose inflorescence and flowers with 2 fertile stamens only but differing by its distinctive, curved, ventricose corolla.

1. M. indica (Nees) Ensermu; *Eranthemum indicum* (Nees) Clarke, *Pseuderanthemum indicum* (Nees) A. & J. Cowan, *Odontonomella indica* (Nees) Lindau

Shrub 1–2.5m. Stem erect, brown, glabrous. Leaves elliptic, 3–21 × 1.5–8cm, shortly acuminate, cuneate at base, glabrous, paler beneath, often deciduous on higher altitude plants at flowering time; petiole 0.5–6cm. Inflorescence of lax terminal and axillary racemes, those in leaf axils shorter and always solitary; racemes 1–19cm, flowers usually in opposite pairs, internodes up to 1.5cm below, less above; rhachis glabrous or puberulent; pedicel 0–5mm. Bracts linear-triangular, c 2mm. Calyx c 4mm, puberulent, lobes lanceolate, acuminate. Corolla white or pink with darker red veins, 2.5–3cm, finely glandular-pubescent, tube ventricose from near base, 5-lobed but weakly 2-lipped. Stamens included, anthers muticous. Style persistent long after the corolla falls. Capsule 2.5–3cm, glabrous.

Bhutan: S – Samchi, Phuntsholing, Sarbhang, Gaylegphug and Deothang districts, **C** – Punakha and Tongsa districts; **Darjeeling:** Lebong, Dumsong. In moist subtropical forest especially in wooded gullies, 500–1700m. December–April.

17. CODONACANTHUS Nees

Small trailing herb similar to *Mackaya* and *Pseuderanthemum* in having only 2 fertile stamens but differing in its short strongly ventricose corolla, very short staminodes and distinctive smooth pollen.

1. C. pauciflorus Nees

Small wiry herb up to 30cm. Stem pubescent. Leaves elliptic, 2–9 × 1.5–4.5cm, acute or shortly acuminate, cuneate at base, glabrous except on veins below, dark above, paler below; petiole 0–1cm. Inflorescence of branched racemes forming a terminal panicle, with or without single racemes in leaf axils below; racemes 3–8cm, lax, internodes 3–13mm, flowers solitary or (rarely) in clusters, rhachis puberulent; pedicels 1–3mm. Bracts lanceolate, 1.5mm. Calyx 2–2.5mm, puberulent, lobes lanceolate. Corolla white or pale pink, campanulate, c 1 × 1cm, glabrous. Stamens included, very short, anther cells muticous. Style persistent. Capsule narrowly clavate, c 1.5cm, puberulent.

Bhutan: S – Sarbhang district (Lam Pati, Sarbhang High School), Gaylegphug district (Taklai Khola). In shade in subtropical forest, 370–600m. February–March.

18. PHLOGACANTHUS Nees

Small shrubs. Leaves petiolate, entire, usually dark shiny green above and noticeably paler beneath. Flowers in small cymes, sometimes axillary but usually arranged in a dense raceme-like panicle, known as a thyrse. Bracts small, inconspicuous; bracteoles 0. Calyx deeply 5-lobed, segments very narrow. Corolla tubular, curved, shortly 2-lipped. Stamens 2; anthers 2-celled, cells oblong, muticous. Ovary glabrous. Capsule elongate, many-seeded from base; seeds lens-shaped, glabrous or hairy.

1. Flowers in axillary cymes; leaves hairy on veins .... **4. P. pubinervius**
+ Flowers in terminal thyrses; leaves completely glabrous 2

2. Corolla c 1cm, pinkish-purple, both lips equal in length, spreading; leaves all less than 15cm .... **3. P. vitellinus**
+ Corolla 1.5–2.5cm, orange-brown, red or yellow, upper lip clearly longer than lower, often reflexed lip; leaves mostly more than 15cm 3

3. Corolla yellow; anthers 2–3mm; thyrse interrupted below with internodes up to 2cm .... **2. P. guttatus**
+ Corolla orange-brown; anthers c 4mm; thyrse dense, not interrupted **1. P. thyrsiformis**

1. P. thyrsiformis (Hardwicke) Mabberley; *P. thyrsiflorus* (Roxb.) Nees. Nep: *Chua* (34).

Shrub up to 3m. Stems erect, glabrous. Leaves often crowded near branch tips, elliptic(-obovate), 11–28 × 2.5–8cm, shortly acuminate at both ends, glabrous; petiole 15–35mm. Flowers in a dense, uninterrupted terminal thyrse, 4–23 × 4–5cm, usually solitary, rarely 2–3; rhachis pubescent. Bracts linear, c 10mm, pubescent. Calyx c 6mm, pubescent, lobes linear. Corolla orange-brown, tubular, 20–25mm, c 8mm wide at mouth, pubescent; upper lip 6–7mm, spreading, about twice as long as deflexed lower lip. Anthers 4–5mm, shortly exserted. Capsule narrowly clavate, 2–3cm.

Bhutan: S – Samchi, Phuntsholing, Chukka, Sankosh, Sarbhang, Gaylegphug and Deothang districts, **C** – Punakha and Tongsa districts; **Darjeeling:** unlocalised (Cowan collection); **Sikkim:** Rangpo; **W Bengal Duars:** Muraghat Forest, Jalpaiguri etc. In subtropical forest and in secondary scrub, particularly in gullies and near streams, common, 200–1100m. February–March.

2. P. guttatus (Nees) Nees

Similar to *P. thyrsiformis* but leaves 11–20 × 4–7cm; thyrse 7–14 × 3–3.5cm, lax and interrupted especially below with internodes up to 2cm; corolla yellow, 15–18mm, 5mm wide at mouth, lower lip strongly bent downwards, slightly shorter than upper lip; anthers 2–3mm.

Bhutan: S – Deothang (315).

3. P. vitellinus (Roxb.) Anderson; *P. quadrangularis* (Hooker) Heine, *P. asperulus* Nees

Similar to the two previous species but leaves only 8–16 × 2–4cm; thyrse 5–12 × 2–2.5cm, rather lax, sometimes interrupted below; corolla pinkish-purple, often with yellow markings, c 10mm, tube widened from c 2mm at base to c 8mm at mouth, lips equal, spreading; anthers c 2.5mm.

Bhutan/Darjeeling: unlocalised Griffith collection (316).

4. P. pubinervius Anderson

Shrub up to 6m, much branched and spreading. Stems grey, puberulent. Leaves broadly elliptic, 4–13 × 2–8cm, acute at both ends, glabrous except for pubescent veins; petiole 0.5–3cm. Flowers in small, subsessile, axillary cymes c 2–3cm. Bracts minute, c 1mm. Calyx 5–8mm, thinly hairy except for conspicuous grey-pubescent margins of lobes. Corolla orange-yellow, strongly curved, shortly pubescent, c 15mm, 7–8mm wide at mouth; tube inflated from base; upper lip longer than lower. Anthers c 3mm, exserted. Capsule c 3cm.

Bhutan: S – Samchi district (Dorokha), Phuntsholing district (Sorchen), Gaylegphug district (Sham Khara), Deothang district (Deothang), **C** – Tongsa district (Pertimi, Khosela–Shemgang); **Darjeeling:** Sitong, Rongsong, Tista Valley and in the terai. Scattered in subtropical forest and scrub, 800–1500m. December–February.

19. LEPIDAGATHIS Willdenow

Herbs or undershrubs. Leaves entire or obscurely toothed. Flowers in dense, sessile, bracteate, 1-sided heads or spikes. Bracts usually scarious; bracteoles similar but smaller. Calyx 4–5-lobed (5-lobed in our species), lobes slightly unequal in size. Corolla usually small, 2-lipped; stamens 4, included; anthers

2-celled, cells at same height, muticous. Capsule 2–4-seeded; seeds flattened, hairy.

1. L. incurva D. Don; *L. hyalina* Nees

Short-lived perennial herb. Stems glabrescent, ascending to 50cm. Leaves petiolate, 1.5–16 × 1–5.7cm, glabrous or shortly glandular-pubescent especially on veins below; petiole 1–3cm; lower leaves commonly ovate and rounded at base; upper leaves larger, elliptic, acuminate at both ends. Flowers in dense, 1-sided, subsessile heads, 0.5–4cm, mostly in a terminal cluster but also with a few in axils of upper leaves. Bracts lanceolate-elliptic, scarious, c 7mm, glandular-puberulent, ciliate-margined; bracteoles similar, c 5mm. Calyx 5-lobed, green, 7–9mm, lobes linear-lanceolate, acute, ciliate. Corolla white, streaked with purple, c 1cm, glabrous. Capsule c 5mm, glabrous except at the tip, 4-seeded.

Bhutan: S – Samchi, Phuntsholing, Chukka, Sankosh and Sarbhang districts; **Darjeeling:** Sureil, Rayong N, Munsang etc.; **W Bengal Duars**. Weed of cultivated land and well-drained disturbed ground in orchards and secondary scrub, 300–1400m. November–April.

L. trinervis Nees is recorded from Darjeeling in F.B.I. but the plant was probably incorrectly labelled. It seems very improbable that this species, which is common on dry hills in C India, would occur in our area.

20. JUSTICIA L.

Herbs or shrubs. Leaves entire. Inflorescence in our species spicate or racemose. Bracts broad and conspicuous in some species but in others linear and inconspicuous; bracteoles linear or absent. Calyx subequally 5-lobed (except in *J. simplex* where one lobe is much smaller or absent); lobes linear-subulate. Corolla 2-lipped, tube short, upper lip notched, lower lip 3-lobed. Stamens 2; anthers 2-celled, one usually distinctly higher than other, spurred at base (except in *J. adhatoda* where cells are acute). Capsule clavate, 4-seeded; seeds glabrous in our species.

1. Bracts conspicuous, elliptic to ovate, at least 5mm wide and concealing calyx 2
\+ Bracts inconspicuous, linear, 3mm wide or less, not concealing calyx 4

2. Bracts obtuse, purple-margined; corolla c 2cm .... **2. J. atkinsonii**
\+ Bracts acute, without distinct marginal colour; corolla c 3cm 3

3. Bracts tomentose, often suffused with red; spikes sessile or nearly so; anthers spurred .... **6. J. brandegeana**
\+ Bracts minutely puberulent or glabrous; spikes borne on peduncles 2–15cm; anthers acute but not spurred at base .... **1. J. adhatoda**

4. Trailing herb; corolla pink, scarcely exceeding calyx .... **5. J. simplex**
\+ Small shrub; corolla white or yellow-brown, at least 3 × as long as calyx 5

5. Corolla white; leaves lanceolate to narrowly elliptic, not more than 2cm wide .... **3. J. gendarussa**
\+ Corolla yellow-brown; leaves broadly elliptic, 5–10cm wide **4. J. vasculosa**

1. J. adhatoda L.; *Adhatoda vasica* Nees, *A. zeylanica* Medikus. Nep: *Asuro, Kalo vashak.*

Shrub 1–1.5m. Stem greenish-brown, glabrescent. Leaves elliptic, 5–25 × 2–9cm, shortly acuminate, attenuate at base, glabrescent but sometimes persistently pubescent on veins below, petiole 0.5–3cm, puberulent. Inflorescence of short, dense, terminal and axillary bracteate spikes 2–8cm; peduncle 2–15cm. Bracts ovate to elliptic, acute, 10–20 × 5–12mm, glabrous to puberulent; bracteoles oblong-lanceolate, similar to calyx lobes. Calyx 5-lobed, c 8mm, puberulent; lobes oblong-lanceolate, caudate, pale-margined. Corolla white with purple lines, c 3cm, glabrous or pubescent, tube short and broad, upper lip hooded, lower lip with 3 ovate lobes c 1.5cm. Anthers not exserted, cells acute at base but not spurred. Capsule woody, 2.5–4cm, pubescent.

Bhutan: S – Phuntsholing, Sankosh, Sarbhang, Gaylegphug and Deothang districts, **C** – Punakha, Tongsa, Mongar and Tashigang districts; **Darjeeling:** Rakti; **W Bengal Duars**. Secondary scrub and broad-leaved forest, often in dry valleys and near settlements, locally abundant and possibly introduced in some places, 200–1610m. January–April.

Flowering shoots sold in markets and used medicinally. They are also reputed to be insecticidal.

2. J. atkinsonii Anderson

Shrub c 1m. Stem glabrous below, thinly bifariously pubescent above. Leaves ovate, 8–22 × 4–12.5cm, shortly acuminate, attenuate at base, nearly glabrous except on veins below; petiole 1–5cm. Inflorescence of dense, simple or few-branched, terminal spikes 4–12cm; peduncle 1–3cm. Bracts suborbicular-ovate, 7–14 × 5–10mm, rounded at apex, glandular-pubescent, purple-margined; bracteoles green, subulate, 10–13mm. Calyx 5-lobed, green, puberulent, 7–9mm, lobes linear. Corolla yellowish-white with rose markings, c 2cm, pilose. Anthers not exserted, anther cells spurred at base. Capsule 14–17mm.

Bhutan: locality unknown, eastern Bhutan (F.B.I.); **Darjeeling:** Pomong, Kalimpong, Pedong, Mahaldiram–Kali Khola. Subtropical forest, 600–1550m. September–November.

3. J. gendarussa L.f.; *Gendarussa vulgaris* Nees

Shrub 1–1.5m. Stem purplish, glabrous or obscurely bifariously pubescent. Leaves lanceolate or narrowly elliptic, 3–12 × 0.7–2cm, tapering to an obtuse apex, attenuate at base, glabrous, veins below prominent, purplish; petiole 0.5–1cm. Inflorescence of terminal and axillary spikes, often interrupted below, 2–11cm, flowers in clusters along rhachis, each cluster subtended by a narrowly lanceolate leaf-like bract 13mm below but shorter above; peduncle 0–4cm; floral bracts similar to calyx lobes. Calyx 5-lobed, brown, c 5mm, thinly pilose, lobes lanceolate. Corolla white with purple markings, c 15mm, glabrous, lower lip with narrowly elliptic lobes. Anthers exserted, lower anther cell white-spurred. Capsule c 12mm, glabrous, very rarely seen.

Bhutan: S – Phuntsholing; **Darjeeling:** terai (locality unknown); **W Bengal Duars:** Jaldapara. Cultivated in subtropical gardens, 200m.

4. J. vasculosa (Nees) Anderson

Perennial undershrub 0.6–1m. Stem glabrous. Leaves ovate-elliptic, 5–20 × 2–8cm, shortly acuminate, attenuate at base, glabrous; petiole 1–5cm, glabrous. Inflorescence of axillary and terminal, solitary or panicled spikes, often interrupted below, 2–12cm; flowers arranged in pairs along rhachis; peduncle 0–3cm. Bracts lanceolate, c 3mm. Calyx 5-lobed, c 4mm, pubescent, becoming scarious in fruit, lobes lanceolate, purple-tipped. Corolla yellow-brown with purple markings, c 13mm, pubescent, lower lip bent downwards, broadly elliptic, c 5mm. Anthers not exserted, lower anther cell indistinctly spurred. Capsule c 18mm, glabrous.

Bhutan: S – Gaylegphug district (3km N of Shershong Bridge). Streamsides in moist, subtropical forests, 700m. January–April.

Also collected south of Deothang, just outside our area.

5. J. simplex D. Don; *J. procumbens* L. var. *simplex* (D. Don) Yamazaki

Wiry perennial herb. Stems decumbent and rooting at the nodes, thinly bifariously pubescent. Leaves ovate-elliptic, 1.5–3 × 0.7–1.7cm, acute, attenuate at base, thinly pilose on both surfaces; petiole 0.3–1.2cm. Inflorescence of short, simple, terminal, hairy spikes, sometimes slightly interrupted below, 1.5–5 × 0.5–1cm, peduncle 2–4cm. Bracts similar to and equalling calyx lobes. Calyx c 6mm, 5-lobed with one lobe reduced in size, lobes linear-lanceolate, scarious-margined with green midrib. Corolla pink, c 6mm, glabrous except for a few hairs on lip, lower lip bent downwards, very shallowly lobed. Anthers not exserted, cells spurred at base. Capsule c 6mm, glabrous or with few hairs at tip.

Bhutan: S – Sankosh district (Daga Dzong), **C** –Tongsa district (Langthel), Punakha district (Punakha). Open rough grassland in dry valleys, 900–1600m. August–November.

The very similar *J. diffusa* Willdenow occurs in Nepal and might be expected in our area. It can be distinguished by its narrower (less than 5mm wide), almost glabrous spikes and bracts distinctly shorter than calyx.

6. J. brandegeana Wasshausen & L.B. Smith; *Beloperone guttata* Brandegee

Shrub 0.5–2m. Stem dark green, bifariously pubescent. Leaves ovate-elliptic or subrhomboid, dark green, 1.5–7 × 0.8–3cm, acute or obtuse, base cuneate, glabrescent above, softly pubescent below; petiole 0.3–3cm, pubescent. Inflorescence of terminal, uninterrupted, bracteate spikes 3–8cm; peduncle c 1cm. Bracts ovate, acute, green when young, becoming wine-red with maturity, 17–25 × 10–15mm, shortly tomentose; bracteoles broadly lanceolate, oblanceolate or elliptic, 8–13 × 5mm. Calyx 5-lobed, 4–5mm, pilose, lobes subulate, scarious. Corolla white, pilose, c 30mm, rather narrow, strongly 2-lipped for about half its length, lower lip with 3 lanceolate lobes c 5mm. Anthers slightly exserted, cells strongly spurred at base. Capsule not seen.

Bhutan: S – Samchi. Cultivated in gardens for its colourful bracts. February–March.

21. LEPTOSTACHYA Nees

Perennial herb. Inflorescence of panicled spikes, usually in Bhutan reduced to a single terminal spike. Bracts inconspicuous, shorter than calyx; bracteoles minute, linear. Calyx deeply 5-lobed, lobes similar. Corolla 2-lipped with short cylindrical tube. Stamens 2; anthers 2-celled, cells at same level or nearly so, not spurred. Capsule 4-seeded, placentas not arising from base. Very close to *Justicia* but differing in its muticous anther cells and hooded upper lip.

1. L. wallichii Nees; *Dianthera debilis* Clarke, *D. virgata* Clarke

Herb to 50cm. Leaves ovate, lanceolate or elliptic, 2–8 × 2–5cm, acute, base rounded or broadly cuneate, entire, nearly glabrous except on veins, petioles 1.5–4cm. Inflorescence with 1–3 spikes, 4–12cm, rhachis glandular-pubescent. Bracts oblong-lanceolate, 1.2–2mm. Calyx 2.5–3mm, lobes subulate. Corolla white, 5–8mm, tube narrowly cylindrical, 4–5 × c 0.5mm. Capsule clavate, 1–1.5cm, finely pubescent.

Bhutan: S – Phuntsholing district (Rinchending–Sorchen), Gaylegphug district (Gaylegphug (117)); **Darjeeling:** Samsing; **W Bengal Duars:** Lapchakhawa (229). Subtropical forest and regenerating scrub in moist ravines, 800–1200m. August–November.

22. ISOGLOSSA Oersted

Herbs or shrubs. Inflorescence an open panicle. Bracts very small, much shorter than calyx; bracteoles absent. Calyx deeply 5-lobed, lobes similar. Corolla 2-lipped, funnel-shaped. Stamens 2; anthers 2-celled, cells at different levels, not spurred at base. Capsule 4-seeded; placentas not rising from base.

Differs from all related genera in its inflorescence which does not consist of panicled spikes but is a true panicle.

1. I. collina (Anderson) Hansen; *Dianthera collina* (Anderson) Clarke

Herb to 50cm. Leaves ovate, 3.5–11 × 2–5cm, acuminate, base broadly cuneate, obscurely sinuate-margined, nearly glabrous, petioles c 15mm. Inflorescence a terminal panicle up to 18 × 10cm, rhachis glandular-pubescent. Bracts lanceolate, 1.5–2 × 0.5mm. Calyx 5–6mm (extending to 8mm in fruit), the lobes subulate. Corolla white with pinkish spots, 17–25mm, tube cylindrical for ⅓ of its length, then gradually widened; lips 8–10mm long, upper lip 2-lobed, lower lip with oblong-elliptic lobes. Capsule c 12mm, glabrous.

Bhutan: C – Tashigang district (Kari La); locality unknown (Griffith Bhutan collection (316)); **Darjeeling:** Rississum; **Sikkim:** Yoksam. Dry forest, 1600–2150m. September–October.

23. RHINACANTHUS Nees

Small shrubs. Inflorescence a panicle, usually divaricate but sometimes contracted, composed of 1–10 branches; flowers sessile or in small cymes. Bracts very small and inconspicuous, much shorter than calyx. Calyx deeply 5-lobed, lobes similar. Corolla 2-lipped, tube cylindrical, long; upper lip narrowly triangular, acuminate, entire, lower lip with 3 broad lobes. Stamens 2, anthers 2-celled, one above the other, not spurred at base. Capsule 4-seeded; placentas not rising from base.

1. R. calcaratus Nees

Small shrub to 1m. Leaves narrowly elliptic, entire, glabrous, 10–23 × 3–8cm, shortly acuminate, gradually tapering to short petiole 0.3–cm. Inflorescence of up to 10 spike-like branches, sometimes much reduced; branches 2–12cm, rhachis glandular-pubescent. Bracts lanceolate, 1–2mm. Calyx 5–6mm, glandular-pubescent, lobes lanceolate. Corolla 40–50mm, pubescent, tube cylindrical, 30–40 × 1–2mm, greenish-white; upper lip greenish-white, c 10mm; lower lip white, lobes elliptic, c 10 × 8mm. Ovary and style persistent long after corolla has fallen, ovary pubescent. Capsule not known.

Readily recognised by the distinctive cylindrical corolla with its long acuminate, entire upper lip.

Bhutan: S – Gaylegphug district (on escarpment above Aie (Shershong) Bridge), 500–700m. March–April.

24. RUNGIA Nees

Similar to *Justicia* in its entire leaves, dense spicate inflorescence, small 2-lipped flowers and 4-seeded capsules but differing in its placenta which rises elastically from base of ripe capsule. Both species found in Bhutan are prostrate weedy herbs with scarious bracts.

1. Corolla pink or white, c 8mm; spikes pedunculate .... **1. R. himalayensis**
+ Corolla blue, c 3.5mm; spikes sessile .... **2. R. pectinata**

1. R. himalayensis Clarke

Wiry decumbent perennial herb, rooting at nodes. Stem thinly puberulent. Leaves ovate or elliptic, often suffused with red, 10–60 × 5–15mm, acute, shortly cuneate onto petiole, minutely scabrid; petiole 3–14mm. Inflorescence of 1-sided terminal and axillary spikes, 1–4cm; peduncle (0–)1–3cm. Bracts ovate, c 6mm, ending in a stiff bristle-like point, pubescent, green with broad, scarious, often pinkish margins. Calyx 5-lobed, pale green, c 4mm, pubescent, lobes linear. Corolla white with pink lower lip, c 8mm, thinly pubescent. Anthers muticous at base. Capsule oblong, c 4mm, mucronate, pubescent.

Bhutan: S – Sarbhang district (above Sarbhang High School), **C** – Tongsa district (frequent around Tintibi); **Sikkim:** Bam, Dikling Khola. Open forest, shaded rocky slopes, roadsides in seasonally dry subtropical regions, 500–1500m. September–April.

2. R. pectinata (L.) Nees; *R. parviflora* Nees

Annual or short-lived perennial herb, sometimes erect but usually prostrate and rooting at nodes. Stem much branched, minutely pubescent. Leaves oblong-elliptic, 10–40 × 4–14mm, acute, cuneate at base, glabrous except for few hairs on veins; petiole 0–5mm. Inflorescence of short sessile, 1-sided axillary and terminal spikes, 0.5–2cm. Bracts dimorphic, outer sterile bracts elliptic, green, c 4mm, acute, glabrous; inner flower-bearing bracts broadly elliptic, 4–5mm, apiculate, pubescent, with prominent scarious margins. Calyx colourless, c 2mm, pubescent, lobes linear. Corolla blue, c 3.5mm, pubescent. Anthers spurred at base. Capsule ellipsoid, c 2.5mm, glabrous.

Bhutan: S – Samchi, Phuntsholing, Sarbhang and Gaylegphug districts, **C** – Punakha district; **Darjeeling:** Sitong, Rakti (Kurseong), Ryang; **W Bengal Duars**. Common weed in hot country growing in gardens, on roadsides and in disturbed bushy places, 220–1100m. November–May.

25. DICLIPTERA Jussieu

Herbs or undershrubs with obscurely 6-angled stems. Leaves entire. Inflorescence of small cymes in the leaf axils, sometimes aggregated into a leafy panicle; flowers paired (one usually sterile) and enclosed within a pair of slightly unequal bracts which conceal the calyx and corolla tube. Calyx 5-lobed, small, the lobes equal. Corolla pink or purple (rarely white) with long slender tube, 2-lipped, the upper lip entire or nearly so, the lower lip weakly 3-lobed. Stamens 2; anthers 2-celled, one above the other, muticous. Capsule 4-seeded, seeds flattened, glabrous.

Differs from *Peristrophe* and *Hypoestes* by having the placenta rise from the base when the fruit is ripe.

1. D. bupleuroides Nees; *D. roxburgiana* Nees var. *roxburghiana* sensu F.B.I. non Nees, *D. roxburgiana* var. *bupleuroides* (Nees) Clarke

Perennial herb or undershrub with long, scrambling, ascending stems to 2m. Leaves elliptic, 2–15 × 1–8cm, shortly acuminate at both ends, entire or obscurely sinuate, dark green, glabrous or thinly pubescent on the undersurface, petiole 0.5–7cm. Flowers in sessile or shortly pedunculate axillary and terminal clusters. Bracts linear to oblong (to oblanceolate in var. *nazimii* Malik & Ghafoor from the Indian plains), apiculate, prominently 3-veined, ciliate-margined, one larger than other, 8–14mm. Calyx c 6mm, minutely pubescent, lobes pale-margined, subulate. Corolla pink with white tube (rarely entirely white), 15–22mm, lobes longer than tube. Capsule clavate, pubescent, 7–8mm.

Bhutan: S – Samchi, Phuntsholing, Chukka, Sarbhang, Gaylegphug and Deothang districts, **C** – Punakha and Tongsa districts; **Darjeeling:** Rakti, Mongpu, Kurseong, Mahaldiram–Kali Khola etc.; **Sikkim:** Tista River; **W Bengal Duars**. Abundant in secondary scrub and along forest margins, particularly in rocky ravines, 200–1100m. September–May.

26. PERISTROPHE Nees

Herbs or shrubs identical to *Dicliptera* except placenta does not rise from base of ripe capsule. Our species are readily distinguished from *Dicliptera* and *Hypoestes* by their large, showy corollas.

1. Bracts linear, less than 3mm wide .... **1. P. speciosa**
+ Bracts ovate-elliptic, 5–12mm wide .... **2. P. fera**

1. P. speciosa (Roxb.) Nees

Perennial undershrub with long, leggy, ascending, glabrescent stems 1–2m. Leaves ovate-elliptic, dark green above, 2–11 × 1–5.5cm, acute, cuneate at base, thinly pubescent or glabrescent on both surfaces; petiole 1–2.5cm. Flowers in small, shortly pedunculate axillary and terminal clusters. Bracts linear (-oblanceolate), 12–16 × 1–3mm, acute, pubescent, green with pale, usually ciliate margin; bracteoles similar, c 9mm. Calyx c 6mm, puberulent, lobes linear, white-margined. Corolla pink, 4–4.5cm, pubescent, upper lip ovate, lower lip elliptic, 3-toothed, lips almost as long as tube. Capsule clavate, 15–18mm, pubescent.

Bhutan: S – Samchi district (Dorokha, Charmurchi River), Sankosh district (Dagapela–Daga Dzong), Deothang district (Deothang), **C** – Punakha district (Sankosh River below Wangdiphodrang); **Darjeeling:** Rongsong, Kurseong. Rocky, scrub-covered slopes and gullies in river valleys, 800–1600m. October–April.

2. P. fera Clarke

Very similar to *P. speciosa*, differing principally in its ovate-elliptic bracts, 12–20 × 5–12mm, uniformly green, never ciliate-margined; calyx c 3mm, uniformly whitish-green; corolla 3–3.5cm.

Bhutan: S – Gaylegphug district (Surey), **C** – Tongsa district (frequent around Tintibi), Mongar district (Namning–Lingmethang), Tashigang district (Yadi curves). On scrub covered slopes and in gullies, usually near streams, 800–1800m. February–April.

Peristrophe nodosa Griff. was described from Bhutan but the description is inadequate and it is impossible to decide what species it refers to.

27. HYPOESTES R. Brown

Herbs or small undershrubs similar to *Dicliptera* and *Peristrophe* but differing from *Dicliptera* in the placenta not rising from base of ripe capsule, from *Peristrophe* in its smaller flowers and from both in its 1-celled anthers.

1. H. triflora (Forsskål) Roemer & Schultes; *Dicliptera roxburghiana* sensu Yamazaki (69) non Nees

Much-branched decumbent or ascending herb. Stem glabrescent. Leaves ovate, 1–6 × 0.7–4cm, acute, broadly cuneate at base; sparsely pilose on both sides, crenate; petiole 0–1.8cm. Flowers in small axillary and terminal clusters, each cluster having 1–5 flowers. Bracts obovate to oblanceolate, 9–12mm, obtuse, pilose, one slightly larger than the other; bracteoles linear-lanceolate, nearly as long as bracts. Calyx c 3mm, glabrous, colourless, lobes linear. Corolla pink, 12–15mm, pubescent, lips about ⅓ as long as very slender tube. Capsule clavate, glabrous but with few hairs at tip, 9–10mm.

Bhutan: S – Chukka district (Chimakothi), **C** –Punakha district (Lomitsawa, Punakha–Sinchu La, Tinlegang), Tongsa district (Tongsa), Tashigang district (Khaling); **Darjeeling:** Darjeeling, Takdah, Tindaria, Dow Hill, Chimli etc.; **Sikkim:** locality unknown. Abundant in and characteristic of moist, broad-leaved hill forest persisting in disturbed grassland after forest clearance but under-collected. Probably frost sensitive and so absent from inner valleys, 1600–2600m. September–November.

Often confused with *Dicliptera bupleuroides* but readily distinguished by its crenate leaves, obtuse bracts and placenta which does not rise from base as capsule ripens.

Family 176. PEDALIACEAE

by E. Aitken

Annual or perennial herbs, rarely shrubs. Leaves simple, opposite or upper leaves alternate, exstipulate, entire, toothed or divided. Flowers bisexual, zygomorphic, axillary, solitary or in few-flowered racemes or fascicles. Prominent glands at base of pedicels. Flowers 5-merous. Calyx divided almost to base. Corolla tubular-ventricose or obliquely campanulate, bilabiate. Fertile stamens 4 (occasionally 2), epipetalous, one pair longer than the others, 5th stamen often reduced to a small staminode or absent. Anthers 2-locular, dehiscing longitudinally. Ovary superior, of 2 or 4 fused carpels. Style filiform, stigma 2- (rarely 4-)lobed. Fruit a capsule or nut, beaked, winged or with hooks or spines. Seeds 1–many in each locule, smooth or rugose.

1. SESAMUM L.

Erect or decumbent annual or perennial herbs. Leaves sessile or petiolate, opposite, often alternate on upper part of plant. Flowers axillary, solitary or in few-flowered fascicles. Calyx small, persistent or deciduous. Corolla tubular-ventricose, narrowing abruptly at base, decurved, bilabiate, lobes of limb patent, rounded. Stamens attached at base of corolla tube; ring of hairs at base of filaments; anthers sagittate, dorsifixed. Ovary 2-celled, appearing 4-celled due to false septum. Capsule oblong or ovoid, obtusely quadrangular, 4-grooved, beaked at apex, dehiscing longitudinally. Seeds small, numerous, oblong or obovate with or without wings, smooth or rugose.

1. S. indicum L.; *S. orientale* L. Eng: *Sesame*. Fig. 106a–d.

Annual, erect, branched herb up to 1m. Stems 4-angled, pilose. Leaves sparsely pilose above, densely glandular pubescent beneath, very varied even on the same plant. Lower leaves opposite, ovate or elliptic, blade 2.8–9 × 1.2–4cm, apex acute, base attenuate or rounded, margins coarsely toothed or lobed; upper leaves alternate, oblong or linear, blade 1.8–6 × 0.2–1cm, margins entire or crenulate; petioles pubescent, 0.2–6cm. Flowers axillary, solitary or in few-flowered fascicles, densely glandular-pubescent. Pedicels 1–5mm. Calyx lobes lanceolate to linear, 3–5 × 1mm, persistent. Corolla bilabiate, tubular-ventricose, narrowing abruptly at base, 1.5–2.5 × 0.7–1cm, white or pink with darker markings. Stamens 7–10mm, glabrous, staminode absent. Ovary densely pubescent, 1–3mm; style 6–7mm; stigmas 1.5–2 × 0.5mm. Capsule oblong, 1.5–2.5 × 0.5–0.7cm. Seeds small, obovate, ridged, smooth.

Darjeeling: unlocalised (Cowan collections); **Sikkim:** Melli. On wasteground. August–September.

FIG. 106. **Pedaliaceae and Gesneriaceae. Pedaliaceae.** a–d, *Sesamum indicum*: a, inflorescence (× ⅔); b, lower leaf (× ⅔); c, dissected calyx with gynoecium (× 1); d, dissected corolla (× 1). **Gesneriaceae.** e–g, *Aeschynanthus hookeri*: e, flowering shoot (× ½); f, dissected calyx and corolla (× 1); g, gynoecium (× 1). h–k, *Lysionotus serratus*: h, flowering shoot (× ½); i, dissected calyx with gynoecium (× 1); j, dissected corolla (× 1); k, seed (× 8). l–o, *Loxostigma griffithii*: l, flowering shoot (× ½); m, dissected calyx with gynoecium (× 1); n, dissected corolla (× 1); o, seed (× 8). Drawn by Mary Bates.

a
b
c
d
e
f
g
h
i
j
k
l
m
n
o
MB

Very little material seen from our area; dimensions of additional specimens from Bengal and C Nepal incorporated in the description above. Native country doubtful. Cultivated world-wide in tropical areas for seeds used on bread and in confectionery and for oil used for cooking and in the manufacture of margarine and soaps. Leaves, seeds and oil also used in herbal medicine.

Family 177. GESNERIACEAE

by O.M. Hilliard

Herbs or shrubs. Leaves simple, opposite, alternate or whorled, those of a pair equal or unequal; plant sometimes with only 1 leaf. Inflorescence usually cymose, rarely racemose, or flowers solitary. Flowers hermaphrodite, often protandrous, zygomorphic, sometimes cleistogamous with reduced corolla. Calyx 5-lobed or divided to base. Corolla with a distinct tube, limb often 2-lipped. Stamens 2 or 4, inserted on corolla tube, staminodes 1 or 3. Disc annular or cupular. Ovary superior, 1-celled with 2 parietal bilamellate placentae, usually inrolled. Ovules many. Fruit often a linear capsule, valves straight or twisted, rarely a berry.

The New World genus, *Achimenes*, which is distinguished from all other gesneriads in our area by its inferior ovary, has been recorded in Sikkim, at Gangtok, as a garden escape on walls at c 1700m (408).

1. Fertile stamens 4 2
+ Fertile stamens 2 9

2. Stems trailing or pendent, sometimes bushy; leaves thick and fleshy; anthers usually far-exserted in male phase, if included stems trailing **1. Aeschynanthus**
+ Plant either stemless or with erect or decumbent stems; leaves herbaceous or leathery; anthers either included or visible in mouth but not far-exserted 3

3. Corolla tube at least as conspicuous as the limb; anthers either cohering in pairs or all 4 cohering 4
+ Corolla tube very short, limb more conspicuous than tube; anthers not cohering 6

4. Leaves rosulate; corolla blue or blue and white, palate bearded **7. Corallodiscus**
+ Either leaves distributed up a stem, or rosulate and then corolla yellow or greenish, palate not bearded (though corolla may be glandular inside) ... 5

5. Seeds with appendages c 1.5–2mm long; 2 elongated invaginations in floor of corolla tube .... **3. Loxostigma**
\+ Seeds not appendaged; no invaginations in corolla tube .... **4. Briggsia**

6. Herb with slender erect stem terminating in 1 leaf (very rarely a second, much smaller, leaf present) .... **8. Platystemma**
\+ Thick-stemmed perennial herbs, undershrubs or shrubs 7

7. Inflorescences several clustered at nodes; fruit globose, white and somewhat fleshy when fresh .... **11. Rhynchotechum**
\+ Inflorescences solitary in leaf axils; fruit linear or oblong-elliptic, brown, not fleshy 8

8. Leaves opposite .... **9. Leptoboea**
\+ Leaves alternate .... **10. Boeica**

9. Flowers crowded in a curled, 1-sided inflorescence, corolla c 5mm long (but flowers often cleistogamous); capsule globose, circumscissile (the upper half detaching like a lid) .... **14. Epithema**
\+ Flowers variously arranged but never crowded in a 1-sided inflorescence, corolla c 7–75mm long; capsule linear, elliptic or ellipsoid, opening lengthwise 10

10. Corolla c 7mm long, widely campanulate, tube shorter than the spreading white limb; capsule spirally twisted .... **12. Paraboea**
\+ Corolla 10–75mm long, tube mostly longer than lobes, if roughly equalling it, corolla violet-blue; capsule not spirally twisted 11

11. Flowers in long racemes; mouth of corolla closed .... **13. Rhynchoglossum**
\+ Flowers cymose; mouth of corolla wide open 12

12. Calyx divided nearly to base; 2 keels on floor of corolla tube; seeds with a hair-like appendage at each end .... **2. Lysionotus**
\+ Calyx variously lobed, cut to base in only one species and then no keels on floor or roof of corolla tube; seeds not appendaged 13

13. No keels on corolla tube; filaments thread-like, straight; stigma either capitate or 2-lipped .... **5. Didymocarpus**
\+ Corolla tube usually with 2 keels on floor or roof, rarely without; filaments variously swollen or geniculate; stigma with dorsal lobe aborted or minute, lower one flat, expanded, entire or bilobed ('fish-tailed') .... **6. Chirita**

1. AESCHYNANTHUS Jack

Undershrubs, epiphytic or lithophytic. Leaves usually opposite, rarely in whorls of 3–4, fleshy or leathery, entire or rarely obscurely toothed. Peduncles terminal, axillary or wanting and then pedicels sub-umbellate; peduncles, when present, 1–2-flowered or rarely cymose, solitary or clustered. Bracts often small, sometimes large and coloured, deciduous. Calyx shallowly to deeply 5-lobed. Corolla tubular-ventricose, limb 2-lipped. Stamens 4, anthers cohering in pairs, in mouth or well exserted. Disc cupular or annular. Ovary stipitate, oblong or linear; stigma peltate. Capsule long-linear, loculicidally 2-valved. Seeds minute, ellipsoid, rugose or papillate, with 1 or more hairs near hilum, 1 at apex.

1. Stems and leaves hairy 2
+ Stems and leaves glabrous (or rarely some hairs on stem only) 3

2. Leaves glabrous above, hairy below; corolla red; gynoecium glabrous except for minute sessile glands on ovary .... **1. A. gracilis**
+ Leaves hairy on both surfaces (hairs sometimes sparse above); corolla white; stipe, ovary and style densely pubescent .... **2. A. chiritoides**

3. Calyx divided to base or very nearly so 4
+ Calyx with a distinct tube 8

4. Flowers in terminal and axillary clusters (1–5 flowers) lacking a peduncle; bracts inconspicuous .... **4. A. micranthus**
+ Flowers in terminal and/or axillary pedunculate cymes, these either expanded or congested, or the flowers sometimes solitary; bracts conspicuous (but sometimes caducous) 5

5. Bracts less than 1cm long; cyme very open; corolla c 15mm, green with red lobes .... **3. A. acuminatus**
+ Bracts 1.5–6cm long; cyme congested or flowers ± solitary; corolla (also calyx and bracts) red 6

6. Bracts 3.7–6cm long; corolla 6.5–9cm .... **5. A. superbus**
+ Bracts 1.5–3cm long; corolla 3.5–5cm 7

7. Leaves broadest near base; peduncles 1–5cm; corolla 3.5–4cm **6. A. bracteatus**
+ Leaves mostly broadest in middle or upper half; peduncles 5–12cm; corolla 4.3–5cm .... **7. A. peelii**

8. Calyx and pedicels clad in spreading hairs .... **12. A. maculatus**
\+ Calyx and pedicels either glabrous (or rarely a few hairs tufted at tips of lobes) or clad in minute nearly sessile glands 9

9. Calyx lobes very obtuse, mostly about ⅓ of the length of the tube **8. A. hookeri**
\+ Calyx lobes acute (rarely subacute) to acuminate, either much longer than tube or very roughly equalling it 10

10. Corolla 3.5–5cm long, with remarkably coarse hairs inside lower part of tube .... **9. A. parasiticus**
\+ Corolla 2–3cm long, without (or rarely 2–3) coarse hairs inside lower part of tube 11

11. Calyx glabrous, tube 1.8–4mm long, lobes 6–12mm, roughly 2–4 times as long as tube .... **10. A. parviflorus**
\+ Calyx covered in minute almost sessile glands, tube 2–5mm long, lobes 2.5–4.5mm, about equalling to slightly shorter than tube **11. A. sikkimensis**

1. A. gracilis C.B. Clarke

Subshrub, stems slender, creeping, rooting at nodes, then trailing or pendulous, loosely branched, hairy. Leaves opposite, thick-textured, veins invisible above, narrowly to broadly elliptic, 12–40 × 5–13mm, subacute to acute, base rounded or cuneate, margins entire or rarely obscurely toothed, lower surface and petiole hairy; petiole 2mm. Flowers solitary in leaf axils, pedicels 8–12mm long, hairy. Calyx divided to base, segments lanceolate, acuminate, 4.5–6.5 × 1–1.8mm, hairy. Corolla 2.3–3.5cm long, mouth very oblique, upper lip elongate, almost hooded, shallowly 4-lobed, lateral margins reflexed, lower lip 1-lobed, corolla hairy outside, light scarlet-crimson outside, inside tube dull yellow, margins of 2 upper lobes and whole of other 3 lobes crimson with darker streaks and blotches, light yellow between streaks. Stamens exserted, filaments pubescent. Disc cupular, oblique, 2-toothed on ventral side. Ovary very minutely gland-dotted, few minute hairs below stigma. Capsule 8–18cm long excluding persistent style. Seeds c 1mm, 1 long hair at each end.

Bhutan: **S** – Sarbhang district (Singi Khola), Gaylegphug district (Gaylegphug, Rang Khola, Karai Khola); **Darjeeling:** Sivok, Kalimpong, Chhota Rangit, Jaldhaka Valley; **Sikkim:** Thinglen, Pakhyong, Singhik, Singtam–Gangtok, Dikchu. Epiphytic in subtropical and terai forest, 400–1500m. January–March.

2. A. chiritoides C.B. Clarke

Resembles *A. gracilis* in habit and hairs on stems, calyx and corolla, but leaves often in whorls of 3, elliptic or oblanceolate, 10–22 × 4.5–9cm, both surfaces hairy (hairs sometimes sparse on upper surface). Pedicels 3–6mm, solitary in upper leaf axils. Calyx segments lanceolate-acuminate, 5.5–9 × 1–1.5mm. Corolla 3.5–4cm long, narrowly tubular below, suddenly expanding to broadly funnel-shaped and 10–15mm across, obscurely bilabiate, lobes small, rounded, white, lower lip with 3–5 purple-crimson lines running down floor of tube, pale yellow between the lines. Stamens included, filaments pubescent at apex. Ovary and style pubescent. Capsule not seen.

Bhutan: S – Gaylegphug district (near Betni Bridge (117)). On trees or rocks in forest, 600–1050m. December.

No material of this species has been seen from E Himalaya and the Bhutan record requires confirmation.

3. A. acuminatus A. DC.; *A. chinensis* Gardner & Champion

Shrublet, well branched, stems woody, bark pale grey. Leaves opposite, leathery, elliptic, penninerved, 5.5–9 × 2.2–3.7cm, acuminate, base cuneate, margins entire or obscurely serrulate; petiole 3–9mm. Flowers in open, 1–4-flowered cymes. Bracts paired, broadly ovate, 5–8 × 5–10mm (first pair). Peduncles 10–23mm, 2–7 at tips of branches; pedicels 7–23mm. Calyx divided nearly to base, segments oblong-elliptic to ovate, 2.5–6 × 2–3.5mm. Corolla glabrous outside, glandular-puberulous inside, 12–15mm long, tube c 8mm broad, mouth very oblique, upper lip ± erect, shallowly bilobed, lower lip 3-lobed, tube and upper lip green, lower lip yellowish or red. Stamens far-exserted, glandular-puberulous. Disc annular. Ovary glabrous; style minutely glandular. Fruit 7.5–12.5cm long, excluding persistent style. Seeds 0.8–1mm long, 1 long hair at each end.

Darjeeling: terai (unlocalised Hooker collection), Rayeng. Epiphytic in mixed forest, festooning trees, 400–500m. October–January.

4. A. micranthus C.B. Clarke

Shrub, stems to 1m or more, laxly branched, often rooting at nodes, glabrous or with a few hairs at nodes. Leaves opposite, fleshy, elliptic, 3–7.5 × 1.2–3cm, acute to shortly acuminate, base cuneate, margins entire; petiole 3–8mm. Flowers 1–5 on short terminal and axillary spurs. Bracts linear-lanceolate, c 1.5–3mm, hairy. Pedicels 4–8mm, hairy or glabrescent. Calyx lobed to base, lobes triangular, very acute, 3.5–6mm long. Corolla pubescent or glabrous except for hairs fringing lobes, 2.4–2.8cm long, narrow (4–7mm), all 5 lobes projecting forwards, subequal, mouth round, crimson, margins of lobes purplish. Stamens far-exserted, filaments glandular pubescent. Disc cup-shaped. Ovary minutely gland-dotted; style glandular-pubescent. Fruit c 9–18cm long. Seeds c 1mm long, 1 long delicate hair at each end.

Bhutan: S – Sarbhang district (Singi Khola); **Darjeeling/Sikkim:** unlocalised

(Hooker collection). Epiphyte in subtropical terai forest on river bank, 390m. Flowering time unknown.

5. A. superbus C.B. Clarke

Laxly branched shrub, stems eventually 3 or more metres long, woody, pendent. Leaves opposite, leathery, elliptic, faintly penninerved, 11–18 × 3.8–10cm, abruptly acuminate, tip obtuse, base rounded or broadly cuneate, margins entire; petiole 1–2cm. Peduncles 1–2.2cm long, terminal and on short (5–10mm) stout axillary spurs, pedicels 0–18mm long, up to c 10 flowers, sub-umbellate. Bracts ovate, 3.7–6 × 2.3–5cm, red, caducous. Calyx divided to base or nearly so, lobes oblong-elliptic, obtuse to subacute, 2.3–3.9 × 0.5–1.3cm, red. Corolla glabrous outside, glandular-puberulous inside, 6.5–9cm long, mouth scarcely oblique, upper lip erect, shallowly bilobed, lower lip 3-lobed, tube rich crimson-scarlet to brick-red, upper part and mouth dirty yellow, all lobes marked with crimson blotches and streaks running back down tube as stripes. Stamens far-exserted, connective and upper part of filaments glandular-puberulous. Disc annular. Ovary and stipe glabrous; style glandular-puberulous. Fruit 30–40cm long. Seeds not seen.

Bhutan: C – Mongar district (Shongar). Epiphytic or on rocks, in forest, 760–1200m. August–September.

6. A. bracteatus DC.; *A. paxtonii* Lindley

Shrub, laxly branched, stems up to 1–2m long, woody. Leaves opposite, fleshy, ovate, faintly penninerved, 5–10 × 1.8–4.3cm, abruptly acuminate, tip rounded, base rounded, margins entire; petiole 10–18mm. Cymes mostly 3–5-flowered, terminal but overtopped by new growth leaving old peduncles apparently axillary. Bracts paired, ovate, 16–30 × 8–14mm, red, caducous. Peduncles 1–5cm long, pedicels 1.3–2cm. Calyx divided to base or nearly so, lobes lanceolate-elliptic, 14–28 × 3–6mm, red. Corolla glabrous outside, glandular-puberulous inside, 3.5–4cm long, mouth scarcely oblique, upper lip erect, shallowly bilobed, lower lip 3-lobed, marked with purplish blotches, corolla crimson or scarlet outside, probably yellowish inside. Stamens far-exserted, filaments glandular-puberulous. Disc annular. Ovary very minutely gland-dotted; stipe and style glandular-puberulous. Fruit 11.5–14cm long. Seeds 1mm long, papillose, 1 short (2–2.5mm) stout hair at each end.

Bhutan: S – Chukka district (Tala–Gedu), **C** –Mongar district (Latun La); **Darjeeling:** above Mongpu; **Sikkim:** Mamring Forest. Epiphytic or sometimes on rocks, mixed and pine forests, 1550–2400m. July–October.

Also collected east of our area at Rupa (Arunachal Pradesh).

7. A. peelii Hook.f. & Thomson; *A. bracteatus* var. *peelii* (Hook.f. & Thomson) C.B. Clarke

Distinguished from *A. bracteatus* by its stems rooting at the nodes, glandular hairs c 0.5mm long present or not on stems, leaves elliptic to obovate, 4–8.5 ×

2.3–3.5cm; petiole 5–12mm; cymes mostly 1–2-flowered; bracts lanceolate, 17–30 × 3–12mm; peduncles 5–12cm; corolla 4.3–5cm long.

Bhutan: C – Punakha district (Monle La, Sinchu La), Mongar district (Sawang), **N** – Upper Kuru Chu (Denchung, Khoma Chu); **Darjeeling:** Darjeeling, Lepcha Jagat, Rungbool road. Epiphytic or on rocks in forest, 1500–2400m. July–September.

8. A. hookeri C.B. Clarke. Fig. 106e–g.

Shrub, laxly branched, stems up to c 1m long, slender, pendent, rooting mainly at lower nodes. Leaves opposite, fleshy, elliptic, 6–11 × 1.2–2.5cm, long-acuminate, base cuneate, margins entire; petiole 7–13mm. Flowers several, clustered at tip of stem, later overtopped by new growth. Bracts linear-lanceolate, 3.5–10 × 1–1.5mm. Pedicels 12–20mm. Calyx tube 8–14mm long, lobes mostly ⅓ length of tube, 3–6mm, oblong or deltoid, obtuse, reddish. Corolla glandular-pubescent outside, 3–3.6cm long, tube curved near apex, there inflated, all 5 lobes projecting forwards, mouth almost round, throat yellow, rest of corolla bright scarlet or orange-scarlet, each lobe with a conspicuous dark purplish median line running back halfway down tube. Stamens far-exserted, anticous filaments glandular-puberulous. Disc cupular. Ovary very minutely gland-dotted; style glandular-puberulous. Fruit c 30cm long. Seeds c 1mm long, with 2 long delicate hairs at one end, 1 at the other.

Darjeeling: Darjeeling, Senchal Hill, Phalut–Dentam, Lepcha Jagat, Manibhanjan, Mahalderam; **Sikkim:** Damthang–Tendong (71), Mamring. Also E Nepal and upper Myanmar, so probably in Bhutan. Temperate evergreen forest, 1500–2400m, epiphytic. May–October.

9. A. parasiticus (Roxb.) Wall.; *Incarvillea parasitica* Roxb., *Trichosporum grandiflorum* D. Don, *Aeschynanthus grandiflorus* (D. Don) Sprengel

Shrub, laxly branched, stems slender, pendent, rooting at the nodes. Leaves opposite, fleshy, elliptic, 8.5–18 × 1.2–2.5cm, long-acuminate, base cuneate, margins entire; petiole 8–12mm. Flowers several, clustered at tip of stem, later overtopped by new growth. Bracts linear-lanceolate, c 3 × 1mm. Pedicels 10–17mm. Calyx tube (3.6–)8–17mm long, lobes triangular, acute, 4.5–11mm long, about equalling to half as long as tube, glabrous. Corolla glandular-pubescent outside, inside with very coarse gland-tipped hairs up to c 1.8mm long near base of tube, 3.5–5cm long, tube curved and inflated in upper half, slightly constricted at mouth, all 5 lobes projecting forwards, mouth nearly round, throat yellowish, rest of corolla orange-red or brick-red, dark purplish median patch on each lobe. Stamens far-exserted, filaments glandular-puberulous. Disc annular. Ovary very minutely gland-dotted; style glandular-puberulous. Fruit c 30cm long. Seeds c 1mm long, 2 long delicate hairs at one end, 1 at the other.

Bhutan: S – Phuntsholing district (above Rinchending), **C** – Tashigang district

(Tashigang–Tashi Yangtsi); **Sikkim:** Chakung Chhu. Epiphyte in forest, 750–1200m. July–October.

Only poor material seen.

10. A. parviflorus (D. Don) Sprengel; *Trichosporum parviflorum* D. Don, *A. ramosissimus* Wall.

Shrub c 1m, stems spreading, laxly branched. Leaves opposite, fleshy, elliptic, faintly penninerved, 7–11.5 × 1.3–4.5cm, long-acuminate, base cuneate, margins entire; petiole 6–12mm. Flowers several, clustered at tip of stem, later overtopped by new growth. Bracts linear-lanceolate, 2–5 × 0.8–1.5mm. Pedicels 10–13mm. Calyx light orange-yellow, tube 1.8–4mm long, lobes triangular, very acute to acuminate, 6–8mm long, roughly twice as long as tube, glabrous. Corolla glandular-pubescent outside, 2.5–3cm long, tube curved and inflated in upper half, mouth scarcely constricted, all 5 lobes projecting forwards, mouth nearly round, throat yellowish or light orange, rest of corolla bright orange or orange-scarlet, paler at base, crimson median patch on each lobe, this papillose on inner face. Stamens far-exserted, anticous filaments glandular-pubescent. Disc annular. Ovary glabrous; style glandular-pubescent. Fruit c 10–20cm long. Seeds 1–1.5mm long, 2 long (25–30mm) delicate hairs at one end, 1 at the other.

Bhutan: S – Phuntsholing district (above Kharbandi, Kamji); **Darjeeling:** Lepcha Jagat, Darjeeling. Evergreen(?) forest, epiphytic or on cliff-faces, 1400–2100m. July–August.

11. A. sikkimensis (C.B. Clarke) Stapf; *A. maculatus* sensu C.B. Clarke in F.B.I. p.p., *A. maculatus* Lindley var. *sikkimensis* C.B. Clarke

Distinguished from *A. parviflorus* by the calyx: tube 2–5mm long, lobes 2.5–4.5mm about equalling to slightly shorter than tube, clad outside in minute almost sessile glands, and possibly by the scarlet colour of the flowers, which may appear earlier.

Bhutan: S – Deothang district (Morong–Narfong), **C** – Mongar district (Shersing Thang–Namning), **N** – Upper Mo Chu district (Khosa–Tamji); **Darjeeling:** Darjeeling, Balasun River, Lebong; **Sikkim:** Chunthang, Lingmo, Yoksam, Lachen, Lachung. Epiphytic or on rock-faces in broad-leaved and mixed deciduous–evergreen forest, 1200–2000m. May–July.

12. A. maculatus Lindley. Nep: *Rati harchim.*

Similar to *A. sikkimensis* but easily distinguished by the spreading hairs on pedicels and calyx.

Bhutan: C – Sakden district (Sakden, Gamri Chu); **Darjeeling:** Darjeeling; **Sikkim:** Relli Chhu; **Arunachal Pradesh:** Nyam Jang Chu. Epiphyte in forest, 1500–1800m. April–July.

2. LYSIONOTUS D. Don

Herbs or subshrubs, stem creeping, flowering stems erect. Leaves opposite or in whorls of 3, often unequal in size. Flowers in axillary cymes. Calyx divided nearly or quite to base. Corolla tubular, abruptly inflated in upper half, bilabiate, lobes ± rounded or ovate, lower lip 2-keeled. Stamens 2, included; anthers confluent at apex and there cohering face to face, connective appendaged or not. Ovary stipitate, linear, disc annular; style simple; stigma obscurely bilobed. Capsule linear, 2-valved, each valve eventually splitting into 2, margins of valves inflexed. Seeds minute with a long hair at each end.

1. Stem hairy (also leaves, mainly on lower surface) .... **4. L. pubescens**
+ Stem and leaves glabrous 2

2. Peduncles axillary, filiform, ± 0.5mm in diameter; corolla glabrous; anther connective not appendaged .... **3. L. atropurpureus**
+ Peduncles usually terminal or subterminal, 1–2mm in diameter; corolla hairy; anther connective with an appendage on the dorsal surface 3

3. Calyx divided nearly or quite to base, lobes 6–10 × 2mm; leaf margins serrate .... **1. L. serratus**
+ Calyx with a distinct tube 2–6mm long, lobes 11–14 × 3–4mm; leaf margins entire .... **2. L. kingii**

1. L. serratus D. Don; *L. ternifolia* Wall., *Calosacme polycarpa* Wall., *Hemiboea himalayensis* Léveillé, *Lysionotus himalayensis* (Léveillé) Wang & Li. Nep: *Kolojhan*. Fig. 106h–k.

Somewhat fleshy glabrous herb, flowering stems 10–30cm, mottled purple. Leaves elliptic or oblong-elliptic, 4–17 × 1.5–6cm, apex acuminate, base cuneate, slightly oblique, margins serrate; petiole 0.5–2cm. Flowers few to many in each cyme, peduncles terminal or subterminal, 5–9cm long, 1–2mm in diameter. Calyx divided nearly or quite to base, lobes lanceolate, acute, 6–10 × 2mm. Corolla white or mauve, veined purple, 2 yellow bars on lower lip in addition to keels in throat, hairy outside, 3–5.5cm long, upper lip shorter than lower, lobes rounded or broadly ovate. Anther connective with conspicuous lateral appendage. Capsules ± 9–11cm long, 2mm in diameter.

Bhutan: S – Samchi district (Samchi, Tamangdhanra Forest), Chukka district (Kalikhola), Gaylegphug district (Karai Khola), **C** – Punakha district (Neptengka, Ngawang, Rinchu), Mongar district (Lhuntse); **Darjeeling:** Darjeeling, Kurseong; **Sikkim:** Gongchung, Pemayangtse. Mossy rocks, banks, tree trunks, in subtropical forest, 500–2100m. July–September.

2. L. kingii (C.B. Clarke) Hilliard; *Aeschynanthus kingii* C.B. Clarke

Similar to *L. serratus* but easily distinguished by its entire leaves and calyx with a distinct tube c 2–5mm long and lobes c 10–15 × 2.5–4mm.

Darjeeling: Mongpu; **Sikkim:** Rangyong Chhu and unlocalised, c 1500m (King collection).

This species is ill-known; it was collected in flower in July and October, but no other information is available.

3. L. atropurpureus Hara

Similar to *L. serratus* (and confused with it in F.B.I.), but peduncles filiform (c 0.5mm in diameter), axillary, 3–8.5cm long, 1–5-flowered, calyx lobes 5–7 × 1mm, corolla red-purple, glabrous, 2–3cm long, anther connective not appendaged, capsules 3.5–8cm long.

Darjeeling: Birch Hill, Rimbi Chhu, Shiri Khola, Rimbik, Budhwari, Kurseong; **Sikkim:** Ratharg Chhu Valley, Mintok Khola–Paha Khola. Steep banks, or epiphytic in forest, 1500–2500m. July–August.

4. L. pubescens C.B. Clarke; *L. wardii* W.W. Smith

Subshrub, stems up to 2.4m, well branched, lax, hairy. Leaves elliptic, 1.7–10.5 × 0.8–2.5cm, apex acute or subacute, base cuneate, margins entire to serrate, upper surface glabrous, lower appressed-hairy; petiole 2–7mm. Peduncles axillary, filiform, hairy, 6.5–9cm, 1–5-flowered. Calyx lobes elliptic, 3–6 × 1.2–2.5mm. Corolla hairy, violet-blue, 2 keels on lower lip strongly raised, yellow(?), 3.5–4cm long. Anthers not appendaged. Capsules 4.5–6cm long.

Bhutan: C –Punakha district (Rinchu, Neptengka). Mossy rocks and tree trunks in forest or in open but sheltered places, 1525–2135m. August–September.

3. LOXOSTIGMA C.B. Clarke

Perennial herbs or undershrubs, stem decumbent at base, rooting. Leaves opposite, often markedly unequal. Peduncles solitary in upper leaf axils, flowers several, cymose. Bracts small. Calyx 5-partite to base. Corolla tubular, ventricose midway, spotted within, 2 elongated invaginations in floor of corolla tube, 2 rows of pouched tissue at base of filaments, limb bilabiate, lobes 5. Stamens 4, included, filaments curved, anthers confluent, all 4 cohering at their tips and inner margins, opening downwards; staminode 1. Disc cupular. Ovary linear, passing ± abruptly into style; stigma equally 2-lipped. Capsule linear-oblong, tapering at base, crowned with persistent style, somewhat flattened, loculicidally 2-valved, placentae inrolled, concealing the seeds. Seeds very small, ellipsoid, appendaged at both ends.

1. L. griffithii (Wight) C.B. Clarke; *Didymocarpus griffithii* Wight, *Dichrotrichum griffithii* (Wight) C.B. Clarke. Fig. 106 l–o.

Coarse herb or undershrub 40–200cm tall, stem decumbent at base, rooting, simple or loosely branched, sparsely puberulous. Leaves broadly elliptic to ovate, largest of each pair 8.5–23 × 5–12cm, apex shortly acuminate, base rounded, oblique or not, margins serrate or crenate-serrate, upper surface sparsely pubescent, hairs nearly confined to veins on lower surface; petioles 0.5–4cm. Peduncles 3–10cm long, pedicels up to 1–3.5cm, both pubescent, hairs mixed glandular and eglandular. Bracts linear-lanceolate, c 4–6 × 1–2mm. Calyx segments elliptic or ovate-elliptic, c 6.5–12 × 2–3.7mm, margins entire or irregularly serrulate, hairy outside. Corolla glandular pubescent outside, inside glandular puberulous on roof of tube in mouth, pale yellow outside, dull red- or brown-spotted inside, 4–5cm long, upper lip much shorter than the lower, bilobed. Ovary glabrous; style glandular puberulous. Capsules c 65–95 × 4mm. Seeds c 1mm long, appendages 1.5–2mm.

Bhutan: S – Phuntsholing district (Gedu–Kharbandi), Chukka district (E of Jumudag), Deothang district (Satsalor), **C** – Tashigang district (Tashi Yangtsi Valley, Shali); **Darjeeling:** Lebong Cant, Mongpu, Rishap, Rambi, Takdah; **Sikkim:** Chakung, Gangtok, Mamring. Epiphytic or on moist shady rocks and banks in forest or scrub, 900–2100m. September–November.

4. **BRIGGSIA** Craib

Rhizomatous herbs, often stemless. Leaves opposite or alternate, often rosulate, petiolate or nearly sessile. Peduncles axillary, flowers solitary or in few-flowered cymes. Bracts small. Calyx 5-lobed almost to base. Corolla large, tube ventricosely inflated from a very short cylindric base, limb relatively small, 2-lipped, upper lip obscurely to distinctly bilobed, smaller than lower lip, lower lip 3-lobed, tube and lower lip characteristically spotted and blotched. Stamens 4, inserted near base of tube, included, filaments long, strongly curved, anthers 2-celled, cells confluent, each pair of anthers lightly cohering at their tips; staminode 1. Disc a shallow 5-lobed ring. Ovary linear-oblong, placentae deeply inflexed, margins revolute, enfolding ovules; stigma shortly and equally 2-lipped. Capsule oblong-elliptic, crowned with persistent style, loculicidal. Seeds very small, ellipsoid, reticulate.

1. Plant stemless; leaves all radical, blade densely silvery-pubescent **1. B. muscicola**
+ Plant with a well-developed stem; leaves cauline, mostly apical, blade thinly pubescent, not silvery .... **2. B. kurzii**

FIG. 107. **Gesneriaceae.** a–c, *Briggsia muscicola*: a, habit (× ½); b, dissected calyx with gynoecium (× 1½); c, dissected corolla (× 1½). d–f, *Didymocarpus podocarpus*: d, flowering shoot (× ½); e, dissected calyx with gynoecium (× 1); f, dissected corolla (× 1). g–j, *Chirita lachenensis*: g, habit (× ½); h, dissected calyx (× 1); i, dissected corolla (× 1); j, gynoecium (× 1). k–n, *Corallodiscus kingianus*: k, habit (× ½); l, dissected calyx (× 2); m, dissected corolla (× 2); n, gynoecium (× 2). Drawn by Mary Bates.

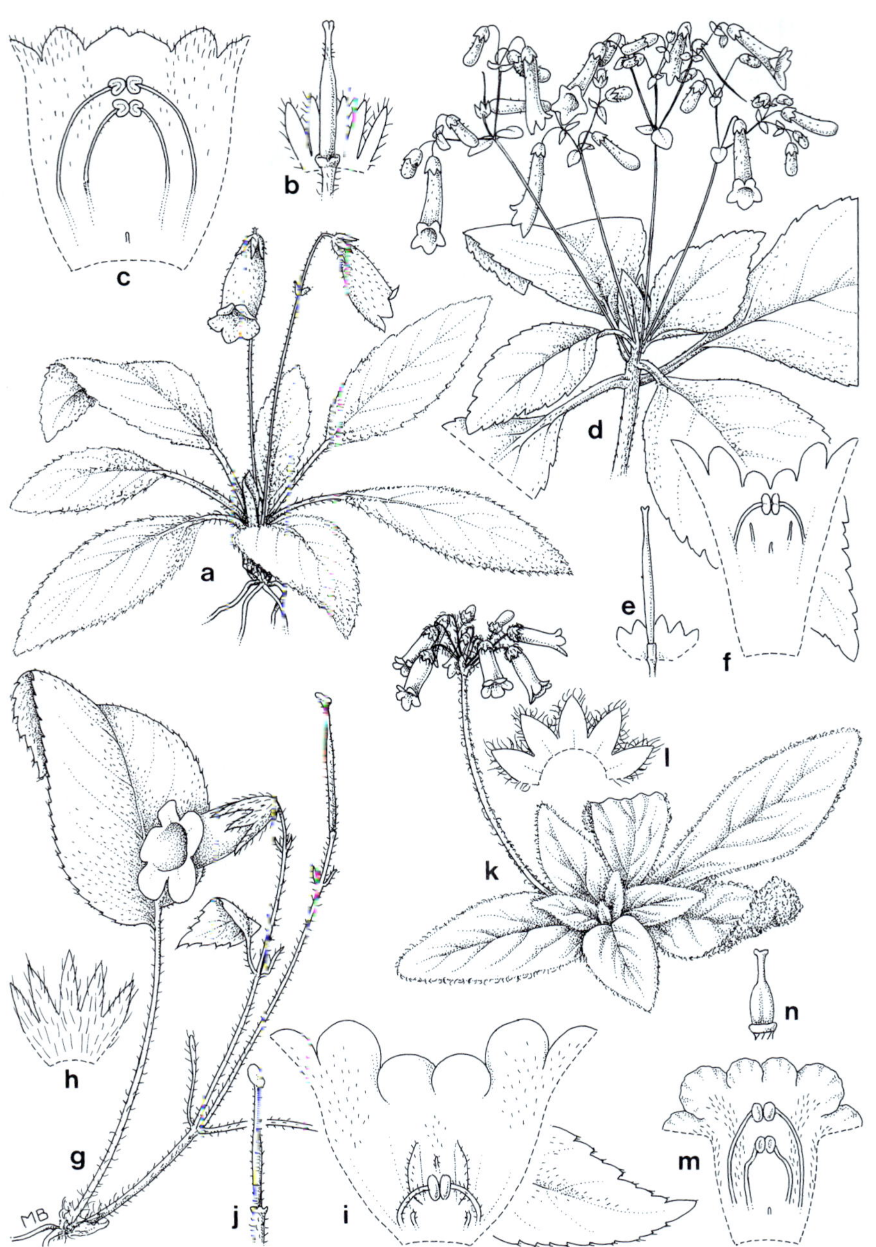
c
b
d
a
e
f
l
k
h
n
g
m
MB
j
i

1. B. muscicola (Diels) Craib; *Didissandra muscicola* Diels, *B. penlopi* C.E.C. Fischer. Fig. 107a–c.

Stemless herb, rhizome c 3.5–4 × 0.8–1cm, knobbly with leaf scars, leaves and peduncles all radical. Leaves several, elliptic, 5.5–13 × 1.8–3.4cm, apex acute, base cuneate, margins crenate-serrate, both surfaces silvery appressed pubescent, veins on undersurface shaggy with long (to 6mm) coarse red-brown acute hairs, these also on petioles; petioles 3.5–7.5cm. Inflorescence a 1- to several-flowered cyme, peduncles 6.5–8cm long, pedicels 0.7–3.5cm, both villous with red-brown acute and glandular hairs. Bracts narrowly ovate, c 4–8 × 1.2–2mm. Calyx tube 0.5–1.5mm, lobes triangular, 5–8 × 1.2–2mm, acute, villous. Corolla pubescent outside and on upper surface of lower lip, glandular puberulous inside on roof of tube, pale yellow or greenish-yellow, purple-spotted, 2.5–3.4cm long, upper lip very short, scarcely divided, lower much larger, 3-lobed. Ovary and style glabrous except for a few hairs at base of style. Capsules 32–65 × 4–7mm excluding persistent style.

Bhutan: C – Thimphu district (near Changri Monastery), Punakha district (Lao La, Pele La–Ritang), Tongsa district (ridge E of Tongsa, Yuto Ridge), Bumthang district (Penge La), Sakden district (Sakden, Gamri Chu), **N** – Upper Bumthang Chu district (Tsampa). Mossy tree trunks in forest, 2600–3500m. June–August.

2. B. kurzii (C.B. Clarke) W.E. Evans; *Chirita kurzii* C.B. Clarke, *Loxostigma kurzii* (C.B. Clarke) B.L. Burtt

Perennial herb, rhizome 1.2–3 × 0.7–1cm, stem solitary, erect, 18–50cm tall, usually simple, sparsely hairy, leafy mostly at apex. Leaves few, main ones crowded (or rarely somewhat distant) at apex of stem, few very small ones lower down, alternate below, opposite above, main leaves oblong-elliptic to obovate, 4–37 × 2–11cm, apex acuminate, base cuneate, margins serrate, upper surface thinly hairy, lower surface with hairs confined to veins, sometimes purplish; petioles 0.3–4(–9)cm. Inflorescences 1–3(–9)-flowered cymes in upper leaf axils, peduncles 3–7.7cm long, pedicels 0.4–1.2cm, both pubescent with glandular and acute hairs. Bracts narrowly ovate, c 4–9 × 1.5–3mm. Calyx tube c 1mm long, lobes elliptic or oblong-elliptic, 9–17 × 2–3(–6.5)mm, acuminate, hairy. Corolla pubescent outside, inside with long (to 3mm) hairs on anticous side, short big-headed glandular hairs as well on both anticous and posticous sides, yellow, spotted orange-or red-brown, 4–6cm long, upper lip much shorter than lower, bilobed. Ovary and style pubescent. Capsules c 40–75 × 5–6mm excluding persistent style.

Bhutan: C – Ha district (Ha Dzong), Mongar district (Latun La); **Darjeeling:** Tonglu, Manibhanjan, Kurseong, Phalut; **Sikkim:** Kalipokhri, Phusum, Dikchu. Moist shady rocks and banks, 1800–3500m. June–September.

5. DIDYMOCARPUS Wall.

Sappy perennial herbs with or without a well-developed stem arising from a small stock. Leaves opposite or alternate, sometimes 3–4-whorled. Peduncles solitary in leaf axils, flowers few to many, cymose. Bracts and bracteoles paired, sometimes connate at base, persistent or caducous. Calyx usually 5-lobed, 5-fid in one species. Corolla tube cylindric or narrowly funnel-shaped, limb strongly to weakly bilabiate, lobes suborbicular or ovate. Stamens 2, included, staminodes 3, anthers confluent, cohering face to face, glabrous or bearded. Ovary stipitate or not, linear, disc shallowly to deeply cupular or cylindric, apex often oblique; stigma obliquely 2-lipped. Capsule linear-oblong or oblong-elliptic, straight or curved, loculicidally 2-valved, valves eventually splitting, placentae on the valves, revolute, concealing the seeds. Seeds ellipsoid, minute, reticulate.

1. Calyx divided to base into 5 separate glandular-puberulous lobes; ovary and capsule glandular-puberulous .... **1. D. mortonii**
\+ Calyx usually with a distinct tube, rarely divided to base and then lobes glabrous or with few hairs; ovary and capsule either glabrous, gland-dotted, or with a very few glandular hairs 2

2. Corolla glabrous 3
\+ Corolla hairy 9

3. Leaves (also stems, which are very short) covered with sessile globose glands, hairs wanting .... **4. D. pedicellatus**
\+ Stems and leaves hairy, globose glands often present as well 4

4. Calyx lobes triangular, ± acute 5
\+ Calyx lobes oblong to suborbicular, rounded 6

5. Stems and leaves without closely felted hairs; corolla orange-red; ovary gland-dotted .... **2. D. aurantiacus**
\+ Stems and leaves closely felted with short appressed hairs (spreading acute hairs as well); corolla purple; ovary glabrous .... **3. D. triplotrichus**

6. Calyx often lobed nearly or quite to base, anthers bearded; stem and leaves without sessile globose glands .... **8. D. albicalyx**
\+ Calyx lobed less than halfway, anthers glabrous; stem and leaves dotted with sessile globose glands 7

7. Plant almost stemless; corolla c 3.5–4cm long .... **5. D. andersonii**
\+ Plant with a well-developed stem; corolla c 1.2–2cm long 8

8. Calyx tube 2–3.5mm long, corolla 1.2–1.4cm long .... **6. D. oblongus**
\+ Calyx tube 3.5–4mm long, corolla c 2cm long .... **7. D. bhutanicus**

9. Lower leaves alternate or subalternate; ovary and capsule without a well-developed stipe (up to c 3mm at fruiting); corolla 1.5–2.5cm long...... 10
\+ Leaves usually opposite or whorled, rarely 1–2 lower leaves alternate; ovary and capsule with a well-developed stipe; corolla 2.5–4cm long 12

10. Hairs on upper surface of leaves spreading .... **11. D. curvicapsa**
\+ Hairs on upper surface of leaves appressed 11

11. Bracts 2–4mm long, pedicels 7–11mm, corolla dark purple or dark violet **9. D. aromaticus**
\+ Bracts 6–9mm long, pedicels 1–5mm, corolla pink to white **10. D. denticulatus**

12. Inflorescences well overtopped by leaves with very long petioles arising near base of plant .... **12. D. cinereus**
\+ Inflorescences either overtopping at least the lowermost leaves or all leaves terminal 13

13. Leaves all terminal, petioles well developed (up to 26mm long); inflorescence axes glabrous, bracts and calyx glabrous except for sessile globose glands, or rarely bracts strigose on backs .... **13. D. podocarpus**
\+ Leaves both cauline and terminal, terminal leaves sessile or subsessile (petiole up to c 4mm long); inflorescence axes, bracts and calyx normally all with spreading glandular hairs, occasionally hairs wanting on bracts and calyx (Note: specimens with characteristics of both *D. podocarpus* and *D. pulcher* are known: see under *D. podocarpus*) .... **14. D. pulcher**

1. D. mortonii C.B. Clarke

Plant up to c 25cm tall, stem 3.5–15cm, glandular-pubescent, spreading acute hairs to 1.5mm as well. Leaves broadly ovate to suborbicular, first pair often unequal in size (or reduced to 1 leaf), often 1 (occasionally 2) pair(s) of smaller leaves above separated by short internodes, 2–15.5 × 1.5–12cm, apex obtuse to subacute, base cordate or cordate-cuneate, margins doubly crenate or serrate, upper surface with acute hairs to 1mm plus shining globose glands, lower surface with hairs nearly confined to veins; petioles 0.4–4.5cm. Flowers few to many in spreading cymes 10–18cm long, peduncles solitary in axils of either one or both primary leaves and/or upper leaves, occasionally upper axils only, pedicels 5–15mm, all axes glandular-puberulous. Bracts 3.5–7 × 2.5–10mm, ovate to suborbicular, hairy as leaves, soon caducous. Calyx divided to base, segments 4–7 × 1–2mm, ± elliptic, glandular-puberulous. Corolla puberulous, rich reddish-purple, throat white, 2.3–3cm long, limb large, very oblique, lobes rounded. Anthers glabrous. Ovary glandular-puberulous, disc cupular, 1mm long. Capsule without a stipe, 17–25 × 2–2.5mm. Seeds c 0.4mm long.

Darjeeling: Aloobari, Darjeeling, Kurseong, Mongpu, Peshok, Pomong, Rimbi Chhu, Senchal; **Darjeeling/Sikkim:** unlocalised (Hooker and Gamble collections). Mossy rocks, 915–2135m. June–September.

2. D. aurantiacus C.B. Clarke

Resembles *D. mortonii* in habit, foliage and indumentum, but distinguished by its calyx (campanulate, tube 2–4.5mm long, lobes 1.3–2.5mm, triangular, gland-dotted outside, sometimes a few hairs as well), corolla orange-red, 3–4.5cm long, glabrous, lobes broadly ovate, ovary and capsule gland-dotted, capsule stipitate, 30–50 × 1.5–2mm.

Darjeeling: Badamtam, Kalimpong, Manjitar Bridge, Chota Rangit; **Sikkim:** Tista Valley. Shady rock-faces, 600–1220m. July–August.

3. D. triplotrichus Hilliard

Resembles *D. mortonii* in habit, foliage and flower-colour, but distinguished by its different indumentum particularly on stem and petioles (felted with short appressed hairs, acute spreading hairs 0.5–1.5mm long as well), calyx campanulate with short triangular teeth, corolla glabrous, c 4cm long, ovary glabrous, capsule stipitate, c 4cm long.

Darjeeling: Pankhabari, Kurseong, Makaibari. 600–1370m. August.

4. D. pedicellatus R. Brown; *D. macrophylla* auct. non D. Don

Stem 0.7–1.5cm. Leaves often only 1 pair with rudimentary second pair, broadly ovate to suborbicular, 4–22 × 4–17cm, apex obtuse, base cordate to cuneate, sometimes oblique, margins crenate-serrate, both surfaces (also stem, peduncles, pedicels, bracts) covered in reddish to colourless globose glands, otherwise glabrous; petioles 1–13.5cm. Flowers few to many in cymes 10–20cm long. Bracts 5–10 × 3–10mm, broadly ovate, partly connate, persistent. Calyx deeply campanulate, tube 6–10mm long, lobes 1–1.5mm, obtuse to subacute. Corolla purple, glabrous, 2.8–4cm long, limb large, oblique, lobes rounded. Anthers bearded. Ovary glabrous, disc cupular, c 2.4–3mm long. Capsule stipitate, c 35–50 × 2–2.5mm.

W Bengal Duars: Buxa. Shady moist banks and rocks, 600–1500m. June–September.

5. D. andersonii C.B. Clarke

Resembles *D. pedicellatus* in habit and the form of leaves, bracts, calyx and corolla, but distinguished by the spreading hairs on stem, leaves, peduncles and first pair of bracts, corolla rose-red, anthers glabrous and capsule without a stipe.

Bhutan: S – Chukka district (Giengo); **Darjeeling:** Kalimpong, Kurseong, Rinchinpong; **Sikkim:** Jarlioke, Martam–Rumtek (71); **W Bengal Duars:** Buxa. Crevices of shady rocks and cliffs, 600–1700m. June–September.

6. D. oblongus D. Don; *D. verticillatus* Wall., *Henckelia oblonga* (D. Don) Sprengel

Stem 2.5–17cm long clad in ± retrorse to spreading acute hairs 1–1.5mm long, colourless to red globular glands as well, usually simple, terminating in a pseudo-whorl of 2 pairs of close-set leaves, otherwise leafless, or occasionally a pair of rudimentary leaves near base, and then rarely the axillary bud producing an exact replica of the primary stem. Leaves elliptic or elliptic-lanceolate, sometimes unequal in size, 1.5–12.5 × 1.2–4.5cm, apex acute, base cuneate or rounded, margins serrate, upper surface with acute hairs to 0.5mm, lower surface with acute hairs to 1–1.5mm mainly on the veins, globose glands on both surfaces; petioles 0.2–1.5cm. Flowers few to many in cymes 4.5–10cm long, pedicels 3–8mm, acute hairs 0.5–1mm and globose glands on peduncles, other axes often glabrous. Bracts c 3–4 × 4–6mm, broadly ovate, partly connate, either glabrous or with a few hairs on the back, eventually deciduous. Calyx glabrous, campanulate, tube 2–3.5mm long, lobes 1–2mm, rounded. Corolla glabrous, shades of purple, 1.2–1.4cm long, limb very oblique, lobes rounded. Anthers glabrous. Ovary glabrous, disc c 2mm long, cupular, strongly oblique. Capsule without a stipe, 9–19 × 1.5–2mm.

Bhutan: C – Tashigang district (Gamri Chu); **Sikkim:** Lachen, Chunthang–Lachen. Damp rock-faces, 1525–3300m. May–September.

7. D. bhutanicus W.T. Wang

Similar to *D. oblongus* but distinguished by its broader leaves (up to 7.5cm broad) often with longer petioles (up to 3.7cm), calyx tube 3.5–4mm long, lobes 1–2.5mm long, corolla c 2cm long, white to pale lilac, capsules 1.4–2.5cm long.

Bhutan: N – Upper Kulong Chu district (Tobrang). Mossy rocks in dense mixed forest, c 2500–2745m. July–August.

A plant from Lachen in Sikkim, with similar long-petioled leaves in pairs on the stem as well as at the apex, needs investigation. This specimen is mentioned in F.B.I. 4: 347.

8. D. albicalyx C.B. Clarke; *D. leucocalyx* C.B. Clarke

Stem 2–32cm long clad in ± retrorse to spreading acute hairs up to 1.5–2mm long, often gland-tipped hairs as well. Leaves broadly elliptic to ovate or suborbicular, sometimes unequal in size, largest pair (or rarely 3) usually well above base, 1–3 pairs of smaller leaves above, often only lowermost internode elongating, 2.7–20 × 2.4–14cm, apex acute to obtuse, base cordate or rounded, sometimes oblique, margins doubly crenate or serrate, upper surface with acute hairs up to 1.5mm long, lower surface with acute hairs confined to veins or not, often mingled with gland-tipped hairs; petioles 0.4–10cm. Flowers usually many in spreading cymes 5–13cm long, pedicels c 6–15mm long, spreading hairs to 1–1.5mm long each terminating in an oblong gland on both peduncles and pedicels. Bracts broadly ovate to suborbicular, c 3–9 × 3–8mm, pubescent, quickly deciduous. Calyx usually pale, rarely purplish and then leaves purplish below, glabrous or nearly so, 2–3mm long, campanulate, lobed nearly or quite to base, lobes rounded. Corolla glabrous, shades of purple, 1.4–2cm long, limb very oblique, lobes rounded. Anthers bearded. Ovary glabrous, disc c 2mm long, very oblique. Capsule without a stipe, 13–22 × 1.8–2.5mm.

Bhutan: S – Phuntsholing and Gaylegphug districts, **C** – Thimphu, Punakha, Tongsa and Tashigang districts, **N** – Upper Mo Chu district; **Darjeeling:** Darjeeling, Ghum, Ghumpahar etc.; **Sikkim:** Rate Chhu N of Gangtok, Bitu, Karponang etc. Moist shady mossy rock-faces and earth banks, occasionally tree trunks in broad-leaved and bamboo forests (not clear if always in forest), 1200–2745m. June–September.

9. D. aromaticus D. Don non Wall.; *D. subalternans* R. Brown

Stem 6.5–23cm tall, clad in acute, upward-pointing appressed hairs to 0.4mm long and shining globose glands. Leaves distant and alternate on lower part of stem, crowded, opposite or subopposite terminally, broadly ovate to elliptic, 2–6.7 × 1.2–5.7cm, apex acute to obtuse, base rounded or cuneate, sometimes oblique, margins crenate, serrate or serrulate, upper surface with appressed acute hairs to 0.8mm long and globose glands, lower surface gland-dotted, hairs confined to veins or scattered; petioles 0.8–3.7cm long. Cymes 3.5–5.5cm long from axils of uppermost leaves, few-flowered, pedicels 7–11mm long, all inflorescence axes glandular-pubescent. Bracts ovate, 2–4 × 1.5–2.5mm, glandular pubescent and gland-dotted. Calyx campanulate, sparsely glandular-pubescent, tube 1.8–2.8mm long, lobes triangular, acute, 1–2.2 × 1.2–2.2mm. Corolla glandular-pubescent, dark purple or dark violet, 2–2.5cm long, limb oblique, lobes ovate. Anthers bearded. Ovary glabrous, disc cupular, 1.5–3mm long. Capsule sessile or with a stipe to 3mm long, often curved, 20–33(–42) × 1.5–2mm.

Bhutan: C – Punakha district (Nobding–Wache). Wet mossy rocks and cliff-faces in forest, 1900–2100m (to c 3000m in Nepal). July–September.

10. D. denticulatus W.T. Wang

Similar to *D. aromaticus* but distinguished by its larger leaves (up to 14 × 6cm) with longer petioles (up to 8cm), longer cymes (to 8cm), larger bracts (6–9 × 6–9mm), shorter pedicels (1–5mm), larger calyx (tube 3mm, lobes c 3 × 2.5mm), and paler flowers, varying from pink to white.

Bhutan: N – Upper Kuru Chu district (Denchung). Wet mossy rocks in forest, c 2300m. July.

11. D. curvicapsa Hilliard; *D. subalternans* R. Brown var. *curvicapsularis* C.B. Clarke, *D. subalternans* R. Brown var. *curvicapsa* C.B. Clarke

Similar to *D. aromaticus* but distinguished by patent (not strongly appressed) hairs on upper surface of leaf and by axillary as well as terminal cymes.

Sikkim: Lachen, Lema, Rabom. Wet shady cliffs in mixed evergreen forest, 2135–2440m. July–August.

12. D. cinereus D. Don; *Henckelia cinerea* (D. Don) Sprengel, *Didymocarpus obtusus* R. Brown

Stem (below primary leaves) 1–6cm long, stout, terminating in 2 (or rarely 3) opposite leaves with remarkably long petioles, 1–4 slender stems from apex of primary stem each terminating in a few greatly reduced almost sessile leaves with a peduncle arising in each axil, inflorescences much shorter than primary leaves, stems felted with tiny acute appressed upward-pointing hairs, plus a few acute spreading hairs and sessile globose glands. Primary leaves broadly elliptic, ovate or subrotund, 4.5–18 × 4.5–13cm, apex subacute to obtuse, base cordate or cordate-cuneate, sometimes oblique, margins crenate to serrate, upper surface clad in acute appressed hairs and shining globose glands, lower surface gland-dotted, hairs mainly confined to veins; petioles 6.5–23cm long. Flowers many in spreading cymes, flowering stem and cyme 6–18cm long, pedicels 3–9mm long, peduncles and pedicels with spreading glandular hairs. Bracts ovate, 3–6 × 2–8mm, subentire or irregularly toothed, hairy. Calyx glandular-hairy, campanulate, tube 3–4mm long, lobes triangular, 1.5–3mm long, acute or subacute. Corolla glandular pubescent, blue or purplish-lilac, throat striate, c 3–3.5cm long, limb small, oblique, lobes ovate. Anthers bearded. Ovary glabrous, disc c 2mm long, cylindric. Capsule stipitate, c 30–43 × 2.2mm, falcate.

Bhutan: S – Sarbhang district (Loring Falls: presence here needs confirmation on flowering material), **C** – Tashigang district (Tashi Yangtsi), **N** – Upper Kulong Chu district (Tobrang). Mossy rocks in forest, 1800–2440m. July–September.

13. D. podocarpus C.B. Clarke. Fig. 107d–f.

Stem 3.5–28cm long clad in short (to 0.3–0.4mm) acute, upward-pointing appressed hairs and shining globose glands, terminating in a pseudo-whorl of 4 main leaves plus a few smaller ones. Leaves ovate to elliptic, often purplish below, 3–16 × 1.5–9.5cm, apex subacute, base rounded or cuneate, margins doubly serrate, upper surface with acute hairs to 0.5–0.8mm, globose glands as well, lower surface gland-dotted, hairs confined to veins; petioles 0.4–2.6cm. Flowers often many, in spreading cymes 4.5–13cm long, pedicels 7–20mm long, peduncles and pedicels glabrous. Bracts broadly ovate to suborbicular, connate at base, 4–7 × 5–8mm, usually gland-dotted on back near tip, rarely strigose as well. Calyx glabrous or 2 lobes gland-dotted on back, campanulate, tube 2–4mm long, lobes 1.5–3mm long, triangular, acute or subacute. Corolla glandular-pubescent, dark red-violet, white lines in throat, 2–3.5cm long, limb small, slightly oblique, lobes ovate. Anthers bearded. Ovary glabrous, disc c 2–3mm long, cylindrical. Capsule stipitate, 30–43 × 2–2.5mm, usually falcate.

Bhutan: S – Deothang district (Ngangshing–Narfong), **C** – Thimphu district (above Dotena, near Changri Monastery), Tongsa district (Chendebi, Tsanka–Chendebi), Mongar district (Sengor, Reb La); **Darjeeling:** Ghumpahar, Tonglu, Sandakphu, Senchal; **Sikkim:** Bakkim–Yoksam, Lachen, Lachung etc. Mossy rocks and banks, in *Tsuga* or broad-leaved forest or on open slopes, 2285–3600m. June–July.

Several specimens from Bhutan and Nepal are intermediate in character between *D. podocarpus* and *D. pulcher* (apical leaves petiolate, calyx and bracts glabrous as in *D. podocarpus*, leaves not all apical, peduncles and pedicels glandular-hairy as in *D. pulcher*). The Bhutanese specimens came from Bumthang district (Kamephu, Shabejethang) and Upper Mo Chu district (Gasa

Dzong), between 2590 and 4875m, flowering in June and July; also Chiong in Nepal, 2440–2745m. Field investigation is needed.

14. D. pulcher C.B. Clarke

Stem 2–45cm long, clad in short (up to 0.4mm) acute upward-pointing appressed hairs and shining globose glands, usually with a whorl of 3–4 petiolate leaves (occasionally only 2) separated by a long internode from a terminal whorl of smaller sessile or subsessile leaves; a small flowering branch may develop from axil of a lower leaf. Leaves (excluding apical ones) ovate, elliptic or oblong-elliptic, often unequal in size, 2.4–15.2 × 1.2–9.5cm, apex acute or obtuse, base cordate to cuneate, sometimes oblique, margins serrate to subentire, upper surface with acute appressed hairs and gland-dots, lower similar but hairs soon nearly confined to veins; petioles 0.4–8(–19)cm. Flowers few to many in spreading cymes 3.5–11cm long, often in axils of terminal leaves only, peduncles and pedicels with patent glandular hairs. Bracts broadly ovate to suborbicular, connate at base, margins irregularly toothed or subentire, sometimes deciduous, 4–7 × 5–9mm, gland-dotted, usually with glandular hairs as well, also strigose hairs. Calyx usually with few to many glandular hairs to 1.5mm long, rarely hairs wanting, gland-dotted, sometimes strigose, campanulate, tube 3–6mm long, lobes 1.5–4mm long, triangular, acute or subacute. Corolla glandular-pubescent, usually shades of purple striped white, rarely greenish striped purple, 2.5–4cm long, limb small, slightly oblique, lobes ovate. Anthers bearded. Ovary glabrous or with a few glandular hairs, disc c 2mm long, cylindrical. Capsule stipitate, c 38–42 × 2mm, often falcate.

Bhutan: S – Phuntsholing district (Gedu–Kharbandi), **C** – Thimphu district (Thimphu), Punakha district (Rinchu, Tinlegang), Tongsa district (Chendebi, Tashiling–Chendebi), **N** – Upper Mo Chu district (Tamji); **Darjeeling:** Singtam, Rishap, Darjeeling etc.; **Sikkim:** Chunthang, Padamchen, Rathang Chhu Valley etc. Shady moist rock-faces, often in forest, 900–2590m. July–September, peak in August.

D. pulcher usually has glandular hairs on peduncles, pedicels, bracts and calyx; their density varies but they appear to be absent from bracts and calyx only in some specimens from the vicinity of Chendebi. There is also variation in flower colour, in specimens from Bhutan as well as in the original material from Sikkim.

6. CHIRITA D. Don

Annual or perennial herbs, sometimes stemless, rarely shrubby. Leaves opposite, rarely alternate. Peduncles solitary in leaf axils, sometimes apparently terminal by suppression of apical bud, flowers either solitary or in few-flowered cymes. Bracts paired, deciduous. Calyx 5-lobed. Corolla funnel-shaped, tube often pouched, often with flanges on roof or keels on floor, limb 2-lipped, lobes 5, rounded, subequal. Stamens 2, included, staminodes 3, anther cells confluent, cohering face to face, glabrous or bearded, filaments weakly to strongly geniculate. Ovary linear, disc a fleshy ring, more or less 5-lobed, stigma with one much-expanded entire or bilobed lip, the other lip aborted or minute. Capsule long-linear, loculicidally 2-valved, placentae on the valves, revolute, concealing the seeds. Seeds ellipsoid, minute, reticulate.

1. Flowering plant with only 2 leaves, one usually much smaller than other **9. C. bifolia**
+ Flowering plant with several leaves 2

2. Calyx glabrous or with only a few scattered hairs 3
+ Calyx pubescent or villous 5

3. Stem glabrous; leaves hairy only on margins, margins entire or subentire **6. C. calva**
+ Stem hairy (sometimes glabrescent); leaves hairy at least on upper surface, margins serrate 4

4. Plant without long-petioled radical leaves; pedicels very short (up to 1cm), flowers paired in crowded, bracteate helicoid cymes .... **4. C. dimidiata**
+ Plant with 1–2 remarkably long-petioled radical leaves; pedicels long (up to 3.6cm), flowers terminal, either solitary or few .... **5. C. macrophylla**

5. Ovary pubescent or glandular-puberulous 6
+ Ovary so minutely gland-dotted as to appear glabrous 7

6. Leaves, peduncles, pedicels and calyx velvety with pale silky hairs; flowers white, without flanges on roof of tube .... **1. C. primulacea**
+ Leaves, peduncles, pedicels and calyx with rather harsh red-brown hairs, these nearly confined to veins on lower leaf surface; flowers shades of blue-violet (white recorded in Myanmar), with 2 flanges on roof of tube **2. C. lachenensis**

7. Red-brown globose glands on stems, leaves and calyx, ovary and capsules pubescent .... **7. C. oblongifolia**
+ Stems, leaves and calyx without globose glands, ovaries so minutely gland-dotted as to appear glabrous (styles glandular-puberulous) 8

8. Calyx c 1.2–2cm long; corolla 3–4.7cm; anthers bearded; lower surface of leaf without scales .... **3. C. pumila**
+ Calyx c 2.3–3.7cm long; corolla 5.5–6.5cm; anthers glabrous; lower surface of leaf covered in minute white scales (long hairs as well, mainly on the veins) .... **8. C. urticifolia**

1. C. primulacea C.B. Clarke

Perennial herb, stem either wanting or up to c 2cm long and then enveloped in old petiole-bases or their scars. Leaves many, all radical, oblong-elliptic to narrowly ovate, 5–17.5 × 3.5–7cm, apex acute, base very oblique, margins coarsely doubly serrate, densely silky-villous, also petioles, peduncles, pedicels and calyx; petioles c 4–15cm long. Inflorescences 1–6-flowered, many. Peduncles 3–6cm long. Bracts linear, up to c 12 × 1.5mm. Pedicels 1.5–4cm long. Calyx deeply divided, c 11–13mm long, lobes triangular, c 9–10mm, very acute to acuminate. Corolla puberulous, c 4.5–5cm long, tube narrowly cylindric in lower ⅓, abruptly expanded and somewhat pouched above, white. Anthers bearded, filaments with minute sessile glands. Ovary and capsule pubescent; stigma c 3 × 3.5mm, entire. Capsule c 60–70 × 1mm.

Darjeeling: Lebong, Mongpu, Rangit Valley, Puttabong, Rishap, Senchal. Damp rock-faces, 600–1675m. March–June.

2. C. lachenensis C.B. Clarke; *C. clarkei* Hook.f., *C. umbricola* W.W. Smith. Fig. 107g–j.

Perennial herb with small rhizome, slender aerial stem up to 7cm tall sometimes present bearing 1–2 pairs of leaves, 1 leaf developed, 1 rudimentary, stem, leaves, peduncles, pedicels, calyx clad in spreading red-brown hairs up to 2–3mm long. Leaves often few, ovate, 3.3–15.5 × 2–8.5cm, apex acute, base oblique, margins serrate; petioles 2–10cm long. Inflorescences 1–5-flowered, few. Peduncles 2.5–14cm long. Bracts linear-oblong, c 4–6 × 0.5–1mm. Pedicels 1.5–6.5cm long. Calyx deeply divided, c 10–17mm long, lobes triangular, c 5–13mm, acute. Corolla densely pubescent, 3–4.5cm long, tube broadly funnel-shaped, shades of blue-violet (once recorded as white), 2 glandular flanges on roof of tube. Anthers bearded, filaments glandular-puberulous. Ovary and style densely glandular-puberulous; stigma c 3 × 7mm, entire. Capsule c 80 × 1.5mm.

Bhutan: S – Deothang district (Tshilingor), **C** –Mongar district (Saling, Namning), **N** – Upper Kulong Chu (Tashi Yangtsi Chu, Tobrang); **Sikkim:** Lachen. Damp rock-faces and boulders, in mixed and broad-leaved forest, 1980–2730m. April–August.

3. C. pumila D. Don; *C. edgeworthii* DC., *C. diaphana* Royle *nom. nud.*, *C. polyneura* Miquel var. *thomsonii* C.B. Clarke, *Calosacme flava* Wall., *Bonnaya pumila* (D. Don) Sprengel, *Henckelia pumila* (D. Don) Sprengel

Annual herb 4–60cm tall. Stem either wanting or up to 45cm long, base decumbent, rooting, spreading hairs to 2mm long either present (few to many) or absent, often a few small leaves at base of stem, 2(–5) pairs higher up, often unequal in size, leaves oblong-elliptic to ovate-elliptic, 1.5–24 × 0.8–11cm, apex acute, base cuneate or rounded, sometimes oblique, margins serrate, scattered acute hairs to 2mm on upper surface, nearly confined to veins on lower surface, this sometimes blotched or wholly purple; petioles 0.3–4.5cm long. Inflorescences 1–3-flowered, mostly in axils of uppermost leaves, sometimes from lower axils as well. Peduncles 1.8–12cm long. Bracts narrowly ovate to lanceolate or elliptic, c 3–10 × 1–6mm. Pedicels 1.5–3cm long. Calyx deeply divided, tube 3.5–6mm, lobes triangular, 9–15 × 2–2.5mm, acute to acuminate, glandular and acute hairs to 3mm outside. Corolla glandular pubescent outside, 3–4.7cm long, narrowly funnel-shaped, slightly pouched, tube white, limb blue to lilac, broad yellow stripe on floor of tube, 2 glandular flanges on roof. Anthers bearded, filaments minutely gland-dotted. Ovary minutely gland-dotted; style glandular-puberulous; stigma c 1.5 × 3.5mm, bilobed. Capsule c 100–145 × 1.5mm.

Bhutan: S – Phuntsholing district, **C** – Punakha, Tongsa and Mongar districts, **N** – Upper Mo Chu and Upper Kuru Chu districts; **Darjeeling:** Darjeeling, Kurseong, Lopchu etc.; **Sikkim:** Lachen, Lachung, Bitu. Damp rock-faces and banks in broad-leaved forest, 1200–2440m. July–October.

4. C. dimidiata C.B. Clarke; *C. polyneura* Miquel var. *amabilis* C.B. Clarke, *Calosacme dimidiata* Wall. *nom. nud.*

Resembles *C. pumila* in habit and foliage but differs markedly in its inflorescence (crowded, many-flowered helicoid cymes, short peduncles (0.5–3.5cm), short paired pedicels (0.3–1cm) hidden by large (c 10–15 × 8–16mm) ovate persistent bracts), calyx (tube c 11–14mm, lobes 9–12 × 3–5mm, sparsely hairy to glabrous), corolla (c 5–5.5cm long, glabrous outside) and puberulous ovary.

Bhutan: S – Phuntsholing district (escarpment above Kharbandi); **Darjeeling:** Darjeeling, Dumsong, Kurseong, Pomong, Jhamchi Hill; **Sikkim:** Pangthang. Wet cliffs, 760–1525m. August–October.

5. C. macrophylla Wall.; *Calosacme macrophylla* Wall. *nom. nud.*, *Henckelia grandifolia* Dietrich

Perennial herb with small rhizome, aerial stem 7–25cm long, lower part decumbent and rooting, simple or 1 (rarely 2) simple branches, sparsely hairy. Leaves: 1–2 radical, ovate, 6–18 × 3.5–12cm, acute, base rounded or cordate, sometimes oblique, margins serrate or rarely subentire, upper surface thinly hairy, on lower surface hairs confined to veins; petioles 5–25cm long; cauline leaves similar but smaller, petioles shorter, 1 (rarely 2) pairs, one of each pair rudimentary. Inflorescence terminal, 1–few-flowered. Peduncles suppressed or up to 2.5cm long. Bracts ovate, c 1.5–1.7 × 1.2–1.8cm. Pedicels c 1.7–3.6cm. Calyx deeply divided, tube 6–15mm long, lobes 7–20 × 3–5mm, triangular, acuminate, glabrous. Corolla minutely puberulous, 5.5–7.5cm long, tube cylindric, curved, slightly inflated above, tube white or yellowish, lobes yellow, throat orange veined red (Sikkim, Darjeeling), pale to deep violet blue, 2 yellow stripes on floor of tube (Bhutan), 2 glandular flanges on roof of tube. Anthers bearded, filaments glandular-puberulous above, gland-dotted below. Ovary and style puberulous; stigma c 3.5–5 × 5–8mm, bilobed. Capsules not seen.

Bhutan: S – Deothang district (Keri Gompa, Riserboo), **C** – Mongar district (Sengor); **Darjeeling:** Darjeeling, Ghum–Tiger Hill, Labha, Rambi Chhu,

Rungirun; **Sikkim:** Yoksam–Bakhim, and unlocalised (Griffith collection). Moist shady banks and rocks in forest. 600–2460. May–August.

6. C. calva C.B. Clarke; *C. glabra* auct. non Miquel

Annual herb; stem 10–25cm long, base decumbent, rooting, either simple or a simple branch or inflorescence from axil of larger of each pair of leaves, all vegetative parts glabrous except for ciliate leaf margins. Leaves: up to 5 pairs, very unequal in size, distant, oblong or oblong-elliptic, 6.5–20 × 2.5–7.5cm, apex acute to acuminate, base remarkably unequal, margins entire or subentire; petioles 1–4cm long. Inflorescences few-flowered. Peduncles 2–3.5cm. Bracts linear, c 3–12 × 0.5–1mm. Pedicels 1–2cm. Calyx tube 4–11mm long, lobes triangular, acute, 6–15 × 4–5mm, glabrous. Corolla minutely puberulous outside, 4.5–6cm, narrowly funnel-shaped, slightly pouched, tube whitish, lobes purple-violet, mouth and throat yellow, 2 glandular flanges on roof of tube. Anthers glabrous or almost so, filaments minutely gland-dotted. Ovary minutely gland-dotted; style glandular-puberulous; stigma c. 3.5 × 12mm, bilobed. Capsules not seen.

Bhutan: S – Sankosh district (Daga Dzong–Daga La); **Darjeeling:** Darjeeling, Takvar Road; **Sikkim:** Yoksam–Paha Khola, Lachen. Moist shady rocks and banks in forest, 1050–1950m. July–August.

7. C. oblongifolia (Roxb.) Sinclair; *Incarvillea oblongifolia* Roxb., *Babactes oblongifolia* (Roxb.) DC., *Calosacme acuminata* Wall. *nom. nud.*, *Chirita acuminata* R. Brown, *Aeschynanthus oblongifolius* (Roxb.) G. Don

Perennial herb eventually shrubby, stem up to c 2m tall, rooting at base, densely pubescent, hairs brownish, acute, small red-brown globose glands as well, also on leaves, peduncles, pedicels, calyx and ovary. Leaves opposite, ovate or ovate-elliptic, 9–38 × 4–14cm, apex acute to acuminate, base cuneate, often oblique, margins serrulate, upper surface more densely pubescent than lower, globose glands most conspicuous there; petioles 1.7–9.5cm. Inflorescences in upper leaf axils, flowers solitary or in several-flowered cymes. Peduncles 1–4cm. Bracts 5–10 × 1–2mm, lanceolate-acuminate. Pedicels 0.5–3cm. Calyx tube 5–10mm long, lobes triangular, acute, 5–10 × 4–5mm, densely pubescent and gland-dotted. Corolla glandular puberulous, 4–5cm long, cylindric in lower part, narrowly funnel-shaped above, creamy white, 2 dark yellow warty keels at base of lower lip. Anther connective glandular, filaments very minutely glandular above. Ovary and capsule densely pubescent and gland-dotted; stigma c 3 × 3mm, shallowly notched Capsules 45–90 × 2.5–3mm.

Bhutan: C – Mongar district (Shongar Chu near Zimgang). Shady banks and rock-faces in forest, 1200–1800m. August–December.

The red-brown glands are unique to this species of *Chirita* (they are commonplace in *Didymocarpus*).

8. C. urticifolia D. Don; *Henckelia urticifolia* (D. Don) Dietrich, *Chirita grandiflora* Wall., *Calosacme grandiflora* Wall. *nom. nud.*, *Henckelia wallichiana* Dietrich, *Gonatostemon boucheanum* Regel.

Perennial herb, stem 5.5–50cm long, pubescent, lower part decumbent, rooting, apparently stoloniferous. Leaves opposite, distant, elliptic to ovate, 3.7–17 × 1.8–9cm, apex acute, base cuneate, oblique, margins serrate, upper surface thinly hairy, lower surface covered in minute white dots, hairs confined to veins; petioles 1–9cm long. Inflorescences in upper leaf axils, flowers solitary or occasionally in 2–3-flowered cymes. Peduncles 1–4.2cm long. Bracts oblong to broadly ovate, c 8–22 × 1–15mm. Pedicels 0.8–2.1cm long. Calyx tube 10–15mm long, lobes 10–20mm, triangular, acuminate, acute hairs up to 4–5mm long outside, minutely glandular inside. Corolla with long acute and shorter gland-tipped hairs outside, minutely glandular inside, 5.5–6.5cm long, tube cylindric below, abruptly expanded above, somewhat pouched, limb shades of reddish-violet, tube paler, yellow inside with purplish lines. Anthers glabrous, filaments glandular-puberulous. Ovary minutely gland-dotted; style glandular puberulous; stigma c 6 × 12mm, bilobed. Capsules c 11cm long.

Bhutan: S – Phuntsholing, Sarbhang and Gaylegphug districts, **C** – Mongar and Tashigang districts, **N** – Upper Kuru Chu district; **Darjeeling:** Darjeeling, Mongpu, Nimbong etc.; **Sikkim:** Gangtok, Lachung. Wet shaded cliffs, rocks and banks in forest, 900–2440m. August–November.

9. C. bifolia D. Don; *Henckelia bifolia* (D. Don) Dietrich, *Calosacme amplectens* Wall. *nom. nud.*

Perennial herb with small scaly rhizome, aerial stem c 2–20cm tall, rooting at base, hairy, 1–2 very small, rudimentary leaves near base, apical pair of unequal leaves. Largest leaf sessile, ovate to orbicular, 2.3–15 × 2.7–11.7cm, apex obtuse, base cordate-clasping, margins serrate to subentire, hairy on both surfaces; second apical leaf similar but smaller and shortly petiolate. Inflorescences terminal, usually 1, rarely 2, but rudimentary ones often present, 1–2(–3)-flowered. Peduncles 0.7–4.5cm. Bracts narrowly to broadly ovate, 4–14 × 1.2–7mm. Pedicels 0.5–2.6cm. Calyx tube 5–10mm long, lobes triangular, acute, 4.5–11 × 2.5–4mm, very hairy. Corolla almost glabrous, 3–5.5cm long, tube narrowly funnel-shaped, slightly pouched, whitish outside, broad yellow band on floor inside, 2 glandular flanges on roof, limb pale blue to bright blue-purple. Anthers bearded, filaments sparsely glandular-puberulous. Ovary and capsule pubescent; stigma c 2.5 × 6mm, entire or shallowly lobed. Capsule c 65 × 1mm (not fully ripe).

Bhutan: S – Deothang district (Khomanagri), **C** –Punakha district (Tinlegang, Wangdu Phodrang), Mongar district (Shongar, Lhuntse, Saling–Mongar etc.), **N** – Upper Kuru Chu district (Kurted). Shady moist rocks and banks in scrub, forest and ravines, 900–2285m. June–August.

7. CORALLODISCUS Batalin

Perennial stemless rosulate herbs, stoloniferous or not. Leaves thick-textured, base tapering into a broad flat petiole. Peduncles axillary, flowers 1–many, cymose. Bracts wanting, bracteoles occasionally present, very small. Calyx deeply divided into 5 lobes. Corolla tubular, 3 bands of relatively long hairs on floor of tube, limb 2-lipped, 5-lobed, upper lip smaller than lower. Stamens 4, one long and one short pair, all included, anther cells confluent, cohering in pairs at apices, filaments inserted in lower half of tube, eventually coiled; staminode 1. Disc annular, bright orange. Ovary conical, passing abruptly into style; stigma equally and shortly bilobed. Capsule oblong or linear-oblong, either septicidal, or loculicidal and eventually septicidal as well. Seeds very small, ellipsoid, reticulate.

1. Peduncles, pedicels and calyx persistently woolly .... **1. C. kingianus**
\+ Peduncles, pedicels and calyx glabrous or glabrescent 2

2. Leaf blade roughly 2–3 times as long as broad, glabrous above even when immature, 1–2(–3) pairs of lateral veins (visible on lower surface); plant stoloniferous .... **3. C. cooperi**
\+ Leaf blade roughly as long as broad to twice as long, either hairy above or glabrous, 3–5 pairs of lateral veins; plant stoloniferous or not 3

3. Leaves persistently hairy above (hairs may be sparse); plant not stoloniferous .... **2. C. lanuginosus**
\+ Leaves either hairy above or glabrous; plant stoloniferous **4. C. × bhutanicus**

1. C. kingianus (Craib) B.L. Burtt; *Didissandra kingiana* Craib, *D. rufa* Hook.f. Fig. 107k–n.

Plant not stoloniferous, caudex up to 7 × 1cm. Leaf blade rhomboid-elliptic, 2.5–5.5 × 1.2–2.8cm (roughly twice as long as broad), apex subacute, base cuneate, margins entire, 3–5 pairs of side veins deeply impressed above even in young leaves, upper surface glabrous, margins and lower surface thickly brown-woolly; petioles 1.7–3cm. Peduncles 4–11cm long, many-flowered, they, the pedicels and calyx all persistently thickly brown-woolly. Pedicels 0.5–1.3cm. Calyx tube 1–1.2mm long, lobes oblong-elliptic or elliptic, 2–3.2 × 1.5–2mm, obtuse to subacute. Corolla 1.4–1.6cm long, violet-blue, 2 dull yellow lines on floor of tube. Capsule 6–12 × 2.8–4mm.

Bhutan: C – Sakden district (Sakden). Cliffs and rock-faces, 2895–4800m (in Tibet). June–August.

2. C. lanuginosus (DC.) B.L. Burtt; *Didymocarpus lanuginosus* DC., *Didissandra lanuginosa* (DC.) C.B. Clarke

Plant not stoloniferous, caudex up to 2 × 1cm. Leaf blade broadly rhomboid, 1.4–6.7 × 1.2–4.8cm (roughly as long as broad to 1½ times as long), apex obtuse, base cuneate, margins entire or subcrenate, 4–5 pairs of side veins, shallowly impressed or not above, upper surface thinly to thickly hairy, hairs white, up to 2–7mm long, lower surface thickly brown-woolly over veins, more sparsely woolly to glabrous between veins; petioles 1–4cm long. Peduncles 4.5–15cm long, 1–15-flowered, they, pedicels and calyx thinly woolly, glabrescent. Pedicels up to 1.2–3.6cm. Calyx tube 0.1–0.5mm long, lobes triangular, 2.5–3.7 × 0.8–1.3mm, acute. Corolla c 1–1.5cm long, varying from whitish to violet-blue, or tube blue, lobes white, 2 yellow lines on floor of tube. Capsule 12–25 × 2–2.5mm.

Bhutan: only doubtful records; see under *C.* × *bhutanicus*; **Darjeeling:** Shiri Khola; **Sikkim:** Chunthang, Lachen, Lachung, Kaydoong Mt. Cliffs and rock-faces, 1800–2440m. May–August.

3. C. cooperi (Craib) B.L. Burtt; *Didissandra cooperi* Craib

Differs from *C. lanuginosus* in its stoloniferous habit, leaf blade elliptic to rhomboid-elliptic, 0.5–2.8 × 0.3–1cm (roughly 2–3 times as long as broad), apex subacute, 1–2(–3) pairs of side veins, scarcely or not visible above, upper surface glabrous, lower brown-woolly, wool soon confined to veins, petioles 0.5–1.6cm long, peduncles 3–8cm long, 1- (very rarely 2-)flowered.

Bhutan: C – Thimphu district (Dotena Chu), Tongsa district (Chendebi, Tongsa to Changkha), Bumthang district (near Bumthang, Dhur Valley, Shabejetang), Tashigang district (Tashi Yangtsi). Cliffs and rock-faces, 1980–3200m. June–July.

4. C. × bhutanicus (Craib) B.L. Burtt; *Didissandra bhutanica* Craib

Differs from *C. lanuginosus* in usually having stolons, leaf blade roughly as long as broad to twice as long, 3–5 pairs of side veins often impressed above, upper surface soon glabrous or nearly so in eastern Bhutan (but persistently white-hairy above in western Bhutan), peduncle 1–6-flowered (possibly never more).

Two forms occur:

Plant stoloniferous, upper leaf surface glabrous or nearly so:

Bhutan: S – Chukka district (2km N of Takhti Chu), Sankosh district (Daga Dzong–Daga La), Deothang district (Deothang–Tashigang Dzong, Kheri Gompa), **C** – Thimphu district (Dotena Chu), Tongsa district (Shemgang), Mongar district (Khoma Chu).

Plant either stoloniferous or possibly sometimes not, upper leaf surface hairy:

Bhutan: S – Chukka district (Pakodyam), **C** – Thimphu district (Tashi Cho

Dzong), Punakha district (Tinlegang, Wangdu Phodrang); **Darjeeling:** Rimbik. Cliffs and rock-faces, 1400–2440m. July–August.

C. × *bhutanicus* possibly represents a range of hybrids between *C. lanuginosus* and *C. cooperi*; this needs further investigation. From the material available to me, true *C. lanuginosus* ranges from the western Himalayas to 88°45′E, *C. cooperi* is known only from Bhutan (89°38′–91°29′E) and *C.* × *bhutanicus* from 88°06′E (Darjeeling) to 89°54′E in Bhutan (leaves hairy above), Bhutan 89°36′–91°34′E (leaves glabrous above).

8. PLATYSTEMMA Wall.

Herb. Leaves 1–2 at apex of stem. Flowers solitary or few in a pedunculate terminal cyme. Calyx divided nearly to base. Corolla ± rotate, limb 2-lipped, lobes 5. Stamens 4, clustered around style; anther lobes confluent. Ovary sessile, disc wanting; style simple; stigma minutely bilobed. Capsule ± elliptic, 2-valved, opening at apex. Seeds minute, ellipsoid.

1. P. violoides Wall.; *P. majus* Wall. Fig. 108a–b.

Slender erect unbranched pubescent herb, stem 3–15cm. Leaf 1 (rarely 2, second much smaller) at apex of stem, broadly ovate to suborbicular, 2–7 × 2–7cm, apex acute or obtuse, base cordate, ± clasping, margin coarsely serrate or occasionally subentire. Flowers solitary or up to 6 on peduncles 2–6.5cm, pedicels 1.2–2cm. Calyx divided nearly to base, lobes triangular, 1.5–2 × 1.2mm. Corolla ± rotate, mauve, limb 10–17mm across, 2-lipped, upper lip shallowly bilobed, 5–7 × 6–7mm, lower lip deeply 3-lobed, 7–10 × 14–16mm. Anthers 4 on short filaments surrounding style base. Capsule ± 6 × 3mm. Seeds minute, ellipsoid.

Bhutan: C – Tongsa district (Chendebi–Rukubji). Shady mossy rocks on earth banks, ± 2700m. July–October.

9. LEPTOBOEA Bentham

Shrubs. Leaves opposite, decussate, simple, petiolate, often crowded. Flowers in axillary few-flowered cymes, peduncles and pedicels filiform. Calyx deeply 5-lobed. Corolla shallowly campanulate, obscurely bilabiate, lobes 5, rounded. Stamens 4, clustered around base of style; anther cells confluent at apex. Ovary sessile, disc wanting; style simple; stigma capitate. Capsule linear, septicidally 2-valved, margins of valves inflexed. Seeds minute, ellipsoid.

1. L. multiflora (C.B. Clarke) Gamble; *Championia multiflora* Clarke, *Leptoboea pubescens* Clarke *nom. nud.* Nep: *Patpati*. Fig. 108c–e.

Shrub to 3m. Leaves crowded at tips of short lateral branches, blade elliptic, 3.5–15 × 1.5–6.5cm, apex acute to acuminate, base cuneate, sometimes slightly oblique, margins minutely serrate, appressed pubescent; petioles up to 2.5cm. Peduncles 2–3cm; pedicels 1–2cm. Calyx lobes narrowly triangular, 1.8–2 × 0.5–0.8mm. Corolla shallowly campanulate, white, red marks at base of upper lip, limb ± 12mm across, tube 4mm, 2 upper lobes ± 4 × 6mm, 3 lower lobes ± 5 × 7mm. Capsule 12–24 × 2mm, excluding persistent style.

Bhutan: S – Sarbhang district (Sarbhang), Deothang districts (Deothang Hills), **C** – Tongsa district (Shemgang); **Darjeeling:** Pedong, Pankhabari, Rambi Chhu, Kalimpong, Rongsong; **Sikkim:** Sirong, Tista Valley. Hot forest and scrub, rock-faces, 600–1300m. June–August.

10. BOEICA Clarke

Undershrubs or woody perennials. Leaves alternate, petiolate. Cymes in upper leaf axils, many-flowered, lax. Bracts small. Calyx deeply 5-lobed. Corolla small, limb shallowly campanulate from a broad shallow tubular base, limb 5-lobed, 2-lipped, upper lip smaller than lower. Stamens 4, inserted near base of tube, included, filaments short, anthers not cohering face to face, 2-celled, cells confluent, minutely glandular at apex, dehiscence valvular; staminode 1. Ovary conical, disc annular, obscure, placentae deeply inflexed, margins revolute enfolding ovules; stigma truncate. Capsule short, oblong-elliptic, crowned with persistent style, loculicidal then septicidal. Seeds very small, ellipsoid, reticulate.

1. B. fulva Clarke. Fig. 108f–g.

Woody perennial at least 40cm tall, sparingly branched, stem decumbent at base, rooting, young parts felted with closely appressed silky-woolly reddish-brown hairs. Leaves crowded at tips of branches, elliptic, 9–25 × 2–7cm, apex acute to acuminate, base cuneate, margins obscurely to distinctly serrulate, upper surface initially with thin covering of hairs, soon glabrescent, lower surface at first felted with silky-woolly reddish-brown hairs, later hairs thinly spread except on veins; petioles 1–5cm long. Peduncles 11–15cm long. Pedicels 0.7–2cm long. Bracts lanceolate, 4–10 × 1.5–2mm. Calyx 2–4mm long, deeply 5-lobed. Corolla shallowly campanulate, purplish-red, c 8mm long, minutely puberulous outside. Capsules 4–9 × 1.5–2mm, excluding persistent style.

Bhutan: S – Deothang district (Tsalari Chu), **C** – Mongar district (Shongar). Mossy rocks in forest, c 900m. September.

Ill-known, last collected in 1915.

FIG. 108. **Gesneriaceae.** a–b, *Platystemma violoides*: a, habit (× 1); b, flower (× 2). c–e, *Leptoboea multiflora*: c, flowering branches (× 1); d, dissected corolla (× 2); e, capsule (× 2). f–g, *Boecia fulva*: f, flowering shoot (× ⅓); g, capsule (× 3). h–i, *Rhynchotechum ellipticum*: h, fruiting shoot (× ⅓); i, calyx and fruit (× 4). Drawn by Mary Bates.

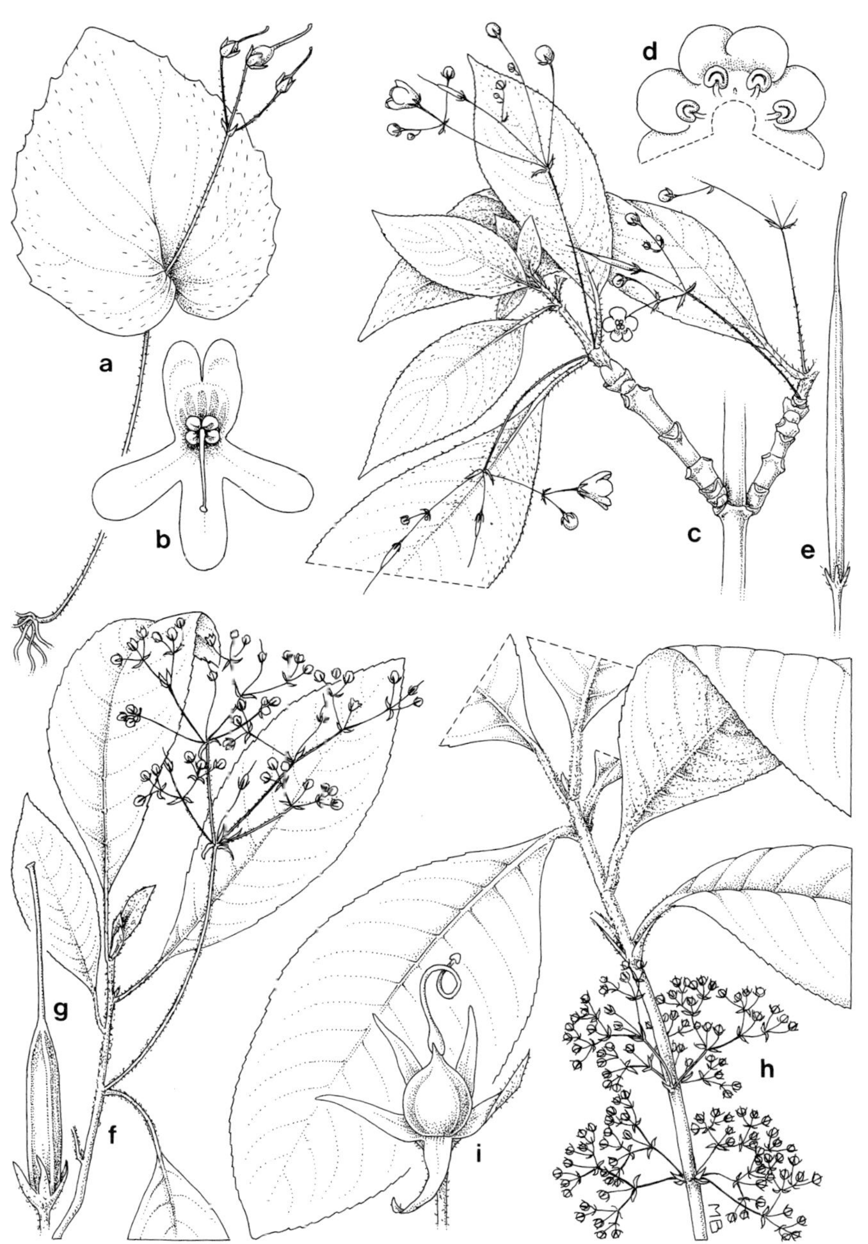
a
b
c
d
e
f
g
h
i
MB

11. RHYNCHOTECHUM Blume

Undershrubs, stem often simple, occasionally sparingly branched, stout, brittle, bark white, spongy, young parts silky-woolly or hispid. Leaves opposite or rarely alternate, petiolate. Cymes in lower leaf axils, many-flowered, compound. Bracts elliptic. Calyx very deeply 5-lobed, persistent. Corolla small, limb shallowly campanulate from a broad shallow tubular base, limb 5-lobed, 2-lipped, upper lip smaller than lower. Stamens 4, inserted near base of tube, filaments short, curved, anthers not cohering face to face, 2-celled, cells confluent, minutely glandular at apex, dehiscence valvular; staminode 1. Ovary conical, disc obscure or annular, placentae deeply inflexed, enfolding ovules; stigma small, capitate. Fruit globose, outer wall somewhat fleshy, white, translucent, thin-walled at base and there separating from receptacle. Seeds very small, ellipsoid, reticulate.

1. Young parts of stem and leaves hirsute, hairs spreading .... **1. R. vestitum**
\+ Young parts of stem and leaves silky-woolly, hairs appressed **2. R. ellipticum**

1. R. vestitum C.B. Clarke; *Corysanthera vestita* Griff. *nom. nud.* Nep: *Ashiney jhar.*

Stem up to 1m tall, 1cm diameter, simple or occasionally sparingly branched, young parts hirsute, hairs spreading. Leaves deciduous on flowering part of stem, elliptic, 10–22 × 3.7–10.5cm, apex shortly acuminate, base cuneate, margins serrulate, both surfaces hirsute with spreading hairs, more densely so below; petioles 2.5–5cm long. Cymes 3–5cm long, fascicled, interspersed with much reduced leaves and bracts. Bracts lanceolate-elliptic, c 6–9 × 1.5–3mm. Peduncles 0.7–3cm long, pedicels up to 1.5cm, both hirsute, hairs spreading. Calyx tube up to c 1mm long, lobes triangular, gland-tipped, c 5–7.5 × 1.5–2.8mm, hirsute, pink or white. Corolla bright pink, glabrous, c 6mm long. Fruit 3–4.5 × 3–4.5mm (when dry).

Bhutan: S – Gaylegphug district (Betni Bridge (38)), Deothang district (Deothang); **Darjeeling:** Darjeeling, Rishap, Rambi Chhu, Lingtam, Rongsong, Sureil, Sureil–Mongpu; **Sikkim:** unlocalised (Gammie collection). Scattered in evergreen forest, 600–1500m. July–September.

2. R. ellipticum (D.F.N. Dietrich) DC.; *Corysanthera elliptica* Wall. *nom. nud.* Nep: *Galeto*. Fig. 108h–i.

Similar to *R. vestitum* but hairs on stems and leaves appressed silky-woolly, leaves often obovate, acute, petioles shorter (1.2–3cm), calyx lobes shorter (c. 3–5 × 1–1.5mm), corolla lobes glandular-puberulous outside, crimson blotch at base of upper lip.

Bhutan: S – Samchi district (Samchi Hill (117)), Gaylegphug district (Gaylegphug (117)), Deothang district (Deothang–Narfong (117)); **Darjeeling:**

Sureil–Mongpu, Sivok, Chota Rangit; **Sikkim:** Tista Valley. Evergreen forest, 300–1500m. July–September.

Used as a fodder plant.

12. PARABOEA (C.B. Clarke) Ridley

Rosulate or stemmed perennial herbs, at least young stems woolly. Leaves opposite or spiral, upper surface usually thinly cobwebby, glabrescent, lower densely woolly, petiolate. Cymes axillary, many-flowered. Bracts small. Calyx very deeply 5-lobed. Corolla small, obliquely campanulate or almost rotate, limb somewhat bilabiate. Stamens 2, inserted near base of tube, anthers facing each other but not cohering, 2-celled, cells confluent. Ovary conical or cylindrical, disc a very shallow ring; stigma small, capitate. Capsule short, in our area dehiscent along both margins, spirally twisted. Seeds very small, ellipsoid, reticulate.

1. P. multiflora (R. Brown) B.L. Burtt; *Didymocarpus multiflorus* Wall. *nom. nud.*, *Boea multiflora* R. Brown, *B. flocculosa* C.B. Clarke. Fig. 109a–b.

Perennial herb, stem stout, 1–12cm long, simple, rooting, young parts brownish-woolly. Leaves crowded at stem apex, elliptic, 5–14.5 × 2–6.5cm, apex subacute to obtuse, base cuneate, margins crenate, upper surface very sparsely pubescent, lower thickly woolly-felted, hairs reddish-brown; petioles 2–5cm long. Peduncles 8–15cm long, pedicels up to 0.6–1cm, both woolly-felted. Bracts linear-lanceolate, c 4 × 1mm. Calyx tube c 0.3mm long, lobes triangular, c 2 × 0.8mm, woolly-felted on backs. Corolla widely campanulate, c 7mm long, limb almost rotate, lobes subequal, white, gland-dotted outside. Stamens lying on lower lip, filaments stout, anthers c 2mm long, covered in globose glistening glands. Ovary clad in glistening globose glands; style and stigma projecting straightforward. Capsules c 5–13 × 1.5–2mm, spirally twisted, glandular as ovary.

W Bengal Duars: Buxa. Damp rocks and banks in shady places, c 500–600m. June–July.

13. RHYNCHOGLOSSUM Blume

Annual or perennial herbs. Leaves alternate, unequal-sided. Flowers in long terminal or axillary racemes. Calyx campanulate, lobes 5, longer or shorter than tube. Corolla tube cylindric, mouth usually personate, limb 2-lipped, upper lip small, bilobed, lower much larger, entire or 3-lobed. Stamens either 4 or 2, included; anthers opposed in pairs, cells confluent. Ovary ovoid, disc shallowly cupular, developed on one side only (?always); style linear; stigma obscurely bilobed. Capsule included within calyx, ellipsoid, loculicidally 2-valved. Seeds ellipsoid, minutely rugulose.

1. R. obliquum Blume; *Wulfenia obliqua* Wall., *Loxotis obliqua* (Wall.) Bentham, *L. intermedia* Bentham, *Rhynchoglossum obliquum* (Wall.) DC. non Blume, *R. blumei* DC., *R. rheedii* DC., *R. obliquum* (Wall.) DC. var. *intermedium* (Bentham) DC., *R. zeylanicum* Hooker, *R. obliquum* Blume var. *parviflorum* C.B. Clarke. Fig. 109c–e.

Annual herb, stem 3–60cm, scabridulous in a broad longitudinal band. Leaves membranous, 1.8–15 × 0.8–9cm, apex acute to acuminate, base very oblique, often cordate-rounded on broad side, ± cuneate on narrow side, margins entire, lower surface thinly scabridulous, upper with coarser scattered minute hairs; petioles 0.2–4cm. Flowers in terminal or axillary racemes and then 1–several often displaced onto petiole, racemes 2–30cm long. Calyx tube 2–3mm, lobes triangular, 2.2–3 × 1–1.7mm, scabridulous on margins and mid-line. Corolla violet-blue, 10–15mm long, upper lip minutely bilobed, lower much larger, ovate to elliptic, entire or with varying development of 2 small lateral teeth, lips pressed together. Stamens 2, included. Capsule ovoid-ellipsoid, 4–5 × 3.5–4mm, included in persistent calyx. Seeds c 0.2mm long.

Bhutan: S – Chukka, Gaylegphug and Deothang districts, **C** – Punakha, Tongsa and Tashigang districts. **Darjeeling:** Kodabari, Ging, Darjeeling, Kurseong etc.; **Sikkim:** Zalipur (69), Chakung. Moist ground in forest or in open places, tends to be a weed, 600–1800m. August–November.

14. EPITHEMA Blume

Short-lived herbs with sappy stems. First foliage leaf always unpaired, upper paired or unpaired, all petiolate or paired upper ones sessile. Peduncles terminal or axillary. Bract solitary, sometimes enfolding the flowers. Flowers many in a dense cincinnus, flowers opening successively, or sometimes cleistogamous. Calyx campanulate, 5-lobed. Corolla small, 2-lipped. Stamens 2, staminodes 2, anthers cohering, confluent. Disc cup-shaped, almost enclosing ovary, often split dorsally or divided into 2 parts, absent in cleistogamous flowers. Ovary globose, unilocular, placentae 2, discoid, stalked, ovules many, on long funicles; style simple; stigma capitate. Capsule globose, thin-walled, circumscissile. Seeds minute, ellipsoid.

FIG. 109. **Gesneriaceae and Orobanchaceae. Gesneriaceae.** a–b, *Paraboea multiflora*: a, habit (× ⅓); b, capsule (× 3). c–e, *Rhynchoglossum obliquum*: c, flowering shoot (× ⅓); d, dissected calyx (× 3); e, dissected corolla with gynoecium (× 3); f–h, *Epithema carnosum*: f, flowering plant (× ½); g, half calyx with capsule (× 4); h, dissected corolla (× 4). **Orobanchaceae.** i–k, *Orobanche solmsii*: i, habit (× ½); j, calyx segments (× 1½); k, dissected corolla with gynoecium (× 1½). l–m, *Aeginetia indica*: l, habit (× ⅔); m, dissected corolla with gynoecium (× 1). Drawn by Mary Bates.

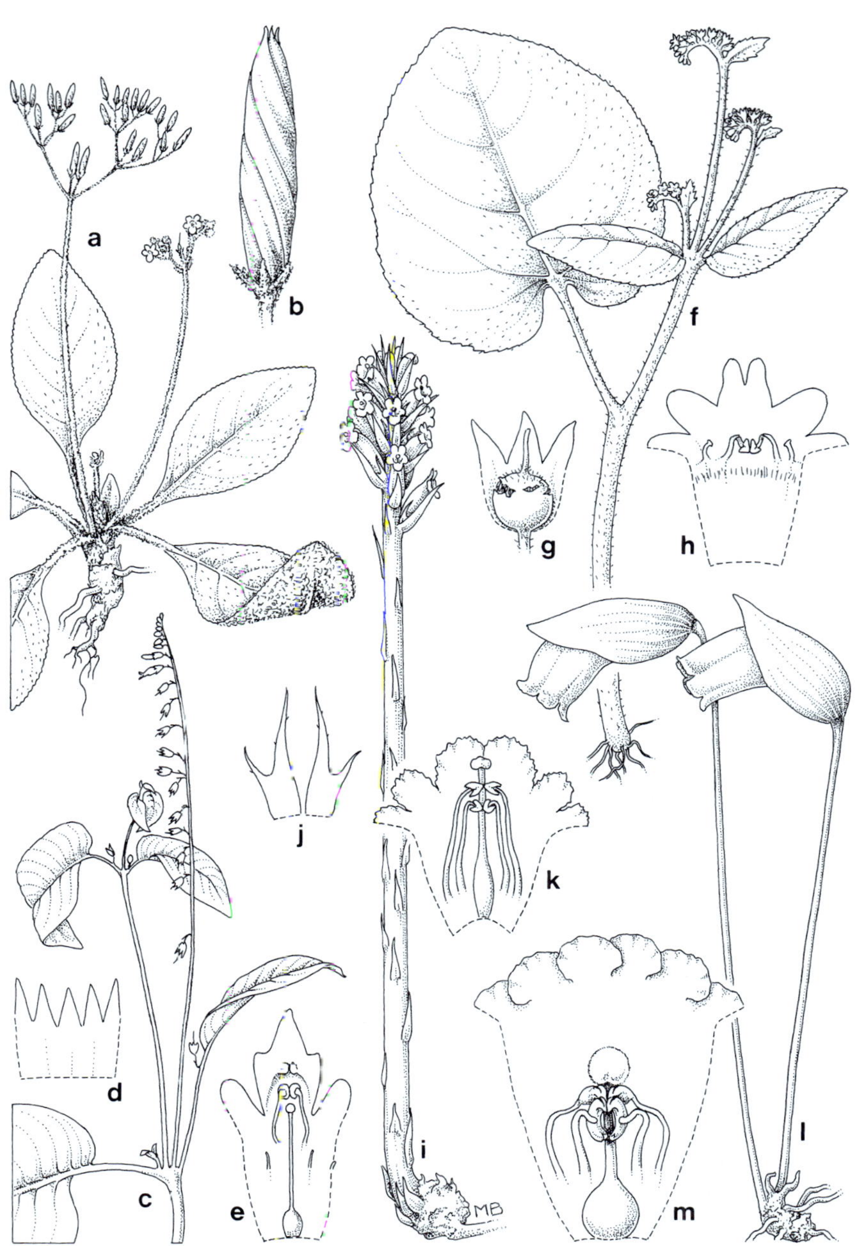
a
b
f
g
h
i
j
k
d
c
e
m
l
MB

1. E. carnosum Bentham; *Aikinia carnosa* (Bentham) G. Don [var. *dentata*, var. *hispida*, var. *zeylanica*, var. *pusilla* in F.B.I. excluded]. Fig. 109f–h.

Herb 5–40cm tall, stem to first leaf 2–18cm. First leaf unpaired, petiolate, upper leaves paired, sessile, all leaves ovate to broadly elliptic, first leaf 5.5–15.5 × 4–13cm, petiole 1.5–7.5cm; leaves below inflorescence 2–10 × 1.5–8cm, all leaves apex blunt, base cordate or subcordate, often unequal, margins entire to serrate, both surfaces pubescent with straight and hooked hairs. Peduncles 0.8–7.5. Bract 5–10 × 6–10mm, margin coarsely toothed. Calyx tube 2–2.5mm (1–2mm in cleistogamous flowers), lobes triangular, 1.75–2.25 × 1–1.25mm (0.75–2 × 0.75–1mm in cleistogamous flowers), pubescent with straight and hooked hairs. Corolla tube 4.5–5 × 2mm, hairy inside, limb mauve. Disc a split ring or 2 separate glands, absent in cleistogamous flowers. Ovary with hairy operculum, hairs straight. Capsule 2.5 × 2.5mm.

Bhutan: C – Mongar district (Shongar); **Darjeeling:** Rishap Ravine, Mongpu. Crevices and shelter of mossy rocks in ravines, 600–1200m. July–October.

Family 178. OROBANCHACEAE

by S.J. Rae

Annual or perennial, herbaceous root parasites, leafless and without chlorophyll. Stems usually simple, often short, scaly. Flowers solitary on slender shoots (scapes), or in axils of scales, or in spikes or racemes, bisexual, zygomorphic. Calyx 4(–5)-lobed or sheath-like and unlobed. Corolla 2-lipped, tube curved, upper lip simple or 2-lobed, lower lip 3-lobed. Stamens 4, in 2 pairs, often coherent at anthers; anthers opening by slits. Ovary superior, 1-celled, ovules numerous, placentation parietal; style 1, stigma 2-lobed. Fruit a capsule, opening by 2–3 valves; seeds numerous.

1. Flowers numerous, sessile or on short pedicels, in spikes or racemes on erect shoots 2
\+ Flowers solitary on long shoots or few on short pedicels arising from short branched stems 4

2. Flowering shoot scaly only at base; calyx glandular; corolla 1–1.5cm **1. Lathraea** (*L. squamaria*)
\+ Flowering shoot scaly throughout; calyx not glandular; corolla 1.5–2.5cm 3

3. Flowering shoots glabrous; corolla lobes ciliate; upper lip 2-lobed, lower lip 3-lobed; stamens exserted .... **3. Boschniakia**
\+ Flowering shoots pubescent; corolla lobes subentire; upper lip hooded, lower lip very short; stamens included .... **2. Orobanche**

4. Flowers solitary on long scaleless shoots; corolla lobes fimbriate **4. Aeginetia**
\+ Flowers few on short pedicels; corolla lobes not fimbriate 4

5. Flowering shoots 7–10cm; calyx truncate; corolla 2.5–3cm **1. Lathraea** (*L. purpurea*)
\+ Almost stemless; calyx lobed; corolla 5–6cm .... **5. Christisonia**

1. LATHRAEA L.

Perennial; rhizome creeping with fleshy scales; stems short, erect with few scales. Flowers few, borne singly in the axils of scales, or numerous in erect bracteate racemes. Calyx equally 4-lobed. Corolla 2-lipped, upper lip strongly concave, simple, lower lip smaller, 3-lobed.

1. Flowering shoots 8–30cm; flowers numerous in racemes; corolla 1–1.5cm; calyx 4-lobed .... **1. L. squamaria**
\+ Flowering shoots 7–10cm; flowers few in axils of scales; corolla 2.5–3cm; calyx almost entire .... **2. L. purpurea**

1. L. squamaria L. Eng: *Toothwort.*
Flowering shoots 8–30cm, white or pale pink. Inflorescence a 1-sided bracteate raceme, pedicels 5mm. Calyx glandular, lobes triangular. Corolla white tinged purple, 1–1.5cm, longer than calyx, lobes short.
Sikkim: Kupup. July.
Hosts unknown.

2. L. purpurea Cummins
Flowering shoots 7–10cm, purple. Flowers borne singly in axils of scales, pedicels 0.5–1cm. Calyx almost entire, purple. Corolla purple, 2.5–3cm, twice as long as calyx, upper lip hooded, subacutely toothed below the apex on both sides, lower lip 3-lobed, purplish-white with dark purple veins.
Sikkim: Dichu Valley, Singalila. 3660–3970m. July.
Hosts unknown.

2. OROBANCHE L.

Erect annual to ?perennial herbs. Stem short, swollen, often scaly. Flowering shoots simple or branched, scaly pubescent. Flowers in dense or lax bracteate spikes, with or without bracteoles. Calyx 4(–5)-lobed. Corolla 2-lipped, upper lip 3-lobed.

1. Stems simple; spikes dense; bracteoles absent; flowers yellow ... **1. O. solmsii**
\+ Stems often branched; spikes lax; bracteoles present; flowers blue **2. O. aegyptiaca**

1. O. solmsii Hook.f. Fig. 109i–k.

Robust perennial (?biennial) 23–57cm, pubescent. Stem base swollen, scaly. Flowering shoots stout, bearing numerous lanceolate scales 2–3cm. Spikes 2.5–15cm, dense; bracts ± as long as flowers, scale-like, bracteoles absent. Calyx 10–15mm, divided to base into 2 segments, each segment divided ± to the middle into 2 unequal, toothed or subentire lobes. Corolla yellow, 1.5–2cm, tube broad, curved, lobes crenulate. Filaments glabrous or with a few hairs at the base. Capsule 5–6mm, oblong-ovoid.

Bhutan: C – Thimphu district (Paga); **Sikkim:** Kupup, Galing; **Chumbi:** Chole-la, Chumbi, Le-ra-oong etc. *Rhododendron* forests, 2130–3960m. July–September.

Parasite on *Rhododendron* and probably other hosts.

2. O. aegyptiaca Persoon; *O. indica* Roxb., *Phelipaea aegyptiaca* (Persoon) Walpers, *P. indica* (Roxb.) G. Don

Annual. Flowering shoots erect, branching, 15–25cm, pubescent, scaly. Spikes lax, 5–11cm; bracts ± size of calyx, scale-like; bracteoles 2, shorter than calyx. Calyx 8–10mm, unequally 4(–5)-lobed. Corolla tube slightly curved, blue, 15–20mm, lobes ciliate. Anthers woolly. Capsule c 10mm, oblong to subglobose.

W Bengal Duars: Jaldhaka River. March.

Parasite on tobacco. Used to cure boils in cattle and to stop diarrhoea.

A third species probably occurs in Bhutan (Simu Dzong) with corollas similar to those of *O. aegyptiaca*, but differing in unbranched stems, denser spikes and absence of bracteoles. The specimen is inadequate for identification.

3. BOSCHNIAKIA Meyer

?Perennial, stem base tuberous. Flowers in dense, erect, glabrous, bracteate, spike-like racemes. Calyx shallowly to deeply 5-lobed. Corolla tube curved, upper lip hooded, simple, lower lip short, 3-lobed. Stamens exserted from corolla tube. Capsule 3-valved.

1. B. himalaica Hook.f. & Thomson; *Xylanche himalaica* (Hook.f. & Thomson) Beck

Tuber subglobose, 2.5–4cm diameter, not scaly. Flowering shoots 15–45cm, pale brown; stems stout, bearing numerous ovate-lanceolate scales 1–2cm. Racemes 7–25cm, dense, bracts scale-like, half as long as flowers, pedicels c 5mm. Calyx lobes 2–5mm. Corolla 1.5–2.5cm, yellow to purplish-pink or brown.

Bhutan: C – Thimphu district (Chelai La), Tongsa district (Yuto La), Bumthang district (Towli Phu), **N** – Upper Bumthang Chu district (Pangotang); **Sikkim:** Kupup, Meguthang, Dzongri, Yampung etc.; **Chumbi:** Yatung. *Abies*/*Rhododendron* forests and juniper/*Rhododendron* scrub, 3200–4700m. June–August.

Parasitic on roots of *Rhododendron* species. Hooker (80) described the calyx limb as being entire but this is not so.

4. AEGINETIA L.

Herbs, stems short, scaly. Flowers solitary, borne on long, simple, naked or shorter, branching shoots. Calyx an unlobed spathe, split nearly to base on lower side. Corolla curved, 2-lipped, 5-lobed.

1. Flowering shoots simple, 10–28cm; corolla pale purple or white **1. A. indica**

\+ Flowering shoots branching (2–)4–15cm; corolla yellow with bluish lobes **2. A. pedunculata**

1. A. indica L. Fig. 109l–m.

Flowering shoots solitary or several, 10–28cm, brown streaked purple, scaly at base. Calyx purple, 3–4cm, acute. Corolla tube broad, curved, usually pale purple outside and darker inside, 2.5–5cm; lobes fimbriate. Capsule c 2cm.

Bhutan: S – Phuntsholing district (Phuntsholing), Deothang district (117), **C** – Punakha district (Chusom, Rinchu, Bhotokha), Tashigang district (Tashi Yangtsi); **Darjeeling:** Gok, Kali Khola, Tista etc.; **Sikkim:** Machong Gompa (298); **W Bengal Duars:** Jalpaiguri. Warm broad-leaved forests, 300–1800m. May–October.

Hosts unknown. Red or even 'red or orange' flowers were recorded for several of the collections, possibly referring to the calyx rather than the corolla. White corollas were recorded for the collections from Tashi Yangtsi and Machong Gompa, the latter at least with the whole plant yellowish or straw-coloured (298). Such plants may be separated as var. **alba** Santapau, with pigmented plants referred to var. **indica**.

2. A. pedunculata (Roxb.) Wall.; *Orobanche pedunculata* Roxb.

Flowering shoots branched, fleshy, 4–15cm, scaly at base. Calyx reddish to yellowish, 3–6cm, fleshy, shortly beaked, filled with liquid. Corolla 4–7cm, tube broad, slightly curved, yellowish; lobes violet or bluish, crenulate. Capsule c 2cm.

Darjeeling. Terai, 150m. June.

Parasitic on grass roots, e.g. *Vetiveria zizanioides* (L.) Nash.

5. CHRISTISONIA Gardner

?Annual; stems usually short, erect, scaly. Flowers few, borne on short pedicels in axils of scales. Calyx 5-lobed. Corolla tube slightly curved, obscurely 2-lipped, 5-lobed.

1. C. hookeri Hook.f.

Almost stemless herb 8–10cm; stems short stout, scales c 1cm with ragged edges. Calyx 2.5cm, pale violet with red-brown lobes. Corolla 5–6cm, tube white, inflated, throat contracted, limb 2cm across, pale violet.

Darjeeling: lower valleys (80). 600–1200m.

Hosts not known.

Family 179. LENTIBULARIACEAE

by H.J. Noltie

Insectivorous plants of wet places (aquatics, epiphytes or lithophytes). Leaves well developed and flat, in a rosette, or filamentous and much divided and inserted along stolons, sometimes bearing bladders. Flowers bilaterally symmetric. Calyx 2-lobed. Corolla 2-lobed, the lower lobe with a basal spur. Stamens 2. Ovary superior, 1-loculed. Fruit a capsule.

1. Leaves flat, in a rosette; bladders absent; flower single on scape; upper lip of calyx lip 3-lobed, lower lip 2-lobed .... **1. Pinguicula**
\+ Leaves filamentous; bladders usually present; flowers commonly several per scape; upper and lower lips of calyx lip not divided .... **2. Utricularia**

1. PINGUICULA L.

Rosette, marsh herb. Leaves flat, surface sticky, margins inrolled. Scape simple, single-flowered. Calyx 2-lipped, the upper lip 3-lobed, the lower 2-lobed. Corolla funnel-shaped, narrowed into basal spur, 2-lipped, the upper lip 2-lobed, the lower 3-lobed. Fruit splitting at apex into 2 valves.

1. P. alpina L. Fig. 110a.

Leaves 1.3–2.3 × 0.7–1cm, ± oblong, blunt, margins inrolled. Scape 3–9.5cm. Calyx glandular, upper lobes c 3mm, lower lobes c 2.5mm. Corolla white, with yellow spot at base of central lobe of lower lip, tube and spur 6–7mm, central lobe of lower lip 5.5–6.5 × 4–5mm, basal spot with long, white hairs; lobes of upper lip c 3.5 × 2.8mm. Style short, c 0.5mm, stigma punctate. Filaments c 1mm; anthers c 1mm, circular, opening by a pore. Capsule c 5.5 × 3.5mm, ovoid, eventually splitting in two at apex, the lobes shortly recurving.

FIG. 110. **Lentibulariacae, Phrymaceae and Plantaginaceae. Lentibulariacae.** a, *Pinguicula alpina*: habit (× 1). b, *Utricularia recta*: habit (× 3). **Phrymaceae.** c–d, *Phryma leptostachya* var. *oblongifolia*: c, apex of flowering shoot (× ½); d, flower (× 5). **Plantaginaceae.** e–f, *Plantago erosa*: e, habit (× ½); f, flower (× 8). Drawn by Glenn Rodrigues.

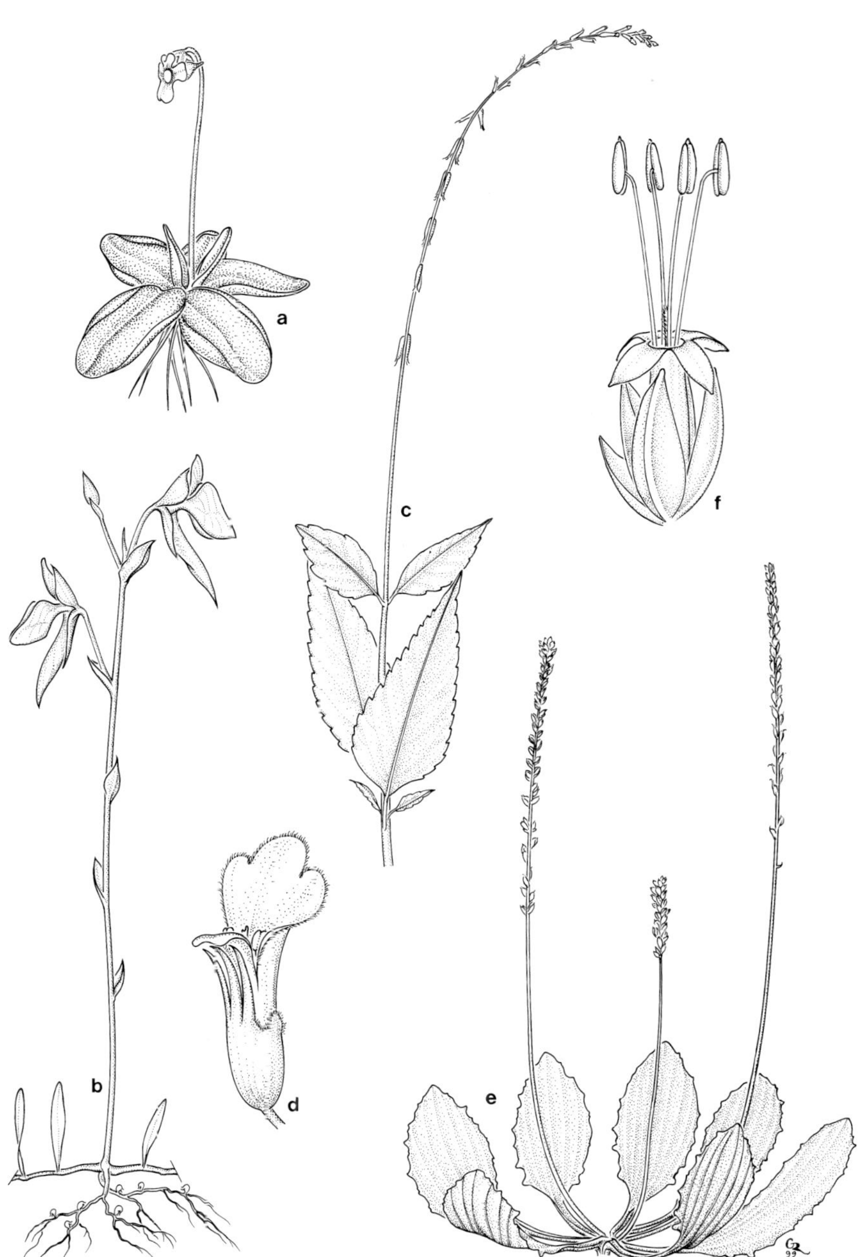
a
b
c
d
e
f

Bhutan: **C** – Thimphu district (Taglung La), Punakha district (Tang Chu), **N** – Upper Mo Chu district (above Naha), Upper Bumthang Chu district (Tolegang, Tsampa, Pangotang), Upper Kuru Chu district (Singhi), Upper Kulong Chu district (Me La); **Sikkim:** Lasha Chhu, Kalaeree, Lachen, Thanggu, Lhonak, Tensutang. Grassy flush; moss on moist gravel; scree, by stream; rocks in spray of waterfall, 3200–4880m. May–July.

2. UTRICULARIA L.

Epiphytic, lithophytic or marsh herbs, or aquatics (floating or anchored); perennial or annual. Rhizoids present; stolons present, bearing finely dissected leaves along their length, or leaves at base of scape, linear or differentiated into blade and petiole. Inflorescence borne on a scape, leafless or with tiny bract-like scales. Inflorescence a raceme or sometimes branched or reduced to a single flower. Bracts medifixed or basifixed. Calyx 2-lipped, lips entire, persistent, accresent in fruit. Corolla 2-lipped, the upper commonly smaller, the lower with a basal spur and expanded, often deflexed limb, with a basal 'palate' often differing in colour, surround by marginal rim. Stamens 2. Fruit dehiscing by a lateral slit or circumscissile.

Scape measurements are to the insertion of the lowest flower. Under-collected in our area and additional species may well occur. These delicate plants are best preserved in spirit, but if pressed, care should be taken to dissect out a few plants from their substrate moss so that the scape-base can be seen. This is necessary to see whether or not a tuber or persistent seed testa is present.

1. Flowers yellow 2
+ Flowers mauve or whitish (sometimes with a yellow spot at the base of the lower corolla lip) 5

2. Floating aquatic 3
+ Terrestrial 4

3. Spur of corolla saccate, wider than long; scape with scales .... **6. U. minor**
+ Spur of corolla longer than wide; scape lacking scales .... **7. U. aurea**

4. Upper lip of corolla shorter than upper calyx lobe; calyx lobes finely acuminate; pedicels very slender, ascending in fruit; plant temperate to alpine .... **8. U. recta**
+ Upper lip of corolla exceeding upper calyx lobe; calyx lobes rounded at apex; pedicels stouter, recurved in fruit; terai plant .... **9. U. bifida**

5. Plants arising from a small tuber; flowers larger, lower corolla lip over 8mm wide 6
+ Plants annual, tuber absent; flowers smaller, lower corolla lip under 8mm wide 7

6. Spur long, over 4mm; lateral lobes of corolla lip narrow, elongate **4. U. brachiata**
\+ Spur short, under 2mm; lobes of corolla lip ± equal, broad **5. U. christopheri**

7. Spur short, saccate; seeds not glochidiate .... **3. U. multicaulis**
\+ Spur slender; seeds glochidiate 8

8. Lower corolla lip 5-lobed .... **1. U. striatula**
\+ Lower corolla lip 4-lobed .... **2. U. furcellata**

1. U. striatula Smith; *U. orbiculata* A. DC.

Delicate lithophytic annual; tuber absent. Stolons filiform, bearing leaves, rhizoids and bladders; leaf blades small, orbicular to transversely elliptic, c 2 × 1.8–2.4mm, base cordate, petioles 2–3mm; bladders 0.5–1.2mm. Scapes 1.5–10cm, filiform, bearing 2–6 flowers, occasionally with a single branch. Lowest pedicel 3–7mm. Bracts medifixed, purplish, 2-lobed. Calyx lobes punctate, the upper transversely elliptic, emarginate, 1.7–2.6 × 2.2–3mm, the lower oblong, c 1 × 0.9mm. Corolla mauve with yellow spot at base of lower lip, upper lip shorter than upper calyx lobe; lower lip wider than long, 4–8mm wide, crenately 5-lobed, lobes ± equal; rim of palate hairy; spur slender, acute, 3–4mm. Capsule subglobose, c 1.7mm in diameter, dehiscing by a lateral slit; seeds c 0.3mm, brown, pear-shaped, with slender, whitish, glochidiate spines.

Bhutan: S – Samchi district (W bank of Torsa River, Phuntsholing), Chukka district (c 2km below Chimkakothi), **C** – Punakha district (Wangdi Phodrang–Wacha); **Darjeeling:** Kalijhora, Damsong Forest, Sivok, Tonglu, Kodabary, Lopchu; **Sikkim:** Phodong Gompa, Lachung, Phadonchen. Wet rocks, including mossy tufa, 200–2280m. February–September.

2. U. furcellata Oliver

Differs from *U. striatula* as follows: plant usually smaller; corolla lip 4-lobed

Darjeeling: Rungbool, Darjeeling; **Sikkim:** Lachen. Wet rocks, 2130–2800m. August–September.

It is impossible to see the shape of the lower corolla lip in some of the badly pressed, old material. Only specimens determined by P. Taylor are included here and none of these is recent.

3. U. multicaulis Oliver

Delicate lithophytic annual, 1.2–5.5cm; tuber absent, seed husk persistent at base of scape. Plants very tightly clumped, so scapes appearing tufted. Stolons filiform, bladders c 0.8mm. Leaves 1–2, at base of scape, blades 0.7–1.3 ×

0.5–1.5mm, orbicular or oblanceolate, base cuneate, petioles 2.4–3mm. Scapes 0.8–4cm, filiform, bearing 2–4 flowers. Pedicels 3–6mm. Bracts c 0.5mm, medifixed, irregularly lanceolate, hyaline. Calyx: upper lobe transversely elliptic, shallowly emarginate, c 1.2 × 1.6mm; lower lobe oblong, c 0.9mm. Corolla mauve, with yellow spot at base of lower lip; upper lip shorter than upper calyx lobe; lower lip ± 3-lobed, 4.5–6 × 4.5–6mm, central lobe 2.2–3.5 × 1.5–2.2mm, blunt, sometimes emarginate, lateral lobes downward-pointing, triangular, acute, sometimes with a lateral tooth; rim of palate hairy; spur short, stout, 1–1.5mm. Capsule globose, c 2mm diameter, dehiscing by a lateral slit; seeds c 0.5mm, dark brown, pear-shaped, covered in rows of subacute papillae, with a tuft of hairs at the narrow end.

Bhutan: C – Ha district (Damthang–Charithang), Bumthang district (Shabjethang), Tashigang district (Roch Chu Valley), **N** – Upper Kuru Chu district (Narim Thang); **Sikkim:** Chamnago. Wet, mossy rocks, 3050–4115m. July–August.

A photograph taken at Darkey Pang Tso (Thimphu district, 3980m) is probably of this species. The specimens from Shabjetang (*Ludlow, Sherriff & Hicks* 19508, K, BM, E) were determined by Taylor as *U. kumaoensis* Oliver and included under that species in Taylor (415). This species differs in having seeds with a tuft of hairs at each end, and less acute papillae. The specimens are in flower, and the only seeds that can be seen are those persisting at the base of the scapes: it is therefore impossible to see if a tuft of hairs has been present at the end at which germination has taken place; however, the papillae look too acute for *U. kumaonensis*, and the specimens have been re-determined.

4. U. brachiata Oliver

Delicate lithophytic or epiphytic perennial. Tuber 1.5–2.3mm, subglobose or elongate. Stolons filiform, bladders 0.7–1mm. Leaves 1–2, at base of scape, blades 1.5–4.5 × 2.5–8mm, transversely elliptic or reniform, base truncate or shallowly cordate, petioles 7–22mm. Scapes 3.2–6.2cm, filiform, bearing 1–2 flowers. Pedicel(s) 4–6.5mm. Bracts 1.7–2mm, medifixed, irregularly oblong, truncate-erose, hyaline. Calyx: upper lobe transversely elliptic, shallowly emarginate, 2–2.4 × 3–3.5mm; lower lobe oblong, c 1.8 × 1mm. Corolla white, with yellow spot at base of lower lip; upper lip shorter than upper calyx lobe; lower lip 5-lobed, 6–9 × 8–10.5mm, central lobe short, blunt, 2–3.5 × 3–5mm, lateral lobes much narrower; rim of palate hairy; spur slender, blunt, 4.5–5mm. Capsule subglobose, c 3mm diameter, dehiscing by a lateral slit; seeds c 0.5mm, brown, oblong in outline, with a tuft of hairs at each end.

Bhutan: C – Thimphu district (Hinglai La–Tsalimaphe, Dechencholing–Punakha, Barshong), Tongsa district (Rinchen Chu), Bumthang district (Yuto La, Dhur Valley), **N** – Upper Mo Chu district (Pari La), Upper Kulong Chu district (Lao); **Darjeeling:** Sandakphu, Tonglu; **Sikkim:** Phedang–Tsoka, Yampung, Lachen. On wet mossy rocks or trees, especially in fir forest, 2740–4270m. July–September.

5. U. christopheri P. Taylor

Differs from *U. brachiata* as follows: spur of corolla shorter, c 1.5mm, conical, lobes of lip ± equal, the lateral not narrow and elongate.

Sikkim: just above Changu. (Presumably on mossy rocks), 3960m. July.

6. U. minor L.

Submerged aquatic perennial; some stolons colourless and buried in substrate; floating stolons long (to 13cm or more), bearing green leaves. Leaves ± circular in outline, finely several times dichotomously divided, primary segments c 3, 4–8mm × c 0.5mm, margins not setulose, ultimate segments with a minute apical seta; bladders borne on margins, ovoid, 1–2mm. Scapes to 7.5cm, stout, emerging above water, bearing several small, bract-like scales, flowers c 6. Pedicels to 4mm. Bracts c 2mm, basifixed, oblong, truncate, base auriculate. Calyx lobes subequal, orbicular, the upper c 2.3 × 2.5mm, the lower c 2 × 2mm. Corolla yellow; upper lip c 4mm; lower lip c 7.5 × 6.5mm, ± oblong; palate streaked purple, rim glandular; spur broadly conical, wider than long, c 0.8 × 2mm. (Fruit not seen; pedicels recurved; capsule 2–3mm diameter, globose, circumscissile; seeds lenticular-prismatic, c 1mm wide (415).)

Bhutan: C – Bumthang district (near bridge over Bumthang Chu, Jakar), **N** – Upper Kulong Chu district (Me La). Shallow pools, 2700–3960m. June.

7. U. aurea Loureiro; *U. flexuosa* Vahl

Differs from *U. minor* as follows: suspended aquatic (i.e. not anchored to substrate); scape lacking bract-like scales; flowering pedicels longer (the lowest often over 1cm); spur cylindric, longer than wide (4–4.5 × 1.7–2mm); capsule larger (4–5mm diameter), persistent calyx lobes coriaceous, spreading horizontally, persistent style to 2mm.

Darjeeling: Siliguri; **W Bengal Duars:** Jalpaiguri. Terai, (tanks and ditches), 150m. August–December.

No recent collections seen.

An inadequate, sterile specimen of a floating species from flooded rice-fields at Chapcha appears to be neither of the two previous species; it is possibly *U. australis* R. Brown, but flowering collections are required. This latter species resembles *U. minor* in having scales on the scape, but differs in having a longer spur and lateral setulae on the leaf segments.

8. U. recta P. Taylor; *U. scandens* Benjamin var. *firmula* (Oliver) Subramanyam & Banerjee, *U. wallichiana* Wight var. *firmula* Oliver. Fig. 110b.

Slender ?perennial, 1.5–13.5cm, usually terrestrial (marsh), sometimes lithophytic. Stolons short, filiform; bladders c 0.7mm, very scarce, inserted at base of stolons and leaf margins; leaves narrowly linear, inserted on stolons, blades to 11 × c 0.5mm. Scapes 0.7–9cm, filiform, bearing 1–4(–8) flowers. Pedicel 2.5–5.5mm, slender, narrowly winged, ascending in fruit. Bracts to 2mm, basifixed, lanceolate to ovate, finely acuminate, hyaline. Calyx lobes subequal, lanceolate to ovate, finely acuminate, the upper wider, 2.5–3.2 × 2.2–2.4mm, the lower 2.5–2.8 × 1.4–1.8mm. Corolla yellow; upper lip shorter than upper calyx lobe; lower lip deflexed, transversely elliptic, blunt, 2.5–5.5 × 3.4–6mm, conspicuously swollen at junction with palate; rim of palate minutely ciliate; spur 3.5–6mm, curved, acute. Capsule ellipsoid, 2.5–2.8 × 1.7–1.9mm, dehiscing by a lateral slit; seeds c 0.3mm, brown, oblong in outline, ± smooth.

Bhutan: C – Thimphu district (Olakthang Hotel, Paro–Chelai La, Paro Bridge), Bumthang district (Lami Gompa, Jakar, Shabjetang, Gyetsa), Tashigang district (Tashi Yangtse Dzong), **N** –Upper Mo Chu district (Gasa); **Sikkim:** Gangtok, Singhik, Lachung. Shallow pond; moss on boulders in fir forest; marshy, peaty turf; edge of stream, 1400–3350m. June–October.

9. U. bifida L. Nep: *Sunakhan jat.*
Differs from *U. recta* as follows: upper lip of corolla exceeding upper calyx lobe; calyx lobes rounded at apex; pedicels stouter, recurved in fruit.

Darjeeling: Katambari. Terai, (marshes; paddy-fields, 150m). November.

No recent collections seen.

Doubtfully recorded species:

U. caerulea L. (syn. *U. nivea* Vahl)

There are old records of this species from the terai, but no specimens have been seen. A mauve-flowered, terrestrial species, with small, subsessile flowers with an unlobed lower corolla lip.

U. gibba L. (syn. *U. exoleta* R. Brown)

A single specimen seen, with the locality 'probably Jalpaiguri District'. The species occurs in Nepal, so is likely to occur in our area at low altitudes. It is a slender, marsh plant, with slender stolons bearing very reduced filamentous leaves; flowers yellow, few, on long pedicels.

Family 180. PHRYMACEAE

by L.S. Springate

Erect herbaceous perennial, sparsely pilose; stems 4-angled. Leaves opposite, simple, ovate, ± acuminate, subtruncate to cuneate at base, serrate-dentate. Inflorescence of terminal and axillary elongate spikes. Flowers bisexual, zygomorphic, spreading, completely deflexed in fruit. Calyx tubular below, very unequally 2-lipped; upper lobes 3, adjacent, linear, apex hooked in flower, coiled in fruit; lower lobes 2, very obscure. Corolla white or pink, often marked purple, tubular below, 2-lipped; upper lip emarginate; lower lip larger, 3-lobed. Stamens 4, inserted on corolla tube, briefly exerted. Ovary superior, 1-ovulate; style 1, slender, briefly exerted; stigma bifid. Fruit an achene retained in enlarged persistent calyx.

1. PHRYMA L.

Description as for family.

1. P. leptostachya L. var. **oblongifolia** (Koidzumi) Honda. Fig. 110c–d.

Plant c 0.5–1m. Leaf blades 30–110 × 15–50mm; petiole 3–25mm. Calyx tube c 3mm in flower, c 6mm in fruit; upper lobes c 2mm. Corolla tube c 4–6mm; lower lip 3–3.5mm, ciliate. Achenes c 4mm.

Bhutan: S – Chukka district (Chukka), **C** – Punakha district (Lumichawa (117)), Tashigang district (Gomchu (117)). Streamsides etc., 1400–2550m. July–September.

Details of specimens from Nepal included in description above. The species is found in E Asia (var. *oblongifolia*) and eastern N America (var. *leptostachya*).

Family 181. PLANTAGINACEAE

by L.S. Springate

In our area one genus of rosulate scapose perennial herbs. Leaves simple, entire, dentate or sometimes lobulate towards base, with 3–5 parallel veins and broad channelled petiole. Inflorescence spicate, dense during anthesis, elongating and sometimes lax in fruit. Flowers bisexual, regular, 4-merous, sessile or subsessile. Bract and sepals with green keel and broad scarious margin; bract 1; sepals free or abaxial pair fused. Corolla gamopetalous, tubular, 4-lobed. Stamens inserted below middle of corolla tube; anthers versatile, long exserted, hanging inverted, ovate, apiculate, sagittate or cordate at base. Ovary superior, 2-celled, though septum free from ovary walls by anthesis, rarely with a false septum dividing 1 or both locules; style filiform, exserted from calyx before corolla lobes and anthers. Capsule ovoid, circumscissile near base; seeds 2–c 27, angular or dorsi-ventrally compressed, peltately attached, partly sunk into thick septum, mucilaginous.

1. PLANTAGO L.

Description as for family.

1. Ovules and seeds 2; bract long scarious-cuspidate; 2 abaxial sepals largely fused .... **4. P. lanceolata**
+ Ovules and seeds 4–c 27; bract very short herbaceous cuspidate; 2 abaxial sepals free 2

2. Ovules and seeds (4–)5; sepals c 1.7mm, barely exceeding c 1.3mm deep capsule base .... **3. P. depressa**
+ Ovules and seeds 6–c 27; sepals 2.3–2.8, much exceeding c 0.8mm deep capsule base 3

3. Ovules and seeds (6–)11–c 27, 0.8–1.2mm; corolla tube 2–3mm; capsule not exceeding 4mm .... **1. P. erosa**
\+ Ovules and seeds (6–)7–9(–11), 1.2–2.3mm; corolla tube 3–3.5mm; capsule often exceeding 4mm .... **2. P. sp. A**

1. P. erosa Wall.; *P. major* sensu F.B.I. p.p. non L., *P. asiatica* auct. non L., *P. major* var. *asiatica* auct. non (L.) Decaisne. Fig. 110e–f.

Leaf blade elliptic, 30–90 × 15–50mm, subacute, attenuate at base, subentire or shallowly dentate above, often coarsely dentate below or bearing 1–2 pairs of oblong to oblanceolate lobes towards base, subglabrous to puberulous; petioles 15–70(–140)mm. Peduncle (25–)70–260mm, ribbed; flowering head 22–200mm up to late anthesis. Flowers subsessile; bracts suborbicular, ovate or elliptic, concave, 1.8–2.5mm, subobtuse to acute. Sepals 2.2–2.8 × 1.3–2mm, broadly obovate (outer 2) or oblong-elliptic (inner 2), with keel briefly projecting as obtuse cusp. Corolla whitish, tube 2–3mm, lobes ovate, obtuse, 0.7mm to narrowly triangular, acute, 1.2mm, soon strongly reflexed. Capsule 2.8–4mm, splitting c 0.8mm above base; seeds (6–9), 11–27, ± elliptic, angular or compressed, 0.8–1.2mm, brown.

Bhutan: S – Samchi, Chukka, Gaylephug and Deothang (117) districts, **C** – Thimphu, Punakha, Tongsa and Mongar districts; **Darjeeling:** Darjeeling, Kurseong, Sitong etc.; **Sikkim:** Gangtok (97a), Phusum, Thanggu etc. Disturbed sites, wasteground, clearings, forest, field margins, grassland, riverside sands, 600–2500m. March–August.

Probably in all administrative districts of Bhutan above 1000m (272). The collection from Thanggu has only 6–9 seeds per capsule and, in one case, scattered flowers almost to base of scapes. No *Plantago* specimens from Chumbi were seen, but Pai (in 421) has recorded *P. asiatica* L. there (as well as *P. depressa*). Pilger (387) distinguished the former from *P. erosa* by the keel of its sepals not protruding as an apical cusp, but being overtopped by a narrow scarious apex, on which basis no reliable example of *P. asiatica* was seen from elsewhere in our area.

2. P. sp. A. Dz: *Casoma*.

Similar to *P. erosa* but small plants with large flower parts: leaf blade 30–60 × 17–35mm; peduncle 50–120mm; flowering head 15–60mm; corolla tube 3–3.5mm, lobes 1–1.4mm; capsules 3.4–4.8mm; seeds (6–)7–9(–11), 1.2–2.3mm.

Bhutan: C – Thimphu district (Barshong, Dochong La, Motithang, Pajoding). Roadside banks etc., 2450–3800m. June–August.

A localised race of rather uniform appearance, provisionally separated from *P. erosa* here, even though most of its potentially distinguishing characters occur exceptionally in that species. A largely dehiscent collection from Thimphu district, Hasimara–Thimphu, 915m, may also belong here. Two flowering collections of uncertain status from the Darjeeling/Sikkim border (Chiya Bhanjang–Phalut, Chiya Bhanjang–Dentam) are similar, but have more ovules (usually 11–14, more on one plant where false septa are present).

3. P. depressa Willdenow; *P. tibetica* Hook.f. Dz: *Tasoma.*

Similar to *P. erosa*, differing most obviously by the fruit. Leaf blades 50–70 × 20–30mm, remotely dentate, not lobulate, petioles c 25mm or 40–60mm (?when etiolated); peduncles 50–140mm in fruit; fruiting head 25–100mm; sepals to 1.7mm, scarcely exceeding capsule base; corolla lobes ± ovate, less than 0.5mm, obtuse; capsule dehiscing c 1.3mm above base; seeds 5, compressed, all erect, 2 pairs side-by-side with one above in 1 locule, 1.4–1.9mm.

Bhutan: C – Thimphu district (Paro), **N** – Upper Mo Chu district (Zambuthang); **Sikkim:** Nathang (113), Thanggu (112); **Chumbi:** without locality (421). Wasteground near habitation, 2300–4020m. In fruit August–October.

Only two fruiting specimens seen. Exceptional capsules with 4 seeds reported outside our area (387).

4. P. lanceolata L.

Differs from species 1–3 by its narrowly elliptic leaf blade, c 100–120 × 20mm, remotely denticulate; peduncle 300–400mm; flowering head very dense, ovoid at first, later cylindric, 7–25 × c 8mm, remaining very dense through fruiting; bracts 3.5–4mm, with long scarious cusp; 2 abaxial sepals largely fused to form ± orbicular, faintly 2-keeled, emarginate blade; 2 adaxial sepals free, with prominent green keel long-ciliate towards apex; corolla lobes ovate, 2mm, short-acuminate, scarious with brown keel; ovules and seeds 2.

Bhutan: C – Bumthang district (Bumthang). Roadsides, 2600m. August.

A weedy species of Europe, N Africa and Asia to W Himalaya, frequent at the site of the only collection seen and probably introduced. Vigorous plants with much longer but equally dense flowering and fruiting heads are known outside our area.

Family 182. CAPRIFOLIACEAE

by R.A. King, *Lonicera* by P.W. Meyer

Herbs, shrubs or small trees, sometimes climbing, usually with pith. Leaves simple or compound, opposite. Stipules present or absent. Flowers hermaphrodite, usually in corymbs or short spikes, rarely paniculate. Calyx adnate to ovary, usually 5-lobed. Corolla gamopetalous, actinomorphic or zygomorphic; lobes usually 5, imbricate in bud. Stamens 5, rarely 6, inserted on corolla tube, alternating with corolla lobes. Ovary inferior, (1–)2–8-locular; style long or short; stigma capitate or lobed. Ovules solitary, few or many. Fruit a drupe or berry.

1. Leaves compound .... **4. Sambucus**
\+ Leaves simple 2

2. Flowers in many-flowered often umbelliform corymbs or panicles **3. Viburnum**
\+ Flowers in short spikes or in pairs 3

3. Herb; leaves obovate to oblong, each pair joined at base around stem **5. Triosteum**
\+ Shrubs or small trees; leaves various, not joined around stem 4

4. Seeds numerous, minute; flowers regular; ovary 5- or 8-locular **2. Leycesteria**
\+ Seeds 2–several, flowers regular or irregular; ovary 2–3-or rarely 5-locular **1. Lonicera**

1. LONICERA L.

by P.W. Meyer

Deciduous or evergreen shrubs, often climbing, rarely tree-like. Branchlets fistulose or solid; winter buds with 2–several scales. Leaves opposite, simple, entire (rarely lobed), short-petioled, sometimes sessile, exstipulate, often glandular. Flowers axillary, sessile or subsessile, in pairs, free or partly to almost completely fused along ovaries; each pair subtended by 2 bracts and 4 bracteoles; bracteoles usually fused below in pairs, sometimes all 4 connate in a cupula sheathing ovaries, rarely absent. Calyx 5-lobed, deciduous or persistent. Corolla 5-lobed, tubular, often gibbous on outer side at base with nectary glands, 2-lipped or nearly equally 5-lobed. Stamens 5, inserted on corolla tube. Ovary inferior, 2-, 3-, or rarely 5-locular; style slender with capitate stigma, usually exserted. Fruit fleshy with 2–several seeds.

L. hypoleuca Decaisne and *L. caucasica* Pallas subsp. *govaniana* (DC.) Hara are reported in Bhutan or Sikkim (135). The validity of these reports is in doubt and no documented specimens from our area have been located.

1. Pith solid 2
\+ Pith hollow 20

2. Leaves on fertile branchlets shorter than 1.4cm 3
\+ Leaves on fertile branchlets longer than 1.4cm 8

3. Corolla not gibbous at base; leaves often linear to oblanceolate 4
\+ Corolla gibbous at base; leaves broadly obovate 7

4. Ovaries almost wholly connate 5
\+ Ovaries connate only at base 6

5. Pedicels 1–5mm .... **1. L. myrtillus**
\+ Pedicels 10–25mm .... **2. L. angustifolia**

6. Stamens shorter than corolla tube; leaves usually hairy beneath **4. L. rupicola**
\+ Stamens exserted; leaves glabrous .... **5. L. spinosa**

7. Flowers and fruits subtended by 2 small bracts and 2 linear bracteoles; flowers emerging after leaves .... **7. L. obovata**
\+ Flowers and fruit subtended by 2 ovate to lanceolate leaf-like bracts; bracteoles absent; flowers emerging together with leaves .... **8. L. litangensis**

8. Leaves glabrous beneath 9
\+ Leaves hairy beneath 13

9. Leaves leathery .... **10. L. ligustrina**
\+ Leaves not leathery 10

10. Bractlets connate, forming a cupula 11
\+ Bractlets distinct 12

11. Pedicels 10–25mm; leaves oblong to lanceolate .... **2. L. angustifolia**
\+ Pedicels 4–7mm; leaves ovate to ovate-oblong .... **3. L. tomentella**

12. Leaves with up to 7 pairs of prominent veins; calyx shallowly lobed; fruits red .... **15. L. webbiana**
\+ Leaves with 9 or more pairs of prominent veins; calyx deeply lobed; fruits black .... **16. L. lanceolata**

13. Bractlets connate, forming a cupula .... **3. L. tomentella**
\+ Bractlets in separate pairs, not forming a cupula, or absent 14

14. Bractlets absent 15
\+ Bractlets present 18

15. Calyx lobes well developed (ovate, c 0.7mm), stiffly and persistently ciliate **13. L. asperifolia**
\+ Calyx lobes eciliate or obsolete 16

16. Fruit red; corolla pilose outside; winter buds tan to brown .. **14. L. hispida**
\+ Fruit blue; corolla glabrous outside; winter buds blue-black 17

17. Corolla yellow .... **11. L. cyanocarpa**
\+ Corolla deep purple .... **12. L. porphyrantha**

18. Leaves broader than 15mm, base rounded to obtuse ...... **15. L. webbiana**
\+ Leaves narrower than 15mm, base attenuate 19

19. Fruit connate only at base; bracts longer than 1mm; corolla wine-red **6. L. purpurascens**
\+ Fruit connate; bracts 0.5mm; corolla pale green .... **9. L. saccata**

20. Erect shrub .... **17. L. quinquelocularis**
\+ Scandent or twining 21

21. Leaves nearly glabrous, not ciliate .... **19. L. glabrata**
\+ Leaves hairy, ciliate 22

22. Leaves pilose beneath; corolla tube shorter than 20mm .. **18. L. acuminata**
\+ Leaves tomentose beneath; corolla tube longer than 60mm **20. L. macrantha**

1. L. myrtillus Hook.f. & Thomson; *L. parvifolia* Hayne var. *myrtillus* (Hook.f. & Thomson) Clarke, *L. myrtillus* var. *depressa* Rehder

Shrub to 4m. Branchlets tan, hirsute, bark peeling off in narrow shreds as it matures. Leaves ovate to oblong, 5–13 × 3–7mm, rounded to acute, base rounded to acute, pilose above when young, glabrescent, grey and glaucous beneath; petiole 0.5–1mm, glabrous. Peduncle 1–5mm, glabrous; bracts leaf-like, obovate, 4–5mm; bractlets connate, 1–1.5mm. Flowers actinomorphic. Corolla white to pinkish, tube 5–8mm; stamens shorter than tube. Fruit orange-red, connate.

Bhutan: C – Thimphu and Punakha/Tongsa districts, **N** – Upper Mo Chu, Upper Pho Chu, Upper Mangde Chu and Upper Kulong Chu districts; **Sikkim:**

Bikbari, Dik Chhu, Thangshing etc. On rocky screes with dwarf *Rhododendron* scrub and on peaty hill slopes, 3050–4400m. May–July.

L. myrtilloides Purpus has been reported from Singalila and Sandakphu (both 3900m, Darjeeling district (73)). The species was based on cultivated material (now lost) and may be of hybrid origin (398). The name is often applied to specimens intermediate in character between *L. angustifolia* and *L. myrtillus* (as in 73).

2. L. angustifolia DC.

Shrub to 4m. Branchlets brown to black, glabrous (young shoots rarely sparsely (red-) glandular-stipitate), pith solid. Leaves oblong to lanceolate, 1.5–4 × 0.6–1.5cm, acute to mucronulate, base attenuate, glabrous above, pilose or sometimes glabrous beneath, silvery (sometimes sparsely glandular-ciliate); petioles 3–5mm, glabrous, leaving persistent raised leaf scar. Peduncle 1–2.5cm; bracts linear to lanceolate, 4–6mm, sometimes glandular-ciliate; bractlets connate forming a cupula, 1.5–2mm. Flowers actinomorphic. Corolla white, greenish, pale pink or pale mauve, tube 8–10mm, glabrous; stamens shorter than tube. Fruit red, connate, 4–5mm.

Bhutan: C – Ha, Thimphu, Punakha, Tongsa, Bumthang, Tashigang and Sakden districts, **N** – Upper Kulong Chu district; **Sikkim:** Boktak, Dik Chhu, Gopethang etc. Among *Rhododendron* in fir forest along stream banks, and in hemlock and oak forests, 2200–3960m. May–August.

3. L. tomentella Hook.f. & Thomson

var. **tomentella**

Shrub to 4m. Branchlets tan, villous when young, pith solid. Leaves obovate to lanceolate, 1.5–2.5 × 1–1.5cm, acute to acuminate, base broadly rounded, glabrous above but sometimes hairy on midrib, villous beneath; petiole 1–2mm, villous with dense tuft of hairs at junction with stem. Peduncle 2–10mm, pilose; bracts linear to lanceolate, 5–6mm; bractlets connate forming a cupula, 2mm. Flowers actinomorphic. Corolla white or flushed pink, tube 8–10mm; stamens shorter than tube. Fruit blue-black with glaucous bloom, 1cm, connate at base.

Bhutan: C – Thimphu district (Barshong, Shanosam), Sakden district (Sakden), **N** – Upper Mo Chu district (Kohina, Pya La area), Upper Kulong Chu district (Me La); **Sikkim:** Phune–Yakche, Zemu (113) etc.; **Chumbi:** Chumbi, Galing. On sandy flats along rivers and in fir forest, 3200–4100m. June–July.

var. **tsarongensis** W.W. Smith

Differs from var. *tomentella* by its leaves which are glaucous, nearly glabrous beneath, and with more conspicuous veins.

Bhutan: C – Tashigang district (Tashi Yangtsi), **N** – Upper Kulong Chu district (Me La).

4. L. rupicola Hook.f. & Thomson; *L. thibetica* Bureau & Franchet

Shrub to 1.3m. Branchlets tan to dull brown, sparsely pilose with short glandular hairs, bark shredding in narrow strips, pith solid. Leaves linear to oblong-lanceolate, 10–15 × 3–5mm at flowering, acuminate, base rounded to subcordate, scattered glandular-hairy above, glabrous to white tomentose beneath; petioles 0.5–1mm, with short-stalked glandular hairs. Peduncle ± absent; bracts linear, 4–5mm, with short glandular hairs; bractlets connate below in pairs, 1.5–2mm. Flowers actinomorphic. Corolla pale yellow to rose-lilac, tube 5–6mm. Fruit bright red, free, 6–8mm, crowned by prominent calyx lobes 2–4mm.

Bhutan: C – Thimphu district, **N** – Upper Mo Chu, Upper Pho Chu, Upper Bumthang Chu and Upper Kulong Chu districts; **Sikkim:** Chemathang–Thangshing, Gurudongmar, Thanggu. Along rocky stream edges, open valleys, and on screes among dwarf *Rhododendron* and *Juniperus*, 3800–4900m. May–June.

L. syringantha Maximowicz has been reported in Bhutan (71) but the examples seen do not match the Chinese type, which has much broader leaves and a more rounded leaf apex. The character of leaf hairiness which is sometimes used to distinguish it from *L. rupicola* is quite variable throughout this group. *L. syringantha* and *L. rupicola* are closely related, each being variable with many intermediate forms. Further work is needed to define their precise relationships. *L. thibetica* has been described as distinct from *L. rupicola* based on minor variations in leaf shape, indumentum, glossiness, and degree of fusion of the bracts. Examination of numerous collections has shown all of these characteristics to be variable, with a full range of intermediate forms.

5. L. spinosa (Decaisne) Walpers; *Xylosteum spinosum* Decaisne

Shrub to 1m. Branchlets tan, often tapering to spiny point, sparsely pilose, bark shredding in fine strips, pith solid. Leaves linear, 8–10 × 1–2mm, apex rounded, base tapered, glabrous above, glaucous beneath; petioles 0.5–1mm, glabrous. Peduncle 1mm; bracts 3–5mm, linear; bractlets connate below in pairs, 0.5–1mm. Flowers zygomorphic. Calyx lobes prominent, 0.5–1mm. Corolla pink, tube 7–12mm; stamens exserted, fused to tube just within throat. Fruit scarlet, translucent, connate at base.

Sikkim: Lhonak Valley; **Chumbi:** Chomo Lhari. On open hillsides and stony flats, 4000–5000m. June.

6. L. purpurascens (Decaisne) Walpers; *Xylosteum purpurascens* Decaisne

Shrub to 5m. Branchlets tan, pilose, bark shredding in narrow strips after 2–3 years, pith solid. Leaves ovate to oblong-lanceolate, 2.5–4 × 0.8–1.2cm, acuminate, rounded to attenuate, ciliate, lightly pubescent above, pilose beneath with glandular hairs on midrib and near base. Peduncle slender, 9–12mm; bracts linear, pilose, 5–8mm; bractlets small, rounded, glandular, 1–1.5mm. Flowers zygomorphic. Corolla wine-red, tube 9–14mm, pilose outside; stamens just equal to length of tube. Fruit blue-black, connate at base.

Sikkim: Yeumtang. Mixed forest among *Abies* and *Betula*, 3350m. May–June.

7. L. obovata Hook.f. & Thomson; *L. parvifolia* Edgeworth non Hayne. Fig. 111a–d.

Shrub to 1.2m. Branchlets brown to tan, glabrous, bark shedding in thin strips, pith solid. Leaves obovate, 6–12 × 4–8mm, obtuse to acute, base cuneate, glabrous above, white-glaucous beneath; petiole purplish, glaucous 1mm. Peduncle purplish, glaucous, 1–2mm; bracts linear, 7–9mm; bractlets connate below in pairs 0.5–2mm. Flowers zygomorphic. Corolla greenish-yellow, tube 6–7mm; stamens barely exserted. Fruit red, connate.

Bhutan: C – Thimphu district (Upper Thimphu Chu), **N** – Upper Mo Chu district (Laya), Upper Pho Chu district (Chojo Dzong), Upper Bumthang Chu district (Marlung, Damakura); **Sikkim:** Dzongri, Pheonp, Yangshong La etc. On stony open hillsides with *Rhododendron*, *Berberis* and *Potentilla*, 3000–4600m. May–June.

8. L. litangensis Batalin

Shrub to 1m. Branchlets dark brown, glabrous, bark peeling in narrow shreds, pith solid. Leaves elliptic to obovate, 6–9 × 3–4mm, acute, base obtuse, glabrous above, scattered glands below; petiole 1mm, glabrous. Peduncle c 3mm; bracts linear-lanceolate, c 5mm; bracteoles connate in 2-lobed pairs, c 1mm. Flowers zygomorphic, appearing before leaves. Corolla pale yellow, tube 7–9mm; stamens exserted. Fruit scarlet, connate, 6–8mm.

Bhutan: C – Thimphu district (Shodug–Barshong, Taglung La), **N** – Upper Mo Chu district (Laya–Laum Thang, Nyeli La); Upper Mangde Chu district (Mangde Chu), Upper Bumthang Chu district (Pangotang), Upper Kulong Chu district (Me La); **Sikkim:** Prek Chhu, Yumthang (409), Dzongri–Thangshing etc. High on rocky slopes among *Rhododendron*, 3900–4500m. May–June.

9. L. saccata Rehder

Shrub to 3m. Branchlets brown, with short bristle hairs persisting even on older branches, bark sometimes shreddy with age, pith solid. Leaves lanceolate, 2.2–4 × 0.5–1cm, rounded to acute, base narrowly attenuate, sparsely pilose above, velutinous beneath with hairs denser at base and on veins; petiole 1–1.5mm, with long pilose hairs. Peduncle 9–13mm; bracts linear, 6–8mm; bractlets small or absent. Flowers zygomorphic. Corolla greenish-white to pale yellow, tube 7–9mm, conspicuously saccate at base; stamens slightly exserted. Fruit scarlet, connate.

Bhutan: C – Thimphu district (Barshong–Nala, Nala–Tzatogang), Punakha district (Ritang), Sakden district (Merak), **N** – Upper Bumthang Chu district

(Lhabja). In *Abies* forest among *Rhododendron hodgsonii* and bamboo, 3400–4100m. May–June.

10. L. ligustrina Wall.; *Xylosteum ligustrinum* (Wall.) D. Don

Shrub to 2.5m. Branchlets brown, strigose, smooth, light tan when older, pith solid. Leaves oblong-lanceolate to ovate, 3–5 × 1–1.5cm, acuminate, base obtuse to truncate, often with pairs of much smaller ovate to elliptic leaves 6–12 × 4–7mm present on same branch, all leaves shiny, nearly glabrous above with strigose hairs on veins, with scattered strigose hairs beneath, denser on veins; petiole 1mm, strigose. Peduncle 1.5–6mm; bracts linear, 2–2.5mm, ciliate; bractlets connate forming a tight cupula. Flowers zygomorphic. Corolla pale green, tube 7mm, gibbous at base; stamens exserted. Fruit free but tightly enclosed by bright red rather fleshy cupula, 5–7mm.

Bhutan: N – Upper Kuru Chu district (Dunkar). Along streams, 2100m. May.

11. L. cyanocarpa Franchet

Shrub to 1m. Branchlets dark brown becoming paler with age, scabrous, pith solid. Leaves elliptic, 15–30 × 7–12mm, acute, base cuneate, scabrous above, more densely so towards margins, strigose beneath, more densely so on veins, ciliate; petioles 1–2mm, strigose, base wrapping nearly halfway around stem. Peduncle 1–10mm; bracts ovate, 6–13mm; bractlets absent. Flowers zygomorphic. Corolla cream-yellow, tube 15–17mm, glabrous outside, pilose within; stamens equalling tube. Fruit free, black with whitish bloom, 1cm; calyx obsolete.

Bhutan: N – Upper Pho Chu district (Tranza, Gafoo La), Upper Bumthang Chu district (Waitang). On screes, among rocks, and on open hillsides, 4300–4700m. June–July.

12. L. porphyrantha (Marquand & Shaw) Nayar & Giri; *L. magnibracteata* Nayar & Giri, *L. cyanocarpa* var. *porphyrantha* Marquand & Shaw

Similar to *L. cyanocarpa* but corolla wine-red to deep purple with broader tube; peduncles 5–25mm; bracts 13–27mm.

Bhutan: C – Tongsa district (Dungshinggang), **N** – Upper Kulong Chu district (Me La); **Sikkim:** Dzongri–Prek Chu, Chemathang–Thangshing etc. On open hillsides among *Rhododendron*, 3040–4260m. June–July.

FIG. 111. **Caprifoliaceae.** a–d, *Lonicera obovata*: a, flowering shoot (× 1); b, flower pair (× 2½); c, dissected corolla (× 2½); d, ovary and style (× 2½). e–g, *Lonicera acuminata*: e, flowering node (× ½); f, flower pair (× 2); g, fruit (× 1½). h–i, *Viburnum nervosum*: h, flowering spur (× ⅔); i, flower (× 4). j, *Leycesteria formosa*: branch of inflorescence (× ⅔). Drawn by Glenn Rodrigues.

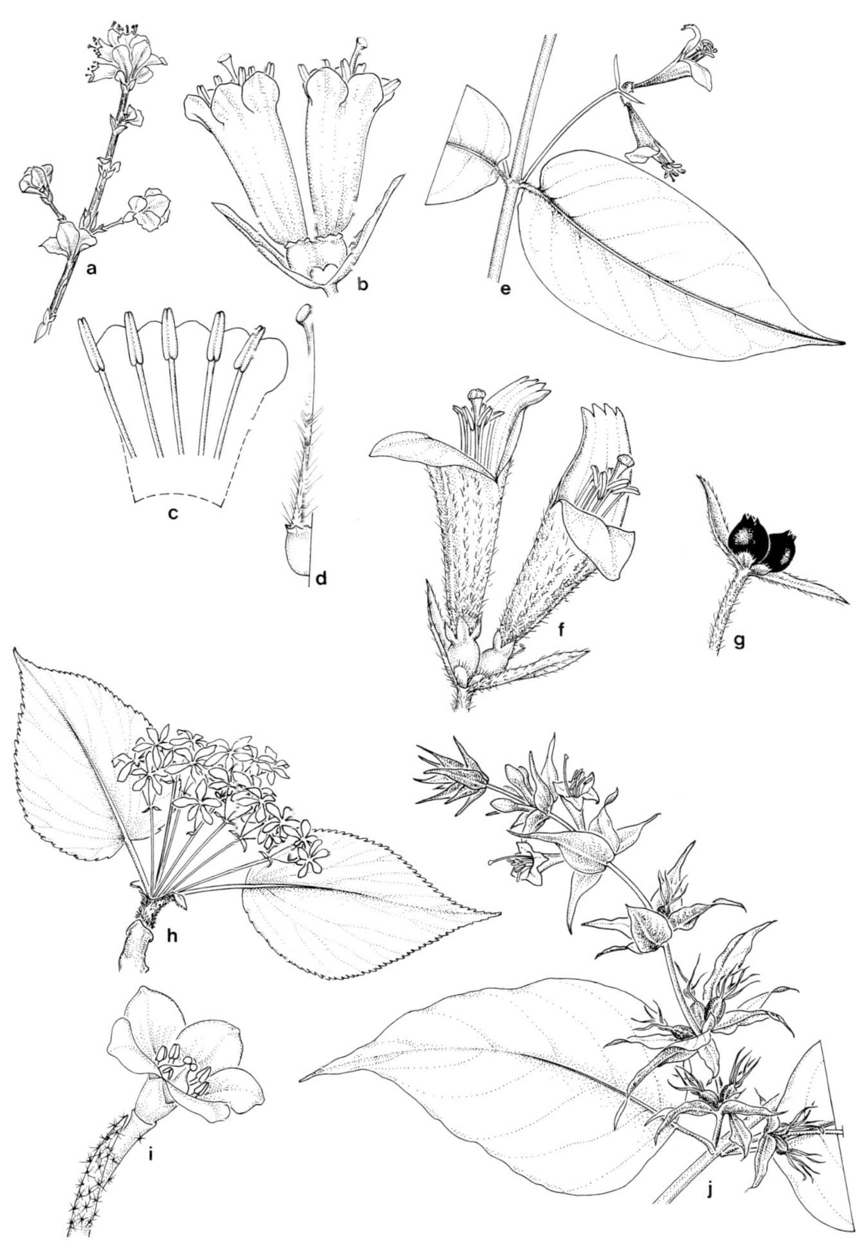
a
b
c
d
e
f
g
h
i
j

13. L. asperifolia (Decaisne) Hook.f. & Thomson; *Xylosteum asperifolium* Decaisne

Shrub. Branchlets brown to tan, scabrous when young, bark shredding in narrow strips when older, pith solid. Leaves elliptic to broad-ovate, 1.5–3 × 1–2cm, acute, base cordate to rounded, sparsely pubescent to scabrous above, white-pilose beneath, more densely so on veins, margins hispid; petioles 4–6mm, with scattered glands and hairs. Peduncle 2–4mm, scabrous; bracts ovate, 6–8mm, acuminate, fringed with scabrous hairs; bractlets absent. Flowers zygomorphic. Corolla white, tube 12–15mm; stamens exserted. Fruit free, elliptic, with persistent lobes.

Bhutan: C – Thimphu district (Thimphu (113)), **N** – Upper Mo Chu district (Shingche La (71)). 2700–3900m. May–June.

The occurrence of *L. asperifolia* is based only on literature reports which require confirmation. Otherwise, this species is restricted to W Himalaya.

14. L. hispida Willdenow var. **setosa** Hook.f. & Thomson

Shrub to 1m. Branchlets brown, with long hispid hairs, denser at nodes, pith solid, bark on second year branches peeling in broad strips. Leaves ovate-elliptic, 20–25 × 9–12mm, acute, base rounded to obtuse, hispid above and more densely so beneath, ciliate; petioles 1–2mm, densely hispid. Peduncle 4–5mm, hispid; bracts obovate-orbicular, 11–13mm; bractlets absent. Flowers zygomorphic. Corolla greenish-yellow, pilose outside, tube 1.5–2cm. Fruit free, scarlet, usually with scattered long-stalked glands, pilose, with persistent calyx lobes.

Bhutan: C – Tongsa district, **N** – Upper Mo Chu, Upper Mangde Chu, Upper Bumthang Chu and Upper Kulong Chu districts; **Sikkim:** Lachen, Lachung etc.; **Chumbi:** Chugya. Among rocks in dry stream beds and with dwarf juniper/*Rhododendron* scrub on open slopes, 4000–4500m. May–June.

Var. *bracteata* (Royle) Airy Shaw has been reported from Bhutan (73) but no specimen has been traced. It differs from var. *setosa* in leaf shape, which is more oblong and narrower.

15. L. webbiana DC.; *L. oxyphylla* Edgeworth, *L. alpigena* sensu F.B.I. non L., *L. adenophora* Franchet

Shrub to 4m. Branchlets brown, tan with age, glabrous or with few glandular hairs especially at nodes, pith solid. Leaves broadly lanceolate, 6–11 × 2–3.5cm, acuminate, base cuneate to subcordate, with scattered scabrous hairs and short-stalked glands especially on veins, both denser below; petiole 5–10mm, ciliate, with short, dense glandular hairs. Peduncle 2–4cm, pilose, glandular; bracts linear, 4–6mm; bractlets free, triangular-oblong, c 1.5mm. Flowers zygomorphic. Corolla crimson, 5–6mm, glandular hairs often present. Fruit red, free, 6–9mm.

Bhutan: C – Thimphu district (Thimphu (71), Barshong–Nala), ?Bumthang district (Bumthang Valley), **N** – Upper Mo Chu district (Laya). In damp ground along streams and in *Betula*, *Rhododendron* and mixed conifer thickets, 2450–3840m. April–June.

16. L. lanceolata Wall.; *L. decipiens* Hook.f. & Thomson

Shrub to 4m. Branchlets brown, becoming light tan, pilose, pith solid. Leaves ovate-lanceolate, 6.5–9 × 2–3.5cm, acuminate, obtuse (subacute) at base, pilose above, pilose with stalked glands beneath, ciliate; petioles 4–5mm, pilose, glandular. Peduncle 11–13mm, villous, glandular; bracts 2.5–3mm, linear; bractlets ± free, broadly ovate, 1–2.5mm, glandular on margins. Flowers zygomorphic. Corolla rose-pink to wine-red, tube 3–5mm, gibbous at base; stamens exserted. Fruit black, free, glabrous.

Bhutan: C – Thimphu, Punakha and Bumthang Sakden districts, **N** – Upper Bumthang Chu and Upper Kulong Chu districts; **Sikkim:** Lachen (80), Phune, Zemu (113) etc.; **Chumbi:** Chumbi, Dotag, Tangka Chu. Understorey shrub in *Abies* and *Betula* forests, 2400–3800m. June–July.

The collection from Dotag is exceptional, having ovate leaves to 5 × 2.5cm and flowers described as blue in September.

17. L. quinquelocularis Hardwicke; *L. diversifolia* Wall.

Shrub to 2.5m. Branchlets purple when young, becoming tan, pilose, pith hollow. Leaves ovate to elliptic, 4–5.5 × 2–2.5cm, acuminate, cuneate to obtuse, pilose above, pale villous beneath; petioles 6–8mm, villous. Peduncle 1–2mm; bracts linear, papery, 4–5mm, soon deciduous; bracteoles connate in pairs, rounded, 2mm, ciliate. Flowers zygomorphic. Corolla yellow to white, tube 5–6mm; stamens exserted. Fruit free, whitish, translucent, ovoid, 2.5mm; calyx persistent, lobes 1mm; seeds blackish-violet.

Bhutan: C – Thimphu district (common in Paro and Thimphu Valleys). In roadside scrub, in disturbed areas and along streams in blue pine forests, 2100–2600m. April–June.

18. L. acuminata Wall.; *L. loureiroi* DC., *L. henryi* Hemsley. Fig. 111e–g.

Climbing shrub to 4m. Branchlets tan turning brown, tomentose, pith hollow. Leaves lanceolate, 6.5–9 × 3–4cm, long acuminate, base subcordate, scattered hairy above with pilose hairs on veins, scattered hairy beneath with veins velutinous, ciliate; petiole 5–9mm, tomentose, union with stem swollen, covered with dense long hairs. Peduncle 9–12mm, tomentose; bracts linear, 4–5mm, ciliate; bractlets free, ovate-obtuse 2–3mm, ciliate. Flowers zygomorphic. Corolla rose-pink to yellow tinged with purple, tube 10–15mm; stamens exserted. Fruit black, free, 5mm, roundish; calyx persistent, lobes 2mm.

Bhutan: C – Thimphu, Punakha, Tongsa, Bumthang, Mongar and Tashigang districts, **N** – Upper Bumthang Chu and Upper Kulong Chu districts; **Darjeeling:** Tonglu area; **Sikkim:** Chiya Bhanjang, Phune–Yakche; **Chumbi:** Sakkurgong. Climbing over shrubs in fir/rhododendron forest, 2400–3700m. June–September.

19. L. glabrata Wall.

Scandent shrub. Branchlets shiny brown, glabrous, pilose when young, pith hollow. Leaves ovate, 5–8 × 3–4.5cm, acuminate, base subcordate, glabrous above, glaucous beneath with scattered hairs, more densely hairy on veins; petioles 6–12mm, velutinous, forming prominent raised ridge at nodes. Fertile axillary and terminal shoots often compressed and relatively leafless. Peduncle 3–14mm, velutinous; bracts linear, 1–1.5mm, ciliate; bractlets free, ovate, 1mm, ciliate. Flowers zygomorphic, very fragrant. Corolla cream to yellow, tube 12–15mm, nearly glabrous outside; stamens exserted. Fruit blue-black, free, 4–5mm, glaucous.

Bhutan: C – Punakha district (Kabjisa, Ritang, Samtengang (71)), **N** – Upper Mo Chu district (Kencho, Tamji), Upper Kulong Chu district (Me La); **Darjeeling:** Baghora, Darjeeling, Rambi etc. Growing over large rocks on forest margins, 1700–2300m. August–September.

20. L. macrantha (D. Don) Sprengel; *Caprifolium macranthum* D. Don

Large scandent shrub with pendulous branches. Branchlets brown, villous, pith hollow. Leaves oblong to oblong-lanceolate, 5–8 × 2–4cm, acuminate, base subcordate to truncate, nearly glabrous above except for pilose hairs on midrib and major veins, villous beneath, whitish when young, less hairy and darker when mature, ciliate; petioles 2–5mm, covered with long villous hairs, hairs longer and denser at union with stem. Peduncle 5–10mm, villous; bracts linear, 9–12mm, ciliate; bractlets free, ovate, 1mm, ciliate. Flowers zygomorphic. Calyx lobes 3–4mm, ciliate, persistent. Corolla creamy white, tube 6–7cm, pilose, glandular-hairy outside. Fruit black, free, 5–6mm, glabrous; pedicel 12–25mm.

Bhutan: S – Deothang district (Wamrung), **C** – Tongsa district (near Tongsa), Bumthang district (Kyertsa), Mongar district (Shersing Thang); **Darjeeling:** without localities (34); **Sikkim:** Namchi (113), Pemayangtse–Thinglen (69), Sang (112). Climbing on trees in warm, wet, broad-leaved forests, 1200–2500m. June–July.

2. LEYCESTERIA Wall.

Shrubs. Stems solid or hollow. Leaves simple, entire or toothed. Interpetiolar stipules present or absent. Flowers in short terminal or axillary spikes, ± regular. Calyx 5-lobed. Corolla 5-lobed, gibbous at base, lobes subequal. Stamens 5. Ovary 5- or 7–8-locular. Fruit a berry. Seeds numerous, minute.

1. Interpetiolar stipules present; petiole 4mm or less 2
\+ Interpetiolar stipules absent; petiole 5mm or more 3

2. Leaves densely woolly on lower surface, usually 4cm or more wide; style glabrous or with few glandular hairs at base .... **2. L. stipulata**
\+ Leaves puberulent with some long hairs on lower surface, usually less than 4cm wide; style hairy .... **3. L. glaucophylla**

3. Bracts conspicuous, 10–30mm, suffused dull red or purple; flowers in whorls of up to 6 .... **1. L. formosa**
\+ Bracts inconspicuous, 2.5–4.5mm; flowers in pairs on spike .. **4. L. gracilis**

1. L. formosa Wall. Fig. 111j.

Stems arching, hollow, 2–5m. Leaves ovate-acuminate, (3.7–)6–13.5 × (1.8–)2.5–6.5cm, usually ± entire, rarely toothed or lobed, lower surface whitish with few hairs on veins; petiole 5–12mm; stipules absent. Bracts ovate-acuminate, 10–30mm, suffused dull red or purple, enlarging in fruit. Calyx densely glandular; lobes ± linear, unequal. Corolla funnel-shaped, 13–16mm, white to pink. Berry subglobose, 6–7mm, red.

Bhutan: C – Thimphu, Punakha, Bumthang and Tashigang districts, **N** – Upper Mo Chu and Upper Kulong Chu districts; **Darjeeling:** Senchal (71), Tonglu etc.; **Sikkim:** Cheungthang, Lachen, Phalut etc. Gravel or loamy slopes, among shrubs in hemlock or fir forest, 1524–3658m. June–August.

2. L. stipulata (Hook.f. & Thomson) Fritsch; *Lonicera stipulata* Hook.f. & Thomson, *Pentapyxis stipulata* (Hook.f. & Thomson) Clarke

Stems arching, ± solid, 2–2.6m. Leaves ovate-oblong, 11.5–17.8 × 3.5–7.5cm, serrate, teeth sometimes very reduced, lower surface densely woolly, coarse-textured; petiole 2–4mm; stipules semicircular to suborbicular, 7–17mm. Bracts ovate to narrowly ovate, c 7mm, woolly. Calyx woolly, somewhat glandular; lobes narrowly ovate, ± equal. Corolla tubular-campanulate, 13–17mm, white. Berry ellipsoid, c 18mm.

Bhutan: S – Puntsholing district (near Jumudag), Chukka district (Gedu, Pasikha–Puntlihir); **Darjeeling:** Ghoom, Batasia, Senchal etc. Rocky slopes and oak forest, 2000–2438m. February–May.

3. L. glaucophylla (Hook.f. & Thomson) Clarke; *Lonicera glaucophylla* Hook.f. & Thomson, *Leycesteria belliana* W.W. Smith

Stems ± hollow, c 2m. Leaves narrowly ovate, 4.5–14 × 0.9–4.7cm, acuminate, subentire to serrate; lower surface papillose with some long hairs, whitish; petiole up to 2mm; stipules semicircular, 4.5–6mm. Bracts ovate to narrowly ovate, up to 18mm, acuminate. Flowers in pairs on spike. Calyx glandular with some eglandular hairs; lobes narrowly ovate, ± equal. Corolla tubular-campanulate, 15–20mm, greenish-white to cream, tinged purple on outside. Ovary 7–8-locular, berry not seen.

Bhutan: C – Thimphu district (Mara Chu), Tongsa district (Changkha, Tongsa), Mongar district (Namning); **Darjeeling:** Pankashari (34) etc.; **Sikkim:** Bakhim, Karponang, Yoksam etc. An undershrub of temperate wet forests, 2400–2896m. May–July.

4. L. gracilis (Kurz) Airy Shaw; *Lonicera gracilis* Kurz

Stems arching, hollow, 1.5–4m. Leaves ovate-elliptic, 7.5–13 × 3–5cm, acuminate, subentire with minute teeth; lower surface whitish, papillose, with few long adpressed hairs on veins; petioles 5–8mm; stipules absent. Bracts narrowly ovate, 2.5–4.5mm, shorter than ovary. Flowers in pairs on spike. Calyx glabrous except for few small glandular hairs, lobes linear, ± equal. Corolla tubular-campanulate, 11–16mm, white to pale yellow. Berry ellipsoid, c 12 × 7.5mm, purple.

Bhutan: S – Phuntsholing district (Kamji, Suntolakha), Chukka district (Gedu, Puntlibhir), Gaylephug district (Chabley Khola); **Darjeeling:** Chhota Rimitti, Labha, Pankabari; **Sikkim:** Karponang, Phedonchen, Yoksam etc. Evergreen oak forest, 1600–2743m. September–October, ?May.

3. VIBURNUM L.

Deciduous or evergreen shrubs or small trees. Leaves opposite, entire, toothed or rarely lobed; stipules present or absent, small, often caducous. Flowers in terminal corymbs or panicles, bracteolate, hermaphrodite. Calyx 5-lobed, lobes equal, persistent. Corolla ± regular, rotate or campanulate to tubular, 5-lobed. Stamens 5, inserted on corolla. Disc absent. Ovary 1–3-locular, style 3-lobed. Ovule solitary, pendulous. Drupe fleshy, 1-seeded, often compressed, red, purplish-black or black when mature.

Records of *V. sambucinum* Blume and *V. corylifolium* Hook.f. & Thomson from Bhutan (117) have not been substantiated.

1. Leaves evergreen, lower surface gland-dotted and glabrous except for tufts of hairs in vein axils; corolla tubular; calyx without lobes **5. V. cylindricum**
+ Leaves deciduous, lower surface glabrous or hairy; corolla tubular-campanulate, campanulate or rotate; calyx lobed 2

2. Corolla tubular-campanulate to campanulate, tube 3.5mm or more 3
+ Corolla rotate or campanulate, tube less than 2mm 5

3. Leaves ovate to broad-ovate, usually 4cm or more wide, densely stellate-tomentose below, margin obscurely crenulate-dentate ... **1. V. cotinifolium**
+ Leaves elliptic to obovate, usually 5cm or less wide, sparsely stellate on veins only below, margin serrulate 4

4. Flowers occurring before leaves; anthers less than 1mm ... **8. V. grandiflorum**
+ Flowers occurring with leaves; anthers 1.5mm or more ... **7. V. erubescens**

5. Leaves with all veins on lower surface prominent and densely stellate-pubescent, base usually cordate; corolla 5–7mm; corymbs sessile **6. V. nervosum**

\+ Leaves with only main veins on lower surface prominent and sparsely stellate, base truncate or cuneate; corolla 2–2.5mm; corymbs pedunculate ... 6

6\. Leaves 2.5–4.5cm, ± elliptic with 1–3 pairs of obtuse teeth or lobes; corolla campanulate ... **3. V. foetidum**

\+ Leaves 4–21.5cm, ovate or oblong-elliptic, serrate with more than 5 pairs of acute teeth; corolla rotate ... 7

7\. Leaves ovate, usually long acuminate, 4–10(–11.5)cm; drupe c 7.5mm; occurring above 1500m ... **2. V. mullaha**

\+ Leaves oblong-elliptic, short acuminate, 11–21.5cm; drupe c 5mm; occurring below 1250m ... **4. V. colebrookianum**

1. V. cotinifolium D. Don; *V. polycarpum* Wall.

Deciduous shrub, 2–5m. Leaves ovate to broadly ovate, 4.8–12.5 × 3.8–9.3cm, acute, base truncate to cordate, entire to obscurely crenulate-dentate, upper surface stellate, lower surface densely stellate-tomentose with whitish appearance; petiole 6–17mm. Flowers in ± umbelliform corymbs; peduncle 5–15mm, stout, densely stellate. Calyx 4–4.5mm, lobes 0.5–1mm. Corolla ± campanulate, 5.5–6.5mm, white tinged pink outside, lobes 1–2mm. Drupe oblong-ellipsoid, black, 7–10 × 5–6mm.

Bhutan: C – Ha district (Chelai La–Ha, Ha Valley), Thimphu district (Paro and Thimphu Valleys); **Darjeeling:** near Darjeeling. Blue pine and conifer/*Rhododendron* forests, often at forest margins, 1524–3720m. April–June.

2. V. mullaha D. Don; *V. stellulatum* DC., *V. stellulatum* var. *involucratum* (DC.) Clarke, *V. mullaha* var. *glabrescens* (Clarke) Kitamura

Deciduous shrub, 0.5–5m. Leaves ovate, 4–10(–11.5) × 2–4.8cm, long acuminate, base cuneate to truncate, serrate, usually sparsely stellate on both surfaces; petiole 5–15mm. Flowers in ± umbelliform sessile or pedunculate corymbs; peduncle up to 3.5cm, hairy. Calyx 1.4–1.8mm, lobes 0.4–0.5mm. Corolla rotate, c 2mm, white sometimes tinged pink, lobes 1.2–1.5mm. Drupe globose, black, c 7.5 × 7.5mm.

Bhutan: C – Thimphu, Tongsa, Bumthang and Mongar districts, **N** – Upper Mo Chu district; **Darjeeling:** Senchal, Tonglu etc.; **Sikkim:** Chunthang etc. Broad-leaved/*Tsuga* forest, frequently in open or cleared areas, 1524–3048m. July–September.

3. V. foetidum Wall. Dz: *Aila shing.*

Erect deciduous shrub, 2–3m. Leaves ± elliptic, 2.5–4.5 × 1.2–2.2cm, acute, base cuneate, margin with 1–3 pairs of obtuse teeth or lobes in upper part, both surfaces sparsely hairy with simple, bifurcating or stellate hairs; petiole 4–6mm. Flowers in dense umbelliform corymbs; peduncle (0.5–)1.5(–2)cm, densely hairy. Calyx c 1.5mm, lobes c 0.3mm. Corolla campanulate, 2–2.5mm, white, lobes c 0.7mm. Drupe ellipsoid, red, c 6 × 4mm.

Bhutan: C – Thimphu (Namselling), Punakha district (Wache, Shenganga, Gangtokha etc.), **N** – Upper Mo Chu district (Tamji). Broad-leaved and blue pine forest, 1570–2134m. June.

4. V. colebrookianum DC.

Large-leaved deciduous shrub, 2–5m. Leaves oblong-elliptic, 11–21.5 × 5–12cm, shortly acuminate, base cuneate, regularly serrate-dentate, some stellate hairs on veins otherwise glabrous; petiole 12–22mm. Flowers in umbelliform corymbs on pendent branches; peduncle 2.5–4.5cm. Calyx 1.5–2mm, lobes 0.5–0.8mm. Corolla ± rotate, 2–2.5mm, white, lobes 1.3–1.5mm. Drupe oblong-ellipsoid, purplish-black, 4.5–5.5 × 3–4mm.

Bhutan: locality unknown (80, 117); **Darjeeling:** Badamtam, Sitong, Birik etc.; **W Bengal Duars:** Muragha Forest. Subtropical forests, 610–2134m. January–April.

5. V. cylindricum D. Don; *V. coriaceum* Blume. Sha: *Yumling shing*.

Evergreen shrub, 2–6m. Leaves narrowly oblong-elliptic, 7.5–18 × 2.7–6.2cm, shortly acuminate, base cuneate, ± entire to obscurely toothed, upper surface glossy dark green, lower surface gland-dotted with tufts of hairs in vein axils; petiole 10–22mm. Flowers in umbelliform corymbs; peduncle 15–60mm. Calyx 1.5–2mm, lobes reduced. Corolla tubular, 3.5–6mm, white to cream; lobes 0.7–1mm, not recurved. Drupe oblong to round, black, 4–5 × 3.5–4mm.

Bhutan: S – Chukka district, **C** – Thimphu, Punakha, Tongsa and Tashigang districts; **Darjeeling:** Darjeeling, Kalimpong, Rissisum, Nimbong etc. Moist broad-leaved, blue pine and evergreen oak forest, 1410–2743m. July–August.

In the vegetative state this species is easily confused with *Cornus oblonga*, which differs by the medifixed hairs on its leaves.

6. V. nervosum D. Don.; *V. cordifolium* DC. Dz: *Ola sima*. Fig. 111h–i.

Deciduous shrub, 2–6m. Leaves ovate, 4.6–12.5 × 2.3–6.8cm, shortly acuminate, base cordate to truncate, serrulate, upper surface sparsely stellate, venation prominent on lower surface and densely stellate-pubescent; petiole 7–15mm. Flowers in sessile corymbs. Calyx 2–3.5mm, lobes 0.5–1mm. Corolla rotate, 5–7mm, white, lobes 4–6mm. Drupe ellipsoid, purplish, 7–9 × 4–5mm.

Bhutan: C – Ha, Thimphu, Punakha, Tongsa and Mongar districts, **N** – Upper Mo Chu, Upper Kuru Chu and Upper Kulong Chu districts; **Darjeeling:** Kalapokri, Tonglu etc.; **Darjeeling/Sikkim:** Phallut, Singalila; **Sikkim:** Chhokha, Dzongri, Tari etc.; **Chumbi:** Yatung, Lachen La–Yatung, Sakkurgong. Conifer/*Rhododendron* forest, occasionally degraded areas, 2743–3962m. April–June.

Often in flower when in young leaf.

7. V. erubescens DC.

Deciduous shrub, 1.5–4m. Leaves elliptic to narrowly oblong-elliptic, (2.5–)3.7–9.5 × (1–)2–5.2cm, acute to shortly acuminate, base ± cuneate, serrulate, upper surface glabrous or thinly appressed-hairy especially at margin. Flowers in small, usually pendulous panicles or corymbs; peduncle 8–37mm, glabrous or hairy. Calyx 2–3mm, lobes c 0.5mm. Corolla tubular-campanulate, 8.5–12mm, pink or white tinged pink on lobes, lobes 3–3.5mm. Stamens inserted in upper part of tube. Drupe ± ellipsoid, purplish-black, 6.5–8.5 × 4.5–6mm.

Bhutan: S – Chukka and Deothang districts, **C** –Punakha, Tongsa, Mongar and Tashigang districts, **N**–Upper Mo Chu and Upper Kuru Chu districts; **Darjeeling:** Shiri Khola, Tonglu, Senchal etc.; **Sikkim:** Tari, Yoksam etc. Broad-leaved conifer/*Rhododendron* and evergreen oak forest, sometimes in degraded areas, 1402–3962m. April–May.

This species shows considerable variation in both leaf size and inflorescence, the latter being conferred or relatively lax, long or short peduncled and variable in indumentum. Var. *prattii* (Graebner) Rehder [var. *wightianum* auct. non Fyson] is recorded from Bhutan (73, without localities) and Darjeeling (Bhotia Basti, Budhwari–Rimbik, Chiya Banjang–Dentam, Takdah (all 69)), but no material has been seen. It is described (69) as a tree of lower altitudes with thinner, broader, oval, minutely to coarsely serrate leaves, larger, loose, drooping panicles and yellowish flowers. A specimen collected near the Ghose Nursery in Darjeeling is similar but probably an escape from cultivation.

8. V. grandiflorum DC. fm. **grandiflorum**; *V. nervosum* sensu F.B.I. non D. Don

Deciduous shrub or small tree, 1.5–4m. Leaves obovate or elliptic to narrow elliptic, 4.2–10 × 1.5–5cm, acute, rarely shortly acuminate, base cuneate, serrulate, upper surface sparsely stellate, lower surface stellate on veins only; petiole 12–28mm. Flowers appearing before leaves in sessile corymbs, corymb branches sparsely to densely hairy. Calyx 3.5–4.2mm; lobes c 1mm, usually ciliate. Corolla tubular-campanulate, 12–15mm, pale pink to rose-pink; lobes broadly rounded, 3.5–4.5mm; stamens inserted in 2 rows, 2 near mouth, 3 lower down, anthers included. Drupe ellipsoid, purplish-black, 7–10 × 5–6mm.

Bhutan: C – Ha, Thimphu, Bumthang and Mongar districts, **N** – Upper Kuru Chu district; **Sikkim:** Changu; **Chumbi:** Yatung, Rinchengang. Conifer/*Rhododendron* forest, stream banks and damp grassy hill slopes, 2743–3810m. March–May, September.

Fm. *foetens* (Decaisne) Taylor & Zappi [*V. foetens* Decaisne] occurs in W Himalaya.

4. SAMBUCUS L.

Shrubs, small trees or woody herbs. Leaves imparipinnate; leaflets opposite or alternate. Flowers regular, articulated with pedicel, in large terminal corymbs. Bracts absent. Calyx 5(–6)-lobed. Corolla rotate, 5(–6)-lobed. Stamens 5(–7), attached to base of corolla. Ovary 3–5-locular; each locule with 1 pendulous ovule. Fruit a berry, 3–5-seeded. Seeds compressed.

Corolla measurements given below are the length of tube and lobe.

1. Upper leaflets connate with rhachis; shrubby herb up to 2m; corymbs ± umbelliform .... **1. S. adnata**
\+ Upper leaflets not connate with rhachis; shrub or small tree of 2–6m; corymbs not umbelliform 2

2. Corolla 1.9–2.3mm; leaflets 10.5 × 3.5cm or more .... **2. S. javanica**
\+ Corolla 3.5–4mm; leaflets 7 × 2.4cm or less ... **3. S. nigra** var. **canadensis**

1. S. adnata DC. Fig. 112a–c.

Shrubby herb, 0.3–1.5m. Leaves (12–)15–30cm, with 1–4 pairs of leaflets, upper 1–2 pairs connate with rhachis; leaflets opposite or alternate, oblong, (5–)7–16 × 1.4–4.2cm with terminal leaflet broader than lateral ones, acuminate, serrate, upper surface hairy on midrib and main veins, lower surface similar with sparse brownish hairs. Corymbs ± umbelliform, pubescent. Calyx c 2mm; lobes c 0.7mm, pubescent. Corolla rotate, 2–2.5mm, cream or white to pink, lobes 1.5–2mm. Berry ± globose, c 4.5 × 3mm, red.

Bhutan: S – Chukka district, **C** – Ha, Thimphu, and Bumthang districts, **N** – Upper Kulong Chu district; **Sikkim:** Gnatong, Kalopokri, Lachen, Sandakphu, Tista; **Chumbi:** near Chumbi. Exposed hill slopes and clearings in mixed rain forest and blue pine forest, 305–3962m. June–September.

2. S. javanica Reinwardt; *S. hookeri* Rehder

Shrub or small tree, 2–6m. Leaves 15–36cm, usually with 2 pairs of leaflets, none connate with rhachis; leaflets opposite, ± oblong, 10.5–20 × 3.5–7cm with terminal leaflet slightly broader than lateral ones, acuminate, serrate, glabrous or with scattered white hairs. Corymbs not umbelliform, broad and loose, pubescent; flowers mostly sessile, some sterile and top-shaped. Calyx c 1mm, lobes 0.4mm. Corolla rotate, 1.9–2.3mm; lobes 1.3–1.8mm, tip apiculate. Berry globose, 4 × 3mm, black.

Bhutan: S – Chukka district (Chima Khothi), **C** –Punakha district (Neptengka), Mongar district (above Mongar), **N** – Upper Mo Chu district (Gasa); **Darjeeling:** Manbhanjan, Mongpu, Rangirum etc. Roadsides, secondary forest scrub, 610–2134m. June–October.

S. hookeri is often maintained as distinct species, distinguished mainly by its sessile flowers, in which case our plants would be referred to it. However, Bolli (322) considers it synonymous with *S. javanica.*

FIG. 112. **Caprifoliaceae.** a–c, *Sambucus adnata*: a, leaf (× ½); b, inflorescence (× ½); c, flower (× 7). d–e, *Triosteum himalayanum*: d, habit (× ½); e, flower (× 3). Drawn by Glenn Rodrigues.

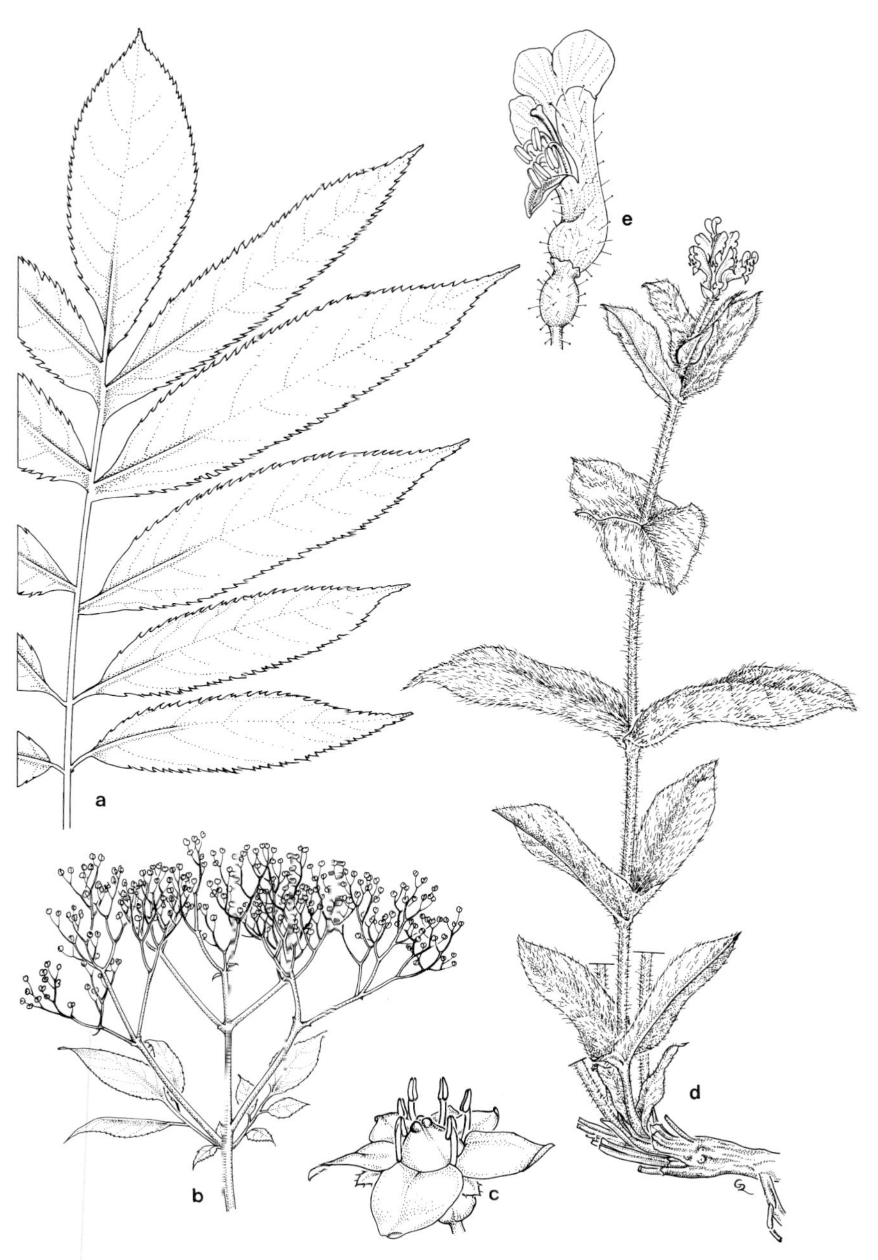
a
b
c
d
e

3. S. nigra L. var. **canadensis** (L.) Bolli; *S. canadensis* L., *S. eberhardtii* Danguy, *S. mexicana* DC.

Shrub or small tree, 3–4m. Leaves 11–17cm with 2–3 pairs of leaflets, never connate with rachis; leaflets opposite, elliptic-ovate, 3.6–7 × 1–2.4cm with terminal leaflet equal to or only slightly broader than lateral ones, acuminate, serrate, upper surface glabrous except for midrib and main veins, lower surface similar with scattered brown hairs. Corymbs c 14cm in diameter, convex, glabrous or minutely papillose. Calyx c 2.5mm; lobes c 0.7mm, glabrous. Corolla rotate, 3.5–4mm; lobes 2.5–3mm, rounded. Berry globose, 4–5 × 4–5mm, black-purple.

Bhutan: S – Phuntsholing district; **Darjeeling:** Mongpu, Takdah (69), Kalimpong (69, 109). Cultivated, rarely escaping to disturbed/waste sites, 200–1600m. February, April–June, August.

A species introduced from N America and frequently cultivated.

5. TRIOSTEUM L.

Coarse erect herb with perennial root-stock. Leaves opposite, joined at base around stem, entire, sinuate or sometimes obtusely lobed. Flowers in short terminal spike, zygomorphic, 2-bracteolate, 5-merous. Corolla tubular-bilabiate, gibbous at base. Fruit a berry, subglobose, scarlet. Seeds 3, oblong.

1. T. himalayanum Wall.; *T. hirsutum* Roxb. Fig. 112d–e.

Stems simple, 45–60cm, pilose with both eglandular and glandular hairs. Leaves obovate to oblong, 5.5–14.5 × 3.1–9cm, indumentum as stem. Flower spike 3–6cm. Calyx lobes 0.5–1mm. Corolla 11–14mm, curved, oblique at apex, greenish, blotched red or purple inside, lower lip deflexed. Berry c 11.5cm. Seeds c 8 × 4mm.

Bhutan: C – Ha, Thimphu and Bumthang districts, **N** – Upper Mo Chu and Upper Bumthang Chu districts; **Sikkim:** Shaitonpokri, Lhonak and Zemu Valleys (113) etc.; **Chumbi:** Galing. Open or cleared *Picea*/*Tsuga* forest, often in damp ground, 3048–3962m. June.

Family 182a. CARLEMANNIACEAE

by D.G. Long & J. Wright

Herbs or subshrubs. Leaves opposite, petiolate, asymmetric, crenate-serrate; stipules absent. Flowers small, in lax or dense terminal cymes. Calyx tube subglobose, adnate to ovary; lobes 4, unequal, spreading. Corolla tubular at base; lobes 4, slightly unequal. Stamens 2, inserted about middle of corolla tube, included; filaments very short; anthers oblong. Ovary inferior, 2-celled; style filiform, stigma bifid. Capsule subglobose, membranous, 4-lobed, loculicidally 2-valved. Seeds numerous, very small, ovoid, rough.

An anomalous family sometimes included in Rubiaceae or Caprifoliaceae.

1. CARLEMANNIA Bentham

Description as for Carlemanniaceae.

1. Leaves glabrous except on veins beneath; cymes lax; calyx lobes linear-lanceolate, c 2mm .... **1. C. griffithii**
+ Leaves whitish pilose on both surfaces; cymes dense; calyx lobes oblong-ovate, 2.5–3mm .... **2. C. congesta**

1. C. griffithii Bentham. Fig. 115a.

Tall herb or subshrub, 1–3m, woody at base, branches slender. Leaves obliquely elliptic, 6–17 × 2–6.5cm, base cuneate, margins serrate, glabrous except on veins beneath; petioles 0.5–2cm. Cymes lax, 3–7cm diameter; peduncles minutely strigose. Bracts very small, bracteoles minute. Calyx persistent on top of capsule, tube 1mm, minutely strigose, lobes linear-lanceolate, 2mm. Corolla white, puberulous, tube 4–6mm, lobes 1–2mm, obtuse. Capsule sparsely puberulous, 3mm diameter. Seeds brown-black, surface honeycombed.

Bhutan: S – Samchi district (Dorokha (117)), Phuntsholing district (Kamji), Chukka district (Sinchula), Gaylegphug (Sureylakha (117)), Deothang districts (Deothang); **Darjeeling:** Labha, Farseng, Rhambi, Sureil, Mongpu, Darjeeling, Kurseong etc. In scrub and in evergreen *Schima* and laurel forests, 300–1820m. July–September.

2. C. congesta Hook.f.

Similar to *C. griffithii* but leaves whitish pilose on both surfaces; petioles longer, 3–6cm; cymes compact, densely-flowered, 2.5–4cm diameter; peduncles hispid; bracteoles oblong, 3–4mm; calyx lobes oblong-ovate, 2.5–3mm; corolla creamy white with yellow or orange spots, larger, tube c 8.5mm, lobes 2.5–3mm, pilose; capsules sparsely long pilose.

Bhutan: S – Gaylegphug district (Gaylegphug–Shemgang); **Darjeeling:** Darjeeling, Ghoom, Birch Hill, Takvar, Lebong etc.; **Sikkim:** Dubde, Rathong Valley, Yoksam. Damp banks and rocks in forest and scrub, 1650–2290m. July–September.

Family 183. VALERIANACEAE

by R.A. Clement

Annual or perennial herbs, often with strong-smelling rhizomes. Leaves opposite or in basal rosettes, simple or compound; stipules absent. Flowers in corymbose cymes that may be paniculate or capitate, zygomorphic (often obscurely so), usually hermaphrodite. Calyx superior, very small in flower, often enlarging in fruit, rarely absent. Corolla superior, 5-lobed, often unequal or spurred at base. Stamens 1–4, inserted on corolla, alternating with corolla lobes. Ovary 1–3-locular, usually only 1 locule fertile, with single pendulous ovule. Fruit an achene crowned by calyx.

1. Stamen 1; flowers and fruit subtended by orbicular bracteole; inflorescence paniculate .... **1. Patrinia**
\+ Stamens 3–4; bracteoles not orbicular; inflorescence capitate at least initially, sometimes becoming lax with maturity 2

2. Stamens 4; leaves simple, linear to narrowly spathulate; calyx with 5 ovate lobes... .... **2. Nardostachys**
\+ Stamens 3; leaves simple or pinnate, not linear or narrowly spathulate; calyx with c 12 setae, rolled in flower, unrolled in fruit .... **3. Valeriana**

1. PATRINIA Jussieu

Perennial herb. Leaves sessile, pinnatifid (rarely simple). Flowers in paniculate corymbs, subtended by scarious bracteole. Calyx lobes obscure or absent, not enlarged in fruit. Corolla 5-lobed, not gibbous or spurred at base. Stamen 1 (in our area). Ovary 3-locular; ovule 1.

1. P. monandra Clarke

Stems erect, 60–130cm, pubescent. Leaves c 8.5 × 5.5cm (c 16 × 2.5cm on one collection (393)), sparsely appressed hairy on both surfaces; lobes (0–)1–2 pairs, lanceolate, irregularly serrate. Panicle up to 30cm. Bracteole orbicular, c 2.5mm diameter, scarious with darker venation, enlarging to c 4.5mm diameter in fruit. Calyx c 1mm when present. Corolla obconical, 3mm, yellow; lobes subequal, 0.6mm, obtuse. Fruit c 2 × 1.4mm.

Darjeeling: Pul Bazar (393); **Sikkim:** Hi Khola, Yoksam. 1050–1850m. Flowering and fruiting in October.

2. NARDOSTACHYS DC.

Perennial herb. Leaves simple, in rosettes at base or opposite on stems. Flowers in bracteate, capitate heads. Calyx 5-lobed, enlarged in fruit. Corolla unequal at base, 5-lobed. Stamens 4. Ovary 3-locular; ovule 1. Fruit obovate.

FIG. 113. **Carlemanniaceae and Valerianaceae. Carlemanniaceae.** a, *Carlemannia griffithii*: habit (× ½). **Valerianaceae.** b–c, *Nardostachys jatamansi*: b, habit (× ½); c, flower (× 3). d–e, *Valeriana hardwickei*: d, habit (× ⅓); e, flower (× 15). Drawn by Glenn Rodrigues.

a
b
c
d
e

1. N. jatamansi (D. Don) DC.; *Valeriana jatamansi* Jones p.p., *Patrinia jatamansi* D. Don, *N. grandiflora* DC. Bengali etc.: *Jatamansi.* Fig. 113b–c.

Rhizome up to 12cm long, aromatic, surrounded by fibrous remains of leaf bases. Basal leaves linear to narrowly spathulate, (2.5–)4–15(–26) × 0.4–1.7(–2.4)cm, glabrous or pilose on main veins and margin, margin ± entire; cauline leaves narrowly ovate to ovate or oblong, 2–5.4 × 0.4–1.4cm, margin entire or serrulate. Flowering stems erect, 9–30cm, sometimes branched above. Flower heads c 2cm in diameter. Calyx 2–2.8mm, entire or toothed. Corolla tubular-campanulate, (4.5–)5.5–13mm, pink to purple, hairy; lobes 2–4mm, obtuse. Fruit c 5 × 2.5mm.

Bhutan: C – Thimphu, Tongsa, and Bumthang districts, **N** – Upper Mo Chu, Upper Bumthang Chu, Upper Kuru Chu and Upper Kulong Chu districts; **Sikkim:** Goecha La, Singalila (393), Yume Chhu etc. Open grassy or gravel areas or among dry scrub, 3810–5155m. June–September.

The rhizome is used for a wide range of medicinal purposes (393). Prakash has treated *N. grandiflora* as a distinct rare species with a locality in Sikkim (393), though it could not be distinguished in our materials.

3. VALERIANA L.

Perennial herbs. Leaves simple or compound or combination of both types, sometimes grouped at base, cauline opposite. Flowers in dense corymbose cymes that become lax with maturity; unisexual flowers sometimes present among bisexual flowers. Bracts persistent. Calyx tightly rolled in flower, unrolled and much elongated in fruit to form pappus of c 10 linear plumose segments. Corolla gibbous or unequal at base. Stamens 3. Stigma 2–3-lobed. Ovary 3-locular with 1 ovule; sterile locules obsolete. Fruit oblong-ovate to elliptic, compressed, 1 rib on one side, 3 on other side, shorter than pappus.

1. Basal leaves numerous, persistent, cordate, 2.5–11.8cm; thick rhizome up to 4.5cm .... **1. V. jatamansi**
\+ Basal leaves few, often caducous, simple or pinnate, if cordate then 2.5cm or less; not forming rhizome 2

2. Mid-cauline leaves more than 3cm with 2–3 pairs of leaflets; terminal leaflet lanceolate to ovate; flowering stems usually 20–130cm .... **2. V. hardwickei**
\+ Mid-cauline leaves 3cm or less, simple or with 1 pair of leaflets; terminal leaflet ovate-oblong to suborbicular; flowering stems usually 3.5–7.5cm **3. V. barbulata**

1. V. jatamansi Jones; *V. wallichii* DC. Eng: *Indian valerian.*

Thick woody rhizome up to 4.5cm with fibrous roots. Basal leaves simple, cordate, persistent, 2.5–11.8 × 1.6–7.8cm, appressed hairy above, pubescent below, margin usually obscurely dentate, sometimes sinuate; petiole 1.5–17.5cm. Cauline leaves few, opposite, upper usually ovate with pair of small lobes at base, up to 3(–5)cm, acuminate, margin obscurely dentate. Flowering stems 4–37(–65)cm. Calyx (rolled) c 0.3mm, unrolling in fruit up to 4mm with 11–13 setae. Corolla obconical, 2.2–3.5mm, white tinged pink or purple; lobes obtuse, 0.7–1mm. Fruit elliptic, c 2.4 × 1mm.

Bhutan: S – Deothang district, **C** – Ha, Thimphu, Punakha, Bumthang and Tashigang districts, **N** – Upper Kulong Chu district; **Darjeeling:** Sureil (393); **Sikkim:** Chunthang–Lachung (393), Rimbi Chhu (393) etc. Moist shady areas near streams or in oak forest, 1200–2900m. February–June.

Rhizome and roots have similar medicinal properties to *Valeriana officinalis* L. and *Nardostachys jatamansi* (393).

2. V. hardwickei Wall. Fig. 113d–e.

Long fibrous roots; stoloniferous; nodes often densely white hairy. Basal leaves on flowering shoots few or absent; on non-flowering rosettes simple, suborbicular to cordate, 0.8–2.5 × 0.4–2.4cm; petiole up to 6cm. Cauline leaves pinnate or pinnatisect with 1–3(–4) pairs of leaflets, 2–10.5(–24)cm; terminal leaflet lanceolate to ovate, 1–5.8(–11) × 0.5–2.2(–5)cm, margin subentire to obscurely dentate-sinuate; lateral leaflets similar, smaller, decreasing in size towards base of leaf, margins ciliate, glabrous or antrorsely appressed-hairy above, glabrous or hairy on veins below. Flowering stems (15–)20–130cm, sometimes branched. Calyx (rolled) c 0.3mm, unrolling in fruit up to 5mm. Corolla obconical, 1.5–2.5mm, white to pink, lobes 0.6–1mm. Fruit ovate, c 2.3 × 1mm, glabrous to pilose.

Bhutan: C – Thimphu district, **N** – Upper Mo Chu, Upper Bumthang Chu, Upper Kuru Chu and Upper Kulong Chu districts; **Darjeeling:** Damsong, Ghoom, Sandakphu etc.; **Sikkim:** Dzongri, Changu, Lagyap, Rathang Chhu etc.; **Chumbi:** unlocalised (Bell collection). Occurring on dry grassy slopes or in damp conditions in open, among shrubs/*Rhododendron* or in fir/juniper forest, 914–4267m. April, June–September.

V. stracheyi Clarke has been reported from Sikkim (393). It is predominantly a species of W Himalaya and Afghanistan with (4–)5–7 pairs of pinnatisect cauline leaves and basal leaves often pinnately divided.

3. V. barbulata Diels

Roots fibrous; long filiform stolons present. Basal leaves few, often simple, oblong to suborbicular, 0.7–1.7 × 0.5–1cm, margin subentire or bearing 2–4 pairs of obscure teeth, subglabrous to sparsely hairy on main veins and petiole margins; petioles up to 1.5cm, flat, often 2–3mm broad. Cauline leaves similar or pinnate or pinnatifid with 1 pair of leaflets or small lobes; terminal leaflet ovate-oblong to suborbicular, up to 2.3 × 0.9cm. Flowering stem 3.5–7.5cm.

Calyx (rolled) c 0.2mm. Corolla obconical, 3(–3.5)mm, pale pink to purplish-pink, lobes 0.8–1(–1.5)mm. Fruit ovate-oblong, c 2.3 × 0.9mm, glabrous.

Bhutan: C – ?Tongsa district (Tampe Tso), **N** –Upper Pho Chu district (Chesha La), Upper Bumthang Chu district (Waitang), Upper Kuru Chu district (Kang La), Upper Kulong Chu district (Shingbe); **Chumbi:** without locality (421). Open hillsides, in grass, on scree or stony banks, 4115–4267m. June–July.

A vigorous collection from Dzongri, Sikkim, 3810m, with larger 5-lobed leaves may belong here, as may one from Zemu Valley, Sikkim, 4550m, reported as *V. himalayanum* Grubov (393), which is otherwise a W Himalayan species similar to *V. barbulata* but lacking stolons and sometimes dioecious.

Family 184. DIPSACACEAE

by R.A. Clement

Herbs. Leaves opposite or whorled, simple, often pinnatifid or lobed, stipules absent. Flowers in cymes or dense heads, hermaphrodite, usually subtended by bracteoles. Calyx superior, very small or with long setae. Corolla, superior, gamopetalous, zygomorphic (often obscurely so), 4–5-lobed. Stamens 4, attached to corolla tube. Style slender. Ovary inferior, unilocular with 1 pendulous ovule. Fruit 1-seeded, enclosed in persistent involucel and crowned with persistent calyx. Seeds with large straight embryo.

1. Flowers in dichotomously branched cymes; conspicuously glandular-hairy in upper part .... **1. Triplostegia**
+ Flowers in capitate heads; without glandular hairs 2

2. Calyx cupular, c 1.5mm; leaves pinnatifid with large terminal leaflet and 2–3 pairs of lateral leaflets .... **2. Dipsacus**
+ Calyx with c 20 setae, c 14mm; leaves linear-oblong, lobed **3. Pterocephalodes**

1. TRIPLOSTEGIA DC.

Erect perennials with small cylindrical tubers. Dark-tipped capitate-glandular hairs at least in upper part. Leaves often lobed. Flowers in dichotomously branched cymes, sessile or shortly pedicellate, obscurely zygomorphic. Ovary enclosed by involucel and group of 4 bracts. Calyx minute, 5-lobed. Corolla 5-lobed. Stamens inserted near base of corolla. Stigma capitate.

1. Corolla 2–3mm; leaves oblong to obovate, if lobed then 1–2 pairs of lobes only .... **1. T. glandulifera**
+ Corolla 5.5–8mm; leaves spathulate to cuneate, 3–5 distinct pairs of lobes **2. T. grandiflora**

1. T. glandulifera DC.; *T. glandulosa* DC.

Leaves mainly grouped near base, variable, oblong to obovate, 1.8–9.5 × 0.7–3.3cm, sometimes pinnatifid with 1–2 pairs of lobes, margin serrate, appressed pilose above, pilose on main veins only below, base attenuate to petiolate. Flowers in few-flowered heads on branched inflorescence that lengthens with maturity. Bracts enclosing base of flower linear, slightly exceeding ovary, mucronate at apex becoming hooked in fruit; involucel urceolate, equalling ovary, obscuring calyx. Corolla obconical, 2–3mm, white with lobe tips tinged pink to purple, tube 1.2–2mm.

Bhutan: C – Thimphu district (Motithang, Pajoding etc.), Thimphu/Punakha district (Dochong La, Dechencholing–Punakha), Bumthang district (Bumthang, Kitiphu, Shabjetang etc.). **N** – Upper Kuru Chu district (Singhi Dzong); **Darjeeling:** Phalut, Rishi La, **Sikkim:** Chiya Bhanjang, Lachen, Tari etc.; **Chumbi:** Chumbi. Open damp areas, hemlock and fir forest, 2286–3962m. July–September.

2. T. grandiflora Gagnepain

Similar to *T. glandulifera* but differing in leaves ± spathulate to cuneate, 3–8 × 1–2.4cm, deeply incised with 3–5 distinct pairs of serrate-crenate lobes, without distinct petiole; corolla 5.5–8mm, tube 3.2–5mm.

Bhutan: C – Thimphu district (Dotena, Tashi Cho Dzong), Punakha district (Wangdu Phodrang, Tilagong), Tongsa district (Chendebi), Tashigang district (Trashiyangtse Dzong). Open hillsides, 1706–2743m. July–September.

2. DIPSACUS L.

Erect herbs. Leaves opposite, basal petiolate, entire or pinnatifid. Involucre of bracts subtending long pedunculate, capitate flower-heads. Bracteoles shorter than flowers, persistent. Involucel scarcely exceeding ovary. Calyx cupular, very small. Corolla 4-lobed. Stamens 4. Stigma elliptic, oblique.

1. Flowers white or cream; involucral bracts linear-oblong, not more than 3.5mm wide .... **1. D. inermis**
+ Flowers blue-black or deep purple; involucral bracts ovate, not less than 3.8mm wide .... **2. D. atratus**

1. D. inermis Wall.; *D. mitis* D. Don, *D. strictus* D. Don, *D. asper* DC.

Perennial. Basal and mid-cauline leaves pinnatifid, 11–21 × 4.5–9cm, margin serrate or crenate-serrate, appressed hairy; terminal lobe elliptic to oblong, larger than lateral ones; lateral lobes 4–6, usually paired; petiole up to 24cm (basal leaves). Upper cauline leaves entire or with 2 small basal lobes, usually lanceolate, serrate. Flowering stems 32–100cm, hirsute, often with small prickles. Involucral bracts linear-oblong, up to 25 × 3.5mm. Flowers in globose heads, hairy. Bracteoles obovate with fine mucronate point at apex, 8–12 × 3.5–4.5mm. Involucel 4-angled with small tooth at each angle. Calyx ± 4-angled, c 1.5mm, apex undulate. Corolla tubular to obconical, 8–11mm, white or cream, lowest lobe longer than others. Fruit (including involucel) c 5 × 1.5mm.

Bhutan: C – Ha district (Ha Dzong), Thimphu (Gunisawa, Motithang, Thimphu etc.), Tongsa district (Tashiling–Chendebi); **Darjeeling:** Damsong, Darjeeling; **?Sikkim:** Barmiak; **Chumbi:** Chubitang, Chumbi, Galing etc. On open hillsides or among shrubs, 1219–3658m. August–September, December.

A rather variable species within which var. *mitis* (D. Don) Nasir has been recognised by some authors.

2. D. atratus Clarke

Similar to *D. inermis* but flowering stems hirsute without prickles; involucral bracts ovate-oblong, 6–12 × 3.8–6mm; bracteoles lanceolate, c 8mm; involucel not angled, toothed at apex; calyx c 1mm, obtusely lobed; corolla narrowly tubular in lower part expanding into campanulate throat and limb, c 9.5mm, blue-black to deep purple, lobes equal.

Bhutan: N – Upper Mo Chu district (Laya), Upper Kulong Chu district (Shingbe); **Sikkim:** Kupup, Phalut etc.; **Chumbi:** Galing. Open areas, bank near cultivation, 3658–3840m. August–September.

3. PTEROCEPHALODES Mayer & Ehrendorfer

Perennial herbs. Leaves opposite, sessile, simple but usually lobed. Flowers in capitate, subglobose heads; heads subtended by involucre of bracts, long pedunculate. Involucel enlarging slightly in fruit. Calyx with plumose setae. Corolla lobes somewhat irregular, usually 5. Stamens 4. Stigma capitate. Fruit ribbed.

FIG. 114. **Dipsacaceae, Morinaceae and Campanulaceae. Dipsacaceae.** a–b, *Pterocephalodes hookeri*: a, habit (× ½); b, flower (× 3). **Morinaceae.** c–f, *Acanthocalyx nepalensis*: c, habit (× ½); d, involucel and flower (× 3); e, dissected involucel (× 5); f, dissected corolla (× 1½). **Campanulaceae.** g, *Campanula pallida*: habit (× ⅔). h–j, *Asyneuma fulgens*: h, habit (× ½); i, flower (× 3); j, dissected corolla and style at female maturity (× 3). Drawn by Glenn Rodrigues.

a
b
c
d
e
f
g
h
i
j

1. P. hookeri (Clarke) Mayer & Ehrendorfer; *Scabiosa hookeri* Clarke, *Pterocephalus hookeri* (Clarke) Diels. Med: *Lugtsedowo.* Fig. 114a–b.

Thick fleshy tap-root. Leaves forming rosette at base, linear-oblong, 4–22 × 1–4.3cm, margin conspicuously lobed, rarely entire, villous on both surfaces, lobes obtuse, Flowering stem 14–36cm, leafless, densely hairy. Involucral bracts ovate-acuminate up to 15mm. Bracteoles linear-oblanceolate, ± equalling flowers. Involucel c 2.5mm, densely hairy, apex ± undulate. Calyx disc-like at base with c 20 setae, disc c 0.5mm, setae up to 10mm. Corolla tubular-obconical, 11–14mm, white or cream, tube 8–11mm; lobes obtuse, subequal. Fruit elliptic, c 4.5 × 1.8mm.

Bhutan: C – Thimphu district, **N** – Upper Mo Chu, Upper Pho Chu and Upper Bumthang Chu districts; **Sikkim:** Gochung, Lhonak, Phamanaga etc.; **Chumbi:** Chumbi, Lingmatang, Phari etc. Among shrubs or on open grassy hillsides, 3048–4877m. August–October.

Family 185. MORINACEAE

by R.A. Clement

Perennial herbs. Leaves opposite or whorled, usually spiny; petioles connate forming sheaths around stem; stipules absent. Bracts conspicuous, at least partially obscuring flowers. Flowers gamopetalous, hermaphrodite, zygomorphic, in verticillate spikes or capitate heads. Involucel present, spiny at apex, persistent. Calyx superior, 2-lipped or oblique at apex. Corolla often curved, 5-lobed, obscurely 2-lipped. Stamens 4, or 2 with 2 staminodes, attached to corolla tube. Nectaries at base of corolla. Ovary inferior, 1-locular. Ovule solitary, pendulous. Style slender, stigma disc-shaped. Fruit an achene, ± obliquely truncate at apex, enclosed in involucel and crowned by calyx.

1. Stamens 4; calyx oblique at apex .... **1. Acanthocalyx**
\+ Stamens 2 with 2 staminodes; calyx 2-lipped 2

2. Corolla up to 1cm; staminodes at base of corolla tube .... **2. Cryptothladia**
\+ Corolla 2–5cm; staminodes in upper half of corolla tube .... **3. Morina**

1. ACANTHOCALYX (DC.) Cannon

Leaves forming rosettes on sterile shoots, linear, lanceolate or ovate-oblong, margin with or without fine spines. Flowers sessile, capitate, sometimes also in 1–2 whorls below head. Bracts free or connate at base. Involucel, tubular, enlarging slightly in fruit. Calyx tubular. Corolla tubular, tube straight or curved, widening at throat, with 2 upper and 3 lower spreading lobes. Stamens 4. Nectary spherical.

1. Basal and mid-cauline leaves linear or linear-lanceolate, usually less than 1cm wide; longer spines on leaf margin 4–8mm, always present; ovary usually glabrous .... **1. A. nepalensis**
+ Basal and mid-cauline leaves oblanceolate to ovate-oblong, usually more than 1cm wide; spines on leaf margin 1–3.5mm, rarely longer, sometimes absent; ovary usually pubescent .... **2. A. delavayi**

1. A. nepalensis (D. Don) Cannon; *Morina nepalensis* D. Don, *M. betonicoides* Bentham. Fig. 114c–f.

Basal and mid-cauline leaves linear or linear-lanceolate, (4–)6–27 × 0.6–1.7cm; upper cauline leaves sometimes narrowly ovate, acuminate, up to 5cm; glabrous or with hairs on margins or upper surface near base, margins always spiny; spines of two general lengths, 2–3mm and 4–5(–8)mm; petiole sheath up to 7cm. Flowering stems up to 50(–72)cm. Bracts ovate-acuminate, spiny on margins, usually obscuring calyx. Involucel not exceeding calyx; spines 12–17, various lengths up to 5mm. Calyx 8.5–10.5mm, with usually 3–5 spinescent teeth. Corolla 20–22mm, pink to rosy purple with darker ring in throat; tube slightly to strongly curved, widening at throat to 3.5–4.5mm; lobes 2–3mm, ± equal. Ovary usually glabrous. Fruit ± oblong c 4.5 × 2.5mm.

Bhutan: C – Thimphu, Tongsa and Bumthang districts, **N** – Upper Bumthang Chu, Upper Mo Chu and Upper Kulong Chu districts; **Darjeeling/Sikkim:** Iskiola, Singalila etc.; **Sikkim:** Bikbari; **Chumbi:** Champitang–Yatung, Trakarpo–Phari etc. In damp or marshy areas and dry ground on cliffs, in clearings in juniper/*Betula* forest and cleared *Picea* forest, 2434–4572m. May–August.

A collection from Chumbi village belongs here, but has flowers recorded as white, which is normally diagnostic of a chinese species, *A. alba* (Handel-Mazzetti) Cannon.

2. A. delavayi (Franchet) Cannon; *Morina delavayi* Franchet

Similar to *A. nepalensis* but basal and mid-cauline leaves oblanceolate to ovate-oblong (basal leaves rarely linear) 4–11 × 1.3–1.8cm, spines on margins usually 1–3.5mm and fewer than previous species (often with 1–2mm between spines). Calyx tubular to tubular-campanulate, 7.5–9mm, tube 2.5–4.5mm. Corolla often sparsely hairy to subglabrous towards base, (17–)20–23(–25)mm. Ovary often pubescent.

Sikkim: Tari. 3960m. July.

2. CRYPTOTHLADIA (Bunge) Cannon

Rhizome covered with fibrous leaf-base remains. Leaves in rosettes on sterile shoots, otherwise in whorls of 4, margins spiny. Flowers in several verticils forming a spike, possibly cleistogamous. Bracts free or connate near base. Involucel tubular to campanulate, enlarging slightly in fruit. Calyx tubular to campanulate with 2 equal lips, lips lobed. Corolla tubular, 2-lipped. Stamens 2, inserted in upper half of corolla tube; staminodes 2, inserted near base of corolla. Nectaries at base of each staminode.

1. Leaf margin incised with spiny teeth; calyx lips equal to or shorter than calyx tube, usually mucronulate .... **1. C. polyphylla**
+ Leaf margin entire with spines; calyx lips at least twice calyx tube, not mucronulate .... **2. C. ludlowii**

1. C. polyphylla (DC.) Cannon; *Morina polyphylla* DC.

Leaves linear to linear-oblong, 5.5–36.5 × 1–4.5cm, margin toothed, each tooth with 2–3 spiny teeth, subsidiary teeth also sometimes present; petiole sheaths up to 2cm. Flowering stems up to 25(–43)cm. Bracts ovate-acuminate, exceeding flowers, margin spiny. Involucel ± tubular, widening slightly to apex, hairy; spiny teeth 5–11, up to 2.5mm; usually shorter than calyx. Calyx 7–10mm, pilose; lips 2-lobed, equal to or shorter than tube; lobes ovate-oblong, 2–3mm, often mucronulate. Corolla ± tubular, 6.5–9mm, white to pink, lips apparently scarcely opening, not exceeding calyx. Fruit ± oblong, c 5.5 × 3mm.

Bhutan: C – Thimphu district (Chekha–Tremo La, Mem La), **N** – Upper Mo Chu district (Nelli La); **Sikkim:** Ningbil; **Chumbi:** Put-lo, Za-ne-gang. Open grassy slopes, 3658–4267m. May–August.

2. C. ludlowii Cannon

Leaves linear-oblanceolate to lanceolate, 5.5–10.5 × 0.7–1.5cm, margin entire with fine spines to 4mm, pilose on margin and main veins; petiole sheaths up to 4.5cm. Flowering stems up to 37cm, densely pilose. Bracts ovate, ± equal to or exceeding flowers, with usually 2–3 pairs of lateral spinose teeth. Involucel obconical, c 11mm at anthesis; spiny teeth 5–9, up to 4mm. Calyx 9–9.5mm, hairy inside, glabrous outside; lips at least twice as long as tube, erose to emarginate. Corolla ± tubular, c 6mm (?in bud), ?yellow, lips shallowly fimbriate, upper lip slightly hooded. Fruit ± oblong, c 4 × 2.3mm.

Bhutan: C – Tashigang district (Preng La), **N** –Upper Mangde Chu district (Ju La), Upper Bumthang Chu district (Kantanang), Upper Kulong Chu district (Me La). On cliff-ledges and rocky areas, 3658–4267m. July.

3. MORINA L.

Rhizome bearing some leaf-base remains. Leaves in pairs or whorls of 3, margins spinose. Flowers in several verticils forming a spike. Bracts free. Involucel tubular. Calyx 2-lipped, lips lobed. Corolla obscurely 2-lipped, upper lip 2-lobed, lower lip 3-lobed, lobes spreading. Stamens 2, inserted in throat of corolla; staminodes 2, inserted in upper half of corolla tube. Nectary 3-lobed.

1. M. longifolia DC.

Leaves linear-lanceolate to lanceolate, 7.5–26.5 × 1.7–3.8cm, margin prominently toothed, teeth spinose in groups of 3, rarely with small spinose teeth, glabrous; petiole sheath up to 4cm. Flowering stem 30–85(–106)cm, pubescent above. Bracts ovate, 3–6.2cm, spiny. Involucel ± tubular, equalling calyx except for 1 longer spine, spines 10, longest up to 8mm at anthesis, villous. Calyx tubular-campanulate, 3.5–10mm, glabrous outside, hairy inside lips; lips longer than tube, emarginate. Corolla narrowly tubular with suborbicular lobes, c 30mm, white becoming pink then red. Staminodes just below throat. Fruit ± oblong, c.5.5 × 3mm.

Bhutan: C – Thimphu district (Bela La, Chapcha), Punakha district (Pele La); **Sikkim:** Thanggu. Open hillsides, among dwarf bamboo, 3200–3810m. August–September.

Family 186. CAMPANULACEAE

by R.A. Clement

Herbs (in Bhutan), usually with milky juice. Leaves simple, alternate, rarely opposite or spirally arranged; stipules absent. Flowers bisexual, actinomorphic or zygomorphic (*Lobelia*), solitary or in racemose inflorescences. Calyx (4–)5-lobed, persistent. Corolla (4–)5-lobed, gamopetalous (occasionally divided virtually to base). Stamens 5, alternate with corolla lobes, filaments free, anthers free or connate, pollen often shed in bud and presented on style when flower opens but before stigmatic surfaces are exposed. Ovary usually inferior, sometimes superior, 2–5-locular. Fruit a capsule or sometimes fleshy and berry-like. Seeds numerous, small.

1. Anthers fused at anthesis 2
+ Anthers free at anthesis 4

2. Corolla actinomorphic with all lobes equal and equally divided; ovary superior, 3–5-locular .... **8. Cyananthus**
+ Corolla zygomorphic with 2 unequal lips; 2-lobed lip divided to base between lobes; ovary inferior, 2-locular .... **11. Lobelia**

3. Flowers (including ovary) to 9mm 5
+ Flowers (including ovary) 10mm or more 7

4. Leaves oblanceolate, to 4mm broad; corolla blue .... **9. Wahlenbergia**
+ Leaves ovate to suborbicular, 5mm or more broad; corolla white, sometimes with pink marking 6

5. Flowers sessile; calyx lobes tridentate; petioles to 4mm .... **4. Heterocodon**
+ Flowers pedicellate, pedicels 10–35mm; calyx lobes triangular; petioles 5mm or more .... **5. Peracarpa**

6. Fruit a capsule dehiscing by pores at base or on sides 8
+ Fruit a capsule dehiscing by valves at apex or fleshy and berry-like 10

7. Corolla divided to base .... **2. Asyneuma**
+ Corolla divided to ¾ or less 9

8. Epigynous disc present; leaves 16mm or more broad .... **3. Adenophora**
+ Epigynous disc absent; leaves to 16mm broad .... **1. Campanula**

9. Ovary 5-locular; fruit dehiscing by 5 valves; corolla 37mm or more; calyx with 10 prominent veins .... **10. Platycodon**
+ Ovary 3-locular; fruit dehiscing by 3 valves; corolla to 36mm; calyx without prominent veins 11

10. Epigynous disc with 5 club-shaped glands; corolla tubular, scarcely lobed; calyx lobes ± cordate, crenate-serrate .... **7. Leptocodon**
+ Epigynous disc without prominent glands; corolla rotate, campanulate or tubular, clearly lobed; calyx lobes not cordate, entire or not **6. Codonopsis**

1. CAMPANULA L.

Biennial or perennial herbs. Leaves alternate. Flowers solitary or in few-flowered racemes, sometimes forming panicles, cleistogamous flowers sometimes present. Calyx 5-lobed, tube adnate to ovary. Corolla 5-lobed, divided to ¾ or less. Stamens 5, free, dehiscing in bud, filaments broadening at base. Epigynous disc absent. Ovary inferior, 3-locular. Style 3-lobed. Capsule dehiscing by small pores at base or near top.

1. Plant slender, glabrous; stem unbranched; flower solitary, terminal....... 2
+ Plant slender or coarse, erect, ascending or decumbent, hairy, branched; flowers 1–6 at apex and on lateral branches 3

2. Capsule cylindrical; calyx lobes usually longer than corolla .. **1. C. aristata**
+ Capsule turbinate (developing shape apparent even at anthesis), calyx lobes usually shorter than corolla .... **2. C. immodesta**

3. Leaves linear, to 3mm wide; stem slender, scabrous; flowers 1–3 grouped at stem apex .... **3. C. sylvatica**
+ Leaves lanceolate, ovate or elliptic, sometimes linear on stem, all or majority more than 3mm wide; stem villous, hirsute or pilose; flowers 1–6 at apex and on lateral branches 4

4. Mid-cauline leaves lanceolate to linear, at least 6 × as long as broad (rarely less); flowers often clustered .... **4. C. benthamii**
\+ Mid-cauline leaves elliptic to ovate, to 5 × as long as broad; flowers never clustered 5

5. All flowers sessile or short pedicellate, pedicels to 7mm; leaves to 14mm **6. C. cana**
\+ At least some flowers long pedicellate with pedicels 7–25mm; leaves 4–50mm 6

6. Plant decumbent, softly villous; calyx lobes entire; leaves rarely more than 15mm .... **7. C. argyrotricha**
\+ Plant erect, hirsute; calyx lobes toothed; leaves usually more than 15mm **5. C. pallida**

1. C. aristata Wall.

Slender glabrous herb. Stems erect or ascending, 9–32cm, unbranched. Basal leaves petiolate, elliptic, ovate or suborbicular, 3–11 × 2.5–10.5mm, apparently caducous; petiole 9–23mm. Cauline leaves sessile, linear, 13–49 × 0.7–2mm, attenuate at base, margin with few minute teeth. Flowers terminal, solitary, deflexed. Calyx lobes linear, 3.5–12mm, usually equal to or longer than corolla. Corolla obconical, 6–13mm, pale lilac to blue; lobes oblong-ovate, 1.5–4.5mm. Ovary cylindrical, 5–14mm. Capsule cylindrical, c.17 × 3.5mm, dehiscing near top.

Bhutan: C – Thimphu and Punakha districts, **N** –Upper Mo Chu, Upper Bumthang Chu, Upper Kuru Chu and Upper Kulong Chu districts; **Sikkim:** Goecha La, Lhonak, Samdong etc. On open grassy hill slopes or among dwarf *Rhododendron*, 3656–4877m. June–September.

2. C. immodesta Lammers; *C. modesta* Hook.f. & Thomson non Schott

Similar to *C. aristata* but stems 7–18cm; basal leaves with petioles up to 45mm; cauline leaves usually smaller and often broader, 9–30 × 1–4.5mm; calyx lobes 3–5.5mm, equal to or shorter than corolla; corolla 6–7.5mm, lobes 1.5–2.5mm; ovary 5–6mm, turbinate shape apparent even at anthesis; capsule ± turbinate, c 12 × 8mm, angular above, base attenuate.

Bhutan: N – Upper Mangde Chu district (Upper Mangde Chu); **Sikkim:** Yume Samdong, Chortenima La, Chapopla etc.; **Chumbi:** Phari. In scree or grass, among shrubs, 4417–5350m. July–September.

3. C. sylvatica Wall.; *C. stricta* Wall. non L.

Scabrous herb. Stems slender, erect, 12.5–31cm, unbranched or branched. Leaves sessile, linear, 15–35 × 1–3mm, margin entire. Flowers in 1–3(–4)-

flowered terminal inflorescences. Calyx lobes triangular-acuminate, 2–9mm, c ½ × as long as corolla. Corolla campanulate, 9–15mm, pale purple to blue; lobes ovate-oblong, 3.5–8mm. Ovary 1.5–2mm. Capsule obovate to subglobose, c 4 × 5.5mm, dehiscing near base.

Bhutan: C – Thimphu district (Drugye Dzong, Chalimarphe and Thimphu Dzong), Punakha district (Ratsoo–Samtengang–Wangdu Phodrand), Mongar district (Khinay Lhakang). Grassy slopes and open pine forest, 1700–2600m. April–July.

4. C. benthamii Kitamura; *C. canescens* A. DC. non Roth

Hirsute herb. Stems erect, up to 50cm; lateral branches spreading-erect, subtended by spreading linear leaf. Leaves sessile, 20–40(–65) × 2–15mm, base attenuate, remotely serrulate, basal ones lanceolate to elliptic, cauline linear to lanceolate. Flowers 4–6 at stem apex and on lateral branches, clustered or ± lax; cleistogamous flowers usually present, c 4mm. Calyx lobes narrowly triangular, 4–8mm, ½–¾ as long as corolla. Corolla campanulate, 6–11mm, blue; lobes ovate-oblong, 3–5.5mm. Ovary 2–3.5mm. Capsule broadly obovate, c 4 × 4.5mm, dehiscing at base.

Bhutan: C – Tongsa district (S of Shamgong); **Darjeeling:** Dumsong, Rishap; **Darjeeling/Sikkim:** unlocalised Hooker collections. On roadsides or banks, 600–1850m. April.

Cephalostigma spathulatum Thwaites may provide an earlier specific epithet for this species (331).

5. C. pallida Wall.; *C. colorata* Wall., *C. ramulosa* Wall., *C. nervosa* Royle. Fig. 114g.

Hirsute suffruticose herb. Stems erect, 18–58cm, lateral branches spreading-erect. Leaves sessile, elliptic, (8–)12–37(–50) × 2–16mm, antrorsely hirsute, veins prominent on lower surface, base cuneate to attenuate, margin serrate-crenate. Flowers 1–2 at apex and tips of branches; pedicel up to 20mm. Calyx lobes triangular to triangular-acuminate, 3.5–6.5mm, usually with small teeth on margin, ½–⅔ as long as corolla. Corolla campanulate to tubular-campanulate, 8–14mm, pale blue or blue to purple; lobes ovate-oblong, 3–5.5mm. Ovary 2–2.5mm. Capsule obovate to subglobose, 3–5 × 4–7mm, dehiscing at base.

Bhutan: S – Deothang district, **C** – Ha, Thimphu, Punakha, Tongsa, Bumthang and Tashigang districts, **N** –Upper Kuru Chu district; **Darjeeling:** Badamtam, Darjeeling, Rississum; **Sikkim:** Domang, Lachen, Thanggu etc.; **Chumbi:** Galing, Yatung, upper Chumbi Valley. Open rocky outcrops, moutain slopes (sometimes among shrubs or ± wet), sandy turf, stream shingles, cultivated areas, 1850–3950m. (April), June–September, (November).

6. C. cana Wall.

Greyish-green, pubescent-hirsute herb. Stems ascending-erect, up to 24cm, lateral branches spreading. Leaves sessile, elliptic to elliptic-oblong, 6.5–14 × 2.5–7mm, margin crenulate-serrulate, lower surface whitish with prominent veins. Flowers 1(–2) at apex and tips of branches. Calyx lobes narrowly triangular, 2.5–5.5mm, ⅓–½ as long as corolla margin toothed. Corolla campanulate, 10–14mm, blue-violet to deep purple; lobes oblong to broad ovate, 3.5–7mm. Ovary 2.5–3.5mm. Capsule obovate to subglobose, c.4.5 × 4.5mm, dehiscing at base.

Bhutan: C – Thimphu district (Thimphu, Tsalimape). On dry banks, 2134–2370m. August–September.

A collection from Takila, Mongar district, Bhutan, 1850m (*Ludlow, Sherriff & Hicks* 21247, BM) differs from typical *C. cana* by its larger ovate-elliptic leaves, 10–24 × 6–13mm, with upper surface pubescent-hirsute and lower surface rather more densely white-tomentose without prominent veins; calyx lobes triangular, 5–6mm, toothed and recurved, ½–⅔ as long as corolla; corolla 11–13mm with lobes 6–8mm. Its status is uncertain.

7. C. argyrotricha A. DC.

Softly villous woody-based herb. Stems many, often decumbent, fine, branched above, up to 24cm, with spreading hairs. Leaves sessile or shortly petiolate, elliptic to ovate, 4–15(–21) × 2–8(–11)mm, margin subentire or remotely serrulate-crenulate, tomentose with long villous hairs, whitish beneath. Flowers solitary at tips of fine branches; cleistogamous flowers usually present, very small (c 2mm) on filiform branches. Calyx lobes narrowly triangular, 4–6mm, ⅓–½ as long as corolla. Corolla tubular-campanulate, 14–17mm, blue to violet; lobes ovate-oblong. 4–5mm. Ovary 1–2mm. Capsule broadly obovate, 2–3 × 3–4mm, dehiscing at base.

Bhutan: C – Thimphu district (Kumar Thang, Mem La, Tzatogang–Dotena), **N** – Upper Mo Chu district (Soe–Shaulung); **Sikkim:** Sherabthang, Tsomogo; **Chumbi:** Chumbi. Dry rocky areas, commonly growing on cliffs, 2800–3962m. May–October.

2. ASYNEUMA Grisebach & Schenk

Perennial or biennial herb. Leaves alternate. Inflorescence relatively few-flowered, spike-like. Calyx tube adnate to ovary. Corolla divided almost to base, lobes linear. Stamens 5, filaments broadening at base, anthers free. Style 3-lobed. Ovary inferior, 3-locular. Capsule dehiscing by 3 pores.

1. A. fulgens (Wall.) Briquet; *Campanula fulgens* Wall. Fig. 114h–j.

Erect herb. Stems 25–38cm, glabrous or pilose. Leaves lanceolate-elliptic, sessile or petiolate (basal leaves), 25–60 × 6–18mm, base attenuate, margin crenulate-serrulate, usually pilose at least on upper surface. Bracts linear, up to

11mm. Calyx lobes linear, 5.5–8mm, straight or recurved at tips. Corolla 6–8.5mm, pale blue; lobes linear-oblong, slightly spreading. Anthers 2.5–3.2mm. Ovary 3.5–4mm. Capsule ± obovoid, 6–11 × 4–5mm, dehiscing laterally near apex.

Bhutan: S – Chukka district, **C** – Thimphu, Punakha, Tongsa, Bumthang and Mongar districts; **Sikkim:** Lachung; **Chumbi:** Chumbi. Open grassy meadows or grassy areas among shrubs or *Pinus wallichiana*, 1219–3048m. June–September.

3. ADENOPHORA Fischer

Perennial with thickened roots. Leaves alternate. Inflorescence spike-like, sometimes branched. Calyx 5-lobed, tube adnate to ovary. Corolla 5-lobed, divided to half or less. Stamens 5, filaments broadening at base, anthers free. Disc cup-shaped, surrounding base of style. Ovary inferior, 3-locular. Style 3-lobed. Capsule dehiscing by lateral pores.

1. A. khasiana (Hook.f. & Thomson) Collett & Hemsley. Fig. 115a–b.

Erect or suberect herb. Stems 40–75cm, subglabrous, branched or not. Leaves sessile, ovate-elliptic to elliptic, 4–6.5 × 1.6–2.7cm, base cuneate or attenuate, margin serrate, subglabrous to puberulent, sometimes pilose on veins below. Flowers subsessile, lax. Calyx lobes linear-lanceolate, 5–8.5mm, margin toothed, glabrous. Corolla obconical-campanulate, 12–19mm, pale blue to lilac; lobes broad ovate to oblong, up to 5.5mm. Capsule subglobose, c 7 × 5mm.

Bhutan: C – Tongsa and Tashigang districts, **N** –Upper Mo Chu, Upper Mo Chu and Upper Kulong Chu districts. On banks on exposed slopes or open woodland, 1981–2591m. July–October.

4. HETEROCODON Nuttall

Trailing herb. Leaves alternate, sessile to shortly petiolate. Flowers axillary, inconspicuous. Calyx 5-lobed, tube adnate to ovary. Corolla 5-lobed. Stamens with filaments broadening at base, anthers free. Ovary inferior. Style 3-lobed. Fruit a capsule.

1. H. brevipes (Hemsley) Handel-Mazzetti & Nannfeldt; *Wahlenbergia brevipes* Hemsley, *Homocodon brevipes* (Hemsley) Hong

Subglabrous, prostrate herb. Stems 25–35cm, usually unbranched above. Leaves broad ovate to suborbicular, 6–9 × 5–8mm, base cuneate to truncate,

FIG. 115. **Campanulaceae.** a–b, *Adenophora khasiana*: a, flowering shoot (× ½); b, flower cut open (× 1½). c–d, *Peracarpa carnosa*: c, flowering shoot (× ½); d, flower (× 3). e, *Leptocodon gracilis*: flowering spur (× ½). f–g, *Codonopsis viridis*: f, section of stem (× ½); g, flower cut open (× ⅔). Drawn by Glenn Rodrigues.

a
b
c
d
e
f
g

margin serrate; petiole up to 4mm. Flowers sessile, 1–2 in leaf axils. Calyx 3–4.5mm; lobes tridentate, 1.5–2.5mm. Corolla campanulate with spreading lobes, c 4mm, white; lobes ± ovate, c 1.8mm. Capsule ± obovate, c 2.5 × 1mm.

Bhutan: N – Upper Kuru Chu district (Denchung). Trailing over boulders, 2134m. July.

5. **PERACARPA** Hook.f. & Thomson

Annual herb. Leaves alternate, petiolate. Flowers axillary, solitary, pedicellate. Calyx (4–)5-lobed, tube adnate to ovary. Corolla (4–)5-lobed, divided to c ⅔. Stamens 5, filaments hairy, anthers free. Ovary inferior. Style 3–4-lobed. Epigynous disc present. Fruit dry, rupturing to release seeds.

1. P. carnosa (Wall.) Hook.f. & Thomson; *Campanula carnosa* Wall. Fig. 115c–d.

Stems weakly ascending to erect, 6–15cm, glabrous. Leaves ovate, 10–30 × 9–20mm, sparsely appressed-hairy on both surfaces, margin serrate-crenate; petiole 5–15mm. Pedicel 10–35mm. Calyx 2.5–4mm; lobes narrowly triangular, 1–2mm. Corolla campanulate, 5–6mm, white with pink marking, lobes ± oblong. Fruit ± elliptic, c 5 × 3mm, with thin papery walls.

Bhutan: C – Punakha district (Dochu La, Menchuang, Ritang–Ratsoo), Tongsa district (Chendebi), Bumthang district (Lami Gompa), **N** – Upper Kulong Chu district (Tobrang); **Darjeeling:** Ghumpahar, Tonglu; **Sikkim:** Chaltakpur etc. Damp shaded areas in broad-leaved or *Tsuga dumosa* forest, 1850–3200m. April–June.

6. **CODONOPSIS** Wall.

Perennials, often with tubers, usually smelling foetid. Stems twining, erect, sprawling or trailing. Leaves alternate and subopposite, petiolate. Flowers solitary, terminal or axillary. Calyx 4–6-lobed, usually with tube adnate to ovary, rarely free. Corolla 4–6-lobed. Stamens usually 5, filaments broadening at base, anthers free. Ovary ± inferior, 3-locular. Style 3-lobed. Capsule somewhat fleshy when young, dehiscing by 3 valves which form beak at apex or apparently indehiscent.

The circumscription of *Codonopsis* and status of *Campanumoea* Blume, *Cyclocodon* Griffith and *Leptocodon* require revision: see (350) and (380).

1. Plant erect, sprawling or trailing 2
+ Plant twining 10

2. Corolla less than 8mm; fruit without beak 3
+ Corolla more than 11mm; fruit with beak 4

3. Calyx free from ovary, separated from fruit by short stipe; leaves 1–2.5cm broad, base rounded-cuneate .... **1. C. parviflora**
\+ Calyx adnate to ovary, up to half adnate to fruit; leaves 1.5–3.4cm broad, base usually rounded to truncate .... **2. C. lancifolia**

4. Leaves about as long as wide, mainly on fine lateral branches near base of stem; upper stem rarely branched, ± leafless 5
\+ Leaves at least 1½ × longer than wide, on main stem or lateral branches; upper stem branched or not, usually leafy 7

5. Leaves less than 5mm; corolla tubular, at least 3 × as long as calyx lobes **3. C. thalictrifolia**
\+ Leaves 5mm or more; corolla campanulate to ± globose, twice as long as calyx lobes 6

6. Corolla tube 20–30mm broad at throat; corolla lobes broad ovate, 6mm or more .... **4. C. foetens**
\+ Corolla tube 10–15mm broad at throat; corolla lobes triangular-ovate, to 6mm .... **5. C. bhutanica**

7. Leaves glabrous or virtually glabrous; corolla red-purple or blue, campanulate, 23mm or more 8
\+ Leaves hairy; corolla green, white or cream with purple marking, campanulate or tubular, if more than 20mm then tubular 9

8. Calyx lobes linear, 7–9mm; corolla lobes 5–7mm; corolla blue **6. C. dicentrifolia**
\+ Calyx lobes narrow triangular, 15–18mm; corolla lobes 10–15mm; corolla red-purple .... **7. C. purpurea**

9. Corolla tubular, green with purple veins, about twice as long as calyx lobes **8. C. benthamii**
\+ Corolla globose-campanulate, white or cream with purple marking, scarcely exceeding calyx lobes .... **9. C. subsimplex**

10. Corolla rotate, divided almost to base, lobes 14mm or more; leaves glabrous **10. C. vinciflora**
\+ Corolla campanulate or tubular-campanulate, divided less than halfway, lobes 10mm or less, leaves glabrous or hairy 11

11. Leaves glabrous or almost glabrous on lower surface 12
\+ Leaves densely hairy on lower surface 13

12. Calyx tube adnate to base of ovary only, about 1mm; calyx lobes 16mm or more .... **11. C. javanica**
\+ Calyx tube adnate to side wall of ovary, 5–8mm; calyx lobes to 12mm **12. C. inflata**

13. Corolla c 10mm, c 10mm broad at throat; calyx lobes to 8mm **13. C. affinis**
\+ Corolla 17–27mm, 20–30mm broad at throat; calyx lobes 9mm or more **14. C. viridis**

1. C. parviflora A. DC.; *Campanumoea parviflora* (A. DC.) Bentham

Stems erect or somewhat sprawling, 60–160cm, glabrous, branching above and below. Leaves opposite, ovate-lanceolate, 3.2–11 × 1–2.5cm, acuminate, base rounded-cuneate, margin serrate, glabrous, lower surface greyish-white; petiole 2–4mm. Peduncle 1–3cm. Calyx 4–6mm, not adnate to ovary, separated from fruit by short stipe, tube absent; lobes 4, linear, serrulate or with small subsidiary lobes. Corolla broad campanulate to rotate, 5–6mm, white or pink, 4-lobed; lobes broad ovate, 3–3.5mm. Fruit ± globular, 6–7mm, apparently indehiscent.

Bhutan: S – Deothang district (Samdrup Jongkar–Deothang (117), Narfong–Deothang (117)); **Darjeeling:** Sitong, Sureil, Birik etc. 610–1676m. September–October, January.

2. C. lancifolia (Roxb.) Moeliono subsp. **celebica** (Blume) Moeliono; *Campanumoea celebica* Blume, *Cyclocodon celebicus* (Blume) Hong

Differs from *C. parviflora* by its ovate-acuminate leaves, 1.5–3.4cm broad, with base usually rounded to truncate; calyx 5–6-lobed, adnate to ovary at base, up to half adnate to fruit; corolla often slightly larger; fruit up to 10mm.

Bhutan: C – Punakha district (Chusom); **Darjeeling:** Rangit, Sureil, Darjeeling etc.; **Sikkim:** Linchyum. In open areas or forest margins, often in damp places, 305–1524m. September, November.

Himalayan specimens are referred to *Cyclodon lancifolius* (Roxb.) Kurz by Hong & Pan (350), rather than to *Cyclodon celebicus.*

3. C. thalictrifolia Wall.

Fleshy tap-root. Stems very slender near base, erect or ascending, 18–40cm, with slender lateral branching near base, ± simple flowering stem above, sparsely pilose. Leaves on lateral branching, subopposite, ovate to ± reniform, 0.2–0.45 × 0.25–0.5cm, base usually truncate, upper and lower surface pubescent, margin entire to crenulate, greyish-green; petiole 1–2mm. Peduncle up to 10cm. Calyx 9–11mm; lobes oblong 5.5–7.5mm, somewhat pubescent. Corolla tubular, flared at mouth, 20–36mm, blue or lavender with purple ring at base; lobes broad ovate, 2.5–6.5mm. Fruit 12–18mm, hemispherical with 7–12mm beak.

Sikkim: Dzongri, Zemu Valley, Bikbari etc. 3810–4267m. August–September.

4. C. foetens Hook.f. & Thomson

Stems ascending, 20–40cm, glabrescent with fine lateral branches near base. Leaves ovate, 0.5–1.1 × 0.4–0.8cm, greyish-green, base truncate to ovate, upper and lower surface pubescent, margin subentire; petiole 1–3mm. Peduncle 2–14cm. Calyx 12–17mm; lobes oblong-elliptic, 9–12mm, margin slightly sinuate. Corolla campanulate or ± globose, 19–26mm, mauve or blue; lobes broad ovate, 6–10mm, extreme tip hooded. Fruit 14–20mm, hemispherical to obconical with 6–9mm beak.

Bhutan: C – Thimphu and Tongsa districts, **N** –Upper Mo Chu, Upper Bumthang Chu, Upper Kuru Chu and Upper Kulong Chu districts; **Sikkim:** Chakung Chhu, Kupup, Lasha Chhu etc.; **Chumbi:** Chumolari etc. In moist peat and turf or scree, 2890–4877m. July–September.

5. C. bhutanica Ludlow

Thick fleshy tap-root. Stems ascending to erect, 15–45cm, pilose to ± glabrous, with fine branches especially near base. Leaves ovate, 0.7–2.3 × 0.5–2.3cm, base cordate to truncate, upper and lower surface ± densely pilose, margin ± entire to crenulate-sinuate; petiole 1–5mm. Peduncle 4–13cm. Calyx 9–13mm, pilose; lobes elliptic-oblong, 6–9mm, margin recurved. Corolla campanulate, 13-19mm, deep reddish-purple or pale blue with red-purple ring at mouth; lobes triangular-ovate, 3.5–6mm, extreme tip often hooded. Fruit 15–20mm, hemispherical with beak 5–7mm.

Bhutan: C – Thimphu, Punakha and Tongsa districts, **N** – Upper Mo Chu, Upper Mangde Chu, Upper Bumthang Chu and Upper Kulong Chu districts. On open grassy hillsides or on scree, 3810–4570m. July–August.

6. C. dicentrifolia (Clarke) W.W.Smith; *Wahlenbergia dicentrifolia* Clarke

Stems decumbent to suberect, 28–60cm, glabrous, branched. Leaves ovate, 1.3–3.9 × 0.8–2cm, glabrescent above, glaucous below, margin ± entire, somewhat sinuate. Peduncle 5–7.5cm. Calyx 14–18mm; lobes linear-oblong, 7–9mm. Corolla campanulate, 23–30mm, blue; lobes broad ovate, 5–7mm. Fruit 20–30mm, obconical with beak 5–8mm.

Darjeeling: Sandakphu; **Darjeeling/Sikkim:** Phalut; **Sikkim:** Gnatong. On sandy banks in shade, 3353–3962m. August.

7. C. purpurea Wall.

Bushy tuberous herb. Stems erect or ascending, up to 60–80cm, glabrous, smooth, jointed at nodes. Leaves opposite, ovate to lanceolate, 3.2–8 × 1.4–3cm, base rounded to cuneate, glabrous, lower surface glaucous, margin entire to sinuate; petiole 3–8mm. Peduncle up to 3.5cm. Calyx 22–25mm, glabrous; lobes narrow triangular, 15–18mm. Corolla campanulate, 23–32mm, red-purple, lobes triangular-ovate, 10–15mm. Fruit c 17mm, hemispherical with beak 8–10mm.

Bhutan: C – Tongsa district (Dotena, Tongsa, Tongsa Bridge etc.); **Sikkim:**

Chunthang. Under low moist forest or in exposed areas among rocks and in grass, 1829–2134m. August–September.

8. C. benthamii Hook.f. & Thomson

Sprawling herb. Stems ± erect with upper part trailing, 20–70cm, sparsely pilose. Leaves ovate to elliptic, 3–7.5 × 1.5–5.4cm, base truncate to cuneate, upper and lower surface hirsute, margin serrate or lobed-serrate; petiole 4–35mm. Peduncle 4–45mm. Calyx 12–18mm; lobes ovate 9–14mm, denticulate. Corolla tubular, 17–26mm, green sometimes with dull purplish veins; lobes triangular, 3–4mm. Fruit 15–22mm, truncate-globose with beak 6–8mm.

Bhutan: C – Thimphu district (Shingkarap, Thimphu Chu), Tongsa district (Omta Tso, Rinchen Chu), Mongar district (Rip La, Sengor–Sheridrang); **Sikkim:** Zemu Valley, Lagyap, Lachen etc.; **Chumbi:** Simdoongtang. On banks in moist *Tsuga*/broad-leaved forest, 1219–3810m. July–August.

9. C. subsimplex Hook.f. & Thomson

Stems ascending, upper part trailing, c 30–60cm, glabrescent. Leaves ovate to ovate-acuminate, 1.8–7.5 × 1.2–5cm, base often slightly unequal, cuneate to rounded, upper surface pilose, lower surface pilose, glaucous, margin usually shallowly serrate-crenate; petiole (2–)6–25(–32)mm. Peduncle 5.5–19cm. Calyx 15–22mm, often glaucous; lobes ovate to ovate-oblong, 8–17mm, denticulate. Corolla globose-campanulate, 12–18mm, white to cream with purple markings inside; lobes broad ovate, 3–6mm. Fruit c 14mm, ± hemispherical with beak c 7mm.

Bhutan: C – Ha, Thimphu, Tongsa and Bumthang districts, **N** – Upper Mangde Chu and Upper Kulong Chu districts; **Darjeeling:** Singalila; **Sikkim:** Changu, Sherabthang, Gnatong etc.; **Chumbi:** Chumbi, Yatung. On grassy slopes and in damp areas among shrubs or in fir forest, 3350–4100. July–September.

10. C. vinciflora Komarov; *C. convolvulacea* Kurz var. *vinciflora* (Komarov) Shen

Stems twining, 2–3m. Leaves ovate to ovate-oblong, 2.5–5.8 × 0.4–2.3cm, base cuneate, truncate or cordate, upper and lower surface glabrous, margin serrately lobed especially near base, rarely obscurely serrate; petiole up to 15mm. Peduncle 2–12cm. Calyx 8–18mm, glaucous; lobes ovate to ± oblong, 4–13mm. Corolla rotate, 15–23mm, blue or purple with red circle near base; lobes ± elliptic, 14–21mm. Fruit c 14mm, obovate with beak c 4mm.

Bhutan: C – Ha (Ha To), Thimphu (Bela La, Cheka, Duke La), Bumthang (Bumthang) districts; **Sikkim:** Galing. Among shrubs and bamboo, 2743–3658m. August–September.

11. C. javanica (Blume) Hook.f.; *Campanumoea javanica* Blume

Stems twining, 2–4m. Leaves ovate, 4–6 × 2.5–5.5cm, deeply cordate at base, upper and lower surface glabrous (?sometimes glaucous), margin obscurely crenate-serrate; petiole c 45mm. Peduncle 1–2cm. Calyx 17–20mm, adnate to base of ovary only; lobes ± oblong, 16–19mm. Corolla campanulate, 22–24mm, cream with purple veins; lobes broadly triangular, c 8mm. Fruit 6mm, hemispherical, purple-black, glaucous.

Bhutan: S – Deothang district (Keri Gompa), **C** –Tashigang district (Shali), **N** – Upper Mo Chu district (Gasa); **Darjeeling:** Rishap; **Darjeeling/Sikkim:** unlocalised (Hooker and Gamble collections). Among scrub in open country, 1050–2438m. August–October.

12. C. inflata Hook.f. & Thomson; *Campanumoea inflata* (Hook.f. & Thomson) Clarke

Stems twining, to 5m. Leaves ovate, 4.3–13 × 2.1–8cm, deeply cordate at base, upper surface sparsely pubescent, lower surface glabrescent and glaucous, margin obscurely dentate; petiole 30–75mm. Peduncle 1.5–3(–7)cm. Calyx 15–17mm, lobes triangular-acuminate, 8–12mm. Corolla broadly tubular-campanulate, 24–26mm, pale sordid yellow, veins purple, lobes triangular, c 10mm. Fruit c 16mm, subglobose, purple-black, glaucous.

Darjeeling: Jorpokri, Tonglu, Kurseong etc.; **Sikkim:** Chhateng. Climber in forest, 1350–2750m. August–October.

13. C. affinis Hook.f. & Thomson

Stems twining. Leaves ovate, 4–7 × 2.2–4.7cm, deeply cordate at base, lower surface densely pubescent-tomentose, upper surface sparsely pubescent, margin usually obscurely crenate-dentate; petiole 10–25mm. Peduncle 1.5–3cm. Calyx 10–15mm; lobes ± oblong, 7–8mm. Corolla campanulate, c 10mm, greenish with apex of lobes reddish; lobes triangular, c 4mm. Fruit 11–12mm, hemispherical, with beak 2–3mm.

Bhutan: C – Tongsa district (Chendebi), **N** – Upper Mo Chu district (Gasa dzong, Gicha); **Darjeeling:** Phalut–Ramman; **Sikkim:** Chiya Bhanjang, Kalej Khola, Lachung. Among bamboo, 2150–3350m. August–September.

14. C. viridis Wall. Fig. 115f–g.

Stems twining, 3–4m. Leaves ovate, 2.7–5.5 × 1.6–3.5cm, base truncate to shallowly cordate, upper surface pubescent, lower surface pubescent-tomentose, greyish-green, margins ± entire; petiole 7–18mm. Peduncle 2.5–5.5cm. Calyx 16–24mm; lobes oblong-elliptic, 9–13mm, recurved, usually folded lengthwise. Corolla broad campanulate, 17–27mm, cream to greenish-yellow suffused reddish-purple at base and on lobes; lobes broad ovate, 7–9mm. Fruit c 17mm, hemispherical, with beak c 7mm.

Bhutan: S – Gaylephug (117) and Deothang districts, **C** – Thimphu and Tongsa districts, **N** – Upper Mo Chu district; **Darjeeling:** Victoria Falls. Climbing over shrubs, 1500–3050m. August–October.

7. LEPTOCODON (Hook.f. & Thomson) Lemaire

Twining herb. Leaves alternate and opposite, petiolate. Flowers solitary, terminal and axillary. Calyx 5-lobed; tube short, adnate to ovary. Corolla shallowly 5-lobed. Stamens 5, filaments broadening at base, anthers free. Ovary semi-inferior. Epigynous disc with 5 prominent club-shaped glands. Style 3-lobed. Fruit a capsule, 3-locular, dehiscing by 3 valves.

Separation from *Codonopsis* dubious: see (380).

1. L. gracilis (Hook.f. & Thomson) Lemaire; *Codonopsis gracilis* Hook.f. & Thomson. Fig. 115e.

Slender, sparsely pilose herb. Stems up to 2.5m. Leaves ovate to broad ovate, 7–20 × 6–20mm, base usually truncate, margin serrate-crenate; petiole up to 28mm, very fine. Pedicel 20–35mm. Calyx 5–7mm; lobes ± cordate, 4–6mm, crenate-serrate. Corolla tubular, 27–35mm, mauve to violet-blue, paler towards base; lobes very broad ovate, up to 1.5mm. Capsule c 14mm, broad ovoid, with beak c 6mm.

Bhutan: C – Tongsa and Tashigang districts, **N** – Upper Mo Chu, Upper Kuru Chu and Upper Kulong Chu districts; **Sikkim:** Karponang, Chunthang, Menshithang etc. Oak forest, 1829–2743m. July–September.

8. CYANANTHUS Bentham

Low annual or perennial herbs. Leaves alternate. Flowers terminal, solitary. Calyx 5-lobed, tube not adnate to ovary. Corolla 5-lobed. Stamens 5. Ovary superior, 3–5-locular. Style 3–5-lobed. Capsule dehiscing by (3–)4–5-valves at apex.

1. Calyx densely black-brown hairy 2
\+ Calyx glabrous or white or yellow-brown hairy 3

2. Leaves cuneate-spathulate, conspicuously lobed .... **1. C. lobatus**
\+ Leaves oblong or oblong-lanceolate, not lobed .... **2. C. pedunculatus**

3. Flowers yellow; leaves spathulate .... **4. C. spathulifolius**
\+ Flowers blue or violet-blue; leaves lanceolate, elliptic, broadly ovate or spathulate 4

4. Perennial with thick fleshy root-stock, tufted .... **3. C. incanus**
\+ Annual, sometimes perennating but never forming a thick root-stock, not tufted 5

5. Flowers 10.5mm or more; calyx 7mm or more, inflated especially after anthesis .... **5. C. inflatus**
\+ Flowers 10mm or less; calyx 7mm or less, not inflated .... **6. C. hookeri**

1. C. lobatus Bentham. Fig. 116a

Perennial. Stems decumbent, 10–40cm, subglabrous to pilose. Leaves cuneate-spathulate, 10–32 × 5–17mm, with 3–5 lobes at apex and sometimes subsidiary lobing, subglabrous to appressed pilose above, pilose to ± lanate or rarely glabrous below. Pedicel 8–20(–30)mm, densely brown-black villous. Calyx broadly tubular, 13–20mm; lobes triangular, 4–5.5mm, densely brownish-black villous. Corolla broadly funnel-shaped, 31–40mm, blue, rarely white; lobes broad obovate, 12–17mm. Capsule ovate-acuminate, ± equal to calyx.

Bhutan: C – Thimphu, Tongsa and Tashigang districts, **N** – Upper Mo Chu, Upper Pho Chu, Upper Mangde Chu, Upper Bumthang Chu, Upper Kuru Chu and Upper Kulong Chu districts; **Sikkim:** Sherabthang, Tsomgo, Dzongri etc.; **Chumbi:** Chaklung La, Chugya, Yatung etc. In meadows, on banks or hillsides, in turf, on peat, sands, gravel or among rocks, in open, among shrubs or in forest glades, 3200–4900m. June–September.

2. C. pedunculatus Clarke

Perennial with woody root-stock. Stems procumbent, 7–26cm, pilose-pubescent, unbranched. Leaves sessile, oblong, sometimes elliptic towards base, 4.5–16 × 2–4.5mm, margin ± entire, pilose-pubescent on both surfaces. Pedicel 15–35mm with spreading brown-black hairs. Calyx tubular, 10–16mm, densely brown-black villous; lobes narrowly triangular, 4.5–6mm. Corolla campanulate, 24–37mm, violet-blue, throat glabrous; lobes obovate, 10–12mm. Capsule ovate, equal to calyx.

Bhutan: C – ?Ha district (Chira: Gould collection); **Sikkim:** Lhonak, Tangka La, Dzongri etc.; **Chumbi:** Dotag, Kupup, Yadong (406) etc. On banks and open hillsides, 3962–5125m. July–September.

A collection from Dzongri, Sikkim is doubtfully referred here. It resembles *C. microphyllus* Edgeworth in its firmer rather coriaceous leaves subglabrous above and short (3.5mm) calyx teeth, but lacks the (densely) hairy corolla throat and much reduced pedicels of that species. It differs from either species by its subglabrous leaf undersurface.

3. C. incanus Hook.f. & Thomson

Perennial with fleshy root-stock. Stems prostrate, 3–16cm, white pilose to tomentose. Leaves sessile to shortly petiolate, lanceolate, elliptic or spathulate, (4–)5–9 × (1–)1.8–3.5mm, appressed-hairy above, white tomentose below, margin ± entire, often recurved. Pedicel up to 6mm, glabrous or yellowish-brown hairy. Calyx tubular, 5–9mm, ± glabrous or densely white or yellowish-brown villous outside; lobes narrowly triangular, 2–3.5mm, always hairy inside. Corolla tubular-campanulate, 13–20(–28)mm, deep blue to violet, rarely white, throat densely villous; lobes ± oblong, 7–10(–14)mm. Capsule ovate, ± equal to calyx.

subsp. **incanus**; *C. incanus* var. *trichocalyx* Franchet

Villous hairs (often sparse) covering whole of calyx outside.

Bhutan: N – Upper Mo Chu district (Kohina and Laya), Upper Mangde Chu district (Passu Sepu); **Sikkim:** Lhonakh, Gochung, Dadong etc.; **Chumbi:** Phari and Yadong districts (349). On banks and open grassy hillsides, 3290–4877m. July–September.

subsp. **orientalis** Shrestha; *C. incanus* var. *leiocalyx* auct. non Franchet

Villous hairs absent from most or all of calyx exterior; when present, sparse, localised, mostly confined to lobes and/or 5 main nerves.

Bhutan: C – Thimphu district, **N** – Upper Mo Chu, Upper Mangde Chu, Upper Bumthang Chu and Upper Kuru Chu districts; **Sikkim:** Bikbari, Lhonakh, Yampung etc.; **Chumbi:** Buchum, Kalaeree, Phari etc. Open grassy areas, 3962–4877m. July–September.

A recent treatment has sunk this subspecies in *C. spathulifolius* (347). A collection from Chumbi village has been identified by Shrestha as *C. leiocalyx* Franchet subsp. *lucidus* Shrestha, a taxon normally distinguished by corollas exceeding 30mm.

4. C. spathulifolius Nannfeldt

Perennial. Stems prostrate to procumbent, 5–14cm, white-pilose to -villous. Leaves subrhomboid-spathulate, 6–11 × 3–6mm, villous to pilose above and below, sometimes glabrescent above, margin entire to crenulate near apex. Pedicel up to 15mm, spreading-hairy. Calyx tubular-globose, 7–11mm, glabrescent with few white or yellow-brown villous hairs on outside; lobes narrowly triangular, 2–3.5mm, villous on inside. Corolla tubular-campanulate, 15–25mm, yellow, throat densely villous; lobes obovate-oblong, 6–8mm. Capsule ovate, slightly exceeding or ± equal to calyx.

Bhutan: C – Bumthang and Tashigang districts, **N** – Upper Mangde Chu and Upper Kulong Chu districts; **Sikkim:** Yumthang, Kangpupchuthang, Bikbari etc.; **Chumbi:** Chomo Lhari, Shamchen. 3350–4900m. June–September.

This taxon has been recently been transferred to *C. macrocalyx* Franchet subsp. *spathulifolius* (Nannfeldt) Shrestha (406).

5. C. inflatus Hook.f. & Thomson

Annual. Stems 1–many, prostrate to procumbent, (6–)10–56cm, thinly pilose, glabrescent, branched. Leaves broadly ovate to semicircular or suborbicular, 3.5–13 × 3.5–12.5mm, margin ± entire to sinuate-dentate, base usually trunc-

---

FIG. 116. **Campanulaceae.** a, *Cyananthus lobatus*: habit (× ½). b–c, *Lobelia nummularia*: b, habit (× ⅔); c, flower (× 6). d–e, *Lobelia erectiuscula*: d, flowering shoot (× ½); e, flower (× 3). Drawn by Glenn Rodrigues.

a
b
c
d
e

ate-attenuate, shortly appressed-hairy above, pilose below. Flowers solitary at apex and on side-branches. Calyx inflated especially in fruit, ovate, 7–11mm, brown or white villous; lobes ovate-oblong, 2–3mm. Corolla tubular with spreading lobes, 10.5–18mm, pale blue to lilac-blue, hairy in throat; lobes oblong-elliptic, 3–5mm. Capsule ovate, 6–12mm, scarcely exceeding calyx.

Bhutan: C – Thimphu, Tongsa and Tashigang districts, **N** – Upper Mo Chu, Upper Bumthang Chu and Upper Kulong Chu districts; **Sikkim:** Bakhim, Nathang, Namgaythang etc.; **Chumbi:** Chumbi, Gowsar, Phari etc. Open areas in thin turf and fir or oak forest, 2850–4722m. August–September.

6. C. hookeri Clarke

Similar to *C. inflatus* but with more compact habit, always branched at base as well as above; stems up to 18cm; leaves 4–8 × 3–4.5mm; flowers usually smaller, 10mm or less; calyx not inflated, 4–7mm, densely brown or white-villous, divided ± halfway; corolla lobes with single bristle at tip; capsule ± equal to calyx.

Bhutan: C – Bumthang district, **N** – Upper Mo Chu, Upper Pho Chu and Upper Kulong Chu districts; **Sikkim:** Gymopchi, Naku La, Sebu Valley (347); **Chumbi:** Chubitang, Dotag, Yadong etc. In turf on exposed hillsides, 3048–3990m. August–September.

9. WAHLENBERGIA Roth

Annual or perennial herbs. Leaves sessile, alternate. Flowers terminal or axillary, pedicillate. Calyx 5-lobed. Corolla actinomorphic, 5-lobed. Stamens free. Ovary inferior. Style 3-lobed. Fruit a capsule, 2–3-locular, dehiscing by 2–3 valves.

1. W. marginata (Thunberg) A. DC.; *W. gracilis* Forster (Schrader) A. DC.

Glabrous to sparsely hairy. Stems erect to procumbent, 8–60cm, branched at base and above. Leaves oblanceolate, 8–20 × 2–4mm, margin undulate, obscurely denticulate, sparsely pilose. Flowers erect. Pedicel up to 7cm. Calyx 3–5mm; lobes ± linear, 1.4–2.5mm. Corolla campanulate, 4–7mm, blue; lobes ovate, 2–5mm. Capsule obconical, 5–7mm; seeds compressed-ellipsoid, 0.5mm.

Darjeeling/Sikkim: unlocalised (Gamble collection).

W. hirsuta (Edgeworth) Tuyn [*Cephalostigma hirsutum* Edgeworth] has been recorded from Bhutan between Gomkhar (? = Gomchu) and Shali, (?Tashigang district), in open forest (117). This species can appear very similar to *W. marginata*, but the seeds are trigonous, c 1mm. Other Himalayan records for it seem confined to the western half.

10. PLATYCODON A. DC.

Herb. Leaves alternate or sometimes whorled, sessile. Flowers terminal, solitary. Calyx 5-lobed, tube adnate to ovary. Corolla 5-lobed, large, ⅓–½-divided.

Stamens 5, broadening at base, anthers free. Ovary inferior. Style 5-lobed. Fruit a capsule, 5-locular, dehiscing by 5 apical valves.

1. P. grandiflorum (Jacquin) A. DC.

Glabrous herb. Stems erect, c 1m. Leaves ovate-elliptic, 4–6.5 × 1.4–2.5cm, upper often lanceolate, base cuneate, margin serrate-dentate. Flowers few, erect. Calyx 15–20mm, prominently ribbed; lobes narrowly triangular, 10–12mm, entire. Corolla obconical, 3.7–4cm, dark purplish-blue; lobes broad ovate, 15–20mm. Capsule obovate with conical apex.

Bhutan: C – Thimphu district (Paro). Cultivated in garden, 2500m. June.

11. LOBELIA L.

Low or tall herbs. Leaves alternate. Flowers zygomorphic, axillary, often resupinate on twisted pedicel. Calyx 5-lobed, tube adnate to ovary. Corolla 2-lipped; 2-lobed lip divided to base of corolla; other lip 3-lobed, not divided to base. Stamens 5; anthers fused, at least shorter anthers with bristles at apex. Ovary inferior, 2-locular. Fruit a capsule, dehiscing by 2 valves at apex or a berry crowned with calyx lobes. Seeds numerous, minute.

Species 5 and 12 below are not closely related, but have previously been segregated in *Pratia* Gaudichaud-Beaupré because of their fleshy fruit.

1. Fruit a dry capsule, dehiscing by 2 valves at apex 2
\+ Fruit fleshy, berry-like 11

2. Corolla to 6mm; calyx to 8mm; plant procumbent, ascending or suberect 3
\+ Corolla 14mm or more; calyx 10mm or more; plant erect 6

3. Shorter 2 anthers only bearded at apex .... **1. L. heyneana**
\+ All anthers bearded or with bristles at apex 4

4. No bracteoles at base of pedicel; corolla 5mm or more; capsule 5mm or more .... **4. L. zeylanica**
\+ Two small bracteoles at base of pedicel; corolla to 5mm; capsule to 5mm 5

5. Plant glabrous .... **2. L. alsinoides**
\+ Plant hairy .... **3. L. terminalis**

6. Corolla 22mm or more; capsule ovoid or oblong-ellipsoid, c 11mm 7
\+ Corolla to 18mm; capsule ovoid or globose, 6–9mm 9

7. Corolla white tinged pink or mauve; found below 1500m .... **9. L. rosea**
\+ Corolla blue or violet; found above 1500m 8

8. Stem and leaves subglabrous to puberulent, leaf veins often pubescent **7. L. seguinii**
\+ Stem and leaves stiffly hairy .... **8. L. tibetica**

9. Stem 15–18mm broad, densely leafy; inflorescence conferted, lanate, with many pendulous bracts; capsule ovoid .... **11. L. nubigena**
\+ Stem 3–8mm broad, not densely leafy; inflorescence lax to fairly dense, ± glabrous to pubescent, with spreading bracts; capsule globose 10

10. Calyx lobes 10mm or more; pedicel 10mm or more; stem glabrous, branched above .... **6. L. pyramidalis**
\+ Calyx lobes to 9mm; pedicel to 8.5mm; stem pubescent above, unbranched **10. L. erectiuscula**

11. Plant creeping; stems hairy; leaves ovate to suborbicular, 1.5cm or less **5. L. nummularia**
\+ Plant erect; stems glabrous; leaves elliptic-acuminate, 6.5cm or more **12. L. montana**

1. L. heyneana Roemer & Schultes; *L. trialata* Hamilton

Glabrous or subglabrous annual herb. Stems ascending to erect, 9–17cm, narrowly winged. Leaves ovate to elliptic, 15–17 × 5–10mm, upper narrower than lower, base cuneate or truncate with lamina decurrent on short petiole, margin subentire to serrulate-crenulate; petiole up to 4mm. Pedicel up to 18mm. Calyx 5–6mm; lobes linear, 1.5–2mm. Corolla 5–6mm, pink or lilac. Shorter anthers only bearded at apex. Capsule c 4.5 × 2.5mm. Seeds ellipsoid, sometimes rounded trigonous, 0.4mm.

Bhutan: unlocalised (Griffith collection); **Darjeeling:** Kalimpong; **Darjeeling/Sikkim:** unlocalised (Hooker collection). 914–1524m.

Material seen from our area belongs to var. **heyneana**.

2. L. alsinoides Lamarck; *L. trigona* Roxb.

Glabrous annual herb. Stems procumbent to ascending, 10–27cm, triangular. Leaves sessile or subsessile, suborbicular to ovate, 7–17 × 5–17mm, upper narrower than lower, base truncate or cuneate, margin usually serrulate. Pedicel up to 25mm with 2 small bracts at base. Calyx 4–5mm; lobes linear, 2.5–3.5mm. Corolla 3.5–4.5mm, blue to lilac. All anthers bearded at apex. Capsule 3.5–5 × 2.5–3mm. Seeds angularly trigonous, 0.5mm.

Darjeeling: unlocalised (Cowan collection).

Material seen from our area belongs to var. **alsinoides**.

3. L. terminalis Clarke; *L. terminalis* Clarke var. *minuta* Clarke

Hirsute herb. Stems procumbent to suberect, 3–28cm, angular. Leaves mostly broad-ovate, 10–16 × 11–17mm, base subcordate or truncate, margin dentate-serrate, hirsute somewhat glabrescent above; uppermost leaves ± oblong, much smaller; petiole 2–5mm. Flowers solitary in upper leaf axils. Pedicels 6–24mm, fine, 2 small bracteoles at base. Calyx hirsute, lobes 1–2.5mm. Corolla 2.5–4mm, blue-violet. All anthers bearded at apex. Capsule ovoid, 2–3.5mm, hirsute. Seeds trigonous.

Sikkim: Phansidown Terai, 150m. December.

4. L. zeylanica L.; *L. succulenta* Blume, *L. affinis* G. Don non Mirbach

Sparsely hairy to subglabrous herb. Stems ascending to suberect, 7–28cm, quadrangular, succulent. Leaves ovate to broad ovate, 2.4–1.5 × 1.1–1.5cm, upper similar proportions to lower, base usually truncate with lamina somewhat decurrent, margin serrulate-crenate; petiole 2–5mm. Pedicel 5–15mm, slightly longer in fruit, no bracts at base. Calyx 5–8mm; lobes linear, 3.5–4.5mm. Corolla 5–5.5mm, purplish-blue. All anthers bearded at apex. Capsule 5–7 × 3–4mm. Seeds angularly trigonous, 0.6mm.

Bhutan: S – Samchi district (Samchi (117)); **Darjeeling:** Kusambing, Phansidown, Sukna; **Sikkim:** Chakung, Lusing. November.

5. L. nummularia Lamarck; *Pratia nummularia* (Lamarck) A. Brown & Ascherson, *P. begonifolia* (Wall.) Lindley. Fig. 116b–c.

Pubescent herb. Stems creeping, up to 30cm long, rooting at nodes. Leaves ovate to suborbicular, 0.8–1.5 × 0.7–1.4cm, base truncate to cordate, margin serrulate; petiole 2–7mm. Peduncle 5–20mm. Calyx ± tubular, slightly constricted at mouth, 6–10mm; lobes linear, 3–5.5mm. Corolla 7–7.5mm, pink, pink streaked white or pink-purple. Short papillose hairs on upper anthers. Berry elliptic to globose, c 9 × 6–8mm, purple.

Bhutan: S – Chukka (Mirichana) district, **C** – Tongsa district (Pertimi, Tama), Tashigang district (Kanglung, Shapang); **Darjeeling:** Pul Bazar River, Mongpu, Sitong etc. In damp or marshy ground in warm wet forest, 610–1400m. February, April–May, August–September.

6. L. pyramidalis Wall.

Glabrous herb except for corolla and anthers. Stems erect, 0.7–2.3m. Leaves sessile, linear-lanceolate to oblanceolate, 7.5–22 × 1.5–2cm, tapering to base, margin serrulate (teeth gland-tipped). Flowers many in a terminal branched raceme, erect to spreading, resupinate or not. Pedicels 10–15mm, narrowly winged, often with 2 small bracteoles, bent in fruit. Calyx 15–18mm; lobes linear, 11–13mm, entire. Corolla 17–18mm, unlobed basal part 5–6mm, white, occasionally suffused pink in upper part. Bristles at apex of shorter anthers only; anthers white-hairy on sutures. Capsule erect or deflexed, globose, c 6 × 5.5mm. Seeds elliptic, 0.6mm.

Bhutan: S – Deothang district, **C** – Thimphu, Punakha, Tongsa and Mongar districts; **Darjeeling:** Darjeeling. Forest edges and clearings, 1676–2134m. March–June.

7. L. seguinii Léveillé & Vaniot var. **doniana** (Skottsberg) Wimmer, *L. pyramidalis* sensu F.B.I. p.p. non Wall.

Sparsely hairy herb. Stems erect, 0.6–1.3m, puberulent. Leaves sessile, lanceolate to linear-lanceolate, acuminate, 7.5–16 × 1–4cm, base ± cuneate, margin serrulate, upper surface glabrous to puberulent, lower surface pubescent at least on veins. Flowers in a terminal, sometimes branched raceme, spreading or somewhat deflexed, sometimes resupinate. Pedicels to 7mm, hairy. Calyx 15–18mm; lobes linear, 10–13mm, denticulate. Corolla 22–24mm, unlobed basal part 10–11mm, blue or violet. Bristles at apex of shorter anthers only; anthers subglabrous to white-hairy on sutures. Capsule erect or deflexed, oblong-ellipsoid, c 11 × 8mm. Seeds elliptic, 0.4mm.

Bhutan: S – Deothang district, **C** – Punakha and Tongsa districts, **N** – Upper Mo Chu and Upper Kulong Chu districts; **Darjeeling:** unlocalised (Cowan and Drummond collections); **Sikkim:** Yoksam–Bakhim, Rambi Ghora. Oak forest margins or moist grassy hillsides, 1981–2591m. August–October.

8. L. tibetica Zheng

Herb 20–50cm, stiffly hairy throughout. Cauline leaves elliptic or elliptic obovate, 23–56 × 10–20mm, obtuse or acute, narrowly cuneate at base, scarcely petiolate, glandular-serrulate. Raceme simple, 9–18cm, with leafy bracts not exceeding flowers. Pedicels 5–7mm, with 2 bracteoles at base. Calyx 9–14mm; lobes narrowly triangular, 5–7mm. Corolla violet, 19–21mm. Bristles at apex of shorter anthers only; anthers glabrescent to stiffly hairy on back and at base. Capsule globose, stiffly hairy, 6–8mm across.

Chumbi: without exact locality (434). Herb zone, 3640m. August–September.

L. colorata Wall. has been reported from Bepa (Mongar district) and Tashigang road (?Tashigang district), C Bhutan (both 117), but no specimens have been seen. It differs from *L. tibetica* by its glabrous state, narrower cauline leaves c 40–160 × 8–26mm and lower bracts exceeding flowers. It differs from *L. seguinii* var. *doniana* by its short calyx lobes (3.5–7mm) and globose capsule. It otherwise occurs in Meghalaya and SW China.

9. L. rosea Wall.

Pubescent herb. Stems erect, 1–3(–4)m. Leaves sessile, lanceolate-oblong, 5–11.5 × 1.5–2cm, base cuneate, margin serrulate, pubescent on both surfaces. Flowers many in a terminal branched raceme, deflexed, resupinate. Pedicel 4–7mm. Calyx 15–18mm; lobes linear, 10–12mm, entire. Corolla 22–26mm, unlobed basal part c 10mm, pink or pale mauve on lobes, whitish towards base. Bristles at apex of lower anthers only; anthers white-hairy on sutures. Capsule deflexed, ovoid, c 11 × 8mm. Seeds oblong-elliptic, 0.4mm.

Bhutan: S – Phuntsholing district (Phuntsholing, above Rinchending), Deothang district (Deothang (117)); **Darjeeling:** Silak, Mongpu; **Darjeeling/Sikkim:** Rungit. Subtropical jungle slopes, 610–914m. February.

10. L. erectiuscula Hara; *Lobelia erecta* Hook.f. & Thomson non de Vriese. Fig. 116d–e.

Partially pubescent herb. Stems erect, 0.3–0.8m, pubescent at least in upper part. Leaves sessile, elliptic to oblanceolate, lower attenuate, 3.7–15.5 × 1–3cm, margin irregularly crenulate-serrate also often minutely denticulate, glabrous except for main veins on upper surface to sparsely pubescent. Flowers in an unbranched raceme, erect, not resupinate. Pedicels 3–8.5mm, hairy. Calyx 10–15mm, ± pubescent; lobes narrowly triangular to ± linear, 5–9mm, minutely denticulate and ciliate on margin. Corolla 14–17mm, unlobed basal part c 5mm, mauve, purple or violet-blue with white marking in throat. Bristles at apex of lower anthers only, otherwise all anthers glabrous. Capsule ± erect, globose, c 9 × 10mm. Seeds elliptic, 0.7mm.

Bhutan: C – Ha, Thimphu, Punakha, Tongsa, Mongar and Tashigang districts, **N** – Upper Mo Chu district; **Darjeeling:** Tonglu; **Sikkim:** Phalut, Phusum, Changu etc.; **Chumbi:** Yatung. In damp grassy places in *Abies/Juniperus* forest or among bamboo, 2286–4267m. July–September.

11. L. nubigena Anthony

Robust biennial herb. Stem erect, up to 1m, unbranched, densely leafy, subglabrous below, puberulous above. Leaves subsessile or sessile, ± entire with glands on margin, subglabrous; basal rosetted, oblanceolate, c 30 × 5cm; cauline oblong-lanceolate, 5–12 × 0.8–2cm. Flowers in a dense, terminal, bracteate spike. Bracts deflexed, c 3 × as long as flowers, densely white lanate. Pedicels 5–10mm. Calyx c 20mm, lobes narrowly triangular, c 15mm, densely white lanate. Corolla c 15mm, blue- or red-purple. Filaments c 1cm. Capsule c 8 × 7mm. Seeds oblong-ellipsoid, 1.5mm, narrowly winged.

Bhutan: C – Tongsa district (Tibdey La, Yuto La). In alpine turf, 3962–4112m. July.

Endemic to Bhutan.

12. L. montana Blume; *Pratia montana* (Blume) Hasskarl

Glabrous herb except for flowers. Stems erect or arching, up to 130cm. Leaves elliptic-acuminate, 5.5–16 × 2–4.3cm, base cuneate, margin serrulate; petiole 5–12mm. Peduncle 12–25mm. Calyx broadly tubular with spreading lobes, 9–10mm; lobes linear, 5–7mm. Corolla 12–16mm, green or white with purple marking or ?violet. Berry globose, 8–14 × 8–14mm, purple.

Bhutan: S – Chukka district (S of Takhti Chu), Deothang district (Deothang–Narfong (117)), **C** – Mongar district (Latun La); **Darjeeling:** Rhambi Chhu, Labha, Ghum etc. On banks, in mixed oak and broad-leaved forest, 1829–2438m. July–August.

Family 187. COMPOSITAE (Asteraceae)

by A.J.C. Grierson & L.S. Springate

Annual, biennial or perennial herbs, sometimes shrubs, rarely trees, tissues sometimes containing milky sap, glabrous, pubescent, tomentose, spinous or variously sessile- or stipitate-glandular. Leaves alternate or sometimes opposite, exstipulate (but sometimes with stipuliform appendages), entire, toothed, lobed or variously dissected. Individual flowers usually numerous, aggregated and ± sessile (briefly immersed to very shortly stipitate) on a common receptacle and surrounded by an involucre of 1–many series of phyllaries (involucral bracts), the whole comprising a capitulum; phyllaries free, rarely connate; capitula solitary to very many, rarely aggregated into secondary capitulum-like glomerules. Receptacle sometimes bearing paleae (scales), hairs or bristles. Flowers (florets) epigynous, bisexual, female, male (at least functionally so) or neuter (sterile). Calyx absent, often replaced on apex of ovary by a pappus of 1 or more series of bristles and/or scales. Corolla gamopetalous, tubular throughout or dilated or 1(–2)-lipped above, variously truncate or 1–5-toothed at apex (apices); rarely corolla absent. Stamens (1–)5, epipetalous, filaments free, anthers laterally connate into a cylinder around style (free in one species). Ovary inferior, 1-celled with one basal ovule; style usually divided above into two branches, sometimes entire on male flowers, emerging through anther cylinder, first collecting and exposing pollen, later exposing stigmatic surfaces if bisexual. Fruit an achene (cypsela) usually bearing a persistent or deciduous pappus; pappus sessile or borne on a beak (rostrum).

Comments without qualification throughout this account refer only to specimens from our area.

Explanation of some terms used here:

Flowers and corollas

Tubular-campanulate: corolla regular, narrowly cylindric below (the *tube*) and (±) dilated above the point of insertion of the filaments (the *limb*, comprising an entire *throat* below and (3–)5 *teeth* or *lobes* above; Fig. 117d, i). Sometimes corollas with tube and throat very weakly differentiated externally are explicitly distinguished as narrowly *infundibular* (Fig. 128i) or *tubular*, then 'tubular-campanulate' only describes corollas with a narrowly campanulate-infundibular to broadly urceolate limb. Bisexual or *male*, the achenes of the latter empty but as long as fertile achenes or variously reduced and the style branches sometimes reduced or absent but a simple style at least is always present to expel pollen.

Eligulate female: corolla narrowly cylindric throughout, truncate at apex or bearing 1–3(–5) usually unequal teeth, usually inconspicuous and tightly sheathing lower style (the flower being *filiform*, Fig. 125e), rarely looser and more conspicuous (*tubular*, Fig. 128d). Flowers with one tooth are here treated as rays if the tooth is spreading or recurved and consistently developed to the outside of the capitulum. *Adenocaulon* is exceptional (Fig. 118c).

Ray: corolla irregular, with narrow *tube* below (rarely absent) and dilated limb above (a *ligule*) directed to the outside of the capitulum. Ligule usually 3-toothed, less often entire, 2- or 4-toothed. Usually female, less often neuter, staminodes sometimes present but flowers never functionally bisexual (Fig. 126f, Fig. 128c).

Ligulate: corolla as ray flower, but limb usually 5-toothed. Always bisexual. Confined to Lactuceae (Fig. 121p).

Bilabiate: corolla tubular below, 2-lipped above; confined to Mutisieae; *bilabiate ray flower*: outer lip enlarged, 3-toothed, inner lip entire, linear, much reduced, female, staminodes present (Fig 118h); *bilabiate disc flower*: lips subequal, 3- and 2-toothed, bisexual (Fig. 118i). Sometimes lips scarcely distinguished and all 5 teeth subequal, then treated here as eligulate female (Fig. 118c) or tubular-campanulate (Fig. 118b, e).

Capitula

Discoid: flowers all of one type and eligulate; usually tubular-campanulate, all bisexual or sometimes male at centre of capitulum; rarely all filiform on monoecious/dioecious plants (then other capitula/plants are male, discoid, with tubular-campanulate flowers). The outer whorl is enlarged and sterile in *Centaurea*.

Disciform: outer flowers eligulate, female; inner flowers tubular-campanulate, bisexual or all or only central ones male.

Radiate: outer 1(–10) rows of ray flowers; inner flowers (comprising the *disc*) tubular-campanulate. Rarely both ray and disc flowers bilabiate (Mutisieae p.p.).

Ligulate: all flowers ligulate. Confined to Lactuceae.

Usually both sexes are represented in each capitulum. *Monoecious* plants bear unisexual capitula with all flowers of the same sex or some neuter, but with capitula of both sexes present on the same plant. *Dioecious* and *subdioecious* plants also rarely occur, the latter with a great predominance of the same sex in each capitulum of the plant.

Phyllaries may be *imbricate*, consisting of succesive series (whorls) of increasing height (Fig. 117b, c, Fig. 118e), may be subequal or the outermost may be largest (Fig. 137g). In the Senecioneae phyllaries of equal length and their involucre are termed *1-seriate*, although consisting of alternate inner and outer elements of distinct form interlocked at the margins (Fig. 117f, g), often with an irregular group of usually smaller bracts inserted below which sometimes extend onto the upper peduncle (a *calyculus*, Fig. 133b, Fig. 134c); *2-seriate* involucres in the Lactuceae comprise similar, tightly appressed but not interlocked inner phyllaries and a distinct set of short outer phyllaries (Fig. 121r).

The receptacle is termed *naked* unless it bears paleae, bristles or long hairs (at least ⅓ length of flower). Paleae and bristles intergrade between deeply dissected paleae and broadly subulate or flattened bristles. Bristles in some species are caducous, sometimes prematurely so when pressed. Receptacles pitted by immersed flower bases are *alveolate* or, when short protrusions from the receptacle surface also develop between the flowers, *fimbrillate*.

Capitula are often grouped into a recognisable general inflorescence (for brevity: the *inflorescence*). For descriptive purposes the capitula are treated as if they were individual flowers, though the supporting axis of each is a *peduncle.* The structure of an inflorescence usually comprises both indeterminate (racemose) and determinate (cymose) elements, but is generally described in terms of the former (*spike*, *raceme*, *panicle*, *corymb*), without implying any sequence of opening of the capitula. If distinctly determinate it is described as *cymose* (broad, branched from top of stem) or *thyrsoid-paniculate* (elongate, branched along stem, obconic-oblong in outline). *Open* inflorescences of few capitula on long flexuous peduncles in no clear arrangement occur particularly in the Lactuceae. *Scape* is loosely applied to any leafless flowering stem; it may be bracteate or naked, the leaves may be basal, undeveloped at flowering or borne on separate shoots, the flowers 1–many.

Flower parts

Stamens and style display a wealth of characters critical to the classification of the Compositae at all levels, but few can be observed satifactorily in the field and are not described in detail here. Further information can be found in (323). Truncate to sagittate anther bases are quite widespread in the Compositae, but tailed or caudate bases (sometimes extensively lacerate and conspicuous when exserted) are only found in Inuleae, Plucheeae, Gnaphalieae, Mutisieae, Cardueae and a few genera of Senecioneae.

The stigmatic surface occurs on the inside of the style branches. On disc flowers it is usually localised, either in a continuous marginal band (e.g. Inuleae) or two lateral bands interrupted by an apical fringe of short hairs (*penicillate* style branches, e.g. Senecioneae) or a sterile broadly triangular to linear appendage (e.g. Astereae, Eupatorieae).

Pappus bristles have a ± regular multicellular structure, are terete to weakly radially compressed (width in upper part not more than 3 × depth) and may be described by one of the following weakly delimited terms:

Capillary: very fine and virtually smooth hairs.

Scabrous: somewhat coarser hairs which are visibly rough due to the projecting tips of the constituent cells.

Barbellate: with the tips of the cells elongated and approximately as long as the breadth of the rachis of the bristle.

Plumose: with the laterally projecting cells elongate, hair-like (Fig. 117d).

Awns are coarser than bristles and sometimes retrorsely barbed (Fig. 135j, k, r–u). *Rims* (continuous) and *ridges* (interrupted) resemble extensions of the achene margin.

Measurements

Involucres distort badly when pressed. Diameters were measured a little above the surface of the receptacle (Fig. 117b, f), preferably on relaxed material. Disc corollas were measured with the tube at maximum development (late/post anthesis) and the teeth or lobes held erect. Corolla ligules often shrink on pressing, some widths in particular may be underestimated here. Ligules usually decrease in length from margin to centre of capitula of the Lactuceae, lengths from marginal flowers are given when possible.

CONSPECTUS OF COMPOSITAE IN THE *FLORA OF BHUTAN* AREA

SUBFAMILY MUTISIOIDEAE

Tribe MUTISIEAE

1. *Ainsliaea*
2. *Gerbera*
3. *Leibnitzia*
4. *Adenocaulon*

SUBFAMILY CICHORIOIDEAE

Tribe CARDUEAE

5. *Arctium*
6. *Cirsium*
7. *Saussurea*
8. *Hemisteptia*
9. *Tricholepis*
10. *Centaurea*

Tribe LACTUCEAE

11. *Crepis*
12. *Youngia*
13. *Stebbinsia*
14. *Soroseris*
15. *Dubyaea*
16. *Taraxacum*
17. *Ixeris*
18. *Ixeridium*
19. *Mulgedium*
20. *Lactuca*
21. *Pterocypsela*
22. *Cicerbita*
23. *Cephalorrhynchus*
24. *Chaetoseris*
25. *Stenoseris*
26. *Notoseris*
27. *Prenanthes*
28. *Syncalathium*
29. *Sonchus*
30. *Launaea*
31. *Hieracium*
32. *Picris*
33. *Tragopogon*

Tribe VERNONIEAE

34. *Vernonia*
35. *Cyanthillium*
36. *Baccharoides*
37. *Elephantopus*

SUBFAMILY ASTEROIDEAE

Tribe INULEAE

38. *Inula*
39. *Pentanema*
40. *Carpesium*
41. *Duhaldea*
42. *Pulicaria*
43. *Blumea*

Tribe PLUCHEEAE

44. *Laggera*
45. *Pseudoconyza*
46. *Blumeopsis*
47. *Sphaeranthus*

Tribe GNAPHALIEAE

48. *Leontopodium*
49. *Anaphalis*
50. *Pseudognaphalium*
51. *Gamochaeta*
52. *Gnaphalium*
53. *Bracteantha*

Tribe CALENDULEAE

54. *Calendula*

Tribe ASTEREAE
55. *Dichrocephala*
56. *Cyathocline*
57. *Grangea*
58. *Myriactis*
59. *Rhynchospermum*
60. *Heteropappus*
61. *Aster*
62. *Callistephus*
63. *Brachyactis*
64. *Erigeron*
65. *Conyza*
66. *Microglossa*
67. *Nannoglottis*
68. *Vittadinia*
69. *Thespis*

Tribe ANTHEMIDEAE
70. *Allardia*
71. *Tanacetum*
72. *Hippolytia*
73. *Chrysanthemum*
74. *Ajania*
75. *Artemisia*
76. *Achillea*
77. *Xanthophthalmum*
78. *Leucanthemum*
79. *Cotula*
80. *Soliva*

Tribe SENECIONEAE
81. *Tussilago*
82. *Petasites*
83. *Parasenecio*
84. *Ligularia*
85. *Cremanthodium*
86. *Doronicum*
87. *Synotis*
88. *Cissampelopsis*
89. *Senecio*
90. *Crassocephalum*
91. *Emilia*
92. *Gynura*
93. *Steirodiscus*

Tribe HELENIEAE
94. *Tagetes*
95. *Gaillardia*

Tribe HELIANTHEAE
96. *Zinnia*
97. *Acmella*
98. *Wedelia*
99. *Eleutheranthera*
100. *Synedrella*
101. *Tithonia*
102. *Helianthus*
103. *Galinsoga*
104. *Tridax*
105. *Acanthospermum*
106. *Sigesbeckia*
107. *Enydra*
108. *Guizotia*
109. *Coreopsis*
110. *Dahlia*
111. *Cosmos*
112. *Glossocardia*
113. *Bidens*
114. *Xanthium*
115. *Parthenium*
116. *Eclipta*
117. *Montanoa*

Tribe EUPATORIEAE
118. *Ageratina*
119. *Mikania*
120. *Adenostemma*
121. *Ageratum*
122. *Chromolaena*
123. *Eupatorium*

ASTEROIDEAE, tribe uncertain
124. *Anisopappus*
125. *Cavea*
126. Centipeda

The following keys allow for some variation in interpretation of difficult characters, e.g. treating the filiform pappus scales of *Allardia* as bristles rather than filiform scales.

GENERAL KEY

Capitula ligulate .... **Key A**
Capitula radiate .... **Key B**

Capitula disciform .... **Key C**
Capitula discoid .... **Key D**

KEY A

1. Pappus bristles plumose 2
\+ Pappus bristles capillary to scabrid 3

2. Plant hispid and/or glandular hairy; flowers overtopping phyllaries **32. Picris**
\+ Plant glabrous; phyllaries longer than flowers .... **33. Tragopogon**

3. Flowers blue, purple, mauve, pink or red 4
\+ Flowers white, yellow or greenish-yellow 18

4. Acaulous; capitula densely crowded, subsessile on plant body **28. Syncalathium**
\+ Caulescent (rarely scapose); capitula more dispersed in pedunculate heads 5

5. Flowers 3–5 per capitula 6
\+ Flowers 7–many per capitula 9

6. Stems tall, scandent, clearly exceeding 1m .... **27. Prenanthes**
\+ Stems low, straggling to decumbent, 60–100cm long or free-standing, erect 7

7. Phyllaries and flowers 3 (very rarely 4); achene body oblong, strongly compressed, with broad thick margin, abruptly contracted to short thick beak ± as wide as long .... **25. Stenoseris**
\+ Phyllaries and flowers 4–5(–6); achene oblong-ellipsoid, moderately compressed or subtrigonous-subterete, with less prominent ribs at margins/angles, gradually tapering at apex without obvious beak or with slender beak 8

8. Lower cauline leaves pinnatisect, narrowly oblanceolate, with 5–6 pairs of leaflets .... **24. Chaetoseris**
\+ Lower cauline leaves simple, triangular to pinnate, oblanceolate in outline with 1(–3) pairs of leaflets 8

9. Phyllaries and flowers 5; achene with 3 unequal faces .... **22. Cicerbita**
\+ Phyllaries and flowers 4; achene with 2 equal faces .... **26. Notoseris**

10. Mid-cauline and upper leaves linear or linear-lanceolate, undivided 11
\+ Mid-cauline and upper leaves broader, often divided or plant scapose .. 12

11. Corolla tube 4.5–5.5mm; achene body much broader than wings, with 3–5 ribs on either face .... **20. Lactuca** (*L. dolichophylla*)
\+ Corolla tube 7–8mm; achene body no wider than either wing, with 1 rib on each face .... **21. Pterocypsela**

12. Inner phyllaries 4.5–5.5mm; corolla tube 2mm, ligule 4mm **20. Lactuca** (*L. dissecta*)
\+ Inner phyllaries 9–19mm; corolla tube 5–10mm, ligule 5–26mm 13

13. Stems trailing or scrambling .... **23. Cephalorrhynchus**
\+ Stems erect or absent 14

14. Achenes subterete, without 2 obvious faces and lateral margins; beak absent **15. Dubyaea**
\+ Achenes somewhat compressed, with 2 faces and lateral margins apparent; beak present (though sometimes short and stout) 15

15. Ligules 4.5–10mm 16
\+ Ligules 13.5–26mm 17

16. Plant 3–30cm; upper leaves subentire, remotely lobulate or absent **19. Mulgedium** (*M. lessertianum*)
\+ Plant 35–100cm; upper leaves closely and acutely denticulate **24. Chaetoseris**

17. Ligules 13.5–15mm .... **19. Mulgedium** (*M. bracteatum*)
\+ Ligules 21–26mm .... **24. Chaetoseris** (*C. macrantha*)

18. All inner phyllaries fused at base for 1–3mm above receptacle 19
\+ Inner phyllaries free 20

19. Flowers 25–37 .... **13. Stebbinsia**
\+ Flowers 4–5 .... **14. Soroseris**

20. Achenes non-rostrate 21
\+ Achenes rostrate 27

21. Ligule of outer flowers clearly exceeding tube or corollas 5.5–6.5mm .. 21
\+ Ligule of outer flowers equalling or shorter than tube; corollas at least 9mm 26

22. Involucres 6–15mm diameter 23
\+ Involucres 1.5–5mm 24

23. Lower leaves simple, dentate to lyrate-pinnatisect, lanceolate to oblanceolate in outline; phyllaries long blackish hairy .... **15. Dubyaea** (*D. hispida*)
\+ Lower leaves simple, subentire, linear-lanceolate to narrowly oblong-elliptic; phyllaries ± glabrous apart from minute cilia .... **31. Hieracium**

24. Achenes ± equally ribbed .... **11. Crepis**
\+ Achenes unequally ribbed with 3–5 stronger ribs .... **12. Youngia**

25. Base of capitula (but not peduncles) becoming thickened and spongy at maturity; caulescent plants .... **29. Sonchus**
\+ Base of capitula not becoming thickened and spongy; acaulous or subscapose plants 26

26. Acaulous; capitula crowded on simple peduncles from centre of leaf rosette **12. Youngia** (*Y. depressa*)
\+ Subscapose; capitula in a racemose to diffusely branched inflorescence **30. Launaea**

27. Acaulous; capitula solitary, on hollow scapes .... **16. Taraxacum**
\+ Caulescent; capitula (few to) many, on solid peduncles 28

28. Achene with 2 or 10 broad wings 29
\+ Achene without wings 30

29. Achene with 2 wings .... **21. Pterocypsela**
\+ Achene with 10 wings .... **17. Ixeris**

30. Perennial; flowers 5–18 .... **18. Ixeridium**
\+ Annual or biennial; flowers 20–30 .... **20. Lactuca**

KEY B

1. Receptacle at least partially paleate or setose 2
\+ Receptacle without paleae or setae 26

2. Leaves alternate 3
\+ Leaves opposite, at least on lower half of stem 10

3. Leaves simple or palmately 3–5-lobed 4
\+ Leaves pinnatifid to bipinnatisect 7

4. Ligules 6–8mm .... **124. Anisopappus**
\+ Ligules 12–70mm 5

5. Receptacles bearing subulate scales c 3mm; ligules red, tipped yellow **95. Gaillardia**
+ Receptacle bearing broad paleae enclosing fertile achenes; ligules yellow or orange 6

6. Leaves usually 3- or 5-lobed; peduncles hollow; pappus persistent **101. Tithonia**
+ Leaves simple, serrate; peduncles solid; pappus deciduous **102. Helianthus**

7. Pappus present 8
+ Pappus absent 9

8. Ray achenes dispersed with 2 paleae and 2 disc flowers attached; disc achenes sterile .... **115. Parthenium**
+ All achenes fertile, dispersed singly and free of paleae or other flowers **112. Glossocardia**

9. Leaves pinnatifid, with 2 pairs of lateral lobes; ligules c 20mm **73. Chrysanthemum**
+ Leaves bipinnatisect with c 20 pairs of primary segments; ligules c 2.2mm **76. Achillea**

10. Leaves simple 11
+ Most leaves 1-pinnatisect to 3-pinnate or 1–2-ternate 23

11. Ligules exceeding 10mm 12
+ Ligules less than 10mm 15

12. Ligules ± sessile and persistent on ray achenes .... **96. Zinnia**
+ Ligules with distinct basal tube, not persistent on ray achenes 13

13. Tree or shrub; ligules white .... **117. Montanoa**
+ Herbaceous, annual or perennial; ligules yellow 14

14. Phyllaries 5; ray flowers female .... **108. Guizotia**
+ Phyllaries many; ray flowers neuter .... **102. Helianthus**

15. Disc pappus of 15–20 fimbriate scales or plumose bristles 16
+ Disc pappus absent or of no more than 2 elements, sometimes with a small coronna 17

16. Perennial; paleae all similar, simple; ligules 3–4.5mm .... **104. Tridax**
+ Annual; paleae dimorphic, inner often 3-fid; ligules c 1.5mm **103. Galinsoga**

17. Ligules white 18
\+ Ligules yellow 19

18. Perennial aquatic or marsh plant; pappus present .... **116. Eclipta**
\+ Annual, weedy; pappus absent .... **107. Enydra**

19. Ligules 4–10mm .... **98. Wedelia**
\+ Ligules 1–3mm 20

20. Only ray achenes fertile, dispersed with enveloping phyllary
105. Acanthospermum
\+ Ray and disc achenes fertile, dispersed without phyllary 21

21. Pappus of 2 stiff awns .... **100. Synedrella**
\+ Pappus absent or of 2 weak bristles c 0.4mm 22

22. All phyllaries glabrous, 1.3–2.5mm .... **97. Acmella**
\+ Outer phyllaries densely stipitate-glandular, 8–10mm..... **106. Sigesbeckia**

23. All flowers fertile; ligules smaller than 9 × 6mm .... **113. Bidens**
\+ Ray flowers sterile; ligules larger than 9 × 6mm 24

24. Achenes 4-sided or terete, narrowed into a beak above .... **111. Cosmos**
\+ Achenes compressed, without a beak 25

25. Slender annual; ligules c 11 × 8mm .... **109. Coreopsis**
\+ Robust perennial or subshrub; ligules c 35–60 × 15–25mm ... **110. Dahlia**

26. Pappus present, entirely of capillary to barbellate bristles 27
\+ Pappus of scales, sometimes including 2 flattened awns, or a coronifom rim or absent 56

27. Phyllaries in 1 series, interlocked at least below (sometimes separating when pressed – examine young capitula; calyculus also often present) 28
\+ Phyllaries in 2–6 series, not interlocked 35

28. Capitula precocious, borne on leafless scapes; leaves borne on separate shoots .... **81. Tussilago**
\+ Capitula borne on leafy stems or on scapes with a basal rosette of leaves 29

29. Petioles of basal leaves at least with thin dilated base sheathing stem .. 30
\+ No petiole bases dilated and sheathing stem 31

30. Capitula 2–many .... see 84. *Ligularia*
\+ Capitulum 1 .... see 85. *Cremanthodium*

31. Plant climbing by basally thickened and twining petioles
88. Cissampelopsis
\+ Plant erect or scrambling, without twining petioles 32

32. Leaves simple, sessile or with simple petiole 33
\+ Leaves simple with remotely lobulate or dentate- or panduriform-winged petiole to bipinnatifid 34

33. Leaves araneous to lanate beneath or rays 1–5 .... **87. Synotis**
\+ Leaves (sub)glabrous beneath; rays c 8–13 .... **89. Senecio**

34. All leaves simple; petiole dentate- or panduriform-winged; ray flowers 1–4; disc corollas 6.5–8mm .... **87. Synotis** (*S. acuminata*)
\+ Cauline leaves 1–2-pinnatifid or rays 6 to c 20 or disc corollas 3.5mm
89. Senecio

35. All corollas bilabiate, ray corollas with small linear lobe opposite main limb 36
\+ Ray corollas with single limb only, disc corollas ± regularly (4–)5-lobed
37

36. Capitula cernous; outer lip of ray corollas 10–18mm .... **2. Gerbera**
\+ Capitula erect; outer lip of ray corollas 4–5mm .... **3. Leibnitzia**

37. Ligules yellow 38
\+ Ligules white, whitish-green, rose, purple or blue 44

38. Flowers trimorphic, with ring of eligulate female flowers inside ray flowers
67. Nannoglottis
\+ Flowers dimorphic, without eligulate female flowers 39

39. Pappus inner series of c 5 bristles, outer series of scales forming a minute cup .... **42. Pulicaria** (*P. petiolaris*)
\+ Pappus entirely of bristles 40

40. Annual; pappus of usually 10 bristles .... **39. Pentanema**
\+ Perennial or shrub; pappus of many bristles 41

41. Phyllaries 3–10-seriate 42
\+ Phyllaries 2-seriate 43

42. Rays 2–3-seriate; ligules 15–45mm .... **38. Inula**
\+ Rays 1-seriate; ligules 1–15mm .... **41. Duhaldea**

43. Pappus well developed on ray flowers .. .. **85. Cremanthodium**
\+ Pappus reduced to c 2 bristles or absent from ray flowers . **86. Doronicum**

44. Flowers trimorphic, eligulate female flowers between ray and disc flowers **64. Erigeron**
\+ Flowers dimorphic, eligulate female flowers absent 45

45. Ligules 3.5–30mm 46
\+ Ligules 0.5–1.5mm 54

46. All leaves bearing 7(–9) dentate lobules; ligules 20–25 × 5–7mm **62. Callistephus**
\+ Leaves entire to remotely and/or shallowly dentate or only some 3(–5)-lobed; ligules shorter than 20mm and/or narrower than 5mm 47

47. Phyllaries 5(–8)-seriate; leaves sparsely but coarsely pilose; ligules white **41. Duhaldea** (*D. nervosa*)
\+ Phyllaries 2–4-seriate; leaves (sub)glabrous or puberulous below or ligules mauve, purple or blue 48

48. Lower leaves decurrent on stem; disc flowers c 7–10; rays c 35–40 **63. Brachyactis**
\+ Leaves not decurrent; disc flowers at least half as many as rays 49

49. Basal part of achene solid, usually attenuate, with tuft of longer (c 0.3–0.4mm) and denser hairs than on remainder of achene .. **68. Vittadinia**
\+ Hairs at base of achene not longer or denser than on remainder of achene 50

50. Disc corollas with 1 tooth more deeply cut; length of ray pappus ¾ of disc pappus .... **60. Heteropappus**
\+ Teeth of disc corollas equal; length of ray and disc pappus ± equal ... 51

51. Capitula 6–many per stem; leaves simple .... **61. Aster**
\+ Capitula 1–5 per stem or some leaves 3-lobed 52

52. Style branch appendages narrowly triangular, length 1.3–2 × width **61. Aster**
\+ Style branch appendages broadly triangular, length less than or ± equal to width 53

53. Leaves densely whitish felted .... **61. Aster** (*A. prainii*)
\+ Leaves sparsely pubescent .... **64. Erigeron**

54. Ray achenes with 1.5mm beak; disc achenes with stout 0.5mm beak; pappus of c 7 weak caducous bristles .... **59. Rhynchospermum**
\+ Achenes without beak; pappus of many ± persistent bristles 55

55. Annual or biennial herb; most flowers bisexual **65. Conyza** (*C. canadensis*)
\+ Shrub, often scandent; nearly all flowers female .... **66. Microglossa**

56. At least lower leaves opposite; achenes blackish, pale and empty at base; pappus including 2 flattened awns longer than achene.......... **94. Tagetes**
\+ Leaves alternate; achenes not markedly paler and empty at base; pappus not including awns longer than achene, sometimes absent 57

57. Ligules 0.8–2mm... **58. Myriactis**
\+ Ligules 6–45mm 58

58. Ligules yellow 59
\+ Ligules white, pink, red or purple 62

59. Ray achenes trimorphic, incurved to coiled .... **54. Calendula**
\+ Ray achenes monomorphic, ± straight or absent........................ 60

60. Phyllaries 1-seriate, ⅔ fused; ligules c 8 × 2mm .... **93. Steirodiscus**
\+ Phyllaries c 3–4-seriate, imbricate, free; ligules 12–20 × 3–5mm 61

61. Basal leaves 3–4-pinnatisect, ultimate segments c 0.2–0.5mm wide **71. Tanacetum**
\+ All leaves 1–2-pinnatisect, ultimate segments 1–3mm wide **77. Xanthophthalmum**

62. Ray flowers neuter, pappus of brown, filiform, bristle-like scales **70. Allardia**
+ Ray flowers female; pappus ± coroniform, usually confined to ray flowers or entirely absent .... **78. Leucanthemum**

KEY C

1. Pappus on fertile achenes of at least 8 bristles or c 5 bristles and an outer series of scales 2
+ Pappus absent or of 1–3 bristles or of ridges or a minute coronna or fimbriate cup 24

2. Leaves remotely denticulate to pinnatisect 3
+ Leaves simple, entire 16

3. Phyllaries 2–6-seriate, unequal, imbricate or outermost largest 4
+ Phyllaries 1-seriate (sometimes with a calyculus) or in 2 subequal series 14

4. Stems not winged; petioles not decurrent 5
+ Stems bearing continuous wings or lines of herbaceous teeth or petioles decurrent as 1–3 pairs of lobes 13

5. Pappus of c 5 bristles and an outer series of scales **42. Pulicaria** (*P. foliosa*)
+ Pappus of many bristles only 6

6. Cleistogamous, corollas concealed by purplish (rarely stramineous) pappus c twice as long; plant scapose .... **3. Leibnitzia**
+ Not cleistogamous; corollas exposed, longer to scarcely shorter than pappus; if pappus purplish, then leafy stems present 7

7. Inner flowers male with undivided style; pappus purple .... **125. Cavea**
+ Inner flowers with divided style, usually bisexual (male in *Thespis*); pappus white, brownish or reddish 8

8. Outer flowers female, filiform, corollas 3–4-lobed; inner flowers of intermediate form, broader, corollas 4-lobed, stamens 1–3; rarely perfect bisexual flowers at centre of capitulum with 5 corolla lobes and 5 stamens **46. Blumeopsis**
+ Flowers of intermediate form absent 9

9. Phyllaries (35–)40 or more; style branches unappendaged, with stigmatic band continuous around apex 10
\+ Phyllaries c 15–30; style branches with lateral stigmatic bands and sterile triangular appendage 11

10. Shrubs; female flowers 1(–2)-seriate, corollas tubular .... **41. Duhaldea**
\+ Annual or perennial herbs or shrubs; female flowers several-seriate, filiform **43. Blumea**

11. Female flowers lacking corollas .... **69. Thespis**
\+ Female flowers bearing corollas 12

12. Annual to perennial herbs; most flowers bisexual .... **65. Conyza**
\+ Shrub, often scandent; nearly all flowers female .... **66. Microglossa**

13. Petioles decurrent as continuous wings or lines of narrow herbaceous teeth **44. Laggera**
\+ Petioles decurrent as 1–3 pairs of lobes only .... **45. Pseudoconyza**

14. Phyllaries c 5; female flowers 1–4; bisexual flowers 3–6 **87. Synotis** (*S. alata*, *S. wallichii*)
\+ Phyllaries and female flowers more than 10; bisexual flowers 1–many .. 15

15. Leaves basal, scarcely deveoped at flowering, reniform, denticulate **82. Petasites**
\+ Leaves cauline, well developed at flowering, obovate-oblong in outline, all deeply pinnatisect or lower ones coarsely dentate *Erechtites* (see 90. *Crassocephalum*)

16. Phyllaries with evident, green, herbaceous keel throughout and scarious margin .... **65. Conyza**
\+ Phyllaries scarious, white, yellow, reddish or brownish, sometimes with green patch at top of claw 17

17. Involucre 10–20mm diameter; longest phyllaries c 13mm; cultivated **53. Bracteantha**
\+ Involucre to 7.5mm diameter; phyllaries to 11mm; not cultivated 18

18. Capitula 1–12(–20) crowded into a head surrounded by radiating star-shaped bracts contrasting with leaves in colour and/or texture **48. Leontopodium**
\+ Inflorescence of 1–many capitula, often more open, not surrounded by radiating bracts 19

19. Phyllaries white above .... **49. Anaphalis**
\+ Phyllaries yellowish or brown above 20

20. Perennial caespitose herbs or compact subshrubs 21
\+ Annual herbs 22

21. Phyllaries yellowish and spreading above, conspicuous .... **49. Anaphalis**
\+ Phyllaries brownish and suberect above, inconspicuous .. **48. Leontopodium**

22. Phyllaries yellowish .... **50. Pseudognaphalium**
\+ Phyllaries brownish 24

23. Pappus bristles connate at base, deciduous in a ring .... **51. Gamochaeta**
\+ Pappus bristles free, separately deciduous .... **52. Gnaphalium**

24. Achenes elongate, beaked (total length at least 5 × width) **40. Carpesium**
\+ Achenes broader, without beak 25

25. All capitula solitary, terminal or axillary 26
\+ Most or all capitula grouped in inflorescences 29

26. Acaulous; capitula clustered at base of plant .... **80. Soliva**
\+ Caulescent; capitula in upper leaf axils or terminal 27

27. Capitula terminal, 10–15mm across .... **71. Tanacetum**
\+ Capitula axillary, 2–5mm across 28

28. Leaves 1–2-pinnatisect; achenes compressed with pale margin or much thickened wing, faces not ribbed or grooved .... **79. Cotula**
\+ Leaves ± simple, 1–3-toothed or -lobed; achenes angular, not compressed, strongly ribbed with grooves in between .... **126. Centipeda**

29. Inflorescence a dense glomerule with only tips of bracts and phyllaries exposed .... **47. Sphaeranthus**
\+ Inflorescence more lax, individual capitula visible 30

30. Achenes c 5.5mm, covered above with stout bristles with red glandular tips **4. Adenocaulon**
\+ Achenes c 0.5–2mm, without gland-tipped bristles 31

31. Phyllaries with scarious margin; receptacle subglobose to clavate or cup-shaped at centre or convex with stipitate flowers; annuals 32
\+ Phyllaries mostly scarious; receptacle concave to convex; mostly herbaceous or suffrutescent perennials, few annuals 34

32. Pappus small, cup-shaped, fimbriate .... **57. Grangea**
\+ Pappus absent, of 2 ridges or of 1–3 weak bristles 33

33. Receptacle subglobose to clavate; corolla of outer (female) flowers tubular to narrowly campanulate, c 0.5mm; inner flowers functionally bisexual **55. Dichrocephala**
\+ Receptacle cup-shaped at centre; corolla of outer (female) flowers narrowly tubular, 1.2mm; inner flowers male .... **56. Cyathocline**

34. Capitula in spike-like racemes or panicles with spike-like ultimate branches; corollas brownish or purplish .... **75. Artemisia**
\+ Capitula in corymbs (exceptionally solitary); corollas yellow ... **74. Ajania**

KEY D

1. Receptacle paleate or setose 2
\+ Receptacle without paleae or setae 11

2. Leaves alternate 3
\+ Leaves opposite 9

3. Capitula unisexual; male rounded, many-flowered, borne at apex of branches; female ovoid, phyllaries fused except hooked spinous tips, borne on lower parts of branches .... **114. Xanthium**
\+ Capitula bisexual; all involucres campanulate; all phyllaries free 4

4. Pappus of plumose bristles, at least in inner series 5
\+ All pappus series of capillary bristles or scales 7

5. Leaf margins spinous; outer phyllaries usually spine-tipped .... **6. Cirsium**
\+ Leaf margins and phyllaries unarmed 6

6. Phyllaries unappendaged; corolla limbs narrowly campanulate **7. Saussurea**
+ Median phyllaries with an upstanding appendage (becoming flattened to one side) in upper half; corollas tubular, limb scarcely dilated ..... **8. Hemisteptia**

7. Phyllaries acicular above, hooked at apex .... **5. Arctium**
+ Phyllaries with hair-like point or scarious appendage 8

8. Phyllaries with long recurved hair-like tips; flowers all alike, bisexual **9. Tricholepis**
+ Phyllaries with scarious appendages apically; marginal flowers enlarged, outcurved, 7-toothed, neuter .... **10. Centaurea**

9. Pappus of 2–4 retrorsely barbed awns 2–3.5mm .... **113. Bidens**
+ Pappus absent, of weak bristles to 0.8mm or a ciliate crown 10

10. Receptacle shallow; flowers 6–9 .... **99. Eleutheranthera**
+ Receptacle conic or elongate; flowers very many .... **97. Acmella**

11. Leaves alternate 12
+ Leaves opposite, at least on lower part of stem 31

12. Pappus present 13
+ Pappus absent 29

13. Phyllaries 2–6-seriate, imbricate or outermost series largest 14
+ Phyllaries in 1 series, interlocked at least below (sometimes separating when pressed – examine young capitula; calyculus also often present) 21

14. Pappus partly or entirely of plumose bristles 15
+ Pappus of capillary, scabrid or barbellate bristles and/or scales 16

15. Inflorescence a narrow, spike-like panicle; flowers 3–4 per capitulum **1. Ainsliaea**
+ Capitula corymbose, concealed in wool or solitary; flowers more than 10 per capitulum .... **7. Saussurea**

16. Capitula in glomerules surrounded by leaf-like bracts (c 20 capitula per glomerule); flowers c 4; pappus of 5–6 bristles .... **37. Elephantopus**
+ Capitula solitary, in corymbs or in panicles; flowers and pappus bristles more than 10 17

17. Perennial, often arborescent 18
+ Annual 20

18. Phyllaries of outermost series largest; flowers all female with tubular corolla or or all functionally male with undivided style apex .... **125. Cavea**

\+ Phyllaries of outermost series shortest; flowers all functionally bisexual with tubular-campanulate corolla and branched style 19

19\. Longest (innermost) phyllaries 20–24mm; corollas bilabiate **2. Gerbera** (*G. maxima*)
\+ Phyllaries not exceeding 12mm; corollas regularly 5-toothed .. **34. Vernonia**

20\. Involucre 7–10mm diameter; middle phyllaries with obtuse herbaceous appendage (base scarious) .... **36. Baccharoides**
\+ Involucre 2.5–3.5mm diameter; all phyllaries simple, acuminate **35. Cyanthillium**

21\. Capitula unisexual, 6–20 borne on bracteate scape, flowering before (basal) leaves develop .... **82. Petasites**
\+ Captitula bisexual or bulbiliferous, 1–many borne on leafy stem 22

22\. Leaves palmately veined; flowers 0 (entirely replaced by bulbils) to 16 .. 23
\+ Leaves pinnately veined; flowers more than 20 25

23\. Climbing plant with twining petiole bases; flowers 9–16 **88. Cissampelopsis** (*C. corifolia*)
\+ Plant not climbing; petioles not twining; flowers 0–7 24

24\. Basal leaves simple to palmatisect, to 18cm across, without dilated petiole base sheathing stem .... **83. Parasenecio**
\+ Basal leaves deeply palmatisect, 30–90cm across, with thin dilated petiole base sheathing stem ....**84. Ligularia** (*L. mortonii*)

25\. Petioles bases thin, dilated, sheathing stem on lower leaves; capitulum 1, cernuous .... **85. Cremanthodium** (*C. discoideum*)
\+ No petiole bases dilated and sheathing stem; capitula 1–many, not cernuous 26

26\. Capitula without calyculus; plants often glaucous .... **91. Emilia**
\+ Capitula with calyculus; plants not glaucous 27

27\. Phyllaries 5; flowers 5–6 ....**89. Senecio** (*S. kumaonensis*)
\+ Phyllaries 10 to c 20; flowers many 28

28\. Subscapose perennial herb from globose tuber or large, sprawling, bushy, perennial herb or subshrub; phyllaries 10–14 .... **92. Gynura**
\+ Erect annual, leafy throughout; phyllaries c 20 .... **90. Crassocephalum**

29\. Involucre c 10–15mm diameter ....**71. Tanacetum**
\+ Involucre 1.5–5.5mm 30

30. Plant thickly lanate; corollas yellow .... **72. Hippolytia**
\+ Plant glabrous to puberulous, sessile-glandular; corollas reddish or purplish .... *Ethulia* (see 35. *Cyanthillium*)

31. Twining climbers; flowers 4; phyllaries 4 (with one adjacent smaller involucral bract) .... **119. Mikania**
\+ Plant erect or scrambling, not twining; phyllaries more than 5; flowers 5–60 + ... 32

32. Phyllaries imbricate; pappus of 10–many bristles 33
\+ Phyllaries subequal; pappus of 3–5(–6) bristles or scales 36

33. Flowers 5–10 34
\+ Flowers c 30–60 35

34. Leaves simple, penninerved; phyllaries prominently sessile-glandular **118. Ageratina** (*A. ligustrina*)
\+ Leaves simple, 3-veined near base or inner phyllaries glabrous (then cauline leaves usually trifoliolate) .... **123. Eupatorium**

35. Flowers c 50–60; inner phyllaries 4.3–5mm; corollas 3.5–4mm **118. Ageratina** (*A. adenophora*)
\+ Flowers c 30; inner phyllaries c 8mm; corollas c 5.5mm **122. Chromolaena**

36. Achenes obovoid; pappus of 3–4 gland-tipped bristles .. **120. Adenostemma**
\+ Achenes linear-oblong: pappus of 5(–6) scales each usually bearing a scabridulous bristle .... **121. Ageratum**

MUTISIEAE

1. AINSLIAEA DC.

Subscapose perennial herbs. Leaves simple, alternate mostly radical. Two distinct flowerings each year: in spring with fully developed flowers, in autumn usually with cleistogamous flowers. Inflorescence narrowly paniculate, often spike-like. Capitula small, subsessile, ± cernuous, borne singly or in small fascicles, discoid. Involucres cylindrical; phyllaries several-seriate, imbricate, after flowering becoming hardened and spreading. Flowers 3–4, bisexual. Corolla limb deeply cut into 5 ± equal lobes. Anther bases caudate. Style branches very short, rounded. Achenes narrowly obovoid, slightly compressed, ribbed, densely silky pubescent; pappus plumose, brownish, rarely almost absent.

1. Petioles winged .... **1. A. latifolia**
+ Petioles unwinged .... **2. A. aptera**

1. A. latifolia (D. Don) Schultz Bipontinus; *A. pteropoda* DC. Fig. 118e–g.

Plant 6–120cm; stems sparsely araneous. Leaf blade ovate or elliptic, 2–11 × 1.2–5.5cm, acute, rounded to tapering at base, weakly crenate, often with linear denticles, sparsely pubescent on both surfaces or araneous beneath; petiole winged throughout or rarely on outer half only, up to 11cm, 0.3–2.2cm wide at apex; sometimes smaller elliptic cauline leaves present. Autumn flowering entirely cleistogamous. Phyllaries ovate to linear-lanceolate, 1–11 × 1–2mm, acute or acuminate, margins scarious. Fully developed corollas white or purplish, tube (3–)5–5.5(–7.5)mm, limb 6–9mm; cleistogamous corollas cylindric, c 4mm. Achenes c 5.5mm; pappus 7–8mm, rarely much reduced (c 0.5mm).

Bhutan: S – Chukka and Gaylegphug (117) districts, **C** – Ha, Thimphu, Punakha and Tongsa districts; **Darjeeling:** Darjeeling, Kurseong, Senchal etc.; **Sikkim:** Bakhim–Zongri, Lachen–Zemu Chhu, Tendong etc.; **Chumbi:** unlocalised (421). In warm broad-leaved forests, 1680–3960m. March–June (fully developed flowers), October–November (cleistogamous generation).

A specimen resembling *A. latifolia* from Dompho Pokri (?NE Darjeeling, Lister collection) produced its last flowers, fully developed, in January.

2. A. aptera DC.

Very similar to *A. latifolia* but leaves ovate, 4–9 × 3–7cm, acute, cordate, more prominently dentate, subglabrous above, sparsely pubescent beneath, petiole unwinged, to 13cm.

Bhutan: S – Chukka district, **C** – Ha, Thimphu, Punakha (71) and Tashigang districts; **Darjeeling:** Rississum; **Sikkim:** Boktak, Pemayangtse, Yakla etc.; **Chumbi:** unlocalised (421). In deciduous and coniferous forests, 2000–3100m. February–June (fully developed flowers), September–December (cleistogamous generation).

Two Bhutan collections of uncertain status from Chukka (Chukka district) and Ghunkarah (Tashigang district) with ovate scarcely dentate leaves, ± truncate at base and unwinged petioles differ from both species in bearing fully developed flowers in November.

2. GERBERA Gronovius

Scapose perennial herbs from short, creeping root-stocks. Leaves radical, simple or pinnatifid, petiolate. Scapes naked or rarely bracteate. Capitula solitary, cernuous, radiate or discoid. Involucres campanulate; phyllaries imbricate, several-seriate. Corollas bilabiate. Ray flowers when present 1–2-seriate; corolla with prominent spreading outer lip and small linear inner lip, functionally female (with staminodes present). Disc flowers bisexual; outer corolla lip 3-toothed;

inner slightly shorter, 2-toothed, often coiled; anther bases caudate; style branches linear, obtuse. Achenes ovoid or elliptic, beaked or not. Pappus scabrid.

1. Capitula discoid .... **1. G. maxima**
\+ Capitula radiate 2

2. Leaves entire .... **2. G. piloselloides**
\+ Leaves pinnatifid .... **3. G. nivea**

1. G. maxima (D. Don) Beauverd; *G. macrophylla* Hook.f.

Leaf blades ovate to oblong, 8–25(–50) × 6–15(–18)cm, acuminate, sagittate, denticulate, glabrous above, appressed whitish tomentose beneath with prominent veins; petioles 15–40cm, tomentose, rarely with few foliose lobules. Scapes 18–70cm, ebracteate, slightly inflated below capitula, loosely brownish tomentose. Capitula discoid. Involucre broadly campanulate, c 15mm diameter; phyllaries lanceolate, mostly glabrous, innermost 20–24mm. Corollas creamy white; basal tube 4.5–6.5mm; limb 9–11mm, divided for 4–6mm. Achenes narrowly elliptic, c 10-ribbed, ± compressed, gradually attenuate into a stout beak, 11–13mm, finely hispid; pappus 14–17mm, stramineous.

Bhutan: C – Tongsa district (Dakpai (117), Shamgong, Tongsa), Tongsa/Bumthang district (Yuto La), Mongar/Tashigang district (Kole La); **Sikkim:** Chunthang. In forests and in the shade of rocks, 1980–2285m. October.

2. G. piloselloides (L.) Cassini. ?Nep: *Pangmento*; Sha: *Laja-mayja*.

Leaf blades oblanceolate, 5–12 × 2–5.5cm, obtuse or subacute, rounded or attenuate at base, entire, sparsely pubescent on upper surface, ± tomentose beneath; petiole 0.5–5(–8)cm. Scape 10–45cm, ebracteate, brownish tomentose, somewhat inflated below capitulum. Capitula radiate. Involucre obconic, c 7mm diameter; phyllaries linear-lanceolate, 7–20 × 1–2mm, sparsely tomentose. Ray flowers 17–25; corolla white to dark crimson or purplish, outer lip 10 × 1mm, basal tube 8mm. Disc corollas 10–13mm, gradually tapered below. Achene body narrowly lanceolate to oblanceolate, 4.5–6mm, scurfy papillose, scarcely compressed, c 5-ribbed; beak slender, 5–7mm; pappus 10mm, reddish-brown.

Bhutan: C – Thimphu district (Pajoding, Tsalimarphe, Thimphu etc.), Punakha district (Pho Chu, Tinlegang, Wangdu Phodrang etc.), Tongsa district (Birti (117), Jirang Chu, Shemgang); **Sikkim:** Chunthang. In coniferous forest and forest clearings, on grassy banks and dry sandy soil, 600–3050m. March–August.

3. G. nivea (DC.) Schultz Bipontinus. Fig. 118h–i.

Leaves deeply pinnatifid, oblanceolate in outline, 2.5–20 × 1–3cm, subacute to abruptly acuminate, glabrous above, appressed whitish tomentose beneath; lateral segments 2–5, decreasing in size towards base, rounded, obiquely apiculate, (sub)entire; petiole 0.5–4.5(–9)cm, dilated and half-sheathing at base.

Scapes 7–15(–30)cm, sometimes with few linear bracts to 1cm, whitish floccose. Capitula radiate. Involucre campanulate, 10–12mm diameter; phyllaries lanceolate, 7–23 × 1.5–5mm, acuminate, glabrous. Corollas cream or pale lemon-yellow. Ray flowers 9–12(–15); corolla basal tube 4–5(–6)mm, outer lip 17–18mm. Disc corollas c 10mm. Achenes oblong, 5mm, several-ribbed, glabrous, erostrate; pappus 10mm, stramineous.

Bhutan: C – Tongsa district, **N** – Upper Mo Chu, Upper Mangde Chu, Upper Bumthang Chu and Upper Kulong Chu districts; **Sikkim:** Dzongri, Kupup; **Chumbi:** localities uncertain (Dungboo collections). In grass, among rocks, on cliffs, 3800–4880m. July–September.

3. LEIBNITZIA Cassini

Similar to *Gerbera* but scapes commonly bracteate, capitula erect, occurring in two distinct forms on the same plant at different seasons:

Spring generation: leaves ovate to weakly lyrate-pinnatifid, lanate or araneous beneath, often not or scarcely developed. Scapes short, capitula rather small. Ray flowers female; outer corolla lip scarcely spreading, enlarged; staminodes present. Disc flowers bisexual, shorter; corolla lips subequal; anthers caudate. Achenes elliptic, with 5–8 ribs, sparsely pubescent, tapering above or beaked.

Autumnal generation: leaves fully developed, lyrate-pinnatifid, araneous to glabrous beneath. Scape long, erect; capitula relatively large. Flowers all tubular, overtopped by pappus before involucre opens, of two types: outer flowers female, corolla longer, with short tridentate outer lip, inner lip reduced, staminodes absent; inner flowers bisexual, cleistogamous, corolla very narrow, shorter, distinctly bilabiate. Achenes as in spring generation.

Sometimes included in *Gerbera* as sect. *Anandria* (Lessing) Schultz Bipontinus. *L. anandria* (L.) Turczaninov [*Gerbera anandria* (L.) Schultz Bipontinus, *G. bonatiana* (Beauverd) Beauverd] may also occur in our area. Specimens of that species from E Nepal differ from *L. nepalensis* by their linear-lanceolate inner phyllaries, stramineous pappus, larger corollas of the vernal generation (ray flowers: basal tube 4.5mm, outer lip 4.5mm; disc flowers: 6–7mm) and thinly araneous to glabrous leaves in autumn. The achenes have no beak.

Most of our specimens of the vernal generation lack fruit and are doubtfully identified. These are marked with an asterisk below. Further vernal specimens requiring confirmation of identity have been collected in **Bhutan: C** – Thimphu district (Bela, Shodug–Barshong (71)), **N** – Upper Mo Chu district (Chamsa–Yabuthang, Chebesa–Lingshi (both 71)), the last three cited as *L. nepalensis.*

1. Achenes not beaked; most inner phyllaries of autumn generation ovate-lanceolate, acuminate, glabrous .... **1. L. nepalensis**
\+ Achenes distinctly beaked; inner phyllaries of autumn generation linear-lanceolate, araneous .... **2. L. ruficoma**

1. L. nepalensis (Kunze) Kitamura; *Gerbera kunzeana* A. Brown & Ascherson. Fig. 118j–k.

Spring generation: leaves ovate or oblong, rarely weakly lyrate-pinnatifid, petiolate, 1–7 × 0.3–1.3cm, apiculate, entire or dentate, glabrous above, araneous below. Scapes 2–4.5cm at anthesis. Involucre 3–4mm diameter; phyllaries acuminate, glabrous or sparsely araneous; outer narrowly triangular; inner ovate to oblong, c 7–8.5 × 2–3mm. Corollas white or pink, sometimes purple above. Ray corolla basal tube 2–3mm, outer lip 4–5mm. Disc corollas 4–6.5mm. Achenes (346) not beaked, 2.5–3mm (?immature); pappus purplish.

Autumnal generation: leaves 4–19 × 1–3.2cm, lyrate-pinnatifid, glabrous above, araneous beneath; terminal segments oblong or ovate, 1.5–5.5cm, with broad, shallow, often obliquely apiculate teeth: lateral segments up to 3 pairs, rounded, to 6mm deep and 13mm wide, obliquely apiculate; petioles up to 8cm, narrowly winged above or throughout. Scapes to 30cm. Involucre 4–7mm diameter; phyllaries mostly ovate-lanceolate, acuminate, outer partly araneous, inner c 14 × 4mm (to 20mm in fruit), glabrous. Outer corollas 4mm, inner 3.5mm. Achenes lanceolate, 6–7mm, rather compressed, narrowed above but not beaked; pappus 9–12mm, purplish.

Bhutan: N – Upper Mo Chu district (Yabu Thang–Laya*); **Darjeeling:** Sandakphu*, Tanglu–Sandakphu–Phalut (69, 101); **Sikkim:** Phalut, Samdong, Yakla etc.; **Chumbi:** Yatung (346). In close turf, 2500–3650m. April*–May (spring generation), autumn generation in fruit in October.

Two unlocalised collections from Chumbi are cited in the description of *Gerbera* (sect. *Anandria*) *lijiangensis* Tseng, distinguished from *L. nepalensis* in the autumnal generation by all phyllaries being ovate-lanceolate with a distinct, elongate apical appendage (none lanceolate, unappendaged), corollas only ⅓ length of pappus, anthers caudate (not truncate) and achenes somewhat narrowed at apex. The relevant organs are often missing from our material, but two collections from Sikkim show some of these characters.

2. L. ruficoma (Franchet) Kitamura; *Gerbera kunzeana* auct. non A. Brown & Ascherson

Very similar to *L. nepalensis*, but differing as follows:

Spring generation: leaves usually absent; phyllaries araneous, inner narrower; achene body ?5mm, beak 2mm.

Autumnal generation: lateral leaf segments 1–4 pairs; phyllaries linear-lanceolate, all araneous, inner to 23 × 2.2mm in fruit; achenes narrowly ellipsoid 4–6mm, beak 2–4mm; pappus 7mm, purple, rarely stramineous.

Bhutan: C – Ha district (Chelai La*), Thimphu district (Barshong), Bumthang district (Ura), **N** – Upper Mo Chu district (Lingshi, Tharizam Chu); **Chumbi:** Chumbi. In close turf, 3350–4300m. April* (spring generation), August–September (autumnal generation).

Details of spring generation from Hansen (346).

4. ADENOCAULON Hooker

Erect perennial herbs from creeping rhizomes. Leaves simple, alternate. Capitula small, few to many in glandular terminal panicles, disciform. Involucre broadly campanulate; phyllaries 5–7, subequal, 1–2-seriate, herbaceous. Receptacle convex. Outer flowers female; hypanthium covered in stout stipitate glands persistent in fruit; corolla very small, infundibular, unequally and deeply 5-lobed. Inner flowers male; corolla tube longer, ± cylindric; anthers sagittate at base; style undivided. Achenes obovoid or clavate, obscurely ribbed, covered above with stout red stipitate glands.

Descriptions supplemented with details of specimens collected outside our area.

1. Leaves ovate or reniform, deeply cordate at base, not decurrent on the stem; phyllaries 2–3mm; anthers rounded at apex .... **1. A. himalaicum**
+ Leaves ± triangular, truncate, weakly hastate or shallowly cordate at base, decurrent on the stem; phyllaries 4–5mm; anthers acute at apex **2. A. nepalense**

1. A. himalaicum Edgeworth; *A. bicolor* sensu F.B.I. non Hooker

Plants 15–45cm; stems erect, araneous, intermixed with subsessile glands on inflorescence. Leaf blade broadly ovate or reniform, 5–10 × 5–10cm, subacute or obtuse, deeply cordate at base, irregularly toothed, subglabrous above, whitish araneous beneath; petioles 1.5–13cm. Capitula c 5mm across; phyllaries ovate, c 2.5–3 × 1.5–2.3mm, reflexed in fruit. Flowers white, corollas glabrous. Female flowers 6–12; corollas 1.1–1.2mm. Male flowers 4–13(–22); corollas 1.7–2.1mm; anther connective rounded at apex, scarcely projecting. Achenes clavate, c 5.5mm, covered on upper parts with gland-tipped bristles c 0.4mm, glabrous at base.

Sikkim: Lachen. Woods, 2300–2800m. August.

Also reported from C Bhutan from Punakha district (Ritang (71)) and Tongsa district (71), but requiring verification.

FIG. 117. **Compositae.** a–d, *Saussurea eriostemon*: a, habit (× ⅔); b, capitulum: involucre 6-seriate, diameter measured between bars (× 3); c, inner, middle and outer phyllaries (× 2); d, flower with pappus partly cut away: flower bisexual, corolla tubular-campanulate, anthers exserted, pappus bristles plumose (× 3). e–i, *Cremanthodium thomsonii*: e, habit (× ⅔); f, capitulum: involucre 1-seriate, diameter measured between bars (× 1); g, inner and outer phyllaries (× 2); h, ray flower with pappus partly cut away: flower female, staminodes present, corolla with expanded ligule, pappus capillary (× 3); i, disc flower: flower bisexual, corolla tubular-campanulate, anthers exserted, pappus capillary (× 3). Drawn by Christina Oliver.

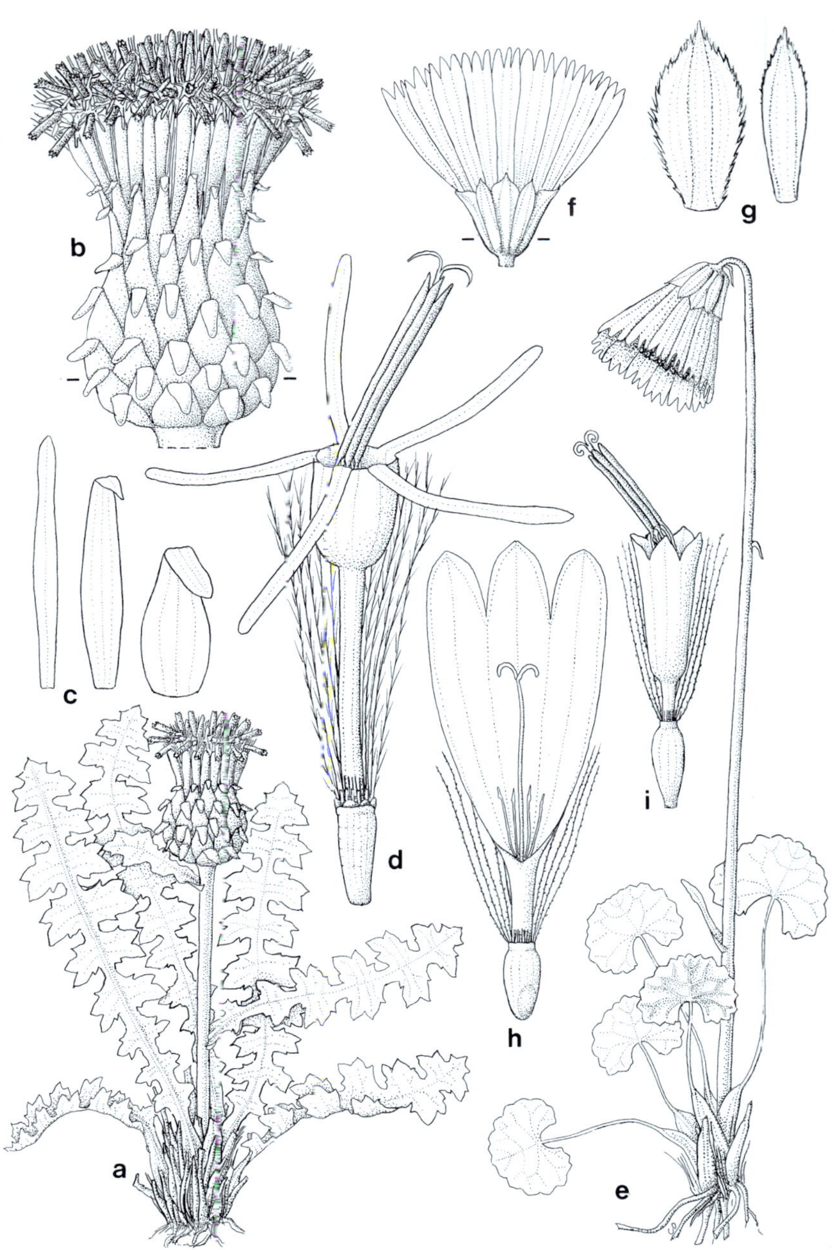
b
f
g
c
d
h
i
a
e

2. A. nepalense Bittmann. Fig. 118a–d.

Similar to *A. himalaicum* but leaves ± triangular, acute, truncate, weakly hastate or shallowly cordate at base, denticulate, petioles decurrent at base; phyllaries (3–)4.5–6.3 × 1.7–2.9mm; female corollas 1.1–1.3(–1.8)mm; male hypanthium sometimes few-glandular; male corollas 2–2.8mm, sometimes hairy outside; anther connective briefly mucronate at apex; achenes often hairy near base.

Sikkim: Tari, Tong, Chiya Bhanjang. 3050–3960m. August.

CARDUEAE

5. ARCTIUM L.

Stout biennial herbs. Leaves alternate, simple. Capitula discoid, subcorymbose. Involucres subglobose; phyllaries multiseriate, stiff, spreading, narrow lanceolate to acicular, hooked at apex. Receptacle flat, setose. Flowers bisexual. Corollas tubular-campanulate, purplish, limb narrowly infundibular, 5-lobed above, shorter than tube. Anther bases caudate. Style branches oblong, somewhat thickened below bifurcation. Achenes oblong-obovate, compressed; pappus short, bristles scabrid, free, deciduous.

1. A. lappa L.

Plant 1–1.5(?–2)m; stems araneous at first. Leaves ovate; blade 10–40 × 8–40, obtuse or acute, truncate or shallowly cordate at base, repand dentate, ± glabrous above, appressed white araneous beneath; petioles to at least 30cm. Involucre 10–15mm excluding hooked spines; phyllaries c 20 × 1mm, ± glabrous. Receptacle bristles 5mm. Corollas tube c 8mm, limb c 6mm including lobes. Achenes incurved, 5mm; pappus 2.5mm, stramineous.

Bhutan: C – Thimphu district (near Thimphu), Tongsa district (Chendebi), Bumthang district (Byakar). Margins of fields and gravel banks, 2440–2560m. July–August.

6. CIRSIUM Miller

The Shachop name *Zombu-zoo* has been applied to species of *Cirsium*.

Perennial herbs, usually caulescent; stems erect, usually branched. Leaves

---

FIG. 118. **Compositae: Mutisieae.** a–d, *Adenocaulon nepalense*: a, habit (× ½); b, male flower (× 12); c, female flower (× 12); d, achene (× 12). e–g, *Ainsliaea latifolia*: e, capitulum (× 3); f, flower, pappus partly removed (× 3); g, anther (× 5). h–i, *Gerbera nivea*: h, ray flower, pappus partly removed (× 3); i, disc flower, pappus partly removed (× 3). j–k, *Leibnitzia nepalensis*: j, habit, vernal phase (× ½); k, habit, autumnal phase (× ½). Drawn by Mary Bates.

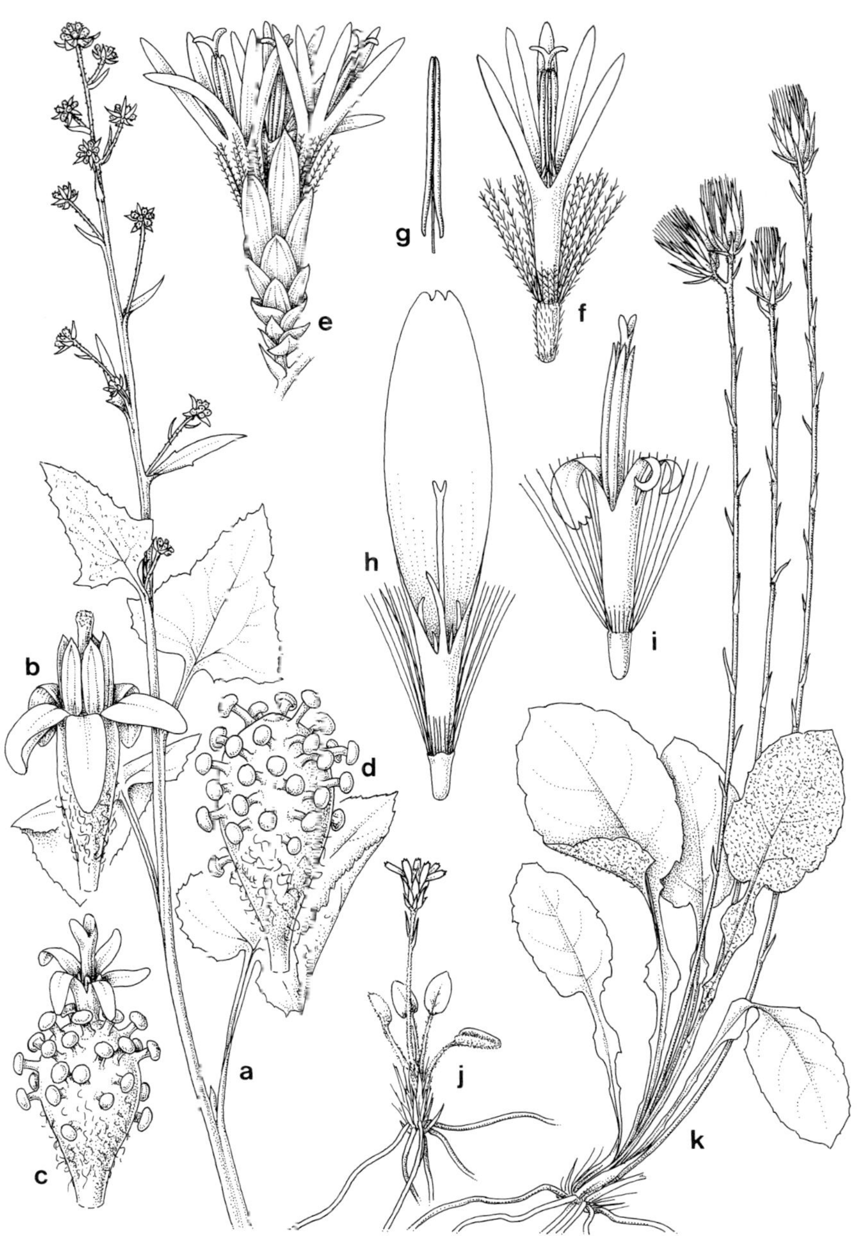

a
b
c
d
e
f
g
h
i
j
k

pinnatifid or pinnatisect, spiny at margins, upper surface setulose or not. Capitula discoid, solitary to several in subcorymbose terminal and axillary heads. Involucres narrowly to broadly campanulate; phyllaries several-seriate, imbricate, appressed or spreading or recurved in upper part, usually spiny, sometimes araneous. Receptacle flat or convex, bearing long simple bristles. Flowers bisexual or rarely unisexual, white to purple or sometimes yellow; corollas tubular-campanulate, limb narrow, with 5 linear lobes. Achenes oblong or obovoid, glabrous, truncate at apex with a cylindrical umbo, attachment sub-basal. Pappus plumose, several-seriate, connate at base, usually deciduous, innermost bristles longest, sometimes thickened at apex.

1. Flowers usually unisexual; corollas lobed almost to base of limb **1. C. arvense**
\+ Flowers bisexual; corollas lobed almost to half length of limb 2

2. Capitula 3–5cm diameter; phyllaries ± concealed by araneous hair, commonly only spines visible .... **2. C. eriophoroides**
\+ Capitula 1–3cm (or more) diameter; phyllaries mostly visible, not or moderately araneous 3

3. Leaves setulose on upper surface 4
\+ Leaves glabrous, cottony or sparsely pubescent on upper surface 6

4. Flowers yellow .... **3. C. lipskyi**
\+ Flowers purple 5

5. Phyllaries uniformly tapered, margin sometimes denticulate but not scarious-dilated near base of spine .... **4. C. falconeri**
\+ Wide, pale, scarious, lacerate appendage on margins of outer and middle phyllaries near base of spine .... **7. C. nishiokae**

6. Acaulescent; phyllaries pinnately spiny .... **5. C. souliei**
\+ Stems 50cm or more; phyllaries terminating in a single spine or acumen . 7

7. Phyllaries c 10-seriate .... **10. C. shansiense**
\+ Phyllaries c 5–6-seriate 8

8. Wide or narrow, pale, scarious, lacerate appendage on margins of outer and middle phyllaries near base of spine .... **6. C. wallichii**
\+ Phyllaries uniformly tapered, margin sometimes denticulate but not scarious-dilated near base of spine 9

9. Leaves moderately spiny (spines 5mm); phyllaries purplish, outer usually ending in spine 2–4mm .... **8. C. tibeticum**
\+ Leaves very spiny (spines 5–12mm); phyllaries greenish, outer with spines 4–8mm .... **9. C. verutum**

1. C. arvense (L.) Scopoli; *Cnicus arvensis* (L.) Hoffmann

Plant 40–70cm; root-stock creeping; stems sparsely araneous. Leaves oblanceolate in outline, 7–15 × 2–3.5cm, with 4–5 pairs of spine-fringed lobes to 1.5 × 1.5cm, ± glabrous above, white araneous beneath. Capitula several, ± cymose. Involucre 8–10mm diameter; phyllaries sparsely araneous, acuminate, outer ovate, shortly spine-tipped, inner linear, soft-tipped, ± appressed, 13–15.5mm. Subdioecious: male and female flowers on separate plants, but few achenes reported to mature sometimes on male plants. Corollas purplish. Achenes ± oblong, c 3mm; pappus white or stramineous. Male corolla tube 9–10mm; limb 6–6.5mm including lobes 4.5–5mm; anthers 5–5.5mm (including tails); pappus 10–10.5mm. Female corolla tube eventually 13–14mm; limb 4–4.5mm including lobes 2.5–3mm; anthers 1.5–2.2mm; pappus c 14mm at anthesis, eventually c 30mm.

W Bengal Duars: Jalpaiguri district (174).

Also collected just north of our area at Gyangtse; description based on specimens from there, Nepal and NE India. Sometimes segregated as *Breea arvensis* (L.) Lessing on account of its unisexual flowers, corolla limb short and 5-partite almost to base and mature pappus much longer than corolla.

2. C. eriophoroides (Hook.f.) Petrak; *Cnicus eriophoroides* Hook.f.

Plant 60–120cm. Leaves ovate-elliptic in outline, acuminate, with numerous lobes tipped with spine to 15mm, araneous on both surfaces but more densely so beneath, upper surface occasionally sparsely setulose; lower leaf blades to 45 × 15cm, narrowed to petiole to 15cm; upper ones 9 × 4cm, rounded at base, sessile. Capitula solitary at branch ends. Involucre 30–50mm diameter; phyllaries acuminate, 40–50 × 2–3mm, densely cottony. Corollas 30–40mm, dark purple, limb 15mm including lobes 5.5mm. Achenes obovoid, 5–6mm; pappus c 25mm, brownish.

Bhutan: C – Thimphu district (Geyche rest-house, Shodug, Somana), Tongsa/Bumthang district (Tibdeh La), **N** – Upper Mo Chu district (Phoudingi), Upper Mangde Chu/Upper Bumthang district (Ju La); **Sikkim:** (Chiya Bhanjang, Lachen, Lachung). Hillsides, grassland, screes and river banks, 3100–4730m. June–October.

3. C. lipskyi Petrak; *Cnicus griffithii* Hook.f. non *Cirsium griffithii* Boissier

Stems 2.5–4m, sparsely araneous. Leaves pinnatisect, auriculate at base with several spiny lobes, setulose above, white araneous beneath; lowest ones ± oblanceolate in outline, to 55 × 24cm, with c 9 pairs of leafy segments, lanceolate, to 12 × 1cm, acuminate to spine 10–15mm, each with 3–4 basal lobes with spines to 25mm; upper leaves smaller, similar, oblong-lanceolate. Capitula solitary or several, racemose or cymose. Involucre 20–25mm diameter;

phyllaries lanceolate to linear, outermost ending in short pale spine, inner c 28–30 × 1.5–2.2mm, cartilaginous above, briefly acuminate, scarcely pungent. Flowers yellow; corolla tube 16–20mm, limb 12.5–13.5 including lobes 5.5–6.5mm. Achenes obovoid, 5mm; pappus 22–27mm, dirty white.

Bhutan: C – Tongsa district (Shemgang), Mongar district (Rudong La), Tashigang district (Kole La). Open hillsides and forest margins, 1950–2440m. August–October.

4. C. falconeri (Hook.f.) Petrak; *Cnicus falconeri* Hook.f., *Cnicus involucratus* sensu F.B.I. non (DC.) Hook.f., *Cirsium verutum* sensu Kitamura (69 etc., not 135) non (D. Don) Sprengel. Fig. 119a.

Plant 0.5–1.5m; stems sparsely pubescent at least above and often araneous. Leaves densely setulose to almost smooth above, white araneous beneath, broadly auriculate at base; lowest 1–2-pinnatisect, oblanceolate in outline, c 20–40 × 5.5–20cm, with 7–12 pairs of leafy primary segments, spinous at tip and margin; upper leaves smaller, ovate-lanceolate in outline, ¼–⅘-pinnately incised; uppermost sometimes involucrate. Capitula at branch ends, usually solitary, sometimes in cymes of 2–4. Involucre 15–25mm diameter; phyllaries lanceolate to linear, narrowly acuminate, sparsely araneous; outer and often middle ones spiny, suberect to recurved; inner ones cartilaginous, 18–33 × 1–3mm. Flowers pale to dark purple; corolla tube 10.5–19mm; limb 10.5–18.5mm including lobes 5–7.5mm, tapered at base to abruptly campanulate. Achenes obovoid, dark brown, 6–7mm; pappus 15.5–23mm, brownish.

Bhutan: C – Thimphu, Tongsa and Bumthang districts, **N** – Upper Mo Chu, Upper Bumthang Chu and Upper Kulong Chu districts; **Darjeeling:** Kalipokhri, Tonglu–Sandakphu; **Sikkim:** Sebuteng, Tsomogo, Yakla etc. Open hillsides and forest clearings, 2745–4265m. July–October.

Extremely variable; in our area larger specimens also differ from *C. lipskyi* by their tapered 10–11mm corolla throats and more spiny phyllaries with longer spines.

5. C. souliei (Franchet) Mattfeld

Acaulescent. Leaves rosulate, pinnatisect, oblanceolate in outline 10–15 × 2–6cm, ± glabrous above, sparsely pubescent beneath; segments 6–8 pairs, ovate in outline, 3 × 1.5cm, bearing few coarse teeth ending in yellowish spines 2–4mm and many finer spinules. Capitula 4–7, ± sessile, clustered at base of leaves. Involucre 15–20mm diameter; phyllaries ovate, 20–30 × 3–5mm, long acuminate to 10mm yellow spine, fringed with yellowish spines to 5mm. Flowers

---

FIG. 119. **Compositae: Cirsieae.** a, *Cirsium falconeri*: apex of flowering shoot (× ⅔). b, *Saussurea graminifolia*: habit (× ⅔). c, *Tricholepis furcata*: apex of flowering shoot (× ⅔). Drawn by Margaret Tebbs.

a
b
c
M.Tejas

purple, corollas 20mm, limb 7mm, lobes 3mm. Achenes obovoid, 4.5mm; pappus 18–20mm, dirty white.

Sikkim: Gyong; **Chumbi:** Phari. 3960m. July–September.

6. C. wallichii DC.; *Cnicus wallichii* (DC.) Hook.f. var. *glabrata* Hook.f., *Cnicus wallichii* (DC.) Hook.f. var. *wightii* Hook.f. p.p.

Stems 2–3m, sparsely pubescent. Leaves pinatisect, elliptic-oblanceolate in outline, with sparse crisped-puberulous hair above and throughout or confined to midrib beneath; mid-cauline c 30 × 8cm excluding spines, segments c 9 pairs, ovate, ending in 1cm yellowish spine, margins with 1–3 large spines borne on coarse teeth and several spinules on each side. Capitula in racemes or panicles. Involucre 12–17mm diameter; outer phyllaries lanceolate, spine-tipped, with ovate, lacerate, scarious appendage at base of spine; inner ones with elongate, acuminate, scarious appendage and obsolete spine, c 17 × 2mm. Corolla tube 8mm, throat 4mm, lobes 4mm. Achenes obovoid, 5mm; pappus 10–12mm, brownish, bristles slightly thickened at apex.

Bhutan: C – Tongsa district (Tashiling (71), Tongsa); **Darjeeling/Sikkim:** unlocalised (Hooker collection). On wasteground, 2300–2400m. September.

The two collections seen are quite divergent. Hooker's, with leaf lamina glabrous beneath, can be referred to var. **glabrata** (Hook.f) Wendelbo, which has also been treated as a distinct species and united with *C. glabrifolium* (Winkler) O. & B. Fedtschenko from C Asia, with the corollas described as purple-red (404). The status of the Tongsa collection, with leaves puberulous but not araneous beneath, is less certain.

7. C. nishiokae Kitamura

Similar to *C. wallichii*, differing most obviously by its spinulose upper leaf surface. Leaves white-tomentose beneath. Only upper leaves seen: segments lanceolate, only basal spines borne on a coarse tooth, middle and upper spines and spinules borne on ± untoothed margin.

Darjeeling: below Tonglu (358), Sandakphu–Tonglu. 2900m. September.

Only one specimen seen from our area, without mature capitula.

8. C. tibeticum Kitamura; *Cnicus wallichii* (DC.) Hook.f. var. *wightii* Hook.f. p.p.

Stems 45–200cm, sparsely pubescent. Leaves pinnatifid, lanceolate in outline, 5–13 × 1–3cm, acuminate to 5mm yellowish spine, sparsely pubescent on both surfaces; lateral segments 4–5 pairs, coarsely toothed and moderately spiny at margins. Capitula in racemes, rarely solitary. Involucre 1.5cm diameter; phyllaries lanceolate, 8–15 × 1.5–3mm, outer acuminate to spine c 2–4mm, sparsely araneous, inner with long-acuminate scarious apex. Flowers purple, corolla 14mm, limb 8.5mm including lobes 2.5mm. Achenes oblong, 3mm; pappus 12–15mm, brownish.

Bhutan: C – Ha district (Sele La (358)), Thimphu district (Paro, Lamnakha,

Taba etc.), Punakha district (Lobeysa), Tashigang (Tobrang, Khaling); **Sikkim:** Lachung. Grassy banks and roadsides, 1450–2750m. June–September.

9. C. verutum (D. Don) Sprengel; *Cnicus argyracanthus* (DC.) Clarke

Similar to *C. tibeticum* but more spiny; leaves lanceolate-elliptic in outline, 12–20 × 4–5cm, acuminate to spine 8–12mm; lateral segments 8–12 pairs, narrowly ovate, acuminate to spine 5–10mm, spiny dentate; capitula usually in tight racemes; involucre 1–1.5cm diameter; phyllaries greenish, 8–15 × 1–2mm, outer acuminate to spine 4–8mm; corollas 17mm, limb 10mm including lobes 4mm; achenes obovoid, 4.5mm; pappus 16mm, dirty white.

Bhutan: C – Thimphu/Punakha district (Simtokha–Yuwak), Punakha district (Lobeysa, Ratsoo–Samtengang). Irrigation ditch, 1500–2100m. April.

10. C. shansiense Petrak; *Cnicus chinensis* sensu F.B.I. non (Gardner & Champion) Clarke

Stems ?c 1.5m, slender, pilose, little-branched above. Leaves pinnatifid, with many fine forward-pointing spinules on margins, subglabrous above, white-tomentose beneath lateral lobes remote, (ovate-)triangular, spine-tipped; mid-cauline c 120 × 35mm with 4 pairs of lateral lobes. Capitula solitary, ?nodding, on slender peduncles 100–300mm, 15–20mm diameter. Phyllaries c 10-seriate, linear-lanceolate; outer with spine c 1mm; inner to 18mm, with apex scarious and somewhat dilated and eroded. Corollas mauve, tube c 11mm, limb c 10mm, divided to unequal depths by linear lobes 5–7mm. Pappus pallid, c 18mm.

Bhutan: C – Tashingang district (Shapang). Open hillside, 2150m. September.

There has been little agreement on the delimitation of species related to *C. lineare* (Thunberg) Scopoli. We have tried to follow Shi (404) here.

7. SAUSSUREA DC.

Acaulescent to medium-sized perennial herbs, rarely monocarpic. Leaves alternate, simple, entire to pinnate or bipinnatifid, without prickles. Capitula 1–many, discoid, sometimes enveloped in long wool or surrounded by leaf-like, coloured bracts. Involucre campanulate to ovoid; phyllaries several-seriate, imbricate. Receptacle flat or convex, usually densely setose, rarely naked or bearing separate linear paleae overtopping corollas. Flowers bisexual. Corollas purplish or violet (rarely white), narrowly tubular-campanulate, limb deeply 5-lobed. Anther bases caudate, lacerate or woolly; filaments glabrous. Style branches filiform, obtuse. Achenes oblong or obovoid, quadrangular or ribbed, smooth or rugose, sometimes coronulate at apex. Inner pappus bristles plumose, connate at base, outer pappus often present, of plumose persistent bristles usually a little shorter than inner bristles or of scabrid bristles usually caducous, much shorter and often fewer than inner bristles, rarely with few plumose hairs as well.

1. Stem apex clavate, hollow; upper stem, upper leaf bases and sometimes involucres with a continuous covering of long whitish lanate hairs 2
\+ Stem apex tapered, solid; upper stem and upper leaf bases rarely with a continuous covering of long lanate hairs (*S. graminifolia*) 8

2. Capitula solitary .... **7. S. sp. A**
\+ Capitula few to many 3

3. Phyllaries chartaceous or achenes with stipitate glands 4
\+ Phyllaries almost membranous to hyaline; achenes glabrous 5

4. Recetacle with bristles; achenes usually glabrous .... **4. S. gossipiphora**
\+ Receptacle without bristles; achenes glandular-stipitate ..... **5. S. nishiokae**

5. Corolla limb 10mm including lobes; outer pappus plumose **6. S laminamaensis**
\+ Corolla limb 6–7mm including lobes; outer pappus scabrid 6

6. Capitula borne on sides and apex of conical upper stem, concealed by leaves and wool .... **2. S. spicata**
\+ Capitula borne only on dilated rounded to flattened stem apex, not concealed 7

7. Cauline leaves denticulate to pinnatisect on upper half; phyllaries lanate at least above .... **1. S. simpsoniana**
\+ Cauline leaves 3–6-lobed at apex; phyllaries sparsely pilose .. **3. S. tridactyla**

8. Upper leaves whitish or coloured, sometimes membranous 9
\+ Upper leaves similar to others in colour and texture 11

9. Upper leaves whitish, completely enveloping capitula ..... **10. S. obvallata**
\+ Upper leaves purplish, not completely enveloping capitula 10

10. Lower leaves linear-lanceolate, grass-like, entire .... **12. S. hookeri**
\+ Lower leaves oblanceolate-elliptic, denticulate .... **11. S. uniflora**

11. Stems 30–200cm 12
\+ Stems 0–30cm 18

12. Capitula solitary or several, well separated; peduncles of mature capitula exceeding 10mm 13
+ Capitula several to many, densely clustered; peduncles of most mature capitula 1–10mm 16

13. Lower leaves simple, serrate 14
+ Lower leaves pinnately divided or lobed 15

14. Leaves elliptic, finely serrate; phyllaries blackish-margined **14. S. fastuosa**
+ Leaves ovate, coarsely serrate; phyllaries greyish .... **21. S. deltoidea**

15. Capitula usually solitary at branch ends, with exposed bearded paleae; phyllaries dark .... **20. S. auriculata**
+ Capitula in elongate racemes of 6 or more, with concealed unornamented receptacle bristles; phyllaries pale .... **22. S. heteromalla**

16. Leaves simple, serrate .... **39. S. candolleana**
+ Leaves pinnately divided 17

17. Terminal leaf segment larger than lateral segments .... **13. S. laneana**
+ Terminal leaf segment scarcely larger than lateral segments **24. S. przewalskii**

18. Capitula 3 or more on some shoots 19
+ Capitula 1(–2) on each shoot (each shoot developed from a separate whorl of senesced leaves) 24

19. Leaves entire or remotely denticulate 20
+ Leaves dentate to pinnatisect 21

20. Acaulous; leaves linear, glabrous beneath .... **8. S. stella**
+ Caulescent; leaves linear-oblanceolate, tomentose beneath **40. S. pantlingiana**

21. Leaf blades green beneath with subsessile glands and few araneous hairs **9. S. kingii**
+ Leaf blades white-araneous and eglandular beneath 22

22. Acaulous; phyllaries conspicuously araneous; outer pappus bristles plumose .... **27. S. fibrosa**
+ Acaulous or caulescent; phyllaries subglabrous, sparsely strigillose or sparsely araneous; outer pappus absent or of ± scabrid bristles 23

23. Acaulous; longest phyllaries 17–28mm .... **28. S. yakla**
+ Stem usually present; longest phyllaries 12–15mm .... **25. S. nimborum**

24. Leaves entire, remotely denticulate or finely serrate 25
\+ Leaves deeply dentate or pinnately divided 30

25. Upper stem, upper leaf bases and involucre with a continuous covering of long whitish lanate hairs .... **23. S. graminifolia**
\+ Acaulous or upper stem and upper leaves glabrous or much shorter-haired 26

26. Leaves linear, blades revolute, entire 27
\+ Leaves linear-oblanceolate to ovate, blades flat or margin recurved and dentate 28

27. Leaves glabrous above, sparsely pilose beneath .... **17. S.** cf. **columnaris**
\+ Leaves covered by long, appressed, interwoven hairs ..... **18. S.** cf. **sericea**

28. Leaf blades 7–40 × 2–6mm 29
\+ Basal leaf blades longer than 40mm or wider than 6mm 30

29. Leaves green and glabrous above .... **19. S. wernerioides**
\+ Leaves greyish, glandular and sparsely araneous above **16. S. cochlearifolia**

30. Acaulous; leaves finely and closely serrate ....**15. S. katochaete**
\+ Plant 80–300mm; leaves subentire to remotely repand-denticulate **41. S. hieracioides**

31. Leaves green beneath and subglabrous or sparsely araneous (then achenes (5.5–)6.5–8mm even at anthesis) 32
\+ Leaves whitish beneath and araneous or appressed tomentose 34

32. Outer pappus absent or of short scabrid bristles .... **38. S. andersonii**
\+ Outer pappus present, plumose, c 2–3mm shorter than inner pappus **31. S. eriostemon**

33. Outer pappus absent or of plumose bristles more than half length of inner pappus, ± persistent 34
\+ Outer pappus of setose or very sparsely plumose, bristles less than half length of inner pappus, caducous at least when pressed 40

34. Phyllaries pinnatifid .... **32. S. pinnatiphylla**
\+ Phyllaries simple, entire or outermost rarely with 1–4 linear teeth 35

35. Receptacle naked or rarely with a few bristles......... **37. S. polystichoides**
\+ Receptacle densely setose 36

36. Acaulous; leaf blades oblanceolate in outline, 15–90mm broad, undivided part often much wider towards apex 37
\+ Stems often present; leaf blades linear-elliptic to -oblong or rarely-oblanceolate in outline, 3–22mm broad, width of undivided part ± uniform throughout 38

37. Capitula always solitary; phyllaries mostly glabrous outside, sometimes weakly woolly towards tip; outer pappus distinct with finer and sometimes shorter bristles than inner pappus .... **29. S. sughoo**
\+ Capitula 1–7; phyllaries finely and densely scabridulous outside; pappus bristles overlapping but fused at base, distinct whorls not readily apparent **28. S. yakla**

38. Leaves thinly araneous above or lateral lobes with 3 or more teeth **30. S. donkiah**
\+ Leaves not araneous above; lateral lobes with 0–1(–2) teeth 39

39. Leaves deeply pinnatisect; outer phyllaries linear-lanceolate, erect **35. S. taraxacifolia**
\+ Leaves less than ⅔-divided or outer phyllaries ovate with ligulate, reflexed appendage .... **34. S. purpurascens**

40. Leaves crisped-pubescent and more thinly araneous above **36a. S. leontodontoides** var. **leontodontoides**
\+ Leaves subglabrous to crisped-pubescent or scabridulous above without araneous hair 41

41. Leaf blades up to 22 × 10mm .... **26. S. obscura**
\+ Leaf blades 30–250 × 10–40mm 42

42. Involucre 13–25mm across; upper half of outer phyllaries reflexed **33. S. pachyneura**
\+ Involucre 5–11mm across; outer phyllaries erect or recurved at tip 43

43. Leaves broadly oblanceolate to narrowly oblong-elliptic, less than ⅔-divided at widest point .... **25. S. nimborum**
\+ Leaves linear-elliptic or -oblanceolate, deeply pinnatisect throughout **36b. S. leontodontoides** var. **filicifolia**

1. S. simpsoniana (Fielding & Gardner) Lipschitz; *S. sacra* Edgeworth

Plant 7–13cm with tap-root and often stout squamate elongate stolons; stems clavate, hollow, densely clothed above in whitish wool and woolly leaves, rounded at apex. Cauline leaves narrowly oblanceolate, 2–6 × 0.4–1cm, acuminate (acute), attenuate at base, denticulate to pinnatisect on upper half, lower ones thinly pubescent above, tomentose beneath, upper ones more woolly; apical leaves linear, ± entire, densely woolly, purplish to dark violet, surrounding and sometimes interspersed among but not concealing capitula. Capitula sessile, crowded around stem apex in inflorescence 2–5cm across. Involucre 3–5mm diameter; phyllaries linear-lanceolate, ± membranous, 10–13 × 1.8–2.7mm, acuminate, woolly at least above. Receptacle setose. Corolla tube 3.7–5.3, limb 6–7mm including lobes 2–3mm. Achenes 3–4mm, glabrous; inner pappus brownish, 9–10mm, outer scabrid, 2–3mm, caducous.

Bhutan: N – Upper Mo Chu (Lingshi La); **Sikkim:** Khungme La, Pauhunri, Yume Samdong etc.; **Chumbi:** Chumbi. 3650–5400m. August–October.

A polymorphic species; our more common form is described above. The Lingshi La collection and one unlocalised Cave collection from Sikkim differ considerably by their longer leaves, involucres to 8mm diameter, phyllaries 11–19mm and receptacles sometimes sparsely short-setose only.

2. S. spicata Kitamura

Similar to *S. simpsoniana*, but plant 20–28cm; upper stem conical, bearing capitula on sides and apex, concealed by leaves and hair; leaf hair floccose; involucres c 5mm diameter; phyllaries to 19mm.

Bhutan: N – Upper Mo Chu district (Lingshi La); **Sikkim:** Theu La. Among boulders, 4570–5180m. Flowers in bud in August.

3. S. tridactyla Hook.f.

Similar to *S. simpsoniana*, but plant 2.5–16cm; cauline leaves spathulate, oblanceolate or obcuneate, 1.5–4(–8) × 0.3–0.7cm, apically 3–7 crenately lobed, narrowed towards base, densely whitish or brownish lanate on both surfaces; involucre 3–4mm diameter; phyllaries oblong-oblanceolate, broadly hyaline-margined, 10–12.5 × c 2.4mm, acute, dark violet and sparsely pilose at apex; corolla tube 6.5–7mm, limb 6.5–7 including teeth 2.5–3.5mm; achenes 4.5mm; inner pappus 12mm, outer 1.5–4mm.

Bhutan: N – Upper Mo Chu district (Kangla Karchu La W), Upper Pho Chu district (Gyophu La), Upper Bumthang Chu district (Marlung, Pangotang, Waitang), Upper Kuru Chu district (Gong La); **Sikkim:** Lhonak, Onglakthang, Yume Samdong etc.; **Chumbi:** Chomp Lhari, Chumighata, Phari etc. Mostly sandy or stony screes, sometimes peaty turf, 4250–5450m. July–October.

4. S. gossipiphora D. Don. Med: *Jagoid sugpa.*

Monocarpic herb 7–40cm from one (rarely several, fasciculate) long fleshy tap-roots; stem solitary, hollow, club-shaped, 1.5–5cm broad at apex, thickly covered above by long whitish or tawny wool borne on stem itself and basal part of leaves, with single opening to exterior at apex. Cauline leaves linear-oblanceolate, 7–16 × 0.3–3cm, acuminate, sessile, remotely denticulate to deeply pinnatifid with small, acuminate, entire or dentate segments rarely extending beyond halfway to midrib, glabrous in upper half, woolly on both surfaces in lower half; uppermost leaves linear, entire, extending 2–6cm above capitula, woolly throughout. Capitula numerous (rarely few), densely crowded at stem apex. Involucre 5–6.5mm diameter; phyllaries chartaceous below, with softer, acute, dark violet, short-pubescent tips, mostly oblong, 7–10.5mm, outermost ovate or triangular, shorter. Receptacle setose, bristles 1–2mm. Corollas 12–13mm including lobes 2mm. Achenes blackish, 4.5–5mm, glabrous (?rarely some stalked glands above); pappus double, inner 10–13mm, outer scabrid, 3–5mm, caducous.

Bhutan: **N** – Upper Mo Chu district (Kangla Karchu La, Lingshi La), Upper Mangde Chu district (Saga La), Upper Bumthang Chu district (Waitang); **Sikkim:** Bikbari, Sebu La, Yakla etc.; **Chumbi:** Chola, Kalaree, Upper Kangbu etc. On screes, hillsides in open or among shrubs, wet gullies, 3950–5000m. August–September.

Three immature collections from Sikkim (Chaunrikhiang, Goecha La and Samiti–Chemthang, 4550–4900m) are ± intermediate between *S. tridactyla* and *S. gossipiphora*, with leaves oblanceolate-spathulate, c 4–10cm, c 4–7-dentate, thinly lanate throughout, phyllaries tending towards either species and glabrous or ?glandular-stipitate achenes. Their status is unclear.

5. S. nishiokae Kitamura

Similar to *S. gossipiphora* but phyllaries narrowly oblong, 7–9mm, obtuse, woolly; receptacle naked, pitted; achenes minutely glandular-pilose.

?Bhutan: Lhonack (Nishioka collection (360)). 4600m. September.

No certain examples seen, though a specimen from Chang Kyetoyakot, Sikkim, 4250m may belong here. It has naked receptacles and achenes with few glandular hairs above. However, specimens from our area with similar achene hair but setose receptacles are referred to *S. gossipiphora* here.

6. S. laminamaensis Kitamura

Similar to *S. gossipiphora*, but plant 8–20cm; cauline leaves 3.5–10cm, deeply pinnatisect with linear segments (cut at least ⅔ to midrib) or bipinnatisect; receptacular paleae reduced or absent; phyllaries ± oblong, thin, hyaline, 9–12 × 1.5–3.2mm, obtuse or acute, lacerate at apex, glabrous; corolla tube 6mm, limb 10mm including lobes 3.2mm; outer pappus of plumose persistent bristles; achenes 5.5–7mm, glabrous.

Sikkim: Goecha La, Kankola, Zongri–Bikbari etc. 4250–4900m. September.

7. S. sp. A. Med: *Sugda.*

Similar to *S. gossipiphora*, but plant c 7–12cm; leaves remotely denticulate or sinuate-dentate, lanate-ciliate at margin, basal to 20cm, cauline 3–8cm with some araneous hair throughout; capitula solitary at stem apex; involucre c 10mm diameter; phyllaries herbaceous-membranous, similar in shape, texture and vestiture and gradually passing into linear-lanceolate uppermost leaves, inner series c 10–20mm; receptacle apparently lacking bristles.

Bhutan: **N** – Upper Mo Chu district (Chhew La), Upper Mangde Chu district (Saga La). Sandy and gravelly screes, 4700m. Flowers in bud July–August.

Mature flowers and fruit not seen (though short, scabrid outer pappus apparent). This may be the taxon described from SE Tibet as *S. gossipiphora* subsp. *conaensis* Liu.

8. S. stella Maximowicz

Monocarpic, acaulous. Leaves rosulate, linear, sessile, 5–10 × 0.6–1.2cm, gradually acuminate from ovate glossy purplish base, glabrous, entire. Capitula 5–30, crowded, subsessile, in sessile corymbose head. Involucre 5–7mm diameter; phyllaries ± oblong, 9–13 × 2–4mm, glabrous, purplish tipped. Receptacle short setose. Corollas 15–20mm including lobes 4–5mm. Achenes oblong, 3mm, ribbed, with corona of c 5 thick stiff scales 2mm; pappus of 2 ± distinct rows of plumose bristles partly fused at base, 15–17mm, whitish stramineous.

Bhutan: C – Thimphu district (Daga La), **N** – Upper Mo Chu district (Tharizam Chu); **Sikkim:** Khangchenyao, Khora Phu, Naku Chho etc.; **Chumbi:** Kalaeree, Ta-loong, Bhutan border near Phari. On grassy hummocks in marshes and on boggy moorland, 4145–4880m. August–September.

9. S. kingii C.E.C. Fischer

Perennial herb c 3cm, viscid, thinly araneous-tomentose, hairs sometimes from stout bases; shoots from deep stolons, subrosulate with short branches. Leaves pinnatifid, linear-oblanceolate in outline, probably rather fleshy; segments triangular (ovate), apiculate, with 0–3 minutely apiculate teeth or small lobes; basal leaves 4–13 × 0.8–2cm. Capitula ± crowded in terminal corymb. Involucre 4–7mm diameter; triangular bracts at base with small leafy appendages passing into ovate acuminate phyllaries, 8–11 × 3–6mm, with brown (purplish) scarious upper margins. Receptacle naked. Corolla c 12mm including lobes 3.5–4mm. Pappus white, inner bristles 9–10mm, outer ones few, scabrid, 1–4mm, caducous.

Chumbi: Dotag, Phari, Tuna. 4400–4550m. August–September.

Immediately north of our area some populations comprise more widely branched plants with more finely divided, more thickly tomentose leaves and obovoid, scarcely compressed, finely ribbed immature achenes with long fine hair.

10. S. obvallata (DC.) Edgeworth

Rhizomatous, 30–70cm; stems erect, pubescent at least above, surrounded at base by dead leaves. Lower leaves oblanceolate, 15–30 × 2–5cm, ± acute, long-tapering to broad semi-amplexicaul petiole, dentate, puberulous (to sparsely tomentose); uppermost leaves membranous, 8–15 × 3–13cm, concave, loosely surrounding inflorescence, whitish or pale green. Capitula 4–11 in compact corymbs. Involucre 8–15mm diameter; phyllaries mostly lanceolate, purplish-black, 10–20 × 2–4mm, acuminate, sparsely pubescent. Corollas 13–17mm, including lobes 4–5.5mm. Achenes oblong, 5–6mm; pappus double, inner 11–14mm, outer scabrid, 3–5mm.

Bhutan: C – Thimphu and Tongsa districts, **N** –Upper Mo Chu, Upper Pho Chu, Upper Bumthang Chu, Upper Kuru Chu and Upper Kulong Chu districts; **Sikkim:** Lachen, Tangkar La, Thangshing etc.; **Chumbi:** Gautsa, Kalaeree, Phari. On damp screes, rocky hillsides, rich pastures, peaty soils etc., 3660–4880m. July–September.

11. S. uniflora (DC.) Schultz Bipontinus

Plant 7–65cm; stems erect, rather sparsely whitish villous above, surrounded at base by leaf remains. Lower leaf blades oblanceolate (to elliptic), 2.5–18 × 1.5–4cm, ± acute, attenuate (subobtuse) at base, denticulate, sparsely pubescent on both sides, petiole 1–15cm; mid-cauline leaves ± ovate, sessile; upper ones dark red or purplish, membranous, ± concave, barely or incompletely enveloping capitula. Capitula 1–10(–16 elsewhere (360)), 7–18mm diameter, corymbose. Phyllaries ovate to linear, 5–22 × 1–6mm, blackish or purplish, sparsely villous. Corollas 13–16mm including lobes 5mm. Achenes oblong, c 5mm, ribbed; pappus double, inner 9–12mm, outer scabrid, 5–8mm.

Bhutan: C – Thimphu district (Pajoding*, Shodug*, Tanga–Barshong*, Shodug–Barshong), Tongsa district (Padima Tso), **N** –Upper Mo Chu district (Chomp Lhari Chakan* (360), Yale La* (360), Shinkarap, Tharizam Chu), Upper Bumthang Chu (Domchen), Upper Kulong Chu district (Me La*); **Darjeeling:** Singalila; **Sikkim:** Chiya Bhanjang, Thanngu, Yume Samdong etc.; **Chumbi:** Gautsa–Phari, Kangboo, Yatung. On open hillsides, damp peaty meadows, among scrub, in shallow streams, 3050–4550m. July–November.

Extremely variable in habit and number of capitula. The material examined can be divided into two varieties: one with all capitula solitary or in heads of 2–4(?–5) only in part of any population, with the central capitulum usually larger on a stouter peduncle (var. *uniflora*); the other with all heads of 5–10 capitula, all small with short phyllaries (longest c 16(–19)mm), on slender peduncles (asterisked in the distributions given above, ?without a valid name but previously recorded as var. *conica* (Clarke) Hook.f. p.p. (80) and *S. conica* Clarke (360)). However, there were no examples among the types seen of *S. conica* Clarke except perhaps an intermediate specimen from Singalila (*Clarke* 12541, K), the remaining types all being var. *uniflora* in the sense used here.

12. S. hookeri Clarke

Tufted plant 7–30cm; stems ± sericeous, surrounded by leaf remains at base. Lower leaves linear, grass-like, 6–18 × 0.3–0.4cm, subacute, broad sheathing at base, entire, revolute at margin, ± glabrous above, sparsely sericeous beneath; upper leaves lanceolate, purplish, c 3.5–5 × 0.5–1cm, ± plane, uppermost gradually passing into phyllaries. Capitulum solitary. Involucre 1–2cm diameter, sparsely sericeous; outer phyllaries herbaceous, purplish, ovate, erect, c 13 × 8mm; inner lanceolate, c 13 × 1.5–4mm, scarious below. Corolla 11.3–15.3mm including lobes 2.7–4mm. Achenes (immature) obovoid, 4mm, ribbed; pappus double, inner subequal to corolla tube, outer 2–5mm, scabrid.

Bhutan: N – Upper Mo Chu district (Shingche La (360)), Upper Pho Chu district (Gyophu La); **Sikkim:** Chapopla, Khangchenyao, Sebu La, Theu La etc.; **Chumbi:** Chomp Lhari. Grassy banks, 4265–5500. July–November.

13. S. laneana W.W. Smith

Stems 30–90cm, whitish araneous at first, base surrounded by fibrous leaf remains. Lower cauline leaves lyrate, petiolate, 10–20cm, decurrent at base, scabrid-puberulous above, araneous beneath; segments acute or obtuse, dentate or denticulate; lateral segments 2–3 pairs, ± ovate, 1–2.5 × 1–2cm; terminal segment ± ovate, 4–8 × 4–6cm, often hastate or lobed; petiole partly (entirely) winged, sometimes dentate. Upper cauline leaves often sessile and shallowly pinnatifid. Capitula 3–10, tightly corymbose at stem apex, surrounded by several leaves. Involucre 7–15mm diameter; phyllaries lanceolate, 5–20 × 2–4mm, blackish above, sparsely pilose. Receptacle densely setose. Corollas 14–20mm including lobes 4–5mm. Anther tails woolly. Achenes oblong, 4–5mm, obscurely ribbed; pappus double, inner brownish, 11–14.5mm, outer scabrid, 1.5–8mm, caducous.

Sikkim: Cho La, Lagyap, Sherabthang etc. 3350–4250m. September–November.

A sterile plant collected at Somana, Thimphu district, Bhutan, 4370m, also seems to be this species.

14. S. fastuosa (Decaisne) Schultz Bipontinus; *S. denticulata* (DC.) Clarke

Stems 40–130cm, sparsely appressed pubescent. Leaves ovate-elliptic to lanceolate, 6–15 × 1.5–5.5cm, acute or acuminate, rounded at base, shortly petioled (sessile), denticulate or finely serrate, green and subglabrous above, finely white appressed tomentose beneath. Capitula 1–5 at branch ends or loosely racemose, 1–1.5mm diameter. Phyllaries ovate to lanceolate or oblong, 7–15 × 2–5mm, with blackish puckered margin. Paleae linear, as long as pappus, incurved and coloured at apex. Corolla 16mm including lobes 3mm. Achenes linear, 4–5mm, ribbed; pappus single, 11–13mm.

Bhutan: N – Upper Mo Chu district (Chusom, Ganguel); **Sikkim:** Lachung, Trakarpo, Yak La etc.; **Chumbi:** Chumbi. On slopes above rivers, 2150–3670m. August–November.

15. S. katochaete Maximowicz; *S. katochaetoides* Handel-Mazzetti, *S. rohmooana* Marquand & Shaw

Acaulous. Leaves rosulate; blade elliptic-oblanceolate, 2.5–8.5 × 1.5–3cm, acute or acuminate, narrowed into broad petiole, sharply denticulate to finely serrate, green, sparsely pubescent above, white-araneous beneath; leaf base sheathing. Capitulum solitary. Involucre 10–22mm diameter; phyllaries ovate to linear, 8–20 × 1.5–6mm, blackish- or black-margined, ± glabrous. Receptacle short setose. Corollas 15–19mm including lobes 4–5mm. Achenes obovoid, 4mm, ribbed; inner pappus 13–16mm, outer scabrid, 1–4mm, some bristles deflexed over achene.

Bhutan: C – Thimphu district (Thodum Chu, Somana); **Sikkim:** Chumegata, Lhonak; **Chumbi:** unlocalised (421). Among boulders on river bank and among rocks, 4170–4570m. August–October.

16. S. cochlearifolia Chen & Liang

Acaulous; rhizome branched, collars surrounded by remains of dead leaves. Leaves spathulate or oblanceolate, 25–40 × 3–6mm, acute, entire, glandular and sparsely araneous above, densely white tomentose below with prominent midrib; petiole 12–13mm, purplish. Capitula solitary, 8–12mm across. Involucre ovoid, 11mm; phyllaries 3–4-seriate, lanceolate, tawny yellowish-green, dark purple on margins above, sparsely araneous. Corollas purple, 15–16mm including lobes 5mm. Anthers turning purple. Achene 1.5mm, transversely rugose. Pappus 2-seriate, tawny yellow, outer scabrid, 2–3mm.

Chumbi: unlocalised (328). In meadows.

No reliable examples seen; description above based on type description (328). A collection from Chholhamoo, Sikkim, 5250m, flowering in August, might be conspecific, but differs by its linear-elliptic leaves with few obscure teeth beneath the revolute margins, campanulate 16mm involucre and smooth 5mm achenes.

17. S. cf. **columnaris** Handel-Mazzetti

Acaulous, densely caespitose, with numerous complete dead leaves persisting and forming columns 1–2cm thick. Leaves linear, rosulate, blade 10–15 × 1mm, acute, margins strongly revolute, glabrous above, sparsely pilose beneath, base sheathing, castaneous. Capitulum solitary, sessile. Involucre c 10–15mm diameter, phyllaries ovate, acuminate to linear-lanceolate, 12 × 1.8–4mm, purplish-tipped, white pilose, inner thickened and coriaceous below. Receptacle naked. Corollas purple, tube (?not fully developed) 4mm, limb 7mm including lobes 4mm. Achenes glandular-stipitate above; inner pappus brownish, c 8mm, outer scabrid, to 2.5mm, scabrid.

Bhutan: N – Upper Bumthang Chu district (Tolegang). On steep open hillsides, 4730m. October.

S. columnaris was described from W Yunnan where usually plants are larger, capitula pedunculate, receptacles densely set with bristles c 7mm and achenes glabrous. The Bhutan specimens may represent a smaller form of the species, linked by rather intermediate plants from SE Tibet with sessile capitula, dense receptacle bristles c 3mm and achenes glandular-stipitate above.

18. S. cf. **sericea** Chen & Liang

Similar to the previous species, but differing by its more slender stem columns with only leaf bases persisting; leaf blades 10–30(–70)mm, covered by long appressed interwoven silvery hairs; thinner inner phyllaries; short setose receptacles; corolla claw 6.5–7mm; inner pappus c 10.5mm.

Sikkim: Lhonak (113), Lungma Chhu; **Sikkim/Tibet:** Naku La; **Chumbi:** locality uncertain (Dungboo collection). 4550–5200m. August.

Similar collections from elsewhere in southern Tibet have been determined as *S. sericea.* However, the phyllaries of our plants are sparsely to densely silver-villous, the receptacle at least sparsely setose and the achenes glabrous, whereas *S. sericea* was described with densely brown-villous phyllaries, naked receptacle and densely white-villous achenes (328). Sterile specimens from Chho Lhamo, Sikkim, 5450m, and Donkung, Sikkim, 5250m, probably also belong here.

19. S. wernerioides Hook.f.

Acaulous; rosettes with crowded leaf remains. Leaf blades elliptic, 7–15 × 2–4mm, acute, margin revolute with 2–4 small teeth on either side, green glabrous above, white araneous beneath; petiole briefly constricted above dilated brown base. Capitula solitary, sessile, 5–7mm diameter; phyllaries linear-lanceolate, 12 × 3mm. Corollas 8–12mm including lobes 2–4mm. Achenes oblong, 3mm, ribbed; inner pappus brownish, 7–9mm, outer scabrid, 1.5–3.5mm.

Bhutan: N – Upper Bumthang Chu (Marlung); **Sikkim:** Dongkya La, Goecha La, Naku La etc.; **Bhutan/Chumbi border:** near Phari. In boulder scree and on glacial grit, 4400–5200m. July–September.

20. S. auriculata (DC.) Schultz Bipontinus; *S. hypoleuca* DC.

Stems 30–160cm, ± glabrous. Leaves lyrate pinnatisect, 8–15(–23) × 3–8cm, sparsely pubescent above, pale appressed tomentose beneath, sessile or lowest ± petiolate, auriculate at base; segments (obtuse to) acuminate, subentire to remotely denticulate; lateral segments 2–3(–5) pairs, ovate to triangular, 1–3.5 × 0.7–2cm; terminal segment ± triangular, sometimes hastate. Capitula cernuous, solitary at branch ends, 2–3cm diameter. Phyllaries ± lanceolate to linear, 9–18 × 1.5–4mm, glabrous or sparsely pubescent, ciliate, purplish, outer spreading with recurved tip. Some paleae exceeding pappus, plumose. Corollas 9mm including lobes 3mm. Achenes obovoid, 3mm, black, 4-angled, strongly tubercled above; pappus single, brownish, 10mm.

Bhutan: C – Ha district (Ha, Sele La (360)), Thimphu district (Barshong, Dochu La, Bela La–Paro), Bumthang district (Kyi Kyi La); **Darjeeling:** Phalut–Raman (101), Rishi La; **Sikkim:** Lachen, Meguthang, Phedang etc.; **Chumbi:** Chumbi, Pongling, Yatung etc. Forest margins, plantations, grassy banks, swamp margins, gravels etc., 3050–4550m. July–October.

21. S. deltoidea (DC.) Schultz Bipontinus; *S. lamprocarpa* Forbes & Hemsley

Stems up to 2m, whitish lanate at first. Lower leaves ± ovate, 13–30 × 6–22cm, acute, cordate or sometimes hastate at base, (bi)dentate, subglabrous above, white appressed tomentose beneath, petiole c 15cm, with 0–2 pairs of ovate to oblong leaf segments to 2.5 × 1.5cm; mid-cauline leaves 10–18 × 7–9(–15)cm, acuminate, rounded or truncate at base, denticulate or serrate, petiole to 6cm, rarely lobulate; upper leaves often sessile, lanceolate or elliptic. Receptacle with long, concealed bristles. Capitula cernuous, 1.5–2cm diameter, loosely racemose or paniculate; phyllaries mostly linear-lanceolate, 8–15 × 1.5–3mm, sparsely pubescent, reflexed above. Corollas white or pale mauve, 13mm including lobes 3mm. Achenes oblong, 2–3mm, with tubercled ribs; pappus single, whitish, 12mm.

Bhutan: S – Chukka district (°Chasilakha (117)), Gaylegphug district (Rani Camp (38), Rani Camp–Tama (38)), **C** –Tashigang district (Pintsogong, Shapang), Deothang district (Keri Gompa); **Darjeeling:** Darjeeling, Kurseong, Tanglu etc.; **Sikkim:** Chakung, Chiya Bhanjang, Gangtok etc. Open (grassy) hillsides, roadside jungle, 1500–3350m. September–December.

S. crispa Vaniot [*S. deltoidea* var. *nivea* Clarke] has been reported from Darjeeling or Sikkim (80), but no reliable examples were seen. It is highly polymorphic and sometimes quite similar to *S. deltoidea*, but shows at least two of the following differences: coarser vestiture, longer lateral lobes on basal leaves, more, smaller, tightly clustered capitula, ovate outer phyllaries, obtuse outer and middle phyllaries or corollas c 9mm.

22. S. heteromalla (D. Don) Handel-Mazzetti; *S. candicans* (DC.) Schultz Bipontinus

Similar to *S. auriculata* but stems greyish cottony at first; leaves not auriculate; capitula 1–1.5cm diameter, erect, ± racemose; phyllaries linear-lanceolate, 8–16 × 1–2mm, acuminate, thinly cottony, not recurved or coloured above; paleae c 4mm, white, concealed; corollas 13.5mm including lobes 2mm; pappus single, white, 13mm.

Darjeeling/Sikkim: Great Rangit, ?Ramom (poorly labelled Hooker collection). In fields, 600m. March–April.

A poor specimen from Patla Khawa, Darjeeling (*Gamble* 7720, K) keys out here, but has a very reduced corolla throat, smooth achenes and outer pappus of broken off bristles or short scales. It represents a distinct species of uncertain identity.

23. S. graminifolia DC. Fig. 119b.

Stems up to 15cm, thickly whitish lanate above, surrounded at base by dead leaves. Basal leaf blades linear, 7–13 × 0.1–0.2cm, obtuse or acute, revolute at margin, green, glabrous above, white tomentose beneath, from dilated sheathing base; cauline leaves similar, broader, lanate and semi-amplexicaul at base. Capitulum solitary, thickly lanate. Involucre 1.5–2.5cm diameter; outer phyllaries similar to leaves; inner phyllaries linear, tapering from base, to 2 × 0.4cm, coriaceous below, reflexed at apex, lanate. Receptacle naked. Corollas 13mm including lobes 3mm. Achenes squarish, smooth, 5mm; pappus whitish, double, inner 13–14mm, outer scabrid, 1–4mm.

Bhutan: N – Upper Mo/Pho Chu district (Kangla Karchu La), Upper

Bumthang Chu district (Marlung). In peaty turf and on open grassy slopes, 4730m. July–September.

24. S. przewalskii Maximowicz; *S. likiangensis* Franchet

Stems 20–45cm, sparsely pubescent, surrounded at base by few dead leaves. Leaves mostly basal, runcinate-pinnatisect, narrowly elliptic in outline, 8–18 × 2–3.5cm, subglabrous on both surfaces, lateral segments 4–5 pairs, oblong, 1–2 × 0.2–0.8cm, apiculate, with 0–3 teeth, terminal segment small; cauline leaves fewer, similar. Capitula 6–10 in very compact terminal corymb. Involucre 5–7mm diameter; phyllaries ± lanceolate, 10–12 × 1.2–3mm, softly herbaceous above, blackish, pubescent. Receptacle with long bristles. Corolla c 12mm including lobes 3mm. Achenes oblong, 4mm, ribbed; pappus double, brownish, inner 8mm, outer scabrid, 1–3mm.

Bhutan: N – Upper Mangde Chu district (Passu Sepu), Upper Kuru Chu district (Gong La), Upper Kulong Chu district (Me La). On open stony hillsides and among tall herbs, 3960–4570m. August–September.

25. S. nimborum W.W. Smith; *S. sughoo* sensu F.B.I. p.p. non Clarke

Stems 0–15cm, with fibrous leaf remains at base. Leaves usually ± pinnatifid, broadly oblanceolate in outline; lower blades 3.5–12 × 1–3cm, scabrid-puberulous above, white-araneous beneath, lateral lobes 4–5 pairs, ovate-triangular, 3–8(–12) × 5–10(–15)mm, with few apiculate teeth; petiole 1–7cm; sometimes all leaves simple, dentate when lobes are very shallow and teeth and lobe apices are subequal. Capitula 1, sessile or pedunculate, or 2–5 in compact or open corymbs. Involucre 5–11mm across; phyllaries ovate- to linear-lanceolate, acuminate, subglabrous to sparsely araneous, dark-margined, inner ones 12–15mm. Receptacle setose. Corollas 16–18mm including lobes 4–5mm. Immature achenes ± smooth, ribbed; pappus double, inner pale brownish, 14mm, outer very incomplete and irregular, 2–5mm, scabrid-plumose.

Bhutan: N – Upper Bumthang Chu district (Lhabja); **Sikkim:** Giaogang, Sebu Valley, Jelap La etc.; **Chumbi:** ?Lingmatang, unlocalised (421). 3050–3800m. September.

A depauperate collection from Somana (Thimphu district, Bhutan) with corollas 12mm and one from Laya (Upper Mo Chu district, Bhutan) with leaves narrowly elliptic in outline and with sparse araneous hair above towards base of blade may belong here. A collection with leaves c 30 × 5mm has been recorded from Subbu Valley, Sikkim (412).

26. S. obscura Lipschitz; *S. sughoo* sensu F.B.I. p.p. non Clarke, *S. inconspicua* Lipschitz non Handel-Mazzetti

Acaulous, base surrounded by fibrous leaf remains. Leaves obovate or oblanceolate, sinuate-dentate, green, crisped-pubescent above, white-araneous beneath; blade to 22 × 10mm. Capitula 1(–2), sessile. Involucre 5–6mm across; phyllaries subglabrous, dark-tipped, outer phyllaries ovate, acuminate, inner longer, linear-lanceolate, 11–13mm. Receptacle setose. Corollas 14.5–15.5mm including lobes 3.5–4.5mm. Immature achenes rugose and muricate; pappus double, whitish, inner 11mm, outer incomplete, scabrid-plumose, c 3–5mm, caducous.

Sikkim: Samdong and/or Yumthang (mixed Hooker collections), Giagong (374). (?3650–)4900m. September.

27. S. fibrosa W.W. Smith

Acaulous, with many heads from stout tap-root. Leaves rosulate, runcinate-pinnatifid, narrowly oblong in outline, 8–12 × 2–3cm, scabrid-pubescent above, whitish tomentose beneath; lobes triangular; petiole 3cm. Capitula 5–12, crowded among leaf bases, peduncles 5–15mm. Phyllaries herbaceous, ovate below, long attenuate above, araneous; receptacle setose. Achenes oblong, glabrous; pappus double, inner 10mm, outer much shorter, plumose.

Bhutan/NE Darjeeling: Kupchee.

No specimen seen. Description from (375) and (411).

28. S. yakla Clarke; *Jurinea cooperi* Anthony. Laya dialect: *Mido.*

Acaulous, usually without leaf remains. Leaves shallowly to deeply pinnatisect, oblanceolate in outline, 8–25 × 2–9cm, green and crisped-puberulous above, white-araneous beneath with scattered; lateral segments 5–7(–10) pairs, ± ovate in outline, 1.5–3 × 1–2cm, simple or lobulate, dentate-apiculate. Capitula very briefly raised above bases of basal leaf rosette, 1–7 in compact subsessile corymb. Involucre 8–15(–20)mm across; phyllaries linear-lanceolate, erect or reflexed above, 15–28 × 2–4mm, scabridulous outside. Receptacle setose. Corolla 19–28mm including lobes 4–6mm. Achenes oblong, 3–5mm, transversely muricate; pappus plumose, bristles overlapping at base and fused ± into 1 whorl, whitish, 18–23mm.

Bhutan: C – Thimphu district (Somana), Tongsa district (Pele La), **N** – Upper Mo Chu district (Laya, Yale La (22)), Thimphu/Upper Mo Chu district (Somana–Soe); **Sikkim:** Changu, Gnatong, Tarkarpo, Tosa, Yakla etc.; **Chumbi:** Chubitang, Chumbi, Yatung etc. Stony peaty meadows, amongst scrub, in grass or bamboo on hillsides, 3300–5200m. August–October(–November).

29. S. sughoo Clarke

Acaulous. Leaves narrowly oblanceolate in outline, pinnately lobed, deeply on basal part of blade, shallowly on apical part, 10–23 × 1.5–5cm, sparsely pubescent above, white-araneous beneath without projecting multiseptate hairs; lateral lobes 4–8, ± ovate, oblique, 5–20 × 7–25mm, dentate-apiculate; petiole to 5cm, sheathing capitulum at base. Capitulum solitary, base tightly sheathed by bases of all leaves. Involucre 12–20mm across; phyllaries lanceolate, 23–28 × 5mm, apex reflexed Receptacle bristles 3–6mm. Corolla 24mm including lobes 7mm. Achenes obovoid, 5mm, ribbed; pappus double, brownish, entirely plumose, inner 16–20mm, outer bristles uneven, from 7mm to as long as inner ones.

Sikkim: Sughoo (375), Yakla, Yampung. 3650–4880. September–November.

The true species is not described in F.B.I.

30. S. donkiah Springate

Loosely to densely tufted plants, acaulous; shoots borne on short rhizomes crowned with marcescent leaf bases or on slender stolons with appressed scale leaves. Basal leaves deeply pinnatisect, narrowly oblong-elliptic or oblanceolate in outline, 35–c 110 × 8–22mm; segments usually 7–10 pairs, occasionally alternating with large teeth, ovate-triangular to obovate in outline, with oblique apiculate apex and undulate margin with 0–7 broad apiculate teeth, sparsely covered above with multicellular, araneous-tipped hairs and sometimes with subsessile glands, thinly to moderately lanate below with some larger multicellular hairs and sometimes subsessile glands. Capitula solitary. Involucre campanulate, 8–16mm across; phyllaries imbricate, c 5-seriate, purplish above, apiculate, ciliate or rarely subentire, pilose above inside, glabrous to short-pilose above outside, sometimes with very few woolly or glandular hairs; outermost phyllaries ovate with tapered appendage to lanceolate, 13–14.5 × 3.5–5.5mm, reflexed at middle; inner phyllaries linear-lanceolate, 16–16.5 × 2–2.5mm, reflexed near tip; innermost as long, c 1.5mm wide, not reflexed. Receptacle densely setose, bristles c 6–9mm, persistent. Corolla tube 7.5–10mm, limb 6.7–8mm including lobes 3.7–4.5mm. Achenes apparently smooth, with prominent corona of many narrow scales; pappus double, inner 11.5–13mm, outer complete, 8–11mm.

Sikkim: Donkiah, Chemathang; **Chumbi:** Lingshi La. Lateral moraine, July–September. 4550–5200m.

Also flowering in November in Nepal, west of Lhonak.

31. S. eriostemon Clarke; *S. nepalensis* Sprengel *nom. superfl.*, *S. chapmannii* C.E.C. Fischer. Med: *Gnodewa*. Fig. 117a–d.

Stems 0–25cm, surrounded by fibrous leaf remains at base, subglabrous or whitish pubescent above. Basal leaves pinnatisect, oblanceolate in outline, 5–18 × 1–3cm, green on both faces, though paler beneath, subglabrous or crisped pubescent on veins especially beneath; lateral segments 4–8(–10) pairs, ovate and well separated to transversely oblong and slightly overlapping, 3–15 × 3–18mm, acute or rounded, mucronate, ± denticulate or toothed; cauline leaves few, similar, smaller. Capitulm solitary. Involucre campanulate, (6–)8–17mm across; phyllaries subglabrous, dark purplish-margined, outer ones ovate-lanceolate, with upper part foliaceous, recurved, sometimes dentate towards apex, inner ones narrower, purplish, suberect, (9–)11–20mm. Receptacle bristles many, persistent, 5–11mm. Corolla purple, 18–23mm including lobes 3.5–6mm. Achenes oblong, 4–6mm, ribbed, ± smooth; pappus double, brownish, inner bristles 13–18mm, outer ones plumose, c 8–22, many fewer to rather more than the inner ones and c 2–3(–6)mm shorter.

Bhutan: C – Thimphu district, **N** – Upper Mo Chu district, Upper Mangde Chu, Upper Bumthang Chu district and Upper Kulong Chu districts; **Sikkim:** Nathang, Tsomogo Chho, Yume Samdong etc.; **Chumbi:** Natu La–Champitang, Phari, Yatung etc. On open slopes and among dwarf rhododendrons, 3650–4450m. July–October.

An extremely variable complex. Plants in the eastern part of its range, including Chumbi and Bhutan, tend to have more pappus bristles (c 16 or more inner, c 11 or more outer) and longer phyllaries and pappus, but vary considerably in stem and leaf characters. They have been distinguished as *S. chapmannii.* Four divergent collections with leaves sparsely araneous and sessile-glandular beneath are also referred here: from Thanggu (Sikkim) with broadly oblanceolate, lyrate-pinnatifid cauline leaves; from Yampung and Zongri (both Sikkim) with narrower, scarcely foliaceous phyllaries and achenes to 8mm; from Thimphu (Bhutan) with very few short receptacle bristles only.

32. S. pinnatiphylla Grierson & Springate

Acaulous or subacaulous. Leaves interruptedly pinnate, elliptic in outline, 7–14.5 × 2.5–3.5mm, principal leaflets 5–10 pairs, ovate to obovate or oblong in outline, 7–15 × 5–12mm, bearing c 7–9 coarse, irregular, mucronulate teeth, green above and very sparsely crisped-puberulous, with small multiseptate hairs with mostly obsolete araneous tips, whitish tomentose beneath; alternating minor lobes usually 1- or 3-toothed, up to 3 × 3mm, absent from some leaves; leaflets towards leaf apex sometimes joined by narrow strip of lamina. Capitulum solitary. Involucre 15–20mm across, 4–5-seriate; outer phyllaries c 19–25 × 4–8mm, with oblong-lanceolate, coriaceous, entire base sparsely puberulous on back and foliaceous, pinnatifid, reflexed limb puberulous on back and sparsely so on face, with c 6–8 pairs of segments triangular below, linear above, up to 4mm; inner phyllaries c 25–28 × 2–4mm with a longer, narrower base and shorter limb and lateral segments. Receptacle bristles numerous, linear, 3mm. Corolla tube 13–14mm, limb 8–8.5mm including lobes 4.5–5mm. Achenes ± ribbed, rugulose; pappus brownish, double, inner series 15mm, outer plumose, 8–9mm.

Bhutan: C – Tongsa district (Taktse La). Alpine grassy pastures, 4730m. August.

Known only from type collection.

33. S. pachyneura Franchet; *S. kunthiana* Clarke var. *major* Hook.f., *S. bodinieri* Léveillé

Stems 0–30cm, sparsely pubescent, usually surrounded with fibrous leaf remains at base. Leaves deeply pinnatisect throughout, oblanceolate in outline, 6–36 × 1.5–4cm, green and sparsely pubescent above, white-araneous beneath; lateral lobes 7–10 pairs, ± ovate to oblong in outline, 6–20 × 3–15mm, coarsely toothed to pinnatifid, teeth/lobules apiculate. Capitula solitary. Involucre 13–25mm across; phyllaries with prominent dark purple margins and recurved mostly green tips, outer ones ovate, long acuminate, inner ones linear-lanceolate, 17–23mm. Receptacle short setose. Corollas 15–23mm including lobes 5–6mm. Achenes oblong 3–4mm, finely transversely wrinkled, squarish; pappus double, brownish, inner bristles 16–18mm, plumose with conspicuous long hairs mostly 2–3mm, outer bristles scabrid, 1–5mm.

Bhutan: N – Upper Mo Chu district (Tharizam Chu), Upper Kulong Chu district (Shingbe); **Sikkim:** Chakung Chhu, Yeumthang, Yakla etc.; **Chumbi:** Chomo Lhari, Guptende La. Open grassy hillsides, 3650–4880m. July–September.

34. S. purpurascens Chen & Liang

Plant c 5cm; stem erect, pubescent, surrounded by leaf remains at base. Leaves rosulate, linear, runcinate, 40–90 × 3–8mm, glabrous above, densely white tomentose beneath except on midrib; lobes 6–8 pairs, narrowly triangular, mucronate, margins entire, revolute. Capitula solitary, 22mm across. Involucre broad campanulate or globose; phyllaries 4-seriate, outer ovate-lanceolate, 13–14 × 4mm, coriaceous, reflexed, glabrous, inner linear, scarious, purplish above, 17 × 2mm, serrulate, mucronate. Receptacle bristles 2mm. Corollas 18mm including lobes 4–5mm. Anthers sky blue. Achenes cylindric, 2mm, brown, smooth; pappus 2-seriate, yellowish, inner 12mm, outer shorter, apparently plumose.

Chumbi: unlocalised (328). Among shrubs on slopes, 4200m.

Description above based on original description (328) and illustration in (421). Two collections from Upper Mo Chu district, N Bhutan, key out here, but differ in several characters: from Yale La on unstable scree at 4780m, flowering in October, with leaves 12cm, thinly tomentose, c 4 pairs of leaf lobes each often bearing a small tooth on the outer margin, inner pappus 9mm, outer pappus complete, plumose, 7mm; from Laya among rocks on steep slope at 4450m, flowering in September, with leaf blades 35 × 5mm, remotely dentate, inner pappus 11.5mm, outer pappus ± complete, plumose, 6.5mm. Their identity is uncertain.

35. S. taraxacifolia (Royle) DC.; *S. taraxacifolia* var. *depressa* Hook.f., *S. caespitosa* (DC.) Schultz Bipontinus var. *depressa* (Hook.f.) Lipschitz

Stems 0–28cm, usually densely tufted with each sheath of senesced leaf bases surrounding several shoots. Leaves deeply pinnatisect, linear in outline 30–150 × 3–25mm, subglabrous above, white araneous beneath; lateral lobes 5–12 pairs, triangular to oblong, usually retrose, sometimes spreading, 1–13mm, mucronate, entire or bearing 1 tooth. Capitulum solitary. Involucre 6–12mm diameter; phyllaries mostly linear-lanceolate, subglabrous or outer sparsely araneous, inner 16–20mm. Corollas 18–22mm including lobes 4–6mm. Achenes smooth, ribbed; pappus double, brownish, inner bristles 14–17mm, outer ones sparsely plumose, c 7–9mm, rarely apparently absent.

Bhutan: C – Ha district (Kang La–Ha), Thimphu district (Barshong, Pajoding, Somana), Punakha district (Punakha), Bumthang district (Rudong La), **N** – Upper Mo Chu district (Shingche La); **Sikkim:** Kaukola, Patang La, Tankar La etc.; **Chumbi**: Yatung. On open stony soil, peat turf, boggy moorland, 3350–4880m. July–October.

36. S. leontodontoides (DC.) Schultz Bipontinus

Tufted, with fibrous leaf remains at base. Leaves pinnatisect, white tomentose beneath. Capitulum solitary. Phyllaries imbricate, at least outer sparsely araneous above. Receptacle setose. Pappus double, brownish, outer bristles scabrid, short, caducous.

Two very distinct geographically isolated varieties in our area, but almost intergrading elsewhere.

a. var. **leontodontoides**; *S. kunthiana* Clarke var. *kunthiana* sensu F.B.I.

Acaulous. Leaves linear-elliptic or -oblanceolate in outline, 30–80 × 4–11mm, densely pubescent and more sparsely araneous above; lateral lobes c 5–8 pairs, triangular to broadly ovate, sometimes oblique, not overlapping, 1.2–3.5 × 3–6mm, apiculate, margin often strongly revolute, with 0–2 shallow, minutely apiculate teeth. Involucre 8–12mm diameter; phyllaries scarcely reflexed above, outer ones ovate, long acuminate, inner ones linear-lanceolate, c 10–12mm, sometimes subglabrous, ciliolate or entire. Receptacle bristles c 2mm. Corollas 14–17mm including lobes 4–5mm. Immature achenes rugose, angular; inner pappus bristles 11mm, outer ones 2–3mm, caducous.

Chumbi: Phari, Tang La. 4250–4570m. August–September.

Another collection from Phari, flowering in July, differing by its leaf blades broadly oblanceolate in outline, c 20 × 7mm, probably belongs here.

b. var. **filicifolia** (Hook.f.) Handel-Mazzetti; *S. kunthiana* Clarke var. *filicifolia* Hook.f.

Stems 0–12cm, tufted, with leaf remains at base. Basal leaves linear-elliptic or -oblanceolate in outline, deeply pinnatisect, 70–230 × 10–23mm, sparsely pubescent and non-araneous above, white tomentose beneath; lobes c 13–24 pairs, narrowly triangular to oblong-oblanceolate, 5–13mm, entire or with 1 well-developed lobule, rarely with second tooth as well. Capitulum solitary. Involucre 6–10mm diameter; phyllaries erect, apiculate, outer ones often recurved at tip, inner ones linear-lanceolate, 16–19mm. Receptacle bristles 2.5–4.5mm. Corollas 17–19mm including lobes 4.5–5mm. Achenes rugose (375); inner pappus bristles 14–16mm, outer ones 2–4mm.

Sikkim: Gnatong, Onglakthang, Yakla, Yome Samdong, Zongri. 3950–4250m. September–November.

37. S. polystichoides Hook.f.

Stems 0–8cm. Leaves deeply pinnatisect or pinnate, narrowly oblanceolate to linear, 40–130 × 2–12mm, dark green glabrous above, white araneous beneath; lateral segments 8–18 pairs, subquadrate, 2.5–6 × 3–6mm, with few apiculate teeth. Capitulum solitary. Involucre 5–10mm across; phyllaries narrowly lanceolate, erect, glabrous to sparsely araneous, inner ones 14–17mm. Receptacle naked, rarely few bristles present. Corollas 15–18mm including lobes 4–5mm. Achenes oblong, 3mm, often wrinkled, angular; pappus brownish, single, 11–14mm, rarely some short plumose outer bristles present.

Bhutan: N – Upper Mo Chu district (Lingshi La and Singche La); **Sikkim:** Lachen, Samdong, Yumthang etc. On margins of streams, 4000–4730m. July–September.

38. S. andersonii Clarke

Stems 0–12cm, densely covered with leaf remains at base. Leaves linear in outline, dentate or shallowly pinnatifid, 2–11 × 0.4–1.2cm, with 3–7 pairs of narrow acuminate teeth, subglabrous on both surfaces, sometimes glaucous beneath. Capitulum solitary. Involucre 6–13mm diameter; phyllaries ovate, acuminate to linear, 12–18 × 2–5mm, subglabrous, outer with purplish or dark violet recurved tips. Receptacle with rather few short caducous bristles. Corollas 19–21mm including lobes 4–5mm. Achenes oblong, 4mm, wrinkled, angular; inner pappus brownish, sometimes purplish at either end, 13–16mm; outer pappus absent or of 1–2 very short scabrid bristles.

Sikkim: Chakung Chhu, Changu, Yampung etc.; **Chumbi:** without localities (421). In peaty meadows and among low rhododendrons, 3350–4265m. June–October.

S. flavovirens Chen & Liang has also been described from Chumbi. From the type description and an illustration (421), it seems very similar to *S. andersonii*, but may differ by the sparsely strigose hair above its leaves and more complete outer pappus.

39. S. candolleana (DC.) Schultz Bipontinus

Stems 60–100cm, narrowly winged, thinly puberulous. Cauline leaves narrowly ovate-elliptic, 7–16 × 2.5–6cm, acuminate, narrowed and decurrent at base, remotely denticulate or finely dentate, sparsely puberulous or minutely scabridulous above, sparsely araneous beneath. Capitula 10–20 in dense terminal corymb, usually in axillary corymbs as well. Involucre c 5mm diameter; phyllaries obovate, acuminate to narrowly oblong or oblanceolate, acute or obtuse, 5–10 × 1.5–2.5mm, dark violet at apex, outer ones sparsely araneous. Corollas c 11.5mm including lobes 4mm. Achenes oblanceolate, 5mm, glabrous, somewhat compressed; pappus double, brownish, inner series c 8.5mm, outer series 2–4.5mm, scabrid, caducous.

Bhutan: C – Ha/Thimphu district (Chile La (360)); **Sikkim:** Lachen, Lachung

Valley (375), Thanggu, Yumthang, Zemu (113). 3550–3650m. August–September.

The specimens from Thanggu and Yumthang are described above, the one from Lachen is probably conspecific but differs considerably by its much narrower, more prominently decurrent leaves with thicker hair beneath and immature capitula with caudate outer phyllaries.

An unlocalised record of *S. parviflora* (Poiret) DC. from Chumbi (421), a similar but otherwise geographically isolated species, may refer to *S. candolleana*.

S. piptathera Edgeworth has been reported from Sikkim without locality (101), but no examples were seen from our area. It differs from *S. candolleana* by its semi-amplexicaul, non-decurrent leaf bases.

40. S. pantlingiana W.W. Smith

Stems 20–30cm, glabrous, surrounded at base by fibrous leaf remains. Basal leaves linear-oblanceolate, c 15 × 1cm, acuminate, narrowed to ± distinct petiole below, entire or very remotely denticulate, glabrous above, white tomentose beneath; cauline leaves similar, sessile, semi-amplexicaul at base. Capitula 2–10, loosely corymbose. Involucre c 5mm diameter; phyllaries ovate to lanceolate, c 10–15mm, acuminate, sparsely araneous. Corollas 16.5mm including lobes 4.5mm. Achenes oblong, slightly angular. Pappus double, inner 12mm, outer scabrid, 4mm.

Sikkim: Yakthang, Tallum Samdong. 3650m. September–October.

41 . S. hieracioides Hook.f.

Plant 8–30cm; stems white pilose, base surrounded by leaf remains. Leaves mainly basal, subacute (to acuminate), apiculate, obscurely remote denticulate and usually densely pilose at margins, sparsely pilose on both surfaces; basal leaves petiolate, ovate or elliptic, 5–13 × 1.5–3cm, obtuse to long attenuate at base, petiole to 8cm, usually broadly channelled; cauline leaves (1–)3(–4), (sub)sessile. Peduncle dilated upwards; capitula solitary, 1–1.5cm diameter. Phyllaries ovate to lanceolate, 12–20 × 2–6mm, acuminate, blackish, partly pilose. Corollas 20–24mm, including lobes 5–6mm. Achenes oblong, 3–4mm, ribbed; pappus double, inner almost reaching base of corolla lobes, outer 3–7mm, scabrid.

Bhutan: N – Upper Mo Chu district (Lingshi); **Sikkim:** Chugya, Dotha, Teyjep La etc. Among rocks and scrub, 3650–4570m. July–September.

S. glanduligera Schultz Bipontinus has been collected near Kamba Dzong, immediately north of Sikkim. It would key out here, but differs by its linear-oblanceolate leaves with dense glandular hair on both faces of blade, the basal ones 50 × 4–200 × 15mm, cottony inside sheathing base.

8. HEMISTEPTIA Fischer & Meyer

Similar to *Saussurea* but annual; midrib of median phyllaries bearing an erect crest below apex; corolla tubular, limb narrow, deeply 5-lobed; achenes oblong, distinctly 15-ribbed, areole slightly oblique; outer pappus of few, minute, scattered, persistent but fragile scales usually present, inner of 15 plumose hairs connate at base, deciduous.

1. H. lyrata (Bunge) Fischer & Meyer; *Saussurea affinis* Sprengel

Stems c 25cm, finely and sparsely greyish araneous. Leaves lyrate-pinnatisect, oblanceolate in outline, to 10cm, sparsely pubescent above, greyish araneous beneath; segments acute, subentire to serrate or sublobulate; terminal segment ± ovate, to 25 × 20mm; lateral segments 2–5 pairs, obovate to oblong, to 20 × 12mm. Capitula loosely corymbose. Involucre 5–9mm diameter; phyllaries ovate (outer) to linear-lanceolate (inner), 2–10 × 1–2mm; receptacle scales many, linear, 1.5–2.5mm. Corollas rose-purplish; tube to 4mm; throat 0.8mm; lobes to 2mm. Outer pappus 0.1–0.2mm; inner pappus to 6mm, white.

Bhutan: C – Thimphu district (Paro, Thimphu–Chimakothi). In fields, 2250–2300m. May–June.

9. TRICHOLEPIS DC.

Perennial herbs. Leaves simple or lobed, alternate. Capitula solitary at branch ends, discoid. Involucre campanulate; phyllaries numerous in many series, very narrow with long recurved hair-like points. Receptacle convex, with many fine linear scales. Flowers bisexual, slender. Corollas white or yellow (in our species). Filaments shortly pubescent, anther bases caudate, tails dentate-lacerate. Style branches linear, slender. Achenes slightly compressed, deeply wrinkled, hilum lateral near base. Pappus many-seriate, capillary, persistent.

1. T. furcata DC. Fig. 119c.

Stems 60–260cm, minutely scabrid-puberulous or glabrescent. Leaves elliptic to lanceolate, usually simple, 5–15 × 1–4.5cm, acuminate, tapered at base, subsessile, finely serrate, upper surface minutely scabrid, pale tomentose beneath. Involucre 1–2cm across; phyllaries blackish, very narrow, 5–25 × 0.1–1.2mm, barbellate; receptacle scales c 8–10mm. Corolla tube 7.5–8.5mm; throat scarcely dilated, 7.5–8.5mm; lobes 5–6.5mm. Achenes oblong-obovoid, c 6mm, pale, glabrous. Pappus white, to 19mm.

Bhutan: C – Thimphu district (Motithang, Paro), Punakha–Tongsa districts (Pele La (360)), Mongar district (Khinay Lhakang), Tashigang district (Balfi); **Sikkim:** Cheungthang, Lachung etc.; **Chumbi:** upper Chumbi Valley. Fields and dry hillsides, 1525–3650m. August–November.

10. CENTAUREA L.

Our cultivated plant an annual herb. Leaves simple, alternate. Involucre campanulate; phyllaries several-seriate, imbricate with scarious apical appendages. Receptacle flat, bearing scales deeply divided into linear segments. Outer flowers enlarged, sterile; corolla funnel-shaped, outcurved, usually with 7 lobes. Inner flowers bisexual; corolla tubular-campanulate, 5-lobed; filaments glabrous, anther bases shortly caudate; style branches short, swollen at base. Achenes glabrous, laterally compressed, apex truncate, hilum large, lateral near base. Pappus several-seriate, narrowly scaly.

1. C. cyanus L. Eng: *Cornflower.*

Stems 15–50(–80)cm, branched, sparsely araneous. Leaves sessile, sparsely araneous above, pale tomentose beneath; lowest ones lyrate, usually withered at flowering time; cauline ones narrowly lanceolate, 4–7 × 0.2–0.3cm, acuminate, usually with 1–3 subulate teeth on each side, upper ones entire. Involucre c 1cm across; phyllaries oblong-lanceolate, 3–14 × 1–2.5mm, including lacerate scarious apical appendage; receptacle scales to 6mm. Corollas blue; outer ones 25 × 12mm; inner corolla tube 7mm, throat 2.5mm, lobes linear, 4mm. Achenes obovoid, 3.5mm; pappus scales 0.8–3mm, brownish.

Bhutan: S – Phuntsholing district (Phuntsholing), 250m. February.

Cultivated in gardens. Native of Europe and Middle East.

LACTUCEAE

11. CREPIS L.

Our species a perennial subscapose herb. Leaves runcinate-pinnatifid. Capitula ligulate, few, at first in 1–3 compact heads, eventually long-pedunculate. Involucre campanulate, 2-seriate; outer phyllaries ovate to ovate-lanceolate; inner phyllaries linear-lanceolate, much longer. Ligules exceeding tube, yellow. Style branches slender. Achenes not beaked, narrowly ellipsoid, subcompressed, finely ± equally ribbed. Pappus capillary, white, free, ± persistent.

1. C. tibetica Babcock

Plant 30–50cm, from horizontal or oblique rhizome. Leaves sparsely hairy; blade 3.5–7 × 0.5–1.5cm, acute or obtuse, with up to 5 pairs of small lateral segments, tapering into petiole 1.5–5cm. Scapes several, sparsely pubescent at base, glabrous above, bearing 2–5 capitula, rarely with 1 reduced leaf and branch near base; peduncles densely dark hairy. Involucre 3–4.5mm diameter; phyllaries blackish, sparsely darker hairy on back; inner 8–10mm. Corolla tube 3–4mm, ligule 6–7.7 × 1.7–2mm. Achenes 4–5mm, brownish, c 10-ribbed; pappus 4mm.

Bhutan: N – Upper Mo Chu district (Lingshi) and Upper Mangde Chu/Upper Bumthang Chu district (Ju La). On open hillsides, 3800–4110m. July.

12. YOUNGIA Cassini

Annual or perennial herbs; stems elongate and erect, short or absent. Leaves broadly ovate to oblanceolate in outline, ± entire or commonly lyrate or runcinate pinnatifid or pinnatisect. Capitula ligulate, 10–15(–30)-flowered. Involucres cylindrical or narrowly campanulate; outer phyllaries short, usually few; inner phyllaries 8–12 in 1 series, becoming carinate and spongy-thickened at base. Ligules yellow. Style branches slender, filiform. Achenes narrowly ovoid or oblong, not beaked, finely and unequally ribbed with 3–5 stronger often spiculate ribs; pappus usually persistent, finely capillary.

1. Dwarf plants; flowering stems less than 10cm 2
+ Plants caulescent; flowering stems more than 10cm 4

2. Leaves ovate or orbicular, margins undulate, abruptly narrowed to petiole **1. Y. depressa**
+ Leaves oblanceolate, usually pinnatifid, base gradually tapering to petiole 3

3. Inner phyllaries 11–15mm in flower; peduncles up to 1.5cm; achenes columnar .... **2. Y. simulatrix**
+ Inner phyllaries 8mm in flower; peduncles up to 5cm achenes contracted at both ends .... **3. Y. gracilipes**

4. Annuals; inner phyllaries 4–4.5mm in flower .... **6. Y. japonica**
+ Perennials; inner phyllaries 6–11.5mm in flower 5

5. Capitula in narrow spike-like racemes or panicles .... **4. Y. racemifera**
+ Capitula ± loosely corymbose 6

6. Stems leafy; pappus sooty (en masse) .... **5. Y. stebbinsiana**
+ Stems naked; pappus dirty white .... **7. Y. silhetensis** subsp. **bhutanica**

1. Y. depressa (Hook.f. & Thomson) Babcock & Stebbins; *Crepis depressa* Hook.f. & Thomson, *Lactuca cooperi* Anthony. Fig. 120a–b.

Rosulate acaulescent perennial from stout tap-root. Leaves oblong, ovate or orbicular; blade 1.5–5 × 1–4cm, obtuse, rounded or weakly cordate at base, entire or obscurely dentate, rarely lobulate-dentate, sparsely pubescent at junction with petiole; petiole to 5cm, pubescent. Capitula numerous, clustered on top of leaves, c 15-flowered; peduncles 3–10mm, densely pubescent. Involucre

FIG. 120. **Compositae: Lactuceae.** a–b, *Youngia depressa*: a, habit (× ⅔); b, achene (× 6). c–d, *Soroseris hookeriana*: c, habit (× ⅔); d, achene (× 6). e–g, *Chaetoseris macrantha*: e, lower leaf (× ⅔); f, inflorescence (× ⅔); g, achene (× 6). Drawn by Margaret Tebbs.

a
b
c
d
e
f
g
M.Tebbs

cylindrical, 3–5mm diameter, green or crimson; outer phyllaries lanceolate, c 5mm; inner ones 7–8, narrowly oblong-lanceolate, 12–16mm, sparsely pubescent. Corolla tube 8–10.5mm, exceeding ligule; ligule 7–10 × 1.7–2mm. Achenes narrowly oblong-elliptic, usually strongly compressed, sometimes 3-angled, 6–7mm, scarcely beaked, spiculate above; pappus bristles white to dirty white, coarse, brittle, 10–12mm.

Bhutan: C – Thimphu district (Paro Pass, Shanosam, Somana), **N** – Upper Mo Chu district (Gyengatang, Laya, Lingshi), Upper Mangde Chu district (Phage la), Upper Bumthang Chu district (Tolegang), Upper Kulong Chu district (Me La); **Sikkim:** Kupup, Lingmatang, Sherabthang etc.; **Chumbi:** Yatung, upper Chumbi Valley. Grassy slopes, 3050–4700m. August–October.

2. Y. simulatrix (Babcock) Babcock & Stebbins; *Crepis simulatrix* Babcock

Rosulate almost acaulescent perennial from deep stolons. Leaves oblanceolate in outline, 2–6 × 0.6–1.2cm, sinuate-denticulate to runcinate, obtuse, tapering to short winged petiole, glabrous or puberulous above, sparsely pubescent beneath. Inflorescence glabrous; flowering stems very short, branched, bracteate, with 4–7 capitula. Capitula 15–20-flowered. Involucres 3–5mm diameter; outer phyllaries ± ovate, less than half as long as inner; inner ones 8–12, lanceolate, 11–15mm. Corolla tube 5–7mm; ligule 12–16 × 2–3mm. Achenes ± oblong, somewhat compressed or angular, 3.5–4mm, c 14-ribbed, spiculate throughout, eventually blackish; pappus 10–11mm, white or whitish.

Sikkim: Theu La (318). 4850m.

3. Y. gracilipes (Hook.f.) Babcock & Stebbins; *Crepis gracilipes* Hook.f.

Dwarf tufted perennial to 6cm from deep stolons. Leaves almost all radical, crowded, pinnatisect (rarely pinnate), oblanceolate in outline, 1.5–5cm, sparsely pubescent on both sides; terminal segment ovate or oblong, sometimes subhastate, 0.5–1 × 0.5–0.9cm; lateral segments 3–4 pairs, rounded or triangular, to 6 × 3mm; petiole to 2cm. Flowering stems several, 8–40mm, bracteate, tomentose, bearing 1(–4) capitula. Capitula 20–30-flowered on ± tomentose peduncles. Involucres cylindrical, 2–4mm diameter; outer phyllaries c 10, ± ovate, to 3.5mm; inner ones 8–10, narrow oblong-lanceolate, 7–9.5mm. Corolla tube (3–)4(–5)mm, ligule 9–10(–13.5) × 2.2–2.8mm. Achenes ± narrowly oblong, 3mm, somewhat compressed or angular, eventually blackish, spiculate throughout; pappus 6–8mm, whitish.

Bhutan: C – Thimphu district (Barshong); **Sikkim:** Lhonak, Natu La. 3650–4600m. June–July.

4. Y. racemifera (Hook.f.) Babcock & Stebbins; *Crepis racemifera* Hook.f.

Perennial, 15–60cm, from short thick root-stock, ± glabrous. Leaves narrowly ovate; blade 2.5–8(–11) × 1.5–3(–8)cm, acute or acuminate, rounded to weakly cordate at base, sinuate dentate, often purple beneath; petiole slender, 1.5–3.5(–8)cm, often narrowly winged above. Capitula cernuous in spicate racemes or sometimes narrow panicles. Involucres campanulate, 3–5mm diameter; outer phyllaries 8–10, ovate to lanceolate, 2–7 × 1mm; inner ones 8–10, narrowly oblong-lanceolate, 7–11.5mm. Corolla tube 4–5.3mm; ligules 7.5–9 × 2.3–2.8mm. Achenes (319) narrowly oblong or oblanceolate, 4.5mm, abruptly constricted beneath disc, with 12–14 smooth uneven ribs; pappus 8mm, whitish or somewhat reddish-brown.

Bhutan: C – Thimphu district (Barshong, Dotena, Paro), Bumthang district (Kopup), **N** – Upper Mo Chu district (Laya, Lingshi, Chamsa–Kohina etc.); **Sikkim:** Thanggu, Gnatong, Thangsing etc. In broadleaf and coniferous forest, 2300–4250m. September–October.

5. Y. stebbinsiana Hu; *Y. gracilis* (Clarke) Babcock & Stebbins non Miquel, *Crepis fuscipappa* sensu F.B.I. non Thwaites

Perennial, 15–40cm, from very short root-stock; stems usually flexuose, glabrous. Leaves oblanceolate in outline, 4–13 × 1–3.5cm, acuminate, attenuate at base, coarsely dentate, subglabous above, sparsely pubescent and often purplish beneath, ± sessile; lower leaves often more deeply divided in lower part of blade, pinnatifid or runcinate-pinnatisect, with 3–5 pairs of lateral segments. Capitula 2–8 in corymbose clusters from upper leaf axils and stem apex. Involucres campanulate, 3mm diameter; outer phyllaries ovate c 1mm; inner ones c 8, narrowly oblong-lanceolate, 7–9mm. Corollas hairy outside at throat; tube 2.5mm, ligule 9.5–10.5 × 3mm. Achenes (318) narrowly oblong-lanceolate, 4mm, dark brown, gradually contracted beneath disc, with c 15 uneven, spiculate ribs; pappus 3.5–5mm, blackish (en masse).

Bhutan: C – Thimphu district (Lamnakha, Paga), Punakha district (Tinlegang), Tongsa district (Chendebi); **Sikkim:** Gaigon, lower Khora, Thanggu etc. On grassy cliffs and among sand and gravel, 2130–3960m. July–August.

6. Y. japonica (L.) DC.; *Crepis japonica* (L.) Bentham

Annual, 15–80cm, stems glabrous or pubescent below. Leaves mostly basal, very variable, usually lyrate pinnatisect, rarely simple, 4–20 × 1.5–4cm, sparsely pubescent at least beneath; segments obtuse to acuminate, subentire to coarsely dentate; terminal segment (whole blade) ± ovate; lateral segments 0–4 pairs, smaller, oblong to triangular. Inflorescence loosely corymbose; capitula numerous. Involucre c 1.5mm diameter; outer phyllaries c 5, ovate, c 1mm; inner ones c 8, narrowly oblong-lanceolate, 4–4.5mm. Flowers (16–)20–25. Corolla hairy outside at throat; tube c 2.7mm; ligule 3–3.5 × 1mm. Achenes oblong-elliptic, 1.5–2.5mm, usually compressed, gradually contracted beneath disc, brown, with c 12 spiculate ribs; pappus 3–4mm, white.

Bhutan: S – Samchi, Phuntsholing and Chukka districts, **C** – Thimphu, Punakha, Tongsa and Mongar districts, **N** – Upper Mo Chu district; **Darjeeling:** Darjeeling, Dumsong terai, Mongpu etc.; **Sikkim:** Lachen, Singtam. Weedy situations, 300–2500m. January–December.

7. Y. silhetensis (DC.) Babcock & Stebbins subsp. **bhutanica** Grierson & Springate

(Sub)scapose perennial, 15–25cm, from short stout root-stock. Leaves oblanceolate, 8.5–25 × 2–4.5cm, acuminate, attenuate at base, remotely denticulate, weakly undulate to dentate, often triangular lobate-dentate towards base, ± glabrous above, sparsely pubescent on veins beneath, densely brown araneous at base. Inflorescence ± corymbose; scapes branched above, rarely with 1 branch from axil of reduced leaf near base, usually several together forming a single head. Involucre 2–3mm diameter; outer phyllaries ovate, 0.7–4 × 0.5–0.7mm, inner phyllaries 10–13, linear-lanceolate, 6–7.2mm. Flowers 15–19. Corolla tube 2.8–5mm, glabrous; ligule 7.5–9.5mm. Achenes (immature) linear-lanceolate, 4mm, puberulous near apex, ribs ± alternately broad and narrow; pappus dirty white or yellow, 5mm.

Bhutan: S – Deothang district (Narfong–Wamrung), **S/C** – Samchi/Phuntsholing/Ha district (Torsa River). On open cliff-faces, (900–)1200–1600m. April–May.

A Griffith collection of a distinct form has been dubiously recorded for Sikkim. Its leaves are subentire, to 30 × 4cm, tapering to a very narrowly winged petiole, 5–6cm. Its origin and status within *Y. silhetensis* are uncertain (see 342a).

13. STEBBINSIA Lipschitz

Rosulate rhizomatous perennial herb. Stems thick, mainly subterranean with linear-lanceolate scale leaves. Upper leaves ovate or suborbicular, angular, dentate or remotely denticulate; petiole long, very rarely pinnately lobulate near apex. Inflorescence terminal, compact, convex, with many capitula; peduncles simple or branched, bracteate. Capitula ligulate, with 25–43 flowers. Outer phyllaries (3–)5, narrowly triangular or lower ones leaf-like; inner phyllaries 13–19, triangular-lanceolate, fused at base. Receptacle flat (concave at margins), fimbrillate. Ligules yellow. Style branches linear. Achenes ± oblong, terete, smooth, not beaked; pappus capillary, deciduous.

1. S. umbrella (Franchet) Lipschitz; *Soroseris umbrella* (Franchet) Stebbins

Stems 4–12cm. Upper leaves sparsely hirsute, blades 2–9 × 2–7cm, obtuse, rounded to subcordate at base; petiole 2–6(–10)cm. Peduncles 1.2–6cm, bracts 1 or few, small, leaf-like or lanceolate and sessile. Involucres c 10mm diameter; outer phyllaries 3–10mm; inner phyllaries 11–14.5mm, dull green or blackish, moderately hirsute. Corolla tube 5–8mm; ligule 8–10(–12)mm. Achenes 3–4.5mm, brown or blackish; pappus 8–10mm, white, sordid or brownish, sometimes rufous below.

Bhutan: N – Upper Mo Chu district (Yala/Pyala/Gilela, Lale La); **Chumbi:** Trakarpo, Phari, Chomp Lhari etc. Screes, 4000–4900m. August–September.

14. SOROSERIS Stebbins

Dwarf rosulate rhizomatous perennials; stems thick, often hollow, usually mostly underground. Lower leaves reduced to scales sometimes present on subterranean stems; upper leaves simple to pinnatisect, entire to dentate. Inflorescence terminal, dense, convex or elongate, bracteate, with many capitula; peduncles rather short, usally simple. Capitula ligulate, 4(–5)-flowered. Involucres cylindrical; outer phyllaries 2, linear, shorter than or exceeding inner ones; inner phyllaries 4, lanceolate, connate at base, inner pair with broad scarious margins. Receptacle concave, naked. Ligules white, yellow or purplish. Style branches linear. Achenes ± oblong, ± terete, not beaked; pappus capillary, deciduous.

1. Scale leaves numerous; blade of upper leaves 5–25mm; ligules 5.5–7mm **1. S. pumila**
\+ Scale leaves few or none; blade of upper leaves more than 25mm; ligules 7–12mm 2

2. Leaves shallowly to deeply pinnatifid, at least when well developed (when length of blade is more than 4 × width); bracts usually densely hirsute at least at base; inner phyllaries hirsute (very rarely subglabrous) **2. S. hookeriana**
\+ Leaves always entire or remotely sinuate-denticulate; bracts and inner phyllaries glabrous or subglabrous .... **3. S. erysimoides**

1. S. pumila Stebbins; *Crepis glomerata* sensu F.B.I. p.p. non (Decaisne) Clarke.

Stems 2.5–10cm, mostly subterranean; scale leaves several or numerous, linear-lanceolate. Upper leaves purplish, spathulate or oblanceolate, obtuse, tapered at base, coarsely dentate or subpinnatifid, hairy on margins and petiole or throughout; blade 5–25 × 3–9(–12)mm; petiole 1–5cm. Inflorescence convex; peduncles 2–10mm; involucre 3–4mm diameter; phyllaries purplish, white or brown hairy; inner phyllaries 8–10mm. Corolla 9–12.5mm; tube (3–)4.5–6.2mm; ligule abruptly expanded, yellow, 5.5–7mm, teeth distinctly irregular. Anthers dark (?violet). Pappus 8–10mm, bluish above.

Bhutan: C – Thimphu district (Wasa La), **N** – Upper Mo Chu district (Yala/Pyala/Gilela), Upper Bumthang Chu district (Pangotang); **Sikkim:** Thanka La, Bikbari and Prek Chhu Valleys etc.; **Chumbi:** Chumolari, Ta-chey-kung; **Nyam Jang Chu/Tibet:** Cho La. On screes, 4300–4880m. July–September.

A specimen collected immediately north of Sikkim (from Kamba Dzong) differs in its glabrous leaf blades, glabrous inner phyllaries fused in lower ⅖, white ligules with small regular teeth and pale (?pink) anthers. It shows more affinity to *S. glomerata* (Decaisne) Stebbins but cannot be reliably referred to that species. However, *S. pumila* and several other species are sunk in *S. glomerata* by Shih (405).

2. S. hookeriana Stebbins; *Crepis glomerata* sensu F.B.I. p.p. non (Decaisne) Clarke. Fig. 120c–d.

Stems hollow, 2–30cm exposed; scale leaves few or absent. Leaves lanceolate to oblanceolate or linear, 4–13 × 0.3–3.5cm, shallowly pinnatifid to deeply pinnatisect, at least when well developed (when length of blade is more than 4 × width, when shorter sometimes simple, dentate), long attenuate at base; lobes entire (rarely with 1 broad and 2 minute teeth); petiole 0–6cm. Inflorescence convex or elongate (to 16cm). Bracts usually densely hirsute at least at base. Involucre 2–3mm diameter; inner phyllaries 8–12.5mm, hirsute (very rarely subglabrous), innermost 2.6–4mm wide. Corolla 12–18mm; tube 3–6.5mm; ligule yellow, 7–12mm, with small regular teeth. Anthers dark (very rarely yellowish with whole flower lacking cyanic pigments). Pappus white, sooty or stramineous, usually with bluish tips, 6.5–13mm.

Bhutan: C – Ha/Thimphu district (Ha–Kangla), Thimphu district (Somana, Tremo), Tongsa district (Thampe La), **N** – Upper Mo Chu district (Laya, Lingshi, Yale La etc.), Upper Pho Chu district (Kanga Karchu La); **Sikkim:** Changu, Chola, Thanka La etc.; **Chumbi:** Trakarpo, Chomp Lhari,Yatung etc. Among boulders and on screes, 3650–4720m. July–September.

Populations from NE Bhutan (Upper Kuru Chu district: Pu La and Gong La; Upper Kulong Chu district: Me La and Shingbe) and Sikkim/Chumbi (Natu La) are doubtfully referred here. They generally have closely set, abruptly attenuate, mucronate or hair-tipped leaf lobes with some deep lateral teeth, inflorescence not elongating, corolla tubes 5.5–7.8mm and rufous pappi. They seem intermediate with *S. hirsuta* (Anthony) Shih [*S. gillii* (S. Moore) Stebbins subsp. *hirsuta* (Anthony) Stebbins].

Plants with an elongate inflorescence are very distinctive. Some from Chumbi (and Nyalam district, Tibet) with the floriferous part of the stem 5–13 × 4.5cm have been described as *S. teres* Shih. Similar examples are distributed across our area, and tend to have longer leaf lobes, more prominent bracts and thicker crisped hair on the inflorescence but grade into forms of *S. hookeriana* with a convex inflorescence. They are provisionally retained in *S. hookeriana* here.

3. S. erysimoides (Handel-Mazzetti) Shih; *S. hookeriana* subsp. *erysimoides* (Handel-Mazzetti) Stebbins

Differs from *S. hookeriana* by its leaves always entire or remotely sinuate-denticulate and bracts and inner phyllaries glabrous or subglabrous.

Bhutan: C – Thimphu district (Tremo La–Kang La), **N** – Upper Mo Chu district (Sinchu La), Upper Bumthang Chu district (Marlung, Waitang); **Chumbi:** Phari, Dotag, Gora La. Screes, 4000–4600m. June–August.

Few populations in our area entirely correspond with this species. Rather more frequent are populations referred to *S. hookeriana* here, but with some more reduced individuals resembling *S. erysimoides* particularly in leaf shape.

15. DUBYAEA DC.

Perennial herbs. Leaves alternate, simple and repand to pinnate. Capitula ligulate, many-flowered, solitary at branch ends or corymbose to racemose, erect or pendent. Involucre campanulate, imbricate. Corollas yellow, pink or purplish. Style branches filiform. Achenes oblong, a little compressed, striate, gradually narrowed above. Pappus several-seriate, capillary.

1. Plant to 100cm; lower leaves pinnate, upper ones long petiolate; corollas purplish .... **3. D. stebbinsii**
+ Plant 10–50cm; all leaves simple, denticulate to lyrate-pinnatifid, at least upper ones sessile; corollas yellow, rarely pink 2

2. Capitula pendulous; corollas yellow .... **1. D. hispida**
+ Capitula erect; corollas pink .... **2. D. cymiformis**

1. D. hispida DC.; *Crepis hispida* D. Don non Forsskål *nom. illeg.*, *Lactuca dubyaea* Clarke, *C. bhotanica* Hutchinson, *D. pteropoda* Shih

Stems 10–50cm, hirsute at least above, borne on long slender rhizomes. Leaves acuminate, denticulate; lower leaves simple to lyrate-pinnatifid (-pinnatisect) with 1–4 pairs of rounded (angular) lateral lobes, blade lanceolate or oblanceolate in outline, 7–12 × 2–5cm, sessile and auriculate or on petiole to 15cm; upper leaves ovate to linear-lanceolate, to 10 × 4cm, sessile. Inflorescence ± corymbose, with 1–15 capitula. Capitula (6–)8–15mm across, cernuous; peduncles up to 1[illegible](–20)cm, blackish hirsute. Phyllaries lanceolate, 8–20 × 2–5mm, blackish hirsute. Corolla tube 9mm; ligules yellow, 13 × 2.5mm. Achenes brown, 9.5mm; pappus dirty white, 10mm.

Bhutan: C – Ha, Thimphu, Punakha, Bumthang and Mongar districts, **N** – Upper Mo Chu, Upper Pho Chu and Upper Bumthang Chu districts; **Darjeeling:** Sandakphu–Phalut, Sandakphu–Tanglu; **Sikkim:** Lachung, Samdong, Jongri etc.; **Chumbi:** Yatung, Chubitang, Phari etc. Open grassy banks and forests, 2750–4370m. July–October.

Extremely variable in leaf and habit. Segregate species are upheld in (405), but could not be reliably recognised among our specimens solely on the basis of the key provided there to leaf shape.

2. D. cymiformis Shih

Similar to forms of *D. hispida* with lyrate-pinnatifid petiolate lower leaves, but capitula erect and corollas pink. Plant c 45cm; capitula c 15.

Chumbi: without locality (405). 3200m. July.

No specimens seen.

3. D. stebbinsii Ludlow

Stems up to 100cm, ± glabrous. Lower leaves pinnate, lateral segments 3–5, oblong, 1–3 × 0.5–2cm, terminal segment ovate-triangular 15 × 12cm, long acuminate, base hastate, margins repand denticulate; upper leaves similar to terminal segments; petioles 1–7cm. Capitula 1–4, campanulate, c 1–1.5cm across, cernuous; peduncles 15–25mm, ± glabrous. Phyllaries ovate or oblong, 5–16 × 3–5mm, blackish, ± glabrous. Corolla tube 6–8mm; ligules reddish-mauve to blue-purple, 12–16 × 5mm. Achenes (immature) 4mm; pappus brownish, 8–10mm.

Bhutan: N – Upper Kuru Chu district (Singhi Dzong). In forests and open hillsides, 3600m. August.

16. TARAXACUM Weber

Eng: *Dandelion.*

Acaulescent perennial herbs with perpendicular tap-roots. Leaves basal, rosulate, runcinate-pinnatifid to subentire, oblanceolate in outline. Scapes erect, hollow. Capitula many-flowered, ligulate. Involucres 2-seriate, inner phyllaries erect, linear-lanceolate, subequal; outer phyllaries shorter than inner ones, appressed or reflexed; phyllary tips thin, callose (thickened), corniculate (weakly bifurcate) or cornute (prominently bifurcate). Flowers white to yellow or reddish; outer ligules greyish or purplish-green on back. Style branches filiform. Achene body oblong-oblanceolate, unevenly ribbed, squamate or spinulose at least at apex, narrowed to an apical cone bearing a slender beak; pappus capillary, persistent.

Leaves are described from flowering specimens; scapes and phyllaries measured at anthesis. Achene beaks are probably at least 6mm long eventually, but were rarely fully elongated in the specimens seen. *Taraxacum* in our area consists of sexually reproducing and apomictic taxa. The latter in particular are poorly understood at present. Specimens not closely corresponding to the descriptions given here can only be identified to the appropriate section. Lawn weeds and other damaged specimens often produce uncharacteristic growth that cannot be reliably identified. Unlobed leaves are sometimes produced in summer. A specimen that only bears such leaves cannot be reliably identified.

1. Apex of all phyllaries thin, entire .... (Sect. *Ruderalia*) **8. T. insigne**
+ Apex of some phyllaries callose, corniculate or cornute 2

2. Stigmas yellow or pale green (drying yellow, brown or grey-green); no phyllaries cornute; outer phyllaries with clearly defined scarious margins; flowering April–May .... (Sect. *Parvula*) 3
+ Stigmas dark green (usually drying black) or some phyllaries cornute or outer phyllaries immarginate (with scarious margin absent or narrow and indistinct); usually flowering July–October .... (Sect. *Tibetana*) 4

3. Leaf lobes often entire; outer phyllaries 8–13 .... **1. T. parvulum**
\+ Leaf lobes always dentate or denticulate; outer phyllaries 14–19 or more **2. T. mitalii**

4. Some phyllaries cornute .... **7. T. pseudostenoceras**
\+ All phyllaries callose or corniculate 5

5. Achenes reddish-brown 6
\+ Achenes stramineous 7

6. Leaves narrow (length usually more than 6 × width), strongly runcinate; leaf lobes narrow, entire (very rarely with 1 small tooth) .. **3. T. sikkimense**
\+ Leaves broad (length usually less than 6 × width), weakly runcinate; leaf lobes broad, often sparsely dentate or denticulate .... **5. T. eriopodum**

7. Outer phyllaries with scarious margins broad, sharply defined, whitish see 3. *T. sikkimense*
\+ Outer phyllaries immarginate or with scarious margins narrow, weakly defined, brownish when dry 8

8. No phyllaries corniculate; pollen absent .... **6. T.** cf. **apargiiforme**
\+ Some phyllaries corniculate; pollen present 9

9. Leaf lobes broad, obtuse, never strongly retrorse .... **5. T. eriopodum**
\+ Leaf lobes narrow, acute, often stongly retrorse .... **4. T. tibetanum**

SECT. **Parvula** Handel-Mazzetti

Phyllaries green or olive-green, with clearly defined scarious margins, callose or corniculate. Stigmas yellow to pale green. Achene stramineous, body gradually constricted into cone. Pappus white.

1. T. parvulum DC.; *T. officinale* Wiggers var. *parvulum* (DC.) Hook.f. sensu F.B.I. p.p., *T. himalaicum* Soest. Fig. 121a–b.

Leaves 35–90 × 8–13mm; lateral segments (1–)3–5 pairs, entire, with 1 tooth or 1–3 denticles. Scapes 30–80mm. Involucres 7mm diameter. Outer phyllaries 8–11(–13), 4–7 × 1.4–2.8mm, callose; inner phyllaries 9–15mm, callose. Stigmas yellowish. Achene body narrow, very weakly spinulose at apex; body and cone 3.8–5mm; beak 6mm; pappus 6–7mm.

Bhutan: C – Thimphu district (Drugye Dzong); **Sikkim:** Chiya Bhanjang. In short grass, 2300–3150m. April–May.

The specimens from our area belong to a group of apomicts of uncertain status that can only be provisionally included in *T. parvulum* sensu lato (356), although they have also been described as *T. himalaicum* fm. *kuluense* Soest. A population from Ha, 2750m, collected at an early stage of development, was described as *T. bhutanicum* Soest. It differs from the specimens cited above by its consistently denticulate leaf lobes, larger capitula and anthers usually lacking pollen, but the material cannot be adequately evaluated or described.

2. T. mitalii Soest. Fig. 121c–e.

Leaves 40–85 × 6–20mm; lateral segments 2–4 pairs, entire, with 1–2 narrow teeth from a broad base or 1–3 denticles. Scapes 16–55mm. Involucres 7–11mm diameter. Outer phyllaries 14–19 or more, 4–8 × 1.2–2.5mm, callose; inner phyllaries 9.7–15mm, callose or slightly corniculate. Stigmas greyish-green. Achene body squamate-spinulose above; body and cone 4.5–4.7mm; beak 7–8.5mm; pappus 5–5.5mm.

Sikkim: Tonglu, Singalilla Ridge; **Chumbi:** Pipitang. 3050–3500m. April–May.

SECT. **Tibetana** Soest

Phyllaries green, olive-green or almost black, with scarious margin clearly or weakly defined or absent, callose, corniculate or cornute. Stigmas pale green to dark green. Achene stramineous or reddish, body abruptly constricted into cone. Pappus white or dirty white.

3. T. sikkimense Handel-Mazzetti; *T. officinale* Wiggers sensu F.B.I. p.p. Fig. 121f–g.

Leaves deeply incised, 50–130 × 7–13mm; lateral segments 3–5 pairs, narrow, strongly retrorse, entire (very rarely with 1–2 denticles). Scapes 60–90mm. Involucres 6–10mm diameter. Phyllaries olive-green, eventually drying dark brown, callose or corniculate; outer phyllaries 12–17, 4.5–6.7 × 1–2.7mm, obscurely bordered; inner phyllaries 10–12mm. Stigmas dark green. Achene body reddish-brown, prominently spinulose on main ribs above; body and cone c 4.2mm; pappus dirty white, 5.5–6mm.

Sikkim: Lachen; **Chumbi:** Lingshi La. 3650–4250m. July.

Several further summer-flowering collections from Sikkim and Chumbi somewhat resemble this species. Those differing in leaf shape, outer phyllaries with more sharply defined paler broader margins, lighter stigmas or stramineous achenes can only be identified to sect. *Tibetana*. Vigorous specimens from Tungu (Sikkim) with less prominently lobed leaves to 210 × 22mm, exceeded by the scapes and larger capitula may be referred to *T. sikkimense* sensu lato (Fig. 121h–i).

FIG. 121. **Compositae: Lactuceae.** a–r, leaves of *Taraxacum* species (× ½): a–b, *T. parvulum*; c–e, *T. mitalii*; f–g, *T. sikkimense*; h–i, *T. sikkimense* sensu lato from Tungu (Sikkim); j, *T. tibetanum*; k–m, *T.* cf. *apargiiforme*; n–o, *T. eriopodum*; p–q, *T. pseudostenoceras*; r, *T. insigne*. s–u, *T. eriopodum*: s, habit (× ½); t, flower (× 1½); u, achene (× 1½). Drawn by Chen Yo-Jiun.

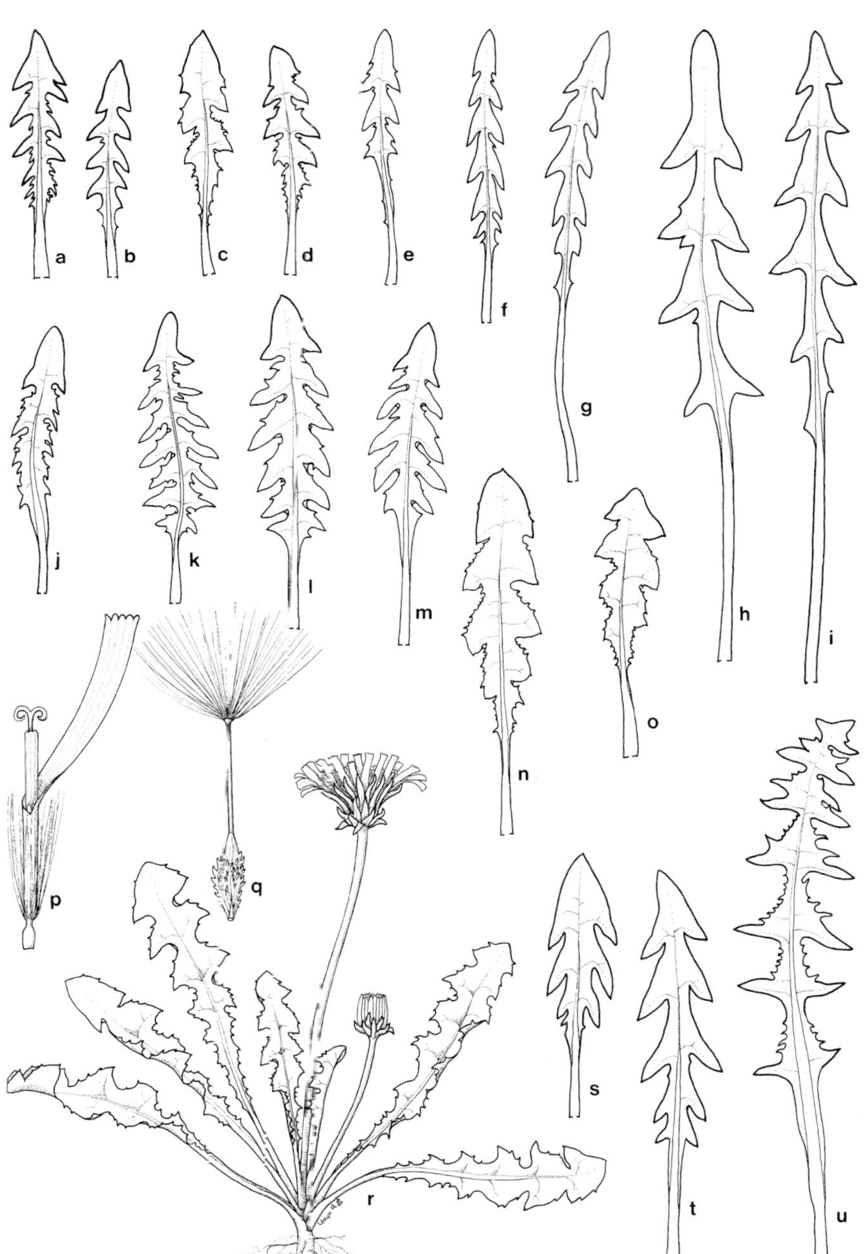
a
b
c
d
e
f
g
h
i
j
k
l
m
n
o
p
q
r
s
t
u

4. T. tibetanum Handel-Mazetti; *T. officinale* Wiggers sensu F.B.I. p.p. Fig. 121j.

Leaves moderately or deeply incised, 55–80 × 10–14mm; lateral segments (2–)3–5 pairs, attenuate, entire to dentate-lobulate. Scapes exceeding leaves. Involucres 10–12mm diameter. Phyllaries dark, eventually drying almost black, mostly corniculate; outer phyllaries c 15–20, 7.5–11 × 1.5–3.5mm, immarginate; inner phyllaries 13–14mm. Stigmas dark green. Achene body stramineous, spinulose above; body and cone c 4.2mm; beak darker; pappus 5.5–6mm.

Sikkim: unlocalised (Hooker collection between 3350m and 5500m).

5. T. eriopodum DC.; *T. officinale* Wiggers var. *eriopodum* Hook.f., *T. wattii* Hook.f. Fig. 121n–o, s–u.

Leaves shallowly or moderately lobed, 55–110(–190) × 13–30(–40)mm; lateral lobes 2–4 pairs, triangular, usually obtuse, entire to sparsely dentate. Scapes 50–110mm. Involucres (5–)7–11mm diameter. Phyllaries callose or corniculate; outer phyllaries 16–24, usually all recurved, dark green suffused purple, drying blackish, 5.5–7.5 × 1.5–2.8mm, immarginate; inner phyllaries olive-green, eventually drying brownish, 11.5–13mm. Stigmas dark green. Achene body reddish-brown or stramineous, spinulose above; body and cone 4.2–5.5mm; beak 6mm; pappus dirty white, 5–7mm.

Bhutan: C – Ha district (Damthang, Ha–Chile La), Thimphu district (Barshong, Motithang, Pajoding), **N** – Upper Mo Chu district (Singche La, Zambuthang, Laya), Upper Bumthang Chu district (Domchen, Pangotang); **Sikkim:** Tsomgo, Nathang, Natu La; **Chumbi:** Chumbi. Pasture, grassy banks, 3000–4250m. July–October.

6. T. cf. **apargiiforme** Dahlstedt. Fig. 121k–m.

Leaves usually deeply incised, 60–120 × 12–26mm; lateral segments (3–)4–7 pairs, narrow, ± oblong, retrorse, usually sparsely dentate at base or between lobes. Scapes 60–130mm. Involucres 8–11mm diameter. Phyllaries callose; outer phyllaries 15–20, dark green, 4.5–7.5mm × 1.2–3mm, immarginate; inner phyllaries olive-green, 11–13mm. Stigmas dark green. Achene body stramineous, weakly spinulose above; body and cone 4.2–4.7mm; beak 6mm; pappus white, 5.5mm.

Nyam Jang Chu: Cho La. On open grassy alp, 3960m. August.

This collection differs from typical *T. apargiiforme* of NW Sichuan only by its lack of pollen.

7. T. pseudostenoceras Soest. Fig. 121p–q.

Leaves deeply incised, 45–70 × c 10mm; lateral segments 5–6 pairs, ± lanceolate, strongly retrorse, entire or 1-dentate between lobes. Scapes (60–)80–90mm. Involucres 7–8mm diameter. Phyllaries eventually drying black; outer phyllaries 20–25, 4.5–7.5mm × 1.2–3mm, with broad, clear margins, cornute; inner phyllaries c 11mm, corniculate. Achene body stramineous, prominently spinulose on ribs above; body and cone 4.7mm; beak 6mm; pappus white, 7.5mm.

Sikkim/E Nepal: unlocalised (Hooker collection between 3350m and 5500m).

SECT. **Ruderalia** Kirschner, Øllgaard & Štěpánek

Introduced weedy section distinguished by their entire, thin phyllary tips.

8. T. insigne Wiinstedt & Jessen. Fig. 121r.

Leaves deeply incised, 80–180 × 20–45mm; lateral segments (3–)5–6(–7) pairs, mostly triangular, spreading, usually with several narrow teeth. Scapes 150–200mm. Involucres 12–13mm diameter. Phyllaries thin and unlobed at apex; outer phyllaries 13–14, reflexed, 10–12 × 2.3–3.3mm; inner phyllaries c 13–15mm. Stigma pale green. Achene body ± stramineous, weakly spinulose and somewhat abruptly contracted above; body and cone 3.5–4.2mm; beak to 7mm; pappus white, 7mm.

Darjeeling: Darjeeling. Grassy fields and roadsides, 2120–2140m. May.

Introduced from Europe. Plants described from specimens cultivated from Darjeeling seed, achene described from European material.

17. IXERIS Cassini

Our species an annual glabrous herb; stems erect or spreading. Leaves oblong to linear and entire to remotely dentate or oblanceolate in outline and deeply pinnatisect with 4–5 pairs of linear segments; basal leaves petiolate (sometimes absent at flowering); cauline leaves sessile, sagittate-auriculate. Inflorescence of 1–many sub-umbellate heads of few capitula. Capitula ligulate, small. Involucres narrow, cylindrical, 2-seriate, inner phyllaries subequal with much smaller, ovate, calyculate outer ones. Flowers up to c 19–27. Ligule yellow. Style branches linear. Achenes prominently and ± equally 10-winged, narrowed into slender beak; pappus simple, of numerous capillary hairs.

1. I. polycephala Cassini; *Lactuca polycephala* (Cassini) Clarke

Stems to 45cm. Lowest leaves oblong-lanceolate with slender petiole to linear with poorly defined petiole, to 13 × 1cm, entire or with few slender teeth especially below, or deeply pinnatisect, to 110 × 35mm, with entire, linear lobes. Cauline leaves lanceolate to linear, 3–13 × 0.3–1.5cm, acuminate, sessile, sagittate-auriculate (auricles to 2cm), margins more often entire. Involucres 2–3mm diameter; outer phyllaries 5–6, ovate, c 0.8mm, acuminate; inner phyllaries 7–8, linear-lanceolate, 5–6mm in flower (6–7mm in fruit). Flowers 19–27. Corolla tube 2–3.5mm; ligules 3.5–4 × 1.4mm. Achene body 2–2.5mm, reddish-brown, beak 1.5mm; pappus 3.5mm, white.

Bhutan: S – Samchi district (Buduni), **C** – Punakha district (Samtengang–Choojom (71)); **Darjeeling:** Rangit, subtropical zone (Hooker collection); **W Bengal Duars:** unlocalised (Haines collection). Wheat-fields etc., c 300–1500m. October–April.

Pinnatisect basal leaves seen only on the (immature) specimen from Buduni, on which the stem has scarcely begun to develop but apparently bore unlobed cauline leaves (except the sagittate base). It can probably be referred to fm. *dissecta* (Makino) Ohwi, which is treated as a distinct species by Shih (405). Some other collections lack basal leaves.

18. IXERIDIUM (A. Gray) Tzvelev

Perennial glabrous herbs; stems usually erect. Leaves simple or hastately or sagittately 3-lobed, entire, denticulate or sparsely slender-toothed, sessile to long petiolate. Capitula ligulate, in loose corymbs or panicles. Involucres cylindrical, 2-seriate. Flowers 5–18. Ligules yellow. Style branches linear. Achenes somewhat compressed, equally 10-ribbed, narrowed into slender beak, scabridulous; pappus simple, of numerous capillary hairs, white or yellowish.

1. Most leaves petiolate and prominently sagittate or hastate; petioles winged above .... **3. I. sagittarioides**
\+ Most or all leaves sessile; if present, petiolate leaves simple, with slender unwinged petioles 2

2. Involucres 4.5–5mm (to 6.5mm in fruit); capitula 5–6-flowered **1. I. beauverdianum**
\+ Involucres 7mm (8mm in fruit); capitula (8–)10–11-flowered .. **2. I. gracile**

1. I. beauverdianum (Léveillé) Springate; *Lactuca gracilis* sensu F.B.I. p.p. non DC., *Ixeris makinoana* (Kitamura) Kitamura. Fig. 122e–g.

Stems usually erect, 15–60cm. Petiolate basal leaves sometimes present, narrowly oblong-elliptic, to 140 × 1.6cm, acuminate, attenuate at base, often with a few slender teeth on lower blade and petiole; petiole slender, to 4cm. Cauline leaves sessile, linear-lanceolate, 3–11(–16) × 0.2–1.5(–2)cm, acuminate, base rounded, semi-amplexicaul and often with few slender teeth. Intermediate leaves with broadly winged, entire or coarsely toothed petioles sometimes present near base of stem. Inflorescence of divaricate corymbose heads. Involucres c 1mm across; outer phyllaries 3(–4), ovate, c 0.4mm; inner phyllaries 5–6, linear-lanceolate, 4.5–5mm in flower (to 6.5mm in fruit). Flowers 5–6. Corolla tube 2.5–3mm, ligule 2.5–3 × 1mm. Achene body 2.5mm, pale brown, beak 0.8mm; pappus 3.5mm, yellowish.

Bhutan: C – Thimphu district (Motithang, Paga, Thimphu etc.), Punakha district (Tashi Choling (73), Tinlegang (73), Wangdu Phodrang–Nahi etc.), Tongsa district (Shemgang, Pele La), Mongar district (Challi); **Darjeeling:** Darjeeling, Labha. Forest roadsides, field margins, grassy slopes and among dry rocks, 1150–2450m. March–August.

2. I. gracile (DC.) Pak & Kawano; *Lactuca gracilis* DC., *Ixeris gracilis* (DC.) Stebbins

Similar to *I. beauverdianum* but involucre c 1.8mm across; outer phyllaries c 6; inner phyllaries 7–8, in flower 7mm (in fruit 8mm); flowers (8–)10–11; ligule 6 × 2.2mm; achenes c 4.5mm (including beak).

Bhutan: C – Punakha district (Gon Chungang–Punakha (71)), Bhutan: unlocalised (Griffith collection). 1600m. May.

3. I. sagittarioides (Clarke) Pak & Kawano; *Lactuca sagittarioides* Clarke, *Ixeris sagittarioides* (Clarke) Stebbins

Plant erect, 20–40cm, usually branched from base. Most leaves basal, narrowly triangular, acute or subacute, prominently sagittate or hastate, entire or distantly denticulate; blade 2–8 × 1.5–5(–8.5)cm; petiole 2.5–8(–13)cm, winged above. Cauline leaves very few or none, narrowly oblong or linear-lanceolate, very weakly lobed or simple, sessile or with short broad petiole, soon passing into scale-like, ovate, acuminate bracts. Inflorescence loosely corymbose. Involucres 2–3mm across; outer phyllaries 5–6, 2–5.5 × c 1mm, inner ones 10–12, linear-lanceolate, 8mm. Flowers c 14–18. Corolla tube 3.5mm; ligule 8 × 2mm. Achenes narrowly ovoid, 3.5mm including beak 1mm; pappus white, inner series 4mm, irregular outer series to 0.2mm present.

Bhutan: C – Punakha district (Ratsoo–Samtengang (71)), Mongar district (Khinay Lhakang), Tashigang district (Kanglung). Dry mountainside, 1800–2000m. March–April.

Exceptional in its leaf shape and white pappus with minute outer series. Transfer to *Mycelis* as *M. sagittarioides* (Clarke) Sennikov has recently been proposed.

19. MULGEDIUM Cassini

Perennial herbs; roots tuberous or slender; stems erect or ± absent. Leaves entire, toothed or pinnatifid. Capitula ligulate, 1–20(–30) in open racemes or panicles. Involucres campanulate, usually narrow. Corolla sometimes hairy outside at throat; ligules pale mauve to blue or violet (?rarely yellow), exceeding tube. style branches linear. Achene body ellipsoid, compressed, with 3–7 weak ribs on each side, gradually tapering to a long beak; pappus single, capillary, white, dirty white or somewhat brownish.

1. Lower leaves often pinnatifid; upper leaves usually entire; ligules 6–7mm **1. M. lessertianum**
\+ All leaves dentate or denticulate; ligules 13.5–15mm .... **2. M. bracteatum**

1. M. lessertianum DC.; *Lactuca lessertiana* (DC.) Clarke

Perennial with evident tuberous root, 3–30cm, subglabrous to sparsely hirsute; stems erect or ± absent. Basal leaves crowded, oblanceolate, 2.5–11(–18) ×

0.5–2.3cm, obtuse or acute, gradually tapering into a winged petiole, entire, dentate or pinnatifid; upper leaves reduced, sessile, narrowly oblong or lanceolate, acute, usually entire. Inflorescence loosely racemose, of 1–10(–25) capitula. Involucres 5–7(–10)mm diameter; phyllaries lanceolate, green (often black when dried), subglabrous to densely hirsute below; inner phyllaries 9–12mm in flower, to 17mm in fruit. Corolla sometimes hairy outside at throat; tube 5–7mm; ligule 6–7 × 1.7mm, blue or purple (one collection described as yellow). Achene 4–6mm, dark brown, 3–5-ribbed on each side; beak pale above, 5–6mm; pappus 5–7mm, white, dirty white or somewhat brownish.

Bhutan: **N** – Upper Mo Chu district (Lingshi); **Sikkim:** Chakung Chu, Onglakthang, Yakla etc.; **Chumbi:** Phari, Gyong. On sandy loam, screes and amongst scrub, 3960–5100m. August–September.

2. M. bracteatum (Clarke) Shih; *Lactuca bracteata* Clarke. Fig. 122a–b.

Perennial, (20–)30–80cm; roots slender; stems erect, hirsute. Leaves obovate-elliptic, 3.5–11 × 1–4cm, acuminate, narrowed and semi-amplexicaul at base, dentate or denticulate, sparsely pubescent. Inflorescence sparse, racemose or paniculate, with prominent ovate (lanceolate), acuminate, denticulate bracts; capitula usually 10–20. Involucre 5–9mm diameter; phyllaries lanceolate, purplish; inner phyllaries 10–13(–16)mm in flower. Corolla tube 5–8mm, ligule 13.5–15 × 2.5–3mm, pale mauve to blue or violet. Achene narrowly ellipsoid, 5.5mm, pale brown, 3–7-ribbed on each side; beak 4mm, brownish; pappus 7mm, white.

Bhutan: **C** – Thimphu district (Barshong, Chapcha, Dotena etc.), Thimphu district (Tashiling, Rukubji), **N** – Upper Mo Chu district (Gasa); **Sikkim:** Lachung, Tsomogo, Yumthang etc.; **Chumbi:** Amo Chu, upper Chumbi Valley, Yatung. In forest, on stony slopes and among scrub, 2130–3650m. September–October.

20. LACTUCA L.

Annual, biennial or ?perennial herbs, glabrous to sparsely crisped-hairy or slightly hispidulous. Lower leaves simple to bipinnatifid, sessile or petiolate; upper leaves sessile, auriculate or sagittate. Inflorescence cymose, corymbose or paniculate. Involucres narrowly campanulate, 2- to several-seriate, dilated below in fruit. Corolla tube hairy above, ligule yellow to blue or purplish. Style branches linear. Achenes obovoid or ellipsoid, compressed, ribs 3–7 on each

FIG. 122. **Compositae: Lactuceae.** a–b, *Mulgedium bracteatum*: a, apex of flowering shoot (× ⅔); b, achene. c–d, *Sonchus wightianus* subsp. *wightianus*: c, apex of flowering shoot (× ⅔); d, achene (× 6). e–g, *Ixeridium beauverdianum*: e, apex of flowering shoot (× ⅔); f, capitulum (× 4); g, achene (× 6). h–i, *Cephalorrhynchus macrorhizus*: h, apex of flowering shoot (× ⅔); i, achene (× 6). Drawn by Margaret Tebbs.

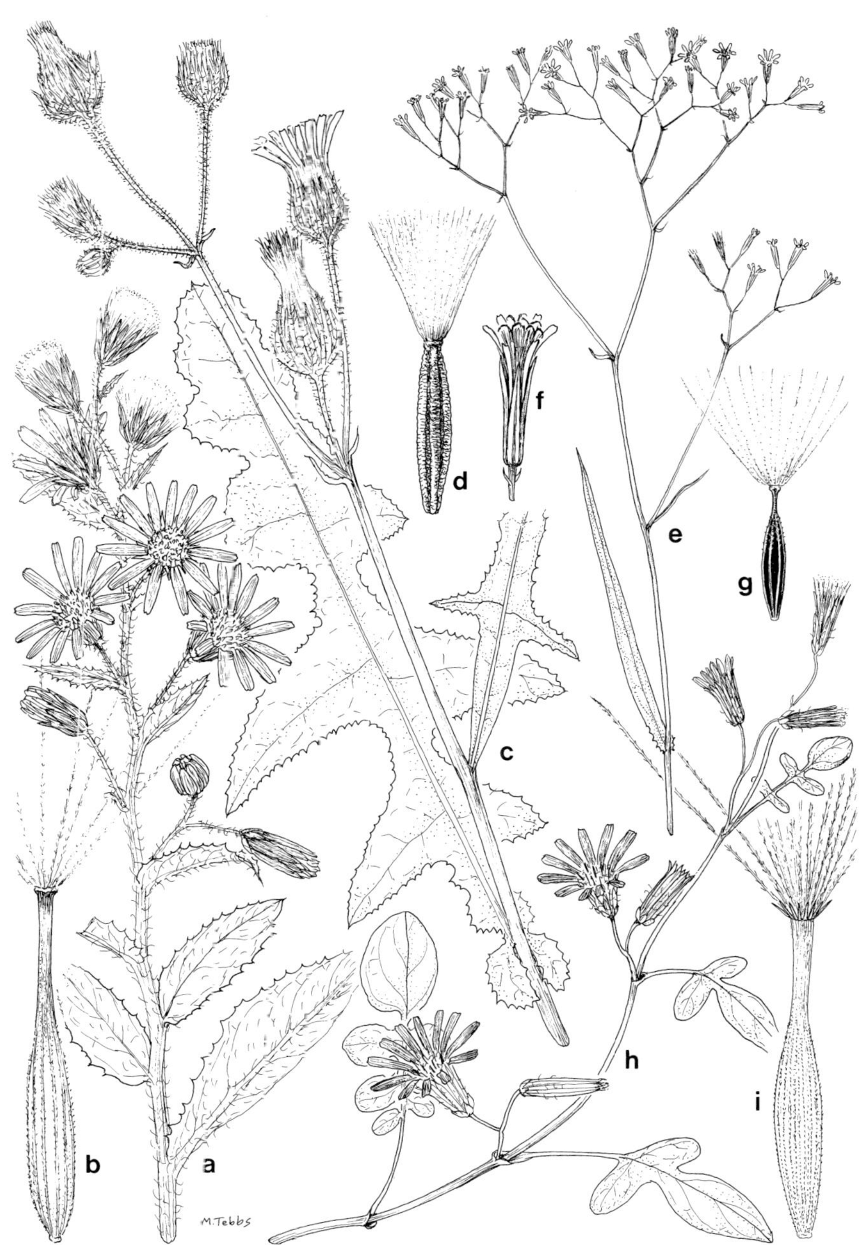
a
b
c
d
e
f
g
h
i
M.Tebbs

side, rather gradually tapering to a long slender beak; pappus single, capillary, white or nearly white.

1. All leaves linear or lower leaves pinnatisect with linear segments **3. L. dolichophylla**
+ At least lower leaves or leaf segments much broader 2

2. Annual to 50cm; lower leaves pinnatifid to bipinnatifid ...... **1. L. dissecta**
+ Biennial 60–100cm; lower leaves simple .... **2. L. sativa**

1. L. dissecta D. Don

Annual, branched above or from base, 9–50cm, subglabrous or sparsely crisped-hairy, sometimes densely hairy at stem base. Lower leaves pinnatifid to bipinnatifid, often lyrate, petiolate, 3–12 × 0.7–3cm; lobes and rachis dentate; upper leaves narrowly oblong-lanceolate, sessile, 2–5 × 0.3–0.5cm, acute, with acute auricles at base, entire. Inflorescence cymose, sparsely to much-branched. Involucres 1.5–2mm diameter, 2-seriate; outer phyllaries (4–)7–8, ovate, to 2.3mm; inner phyllaries oblong-lanceolate, 4.5–5.5mm in flower (to 10mm in fruit). Flowers 20–25. Corolla tube 2mm; ligule 4 × 1mm, pale yellowish-green, pale lilac to blue. Achene body obovoid, 2.3–3 × 1mm, pale brown, 3-ribbed on each side, muricate; beak c 4.5mm, pale; pappus 3–4mm, whitish.

Bhutan: C – Thimphu district (Tanalum Bridge), Tashigang district (near Tashigang (117)); **Darjeeling:** Darjeeling, Badamtam. Weed of tea gardens etc., 850–2000m. April–June.

2. L. sativa L. Eng: *Lettuce.*

Biennial, glabrous; stem simple, erect, 60(–100)cm. Leaves obovate or elliptic, obtuse or acute, entire or denticulate, often shallowly lobed below; lower leaves rosulate, 3–18 × 1.5–10cm, tapered to usually truncate base; upper leaves reduced, auriculate, semi-amplexicaul. Inflorescence usually a large dense corymb. Involucres 3mm diameter; phyllaries imbricate, ovate to lanceolate, 2–8 × 1.5–2.5mm. Flowers 20–30. Corolla tube 3–5mm; ligule 6 × 1.5mm, yellow, sometimes with some mauve. Achene body ellipsoid, 4–5mm, brown, closely 5–7-ribbed on each side, sometimes sparsely hispid at apex; beak c 4mm, pale; pappus 4mm, white.

Bhutan: unlocalised (341). Cultivated as salad vegetable.

No specimens seen from our area. Description from plants cultivated elsewhere in N India.

3. L. dolichophylla Kitamura; *L. longifolia* DC. non Michaux

Herb with stout tap-root; stems erect, simple. Upper leaves linear, to 22 × 1.5cm, acuminate, sessile, sagittate at base with narrow lobes, glabrous. Lower leaves similar or deeply pinnatisect with slender, entire segments, hispidulous. Capitula in open terminal panicle. Involucre 2.7–3.5mm diameter; phyllaries weakly imbricate, inner ones 8–9, lanceolate, c 9.5mm. Flowers c 17. Corolla tube 4.5–5.5mm, densely hairy at top; ligule 11–12.5 × 2–2.7mm, pale blue or lavender. Achenes oblanceolate, compressed, with broad thick wing and 3–5 shallow ribs on each face, c 4.5 × 1.1mm; beak filiform, 2.5–4.5mm; pappus white, c 4.5mm.

W Bengal Duars: Buxa (229). Open rocky slopes, 850m. September–October.

No specimens seen from our area. Description based on specimens from Nepal and NW India.

21. PTEROCYPSELA Shih

Glabrous annual. Leaves linear-lanceolate, acuminate, base tapering, auriculate, margin entire. Capitula in elongate racemose panicles; involucres 2-seriate; outer phyllaries ovate; inner phyllaries oblong. Flowers 10–20. Corollas densely hairy outside at throat; ligule blue. Style branches linear. Achenes oblong-obovate, compressed, acuminate, with thin broad wing as wide as body and single midrib; beak slender; pappus single, capillary, dirty white.

1. P. indica (L.) Shih; *Lactuca brevirostris* Champion

Plant 0.6–1.5m. Leaves 12–30 × 1–2.5cm. Involucres 3–4mm diameter; inner phyllaries c 9–10mm in flower. Corolla tube 7–8mm; ligule 8–10 × 1.5mm. Achenes 5 × 2mm; beak 1mm; pappus 7mm.

Darjeeling: Rangit Valley (80, 329).

The only reports from Himalaya. Description based on (329). Specimens with leaves sparsely dentate, capitula with up to 30 flowers and ligules pale yellow inside (80) occur in NE India.

22. CICERBITA Wallroth

Straggling or decumbent perennials, glabrous or partly hirsute. Leaves simple or pinnate with 1(–3) pairs of lateral leaflets; petiole broadly winged at least towards base. Inflorescences lax, racemose or paniculate, of few to many capitula, from upper stem. Involucres cylindric, 2-seriate; outer phyllaries 3–6 phylls; inner phyllaries 5. Flowers 5. Corolla hairy outside at throat; ligule exceeding tube, pale mauve or purplish. Style branches linear. Achenes oblong-elliptic, rather compressed, weakly several-ribbed on each face, without beak; pappus double, white, capillary; inner series long; outer series c 0.1mm.

1. C. violifolia (Decaisne) Beauverd; *Prenanthes violifolia* Decaisne, *P. sikkimensis* Hook.f., *C. sikkimensis* (Hook.f.) Shih

Plant glabrous or with few large pellucid hairs at top of petiole and beneath leaves; stems 60–100cm. Leaves 4–20cm, blade or terminal leaflet triangular (rarely ovate), 2.5–5(–10) × 2–4(–7)cm, subacute, base truncate, hastate, sagittate or cordate, margins subentire to undulate-dentate; petiole slender, channelled to broadly winged towards base, sometimes with 1(–3) pairs of ± oblong lateral leaflets to 5 × 1.7cm. Inflorescences from stem apex and upper leaf axils, ± forming a single open panicle. Involucre c 2.5mm diameter; outer phyllaries (3–)4(–5) with 1–2 overlapping bracts immediately below (4–6 in total); inner phyllaries narrowly oblong-lanceolate, 10–13(–16)mm, glabrous. Corolla tube 5–7mm; ligule (8–)10–13 × (1.5–)3–3.5mm, pale mauve or purplish. Achenes 5.5–7mm, weakly several-ribbed on each face; pappus 7–8mm.

Bhutan: C – Thimphu district (Barshong, Hinglai La–Tsalimarphe), Tongsa district (Rinchen Chu), Tashigang district (Rocha Chu), **N** – Upper Kuru Chu (Singhi); **Sikkim:** Kankola, Lachen; **Nyam Jang Chu:** Pangchen. Open meadows, forests and scrub jungle, 2300m–3600m. August–November.

The plants from our area are taller and much more variable in leaf form than those of western Himalaya and are often separated as *C. sikkimensis* (Hook.f.) Shih. A record from Thimphu district, 300m (117) with inappropriate description requires confirmation. A species of *Cicerbita* from Meghalaya has been described as *Prenanthes hookeri* Hook.f., distinguished from *C. violifolia* by its more numerous hairs, thicker upper leaves with simple blades exceeding the broadly winged petioles and at least 5 outer phyllaries with several similar bracts immediately below. A poor unlocalised specimen from Bhutan (Griffith collection) might belong with it, though the leaves have few hairs.

23. CEPHALORRHYNCHUS Boissier

Perennial herbs, glabrous or hirsute in part; root tuberous; stems weak, scrambling or trailing. Leaves pinnatifid or pinnate with 2–6 pairs of lateral segments, rarely simple, sessile or petiolate, sometimes auriculate, entire or toothed. Capitula few, in small terminal and axillary clusters. Involucre 2-seriate; phyllaries sometimes ciliate or stiffly hirsute on back. Flowers c 10. Corollas densely hairy outside around throat; ligule exceeding tube, mauve, purple or blue. Style branches linear. Achene oblong-ellipsoid, little compressed, fine-ribbed, with marginal pair scarcely more prominent, finely scabrid, with slender beak shorter than body; pappus double, white, inner series of many long capillary bristles; outer series c 0.1mm.

1. C. macrorrhizus (Royle) Tuisl; *Lactuca macrorrhiza* (Royle) Hook.f., *Cicerbita macrorrhiza* (Royle) Beauverd. Sha: *Karnpaati.* Fig. 122h–i.

Plant glabrous except some phyllaries; stems 30–75cm. Leaves usually pinnate, rarely simple; terminal leaflet (entire blade) ovate or 3-lobed, 1.5–7 × 0.7–4.5cm, obtuse, abruptly rounded or cordate at base, entire or shallowly and distantly toothed; lateral leaflets 2–6 pairs, oblong, 3–30 × 3–18mm; petiole slender, to 110mm, sometimes with base winged and auriculate. Involucres 3–5mm diameter; phyllaries glabrous or sparsely hirsute on the back; outer phyllaries ovate to lanceolate; inner phyllaries linear-oblong, 10.5–13mm in flower. Flowers c

9–11, mauve, purple, or blue. Corolla tube 5–6mm; ligule 9–14 × 2.5–3.5mm. Achenes c 5.8–9mm including beak, dark brown or dark grey; inner pappus 8mm.

Bhutan: **N** – Upper Mo Chu district (Laya), Upper Bumthang Chu district (Pangotang); **Sikkim:** unlocalised (Elwes, Cunningham collections); **Chumbi:** Chumbi. On banks and in rock-crevices, 2750–3650m. August–September.

An April-flowering collection from Khinay Lhakang, Mongar district, Bhutan, with very immature fruit (*Ludlow, Sherriff & Hicks* 20144, BM, E) seems to be an undescribed species of this genus. It differs from *C. macrorrhizus* by its narrow, always pinnatifid, non-lyrate, sessile, auriculate leaves with acuminate lobes and teeth and by its densely hairy peduncles.

24. CHAETOSERIS Shih

Perennial herbs; root tuberous; stems erect. Leaves simple, pinnatifid, pinnate or bipinnatisect, sessile or petiolate, sometimes auriculate, entire or toothed. Inflorescence terminal, racemose, paniculate or thyrsoid; capitula 2–many. Involucre several-seriate; phyllaries entire or ciliate, glabrous or stiffly hirsute on back. Flowers numerous. Corollas hairy outside around throat; ligule shorter or much longer than tube, mauve or reddish-purple to blue. Style branches linear. Achene body ± ellipsoid, at least weakly compressed, with 3–7 ribs on each face, broad thick to rather thin margin and distinct stout or slender beak; ribs and margins finely scabrid; pappus double, white or dirty white, of capillary hairs; inner series numerous, long; outer series to 0.15mm.

Two species are described below; a dubiously localised collection from Darjeeling/Sikkim (?or E Nepal) with all capitula in bud (*Hooker* s.n., K) apparently represents a further species of *Chaetoseris* found also in W and NE Nepal and ?SE Tibet. Its lower leaves are bipinnatisect, inflorescence racemose to narrowly thyrsoid and phyllaries eciliate, finely papillose outside and coarsely hirsute on the keel and its achene beak slender, c 4mm. It is either conspecific with specimens from Kashmir described as *Lactuca decipiens* Clarke var. *multifida* Hook.f. (none seen) or it is unnamed. However, it is neither *Lactuca* in the restricted sense used here nor is it obviously a variety of Clarke's species.

Specimens from Ritang and Samtengang, Punakha district, Bhutan, 2300m (*Cooper* 2030, E; *Cooper* 2357, E, BM) comprise another, ?new species, distinguished by its pinnasect lower cauline leaves, narrowly oblanceolate in outline with 4–6 pairs of leaflets, 5 flowers per capitulum, corolla tubes 8–10mm and ligules c 8.5mm.

1. Phyllaries ciliate; ligules (9–10) 21–26mm .... **1. C. macrantha**
\+ Phyllaries eciliate; ligules 5mm .... **2. C. cyanea**

1. C. macrantha (Clarke) Shih; *Lactuca macrantha* Clarke, *Cicerbita macrantha* (Clarke) Beauverd. Fig. 120e–g.

Plant 45–100cm, sparsely pubescent. Leaves pinnatifid, 12–45 × 4–12cm, acute, margins dentate with apiculate teeth; lateral segments 3–6 pairs, oblong-triangular, acute; terminal segment triangular; basal leaves dissected almost to rachis, long petiolate; cauline leaves sessile, with broad, auriculate, stem-clasping base. Inflorescence terminal, loose, racemose or paniclulate, of few capitula. Involucres greenish-brown, 7–10mm diameter; phyllaries ovate to lanceolate, coarsely pale-ciliate; inner phyllaries 14–19mm in flower. Flowers numerous. Corolla tube 8–10mm; ligule blue (sometimes lavender), 21–26 × 4.5–5mm. Achenes ellipsoid, 5.5–6mm, with c 5 finely scabrid ribs on each face, tapering gradually into a thick beak; beak 2–2.5mm; pappus dirty white, 8mm.

Bhutan: C – Ha district (Damthang–Sharithang), Thimphu district (Barshong, Pajoding, Paro etc.), **N** – Upper Mo Chu (Tharizam Chu), Upper Mangde Chu (Dur Chutsen, Worthang), Upper Bumthang Chu (Marlung, Tolegang), Upper Kulong Chu district (Me La); **Darjeeling:** Sabargam; **Sikkim:** Chomnagu, Phalut, Singalila etc.; **Chumbi:** Gowsar, Trakarpa–Yatung. Alpine turf, rocky hillsides, meadows, clearings in forest, by streams, 2600–4250m. June–October.

A further specimen from Worthang (*Ludlow, Sherriff & Hicks* 17308, BM), of uncertain status, approaches *C. cyanea* in its mostly simple upper leaves, narrow, few (c 15)-flowered capitula, phyllaries with reduced cilia and dorsal hair and short ligules (9–10mm).

2. C. cyanea (D. Don) Shih; ?*Lactuca hastata* DC., *Cicerbita cyanea* (D. Don) Beauverd

Plant 35–100cm, sparsely pubescent. Median leaves usually simple, rarely weakly pinnatifid or pinnate; blade or terminal segment triangular or ovate, often hastate, 4.5–10(–18) × 3–7(–11)cm, acuminate, truncate or cordate at base, dentate or denticulate; lateral leaflets 2–3 pairs, ovate, acute, to 4 × 2cm; petiole usually narrowly winged, to 7(–11)cm; lower leaves simple and petiolate or pinnatifid and sessile with up to 6 pairs of lateral segments. Inflorescence of 1(–several) racemose panicles at stem apex. Involucres 4–5mm diameter; phyllaries ovate to narrowly oblong-lanceolate, dark-tipped, eciliate; outer ones glabrous or sparsely pale hirsute on back; inner ones 10–13mm in flower, glabrous. Flowers 15–35. Corolla tube 7mm; ligules 5 × 1.7mm, purplish-red to dark purple or blue. Achene dark brown; body oblong-ellipsoid, 4mm, 3-ribbed on each face; beak slender, 3mm; pappus white, 7mm.

Bhutan: C – Thimphu district (Paro (117), Shingkarap); **Darjeeling:** Goom, Senchal, Tanglu etc.; **Sikkim:** Chiya Bhanjang, Lachung. On grassy banks, among trees, 2150–3350m. September–November.

Shih (405) segregates plants with reduced vestiture comparable with that of most or all of our plants as *Chaetoseris hastata* (DC.) Shih.

25. STENOSERIS Shih

Erect annual or biennial, often puberulous. Leaves simple or pinnate with 1–2 pairs of lateral segments, petiolate. Inflorescence terminal, paniculate, often narrow and elongate. Capitula ligulate. Involucres narrowly cylindric, 2-seriate; outer phyllaries ovate, short; inner phyllaries and flowers 3(–4). Corolla hairy outside at throat; ligule shorter than tube, deep red or mauve. Style branches linear. Achene body ± oblong, compressed, mid-brown, c 5-ribbed on each side, with broad, thick margin; beak distinct, thick; pappus capillary, dull yellowish-white, with minute outer series.

1. S. graciliflora (DC.) Shih; *Lactuca graciliflora* DC., *L. rostrata* auct. non (Blume) Kuntze

Plant 60–200cm. Terminal leaf segments or entire blade ± triangular or 3-lobed, (3.5–)5–7(–10) × (2–)3.5–6(–8)cm, acuminate, cordate to attenuate at base, dentate; lateral segments elliptic, up to 5 × 2cm, acuminate, subsessile; petiole slender, to 9cm. Involucres 1mm diameter with 3 inner phyllaries and flowers involucre 1.5mm diameter (with 4 inner phyllaries and flowers consistently on one specimen,); phyllaries often purplish, minutely papillate and bearded at apex; inner phyllaries linear-oblong, 7.5–9.5mm. Corolla tube 7.5mm, ligule 4.5–5 × 1.8mm. Achene body 4.5–6mm, brown; beak 0.5–1.2mm, brownish; inner pappus 7mm, outer pappus c 0.1mm.

Bhutan: C – Thimphu district (Shingkarap, Tanga–Barshong (359)), Punakha district (Wangdu Phodrang), Bumthang district (Shimitang), **N** – Upper Bumthang Chu district (Kopub), Upper Mo Chu district (Gasa–Pari La); **Darjeeling:** Kalipokhri, Senchul, Sureil etc.; **Sikkim:** Chhateng, Chiya Bhanjan, Zongri etc. In forest, woodland and scrub, on moist gravel, 1500–3650m. September–December.

Extremely variable in leaf shape. Plants from Chumbi and other districts of S Tibet with all leaves simple, ovate-triangular to hastate-triangular, with blade 5–14 × 5–10cm have been described as *S. tenuis* Shih. Other characters were within the range of *S. graciliflora*. We have seen no certain examples. Specimens with similar upper and middle leaves have been collected infrequently in Sikkim, N Bhutan and E Nepal, but the lower leaves are missing or bear small lateral lobes.

26. NOTOSERIS Shih

Erect herb, sparsely pubescent. Leaves simple or ternate (possibly also pinnatisect or pinnate), long petiolate, linear-denticulate. Capitula ligulate, in terminal, narrow, paniculate inflorescence. Involucres narrowly cylindric; outer phyllaries ovate to lanceolate, short, c 6 in 2 series; inner phyllaries 4, linear-oblong, reddish-purple. Flowers 4. Corolla hairy outside at throat; ligule shorter than tube, blue. Style branches linear. Achene body ellipsoid-oblanceolate, compressed, purple-brown, c 6–8-ribbed on each side with marginal ribs not enlarged, apex truncate; pappus bristles capillary, brownish, all long.

1. N. sp. A

Very similar to *Stenoseris graciliflora* but inner phyllaries 4, 11mm; flowers 4; ligules 5.5–6mm; achenes purple-brown, without beak or broadened or thickened margin; pappus bristles all long, erect.

Bhutan: S – Deothang district (Keri Gompar). On damp bank in forest, 2000m. November.

A single collection seen from our area, quite isolated from the nearest *Notoseris* species in Meghalaya (described as *Prenanthes khasiana* Clarke) and SW China. More collections are required for complete description and exact determination of species, though our plant is quite similar to some Meghalayan material.

27. PRENANTHES L.

Our species a scandent perennial herb, ± puberulous, stems zigzag. Leaves simple, petiolate, ovate to triangular, acuminate, cordate or hastate at base, linear-denticulate. Inflorescences paniculate, from upper leaf axils, usually forming a single, terminal panicle of many capitula; peduncles pubescent. Capitula ligulate. Involucre narrowly cylindric, 2-seriate; outer phyllaries short; inner (4–)5(–6), linear-oblong, sparsely pubescent. Flowers (4–)5. Corolla with few hairs outside at throat; ligule subequal to tube, dull violet. Style branches filiform. Achenes (immature) narrow oblong-lanceolate, gradually contracted beneath rather wider disc but not beaked, somewhat compressed, without broadened or thickened margins, several-ribbed on each face; pappus bristles all long, erect, capillary, stramineous.

1. P. scandens Bentham & Hook.f.

Leaf blades 5–12 × 3–7cm; petiole 1.5–4cm. Involucres 2–3mm diameter; inner phyllaries 11mm. Corolla tube 7mm; ligule 7 × 2–2.5mm. Achenes 4–5mm; pappus 7mm.

Darjeeling: Ghumpahar, unlocalised (Griffith and Gamble collections); **Sikkim:** Kulhait. 910–2100m. November.

28. SYNCALATHIUM Lipschitz

Rosulate perennial herbs from lsender rhizomes; stems short,broadening towards apex, or almost absent. Leaves simple to lyrate-pinnatisect; petiolate. Capitula ligulate, 3–6-flowered, densely crowded on stem apex, ± sessile. Involucres cylindrical, subuniseriate; phyllaries 3–5, subequal. Ligules pink to purple. Style branches filiform. Achenes obovoid, compressed, contracted at apex and very shortly beaked. Pappus many-seriate, of deciduous bristles.

1. S. cf. **souliei** (Franchet) Ling

Leaves often reddish or purplish, often simple, subglabrous; blade or terminal segment elliptic to suborbicular, 7–25 × 5–20mm, obtuse, rounded or cordate at base, dentate; leaf rachis and petiole 1–4.5cm, with 0–4 lobes to 4 × 5mm. Phyllaries (4–)5(–6), lanceolate, 12 × 3mm, glabrous. Flowers (4–)5. Corolla tube 8.5–10.5mm, ligule 6 × 2.5mm, dull purple. Anthers c 2mm. Pappus c 10mm, whitish.

Nyam Jang Chu/Tibet: Cho La. Among scrub, 3650m. August.

A single collection remote from the main range of *S. souliei*, with short anthers, lacking mature achenes. Achenes of *S. souliei* [*Lactuca souliei* Franchet] are strongly compressed with 1 fine nerve on either face, achenes of other species are less compressed with 5 ribs.

29. SONCHUS L.

Eng: *Sow thistle*.

Annual, biennial or rhizomatous perennial herbs. Leaves alternate, simple or pinnatifid (then often lyrate), auriculate at base, without distinct petiole, though lower leaves usually much narrowed below. Inflorescence terminal, open, of 1–several corymbose heads of few capitula. Capitula ligulate. Involucre campanulate, often white tomentose below; phyllaries linear-lanceolate, several-seriate, becoming swollen and spongy at base when mature. Flowers numerous. Corolla hairy outside at throat; ligules yellow, much longer towards margin of capitula but apparently not exceeding tube in our area. Style branches linear. Achenes obovoid to ellipsoid, strongly compressed and distinctly winged to subterete, unbeaked, with 3–5 ribs or 2 grooves on each face, ± smooth or finely rugulose. Pappus white, dimorphic, of soft capillary bristles and fewer coarse scabrid ones united in a ring at base.

1. Annual; achenes strongly compressed, winged, with 3 ribs on each face, ± smooth or remotely scabridulous .... **1. S. asper**
\+ Annual/biennial or rhizomatous perennial; achenes compressed to subterete, not winged, with c 5 ribs or 2 grooves on each face; uniformly rugulose .. 2

2. Annual/biennial; achene compressed, 2-grooved on each side **2. S. oleraceus**
\+ Perennial from elongate rhizome; achene compressed to subterete, ± with 5 broad ribs on each side, most without grooves .... **3. S. wightianus**

1. S. asper (L.) Hill

Annual, 15–75(–120) cm, glabrous or stems above and peduncles glandular-hairy. Leaves simple or pinnatifid, 5–25 × 3–8cm, segments ± triangular, sharply dentate or denticulate, with rounded auricules at base. Involucres 5–7mm diameter, without cottony hair; outer phyllaries sometimes bearing few coarse glandular hairs; inner phyllaries 8–10mm. Corolla tube 6–8mm, ligule 5–6 × 1mm, shorter than tube. Achene obovate-ellipsoid, strongly compressed, 3 × 1.5mm, distinctly winged, ± smooth or remotely scabridulous on wings and ribs, most with 3 ribs on each face; pappus 8–9mm.

Bhutan: C – Thimphu district (Paro, Thimphu, Thimphu–Dochong La (71)), Bumthang district (Nangsephell); **Darjeeling:** Lebong, Rimbik. Roadsides and cultivated land, 1800–2700m. February–September.

2. S. oleraceus L.

Similar to *S. asper* but sometimes biennial; leaves simple or lyrate, with acuminate auricles at base; phyllaries usually cottony when young; corolla tube (4.5–)6–7mm, ligule 4.5–6.3mm, shorter than or equalling tube; achenes narrowly obovoid, less compressed, 2.5–3.75mm, not winged, 2-grooved on each face with obscure ribs, minutely and uniformly rugulose; pappus 7–8mm.

Bhutan: C – Thimphu district (Motithang, Simtokha–Pyemitangka, Thimphu), Punakha district (Tinlegang), Tashigang district (Tashigang); **Darjeeling:** Darjeeling. Roadsides, open ground, gardens, 1950–2450m. March–November.

3. S. wightianus DC.; *S. arvensis* sensu F.B.I. non. L.

Perennial, 25–140cm, from elongate rhizome; stems and leaves glabrous. Leaves usually simple, lanceolate or oblanceolate, sometimes pinnatisect, 6.5–30 × 1–6cm, acute, with rounded auricules at base, denticulate. Involucres (5–)7–10mm diameter, whitish tomentose below at least when young; inner phyllaries (9–)11–12(–13)mm. Corolla tube 7–8(–9.5)mm; outer ligules 5–6.5 × c 1.4mm. Achenes narrowly ellipsoid, compressed to subterete, 2.8–3.7mm, broadly 5-ribbed on each side, minutely and uniformly rugulose; pappus 8mm.

subsp. **wightianus**. Fig. 122c–d.

Peduncles and involucres glandular-hairy.

Bhutan: S – Phuntsholing, Chukka and Deothang districts, **C** – Thimphu, Punakha, Tongsa, Mongar and Tashigang districts; **Darjeeling:** Birik, Darjeeling, Monpu etc.; **Sikkim:** Rate Chhu, Singtam, upper Tista etc. Roadsides, walls, cultivated land, damp grassy banks, forest openings, 50–2350m. February–October, December.

subsp. **wallichianus** (DC.) Boulos (*S. arvensis* L. var. *laevipes* Koch)

Glandular hairs absent.

Bhutan: S – Phuntsholing district (Phuntsholing), **C** – Thimphu district (Namselling, Tsalimarphe–Simo Sampa), Punakha district (Lobesa: with subsp. *wightianus*); **Darjeeling/Sikkim:** unlocalised (Hooker collection); **Sikkim:** Singhik–Gangtok. Roadsides, 750–2500m. April–July.

30. LAUNAEA Cassini

Biennial or perennial subscapose herbs, glabrous. Leaves mostly basal, sometimes with 1–3 reduced leaves at lowest nodes of stem/common peduncle, simple or pinnatifid, narrowed towards base but without distinct petiole. Inflorescence racemose, narrowly paniculate, ± corymbose or divaricately branched. Capitula ligulate. Involucres tubular-campanulate, often apparently double; inner phyllaries linear-oblong. Corolla hairy outside at throat; ligule yellow, subequal to tube. Style branches linear. Achenes narrowly ovoid or oblong, unbeaked, longitudinally ribbed. Pappus silky, capillary, white, deciduous.

1. Leaves pinnatifid .... **1. L. asplenifolia**
\+ Leaves entire or remotely and shallowly dentate .... **2. L. acaulis**

1. L. asplenifolia (Willdenow) Hook.f.

Biennial or perennial, c 10cm. Leaves pinnatifid, 5–9 × 1.2–2.2cm, segments ± triangular, rounded and apiculate at apex, shallowly and distantly dentate or denticulate; lateral segments up to 4 pairs. Inflorescence divaricately branched. Involucre 2.5mm diameter; apex of some phyllaries spurred; inner phyllaries c 8.5mm. Corolla tube 7mm; ligule 6–7 × 2.3mm. Achene oblong, 2mm, with 5 major ribs and 10 minor ribs, all pale, rugulose; pappus 7mm.

Bhutan: unlocalised (Griffith collection).

2. L. acaulis (Roxb.) Kerr; *Crepis acaulis* (Roxb.) Hook.f.

Stems (peduncles) 7–15cm. Leaves linear-oblanceolate, 4–13(–30) × 0.4–1.5(–2.2)cm, acuminate, entire or remotely and shallowly dentate, often sparsely denticulate, rarely ciliate. Inflorescence racemose, narrowly paniculate or corymbose, usually of few capitula. Involucres 3mm diameter; inner phyllaries 11–13mm, apex not spurred. Corolla tube c 8mm, ligules 8–9 × 2–2.5mm. Achene oblong-lanceolate, compressed, 5mm, with 5 pale major ribs and darker minor ribs between; pappus 7mm.

Bhutan: C – Punakha district (Punakha, Samtengang–Wangdu Phodrang etc.); **Darjeeling:** Tanglu (97a). 1500–3000m. March–July.

31. HIERACIUM L.

Medium-sized erect perennial herbs. Leaves alternate, simple, sessile, subentire. Inflorescence terminal, cymose, with c 1–3 capitula. Capitula ligulate, many-flowered. Involucre campanulate; phyllaries glabrous, imbricate. Receptacle fimbrillate, flat (convex in fruit). Corolla sparsely pilose outside at throat. Ligule, anthers and style yellow. Style branches linear. Achenes oblong, not compressed, with 10–15 ribs; pappus capillary, fragile, ± 2-seriate with scattered shorter hairs.

1. H. umbellatum L.

Plant c 40cm, thinly stellate puberulous and scabridulous; stems slender. Leaves numerous, sessile, linear-lanceolate to narrowly oblong-elliptic, to 60 × 8mm, acute, obtuse to cuneate at base, margins densely scabridulous, sometimes with 1–2 small teeth. Involucre c 7mm diameter; phyllaries c 40, dark green, lanceolate, to 12mm, subacute, minutely serrulate; outermost ones recurved. Corolla tube 4.5–6mm; ligule 8–14mm. Achenes dark brown, 3mm, acute at base; pappus dirty white, 6mm.

Bhutan: C – Thimphu district (Simtokha–Talukah Gompa). Eroded slopes with stunted trees, 2600m. August.

32. PICRIS L.

Erect annual or short-lived perennial herb; indumentum hispid, partly consisting of glochidiate bristles. Leaves simple, alternate, lower oblanceolate, obtuse, attenuate towards base but sessile, weakly crenulate. Inflorescence loose, paniculate, often corymbose. Capitula ligulate. Involucres narrowly campanulate; all phyllaries hispid along keel; inner subequal, 1-seriate; outer several-seriate, usually all much shorter. Corolla hirsute outside at throat; ligule exceeding tube. Anther bases caudate. Style branches linear. Achenes ellipsoid, somewhat incurved, subterete and broadly 5-ribbed when fully developed, transversely rugose-squamate, with poorly defined, short, stout beak. Pappus plumose, bristles c 20, deciduous.

1. P. hieracioides L.

Plant 40–100cm; stem densely bristly below. Lower leaves 5–10 × 0.8–2cm, hispid on both surfaces; upper leaves shorter, oblong or lanceolate. Involucres 5–7mm diameter; inner phyllaries 8–12mm. Corolla tube 4–5mm; ligule 7–8 × 1.7–2mm. Achenes 4–5mm, mid-brown; pappus 6–7mm, whitish or dirty white.

Bhutan: S – Chukka district (Raidak Valley), **C** – Thimphu district, Punakha district (Samtengang–Chusom (71)), Tongsa district (Tongsa–Yuto La (71)), Bumthang district (Byakar), **N** – Upper Mo Chu district (Tamji–Gasa (71)); **Sikkim:** Lachen, Sangachoiling, Dentam–Pemayangtse (69) etc.; **Chumbi:** Yatung. Roadsides, sands and gravels, dry slopes, meadows, 1220–3750m. June–August.

Plants from our area are referable to subsp. **kaimainensis** Kitamura with mid-brown achenes.

33. TRAGOPOGON L.

Erect glabrous biennial herb, sparingly branched, from cylindric tap-root. Leaves linear, suberect, dilated at base and semi-amplexicaul, entire, with c 9 parallel veins prominent beneath (at least when dried). Capitula ligulate, few, on long, gradually inflated, leafy peduncles; flowers numerous. Phyllaries 1-seriate, linear-lanceolate, exceeding ligules. Corolla yellow (our species), hairy outside at throat; outer ligules much exceeding tube. Style branches linear. Achene rostrate; body linear, pentagonal, with 5 main ribs, stramineous, squamate throughout; beak as long as body, smooth. Pappus plumose, as long as achene, dirty white, borne on a pubescent annulus.

1. T. dubius Scopoli

Plant c 70cm. Cauline leaves to 15cm, blade to 8mm wide above dilated base. Involucre c 10mm diameter; phyllaries c 35mm, to 70mm in fruit. Corolla tube to 8mm; ligule to 16 × 3mm. Achene c 35mm.

Bhutan: C – Thimphu (Paro). On open grassy plain and orchard weed, 2200–2300m. June, September.

Introduced; native to Europe and NW Asia.

VERNONIEAE

34. VERNONIA Schreber

Herbs, shrubs, sometimes scrambling, or small trees. Leaves alternate, simple. Capitula in terminal panicles, often corymbose, discoid. Involucres oblong or campanulate; phyllaries linear or narrowly ovate in several series, imbricate. Corollas equal, tubular-campanulate, 5-toothed, pink or red to dull purple or bluish (colour rarely recorded). Style branches subulate. Receptacle flat, naked or with a few short hairs. Achenes oblong, 5–10(–15)-ribbed, terete, tapered at base; pappus deciduous, reddish or dirty white, with inner series of scabrid bristles and incomplete outer series of short bristles or narrow scales.

Vernonia should be restricted to species from the New World (323). However, at present in our area only *Baccharoides* and *Cyanthillium* can be completely separated from *Vernonia* sensu F.B.I. and the following species are treated under that genus.

Postscript: Recognition of the following species by Robinson (399a) in a paper preliminary to a generic review of the Vernonieae should be noted: *Acilepis aspera* (Hamilton) Robinson [*V. pyramidalis*], *A. saligna* (DC.) Robinson [*V. saligna*], *A. silhetensis* (DC.) Robinson [*V. silhetensis*], *A. squarrosa* D. Don [*V. squarrosa*], *Gymnanthemum obovatum* Gaudichaud [*V. scandens*], *G. volkameriifolium* (DC.) Robinson [*V. volkameriifolia*].

1. Achenes without eglandular hairs (glandular hairs sometimes present) ... 2
+ Achenes with eglandular hairs (glandular hairs sometimes also present) . 6

2. Tree 3
+ Subshrub or scandent shrub 4

3. Pappus whitish .... **5. V. volkameriifolia**
\+ Pappus reddish .... **6. V. talaumifolia**

4. Scandent; leaves entire .... **10. V. scandens**
\+ Free-standing; leaves remotely toothed at least towards apex 5

5. Involucre 7–10mm diameter; inner phyllaries 8–12mm at anthesis; corolla 13mm .... **3. V. silhetensis**
\+ Involucre 4–6mm diameter; inner phyllaries 5–8mm at anthesis; corolla 9–11mm .... **9. V. saligna**

6. Plant 15–20cm; leaves linear, margins revolute .... **4. V. revoluta**
\+ Plant exceeding 30cm; at least lower leaves ovate to obovate or oblanceolate, margins plane 7

7. Pappus whitish, inner series 6–9mm 8
\+ Pappus reddish, inner series 11–13mm 9

8. Phyllaries more than 100 .... **1. V. squarrosa**
\+ Phyllaries c 35 .... **8. V. attenuata**

9. Leaves obovate; inner phyllaries acuminate .... **2. V. subsessilis**
\+ Leaves oblanceolate; inner phyllaries obtuse .... **7. V. extensa**

A further unidentified taxon with immature capitula has been collected in Bhutan NW of Mongar. It is a summer-flowering shrub with leaves entire, elliptic, acuminate, involucre cylindric in bud, phyllaries ovate to oblong, obtuse, ciliate and flowers 3–4.

1. V. squarrosa (D. Don) Lessing; *V. teres* DC.

Perennial from woody stock, 15–60cm; stems rigid, shortly pubescent. Leaves obovate to oblanceolate, 4.5–9 × 1–5.5cm, acute or acuminate, cuneate at base, subsessile, subentire or remotely serrulate, shortly pubescent especially on midrib beneath. Capitula solitary, terminal and axillary, upper ones subsessile, lower ones on leafy peduncles. Involucre campanulate, 1–1.5cm diameter; phyllaries multiseriate, more than 100, subulate to narrow oblong and long acuminate, 3–10 × 0.5–1.5mm, outer ones spreading, often recurved above, inner ones erect and appressed. Corolla purplish, 14mm. Achenes pale brown, 3mm, pubescent. Pappus whitish, inner series 8–9mm, outer series setose, 1–1.5mm.

Darjeeling: unlocalised (Hooker and Griffith collections). Terai.

V. pyramidalis (D. Don) Mitra [?*V. aspera* sensu DC. non Hamilton, *V. roxburghii* Lessing] has been reported from Bhutan (Rhogma–Wangdu Phodrang, 1300m) (360). It differs from *V. squarrosa* by its distinctly serrate leaves, more numerous capitula arranged in a panicle and fewer (c 60) phyllaries, the inner ones abruptly short acuminate or obtuse and mucronate.

2. V. subsessilis DC.

Stout evergreen shrub. Leaves obovate, 13–30 × 8–12cm, subacute to acuminate, attenuate at base, subentire to serrate-dentate towards apex, subglabrous above, appressed puberulous beneath; petiole 5–10mm. Inflorescence terminal, large, open, paniculate, pubescent; capitula numerous. Involucre campanulate, c 5mm diameter; phyllaries c 4-seriate, lanceolate, acuminate, 1–6.5 × 0.5–1.5mm, outer ones spreading. Corolla 15mm, tube glandular-pubescent. Achenes pubescent, 4mm (immature); pappus rufous, inner series 13mm, outer series of bristles, 1–2mm.

?Bhutan: junction of Dhazi Chu and Panalsampa Chu (Lister collection); **Darjeeling:** Tista Valley. Lower hill forest, 600m. January.

3. V. silhetensis (DC.) Handel-Mazzetti; *V. bracteata* Clarke

Suffrutescent, 1–2m; stems ± unbranched except at inflorescence, shortly pubescent. Leaves rather coriaceous, elliptic to narrowly lanceolate, 4–11.5 × 0.75–3cm, acuminate, narrowed to base, ± sessile, remotely serrulate, finely glandular-punctate, pubescent at least along veins beneath. Inflorescence terminal, paniculate, subcorymbose. Involucre broadly campanulate, 7–10mm diameter; phyllaries ovate to oblong, 3–12 × 1–3mm, acuminate, all ± appressed, pubescent, ciliate. Corollas purplish, 13mm. Achenes 4mm, glabrous; pappus whitish, inner series 8mm, outer series of bristles, 1–2mm.

Bhutan: unlocalised (Griffith collection); **Darjeeling:** unlocalised (Hooker collections). Terai, c 300m. October.

4. V. revoluta Hamilton

Perennial from woody stock, 15–20cm; stems simple, ± erect, sparsely pubescent. Leaves 2–4 × 0.2–0.5cm, acute, tapering to ± sessile base, entire, revolute at margins, minutely glandular-punctate and shortly pubescent. Inflorescence lax, terminal, of few capitula. Involucres campanulate, c 5mm diameter; phyllaries lanceolate, 3–10 × 0.7–1.5mm, acuminate, all ± appressed, pubescent. Corollas 11mm. Achenes hairy; pappus whitish, inner series 9mm, outer series of few bristles 0.5–1mm.

Bhutan: unlocalised (Griffith collection); **W Bengal Duars:** Jalpaiguri, Chokerboo. February–March.

5. V. volkameriifolia DC. Nep *Nundheki* (34), *Nanriki*. Fig. 123f–h.

Small evergreen trees 2–5(–8)m. Leaves oblanceolate or obovate, 13–33 × 3.5–13cm, ± coriaceous, acute or acuminate, long attenuate to base, undulate to irregularly serrate, ± glabrescent above, sparsely pubescent or glabrescent beneath; petiole 0.5–2.5cm. Inflorescence terminal, paniculate; capitula numerous, 4–6(–7)-flowered. Involucre campanulate, 3–4mm diameter; phyllaries ovate to oblong, 1–7.5 × 1–3mm, rounded or subacute, sparsely pubescent. Corollas white to mauve, 10–12.5mm; lobes smooth or papillose at apex.

Achenes oblong, 3.5–6.5mm, glabrous; pappus whitish, inner series 8–10mm at maturity, outer series shorter, mostly 1–3mm.

Bhutan: S – Samchi, Phuntsholing, Chukka and Gaylegphug districts, **C** – Tongsa and Tashigang districts; **Darjeeling:** Kurseong, Rangit, Takdah (69), terai etc. Subtropical forest slopes, 500–1800m. October–March.

One specimen from Darjeeling district approaches the next species in its mostly 10-flowered capitula, occasional apical hair on corolla lobes and 10.5mm immature inner pappus. It is exceptional for our area, but not for *V. volkameriifolia* elsewhere.

6. V. talaumifolia Clarke. Nep: *Nundeki*.

Similar to *V. volkameriifolia*, most readily distinguished by its rufous pappus. Generally larger trees (trunk sometimes 30cm diameter); leaves to 40 × 15cm; capitula 10–12-flowered; corolla lobes weakly bearded; pappus inner series 10–12mm.

Bhutan: C – Mongar/Tashigang district (Mongar–Tashigang (117)), Tashigang district (Jiri Chu); **Darjeeling:** Badamtam, Tarkhola, Rangit Valley etc.; **Sikkim:** Yoksam. Forest, dry hillsides, 300–1350m. September–November.

7. V. extensa DC.

Shrub; stems to 2m, glabrescent. Leaves oblanceolate, 10–22 × 2.5–5cm, acuminate, attenuate to base, finely serrate, ± glabrous above, sparsely glandular and pubescent beneath. Inflorescence terminal, paniculate, corymbose, pubescent; capitula numerous, 6–11-flowered. Involucre ± cylindric at first, c 5mm diameter; phyllaries ovate to oblong, 1–8.5 × 1–3mm, subacute or obtuse, sparsely pubescent at first. Corolla 12–13mm, tube glandular-pubescent. Achenes oblong, 4.5mm, glandular and pubescent; pappus reddish, inner series 11mm, outer series of bristles or narrow scales, 2–3mm.

Bhutan: C – Tongsa district (Tashiling (71)), Mongar district (Lhuntse); **Darjeeling:** unlocalised (Hooker collection). 300–2300m.

8. V. attenuata DC.

Diffuse herb to 1m; stems slender, stiff, suberect. Leaves ovate to obovate, stiff, 8–14 × 3–6cm, acuminate, cuneate to base, shallowly serrate, subglabrous above, puberulous beneath, reticulate veins prominent beneath; petiole 5–8mm. Inflorescence sparse, terminal, ± paniculate, with much narrower oblong or elliptic leaves; capitula few, subsessile to long pedunculate. Involucre campanulate, c 6mm diameter; phyllaries c 30, narrow oblong-lanceolate, 1.5–10.5 × 0.7–1.5mm, aristate. Corollas light red (one reference only), 8mm, sparsely glandular-puberulous outside (densely so in bud). Achenes ± oblong, 3–4mm, pubescent; pappus whitish, inner series 6–7mm, outer series of bristles, c 1mm.

Darjeeling: Gok, Rangit and Tista Valleys etc.; **Sikkim:** Lusing; **W Bengal Duars:** Buxa. Hot valleys, 150–1200m. October–March.

A distinct form from Keri Gompar in E Bhutan differs by its more numerous (c 60), narrower, mucronate phyllaries and longer (15mm) pale mauve corollas.

9. V. saligna DC.

Erect herb 100–130cm. Leaves lanceolate to oblanceolate, 8–16(–25) × 2–4(–8)cm, acute or acuminate, attenuate to base; coarsely serrate at least above, sparsely pubescent on both surfaces, minutely glandular beneath; petiole 2–5mm. Inflorescence large, terminal, paniculate. Involucre campanulate, c 5mm diameter; phyllaries ovate to narrow oblong, 1–8 × 0.7–2mm, rostrate to subacute, araneous at first. Corollas purplish, 9–11mm, sparsely glandular-pubescent outside. Achene ± oblong, c 3.5mm, subsessile-glandular between ribs; pappus whitish, inner series 8mm, outer series of subulate often caducous bristles, c 1.5mm.

Bhutan: S – Samchi (117), Chukka, Gaylegphug (117) and Deothang districts, **C** – Punakha and Tashigang (117) districts; **Darjeeling:** Balasun Valley, Gok, Rangli Rangliot etc.; **Sikkim:** Lingthem. Dry valleys and forest margins, 150–1500m. October–December.

10. V. scandens DC.; *V. vagans* DC.

Scrambling evergreen shrub; young stems pubescent. Leaves ovate or elliptic, thin, 5.5–15 × 2–7cm, acuminate, rounded or cuneate at base, entire, minutely puberulous beneath; petiole c 8mm. Inflorescence large, terminal, leafy, paniculate, pubescent; capitula numerous, 9–17-flowered. Involucre campanulate, c 4mm diameter; phyllaries ovate and acuminate to oblong, obtuse and mucronulate, 1–8 × 0.7–2.5mm, ciliate. Corollas (immature) c 8.5mm, glabrous. Achenes oblong, 3mm, tapered at base, glabrous; pappus reddish or brownish, 7mm, inner series 8.5–10mm, outer series of bristles, 1–3(–5)mm.

Darjeeling: Chhota Rangit, Dalkarjhar, Rishap etc.; **W Bengal Duars:** Buxa Reserve. 450m. January–February.

Outside our area flowers are recorded as reddish-purple and whitish-blue and mature corollas are 10.5–12mm.

35. CYANTHILLIUM Blume

Our species an annual; stems greyish pubescent. Leaves ovate, lanceolate or elliptic, petiolate, rounded to apiculate at apex, narrowly attenuate to base, undulate to serrate, sparsely pubescent above, pale greyish tomentose beneath. Inflorescence terminal, paniculate, usually corymbose. Capitula discoid. Involucre campanulate; phyllaries linear-lanceolate, all ± appressed. Corollas white, blue or mauve. Style branches subulate. Achenes pubescent; pappus whitish, 2-seriate with minute outer series of bristles.

1. C. cinereum (L.) Robinson; *Vernonia cinerea* (L.) Lessing. Fig. 123d–e.

Plant 15–100cm. Leaves to 8 × 3cm; petiole 0–2cm. Involucre c 3mm diameter. Phyllaries 1–4.5 × 0.2–0.8mm. Corollas 4–7.5mm. Achenes 1.5–2mm; inner pappus 5–6mm, outer pappus 0.2mm.

Bhutan: S – Phuntsholing and Sarbhang districts, **C** – Punakha (71), Tongsa (117), Mongar and Tashigang districts; **Darjeeling:** Mongpu, Rayeng (69), Sukna etc.; **Sikkim:** Gangtok (69). Roadsides and disturbed ground, 250–1700m. Flowering all year, peak December–February (97a).

Ethulia conyzoides L.f. subsp. *conyzoides* [*E. megacephala* auct. non Schultes Bipontinus] has been reported from Sikkim and W Bengal (344, 397). It is also an annual species, differing from *C. cinereum* by its covering of sessile glands (as well as pubescence usually), always dentate leaves, short, subacute phyllaries and lack of any pappus.

36. BACCHAROIDES Moench

Our species a sparsely pubescent annual. Leaves elliptic or ovate, acute or acuminate, long attenuate at base, usually without distinct petiole, usually coarsely serrate. Inflorescence of compact terminal heads of few capitula, progressively overtopped by lateral flowering shoots. Capitula discoid. Involucres campanulate; phyllaries loosely imbricate; outer ones linear, herbaceous; middle ones scarious below with oblong or spathulate herbaceous appendage, often exceeding inner; inner ones elliptic, scarious. Corollas mauve. Style branches subulate. Achenes oblong-obconic, blackish, sparsely pubescent. Pappus rufous; inner series of bristles; outer series of narrow scales.

1. B. anthelmintica (L.) Moench; *Vernonia anthelmintica* (L.) Willdenow

Stems 60–120cm. Leaves 6–12 × 1.5–5cm. Involucres c 7–10mm diameter; phyllaries 5–10 × 0.7–2mm. Corollas c 11mm. Achenes 5mm; inner pappus 8mm, outer 0.4–2mm.

Bhutan: C – Tashigang district (Kheri (117)); **?Darjeeling:** unlocalised (Hooker collection). 1000m. September.

37. ELEPHANTOPUS L.

Stiff perennial herb; stems erect, branched. Basal leaves oblanceolate; cauline leaves reduced, alternate. Capitula in terminal glomerules. Glomerules surrounded by leaf-like bracts, of c 20 capitula. Capitula discoid, few- (usually 4-) flowered. Involucre narrow oblong; phyllaries 8–10, imbricate, acuminate. Receptacle small, naked. Corolla tubular-campanulate, limb deeply and unequally cut into 5 secund lobes. Style arms subulate. Achenes oblong-oblanceolate, somewhat compressed, 10-ribbed, pubescent. Pappus of 5 (rarely 6) stiff bristles, dilated and ciliate at base.

1. E. scaber L. Fig. 123a–c.

Rhizomatous; stems 10–60cm, appressed stiffly white pubescent. Basal leaves 6–25 × 2–7cm, obtuse or subacute, base attenuate, sparsely hirsute above, pubescent and glandular beneath, margin crenate-serrate, petiole to 1.5cm, semi-amplexicaul at base; cauline leaves shorter, ovate or oblong, semi-amplexicaul at base. Glomerules 5–7mm diameter, surrounded by 3 cordate carinate bracts, 1–1.5 × 1cm. Involucres 8–10 × 2mm; phyllaries in 2 series, oblong-acuminate, stiff, whitish pubescent. Corollas lilac or white, c 8.5mm, including linear-lanceolate lobes c 3mm. Achenes 4mm, pappus bristles 5mm

Bhutan: S – Samchi district (near Samchi (117)); **Darjeeling:** Balasun (97a), Lebong, Sukna etc.; **W Bengal Duars:** Buxa. In lawns and grassland, 600–1800m. September–December.

INULEAE

38. INULA L.

Coarse perennial herb from thin elongate rhizome. Leaves simple, alternate, usually finely toothed. Capitula few in terminal corymb, sometimes also borne singly on long axillary peduncules, radiate. Involucres broadly campanulate, many-seriate, phyllaries linear, recurved, villous. Receptacle naked, convex. Ray flowers 2–3-seriate, female, ligules long, linear, yellow. Disc flowers bisexual, yellow, tubular-campanulate, 5-toothed at apex. Style branches linear, obtuse. Achenes angular, oblong, glabrous; pappus bristles 1-seriate.

1. I. hookeri Clarke

Plant 0.6–1.5m; stems loosely lanate at first. Leaves elliptic-lanceolate, 8–15 × 2.5–4cm, acuminate, base attenuate to petiole c 5mm, margins minutely denticulate, pubescent and short glandular on both faces but much sparser above. Involucres 18–30mm diameter below; phyllaries to 15 × 1mm, brown villous. Tube of ray flowers 5–7mm; ligule 18–45 × 1mm, with long soft hairs at base. Disc corollas 5–6.5mm. Achenes 1.5mm; pappus whitish, rather shorter or subequal to disc corollas.

Bhutan: S – Deothang district; **C** – Ha (359), Thimphu, Punakha, Tongsa, Bumthang and Mongar districts, **N** – Upper Mo Chu and Upper Kuru Chu districts; **Sikkim:** Chunthang, Lachung etc.; **Chumbi:** upper Chumbi Valley. Hillsides, forest clearings, coniferous and wet broad-leaved forest, riverside scrub on gravel, 2450–4300m. August–October.

39. PENTANEMA Cassini

Erect annual; stem single. Leaves alternate, simple. Capitula in terminal leafy corymb, radiate. Involucre broadly campanulate, 4–5-seriate. Receptacle convex, naked. Ray flowers female, 1–2-seriate; corolla yellow. Disc yellow;

flowers bisexual; corolla 5-toothed. Style branches obtuse. Achenes oblong, scarcely ribbed; pappus bristles few.

1. P. vestitum (DC.) Ling; *Vicoa vestita* DC.

Plant 8(–45)cm; stem simple (or branched above), pilose (at first). Leaves oblong (or ovate in inflorescence), 2–4(–8) × 0.6(–2)cm, acute or obtuse, cordate, semi-amplexicaul at base or lower ones tapered (sometimes ± to a long petiole), (sparsely) pilose (or pubescent) and subsessile glandular on both surfaces, partially and usually shallowly serrate. Involucre 4(–5)-seriate; outer phyllaries herbaceous, linear, to 4.5(–6)mm, inner phyllaries scarious-margined, linear-lanceolate, acuminate, to 6(–7.2)mm. Ray corolla tube c 2.7mm; ligule 4.5(–6) × c 0.6mm, pilose at base. Disc c 0.6(–1)cm diameter; corollas 3.3(–4.5)mm, glandular at apex. Achenes c 0.8mm, brown, sparsely pubescent; pappus bristles fine, usually 10, 2.5(–3.5)mm, white.

Darjeeling/Sikkim: unlocalised (Treutler collection). March.

Only one exceptionally small specimen seen from our area. Description supplemented with details of further specimens from N India and Nepal in brackets.

40. CARPESIUM L.

Rhizomatous perennial herbs. Leaves simple, alternate. Capitula disciform, (suberect-)cernuopus, usually with leafy bracts at base, rarely ebracteate, solitary or in racemes, spikes or panicles. Involucre campanulate, few-seriate. Receptacle flat, naked. Outer flowers female; corolla 3–5-toothed. Inner flowers bisexual; corolla tubular-campanulate, 5-toothed; style branches linear, obtuse. Achenes terete, elongate, with glandular beak at apex; pappus absent.

1. Capitula numerous, ± closely arranged in a panicle of stiffly ascending branches; no spreading leafy bracts at base of capitula **4. C. abrotanoides**
\+ Capitula solitary at end of branches or remotely arranged along usually lax branches; spreading leafy bracts at base of capitula 2

2. Outer scarious phyllaries 2.5–3.5mm, shorter than inner ones; achenes c 3mm .... **3. C. tracheliifolium**
\+ Outer scarious phyllaries 4.5–6mm, a little longer than inner ones; achenes at least 3.5mm 3

3. Leaves ± regularly spaced along stem, usually ovate, (briefly) acuminate, blades not reduced on middle stem; capitula 3–many .... **1. C.nepalense**
\+ Leaves approximate at base of stem, ovate-lanceolate to oblanceolate, rounded to subacute at apex, distant on middle and upper stem with much-reduced blades; capitula 1–2(–6) .... **2. C. scapiforme**

FIG. 123. **Compositae: Vernonieae.** a–c, *Elephantopus scaber*: a, aggregate capitulum (× ⅔); b, single capitulum (× 4); c, achene (× 6). d–e, *Cyanthillium cinereum*: d, upper flowering stem (× ⅔); e, capitulum (× 4). f–h, *Vernonia volkameriifolia*: f, leaf (× ⅔); g, inflorescence (× ⅔); h, capitulum (× 4). Drawn by Margaret Tebbs.

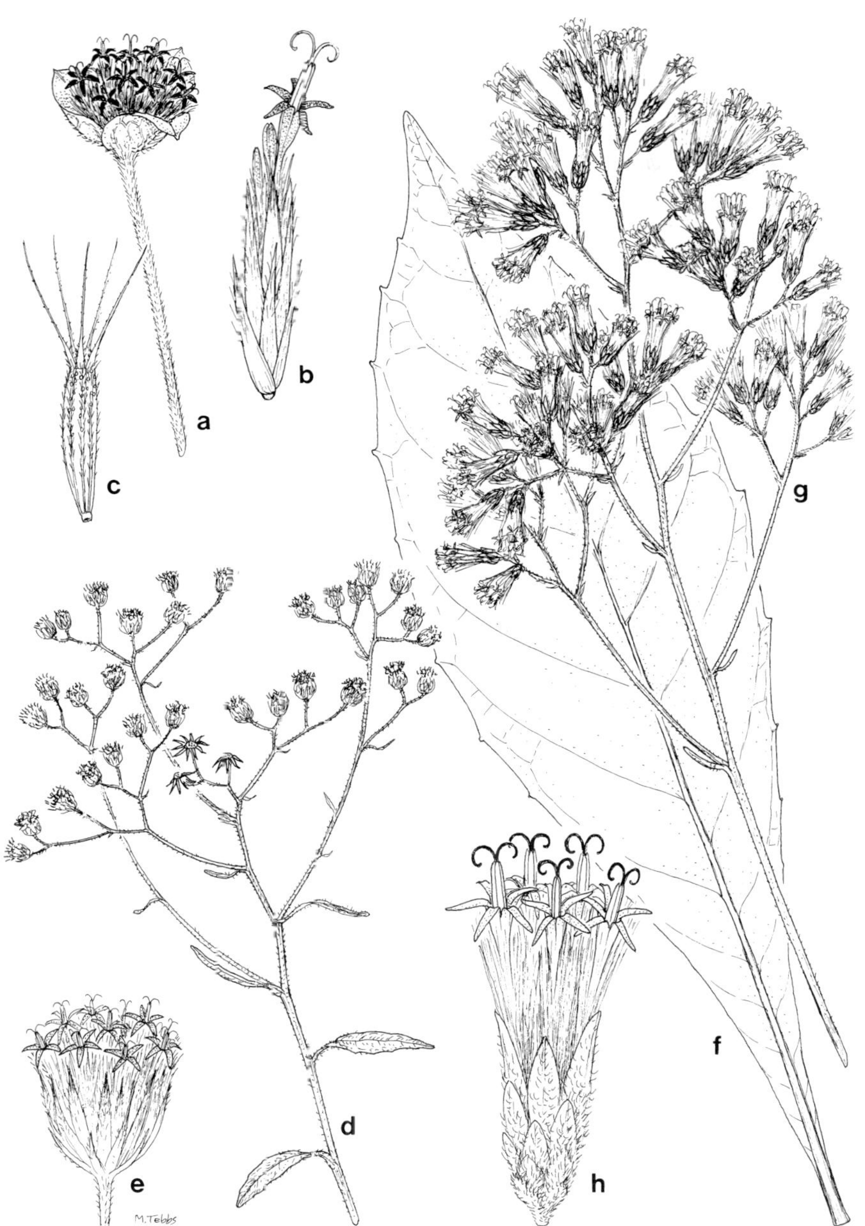
a
b
c
d
e
f
g
h
M.Tebbs

1. C. nepalense Lessing; *C. cernuum* var. *cernuum* sensu F.B.I. p.p. non L. Fig. 124a.

Plant 30–80cm; stems puberulous at first to persistently lanate. Leaves ovate, rarely elliptic, narrower in inflorescence, 4–25 × 1.2–7cm, usually briefly acuminate, truncate and petiolate to narrowly long attenuate at base, entire to sinuate-denticulate, sparsely puberulous to pilose above and to thinly lanate beneath, short glandular hairy on both surfaces. Capitula usually solitary at branch ends, rarely several, well spaced along branches and subsessile, nodding, surrounded by a cluster of elliptic leafy bracts, each 1–5 × 0.3–1cm. Involucre 3–4-seriate; outermost phyllaries herbaceous at least above, oblong, c 6–8 × 2mm, acute or acuminate; inner ones shorter, scarious, oblong, acute or some obtuse. Disc 6–17mm diameter, greenish-yellow; corollas 1.4–3mm, tube glabrous or hairy. Achenes 4mm.

Bhutan: S – Deothang (117) and Chukka districts, **C** – Ha, Thimphu, Punakha and Bumthang districts; **N** – Upper Mo Chu district; **Darjeeling:** Darjeeling, Raman–Rimbik (101), Tanglu (71) etc.; **Sikkim:** Phune, Thanggu, Gangtok (69) etc. Meadows, clearings, pathsides, banks, streamsides in forests, 1200–3650m. July–October.

Extremely variable; plants from Burtung, Lusing and Yoksam in Sikkim with lanate stems, thinly lanate leaves beneath and few large capitula are **C. nepalense** var. **lanatum** (Thomson) Kitamura; plants with small capitula (8–10mm across) are sometimes separated as *C. pubescens* DC. (420).

2. C. scapiforme Chen & Hu; *C. cernuum* L. var. *glandulosum* Clarke, *C. lipskyi* auct. non Winkler

Plant 25–50cm; stems erect, simple, sparsely pilose, rarely villous. Leaves 4–5(–8) per stem below inflorescence, ovate-lanceolate to oblanceolate, rounded to subacute, attenuate at base, subentire, rarely serrulate, pubescent (usually sparsely), with shorter glandular hairs more numerous beneath, rarely villous on veins beneath; lower leaves close together, 7–28 × 2.5–6cm, usually long petiolate; middle and upper leaves distant, much reduced, (±) sessile. Capitula 1–2(–3, rarely –6 elsewhere), surrounded by several whorls of ± regular, spathulate, obtuse, herbaceous, pilose bracts, 7–15 × 1.5–6mm, rarely oblong, acute or few much larger; phyllaries oblong, acute or obtuse, scarious, c 6 × 1.6mm. Disc 8–20mm diameter, yellow; corollas c 2mm, tube hairy. Achenes 4mm.

Bhutan: N – Upper Mo Chu district (Yale La (361), near Shingche La); **?Sikkim:** ?Thanggu and unlocalised (both Hooker collections); **Chumbi:** Chumbi. Streamsides, meadows, 3650–4100m. July–September.

FIG. 124. **Compositae: Inuleae.** a, *Carpesium nepalense*: capitulum (× 2). b, *Duhaldea nervosa*: inflorescence (× ⅔). c–e, *Blumea balsamifera*: c, leaf (× ⅔); d, inflorescence (× ⅔); e, capitulum (× 4). Drawn by Margaret Tebbs.

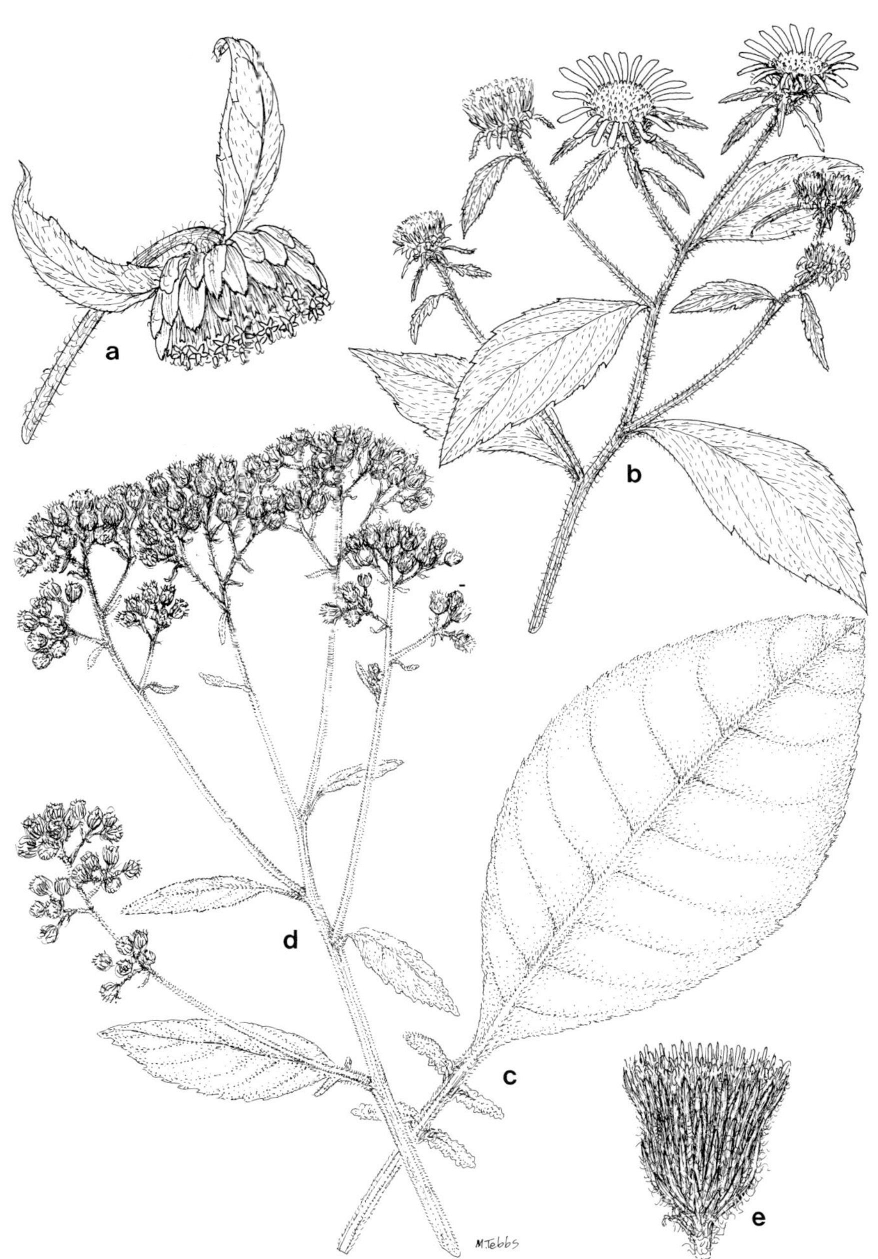
a
b
c
d
e
M.Tebbs

The specimens from Chumbi approach *C. lipskyi* in their small capitula with bracts often oblong and acute.

3. C. tracheliifolium Lessing

Plant 30–50cm; stems usually sparsely pubescent. Leaves ovate, 4–15 × 2–8cm, usually acuminate, subcordate to attenuate at base, subentire to coarsely serrate, sparsely pubescent on both surfaces, with subsessile glands beneath; lower ones broadly (narrowly) petiolate; upper ones sessile or short petiolate. Capitula 1–8 in racemes or spikes at branch ends, suberect to pendulous, surrounded by leafy bracts 4.5–37 × 2–9mm. Involucre 4–5-seriate; phyllaries mostly oblong, obtuse, scarious, to 3.5–5.2mm; outer ones sometimes shorter or herbaceous above or ovate and rounded to acuminate at apex. Disc 2–10mm diameter; corollas c 2mm, tube glabrous or hairy. Achenes 3mm, narrowed above to glandular apex.

Bhutan: C – Punakha district (Rinchu), Mongar district (Sengor), Tashigang district (Tashi Yangtsi); **N** – Upper Kulong Chu district (Tobrang); **Darjeeling:** Senchal, Tonglu–Sandakphu, Budhwari–Bhotia Basti (69) etc.; **Sikkim:** Lachen, Lachung. Streamsides, sandy levels, forest, 1500–3000m. July–August.

4. C. abrotanoides L.

Similar to *C. tracheliifolium* but leaves narrowly elliptic, acuminate or lower ones oblanceolate, subacute, 5–15 × 1–3.5cm, narrowly cuneate at base, entire; capitula numerous in terminal panicle of spicate branches, secund, ebracteate; outermost phyllaries ovate-triangular, c 2.8 × 2.1mm, ciliate; disc 3mm diameter, corolla tube glabrous.

Darjeeling: Shiri; **Sikkim:** Kalhej Khola. 2150m. October–November.

41. DUHALDEA DC.

Perennial herbs or shrubs. Leaves simple, alternate, usually finely toothed. Capitula solitary or few to many in corymbs, racemes or panicles, usually radiate, rarely disciform. Involucres (3-), 5–8-seriate, imbricate, outer phyllaries sometimes herbaceous. Receptacle naked, flat to hemispheric. Marginal flowers 1-seriate, female; ligule yellow or white or absent. Disc flowers bisexual; corolla yellow, tubular-campanulate, 5-toothed at apex; style branches linear, obtuse. Achenes subterete or angular, 4–5-ribbed, oblong; pappus bristles usually 1-seriate, sometimes thickened above.

1. Plant ± herbaceous; ligules white .... **5. D. nervosa**
\+ Shrubby plants; ligules yellow or absent 2

2. Small racemes or corymbs borne in most leaf axils and at stem apex **4. D. rubricaulis**
\+ Inflorescence a terminal corymb or panicle, sometimes additional corymbs in upper leaf axils only 3

3. Leaves persistently lanate beneath; ligules to 1mm or absent (rarely to 3mm) .... **1. D. cappa**
\+ Leaves pubescent or scabrid beneath; ligules at least 3mm (rarely absent) 4

4. Inflorescence ± hispid with spreading hairs at least 1mm long; disc 7–10mm .... **3. D. simonsii**
\+ Inflorescence pubescent with appressed or spreading hairs to 0.5mm; disc 4–6mm 5

5. Leaves 25–30 × 7–8cm, coarsely dentate above .... **6. D griffithii**
\+ Leaves 10–19 × 2–6cm, remotely serrulate or denticulate **2. D. eupatorioides**

1. D. cappa (D. Don) Anderberg; *Inula cappa* (D. Don) DC. Nep: *Taimakhu* (117).

Shrub 50–200cm; stems lanate-tomentose. Leaves thick, rather chartaceous, elliptic, lanceolate or narrowly oblong, 8–23 × 2.5–5.5cm, acute or briefly acuminate, rounded (rarely tapered) at base, remotely serrulate, green and sparsely coarse pubescent above, whitish lanate beneath, obscuring minor veins; petiole c 5mm (rarely to 20mm). Capitula radiate or disciform, in dense corymbs. Involucres 6-seriate; phyllaries lanceolate, tomentose, inner ones 4.5–6mm × c 0.7mm. Flowers yellow; female ones few, marginal, corollas usually 4.5–5.3mm, tubular and erect or outcurved or with short ligule to 1 × 1mm, rarely tube 2.9–3.5mm with ligule 2.3–2.9mm. Disc 4–6mm diameter; corollas 4.7–6mm. Achenes 1.5mm, sericeous; pappus slightly shorter than disc corollas, whitish, sometimes brownish tipped.

Bhutan: C – Punakha district (Ritang, Tang Chu, Wangdu Phodrang), Tongsa district (Tama (117), Tashiling–Tongsa), Tashigang district (Tashigang (117), Shiri Chu); **Darjeeling:** Darogadara, Darjeeling (128), Peshok; **Sikkim:** Chunthang, Yoksam; **W Bengal Duars:** Applachand (175). Arid banks, sun or shade, 450–2400m. All year

2. D. eupatorioides (DC.) Anderberg; *Inula eupatorioides* DC.

Similar to *D. cappa* but leaves tapered at base, sparsely scabridulous above, more densely pubescent at least on veins beneath, minor veins prominent beneath; female flowers ligulate, corolla tube 3mm, ligule 3–4 × 1–1.4mm; disc flowers 4–5.7mm; pappus brownish tipped.

Bhutan: unlocalised (Griffith collections); **Darjeeling:** Kurseong, Kurseong–

Tindharia (97a), Kariabasti (97a); **Darjeeling/Sikkim:** unlocalised (Hooker collection). 1200–1650m. September–December.

An unlocalised Hooker specimen from Darjeeling/Sikkim with leaf indumentum as above but deformed marginal flowers probably also belongs here, though it might represent a hybrid. It was collected by Hooker below 300m, probably in Darjeeling. *D. cuspidata* (DC.) Anderberg has been cited from Sikkim and Bhutan (343), but no specimens were seen from our area. It differs from *D. eupatorioides* by its thin, distinctly acuminate, more regularly serrulate leaves and often longer ligules.

3. D. simonsii (Clarke) Anderberg; *Inula simonsii* Clarke

Shrub c 2m or undershrub; stems sparsely hispid (hairs c 2.5mm, brownish). Leaves elliptic-lanceolate or oblanceolate, 9–20 × 2.5–5cm, acuminate, rounded at base, sparsely hispid on both surfaces, denser on margins, slightly sticky, distantly denticulate, reticulate veins prominent on both faces; petiole c 3mm. Capitula radiate, rather few in terminal corymb or panicle. Involucre 5–6-seriate; phyllaries acuminate or caudate, brownish pilose, outer ones lanceolate, tips recurved, inner ones erect, innermost longest, linear, 5.5–7.5 × 0.7mm. Ray corollas yellow; tube 4mm; ligules 9 × 2mm (apparently 7 × 1mm on a senescent specimen). Disc 7–10mm diameter; corollas 5.5–6.5mm. Achenes oblong, 1.8–2.5mm, sericeous; pappus a little shorter than disc corollas, bristles whitish with brown tips.

Bhutan: S – Sarbhang district (Damphu, Tori Bari–Loring Falls). Warm broad-leaved forest, roadside cliffs, 1200–2000m. October–February.

4. D. rubricaulis (DC.) Anderberg; *Inula rubricaulis* (DC.) Clarke

Shrub 1–3m; stems sparsely grey tomentose at first, later reddish, glabrous. Leaves elliptic-lanceolate, 6–19 × 1–4cm, acuminate, cuneate at base, sessile, serrulate, sparsely pubescent on both surfaces. Capitula radiate, 2–6 in small lateral racemes or corymbs. Involucres 5–6-seriate; phyllaries acuminate, outer ones spreading, lanceolate, inner ones linear-lanceolate, to 8–9mm. Ray flowers few; corolla tube 5mm; ligule 6–8.5 × 1.7–2mm. Disc c 7mm diameter; corollas 6–7mm. Achenes oblong, 2.5mm, sericeous; pappus white, 5–6mm.

Bhutan: unlocalised (329); **Darjeeling:** Darjeeling (128), Kurseong (97a), Pankhabari etc.; **Darjeeling/Sikkim:** unlocalised (Hooker collections). 150–1700m. December–March.

5. D. nervosa (DC.) Anderberg; *Inula nervosa* DC. Fig. 124b.

Plant ± herbaceous with tuberous roots; stems often decumbent, 0.2–1m, usually simple, sparsely long (2.5–3mm) tawny pilose or strigose, sometimes with rounded densely soft-hairy buds at base. Leaves elliptic, 5–13 × 2–4.5cm, acute or acuminate, base narrowed and cuneate, margin distantly and shallowly serrate, sparsely spreading pilose on both surfaces; petiole 0–6mm. Capitula 1 or few, loosely corymbose. Involucre 5(–8)-seriate; inner phyllaries linear-lanceolate, mainly scarious, 7–10 × 0.7–1.2mm, acute to acuminate, pilose, tips sometimes purple; outer ones suberect, similar but smaller or herbaceous, dilated above, sometimes to 22mm, obtuse. Tube of ray corollas 3.5–4.7mm; ligule white, 9–13.5 × c 2.5mm. Disc yellow, 8–15mm diameter; corollas 6–6.7mm. Achenes 2–2.5mm, sericeous; pappus white, as long as disc flowers, rarely shorter.

Bhutan: S – Chukka district (below Chapcha), **C** – Punakha district (Nobding), Tongsa district (Tongsa), Mongar district (Anjor, Sawang), Tashigang district (Tashiling, Tashi Yangtsi); **Sikkim:** Chunthang; **SE Tibet:** Nyam Jang Chu. Rock-crevices, 1800–2285m. July–November.

The species varies considerably outside our area. Taller plants from Chapcha, Chukha district, Bhutan, with narrower lanceolate leaves, numerous capitula and smaller ray corollas (tube 2mm, ligule 7 × 1.7mm) probably belong here.

6. D. griffithii (Clarke) Anderberg; *Inula griffithii* Clarke

Shrub; branches c 10cm, stiffly pubescent at first. Leaves obliquely oblanceolate, 25–30 × 7–8cm, acuminate, base attenuate, ± sessile, margin coarsely denticulate, sparsely scabrid on both surfaces with short stiff hairs with persistent swollen bases. Panicles terminal, 13–18cm, branches densely yellowish pubescent; bracts lanceolate, 15–45 × 6–12mm. Capitula few, c 7mm diameter. Phyllaries 3-seriate, linear-lanceolate, up to 5 × 1mm, pubescent. Corollas in bud yellow, c 5mm. Achenes (immature) c 1.2mm, sparsely pubescent; pappus simple, 5.5mm.

Assam Duars: Dairang. In bud in January.

Apart from noting the flower colour, Griffith gives no information on this species which is known only from the type specimen. The size of leaves give the impression of a large shrub perhaps 2m tall. Ligulate flowers are definitely present at the margins of the capitula but, like the disc flowers, in tight bud. The anther bases, as Clarke states, are caudate.

42. PULICARIA Gaertner

Small plants, annual or woody-based perennial, with long soft hairs intermixed with short-stalked glands. Leaves alternate, simple, entire or dentate, sessile or petiolate. Capitula in small corymbs, radiate or disciform. Involucre broadly campanulate, c 5-seriate. Receptacle convex. Corollas yellow. Outer flowers female, rather few, ligulate or filiform. Inner flowers bisexual, numerous; corollas narrowly infundibular, 5-lobed; anthers with branched tails; style branches linear-oblong, rounded at apex. Achenes oblong, scarcely ribbed; pappus double, outer of short scales, inner of (1–)5 caducous bristles.

1. P. foliolosa DC.

Bushy annual c 20cm; stems pilose at first. Leaves oblanceolate, sessile, 2–3.5 × 0.6–0.8cm, obtuse with apiculus, attenuate at base and semi-amplexicaul, entire, mainly glandular above, more eglandular hairs beneath. Involucre c 5-seriate, to c 5mm across; phyllaries linear-lanceolate, scarious-margined; outer c 2.5mm; inner to 3.7mm. Corollas subequal, 3mm, with glandular hairs at apex. Achenes 0.8mm, sparsely pubescent; inner pappus bristles c 2mm, white, outer scales c 0.2mm.

Darjeeling/Sikkim: locality not known (Treutler collection). March.

P. petiolaris Jaubert & Spach has been reported from Sikkim without locality (344) but no examples have been seen. Specimens from W Nepal differ from *P. foliolosa* by their perennial woody bases, narrowly ovate, acute to acuminate, dentate leaves on long slender petioles, radiate capitula (ligules c 2.7 × 0.5mm), corollas with glandular hairs few and scattered throughout, and pappus bristles subequal to corollas.

43. BLUMEA DC.

Annual, biennial or perennial herbs, shrubs or small evergreen trees, mostly strong-smelling; stems erect or ascending, sometimes scrambling, rarely prostrate. Leaves alternate, simple or pinnately or lyrately lobed, denticulate to biserrate, usually pubescent on one or both surfaces. Capitula disciform, in terminal and sometimes axillary panicles. Involucres cylindrical to hemispheric, multiseriate; phyllaries ovate to linear. Receptacle flat or convex, glabrous or pilose, sometimes fimbrillate. Flowers numerous, marginal ones female, central ones many fewer, bisexual. Corollas often yellow, rarely purple. Female flowers filiform; corolla minutely 2–4-toothed, usually glabrous. Bisexual flowers (3–)5-merous; corolla narrowly infundibular, teeth usually with minute glandular hairs on back and often longer eglandular ones; anthers tailed. Achenes usually oblong, minutely 5-ribbed, pubescent. Pappus of 1 series of simple white or reddish bristles, subequal to corollas.

Considered an artificial genus with many species that probably require transfer to the Plucheeae (323).

1. Plants ± prostrate; leaf margins with sharp subspiny teeth **15. B. oxydonta**
\+ Plants ± erect; leaf margins with softer teeth 2

2. Leaves white to grey cottony beneath .... **7b. B. hieracifolia** var. **hamiltonii**
\+ Leaves subglabrous to appressed pilose or villous to lanate with brownish sericeous hairs beneath 3

3. Leaves villous to lanate beneath with brownish sericeous hairs and either plant arborescent or corollas at least 6mm 4
\+ Leaves subglabrous to appressed pilose beneath or annuals with corollas less than 6mm 5

4. Evergreen undershrub to small tree .... **6. B. balsamifera**
\+ Herbaceous, biennial or perennial ...... **7a. B. hieracifolia** var. **hieracifolia**

5. Annual or biennial 6
+ Herbaceous perennial or shrub 11

6. Receptacles pubescent 7
+ Receptacles hairless, though often fimbrillate 8

7. Most phyllaries recurved from middle at anthesis .... **9. B. fistulosa**
+ Most phyllaries erect at anthesis, often spreading from base later **10. B. sinuata**

8. Leaves simple; phyllaries 4-seriate; corollas purple .... **11. B. axillaris**
+ Some leaves lyrate or pinnatifid or phyllaries 5–6-seriate; corollas yellow 9

9. Capitula many, usually in dense narrow panicles, rarely in more open leafy ones; hairs on c ⅙ of style beneath branches in bisexual flowers **12. B. lacera**
+ Capitula few or in open panicles with reduced leaves; hairs on style only immediately beneath branches in bisexual flowers 10

10. Stems above and inflorescence axes glandular-pubescent and sometimes sparsely eglandular-pilose .... **13. B. membranacea**
+ Stems and inflorescence axes glabrous or sparsely eglandular-pilose **14. B. virens**

11. Leaves pinnatifid .... **5. B. hookeri**
+ Leaves simple 12

12. Scandent shrub; outer 2 series of phyllaries ovate .... **2. B. riparia**
+ Herb or undershrub, free-standing or scandent; outer phyllaries oblong, lanceolate or linear 13

13. Leaves 4.5–12 × 0.7–2.5cm; capitula in narrow often spicate panicles **8. B. clarkei**
+ Most leaves longer than 12cm or wider than 2.5cm; capitula in broad panicles 14

14. Indumentum beneath leaves including short glandular hairs; receptacle hairless; peduncles of individual capitula (1.5–)3–15mm .... **4. B aromatica**
+ Leaves eglandular-hairy or glabrescent; receptacle hairy or capitula subsessile 15

15. Leaves 6–20cm, usually obovate; capitula subsessile (peduncles not exceeding 1mm at anthesis); female corollas 4.5–5.7mm ..... **1. B. procera**
+ Leaves 15–30cm, oblanceolate; peduncles usually exceeding 1mm; female corollas 6.2–7.2mm .... **3. B. lanceolaria**

1. B. procera DC.

Large herb or undershrub; stems 120–250(–350)cm, shaggily tomentose in inflorescence. Leaves obovate or sometimes oblanceolate, 6–20 × 3–7.5cm, acuminate, tapered towards base or sometimes truncate, usually petiolate, denticulate and undulate to coarsely serrate-dentate, thinly appressed puberulous at maturity; petiole poorly defined, broad, almost semi-amplexicaul, to 6mm. Capitula subsessile in dense glomerules in open terminal panicles. Involucres 4–5-seriate; phyllaries 2–7mm, outer ones lanceolate, thickly tomentose, inner ones linear. Receptacle subglabrous to moderately long-haired. Corollas yellow, sometimes purplish above, 4.5–6.2mm; lobes of bisexual ones with few eglandular and often some glandular hairs. Pappus white.

Bhutan: S – Gaylephug district (Chabley Khola); **Darjeeling:** Kalimpong, Kurseong, Darjeeling–Takvar etc. Woods, broad-leaved evergreen forest, 720–1650m. March–November.

2. B. riparia (Blume) DC.; *B. chinensis* sensu F.B.I. non (L.) DC.

Scandent shrub; stems 0.5–2.5m, somewhat pubescent among inflorescences. Leaves ovate-lanceolate, petiolate, 5–13 × 1.5–4cm, acuminate, rounded or sometimes subacute at base, remotely denticulate without coarser teeth, ± glabrous on both surfaces; petioles usually distinct, narrow, up to 6mm, not amplexicaul. Capitula on short peduncles (1–7mm) in tight or loose clusters in axillary and terminal panicles. Involucre c 5-seriate; phyllaries of outer 2 series ovate, c 2.2–3.5mm, pubescent, of inner 2 series narrowly oblong to linear, c 7mm. Receptacle densely long-haired. Corollas yellow, 5.5–7mm; lobes of bisexual ones with glandular and few or many eglandular hairs. Pappus white.

Bhutan: S – Sarbhang district (Sarbhang–Chirang), Deothang district (Tsalari Chu), **C** – Punakha district (Punakha–Bhotokha (71)), Punakha or Upper Mo Chu district (Mishichen–Khosa); **Darjeeling:** Kurseong, Peshok, terai etc. Cool broad-leaved forest, 1200–1810m. March–April.

3. B. lanceolaria (Roxb.) Druce; *B. myriocephala* DC.

Large herb or shrub; stems erect, 1–2m, ± woody, pubescent above. Leaves oblanceolate, ± sessile, 15–30 × 3–8cm, acuminate, long attenuate at base but ± lacking distinct petiole, usually shallowly serrate, glabrous at maturity or with some hairs (eglandular) persisting beneath. Capitula in large terminal panicles, usually pedunculate at anthesis, rarely most subsessile; peduncles to 12mm. Involucres c 5-seriate; phyllaries 2.5–9.5mm, outer ones lanceolate, pubescent, inner ones linear. Receptacle moderately hairy. Corollas yellow, 6.5–7.5mm; lobes of bisexual ones with glandular and often few eglandular hairs. Pappus pale reddish.

Bhutan: S – Samchi district (Sangura (117), Chepuwa Khola), Gaylegphug district (Sham Khara (117)); **Darjeeling:** Badamtam–Rangit, Rishap; **Darjeeling/Sikkim:** Great Rangit. In subtropical forest, 450–1550m. March–April.

4. B. aromatica DC.

Perennial; stems (0.15–)0.4–3m, densely glandular-pubescent interspersed with a few eglandular hairs. Leaves oblanceolate (rarely obovate), rapidly wilting, 6–40 × 1.8–13cm, acuminate, long attenuate, sometimes with a short petiole, ± biserrate or bidentate, glandular-pubescent interspersed with eglandular hairs on both surfaces. Capitula usually crowded in narrow terminal panicles, sometimes panicles broad, lax or axillary as well; peduncles (1.5–)3–15mm at anthesis. Involucres c 5-seriate; phyllaries 2–8mm; outer ones mostly lanceolate or linear-lanceolate, glandular-pubescent, often with longer appressed eglandular hair; inner ones linear. Receptacle hairless. Corollas yellow, 5.7–7.3mm; lobes of bisexual ones with glandular and sometimes few short eglandular hairs. Pappus pale reddish, rarely dirty white.

Bhutan: S – Gaylegphug district (Surelakha (38), Kabare), Chukka district (Chumi Lingu–Tala (359)), **C** – Punakha district (Misichen–Khosa (71), Tinlegang–Gon Chungnang (71), Wangdu Phodrang–Samtengang), Tongsa district (Mangde Chu Valley); **Darjeeling:** Darjeeling, Dumsong, Lebong etc.; **Sikkim:** Lachen Lachong, Rhimbi Chhu. Dense moist subtropical forests, 900–1950m. October–December, March–April.

Very diverse elements are included here. Material seen consists of shorter herbs with basal leaves larger, phyllaries with glandular hair only and corollas to 6.5mm and of portions of longer ?scrambling plants approaching *B. lanceolaria* in habit, with mixed phyllary hair and larger corollas. A further collection differs by its stiffer elliptic leaves, eglandular phyllaries and bisexual corolla tube abruptly stepped at base. It is from an unlocalised Hooker collection and is a type of *B. sikkimensis* Hook.f. Its status is uncertain.

5. B. hookeri Clarke

Coarse perennial herb (80); stems erect, puberulous. Leaves pinnatifid, ± elliptic in outline, ± sessile, 7–38 × 3–18cm, acuminate, long attenuate at base, puberulous on both faces, lobes 3–5 pairs, oblong to lanceolate, acuminate, subentire or remotely serrulate, sometimes also with few coarse teeth. Capitula numerous, in open, leafy, rounded or elongate, terminal panicles. Involucre 5–6-seriate; phyllaries at least 2–6mm, outermost oblong-lanceolate, pubescent, remainder linear. Receptacle hairless, sometimes also sparsely hairy. Corolla lobes of bisexual flowers with minute glandular pubescence and eglandular hairs. Ovaries pubescent; pappus somewhat reddish.

?Bhutan: Nichu La; **Darjeeling:** near Darjeeling; **Darjeeling/Sikkim:** unlocalised Hooker collections. 900–1220m. In bud November–January.

All material seen in bud. Considered possible hybrids between *B. densiflora* DC. and *B. aromatica* (395), but no specimens of *B. densiflora* were seen from our area.

6. B. balsamifera (L.) DC.; *B. densiflora* sensu F.B.I. p.p. non DC. Nep: *Baburi*, *Kapur* (both 34). Fig. 124c–e.

Evergreen undershrub or small tree, (0.5–)3–6m, smelling of camphor, with buff sericeous hairs; stems densely tomentose. Leaves ovate-lanceolate or lanceolate, 6–30 × 1.5–12cm, acuminate, obtuse to attenuate at base, usually with petiole to 3cm, tomentose above, lanate beneath, coarsely or weakly (bi)serrate, often 1(–3) pairs of lateral lobes present, usually confined to petiole or base of blade, oblong-elliptic, oblique, 7–23 × 2–5mm. Capitula in dense, terminal, flat-topped panicles. Involucres 5–6-seriate, c 5mm diameter; phyllaries all linear or outermost lanceolate, 1.5–7mm, lanate-tomentose. Receptacle fimbrillate. Corollas 4.5–6mm, lobes of bisexual ones with glandular and sometimes few eglandular hairs. Pappus reddish.

Bhutan: S – Phuntsholing district (Phuntsholing), Gaylegphug/Tongsa district (Rani Camp–Tama (117)), **C** – Tongsa district (Tama (117)); **Darjeeling:** Tista–Rangpo (69), Chhota Rangit, Rayeng etc.; **Jalpaiguri Duars:** Chel River. Forest clearings, 350–1350m. March–April.

Distributed from Nepal to Malesia, the extratropical elements have been distinguished as var. *microcephala* Kitamura.

7a. B. hieracifolia (D. Don) DC. var. **hieracifolia**; *B. hieracifolia* (D. Don) DC. var. *macrostachys* (DC.) Hook.f.

Biennial or perennial herb, (60–)100–150cm; stems erect, usually simple, villous above. Leaves oblanceolate, 5–12 × 1.5–3.5cm, acute, attenuate below, sessile, serrate- or biserrate-dentate, buff sericeous on both faces at first, persistently so beneath. Capitula in dense clusters, forming narrow spicate inflorescences or broader panicles. Involucre c 4.5mm diameter; phyllaries 5–6-seriate, 3–8.5mm, purplish at tips, outer series lanceolate, villous, remainder linear-lanceolate to -oblanceolate. Receptacle shortly fimbrillate. Corollas yellow, 7–8.5mm, lobes of bisexual ones with few glandular and eglandular hairs. Pappus white.

Darjeeling: Kalimpong (69), Kariabasti (97a), Siliguri etc. Fields, grassland, 350–1200m. September–March.

Also collected outside our area near Deothang. Tall robust plants with well-developed stem leaves and usually distinctly peduncled clusters of capitula are sometimes separated as var. *evolutior* Clarke [var. *macrostachys* (DC.) Hook.f.] (395, 344), which would include the few specimens seen from our area and described above. Var. *hieracifolia* s.s. would then be restricted to shorter slender plants with much-reduced stem leaves and sessile clusters of rather few capitula. However, these are weakly defined taxa.

7b. B. hieracifolia (D. Don) DC. var. **hamiltonii** (DC.) Clarke

Differs from var. *hieracifolia* in stature (40–60cm); leaves finely serrate-dentate, white or grey cottony; involucre 3.5–4mm diameter, 4–5-seriate; phyllaries to 6mm, often stramineous at first, sometimes weakly purplish later;

corollas 4–5.5mm, lobes of bisexual ones with many short marginal eglandular hairs.

Darjeeling: Siliguri. May–June.

8. B. clarkei Hook.f.

Perennial, c 1.2–1.4m; stems erect, pubescent at least above. Leaves elliptic or oblanceolate, 4.5–12 × 0.7–2.5cm, briefly acuminate, attenuate at base, subsessile or sometimes with petiole to 5mm, remotely serrulate, greyish beneath, sparsely pubescent on both surfaces. Capitula in narrow often spicate panicles; peduncles 5–10(–20)mm. Involucre c 5-seriate, c 6mm diameter; phyllaries linear, 2.5–8m, appressed pubescent and sparsely glandular. Receptacle densely pilose. Corollas yellow, 5.5–7mm, lobes of bisexual ones with few glandular and eglandular hairs. Pappus white.

Darjeeling: Mongpu, Panchkilla, Rangit, Tista. Terai, 150–600m. February–March.

9. B. fistulosa (Roxb.) Kurz; *B. glomerata* DC.

?Annual; stems erect, 15–100cm, simple, shaggily pubescent above. Leaves oblanceolate to obovate, 3–15 × 0.5–5cm, acute, narrowly long attenuate at base, bidentate (sometimes sublobulate), pubescent (rarely sparsely pilose) on both surfaces. Capitula in small ± sessile clusters arranged in interrupted spike-like terminal racemes or sparsely branched panicles; involucres 4–5-seriate, c 3.5mm diameter; phyllaries purplish above, mostly recurved from middle by anthesis, 2.5–6mm, pubescent, sparsely glandular, outer series ± lanceolate, remainder linear. Receptacle sparsely short pubescent. Corollas yellow, 4.2–5mm, lobes of bisexual ones with glandular and few eglandular hairs. Pappus white.

Bhutan: S – Sankosh district (Sankosh); **Darjeeling:** Chenga Forest, ?Kalimpong district (395); **Darjeeling/Sikkim:** unlocalised (Hooker collections and (395)). 600–700m. January–February.

10. B. sinuata (Loureiro) Merrill; *B. laciniata* (Roxb.) DC. Dz: *Hudo.*

Biennial, 20–180cm; stems pubescent (rarely subglabrescent or pilose) and sometimes sparsely stipitate-glandular. Leaves simple and ± elliptic to deeply lyrate-pinnatisect, 3–20 × 1–8cm, acute, finely or coarsely serrate or dentate, pubescent on both surfaces (rarely puberulous or long soft pilose); lower leaves long attenuate at base, usually lyrate with 1–3 pairs of acute triangular lateral lobes to 3cm; upper leaves sessile, usually simple, sometimes pinnatifid. Capitula in broad lax or sometimes narrow dense terminal panicles. Involucres sometimes purplish, 6–7-seriate; phyllaries 2.5–7mm, pilose and often stipitate glandular; outer ones ± lanceolate, thickened above and often narrowed into a terete appendage; inner ones linear. Receptacle short pubescent. Corollas 4–6mm, lobes of bisexual ones often with sparse glandular hairs and usually with short eglandular hairs. Pappus white.

Bhutan: S – Samchi district (Daina Khola), Phuntsholing district (Phuntsholing), Sankosh district (Sankosh), **C** – Punakha district (Wangdu Phodrang), Tongsa district (Berthi), **N** – Upper Kuru Chu district (Lingmachey); **Darjeeling:** Mongpu, Pankhabari, Ryang etc.; **Darjeeling/Sikkim:** Great Rangit Valley; **Sikkim:** Namphak. Subtropical forest slopes and riverside shingle, 200–1500m. March–May.

11. B. axillaris (Lamarck) DC.; *B. mollis* (D. Don) Merrill, *B. wightiana* DC.

Strong smelling annual or biennial; stems 5–100cm, simple to short branched throughout, pubescent with white spreading hairs interspersed with shorter glands. Leaves ovate-oblong to obovate, 2.5–12 × 1–5cm, obtuse or acute, attenuate to broad petiole up to 2cm, (bi)serrate-dentate, sparsely to densely pilose on both surfaces with long soft white hairs and shorter glands. Capitula few from most outer leaf axils, often forming corymbose or interrupted spicate panicles; involucres 4-seriate; phyllaries linear or outermost lanceolate, 3–5.2mm, pubescent with white hairs and shorter glands. Receptacle hairless. Corollas purplish, 2.8–4(–5)mm. Bisexual flowers with glandular hairs and sometimes 1–2 long eglandular ones on corolla lobes and hairs on c ⅙ of style beneath branches. Pappus white.

Bhutan: S – Phuntsholing district (Phuntsholing), **C** – Punakha district (near Lobesa); **Darjeeling:** Badamtam, Kurseong, Pankhabari etc. Tea gardens, ditches and paddy-fields, 200–1300m. February–April (September–December (97a)).

12. B. lacera (Burman f.) DC.; *B. subcapitata* DC.

Annual or biennial, 20–100cm; stems glandular-pubescent interspersed with long eglandular hairs, often with villous buds below. Leaves obovate or oblanceolate, 2.5–10 × 1–4.5cm, acute or obtuse, gradually or abruptly attenuate at base, sometimes ± petiolate, serrate or dentate, sometimes lyrately lobed near base or lyrate-pinnatisect, sparsely to densely eglandular-pilose, sometimes with few glandular hairs. Capitula 4–6mm diameter, in compact narrow axillary and terminal panicles; involucres c 5-seriate; phyllaries all linear or outermost ± lanceolate, 2.5–7mm, glandular-pubescent and/or eglandular-pilose. Receptacle hairless. Corollas yellow, 4–5.5mm, lobes of bisexual ones with glandular and few short eglandular hairs. Pappus white.

Bhutan: S – Phuntsholing district (near Phuntsholing); **?Darjeeling:** Rangit, unlocalised (Hooker collections); **W Bengal Duars:** Chel (175), Jalpaiguri. Roadside ditch, roadside in hot forest, 200–800m. February–May.

13. B. membranacea DC.

Annual, 7–100cm; stems usually simple, glandular-pubescent, sometimes with few longer eglandular hairs above; leaves simple, obovate or lyrate, often oblanceolate in outline, with 1–2(–3) pairs of lateral lobes, thinly herbaceous, 2–13 × 1–5cm, acute or obtuse, narrowly attenuate at base, usually petiolate, denticulate (to shallowly dentate), pubescent on both surfaces, sometimes glandular-stipitate above at least in inflorescence; petioles to 3cm. Capitula rather few in narrow terminal panicles. Involucre 4mm diameter; phyllaries purplish tinged, 6-seriate, 2–6.5mm, glandular-pubescent, sometimes with few eglandular hairs, outermost lanceolate, inner ones linear. Receptacle hairless. Corollas yellow; female 3.5–4.5mm; bisexual 4.5–5.5mm, lobes with glandular hairs and sometimes very reduced eglandular ones. Pappus white.

Darjeeling: Kurseong, Rungait, Siliguri; **Darjeeling/Sikkim:** unlocalised (Hooker collections). 300–1200m. February–June.

14. B. virens DC.

Similar to *B. membranacea*, but plant entirely lacking glandular hairs except on bisexual corolla lobes; stems and inflorescence axes glabrous or sparsely pilose; leaves membranous, usually more regularly serrate, glabrescent or sparsely pilose on both faces.

?Darjeeling: Kalimpong district, (395) as 'Bhutan'.

No specimens seen from our area. Described from other Indian collections.

15. B. oxydonta DC.

?Annual, with stout tap-root; stems usually numerous, ± prostrate, 15–30(–40)cm, finely pubescent. Leaves elliptic to obovate, 1–5 × 0.5–1.5cm, acute, cuneate at base, ± sessile, with a few sharp subspinous teeth, sparsely to densely villous on both surfaces and glandular-stipitate. Capitula few in terminal and axillary panicles. Involucre c 4-seriate; phyllaries lanceolate to linear, 2–6mm, with long eglandular hairs and some stalked glands. Receptacle hairless. Corollas yellow, c 4mm; lobes of female flowers with few eglandular hairs, lobes of bisexual flowers with dense minute glandular pubescence and some long eglandular hairs. Anthers usually abortive. Pappus whitish.

Bhutan: C – Punakha district (Choojom and Tinlegang (both 71)). 1450–2450m. April–May.

No specimens seen from our area, description based on material from Nepal.

PLUCHEEAE

44. LAGGERA Koch

Annual to perennial herbs; stems erect, with entire or deeply incised wings. Leaves alternate, simple, sessile or petiolate. Capitula ± cernuous, in large, open, leafy panicles, disciform. Involucre campanulate, 7–8-seriate; phyllaries imbricate, linear-lanceolate to linear. Receptacle naked. Marginal flowers female, numerous, central ones bisexual, few. Corollas pink or mauve. Outer flowers numerous, female; corollas filiform, 3–4-toothed. Inner flowers few, bisexual; corollas tubular with 5 sparsely glandular teeth; anthers sagittate. Achenes oblong, pubescent; pappus 1-seriate, of white capillary bristles.

Our two species are often regarded as conspecific in other parts of their range.

1. Stems bearing lines of narrow uneven herbaceous teeth 2–16mm deep, usually joined by a narrow wing .... **1. L. crispata**
+ Stems bearing entire (rarely denticulate) herbaceous wings 1–5mm deep **2. L. alata**

1. L. crispata (Vahl) Hepper & Wood; *L. pterodonta* (DC.) Oliver & Hiern

Plant (20–)40–100cm; stems bearing lines of narrow uneven herbaceous teeth 2–16mm deep, usually joined by a narrow wing; stem teeth and leaves densely pubescent, mainly with short erect glandular hairs. Lower leaves narrowly obovate, 5–17 × 1.5–6cm, acute, long attenuate at base, sessile or sometimes with winged, deeply toothed petiole to 15mm, subentire to serrate-dentate; upper leaves narrower, little tapered below, ± oblong, sometimes subobtuse. Involucre c 7mm diameter; phyllaries 4.5–9mm, at least outer ones glandular-pubescent. Corollas 6.5–8mm. Achenes c 1mm; pappus c 6mm.

Bhutan: S – Gaylegphug district (Sher Camp), **C** – Tongsa district (Tama (117)); **Darjeeling:** Badamtam, Pankhabari, Ryang etc. Roadside banks, 600–1050m. March–April.

2. L. alata (D. Don) Oliver

Similar to *L. crispata*, differing by its entire (rarely remotely denticulate) herbaceous wings, 1–5mm deep, along the stems; wings and leaves with thicker covering of predominantly longer, spreading multiseptate hairs; all leaves often subobtuse, lower ones less tapered below.

Bhutan: C – Tongsa district (Dakpai (117)); **Darjeeling:** Sepoydhura (97a), unlocalised (Hooker and Cowan collections). Forest clearings, unweeded gardens, 1650–1700m.

45. PSEUDOCONYZA Cuatrecasas

Herbaceous perennial; most parts glandular-pubescent with less persistent eglandular hairs; stems erect or decumbent. Leaves alternate, simple to pinnatisect, serrate-dentate, sometimes petiolate, decurrent on stem as 1–2(–3) pairs of lobes. Capitula disciform, in small loose corymbs on leafy branches from most upper leaf axils. Involucre campanulate, 5–6-seriate; phyllaries imbricate, lanceolate (outer 2 series) to linear. Receptacle naked. central ones bisexual, few. Corollas creamy white or pink. Outer flowers female, numerous; corollas filiform, 3–4-toothed. Central flowers few, bisexual; corollas tubular with 5 minutely glandular-pubescent teeth; anthers caudate. Achenes oblong, pubescent; pappus 1-seriate, of white capillary bristles, subequal to corolla.

1. P. viscosa (Miller) d'Arcy; *Laggera aurita* (L.f.) Clarke

Plant c 30–60cm. Leaves ovate to obovate, sometimes lyrately or pinnately lobed, 2–12 × 1–6cm, acute, cuneate to a narrow base or sometimes lobed to base, dentate; decurrent lobes ± obovate, c 8 × 5mm, rarely much larger, dentate. Involucres 4–6mm diameter; phyllaries 3–7mm, hairs mainly glandular. Flowers 4.5–6mm. Achenes c 1mm, red-brown.

W Bengal Duars: Jalpaiguri (174).

46. BLUMEOPSIS Gagnepain

Slender annuals. Leaves alternate, simple, sessile, dentate. Capitula disciform, in axillary and terminal pedunculate corymbs. Involucre imbricate, c 5-seriate; outer phyllaries ovate, subacute; inner ones linear-oblong, caudate. Outer flowers numerous, female; corolla filiform, minutely toothed. Inner flowers few, subbisexual with partially developed androecium; corolla broader, 4-toothed; anthers 1–4, often unequal, tapered at base into filament, not tailed. Rarely (in our area) some perfect inner flowers present, with 5 corolla lobes and 5 caudate stamens. Achenes oblong, glabrous; pappus 1-seriate, of deciduous bristles barbed at base.

1. B. flava (DC.) Gagnepain; *B. falcata* auct. non (D. Don) Merrill, *Laggera flava* (DC.) Clarke. Fig. 125c–?.

Plant 25–100cm; stems usually simple, glabrous. Leaves 2–11 × 0.5–4.5cm, coarsely or finely toothed, glabrous or sparsely pubescent; lowest ones obovate, narrowed to base; remainder oblong (rarely ovate), acute, truncate at base. Capitula 3–4mm diameter; phyllaries 2.5–7mm, glabrous (rarely few coarse hairs above on inner ones). Flowers 3.5–4.5mm, teeth of all bisexual ones papillose. Achenes 0.5mm, pappus white, 3.5–5mm.

Bhutan: C – Mongar district (Mongar (117)); **Darjeeling:** Chenga, Sivok, Silak etc. Roadsides, 250–600m. November–February.

47. SPHAERANTHUS L.

Overwintering annuals; stems woody at base, ascending. Leaves alternate, simple, decurrent. Capitula disciform, crowded on a globose common receptacle, forming terminal rounded glomerules, each capitulum subsessile with several bracts below. Involucre 2–4-seriate; phyllaries rather similar to and ± equalling bracts; flowers ± included. Outer flowers several, female; corolla filiform, minutely 3-lobed. Central flowers fewer, male; corolla campanulate-infundibular, 5-lobed, basal part becoming suberised and much dilated in fruit; abortive ovary present, style linear, papillose, minutely bifid at apex. Achenes small, ± oblong; pappus absent.

1. S. indicus L. Fig. 125a–b.

Rank smelling; stems 10–45cm with 4 irregularly and sharply toothed wings, minutely stipitate-glandular and whitish pubescent. Leaves oblanceolate or spathulate, 2.5–6 × 0.8–2.5cm, obtuse or acute, apiculate, attenuate, semi-amplexicaul and strongly decurrent at base, rather irregularly biserrate-dentate, minutely stipitate-glandular and white pubescent on both surfaces. Glomerules ovoid-globose, c 12 × 10mm; capitulum bracts linear-lanceolate, c 4–5mm, finely acuminate, ciliate, hispid and stipitate-glandular; phyllaries c 12, linear-oblong to linear-spathulate, more scarious and less glandular than bracts. Female flowers 10–15, male flowers 2–3. Corollas purplish, c 2.4mm. Achenes 1mm, puberulous.

Bhutan: S – Samchi district (Samchi (117), Torsa), Phuntsholing district (Phuntsholing (117)); unlocalised (Griffith collection from Bhutan); **W Bengal Duars:** Jalpaiguri district (174). Wet places, 600m. January.

S. senegalensis DC. has been reported from Sikkim (343). It is very similar to *S. indicus*, but has leaves simply serrate-dentate with sessile glandular hairs.

GNAPHALIEAE

48. LEONTOPODIUM R. Brown

Perennial, pulvinate, tufted or stoloniferous, tomentose herbs; flowering stems erect or absent. Leaves alternate, simple, entire, basal ones sometimes rosulate. Inflorescence terminal, of 1–12(–20) crowded capitula usually surrounded by radiating inflorescence bracts conspicuous by colour and/or texture of tomentum. Flowers unisexual; capitula usually either discoid and dioecious or disciform and of two types borne on separate plants with one sex (usually female) much predominant; however, in *L. himalayanum* discoid and disciform capitula may occur in the same inflorescence. Involucre campanulate, imbricate, several-seriate; phyllaries (pale) dark brown scarious at least above. Receptacle flat or convex, naked, glabrous. Corolla of female flowers filiform, 3–4-toothed at apex. Corolla of male flowers narrowly tubular-campanulate, 5-toothed; styles oblong, bifid. Achenes oblong-ellipsoid, glabrous or papillose-pubescent; male ovaries equally developed in discoid and disciform capitula, papillose pubescence often reduced; pappus simple, white, deciduous, subequal to corolla, bristles coherent at base, thicker towards apex on male flowers.

FIG. 125. **Compositae: Plucheae and Gnaphalieae. Plucheae.** a–b, *Sphaeranthus indicus*: a, shoot with aggregate capitulum (× ⅔); b, single capitulum on bracteolate peduncle (× 8). c–f, *Blumeopsis flava*: c, upper flowering stem (× ⅔); d, capitulum (× 4); e, female flower (× 8); f, perfect bisexual flower (× 8). **Gnaphalieae.** g–h, *Anaphalis contorta*: g, upper flowering stem (× ⅔); h, capitulum (× 6). Drawn by Margaret Tebbs.

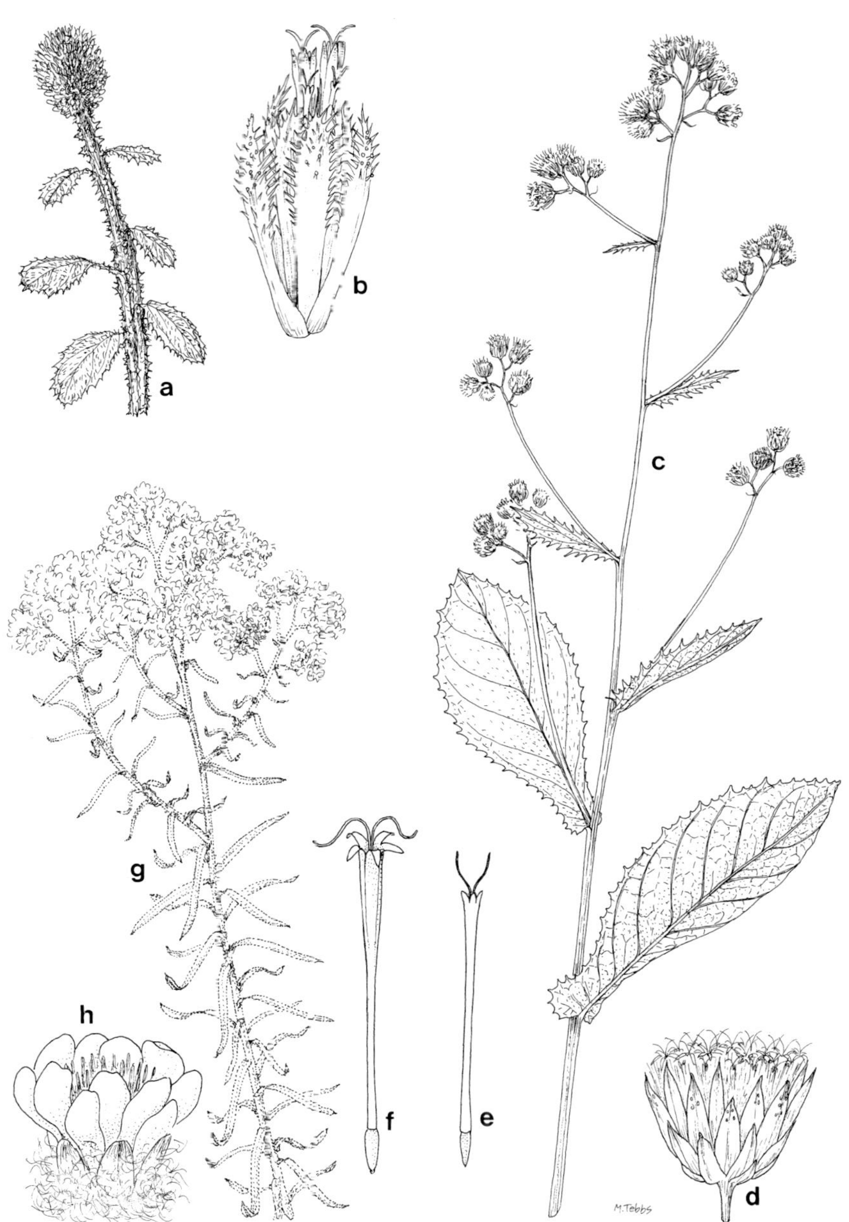
a
b
c
g
f
e
h
d
M.Tebbs

The habit of most of our species is distinctive, but rarely apparent in dwarf specimens. Their identification here is tentative, apart from *L. haastioides.*

1. Stems and leaves bearing glandular hairs .... **1. L. stracheyi**
+ Stems and leaves without glandular hairs 2

2. Inflorescence bracts absent or not forming a conspicuous star and scarcely differing from leaves in vestiture 3
+ Inflorescence bracts present, at least tips spread in a conspicuous star, differing from leaves (except upper cauline sometimes) in colour, length, or texture of indumentum 4

3. New leaves at apex of sterile rosettes 10–25, blades or herbaceous tips triangular to obovate, 1.5–8 × 1–3.2mm .... **2. L. haastioides**
+ New leaves at apex of sterile rosettes 3–10, blades spathulate or oblanceolate, 7–13 × 1.8–5mm .... **3. L.** cf. **nanum**

4. Plant stoloniferous, ± matted; stolons spreading, often elongate, naked or bearing scattered recurved scales, small spreading leaves or rarely scattered appressed petiole-like scales 5
+ Plant tufted, with replacement shoots erect, short, crowded, entirely sheathed by old leaf bases 7

5. Stolons bearing scattered, loosely appressed, scarious, modified petioles without blades, sometimes most stolons filamentous (width hardly exceeding width of main roots) .... **3. L.** cf. **nanum**
+ Stolons naked or bearing scattered recurved scales or small spreading non-sheathing leaves, most or all stolons thicker 6

6. Flowering stems 0.5–6cm; radiating bracts greyish-yellow to brown villous above, oblanceolate or spathulate .... **4. L. monocephalum**
+ Flowering stems more than 6cm or indumentum of radiating bracts whitish (araneous-)tomentose above or radiating bracts lanceolate to elliptic **5. L. jacotianum**

7. Inflorescence diameter across radiating bracts 2–4 × diameter of its cluster of capitula .... **6. L. himalayanum**
+ Inflorescence diameter across radiating bracts 1–2 × diameter of its cluster of capitula .... **7. L. leontopodinum**

1. L. stracheyi (Hook.f.) Hemsley; *L. alpinum* Cassini var. *stracheyi* Hook.f.

Stoloniferous, 30–45cm; stolons thick, bearing scarious scale leaves up to apex; flowering stems brownish, glandular and sometimes crisped pubescent. Leaves narrowly oblong-lanceolate, acute or acuminate, sessile, almost auriculate, green, glandular and sometimes crisped pubescent above, closely white tomentose beneath, upper ones 2–3 × 0.3–0.7cm. Capitula 5–11, crowded, apparently always dioecious; bracts similar to leaves but white tomentose on both surfaces, more densely so above, with some erect hairs beneath. Phyllaries ovate to oblanceolate, 4–4.5 × 1–2mm, with dark brown scarious margin. Corollas 3.5–4mm. Achenes 0.75–1mm, pubescent; male ovary pubescent.

Bhutan: C – Thimphu district (Jakhuthethe), **N** –Upper Mo Chu district (Lingshi, Jari La). Grassy banks, dry hills, 3950–4250m. September–October.

2. L. haastioides Handel-Mazzetti; *Antennaria muscoides* Clarke non *L. muscoides* Handel-Mazzetti, *L. jacotianum* Beauverd var. *haastioides* (Handel-Mazzetti) Srivastava. Fig. 126a.

Forming dense cushions, shoots c 1.5cm, short columnar, with closely imbricate leaves. Leaves spathulate or oblong, 4–8 × 1–2mm; tip spreading, thickened, herbaceous and thickly greyish tomentose (or subglabrescent (345)), rounded at apex; base thin, appressed, scarious and sparsely araneous below. Capitula solitary, 3–4mm diameter, dioecious or predominantly female, ebracteate, immersed among leaves; phyllaries lanceolate to oblanceolate, 4–6 × 0.6–2mm, acute or obtuse, tips scarious, brownish. Corollas 3–4mm. Achenes c 1.2mm, pubescent or glabrous; male ovary glabrous (or sparsely pubescent (345)).

Sikkim: Donkung, Giagong (345), Gayum Chhona (345), Lhonak, Tang La; **Chumbi:** Dongka La (345). Stony ground, bare soil, 4550–4900m. August.

A collection from Tolegang, Upper Bumthang Chu district, Bhutan differs considerably in its distinct, obovate, wider (2.4–3.2mm), greenish leaf blades with yellowish tomentum, larger capitula (6–8mm diameter) and phyllaries with dark brown upper margin and tip. It may represent a distinct taxon.

3. L. cf. **nanum** (Hook.f. & Thomson) Handel-Mazzetti; *Antennaria muscoides* sensu F.B.I. p.p. non Clarke. Fig. 126b.

Pulvinate-caespitose or matted, 2–7cm, greyish tomentose; stems mostly suberect, covered in scarious appressed leaf bases, with rosette of few leaves at apex (sometimes of many leaves on flowering shoots); spreading stolons sometimes present, bearing scattered, loosely appressed, scarious, modified petioles without blades. Leaf blades oblanceolate to spathulate, 7–13 × 1.8–4(–5)mm, usually subequal to sheathing base, obtuse. Flowering stem absent or 1.5–2cm with scattered non-sheathing leaves. Capitula solitary or c 3–7 interspersed with inflorescence bracts; inflorescence bracts inconspicuous, non-radiating and ± similar to rosette leaves in shape and colour but non-sheathing, few large (subequal to leaf blades), sometimes with others much shorter or some transitional to phyllaries. Capitula predominantly female or all male, 2–6mm diameter; phyllaries mostly oblong to oblanceolate, 4–6.5mm, obtuse to acuminate. Corollas 3–3.6mm. Male ovary and achenes usually glabrous.

Sikkim: above Giagong, Kanchenyao; **Sikkim/Tibet:** around Khamba Dzong. Between 4900 and 5400m. In fruit in July.

Only four disparate collections seen from our area, mostly of incomplete specimens. Some have been determined as *L. pusillum* (Beauverd) Handel-Mazzetti (345), including the collection from Kanchenyao with a rosette of many non-sheathing leaves beneath a sessile inflorescence. *L. pusillum* is usually distinguished by its filiform stolons with remote scales and conspicuous radiating bracts. It occurs further north, just beyond our area and possibly at Lhonak, Sikkim (plants with flowering stems c 1cm). An incomplete collection of uncertain identity from Lugnak La, Sikkim shows some similarities to *L. nanum*. It has flowering stems c 5cm with lorate leaves c 18 × 2mm and inflorescences of c 10 capitula surrounded by thickly tomentose lorate ?stellate bracts.

4. L. monocephalum Edgeworth; *L. alpinum* var. *alpinum* sensu F.B.I. p.p. non *L. alpinum* Cassini. Fig. 126c.

Stoloniferous, forming mats, 4–10cm; stolons slender, with scattered recurved scales or small non-sheathing leaves, ± prostrate, terminated by a loose rosette of leaves. Leaves oblong-oblanceolate to spathulate, 7–18 × 2–3.3mm, obtuse, rarely brown-apiculate, dilated at base, yellowish (rarely greyish) tomentose. Flowering stems 0.5–6cm. Capitula 1–10, dioecious (in our specimens), central one 5–7mm across; radiating bracts oblanceolate or spathulate, to 14 × 3(–4)mm, greyish-yellow to brown villous(-sericeous) above, sometimes with concealed apiculus. Phyllaries (ovate–)oblanceolate, 4–6.2 × 1–2mm, with blackish, scarious, deeply lacerate margins. Corollas 3.3–4.2mm. Achenes pubescent; male ovary sparsely pubescent.

Sikkim: Lhonak, Lungia Chur (345); **Chumbi:** Guptende La. 4700–5200m. August.

Description above includes some dubiously localised Hooker specimens from Sikkim that are intermediate with *L. jacotianum*.

5. L. jacotianum Beauverd; *L. alpinum* var. *alpinum* sensu F.B.I. p.p. non *L. alpinum* Cassini. Fig. 126e.

Stoloniferous, forming mats, 6–28cm, greyish (greyish-yellow) tomentose; stolons slender, elongate, usually naked (sometimes remotely squamate), terminated by a loose rosette of leaves. Leaves 7–25mm, subobtuse to acuminate, usually brown-apiculate; rosette leaves linear, lorate-spathulate or oblanceolate, 1–3mm wide; cauline leaves lanceolate to lorate or linear with margins recurved, 1–5mm wide. Flowering stems 2–25cm. Capitula 4–9(–18), all female or predominantly male or female, usually densely crowded; radiating bracts lanceolate to oblong-elliptic, 8–25 × 2.2–7mm, often long acuminate, more densely whitish

FIG. 126. **Compositae: Gnaphalieae.** a–e, *Leontopodium* species (× ⅔): a, *L. haastioides*; b, *L.* cf. *nanum* from Giagong; c, *L. monocephalum*; d, *L. himalayanum*; e, *L. jacotianum*. Drawn by Margaret Tebbs.

a
b
c
d
e
M.Tebbs

(rarely yellowish or brownish) tomentose than cauline leaves. Phyllaries lanceolate to oblanceolate, 3–5mm, brown scarious above. Corollas 3–3.5mm. Achenes pubescent; male ovaries pubescent or subglabrous.

Bhutan: C – Thimphu district (Pajoding, Paro Pass, Somana), Bumthang district (Kyi Kyi La), **N** – Upper Mo Chu district (Kohina, Laya, Singche La etc.), Upper Bumthang district (Pangotang); **Sikkim:** Gochung, Yeumthamg, Zongri etc.; **Chumbi:** Lingmatang, Tangu. Open grassy banks and hillsides, 3050–5150m. July–October.

Two varieties have been recognised in our area (345), but can scarcely be maintained: var. *jacotianum* with flowering stems 3–12cm, leaves araneous above, finely white tomentose beneath; var. *paradoxum* (Drummond) Beauverd, more compact with rosette leaves more slowly marcescent, leaves and bracts villous-lanate on both sides.

6. L. himalayanum DC.; *L. alpinum* var. *alpinum* sensu F.B.I. p.p. non *L. alpinum* Cassini. Fig. 126d.

Plant densely tufted, (3–)7–32cm; offsets sheathed below by many old leaf bases, terminated by loose tuft of few leaves. Offset leaves linear-spathulte, sometimes brown apiculate; cauline leaves ± lorate, 18–70 × 1–4mm, acute or subobtuse, less often apiculate, greyish tomentose. Capitula crowded, (4–)7–12(–20), discoid or disciform; disciform capitula usually all similar, sometimes central one predominantly male and remainder predominantly or entirely female; bracts lorate, usually dilated at or near base or sometimes linear-lanceolate, (obtuse) acute or acuminate, basal half at least of upper face more densely white (yellowish-grey or tawny) tomentose than leaves or very rarely villous, outer ones 12–40 × 2–7.5mm. Inflorescence diameter across bracts 2–4 × diameter across capitula. Phyllaries ± oblong, 3.5–6 × 1–2.5mm, with broad brown scarious margin at least above. Corollas 3.3–4mm. Achenes 1.5–2mm, glabrous or sparsely pubescent; male ovaries ?always glabrous.

Bhutan: C – Thimphu and Tongsa districts, **N** –Upper Mo Chu, Upper Pho Chu, Upper Kuru Chu and Upper Kulong Chu districts; **Sikkim:** Chemathang, Yakla, Yampung etc.; **Chumbi:** Kalaeree, Phari (369). Rocky hillsides and alpine turf, 3800–5500m. June–October.

The plants from Phari have been described as *L. himalayanum* var. *pumilum* Ling, with stems 1–3cm, leaves 8–15 × 1–2mm, inflorescence 2–3cm across bracts, radiating bracts 10–15 × 1–3mm, capitula 1–3.

7. L. leontopodinum (DC.) Handel-Mazzetti; *L. alpinum* var. *alpinum* sensu F.B.I. p.p. non *L. alpinum* Cassini, *L. ochroleucum* Beauverd

Very similar to *L. himalayanum*, but differs by its cauline leaves oblong-lanceolate to spathulate, very obtuse to subacute; diameter across radiating bracts 13–35mm, 1–2 × diameter across capitula; radiating bracts ovate to linear-lanceolate.

Some fragmentary specimens of small plants (3–8cm) from Sikkim, mixed with *L. himalayanum*, have been identified as this species ((345), localities considered unreliable). Very dwarf specimens with flowering stems c 5mm and inflorescence bracts radiating, elliptic, short, with tips exposed and conspicuously tomentose may also belong here, including some from Gayum Chhona (Sikkim).

49. ANAPHALIS DC.

Perennial herbs or subshrubs; stems ± erect, usually cottony lanate-tomentose. Leaves simple, alternate, entire. Capitula small to medium-sized, several to many in terminal corymbs, rarely solitary. Flowers unisexual; capitula discoid and dioecious or disciform and of two types, borne on separate plants, each with one sex much predominant, usually female. Involucre broadly campanulate; phyllaries several- to many-seriate, scarious, inner petaloid, white or yellowish. Receptacle convex, naked. Corolla of female flowers filiform, 3–4-toothed at apex. Corolla of male flowers narrowly tubular-campanulate, 5-toothed; anthers tailed; style branches obtuse. Achenes oblong, usually papillose-pubescent; male ovary abortive in predominately female capitula, but as long as those of female flowers, ± absent in predominately or entirely male capitula (?except *A. cavei*); pappus simple, just exceeding corollas (?longer in *A. deserti*), of separately deciduous bristles often thickened at tips on male flowers.

1. Leaves decurrent on stems for at least 5mm, though often very narrowly so ... 2
\+ Leaves not or scarcely decurrent on stems ... 5

2. Short-lived (?biennial) herbs, without woody crowns or branch systems ... **2. A. busua**
\+ Long-lived plants with woody crowns or branch systems ... 3

3. Well-developed system of persistent aerial branches; longest phyllaries 8–9.5mm ... **10. A. cooperi**
\+ Herbs or sprawling subshrubs; longest phyllaries 5.5–7mm ... 4

4. Weakly suffrutescent stock bearing many woolly buds composed of sessile scale leaves ... **8. A. deserti**
\+ Much-divided woody crown at soil surface bearing non-woolly buds with large erect petiolate leaves often much longer than cauline leaves and with similar indumentum ... **9. A. xylorhiza**

5. Flowering stems persistently glandular hairy 6
+ Flowering stems araneous to tomentose, sometimes glabrescent 7

6. Longest phyllaries 7–8mm .... **12. A. subumbellata**
+ Longest phyllaries to 4mm .... **4. A. hookeri**

7. At least some basal leaves petiolate or plant less than 15cm 8
+ Leaves sometimes narrowed towards base but not petiolate and plant more than 15cm 11

8. Phyllaries 6–7-seriate, longest 5–7.1mm 9
+ Phyllaries 8–9-seriate, longest 7.5–11mm 10

9. Much-divided woody crown without stolons; phyllaries white **9. A. xylorhiza**
+ Plant caespitose but without woody crown; phyllaries yellowish **11. A. cavei**

10. Leaves 1- or 3-veined, ± not amplexicaul, all less than 2.5cm wide; inflorescences with 1–15 capitula .... **13. A. nepalensis**
+ Leaves 3- or 5-veined, semi-amplexicaul at base, either some wider than 2.5cm or some inflorescences with more than 15 capitula .. **14. A. triplinervis**

11. Leaves oblanceolate, flat, whitish tomentose on both faces; phyllaries to 5.2mm, most white below .... **1. A. adnata**
+ Leaves narrowly lanceolate to narrowly oblong or linear, often yellowish to cinnamomeous tomentose below, sometimes glabrescent above or some phyllaries longer than 5.2mm, all brownish below 12

12. Leaf margin often strongly recurved; phyllaries 30–50 in c 5–6 series **6. A. contorta**
+ Leaf margin scarcely recurved; phyllaries 55–120 in c 8–10 series 13

13. Tufted plant, much branched, with slender stems 1–2mm thick below **5. A royleana**
+ Stems simple (rarely few-branched) from long underground runners, c 3–5mm thick below (rarely thinner if branched at base) 14

14. Leaves densely cinnamomeous tomentose beneath, thinly greyish tomentose above; lateral veins absent or 2, concealed .... **7. A. griffithii**
+ Leaves of flowering stems sometimes grey tomentose beneath, usually glabrous above; lateral veins 2 or 4, prominent .... **3. A. margaritacea**

1. A. adnata DC.; *Gnaphalium adnatum* (DC.) Kitamura

Herbaceous; stems erect, simple, 15–45cm. Leaves oblanceolate, 5–9 × 0.5–2cm, acute, base narrowed, semi-amplexicaul, appressed whitish tomentose on both surfaces but thinner on upper surface, distinctly 3- or 5-veined. Capitula small, 2.5–3.5mm diameter, in dense clusters in terminal corymbs, usually with short or rarely long-pedunculate glomerules from most leaf axils; only capitula with predominately female flowers seen from our area. Involucre 5–8-seriate; phyllaries ovate or elliptic, white, subacute, base narrowed, outermost brownish below, outer and middle ones subequal, to 3.7–5.2mm long, inner ones unguiculate. Corollas 2–3mm. Achenes 0.5mm, papillose.

Bhutan: S – Deothang district (Deothang–Raidong), **C** – Thimphu district (Motithang), Punakha district (Wangdu Phodrang–Naki (345)), Tsarza–Samtengang); **Darjeeling:** Kariabasti, Kurseong–Tindaria; **Darjeeling/Sikkim:** unlocalised (Hooker collections). 1200–2450m. August–January.

2. A. busua (D. Don) DC.; *A. araneosa* DC., *A. alata* Maximowicz. Dz: *Wadepusang shing.*

Short-lived herb (?biennial), 20–80cm, sparsely to moderately arachnoid-tomentose, sometimes also short glandular-pubescent; stem erect, usually only branched beneath inflorescence. Leaves linear-lanceolate, 3.5–10 × 0.3–1cm, acuminate, base shortly or strongly decurrent, green and sparsely tomentose above, greyish moderately tomentose beneath. Capitula numerous, in corymbs; flowers all female or predominantly male or female; involucre c 4-seriate, 2.5–5mm diameter; phyllaries ovate or elliptic, obtuse or acute, white above, outer ones somewhat larger, to 3.5–4.7 × 1.3–2.2mm, at least inner ones unguiculate. Corollas 2.2–3mm. Achenes 0.5mm, papillose.

Bhutan: S – Phuntsholing (117), Chukka (117), Gaylegphug (117) and Deothang districts, **C** – Punakha, Tongsa, Mongar (117) and Tashigang districts, **N** – Upper Mo Chu district; **Darjeeling:** Darjeeling, Rayong, Tanglu etc.; **Sikkim:** Lachen. Hillsides, among shrubs, roadsides, 300–4080m. February–November.

3. A. margaritacea (L.) Bentham & Hook.f.; *A. cinnamomea* (DC.) Clarke. Dz: *Daningon.*

Herb 30–80cm, from long subterranean stolons; stems closely greyish or brownish tomentose. Leaves of flowering stems narrowly elliptic or lanceolate, 2.5–12 × 0.2–1.8cm, acuminate, sessile, narrowed at base, sometimes weakly auriculate, non decurrent, thinly tomentose or glabrescent above, greyish or brownish (rarely yellowish) closely tomentose beneath, 3-veined; short sterile stems with oblanceolate, obtuse to cuspidate leaves to 8 × 2.5cm sometimes present. Capitula in usually dense clusters in dense to diffuse corymbs; flowers predominantly female or all male, rarely predominantly male. Involucre c 8–10-seriate, 4–6mm diameter; phyllaries white, brownish at base, middle ones longest, oblong-elliptic, 4.3–6mm, obtuse to acuminate, inner ones unguiculate. Corollas 3–5.3mm. Achenes 0.6mm, papillose.

Bhutan: S – Chukka and Deothang districts, **C** – Thimphu, Punakha, Tongsa, Mongar and Tashigang districts, **N** – Upper Mo Chu; **Darjeeling:** Darjeeling, Rongsong, Tanglu etc.; **Sikkim:** Bakhim, Lachen, Lachung, Takla; **Chumbi:**

Yatung. Cleared forests, roadsides, rock-crevices etc., 900–4500m. July–December.

4. A. hookeri Hook.f.

Herb more than 60cm; stems erect, glandular-pubescent throughout, araneous above. Leaves lanceolate, 5–10 × 0.5–1.3cm, acuminate, cordate-auriculate at base, sessile, 3-veined, sparsely araneose on both surfaces, glandular-pubescent beneath. Capitula numerous, in terminal corymbs; flowers predominantly female or ?all male. Involucre 3–3.5mm diameter; phyllaries white, pale brown at base, outer ones ovate-oblong, middle ones longest, oblong to elliptic, to 3–4mm, obtuse or subacute, inner ones unguiculate. Corollas 2.5mm. Achenes oblong, 0.5mm, papillose.

Sikkim: Lachen and Lachung. 2745m. July–August.

5. A. royleana DC. var. **concolor** Hook.f.

Tufted subshrub, much branched at base; stems slender, 15–35cm, erect or ascending, woody below with dead leaves and prominent adventitious buds at flowering, lanate. Upper leaves narrowly elliptic, 20–60 × 1.5–6mm, ± acuminate, base scarcely narrowed, sessile, 1 vein apparent, thinly yellowish or greyish lanate on both surfaces. Capitula 3–30 in dense terminal corymbs; flowers entirely male or predominantly female. Involucre 5.5–6mm diameter; outer phyllaries ovate, entirely brown; middle ones longest, ovate or elliptic, to 5.8–7mm, usually acute, white above, inner ones unguiculate. Corollas 2.8–4mm. Achenes 0.75mm, papillose.

Bhutan: N – Upper Mo Chu district (Tharizam Chu); **Sikkim:** Dadong, Lachung, Yakla etc.; **Chumbi:** Gautsa. Grassy hillsides, 3950–4570m. June–September.

A small specimen from Tangu (Sikkim) may belong here, but differs by its scarcely brown middle phyllaries and non-unguiculate inner ones.

6. A. contorta (D. Don) Hook.f.; *A. tenella* DC. Dz: *Churkarp*. Fig. 125g–h.

Plant woody at least at base, 15–40cm; stems erect or decumbent, usually with prominent tomentose buds below. Leaves numerous, linear, 1–3 × 0.1–0.5cm, acute, sometimes with black tip, base cordate, margin usually revolute, whitish tomentose on both surfaces but more densely so beneath, 1-veined. Capitula in rounded usually densely crowded corymbs; flowers predominantly or entirely male or female. Involucres 5–6-seriate, 3–4.5mm diameter; outermost phyllaries ovate, brownish or purple-tipped, remainder white or ?yellowish, middle ones longest, ovate to oblanceolate, to 4–5.3mm, inner ones unguiculate or not. Corollas 2–3.3mm. Achenes 0.5mm, papillose.

Bhutan: S – Phuntsholing or Chukka district, **C** – Ha, Thimphu, Punakha, Tongsa (71), Tashigang, Mongar and Bumthang districts, **N** – Upper Mo Chu and Upper Pho Chu districts; **Darjeeling:** Darjeeling, Kurseong, Sandakphu etc.; **Sikkim:** Meguthang, Yakla, Zongri etc.; **Chumbi:** Natu La–Chubithang,

Phari, Yatung etc. Open slopes and screes, 2150–4400m. February, July–November.

An extremely variable taxon. Small densely tufted plants with weak stems persistent and much branched in lower half and non-unguiculate inner phyllaries have been segregated as *A. tenella* DC.

7. A. griffithii Hook.f.

Perennial, 20–50cm; stems erect or ascending, thinly pale brownish tomentose. Leaves numerous, narrowly oblong or elliptic, 2–5.5 × 0.2–0.5cm, subacute to acuminate, with brownish tip, sessile, neither decurrent or auriculate at base, densely pale brownish tomentose beneath, greyish araneous above, 1-veined, margin flat. Capitula rather few in compact heads sometimes grouped in loose corymbs; flowers predominantly female in our specimens. Involucre c 8-seriate, 5–6mm diameter; phyllaries whitish, outer ones at least brown below, ovate, middle ones elliptic, to 5–6mm, acute or acuminate, inner ones unguiculate, to 6.3mm. Corollas 4.4–5.2mm. Achenes c 0.5mm, subglabrous.

Darjeeling: Chhota Rangit, Kurseong, Mongpu; **Sikkim:** Lusing. 300–1350m. October.

8. A. deserti Drummond

Subshrub 15–25cm; stock with short, stout, creeping, woody branches with many woolly buds; flowering stems slender, erect, glabrescent. Leaves lanceolate to narrow-oblong, 1.5–4 × 0.3–0.6cm, obtuse or subacute, decurrent, short glandular-pubescent on both faces with sparse arachnoid tomentum at least beneath uppermost leaves. Inflorescence a compact terminal corymb of c 7–33 capitula. Involucre c 5-seriate, 3.5–5mm diameter; phyllaries c 25–40, white above, mostly elliptic, to 5.5 × 2.8mm, usually obtuse; outer purplish and tomentose below; inner unguiculate, sometimes acute. Flowers all male or all female except 2–3 male. Corollas 3–4mm. Pappus to 5.3mm.

Sikkim: Lhonak. 4500–4700m. August.

9. A. xylorhiza Hook.f. Dz: *Jau metog*

Tufted herb, almost suffrutescent, 2–20cm; tap-root very woody, twisted; stock divided into stout partly exposed woody crowns with erect petiolate leaves usually longer than cauline leaves. Flowering stems loosely greyish tomentose, sometimes glabrescent below. Leaves oblanceolate or spathulate, 1.3–6 × 0.2–0.8cm, attenuate at base, yellowish or greyish arachnoid tomentose and glandular-pubescent on both surfaces or tomentum soon lost from upper face; leaves of crown obtuse, petiolate; cauline leaves ± sessile, sometimes narrowly decurrent, upper ones usually acuminate to a fine point. Capitula in compact corymbs; flowers predominantly female or entirely male. Involucre 6–7-seriate, c 4mm diameter; outer phyllaries brown below, middle ones longest, lanceolate to oblanceolate, to 5–7mm, ± obtuse, inner ones unguiculate. Corollas 2.5(–4)mm. Achenes 0.8–1mm, papillose.

Bhutan: C – Thimphu district (Somana), **N** – Upper Mo Chu district (Lingshi, Laya, Sinchu La etc.), Upper Bumthang Chu district (Marlung); **Sikkim:** Chugya, Giaogang, Lhonak etc.; **Chumbi:** Chomp Lahari, Gautsa–Phari. Grassy hillsides to screes, 3950–4900m. June–September.

10. A. cooperi Grierson

Rounded (sub)shrub; branches up to 40cm, loosely covered with dead leaves; flowering shoots 10–15cm, ± erect, loosely covered by pale brownish tomentum. Leaves narrowly oblanceolate, 4–5 × 0.4–0.8cm, acute or acuminate, attenuate at base, narrowly decurrent on stem, loosely brownish tomentose on both surfaces, 1-veined. Capitula 6–20 in compact corymbs; flowers predominantly female or almost entirely male; involucre 5–7-seriate, c 6mm diameter; phyllaries white above, ovate-oblong, 7–9.5 × 2–3mm, obtuse or subacute, often erose. Corollas 5.5mm. Achenes (immature) 1mm.

Bhutan: C – Thimphu district (Barshong), **N** – Upper Mo Chu district (Laya), Upper Pho Chu district (head of W Pho Chu), Upper Bumthang Chu district (Kantanang). Pebbly shores of glacier lake and grassy ledges of exposed ridge, 3800–4480m. June–September.

11. A. cavei Chatterjee

Caespitose herb, 2–7cm; stems densely brownish tomentose. Leaves of flowering stems ± oblanceolate, 5–8 × 1.5–3.5mm, obtuse or subacute, base attenuate, sessile, densely brownish lanate; leaves of sterile rosettes shorter and wider. Capitula c 4mm diameter, in small dense corymbs of 5–9 capitula. Phyllaries ± oblanceolate, obtuse, yellowish with brown band above weakly defined claw, middle ones longest, to 6.5–7.1mm. Flowers 15–25; corollas 3.8mm. Achenes 1mm, sparsely papillose.

Sikkim: Naku Chho, Goecha La. 5000–5300m. September.

Capitula c ⅔ female on our specimens; in E Nepal specimens have been collected with entirely female or male flowers, the latter with partly developed ovaries, c 0.6mm.

12. A. subumbellata Clarke

Herb to 45cm; stems erect, simple, minutely glandular, with whitish tomentum deciduous except from inflorescence branches. Leaves lanceolate to spathulate, 3–6 × 0.25–1.8cm, subobtuse to acuminate, with a fine blackish tip, obtuse to long attenuate at base, sessile, 3-veined, glandular-pubescent on both surfaces but more densely so above, with whitish tomentum ± persisting at margins and along veins. Capitula rather few in compact corymbs; flowers predominantly male or predominantly female. Involucre c 6–8-seriate; phyllaries lanceolate, acuminate, white, prominently brownish below, middle ones longest, to 7.2–8.5mm, inner ones unguiculate. Corollas c 4mm. Achenes c 0.8mm, papillose.

Sikkim: Gnatong, Lachung; **Chumbi:** Yatung, Natu La–Chubitang. 3050–3650m. July–October.

13. A. nepalensis (Sprengel) Handel-Mazzetti

Herbs, sometimes with arching stolons, 2.5–50cm; stems sparsely or densely whitish tomentose. Lower leaves oblong or elliptic to spathulate, 1–10 × 0.3–2.3cm, subacute to acuminate, with a blackish tip, tapering at base, sometimes petiolate, ± not amplexicaul, 1- or 3-veined, whitish (rarely yellowish) tomentose on both surfaces but more densely so beneath or sometimes yellowish to cinnamomeous beneath, petiole to 4.5cm; upper leaves usually narrower, usually sessile, uppermost sometimes with scarious tip. Capitula 1–15, in corymbs; flowers predominantly male or predominantly female. Involucre 8–9-seriate, 6–7.5mm diameter; phyllaries brownish or blackish below, white above, middle ones longest, ± lanceolate, 6.5–11 × 2–3mm, acuminate, inner ones with green claw. Corollas 3–4mm. Achenes 0.6–0.8(–1.1)mm, papillose.

1. Lower leaves 0.5–2.2cm wide, 1- or 3-veined, lowest usually petiolate; inflorescences of (1–)5–15 capitula .... **a.** var. **nepalensis**

\+ Lower leaves 1–5 × 0.3–0.8mm, often sessile, usually 1-veined; most or all capitula solitary, rarely 2–4 together .... **b.** var. **monocephala**

a. var. **nepalensis**; *A. triplinervis* (Sims) Clarke var. *intermedia* (DC.) Airy Shaw, *A. nubigena* sensu F.B.I. p.p., *A. cuneifolia* (DC.) Hook.f.

Plant 10–50cm.

Bhutan: C – Thimphu, Punakha, Tongsa, Bumthang and Mongar districts, **N** – Upper Mo Chu, Upper Pho Chu and Upper Kulong Chu districts; **Darjeeling:** Senchal, Sandakphu etc. (all 69); **Sikkim:** Chomnagu, Dzongri–Olothang (69) etc.; **Chumbi:** Natu La–Chubitang, ?Patong La etc. Forest margins, open grassy hillsides, peaty slopes and screes, 2400–4265m. June–October.

b. var. **monocephala** (DC.) Handel-Mazzetti; *A. nubigena* sensu F.B.I. p.p., *A. triplinervis* var. *monocephala* (DC.) Airy Shaw

Plant 2.5–15(–18)cm.

Bhutan: C – Thimphu, Tongsa, Bumthang and Tashigang districts, **N** – Upper Mo Chu, Upper Pho Chu, Upper Bumthang Chu and Upper Kulong Chu districts; **Sikkim:** Dzongri, Lasha Chhu, Yume Samdong etc.; **Chumbi:** Phari, Tangkar La, Yatung etc. River sands, peaty sites, in grass, among dwarf shrubs, rocky slopes and crevices, 2750–5200m. June–October.

A poor unlocalised collection from Darjeeling or Sikkim, of uncertain identity, differs from var. *nepalensis* by its narrowly winged petioles, non-unguiculate inner phyllaries and glutinous tomentum with clavately tipped hairs.

A. acutifolia Handel-Mazzetti has been reported from Chumbi (421) but no examples have been seen. Its indumentum, inflorescence and capitula are similar to those of var. *nepalensis* but it is laxly caespitose with a short thick rhizome and no slender leafy stolons, it lacks long-petiolate basal leaves and has many narrowly oblong cauline leaves

14. A. triplinervis (Sims) Clarke

Plant 23–50cm, differing from *A. nepalensis* by its semi-amplexicaul leaves; lower leaves 3- or 5-veined, 4–17 × 1.5–4cm, usually with petiole to 8cm; capitula 9–70 per inflorescence.

Bhutan: C – Bumthang district (Bumthang, Dhur); **Darjeeling:** Darjeeling, Sandakphu, Senchal–Ghum etc.; **Sikkim:** Tonglo, Lachung, Prek Chhu etc. Grassy slopes, open meadows and stony banks, 1800–4250m. July–October.

A number of collections could not be certainly separated from *A. nepalensis* including one from Motithang (Bhutan: Thimphu district) in forest and one from Laya (Bhutan: Upper Mo Chu district). Basal lateral shoots alone are often inadequate for identification.

50. PSEUDOGNAPHALIUM Kirpichnikov

Annual herbs. Leaves simple, alternate, entire. Capitula in corymbs, disciform. Involucres campanulate; phyllaries several-seriate, scarious, yellow, glossy. Receptacle flat, naked. Marginal flowers female, numerous, filiform. Inner flowers bisexual, few; corollas narrowly tubular-campanulate, 5-toothed. Corollas yellow. Achenes ± oblong, papillose; pappus simple, bristles slender, free or partly coherent at base, deciduous.

1. Leaves whitish tomentose on both surfaces .... **1. P. affine**
+ Leaves green glandular on upper surface, white tomentose beneath **2. P. hypoleucum**

1. P. affine (D. Don) Anderberg; *Gnaphalium affine* D. Don, *G. luteoalbum* L. var. *multiceps* (DC.) Hook.f. Dz: *Mito kapa*; Nep: *Hooki phul* (117).

Stems erect or spreading, 10–40cm, often branched from base, whitish tomentose. Leaves lanceolate to oblanceolate, 1.5–6 × 0.3–1cm, obtuse to acuminate, sessile at base, whitish tomentose on both surfaces. Capitula 2–2.5mm diameter, in dense corymbs. Phyllaries 3–4-seriate, ovate-oblong, 1.7–3mm, obtuse or outermost subacute, inner ones usually with broad claw. Corollas 2mm. Achenes 0.5mm, papillae sparse; pappus as long as corollas, bristles slightly coherent at base.

Bhutan: S – Chukka, Gaylegphug and Deothang districts, **C** – Thimphu, Tongsa and Mongar districts; **Darjeeling:** Darjeeling, Jhepi, Kalimpong etc.; **Sikkim:** Lachen, Lhonak, Yoksam; **Chumbi:** Chubitang–Yatung. Common weed on roadsides and cultivated ground, 1050–4900m. March–December.

2. P. hypoleucum (DC.) Hilliard & Burtt; *Gnaphalium hypoleucum* DC.

Plant 20–80cm; stems erect, usually simple, branching above, greyish tomentose at first, later brownish glandular. Leaves 1.5–4 × 0.2–0.5cm, acuminate, sessile, semi-amplexicaul at base, usually undulate, greenish, glandular-pubescent above, greyish tomentose beneath. Capitula 3–3.5mm diameter, in corymbs; phyllaries c 5-seriate, elliptic-oblanceolate, to 3–3.5mm, yellow, obtuse, at least inner ones unguiculate. Corollas 2.5mm. Achenes 0.5mm; pappus as long as corollas, bristles free.

Bhutan: C – Thimphu district (Paro, Tho Cu La), Punakha district (Ritang, Tinglegang), Tongsa district (Tashiling), Bumthang district (Tangphomrong, Ura); **Darjeeling:** Kurseong, Sepoydhura (both 97a); **Sikkim:** Gangtok (97a), Pemayangtse–Thinglen; **Chumbi:** Trakarpo; **W Bengal Duars:** Jalpaiguri district (174). Hillsides, open ground, roadsides, fields, crops and pine forest, 1650–3350m. March–October.

51. GAMOCHAETA Weddell

Annual herb. Leaves spathulate, alternate, entire. Capitula in axillary clusters forming a more or less interrupted leafy spike, disciform. Involucres campanulate; phyllaries 2–3-seriate, scarious, pale brownish. Receptacle naked, glabrous. Corolla tips reddish at anthesis. Marginal flowers female, filiform, numerous; inner flowers bisexual, few, corollas narrowly tubular-campanulate, 5-toothed. Achenes oblong, sparsely papillose; pappus simple, bristles slender, connate at base and deciduous as a ring.

1. G. pensylvanicum (Willdenow) Cabrera; *Gnaphalium purpureum* sensu F.B.I. non L.

Plant 10–40cm; stems simple or branched from base, erect or ascending, greyish tomentose. Leaves 2–4(–6) × 0.5–1(–1.5)cm, obtuse, apiculate, long attenuate at base, sparsely lanate on upper surface, greyish tomentose beneath. Involucre 2–3-seriate; outer phyllaries ovate; inner ones oblong, c 3 × 0.7mm, subacute. Corollas 2mm. Achenes 0.5mm; pappus as long as corollas, bristles of all flowers less than 0.05mm across teeth at apex.

Bhutan: S – Samchi, Phuntsholing and Gaylegphug districts, **C** – Punakha (71), Tongsa and Tashigang districts; **Darjeeling:** Badamtam; **W Bengal Duars:** Jalpaiguri. Weed of roadsides and cultivated ground, 250–1400m. February–June.

Originally a native of N America, now a widespread weed.

52. GNAPHALIUM L.

Annual herb. Leaves oblanceolate, alternate, entire. Capitula in axillary clusters forming a more or less interrupted leafy spike, disciform. Involucres campanulate; phyllaries 2–3-seriate, scarious, pale brownish. Receptacle naked, glabrous. Corollas reddish above at anthesis. Marginal flowers female, filiform, numerous; inner flowers bisexual, few, corollas narrowly tubular-campanulate, 5-toothed. Achenes oblong, sparsely papillose; pappus simple, bristles slightly clavate, free, separately deciduous.

1. G. polycaulon Persoon; *G. indicum* sensu F.B.I. non L.

Similar to *Gamochaeta pensylvanicum*, but differing by its more compact habit; oblanceolate leaves; free, separately deciduous pappus bristles (not always apparent on flowers pressed in bud), more dilated at apex (more than 0.05mm across teeth on bisexual flowers).

Darjeeling: Sepoydhura, Mahanadi Station (both 97a). 1200–1700m. Flowering nearly all year.

No specimens seen. Described from specimens collected in W Bengal. Most records from our area proved to be of *Gamochaeta pensylvanicum*. Records from Puttabong, Darjeeling (218) and Apalchand, W Bengal Duars (175) need confirmation.

53. BRACTEANTHA Anderberg & Haegi

Annual herb. Leaves alternate, narrowly elliptic, simple, entire. Capitula (1–)several, ± corymbosely arranged, disciform; involucre broadly campanulate, white, yellow or reddish; phyllaries 6–7-seriate, stiffly scarious. Corollas yellow. Marginal flowers female, filiform. Central flowers more numerous, bisexual; corollas very narrowly tubular-campanulate, 5-lobed; style branches oblong, obtuse. Achenes oblong, obscurely 4–5-angled; pappus 1-seriate, hairs barbellate.

1. B. bracteata (Ventenat) Anderberg & Haegi

Stems 45–100cm, erect, sparsely tomentose at first. Leaves 4–8 × 0.5–1.0cm, acuminate, base attenuate, ± glabrous. Involucres 10–20mm diameter at base; outer phyllaries ovate, obtuse; middle ones oblong, c 13 × 3–5.5mm, obtuse; innermost linear-lanceolate, acute or acuminate. Corollas c 6mm. Achenes 2.5 × 0.75mm, blackish; pappus whitish or yellow, a little shorter than disc corollas.

Bhutan: C – Punakha district (Wangdu Phodrang), Tongsa district (Shamgong). Cultivated in gardens, 1450–2000m. June.

Originally native of Australia.

CALENDULEAE

54. CALENDULA L.

Annual herb; stems erect. Leaves alternate, simple, ± sessile. Capitula solitary or loosely racemose at branch ends, heterogamous, radiate. Involucre hemispherical; phyllaries 2-seriate, free. Receptacle flat, naked. Corollas yellow or orange. Ray flowers several-seriate, female; ligule 3-toothed. Disc flowers male; corolla tubular-campanulate, 5-toothed; style branches thickened, scarcely divided. Achenes heteromorphic, rather woody, hispid, tuberculate on the back; outer ones incurved, with stout, indistinct beak; intermediate semicircular or almost coiled, with broad thick lateral wing; innermost tightly coiled, with reduced wing; pappus absent.

1. C. officinalis L. Fig. 127a–d.

Plant 30–45cm; stems glandular-pubescent. Leaves oblong or oblanceolate, 3–8 × 0.7–3.5cm, obtuse, apiculate to acuminate, attenuate towards base, semi-amplexicaul, remotely denticulate, glandular-pubescent on both surfaces. Involucre 12–15mm diameter; phyllaries lanceolate or elliptic, 10–11 × 1.5–2.2mm, acuminate, glandular-pubescent. Ray flowers numerous; corolla tube 1.5mm, pubescent; ligule oblanceolate, 18–20 × 3–5mm. Disc corollas 3.5–4mm. Outer achenes c 12mm across arch × 2mm in section, intermediate 8 × 6mm across wing, innermost 5 × 2.5mm.

Bhutan: C – Tongsa district (Chendebi, Shamgong). 1980–2300m. June.

Cultivated in garden; origin of species unknown.

ASTEREAE

55. DICHROCEPHALA DC.

Erect or spreading annual herbs, puberulous to pubescent. Leaves alternate, simple or pinnatifid (often lyrate), dentate, sometimes petiolate. Capitula small, globose, disciform, in loose panicles at branch ends. Involucre c 2-seriate; phyllaries subequal. Receptacle subglobose to clavate. Corollas and achenes bearing some subsessile glands. Outer flowers numerous, in several series, female; corollas tubular, sometimes dilated above, 2–3-toothed. Central flowers fewer, bisexual; corollas tubular-campanulate, 4-toothed; style scarcely divided beyond triangular-lanceolate appendages. Achenes of both flower types laterally compressed, ribbed at margin. Pappus absent or of 2 ridges or 1–3 weak bristles shorter than corolla.

1. Female corollas dilated above, broadly and deeply 3-toothed, similar in shape to bisexual corollas but smaller .... **3. D. chrysanthemifolia**
\+ Female corollas tubular, sometimes bulbous at base, minutely 2–3-toothed, quite different from bisexual corollas 2

2. Leaves lyrately pinnatifid; female corollas loosely tubular, constricted at base .... **1. D. integrifolia**
\+ Most or all leaves simple; female corollas bulbous below, tightly tubular above .... **2. D. benthamii**

1. D. integrifolia (L.f.) Kuntze; *D. latifolia* DC. Fig. 127e–g.

Plant 13–80cm; sparsely puberulous; stems ± erect. Leaves lyrate, 3–10 × 1.5–4cm, acute, attenuate at base and usually petiolate, or truncate and petiolate; lobes serrate or rarely lobulate. Capitula 3.5mm diameter; phyllaries lanceolate to obovate, 1–1.3 × 0.4–0.6mm, margin hyaline, lacerate; receptacle subglobose. Female corollas whitish, c 0.5mm, tubular, usually not appressed to style, 2–3-toothed, suddenly narrowed at base. Bisexual corollas c 1mm, yellowish, ± tubular-campanulate, 4-toothed. Achenes obovoid, c 1mm. Pappus usually absent on female achenes, sometimes 2 ridges or 1 bristle; 2(–3) bristles on bisexual achenes.

Bhutan: S – Phuntsholing, Chukka (117) and Gaylegphug districts, **C** – Thimphu, Punakha and Tongsa (71, 117) districts, **N** – Upper Kulong Chu district; **Darjeeling:** Darjeeling, Siliguri, Tista etc.; **Sikkim:** Gangtok, Talung Chhu, Thinglen–Yoksam. Roadsides, field margins, crops, orchards and clearings in forest, 200–2500m. March–July.

Reported to occur in most administrative districts of Bhutan (272). Infraspecific division of our material is problematic. Most specimens can be referred to subsp. *integrifolia*; several have the more evenly pinnatisect leaves that distinguish subsp. *gracilis* (DC.) Fayed [*D. gracilis* DC.], but otherwise form a very heterogenous group.

2. D. benthamii Clarke

Plant branched from base, 15–20cm, pubescent; stems decumbent. Leaves obovate to ± spathulate, 15–35 × 5–12mm, obtuse, toothed above, rarely also lobed, somewhat narrowed to a broad sessile base; lowest leaves to 60 × 18mm, long attenuate at base. Capitula in small terminal cymes, soon overtopped, 3.5–4mm diameter; phyllaries oblong to obovate, 1.5 × 0.6–1mm, margin ± scarious, erose; receptacle semi-globose. Female corollas 0.5mm, white, bulbous below, narrowed, appresed to style above, minutely 3-toothed. Bisexual corollas 1.2mm, greenish, ± tubular-campanulate, with 4 broad deep teeth. Achenes yellowish, obovoid, c 1.25mm; pappus bristles absent.

Bhutan: S – Chukka district (Putlibhir), **C** – Thimphu district (Thimphu), Punakha district (Tashi Choling–Kyebaka), Tongsa district (Tongsa–Tratang, Neylong–Tashiling etc. (all 71)). Roadsides etc., 1600–2300m. April–August.

FIG. 127. **Compositae: Calenduleae and Astereae. Calenduleae.** a–d, *Calendula officinalis*: a, fruiting capitulum (× ⅔); b, outer achene (× 3); c, middle achene (× 3); d, inner achene (× 3). **Astereae.** e–g, *Dichrocephala integrifolia*: e, apex of flowering shoot (× ⅔); f, female flower (× 14); g, bisexual flower (× 14). h–i, *Aster albescens*: h, apex of flowering shoot (× ⅔); i, capitulum (× 3). j–l, *Myriactis nepalensis*: j, capitulum (× 2); k, ray flower (× 14); l, disc flower (× 14). Drawn by Margaret Tebbs.

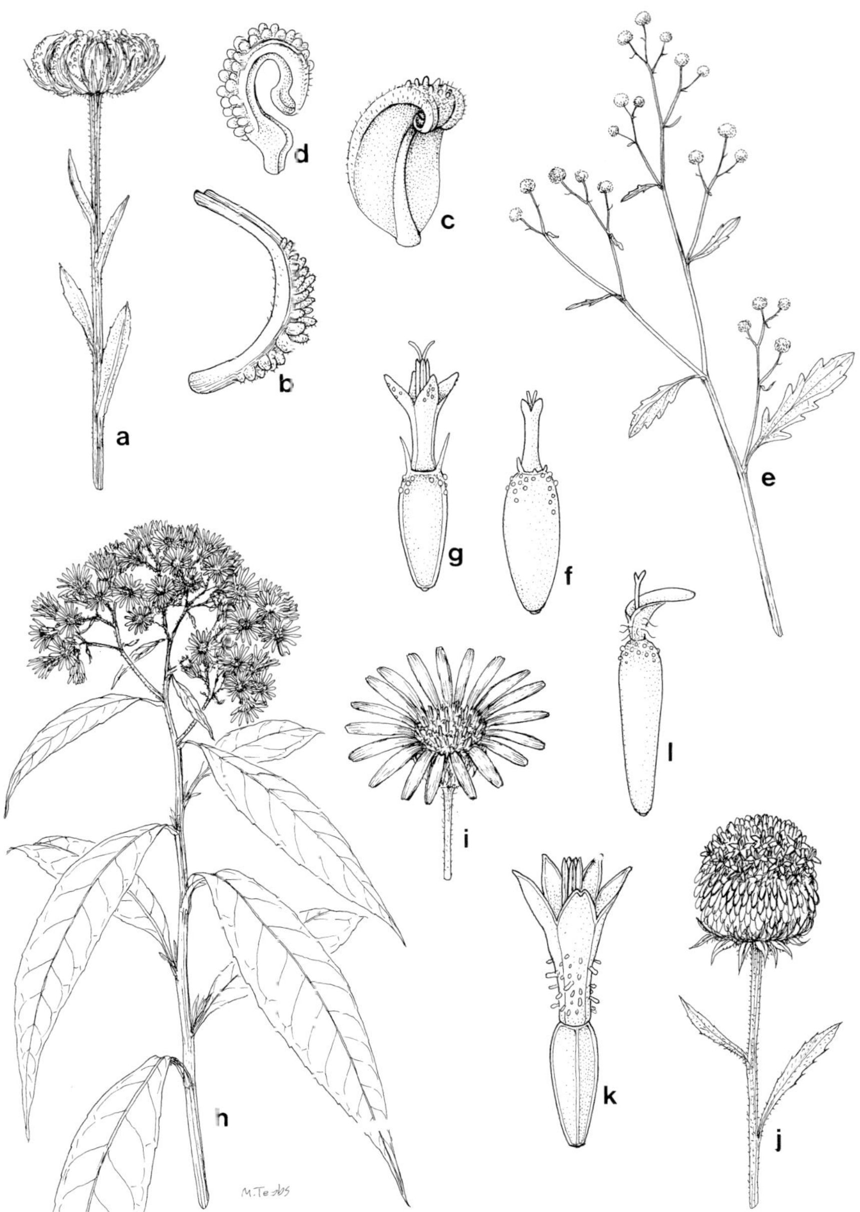

a
b
c
d
e
f
g
h
i
j
k
l
M. Tebbs

3. D. chrysanthemifolia (Blume) DC.

Plant 35–50cm, pubescent; stems erect. Leaves oblong in outline, 2–6 × 0.75–1.5cm, regularly pinnatifid with 3–4 pairs of triangular lateral segments, acute, auriculate and semi-amplexicaul at base. Capitula 6mm diameter, reddish; phyllaries oblong, 1.3–2 × 0.7–1mm including broad hyaline lacerate margin; receptacle clavate. Corollas reddish; female 0.5mm, ± tubular-campanulate, with 3 broad deep teeth; bisexual c 1mm, similar but 4-toothed. Achenes obovoid, 1.5mm; pappus bristles absent.

Bhutan: C – Thimphu district (Paro (117)). On open ground, 2500m. October.

Described from specimens from Nepal; corollas yellowish elsewhere.

56. CYATHOCLINE Cassini

Erect annual aromatic herb, mostly covered in mixture of short glandular hairs and fewer, long, multiseptate, eglandular ones. Leaves irregularly bipinnatisect, alternate, sessile. Capitula in terminal corymbs, disciform. Involucres broadly campanulate; phyllaries narrowly elliptic to oblanceolate, subscarious, c 2-seriate. Receptacle cup-shaped at centre. Corollas purplish. Female flowers numerous, mostly borne outside receptacular cup; corollas narrowly tubular, 2–3-toothed; style shortly divided. Inner flowers male, fewer, borne within cup; corolla tubular-campanulate, irregularly (4–)5-toothed, eglandular hairy above; anthers white, style undivided. Achenes minute; pappus absent.

1. C. purpurea (D. Don) Kuntze; *C. lyrata* Cassini

Stems 12–50cm, glandular pubescent. Leaves ovate to oblanceolate in outline, 2–15 × 1–5cm, with 3–6 pairs of primary segments. Capitula 2.5mm diameter; phyllaries 2.5 × 0.5mm. Female corollas filiform, c 1.2mm. Male corollas c 1.7mm. Achenes ellipsoid, 0.5mm.

Bhutan: unlocalised (80); **Darjeeling:** Pul Bazar. Rocky banks by roadside. April.

Only depauperate material seen from our area, some dimensions of specimens from Nepal included in description.

57. GRANGEA Adamson

Spreading or ascending puberulous annual herb. Leaves alternate, pinnatisect, sessile. Capitula few in compact leafy terminal or axillary corymbs, disciform. Involucre broadly campanulate; phyllaries 2-seriate, subequal, herbaceous, thickened at base. Receptacle convex, naked; flowers stipitate. Outer flowers female, numerous, in c 6 series, filiform; corolla of outermost series minutely 2-toothed, of inner series irregularly 3–4-toothed. Central flowers bisexual; corollas tubular-campanulate, 5-toothed; style branches with short triangular appendages. Achenes oblong-obovoid, 2-ribbed; pappus small, cup-shaped, fimbriate.

1. G. maderaspatana (L.) Poiret

Plant c 15cm. Leaves ± lyrate, ± obovate in outline, 1.5–5 × 0.75–2.5cm, subauriculate at base; lateral lobes up to 4 pairs, ovate or oblong, acute, with few teeth. Capitula c 6mm diameter; phyllaries oblong, 3.5–4 × 1.3mm, obtuse, erose. Corollas yellow; female ones 1.1mm; bisexual ones 1.3mm. Achenes c 1.5mm, sparsely glandular; pappus 0.4mm.

Bhutan: S – Phuntsholing district (Phuntsholing); **W Bengal Duars:** Jalpaiguri (174). Damp streamsides, 200m. May.

58. MYRIACTIS Lessing

Erect annual herbs. Leaves simple, alternate. Inflorescence of rather few capitula, usually in an open leafy panicle. Capitula radiate. Involucre globose or hemispherical; phyllaries 2–4-seriate, the inner often shorter and ± concealed. Ray flowers 2- to many-seriate, female; corolla tube obsolete. Disc flowers bisexual, tubular-campanulate, 5-toothed; style branches with short lanceolate appendages. Receptacle domed or ± conical, naked. Achenes obovate, ± compressed, bordered, umbonate and glandular at apex. Pappus absent.

The following two species show considerable overlap in characters and have sometimes been combined under *M. nepalensis* (71, 135). The literature records require confirmation.

1. Capitula globose; ray flowers c 10-seriate, ligules ovate ... **1. M. nepalensis**
\+ Capitula ± hemispherical; ray flowers 2–5-seriate, ligules ovate to narrow oblong .... **2. M. wallichii**

1. M. nepalensis Lessing. Fig. 127j–l.

Plant (10–)30–60cm; stems finely appressed pubescent. Leaves ovate-elliptic, 5–16 × 1–6cm, acute, obtuse to attenuate at base, often with petiole broadly winged at least above, semi-amplexicaul, usually serrate-dentate, rarely subentire, glabrous or pubescent on both surfaces. Capitula globose, 7–10mm diameter; outer phyllaries lanceolate, 3.5–5.5 × 1mm, acuminate. Corollas puberulous below, yellow or rays whitish, all becoming purplish with age. Ray flowers c 10-seriate; ligules ovate, 0.8–1 × 0.5–0.6mm. Tube and throat of disc corollas 1–1.2mm; teeth ovate, 0.8–1mm. Achenes 2 × 0.75mm.

Bhutan: S – Chukka district (117), **C** – Thimphu (117), Punakha, Bumthang and Tashigang (117) districts. **N** – Upper Mo Chu district; **Darjeeling:** Darjeeling, Panhkabari, Tanglu etc.; **Sikkim:** Lachen. Open scrub, grassy meadows, streamsides, 1525–3050m. April–October.

A collection from Rinchu, Punakha district with narrow-elliptic leaves, long wingless petioles and ray flowers in c 7 series is doubtfully referred to this species.

2. M. wallichii Lessing

Similar to *M. nepalensis* but leaves usually attenuate to a narrow petiolate base; capitula ± hemispherical, sometimes smaller with shorter phyllaries; ray flowers 2–5-seriate; ligules often narrower (down to 0.3mm) or longer (to 2mm), pinkish-white; achenes often 2.5 × 1mm.

Bhutan: C – Thimphu district (Motithang, Taksang, Thimphu); **Darjeeling:** Sandakphu–Sabargam, Sabargam–Phalut, Pul Bazar etc.; **Sikkim:** Lachen, Lachung, Phusum. Forests, clearings, roadsides, 1500–3650m. July–October.

A population from Thimphu with lyrate leaves is provisionally referred here. A record of *M. wightii* DC. from Rani Camp, Gaylephug district, Bhutan (117) may also refer to this species. *M. wightii* is native to southern India and Sri Lanka and is distinguished by its longer ligules (c 3.5mm) and often lyrate leaves.

59. RHYNCHOSPERMUM Reinwardt

Erect perennial herb; stems stiffly branched above. Leaves alternate, simple, petiolate. Capitula racemosely arranged on upper branches, radiate. Involucre campanulate (substellate in fruit); phyllaries 3-seriate; receptacle naked. Ray flowers 2-seriate, female; ligules elliptic, minutely 3-toothed. Disc flowers bisexual, few, tubular-campanulate; style branches flattened, with obtuse tips. Achenes compressed, beaked, subsessile-glandular; pappus of several weak caducous bristles.

1. R. verticillatum Reinwardt

Plant c 90; stems finely pubescent. Leaves elliptic-lanceolate, 5–10 × 1–2.5cm on main stems, acuminate, cuneate at base, denticulate to finely serrate, scabridulous above, sparsely puberulous and densely subsessile-glandular beneath; petiole slender, to 7mm. Involucre 3mm across; phyllaries oblong, 1.5–2mm, obtuse, scarious-margined. Corollas white, tubes subsessile glandular. Tube of ray corollas 0.7mm; ligule c 1.1 × 0.7mm. Disc corollas 2.5mm. Achenes obovoid, sparsely subsessile-glandular; body c 3mm; beak of ray achenes slender, to 1.5mm, of disc achenes very stout, 0.5mm; pappus bristles c 7, whitish, c 2mm.

Bhutan: N – Upper Mo Chu district (Mo Chu); **Sikkim:** Lachung Chhu. Woods and wet jungle, 1600–1850m. October (last flowers).

60. HETEROPAPPUS Lessing

?Annual, rather woody at base; most parts bearing eglandular and glandular hairs; stems prostrate or ascending. Leaves simple, sessile, entire. Capitula radiate, solitary or few at branch ends. Involucre 3-seriate, broadly campanulate at first; phyllaries mostly lanceolate, herbaceous, inner ones with scarious margins. Ray flowers female, ± 2-seriate; ligules bluish. Disc flowers bisexual; corollas tubular-campanulate, yellow, 5-toothed, with 1 tooth more deeply cut than other 4; style branches ± oblong, flattened, appendages triangular. Achenes obovoid, compressed, densely appressed pubescent. Pappus simple, of many free persistent bristles; length of ray pappus ¾ of disc pappus.

1. H. gouldii (C.E.C. Fischer) Grierson

Plant 8–15cm; stems spreading pubescent, intermixed with stipitate glands above. Leaves oblanceolate, 1–2.5 × 0.3–0.5cm, ± obtuse, narrowed at base, appressed hispid pubescent at least above. Phyllaries 5.5–7 × 1.2–2.2mm, mostly glandular pubescent. Corolla tubes sparsely hairy. Ray flowers 20–40; ligules 10–13 × 2mm. Disc corollas 5.5–6mm, teeth pubescent outside below apex. Achenes 2–2.5mm; pappus brownish, those of ray achenes 3–3.7mm, those of disc achenes 4.2–5mm.

Sikkim: Dotha, Kamba. 3655–5340m. August–September.

Pappus of ray and disc achenes usually equal outside our area.

61. ASTER L.

Perennial herbs or shrubs, often rhizomatous, sometimes stoloniferous, rarely tuberous-rooted. Leaves alternate, simple, sometimes in basal rosettes. Capitula 1–many, radiate. Involucre several-seriate; phyllaries generally relatively broad, subequal or imbricate, ± herbaceous. Receptacle flat or convex. Ray flowers female, usually rather few, sometimes to c 120; ligules white, mauve or bluish. Disc flowers bisexual; corolla tubular-campanulate, 5-toothed at apex, usually yellow, sometimes purplish; style branches flattened with a sterile lanceolate or triangular appendage. Achenes oblong or obovoid, compressed, with 2 or more ribs. Pappus bristles scabrous, ± persistent, white or brownish, simple, with all bristles of similar length, as long as or less than half as long as disc corollas, or double, with an outer series of short (c 2mm) often subpaleaceous bristles.

1. Pappus hairs less than half as long as disc corollas .... **16. A. souliei**
+ Pappus hairs, at least the inner ones, as long as disc corollas 2

2. Pappus simple, all hairs ± same length and thickness 3
+ Pappus double, outer elements shorter and sometimes broader than inner 14

3. Capitula c 5–many; involucres corymbose, c 5–8mm diameter 4
+ Capitula 1–3(–5); involucres usually more than 10mm diameter 8

4. Leaves distinctly 3-veined above the base; herbs or subshrubs usually more than 30cm .... **2. A. ageratoides**
+ Leaves pinnately veined or basally 3-veined 5

5. Shrubs (0.3–)1–2m tall .... **3. A. albescens**
+ Herbs, sometimes woody at base, usually less than 1m tall 6

6. Inflorescence ± racemose; peduncles conspicuously several-bracteate; phyllaries with distinct diamond-shaped green apical patch .... **5. A. laevis**
+ Inflorescence corymbose; peduncles with 0–1 small linear bracts; phyllaries diffusely green, not in diamond-shaped patch 7

7. Stems ± straight; ligules 3–4-veined .... **1. A. vestitus**
+ Stems flexuose; ligules distinctly 2-veined .... **4. A. sikkimensis**

8. Pappus hairs whitish 9
+ Pappus hairs brownish 12

9. Phyllaries c 1mm broad; generally slender plants, sometimes stoloniferous, with (1–)2–3(–5) capitula; leaves entire .... **7. A. neoelegans**
+ Phyllaries 2–4mm broad; generally stouter plants with 1 or sometimes 3 capitula; leaves serrate (sometimes entire in *A. tricephalus*) 10

10. Flowering stem borne terminally on rhizome .... **6. A. tricephalus**
+ Flowering stem borne laterally under basal leaf rosette 11

11. Plants not stoloniferous; cauline leaves 1.5–4cm .... **9. A. himalaicus**
+ Plants stoloniferous; cauline leaves seldom more than 1cm **12. A. stracheyi**

12. Dwarf, flowering stems less than 10cm; plants densely white villous **11. A. prainii**
+ Moderate sized, flowering stems 15–30cm; plants not white villous 13

13. Phyllaries 1–2mm broad; ligules c 2mm broad .... **8. A. barbellatus**
+ Phyllaries 4–7mm broad; ligules c 0.7mm broad .... **10. A. heliopsis**

14. Ligules c 2cm; stem surrounded at base by a collar of leaf remains **15. A. diplostephioides**
+ Ligules c 1cm; stems bases not surrounded by leaf remains 15

15. Roots not tuberous; disc corolla teeth yellow or orange; phyllaries ± shaggily pubescent .... **13. A. flaccidus**
+ Roots tuberous; disc corolla teeth yellow or purplish with pale or dark glands or eglandular hairs; phyllaries glandular .... **14. A. asteroides**

1. A. vestitus Franchet; *A. sherriffianus* Handel-Mazzetti

Erect perennial; stems 60–130cm, densely glandular, usually intermixed with spreading eglandular hairs. Leaves elliptic-lanceolate, 2.5–6.5 × 0.45–1.5cm, acute, rounded at base, sessile, with a few small teeth, glandular pubescent on both surfaces, intermixed with eglandular hairs especially beneath. Capitula ±

numerous, in terminal corymbs; involucres c 8mm diameter; phyllaries linear-lanceolate, imbricate, 4.5–5.5 × 0.5–1.2mm, acute, glandular. Ray flowers 20–30; corolla tube 1.5mm; ligule white, mauve or blue-violet, 10 × 2mm. Disc corollas yellow, 4.5mm. Achenes obovate, 2.5 × 1mm, appressed pubescent; pappus white, simple, ± as long as disc corollas.

Bhutan: C – Ha, Thimphu, Tongsa and Bumthang districts, **N** – Upper Mo Chu district; **Chumbi:** Chumbi, Lingmatang, Yatung etc. Open banks, forest margins, 2350–3200m. August–October.

A. nakaoi Kitamura has been described from Chile La, Ha/Thimphu district, Bhutan, 3000m, stated to differ from *A. vestitus* by its more herbaceous phyllaries, with only the margins of the middle and inner series scarious and by the leaves densely covered with short hairs only (glandular and eglandular). No examples have been seen, but the remaining details of the description fall within the range of *A. vestitus* as described here.

2. A. ageratoides Turczaninov; *A. trinervius* D. Don *nom. illeg.* non *A. trinervis* Nees

Perennial suffrutescent herb; stems 0.3–2m, finely pubescent above. Leaves ovate-elliptic, 4.5–10 × 1.5–2cm, acuminate, attenuate at base, with few coarse teeth, pinnately veined but superficially 3-veined with 1 pair of veins near base more prominent than rest, usually finely pubescent and glandular beneath. Capitula in lax corymbs; involucres c 6mm diameter; phyllaries imbricate, 3-seriate, spathulate or linear-lanceolate, 4–5.5 × 0.7mm, acute or obtuse, usually ciliate. Ray flowers 10–20; corolla tube 2–3mm; ligules usually white, sometimes pale pink or mauve, 7–10 × 1.5–3mm. Disc corollas yellow, 5–6.5mm, lobes often glandular. Achenes obovoid, 2.5–3 × 0.75–1mm, sparsely silky pubescent, sometimes glandular; pappus whitish, 4.5–6mm, simple.

Bhutan: C – Thimphu district (Paro, Simtoka, Chapcha etc.), **N** – Upper Mo Chu district (Kohina); **Sikkim:** Chhateng, Lachen. Forest margins, clearings, among shrubs, 1800–3650m. June–November.

3. A. albescens (DC.) Koehne; *Microglossa albescens* (DC.) Clarke. Fig. 127h–i.

Erect shrub; stems 0.5–2m, sparsely pubescent intermixed with glands, rarely white tomentose. Leaves ovate-lanceolate, 3–10 × 0.8–2.5cm, acute or acuminate, cuneate at base, sessile or shortly petiolate, entire or finely serrate, sparsely pubescent beneath and sometimes bearing a few glands. Capitula usually numerous, in terminal corymbs; involucres c 5mm diameter; phyllaries usually c 20, imbricate, linear-lanceolate, 3.5–4.5 × 0.5mm, green or purplish at tip, pubescent or tomentose towards base. Ray flowers 12–30; corolla tube c 2.5mm; ligules blue, mauve or white, 4–4.5 × 0.6–1.2mm. Disc corollas yellow, 4.5mm. Achenes 1.75–2.5 × 0.5mm, sparsely pubescent, often glandular at apex; pappus brownish, 3.5–4mm, simple.

Bhutan: S – Chukka district, **C** – Ha, Thimphu, Tongsa and Bumthang districts, **N** – Upper Bumthang Chu and Upper Kulong Chu districts; **Sikkim:**

Lachen, Lachung, Lower Zemu Valley etc.; **Chumbi:** Chumbi. Clearings, forest margins, rocky scrub, 1750–3650m. May–September.

Leaves of specimens from Changka and Tongsa, Tongsa district, Bhutan and ?Sikkim (specimen (not seen) supporting var. *niveus* Handel-Mazzetti *nom. inval.*) are covered with a fine white or grey tomentum beneath. Of these the specimen from Changka has c 60 ligules and the one from Sikkim only 10–12 phyllaries.

4. A. sikkimensis Hooker

Erect perennial; stems 50–80cm, pubescent, sometimes densely so above. Leaves ovate-lanceolate, 5–12 × 1.5–5cm, acuminate, ± sessile but usually contracted above semi-amplexicaul, somewhat auriculate base, denticulate, very sparsely pubescent, with scattered glands beneath. Capitula numerous, in corymbs; involucre c 8mm diameter, imbricate, 3-seriate; phyllaries lanceolate, 3–4 × 0.5mm, often purplish, sparsely pubescent. Ray flowers 40–60; corolla tube 1.5mm; ligule white, rose or purple, 6–7 × 0.7–1mm. Disc corollas 3.5–4mm; teeth pale at first then purplish on opening. Achenes obovate, 2 × 1mm, pilose; pappus brownish, 3–3.5mm, simple.

Sikkim: Chiya Bhanjan, Kalej Khola, Singalila etc. 3050–3650m. October.

5. A. laevis Torrey & Gray

Erect perennial; stems 30cm or more, glabrous. Leaves oblanceolate, 7–10 × 1.5–2cm, acute or acuminate, attenuate to a broad subpetiolate semi-amplexicaul base, finely serrate, glabrous. Capitula several to numerous, in leafy racemes; pedicels 2–3cm, bearing several ovate bracts, sparsely pubescent; involucre c 8mm diameter, imbricate, 3–4-seriate; phyllaries linear-lanceolate, 3–6 × 1mm, with diamond-shaped green patch at apex. Ray flowers 25; corolla tube 3mm; ligule bluish-purple, 12 × 2mm. Disc corollas yellow, 7mm, teeth glabrous. Achenes obovoid, 2 × 1mm; pappus white or reddish, 5mm, simple.

Bhutan: C – Thimphu district (Taba). Cultivated ornamental, 2400m. July. Native of N America.

6. A. tricephalus Clarke

Erect perennial; stems 30–40cm, simple. Leaves narrowly spathulate-oblanceolate, 3.5–6.5 × 0.5–2cm, upper ones often surrounding capitula, obtuse or acute, attenuate to sessile semi-amplexicaul base, entire or 2–3-toothed on either margin, sparsely pubescent on upper surface, subglabrous beneath. Capitula 1–3; involucres c 20mm diameter, 2–3-seriate; phyllaries subequal, 12–14 × 1.5–2mm, linear-lanceolate, acute, sparsely pubescent. Ray flowers 50–60; corolla tube 2mm; ligule white or blue, 18 × 1.5mm. Disc corollas yellow, 5mm, glandular in lower half. Achenes obovoid, 3.5 × 1mm, at least sparsely pilose and with a few glands at apex; pappus whitish, 5mm, simple.

Bhutan: C – Thimphu district (Paro (117)); **Darjeeling:** Sandakphu–Phalut,

Manibhanjan–Tanglu (101) etc ; **Sikkim:** Tari, Botak, Rathong Chhu etc. 2400–3950m. July–October

7. A. neoelegans Grierson; *A. elegans* Clarke non Willdenow, *A. tricephalus* sensu F.B.I. non Clarke

Erect stoloniferous perennial; stems 40–75(–100)cm, simple. Leaves oblanceolate or narrowly spathulate, 2.5–7 × 0.4–1cm, acute or obtuse, attenuate to sessile base, entire, finely pubescent. Capitula 1–2(–5); involucre c 8–13mm diameter, 3–4-seriate; phyllaries ± oblong to linear, 4–8 × 1–1.5mm, acute or acuminate, sparsely appressed pubescent. Ray flowers 30–40; corolla tube 2–3mm; ligule white, mauve or blue-violet, 7–12 × 1mm. Disc corollas yellow, 4–5mm, teeth glandular. Achenes obovoid, 2.5–3mm, sparsely pubescent, with few glands at apex; pappus whitish, 4–5mm, simple.

Bhutan: C – Thimphu district (Thimphu), Punakha district (Hinglai La–Nahi), Tongsa district (Rukubji, Tashiling–Chendebi, Yuto La), Bumthang district (Byakar, Dhur), **N** – Upper Mo Chu district (Gasa); **Darjeeling:** Pangrizampa; **Sikkim:** ?Lachen, ?Yumthang; **Chumbi:** Yatung. Marshy ground, meadows, forest clearings, open hillsides, 2450–3200m. May–September.

8. A. barbellatus Grierson

Erect perennial; stems 30–60cm, simple, puberulous (hirtellous) and sparsely glandular short-stipitate. Leaves spathulate, 3–6.5 × 1–1.2cm, obtuse or acute, long attenuate to sessile base, entire, sparsely pubescent. Capitulum solitary; involucre 12–20mm diameter, 2–3-seriate; phyllaries linear-oblong, subequal, 10 × 1.5–2.5mm, acute (rounded), ciliate. Ray flowers 30–40; corolla tube 2mm; ligules mauve or blue, 15–20 × 2–2.5mm. Disc corollas yellow, 6.5mm, teeth apparently sometimes purplish at least in bud, glandular. Achenes obovoid, 3mm, finely pubescent and glandular; pappus reddish, 4–5.5mm, bristles finely barbellate above.

Bhutan: N – Upper Pho Chu district (Leji, Choju Dzong), Upper Mangde Chu district (Dur Chutsen); **Sikkim:** Lingmuthang; **Chumbi:** Cho-lih La, Chumbi, Yatung. Open grassy slopes and moist ground, 3050–3950m. June–September.

A. ionoglossus Ling has been described from Chumbi. No examples have been seen, but its characters all fall within the range of those of *A. barbellatus* as described here, apart from a yellowish pappus.

9. A. himalaicus Clarke

Perennial herb; stems (3–)8–25cm, ± erect, arising laterally from base of leaf rosette. Radical leaves usually lacking on flowering stems, on sterile shoots obovate-elliptic, 2–3.5 × 0.8–2.5cm, obtuse or acute, attenuate to 1–4.5cm petiole, entire or remotely denticulate, sparsely pubescent, with scattered glands beneath; cauline leaves oblong-obovate or lanceolate, 1.3–5.2 × 0.5–2cm, acute or obtuse, rounded at base, sessile, semi-amplexicaul. Capitulum solitary;

involucre 12–17mm diameter, 2–3-seriate; phyllaries oblong-lanceolate, 9–12 × 2.5–3.5mm, usually purplish, sparsely pubescent and glandular. Ray flowers 50–70; corolla tube 2mm; ligule purplish-blue, 11–17 × 1.25–2mm. Disc corollas yellow or purplish, 6.7–7.5mm, ± pubescent at base. Achenes obovoid, 2.5–3 × 1mm, sparsely pubescent, glandular near apex; pappus whitish, 5.5mm, usually simple or sometimes with an outer series of bristles.

Bhutan: C – Thimphu district (Pajoding, Paro Pass), **N** – Upper Mo Chu district (Jari La, Laya, Lingshi (101)), Upper Mangde Chu district (Phage La, Passu Sefu), Upper Bumthang Chu district (Domchen); **Sikkim:** Megu, Nathu La, Zongri etc.; **Chumbi:** Cho La, Yatung. On open or scrub or rhododendron-covered slopes, in turf or gravel or on cliffs, 3750–4900m. August–October.

10. A. heliopsis Grierson; *Doronicum latisquamatum* C.E.C. Fischer, *A. platylepis* Chen

Similar to *A. himalaicus* but stems 15–35cm; radical leaves oblong or ovate, 2–4 × 1–3cm, ± obtuse, rounded or truncate at base, entire or denticulate, petioles 2.5–5cm; cauline leaves elliptic, 2–4 × 1.2–2cm, sessile; involucre 1.5–20mm diameter, 2–3-seriate; phyllaries ovate-lanceolate, 15 × 3–6mm, acute, longer than disc flowers; ray flowers 100–120, corolla tubes 2mm, ligules bluish, 15 × 0.5mm; disc corollas yellow, 4–4.5mm, glabrous; achenes obovoid, c 5 × 1.5mm, sparsely pilose and glandular; pappus simple, 5mm, brownish.

Bhutan: C – Tongsa district (Taktse La); **Sikkim:** Onglakthang, Tsomgo Chho. Cliff-faces, rocky shale around lake, 3650–4570m. August.

11. A. prainii (Drummond) Chen; *Chlamydites prainii* Drummond, *Wardaster lanuginosus* J. Small, *A. lanuginosus* (J. Small) Ling non Wendland

Rhizomatous; stems up to 5cm, densely white felted. Leaves mostly broad spathulate, 2–3 × 1–1.5cm, obtuse, attenuate at base, ± entire, densely whitish felted on both sides. Capitula solitary, broadly campanulate or hemispherical; involucre c 17mm diameter; phyllaries 2–3-seriate, lanceolate, 8–11 × 1.2–2mm, densely white pubescent outside. Corollas puberulous below. Ray flowers 1-seriate; corolla tube 3.5mm; ligule oblong, blue-violet, 12 × 2.5mm. Disc corollas brownish, 7.2mm, teeth pilose or subglabrous. Achenes (immature) 3mm, appressed pubescent; pappus simple, brownish, 7mm.

Bhutan: N – Upper Bumthang Chu (Marlung, Waitang). Open sandy screes, crevices, cliff-ledges, 4400–4700m. July–September.

Specimens with violet disc corollas occur immediately north of Sikkim at Khamba Dzong.

12. A. stracheyi Hook.f.

Similar to *A. himalaicus* but bearing stolons up to 30cm. Flowering stems 3–18cm, erect or ascending, springing from under basal leaf rosette, often reddish, sparsely pubescent. Basal leaves spathulate, blade 1–3.5 × 0.5–1.3cm, obtuse or acute, attenuate to petiole 1–6.5cm, entire or serrate with a few teeth mostly in upper half, glabrous or sparsely pubescent; cauline leaves oblanceolate, 9–20 × 1.5–8mm, acute or obtuse, sessile, margin entire or with a few teeth at apex. Capitulum solitary, involucre 10–15mm diameter, 2–3-seriate; phyllaries lanceolate, 8–12 × 2.5–3.75mm, glabrous or ciliate, usually purplish. Ray flowers 30–40; corolla tube 1.5mm; ligule blue or mauve, 9–13 × 1.75–2mm. Disc corolla yellow, 5.5mm. Achenes obovoid, 2.5–3 × 1mm, densely whitish pubescent; pappus simple, 4.5–5mm, whitish.

Bhutan: C – Thimphu district (Barshong), Tongsa district (Takse La), **N** – Upper Mo Chu district (Lingshi), Upper Mangde Chu district (Saga La); **Sikkim:** Kangling, Onglakthang. Grassy rock-crevices etc., 3800–4722m. July–August.

13. A. flaccidus Bunge subsp. **flaccidus**; *A. heterochaeta* sensu F.B.I. non (DC.) Clarke, *A. tibeticus* Hook.f. p.p.

Erect perennial; stems 5–15cm, whitish pubescent and sometimes glandular. Basal leaves spathulate or oblanceolate, 1.7–6.5 × 0.8–1.5cm, acute or obtuse, base attenuate and shortly petiolate, pubescent on both surfaces; cauline leaves few and smaller, narrowly lanceolate. Capitulum solitary, involucre 12–17mm diameter, 2–3-seriate; phyllaries lanceolate, 8–10 × 1.5–2mm, ± shaggily pubescent at least at base. Ray flowers 60; corolla tube 1–2mm; ligule blue or mauve, 10–15 × 1–1.5mm. Disc corollas yellow, 6mm, usually sparsely eglandular pubescent on and below teeth. Achenes narrowly obovoid, 2.5mm, sparsely appressed pubescent and glandular; pappus double, whitish, inner series 7mm, outer subpaleaceous, 2mm.

Bhutan: C – Thimphu district (Ceka); **N** – Upper Paro Chu district (Gafoo La), Upper Bumthang Chu district (Marlung); **Darjeeling:** Phalut–Ramman (101); **Chumbi:** Phari, Chomolhari, Chugya. In peaty turf, on open sandy hillsides, 3750–5025m. July–September.

The specimens cited from Bhutan and Chomolhari belong to fm. **griseobarbatus** Grierson with phyllaries greyish shaggy pubescent throughout.

14. A. asteroides (DC.) Kuntze; *A. heterochaeta* (DC.) Clarke *nom. illeg.*

Erect perennial herb with tuberous roots; stems 3–15cm, pubescent and glandular. Basal leaves ovate-elliptic, 2–4 × 0.8–1.7cm, acute or obtuse, gradually attenuate at base, subpetiolate, usually entire, ± hoarily pubescent on both sides; cauline leaves few, lower ones similar to basal leaves, upper ones linear-lanceolate. Capitulum solitary, involucre 10–17mm diameter, 2–3-seriate; phyllaries oblong-lanceolate, 7–10 × 1.5–2mm, glandular pubescent especially at base. Ray flowers 30–50, corolla tube 1.5mm, ligule blue or purplish, 9–12 × 1.5–2mm. Disc corollas yellow or purplish, 4–5mm, teeth with pale or dark glands or eglandular hairs. Achenes obovoid, 2 × 0.75mm (?immature), sparsely pale pubescent; pappus double, inner bristles whitish, ± as long as disc, outer series paleaceous, 1–1.5mm.

Bhutan: C – Thimphu district (Paro Chu, Tremo La), Tongsa district (Black

Mountain), **N** – Upper Mo Chu district (Lingshi, Shingche La), Upper Pho Chu district (Kangla Karchu La); **Sikkim:** Gochung, Lhonak; **Chumbi:** Dotag, Phari, Tang La. Open marshy ground, steep, rocky slopes, 3650–4760m. May–September.

There are two subspecies: subsp. **costei** (Léveillé) Grierson with disc flowers purplish, teeth glabrous, glandular-pubescent or sparsely white-pilose; and subsp. **asteroides** with disc flowers yellow and sparsely covered with short blackish hairs. Examples of both are known from Bhutan, Sikkim and Chumbi.

15. A. diplostephioides (DC.) Clarke

Erect perennial herb; stems 15–40cm, surrounded at base by layer of leaf remains, glandular and pubescent above. Basal leaves lanceolate or oblanceolate, 6–13 × 1.3–2cm (including petiole 4–7cm), acute, attenuate at base, usually entire, sometimes denticulate, glandular or sparsely pubescent; cauline leaves oblong or linear, decreasing in size above; upper part of stem ± naked. Capitulum solitary; involucre 17–22mm diameter, ± 2-seriate; phyllaries oblong-lanceolate, 10–12 × 3–4.5mm, acuminate, blackish glandular and bearing long white hairs especially towards base. Ray flowers ± numerous, 1–2-seriate; corolla tube 1.5mm; ligules mauve or blue, 20–30 × 1.25–2.5mm. Disc corollas blackish-purple fading to orange, 5mm, teeth glabrous. Achenes obovoid, 3–3.5 × 1–1.5mm, glandular and whitish pubescent; pappus white, double, inner series 5mm, outer series 1–1.5mm, subpaleaceous.

Bhutan: C – Ha and Thimphu districts, **N** – Upper Mo Chu, Upper Pho Chu, Upper Bumthang Chu, Upper Kuru Chu and Upper Kulong Chu districts; **Sikkim:** ?Thanggu; **Chumbi:** Lingmatang, Yatung. Grassy slopes and screes, 3200–4730m. June–September.

16. A. souliei Franchet

Erect perennial; stems 5–15cm, pubescent or ± glabrous. Basal leaves oblanceolate or spathulate, 2–6.5 × 0.7–1.5cm, obtuse or acute, attenuate at base, entire, sparsely pubescent and ciliate with long white hairs; cauline leaves oblong or linear, decreasing in size above. Capitula solitary, involucre c 12mm diameter, 2–3-seriate; phyllaries linear-lanceolate, 7–10 × 1.5mm, acute or obtuse, pubescent and ciliate. Ray flowers 40–60; corolla tube 1.5–2mm; ligule blue or mauve, 10–15 × 1.5–2mm. Disc corollas yellow, 3–4mm. Achenes obovoid, 1.5 × 0.5mm, sparsely pubescent; pappus simple, brownish, 1.2–1.8mm.

Bhutan: C – ?Bumthang district (Bumthang–Tsampa (Upper Bumthang Chu district)), **N** – Upper Mangde Chu district (Saga La), Upper Bumthang Chu district (Pangotang, Tolegang), Upper Kulong Chu district (Me La). On open grassy hillsides, 3800–4722m. May–July.

62. CALLISTEPHUS Cassini

Annual; stem erect, branched above or throughout; branches stiffly ascending. Leaves simple, alternate, mostly petiolate. Capitula solitary at branch ends, hemispherical, radiate. Involucre several-seriate; outer phyllaries herbaceous, exceeding inner scarious ones. Receptacle flat, naked. Ray flowers 2–3-seriate, female; ligules violet, showy. Disc flowers bisexual; corolla tubular-campanulate, yellow, 5-toothed; style branches narrowly oblong with ovate acute appendage. Achenes ± compressed; pappus of many bristles, double, outer very short, inner ± as long as disc corollas, deciduous.

1. C. chinensis (L.) Nees

Plant c 45cm; stems sparsely pubescent. Leaves ovate, 5–12 × 2–5cm, ± acute, obtuse at base, coarsely and irregularly dentate, ± glabrous; petioles up to 6cm. Involucres 3cm diameter; phyllaries oblanceolate, 10–19 × 3–5mm, ciliate. Corollas sparsely hairy below. Tube of ray corollas 3–4mm; ligule 20–25 × 5–7mm. Disc corollas 5mm. Achenes obovoid, 3.5mm, sparsely sericeous; pappus white, outer series 0.2mm, inner ones 4mm.

Bhutan: C – Thimphu district (Thimphu). In garden, 2370m. September.

Highly selected form, cultivated as an ornamental; native to China. Forms with capitula much smaller and ligules smaller, pink or white, or 1- or many-seriate have been grown in India outside our area.

63. BRACHYACTIS Ledebour

Perennial herb; stems erect, usually simple. Leaves alternate, simple. Capitula axillary and in terminal corymb, sometimes forming an open leafy panicle, radiate. Involucre campanulate, soon spreading, imbricate, 2–3-seriate; receptacle flat, naked, fimbrillate. Ray flowers 2–3-seriate, female; ligule small. Disc flowers fewer, bisexual; corolla tubular-campanulate, (4–)5-toothed; style branches flattened, with ovate-lanceolate appendage. Achenes obovoid, flattened, pubescent; pappus of bristles, 2-seriate, outer series much shorter, incomplete.

1. B. anomala (DC.) Kitamura; *B. menthodora* Bentham

Stems 30–50cm, purple, finely pubescent. Basal leaves ovate; blade 2.5–6 × 2–4cm, obtuse, ± truncate at base, crenate, glabrescent; petiole 4–10cm. Cauline leaves lanceolate with broad-winged petiole to panduriform, sessile, 3–8 × 1–2.5cm, acute or subobtuse, shortly decurrent at base, crenate to coarsely serrate, finely pubescent (mostly glandular) on both faces. Involucres c 4mm diameter at base; phyllaries linear-lanceolate, 6–8 × 1–1.5mm, mostly glandular pubescent. Ray flowers c 35–40; corolla tubes 3.5mm; ligule white, later turning purplish, 3.5 × 0.2–0.5mm, simple, deeply divided or irregularly 2–3-toothed. Disc flowers c 7–10; corolla yellowish, 5–6mm. Achenes 3 × 1mm; pappus brownish, inner series c 4.5mm, outer series usually c 1mm.

Bhutan: C – Thimphu district (Pumo La); **Darjeeling:** Phalut, Singalila; **Sikkim:** Boktak, Kupup, Tsomgo etc. Open grassy slopes, 3050–4100m. August–October.

64. ERIGERON L.

Annual or perennial herbs; stems prostrate, ascending or erect, simple or branched. Leaves alternate, simple, entire, toothed or lobed. Capitula radiate, solitary or in loose racemes. Involucres hemispherical; phyllaries 3–4-seriate, imbricate, herbaceous. Receptacle flat or slightly convex, naked. Flowers dimorphic or trimorphic. Ray flowers usually numerous, female; ligules purplish or white, narrow. Disc flowers bisexual; corolla tubular-campanulate, 5-lobed; style branches with short subacute or obtuse appendage. Eligulate female flowers sometimes present between ray and disc flowers, with short filiform corollas (species 2 and 3, sometimes separated under *Trimorpha* Cassini). Achenes obovoid, compressed; pappus ± as long as disc flowers, simple or double with an outer series of shorter bristles.

Species 4 and 5 most closely resemble species of *Aster*, but can be distinguished by their relatively narrower phyllaries and ligules as well as by their shorter style branch appendages.

1. Stems prostrate or ascending; basal leaves 3(–5)-lobed; ligules white or sometimes pink, 5mm .... **1. E. karvinskianus**
+ Stems ± erect; basal leaves unlobed or ligules purplish or blue; ligules 2–4 or 8–15mm 2

2. Eligulate female flowers present between ray and disc flowers 3
+ Eligulate female flowers absent 4

3. Leaves entire, glabrous or sparsely pubescent .... **2. E. acer**
+ Leaves coarsely toothed, white pubescent .... **3. E. sublyratus**

4. Stems 15–45cm; involucres 1.5–2.5cm across; rays numerous, 120–200; ligules 8–15mm .... **4. E. multiradiatus**
+ Stems 10–15cm; involucres 1–1.5cm across; rays fewer, 50–100, ligules c 10mm .... **5. E. bellidioides**

1. E. karvinskianus DC.; *E. mucronatus* DC.

Rhizomatous perennial; stems prostrate or ascending, 20–30(–45)cm, sparsely pubescent. Basal leaves obovate, 3(–5)-lobed, 3–4.5 × 1–2cm, acute, attenuate at base, sparsely pubescent; upper leaves elliptic, 1–2.5 × 0.2–0.3cm, entire. Capitula usually solitary, terminal or axillary on peduncles 5–10cm; involucre c 6mm diameter, 3–4-seriate; phyllaries lanceolate, c 4 × 1mm, acuminate, sparsely pubescent, mid-vein brownish. Flowers dimorphic. Ray flowers 2-seriate; corolla tube 2mm; ligule white, sometimes becoming pink on fading, 5 × 0.4mm. Disc corollas yellow, 3mm. Achenes oblong, 1 × 0.4mm, sparsely minute pubescent; pappus 1-seriate, 3mm.

Bhutan: S – Phuntsholing district (Phuntsholing); **Darjeeling:** Bhotia Basti,

Chimli (97a), Kurseong (97a) etc.; **Sikkim:** Gezing, Yoksam, Gangtok (97a). 200–2050m. All year, peak April–May.

A native of C America; sometimes planted along irrigation ditches to stabilise banks.

2. E. acer L. var. **multicaulis** (DC.) Clarke; *E. alpinus* L. var. *multicaulis* DC.

Perennial; stems 10–30cm, simple, sparsely pubescent, branched above and bearing several capitula. Leaves ovate to oblanceolate, 2–5.5 × 0.4–0.7cm, acute, attenuate to short petiole or sessile, semi-amplexicaul, glabrous or sparsely pubescent. Capitula c 7.5mm diameter, on peduncles up to 9cm; involucre 3–4-seriate; phyllaries linear-lanceolate, 3–7 × 0.7mm, acuminate, often purplish, minutely glandular and sparsely pubescent. Flowers trimorphic. Ray flowers numerous; corolla tube 4–5mm; ligule purplish or white, 3.5–4 × 0.4–0.5mm. Eligulate female flowers present; corollas 3–4mm. Disc corollas yellow, 4.5–5mm. Achenes obovoid, 2 × 0.7mm, pubescent; pappus simple, 5–6mm, white or reddish.

Bhutan: Thimphu/Chukka district (Thimphu–Chima Kothi (71)); unlocalised (Griffith Bhutan collection). 2250–2450m. June.

3. E. sublyratus DC.; *E. asteroides* Roxb. non Besser

Annual; stems up to 45cm, spreading whitish pubescent. Leaves oblanceolate or spathulate, 3–6 × 0.6–2.5cm, acute or obtuse, auriculate and semi-amplexicaul at base, coarsely toothed, spreading white pubescent on both surfaces. Capitula solitary on axillary peduncles or several in loose racemes; involucres 7.5mm diameter, 3-seriate; phyllaries linear-lanceolate, 3–4 × 0.5mm, acuminate, whitish hoary pubescent. Flowers trimorphic. Ray flower corolla tubes 2.5mm; ligules purplish, 2–3 × 0.3–0.4mm. Eligulate female flowers present; corollas 1.5–2mm. Disc flowers 15–20; corollas yellow, 3.5–4mm. Achenes narrowly obovoid, c 0.75mm, puberulous; pappus whitish, 3–4mm.

Darjeeling: Mahanadi. 900–1800m. May.

4. E. multiradiatus (DC.) Clarke. Fig. 128a.

Erect perennial herb; stems (5–)15–45cm, simple or branched, sparsely pubescent and glandular above. Leaves very variable, basal ones when present oblanceolate, 5–11 × 0.7–1.5cm, acute, attenuate to broad petiole, sparsely pubescent on both surfaces; upper leaves lanceolate or oblong, 3–7 × 0.7–3cm, acute or acuminate, narrowed or rounded at base, sessile, semi-amplexicaul, entire or with a few serrate teeth on either side near apex. Capitula usually solitary; involucre 12–22mm diameter, 3-seriate; phyllaries linear-oblanceolate, 10–12 × 1–1.5mm, acuminate, often purplish at tip, usually pubescent, sometimes densely so and glandular above. Flowers dimorphic. Ray flowers 120–200, 2–3-seriate; corolla tube 1.5–3mm; ligule mauve-red to purple-blue, 8–15 × 0.5–1.3mm. Disc corollas yellow or purplish, 4mm. Achenes oblong or narrowly obovoid, 2.5 × 0.5mm, minutely pubescent, compressed, with a brown rib down each edge; pappus brownish, double, inner bristles 3mm, outer series 0.2mm slightly wider than inner ones.

Bhutan: C – Ha, Thimphu, Punakha, Tongsa and Bumthang districts, **N** – Upper Mo Chu, Upper Pho Chu, Upper Bumthang, Upper Kuru Chu and Upper Kulong Chu districts; **Darjeeling:** Sandakphu; **Sikkim:** Kupup, Nathu La, Yume Samdong etc.; **Chumbi:** Chumbi, Chugya, Lingmatang etc. Open grassy slopes, among shrubs or boulders, on cliffs and rocky slopes, in wet pine forest, 3050–4800m. May–September.

5. E. bellidioides (D. Don) Clarke

Similar to *E. multiradiatus* but smaller; stems 10–15(–45)cm; leaves oblanceolate or oblong, 1–4 × 0.2–1cm, acute, attenuate at base, usually entire; capitula usually solitary, sometimes several; involucre 6–9mm diameter; phyllaries linear-oblong or -lanceolate, 5–7 × 1mm, sparsely pubescent; ray flowers 50–100; ligules mauve or sometimes white, 10 × 0.5mm; disc corollas yellow, 3mm; achenes oblong, 2 × 0.5mm; inner pappus 2.5mm, outer 0.4mm.

Bhutan: C – Ha district (Damthang, Ha Dzong), Thimphu district (Chapcha, Paga, Pajoding etc.), Tongsa district (Tongsa–Yuto La (101)), Bumthang district (Bumthang, Byakar); **Darjeeling:** Tanglu–Sandakphu–Phalut (101). Open meadows and hillsides, 2133–3350m. April–August.

65. CONYZA Lessing

Annual or perennial herbs; stems ± erect. Leaves alternate, simple, entire or coarsely toothed, 3-lobed or pinnatifid. Capitula corymbose or paniculate, disciform or with minute ligules. Involucre campanulate, sometimes contracted above, 2–3-seriate; phyllaries narrow, imbricate. Receptacle flat or convex, naked. Outer flowers female, usually numerous; corollas filiform, sometimes with minute ligule. Inner flowers bisexual, tubular-campanulate, usually 5-toothed, usually few; style branches flattened, acute. Achenes compressed; pappus hairs 1-seriate, fused into a ring at base.

1. Stems 10–30(–40)cm; leaves oblong to oblanceolate, acute or obtuse, usually regularly serrate in the upper half; female corollas ± half as long as style .... **1. C. japonica**

\+ Stems usually taller; leaves otherwise; female corollas as long as style or about ¼ as long 2

Fig. 128. **Compositae: Astereae.** a, *Erigeron multiradiatus*: habit (× ⅔). b–e, *Nannoglottis hookeri*: b, capitulum (× ⅔); c, ray flower (× 6); d, eligulate female flower (× 10); e, central flower (× 6). f–i, *Conyza stricta*: f, flowering stem (× ⅔); g, capitulum (× 10); h, female flower (× 20); i, bisexual flower (× 20). Drawn by Margaret Tebbs.

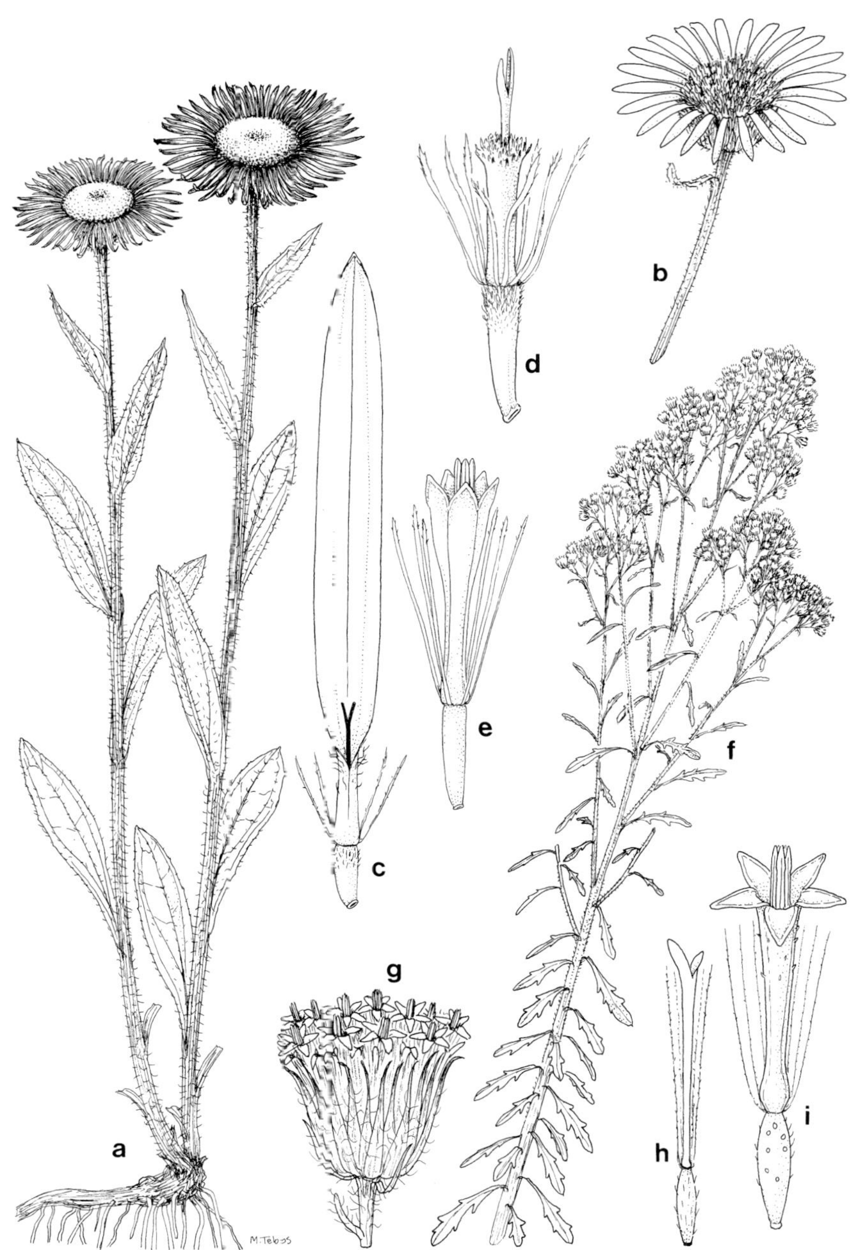

a
b
c
d
e
f
g
h
i
M.Tebbs

2. Capitula 1.5–2mm diameter, eligulate; sometimes most leaves 3-lobed or pinnatifid .... **2. C. stricta**
\+ Capitula 2.5–5mm diameter or female corollas with ligule c 0.7mm; most or all leaves simple 3

3. Indumentum conspicuously glandular especially on inflorescence; female corollas about ¼ as long as styles .... **3. C. leucantha**
\+ Indumentum ± eglandular; female corollas ± as long as styles 4

4. Female corollas with elliptic ligule c 0.7mm .... **6. C. canadensis**
\+ Female corollas with 2–3 small teeth at apex, eligulate 5

5. Panicles pyramidal, usually overtopped by lateral inflorescences; upper stems, leaves and inflorescence dull greyish-green; involucres contracted above .... **4. C. bonariensis**
\+ Panicles cylindrical, not overtopped by lateral inflorescences; upper stems, leaves and inflorescence yellowish-green; involucres scarcely contracted above .... **5. C. sumatrensis**

1. C. japonica (Thunberg) Lessing

?Annual; stems 10–30cm, sparsely spreading-pubescent. Leaves oblanceolate or oblong, 2–6 × 0.75–1.5cm, acute or obtuse, attenuate at base, serrate in the upper half, pubescent on both surfaces; basal leaves with a petiole up to 2cm; cauline leaves sessile, ± semi-amplexicaul. Capitula several in compact terminal corymbs; involucre 3–5mm diameter; phyllaries 2–3-seriate, lanceolate, 3–4 × 0.5–1mm, outer and middle ones pubescent on the back. Flowers yellowish at first, later purple. Female corollas eligulate, 1.5mm; style 2.5mm. Bisexual corollas 2.5mm. Achenes obovoid, 1 × 0.5mm; pappus 2.5mm, white or reddish.

Bhutan: C – Thimphu district (Namselling, Thimphu), Punakha district (Lometsawa, Ratsoo–Samtengang (71), Tinlegang (71) etc.), Tongsa district (Kinga Rapden, Tama (117)); **Darjeeling:** Darjeeling, Jhepi, Mongpu etc.; **Sikkim:** Lachen–Lachoong, Tista. Roadside, open, dry hillside, edge of ditch, 900–2500m. April–November.

2. C. stricta Willdenow. Fig. 128f–i.

Annual; stems 10–70cm, ascending-pubescent. Leaves linear-oblanceolate, simple, to obovate in outline, 3-lobed or pinnatisect, 1–7 × 0.2–1.5cm, acute or obtuse, attenuate at base, ± petiolate, entire, coarsely dentate or entire-lobed, pubescent on both surfaces intermixed with sessile glands beneath. Capitula numerous in dense terminal corymbs; involucre 1.5–2mm diameter; phyllaries 2–3-seriate, lanceolate, 2–2.5 × 0.5mm, acuminate, pubescent and glandular. Flowers yellowish. Female corollas eligulate, c 1.2mm, ± as long as style. Bisexual corollas 1.5mm. Achenes obovoid, 0.7 × 0.3mm; pappus caducous, 2mm, whitish.

Bhutan: C – Thimphu, Punakha, Tongsa (117), Mongar and Tashigang districts, **N** – Upper Mo Chu district; **Darjeeling:** Badamtam, Bom Forest, Ryang Chhu etc. On grassy wasteground and shingle, 600–4300m. All year.

Specimens with pinnatisect leaves may be separated as var. **pinnatifida** (D. Don) Kitamura.

3. C. leucantha (D. Don) Ludlow & Raven; *C. viscidula* DC. *nom. illeg.*

?Perennial; stems up to 2m, minutely pubescent. Leaves elliptic, 3–12 × 1–4.5cm, acuminate, cuneate at base, sharply serrate, pubescent and glandular on both surfaces. Capitula in ± spreading terminal corymbs; involucre 3–5mm diameter; phyllaries 2–3-seriate, narrowly lanceolate, 3–5 × 0.75mm, acuminate, glandular pubescent. Female corollas eligulate, white or purplish, filiform, 0.3–0.4mm; styles 2.5mm. Bisexual corollas 3mm. Achenes ovoid, 0.75 × 0.4mm; pappus 4mm, dirty-white.

Bhutan: C – Mongar district (Namning); **Darjeeling/Sikkim:** unlocalised (Hooker collections). In mixed forest, 1000–2600m. June.

4. C. bonariensis (L.) Cronquist; *C. angustifolia* Roxb.

Annual or biennial; stems 30–100cm, indumentum of few stronger spreading hairs and finer appressed eglandular hairs; upper stems, leaves and inflorescence dull greyish-green. Leaves linear or linear-oblanceolate, 2–8 × 0.2–0.75cm, acuminate, gradually attenuate at base, entire, appressed pubescent; basal leaves sometimes to 12mm wide and coarsely serrate with few teeth. Capitula in pyramidal panicles, the terminal panicle commonly overtopped by lateral ones, involucre 3–5mm diameter, conspicuously contracted above; phyllaries 3-seriate, linear-lanceolate, pubescent, inner ones 4–5 × c 0.7mm, pubescent. Female corollas eligulate, c 4mm, ± as long as styles. Bisexual corollas c 3.7mm, 5-toothed. Achenes oblong, 1–1.5mm, pubescent; pappus often reddish, c 3.5mm.

Bhutan: C – Thimphu district (Paro, Thimphu), Punakha district (Bhotokha, Chusom–Mishina, Tashi Choling), Tashigang district (Khaling); **Darjeeling:** Rimbi Chhu, Pemayangtse–Thinglen (69); **Sikkim:** Namphak, Yantong. On fallow land and roadsides, in ditches, 450–2450m. April–June.

Recorded from Mongar and Thimphu administrative districts of Bhutan at a wide range of altitudes, but probably in most districts (272). Introduced, native to tropical S America.

5. C. sumatrensis (Retzius) Walker; *Erigeron sumatrensis* Retzius, *C. floribunda* Humboldt, Bonpland & Kunth

Very similar to *C. bonariensis* but plant less bushy; stems to 2m; upper stems, leaves and inflorescence yellowish-green; sometimes most leaves remotely serrate; panicles cylindrical, not overtopped by lateral inflorescences; involucres 2.5–4mm across, scarcely contracted above; pappus dirty white, stramineous or brownish only.

Bhutan: **S** – Chukka or Phuntsholing (71) and Gaylegphug districts, **C** – Thimphu (117), Punakha, Tongsa (71) and Tashigang districts; **Darjeeling:** Manjitar, Kalimpong (71); **Sikkim:** Pemayangtse–Thinglen (71). On waste-ground and roadsides, 600–1800m. May–July.

Reported from Chhukha, Lhuntshi, Mongar, Punakha, Sarbang Tashigang and Wangdi administrative districts, mainly below 2000m (272). A record from Thimphu, Bhutan (117) requires confirmation. Introduced, native to S America.

6. C. canadensis (L.) Cronquist; *Erigeron canadensis* L.

Similar to *C. bonariensis* but plant less bushy; upper stems, leaves and inflorescence yellowish-green; sometimes most leaves remotely serrate; panicles cylindrical, not overtopped by lateral inflorescences; involucres c 2mm across, scarcely contracted above; female corollas 3mm, including elliptic ligule c 0.7mm; bisexual corollas 3mm, with 4 teeth; pappus dirty white, stramineous or brownish only.

Bhutan: **S** – Phuntsholing/Chukka district (Phuntsholing–Chimakothi (71)), **C** – Thimphu district (Damgi, Yosepang); **Darjeeling:** Kurseong, Sepoydhura, Tindaria etc. (all 97a); **Sikkim:** Gangtok–Temi (101). Weed of perennial crops and fallow land, 1600–2700m. Flowering ± all year with peak April–June (97a).

Reported from Chhukha, Paro, Tashigang, Thimphu and Wangdi administrative districts, mainly above 2000m (272). Introduced, native to N America.

66. MICROGLOSSA DC.

Shrub, often scandent. Leaves alternate, ovate, acuminate, obtuse at base, minutely and remotely serrulate, petiolate. Capitula in 1–several small rounded heads arranged in a terminal corymb, disciform. Involucre campanulate, several-seriate, imbricate; phyllaries linear-lanceolate. Receptacle flat. Corollas greenish-white. Outer flowers numerous, female; corolla filiform with 1 erect linear tooth barely exceeding pappus. Inner flowers few, bisexual; corolla tubular below, with scarcely dilated limb cut almost to middle by suberect teeth; style branches flattened, with ovate appendage. Achenes oblong-obovoid, pubescent; pappus simple, reddish-brown, of filiform bristles.

1. M. pyrifolia (Lamarck) Kuntze; *M. volubilis* DC.

Plant c 2–2.5m; stems glabrescent. Leaves 6–7 × 2.5–3.5cm, reduced beneath inflorescence, subglabrous above, puberulous beneath; petioles 5–10mm. Involucre 2.5–3.5mm diameter; outer phyllaries puberulous, inner ones 4.5–5.5mm. Female corolla tube 3–3.5mm; tooth c 1.3mm. Bisexual corollas 4–4.5mm. Achenes c 1mm; pappus 4–4.5mm.

Bhutan: **S** – Tashigang (Jirgang Chu), unlocalised (Griffith collection). Dense wet evergreen forest, 1350m. May.

Also reported from Sikkim without locality (343).

67. NANNOGLOTTIS Maximowicz

Erect perennial herbs with thickened rhizomes. Leaves simple, alternate. Capitula solitary or 2–5 in terminal corymbs, radiate. Involucre herbaceous, several-seriate; phyllaries subequal. Flowers trimorphic; corollas yellow. Ray flowers 1-seriate, female. Eligulate female flowers 1–2-seriate; corolla tubular, short, truncate above. Central flowers tubular-campanulate, 5-toothed above, male; style branches erect, broadened, acute. Fertile achenes oblanceolate, somewhat compressed, c 10-ribbed. Pappus of fertile achenes almost barbellate, 1–2-seriate; pappus of sterile achenes sparse.

1. N. hookeri (Hook.f.) Kitamura; *Doronicum hookeri* Hook.f., *Inula macrosperma* Hook.f. Fig. 128b–e.

Stems 13–70cm, simple or branched only in inflorescence, often weakly lanate at first, glandular-stipitate above. Leaves lanceolate to oblanceolate, 7–25 × 1–7cm, acute, attenuate at base, decurrent on stem, sessile or lowest ones with winged petiole to 10cm, denticulate to sharply toothed, lanate or glabrescent on both surfaces. Lateral capitula, when present, considerably overtopping central one. Involucres ± hemispherical, 15–20mm across top; phyllaries lanceolate, 10–15 × 1–4mm, acuminate, glandular-stipitate and often sparsely araneous. Ray corolla tubes 2.3–4.5mm, puberulous above; ligule 10–15 × 2–4mm. Eligulate female corollas 2–3.5mm, puberulous above. Male corollas 4–5.5mm, teeth puberulous outside. Fertile achenes oblong, c 4.5–5.5mm, appressed pubescent above; pappus brownish, 4–5mm. Sterile achenes ± glabrous.

Bhutan: C – Ha, Thimphu, Punakha, Tongsa, Bumthang and Mongar districts, **N** – Upper Mo Chu, Upper Pho Chu and Upper Bumthang Chu districts; **Sikkim:** Giaogang, Lachen, Thanggu etc.; **Chumbi:** Chumbi, Dotag, Yatung etc. Hillsides, in grass, among boulders or sometimes among shrubs, 2600–4550m. May–September.

At flowering our plants vary considerably from precocious, small, densely lanate ones with remotely denticulate leaves only partially expanded and solitary capitula to later, tall, glabrescent plants with broad coarsely dentate leaves and most stems bearing several capitula.

68. VITTADINIA Richard

Annual or perennial herbs, sometimes woody at base, variously covered in long, septate, sometimes araneous, egandular hairs and shorter glandular ones; stems much branched. Leaves simple to doubly trifid, often cuneate. Capitula solitary, teminal, radiate. Involucre campanulate, several-seriate; phyllaries narrowly oblong to elliptic, obtuse to acuminate. Receptacle convex, naked, fimbrillate. Ray flowers female, several-seriate; ligule ± oblong, subequal to tube. Disc flowers fertile; corolla tubular-campanulate, 5-toothed; style branch appendages subulate. Achenes oblanceolate, compressed, finely ribbed on both faces, with basal tuft of hair, pubescent elsewhere. Pappus bristles (1- to) several-seriate, connate at base, subequal to disc corollas.

1. cf. **V. australis** Richard

Darjeeling: Darjeeling, 1900m (69).

No specimen seen. An introduced plant, reported as *V. triloba* (Gaudichaud) DC. [*V. australis* Richard], from New Zealand. However, the two species are distinct, the former a *Brachycome* (*B. triloba* Gaudichaud) (324). Our plant is probably *V. australis* Richard from New Zealand, with white ligules, or perhaps an Australian *Vittadinia* with blue or purplish ligules. Pappus bristles of *Brachycome* are minute and scale-like or absent. Often confused with *Erigeron karvinskianus*, to which Mukherjee (218) refers this collection.

69. THESPIS DC.

Annual, subglabrous; stem simple, erect. Leaves simple, alternate. Capitula small, oblate, subsessile, in rounded clusters forming a corymbose terminal inflorescence, disciform. Involucres 2-seriate; phyllaries herbaceous with scarious margin. Receptacle shallowly convex, naked. Outer flowers female, many-seriate, without corollas. Central flowers very few, male; corollas yellow, tubular-campanulate, 4-toothed, with few stipitate glands; style branches appressed forming a subacute cone. Achenes oblanceolate, subterete, papillose; pappus of c 10 bristles, equal to style; also present on male flowers, incomplete, with some bristles longer.

1. T. divaricata DC.

Plant 30–50cm. Leaves obovate or oblanceolate, 3–5 × 0.8–1.5cm, obtuse, long attenuate at base, ± petiolate on lower stem, serrate, teeth with stiff hooked tip and midrib scabridulous beneath. Capitula c 2mm diameter. Phyllaries oblong-elliptic, to 1.2 × 0.8mm, obtuse. Style of female flowers c 0.4mm. Male corollas c 1mm. Achenes ellipsoid, c 0.6mm; pappus whitish.

Darjeeling: Balasun; **W Bengal Duars:** Chapramari (407). Among grasses and on sandy river bed, 100m. May.

Only one collection seen from our area. Elsewhere in NE India often a weed of paddy-fields, usually more branched, sometimes only 3cm tall.

ANTHEMIDAE

70. ALLARDIA Decaisne

Fragrant tufted alpine herb, sparsely covered with araneous hairs, rarely with conspicuous sessile glands. Leaves alternate, cuneate, cut into ± linear lobes at apex. Capitula radiate, usually solitary at branch ends. Involucre hemispherical, several-seriate; phyllaries blackish or brownish scarious-margined. Receptacle flat, naked. Ray flowers 1-seriate, sterile; ligule pale pink to red or purple, 3-toothed; pappus, ovary and sometimes a simple style ± developed. Disc yellow; flowers bisexual; corolla tubular-campanulate, 5-toothed; style branches with rounded or truncate hairy appendages. Achenes oblong, subterete, 5-angular, conspicuously sessile-glandular. Pappus of brownish, linear, bristle-like scales.

1. A. glabra Decaisne; *Waldheimia tridactylites* Karelin & Kirilow, *W. glabra* (Decaisne) Regel

Plants 8–15cm. Leaves 1–1.5cm on sterile shoots, to 3cm on flowering ones, lobes often 5. Involucre c 1.5cm diameter; phyllaries lanceolate to oblong, c 9 × 3mm, obtuse, sometimes glandular. Ligules oblong-elliptic, 9–12 × 3–6mm. Disc corollas c 4.5mm. Achenes 2mm; pappus 7(–10)mm.

Bhutan: **N** – Upper Mangde Chu district (Passu Sefu), Upper Bumthang Chu district (Tolegang, Mon La Karchung La), Upper Kuru Chu district (Narim Thang); **Sikkim:** Lhonak, Yume Samdong, Yumthang etc. In open screes, sandy soil and rocky streamsides, 4265–5050m. July–September.

71. TANACETUM L.

Erect herbs from fleshy root-stock. Leaves 2–4-pinnatisect, alternate. Capitula radiate, disciform or discoid, solitary or 2–4 in loose raceme. Involucre hemispherical; phyllaries imbricate, c 4-seriate, ovate to linear-oblong, with pale or dark scarious margins. Receptacle convex, naked. Marginal flowers ligulate, female, or tubular-campanulate with some female or ± bisexual with reduced androecium and others undifferentiated from disc flowers. Disc flowers bisexual; corollas tubular-campanulate, 5-toothed; style branch tips truncate, hairy. Achenes oblong or clavate, terete, 5–10-ribbed; myxogenic cells absent. Pappus absent or reduced to c 3 minute teeth in our species.

1. Basal leaves 3(–4)-pinnatisect, 12–30cm; marginal corollas conspicuously ligulate .... **1. T. atkinsonii**
+ Basal leaves 2-pinnatisect, 2.5–9cm; marginal corollas scarcely differentiated **2. T. tatsienense**

1. T. atkinsonii (Clarke) Kitamura; *Chrysanthemum atkinsonii* Clarke

Root-stock thick, fleshy; stems erect, usually simple, 15–42cm, puberulous, lanate beneath capitulum. Basal leaves 3(–4)-pinnatisect, obovate in outline, 12–30cm, sparsely pubescent with long fine hairs, ultimate segments narrowly oblong, acuminate, c 0.5mm wide (0.2–0.4mm wide on 4-pinnatisect leaves). Cauline leaves reduced, mostly 2-pinnatisect. Capitula usually solitary, terminal, rarely 1–3 more on short lateral branches, radiate. Involucre hemispherical, c 1.5cm diameter; phyllaries oblong-lanceolate, 6–10 × 1.5–3mm, obtuse, scarious margin blackish, eroded. Flowers often sparsely glandular below; corollas yellow. Ray flowers 15–25; ligules 12–17 × 3–5mm, densely papillose inside; staminodes exerted. Disc corollas 3–4.5mm. Achenes c 2.5mm, 5-ribbed, minutely glandular; pappus usually absent, sometimes achene rim developed into 3 small lobes.

Bhutan: C – Ha district (Ha, Tare La, Chile La etc.), Thimphu district (Pajoding), Tongsa district (Takse La), Tashigang district (Tashigang); **Sikkim:** Kupup, Tsomgo, Zongri etc.; **Chumbi:** Natu La–Chubitang, Yak La, Yatung etc. Damp meadows, hillsides and cliffs, mainly among grass, 3650–5050m. July–October.

2. T. tatsienense (Bureau & Franchet) Bremer & Humphries var. **tanacetopsis** (W.W. Smith) Grierson; *Chrysanthemum jugorum* W.W. Smith var. *tanacetopsis* W.W. Smith

Root-stock rather thick, fleshy; stems 5–25cm, sparsely pilose. Basal leaves bipinnatisect, 2.5–9cm, ± oblanceolate in outline, sparsely pubescent with long hairs, ultimate segments linear-oblong, c 0.4mm wide, acuminate. Capitula solitary, weakly disciform. Involucre hemispherical, c 1.5cm diameter; phyllaries linear-oblong, 7–9 × 1–2mm, obtuse, scarious margin blackish, eroded, lacerate and sometimes ?yellowish at apex. Disc flowers bisexual; corollas yellow or orange or teeth purplish, c 4.5mm, with very few glands below. Some marginal flowers with reduced (sometimes sterile) androecium and slightly elongated yellow corolla. Achenes c 2.5mm, 5-ribbed; pappus absent.

Bhutan: N – Upper Bumthang Chu district (Pangotang). Among dwarf rhododendrons, 4400m. July.

Matricaria matricarioides (Lessing) Porter will key out to *Tanacetum*. It is a branched annual with discoid capitula, conic receptacles and greenish 4-toothed corollas. It is a widespread weed, recently collected from roadsides and waste places in Bumthang (Bumthang district, Bhutan).

72. HIPPOLYTIA Poljakov

Densely tufted herb, thickly brown or grey lanate. Sterile shoots columnar; leaves overlapping, usually trifid at apex, sometimes more divided. Flowering stems erect, simple, bearing c 10–15 capitula in dense semi-globose cyme at apex; leaves sometimes much more dissected and/or with segments blackish scarious apically. Capitula discoid. Phyllaries lanceolate to linear-oblong, margins blackish scarious. Corollas yellow, sometimes scattered subsessile glands on tube. Achenes narrowly obovoid; pappus absent.

1. H. gossypina (Hook.f. & Thomson) Shih; *Tanacetum gossypinum* Hook.f. & Thomson. Fig. 129a–b.

Plant 4–10cm. Leaves of sterile shoots c 7 × 3mm, appressed and ± scarious below, lobes usually oblong and obtuse. Leaves of flowering stems often larger, spreading from base. Capitula 3.5–5mm diameter. Phyllaries 4.5–6mm, sparsely hairy. Corollas 2.5–4mm. Achenes 2mm.

Bhutan: N – Upper Mo Chu, Upper Phu Chu, Upper Mangde Chu, Upper Bumthang Chu and Upper Kulong Chu districts; **Sikkim:** Bikbari, Donkiah, Lhonak etc.; **?Chumbi:** Chomo Lhari, border E of Phari, Tangkar La. Sandy screes and rocky gullies, 4100–5200m. July–September.

73. CHRYSANTHEMUM L.

Perennial stoloniferous herb; stems somewhat decumbent, simple or branched, leafy. Leaves alternate, pinnatifid with 2 pairs of lateral lobes, petiolate, upper leaves at least with stipuliform segments at petiole base. Capitula few in terminal cymes and solitary in upper leaf axils, radiate. Involucre c 4-seriate; most phyllaries ± oblong with broad scarious margin. Receptacle convex, paleaceous among ray flowers and outer disc flowers, naked within. Ray flowers 2–3-seriate, female. Disc flowers bisexual; corolla tubular-campanulate, unwinged, 5-toothed; anther bases rounded; style branches oblong, truncate. Achenes oblong, ribbed; pappus absent.

Use of the name *Chrysanthemum* L. for this genus was authorised in 1998, previously it was correctly known as *Dendranthema* Des Moulins. Species previously treated under *Chrysanthemum* can be found under 76. *Xanthophthalmum.*

1. C. cf. **indicum** L.

Stems c 45cm, shortly pubescent. Leaves ovate in outline 3–8 × 2–5cm, shortly acuminate, ± truncate at base, coarsely serrate, sparsely appressed puberulous above, paler beneath and more densely appressed pubescent; petioles up to 2cm, with trifid stipuliform segments 2–5mm. Capitula broadly campanulate, 1–1.5cm diameter. Phyllaries sparsely tomentose; 2 outermost herbaceous, c 9 × 1.5mm; remainder c 9–10 × 3.5mm, with scarious border with brownish outer margin. Paleae oblong-obovate, mostly hyaline, c 7 × 2.5mm. Ray corolla tube 3.5mm; ligule oblong, white, 20 × 4.5mm. Disc corollas yellow, 4.5mm, teeth sparsely subsessile-glandular. Achenes (immature) c 1mm.

Bhutan: S – Samchi district (Samchi). Cultivated, 500m. March.

Introduced from China or Japan. Differs from more typical forms grown elsewhere in India by its more deeply divided leaves, longer phyllaries and larger ligules.

74. AJANIA Poljakov

Tomentose perennials, from usually exposed woody stock; stems erect to decumbent, simple to much branched above. Leaves small, 1–3-pinnatisect. Capitula disciform, 1–many in terminal usually corymbose inflorescence. Involucre hemispherical; phyllaries c 4-seriate, mostly obtuse and broadly scarious-margined. Receptacle conic. Some marginal flowers female, remaining flowers bisexual. Corollas yellow, sparsely subsessile glandular, obliquely inserted on ovary. Female corollas ± tubular, with 2–4 often irregular teeth. Bisexual corollas tubular-campanulate, 5-toothed; style branches oblong, apex truncate, hairy. Achenes narrowly obovoid, oblique, with myxogenic hairs; pappus absent.

1. Plant 6–40cm; corollas 3.2–4.2mm .... **1. A. nubigena**
\+ Plant at least 40cm; corollas 2–2.5mm .... **2. A. myriantha**

1. A. nubigena (DC.) Shih; *Tanacetum nubigenum* DC. Med: *Khenkar*. Fig. 129c–e.

Erect to decumbent herb, 6–40cm, silver (sometimes brownish) grey tomentose; stems simple. Leaves 1–3-pinnatisect, to 3 × 3cm, primary lateral segments 1–3(–4) pairs, sometimes subpalmately arranged; secondary lateral segments 1–3 pairs, ultimate segments (obovate to) oblanceolate to linear, 1.5–9mm, ½–1⁄12 as wide, subacute to acuminate. Capitula 1–27 in loose to compact terminal corymbs or panicles. Involucre broadly campanulate, 4–7mm diameter; phyllaries brown or blackish-margined, sparsely to densely tomentose, outermost often linear, inner ones ± oblong, 4.2–6.2 × 2–3mm. Corollas 3.2–4.2mm. Achenes 1.5–2mm.

Bhutan: C – Thimphu district, **N** – Upper Mo Chu, Upper Pho Chu, Upper Mangde Chu, Upper Bumthang Chu and Upper Kulong Chu districts; **Sikkim:** Kongra La, Phaklung, Tangu; **Chumbi:** Chugya, Phari, Tuna etc. Open hillsides and screes, 3650–4720mm. August–September.

Usually considered to comprise two species: *A. nubigena*, taller, usually erect plants with leaves 2–3-pinnatisect, corymbs looser and small capitula (4–5mm diameter) and *A. khartensis* (Dunn) Shih [*Tanacetum mutellinum* Handel-Mazzetti], shorter plants, often decumbent, with leaves 1–2-pinnatisect, compact corymbs and larger capitula (c 6mm diameter). However, most plants from our area show different combinations of these characters.

2. A. myriantha (Franchet) Shih

Subshrub, at least 40cm; stems sometimes much branched above. Leaves bipinnatisect, ovate to oblong in outline, 15–40 × 10–20mm, densely greyish tomentose beneath, sparsely so above; primary lateral segments 3–4 pairs; secondary lateral segments 3–6 pairs, oblong to obovate, 1–3.5mm, c ½ × as wide, subacute. Capitula globose, in small compact cymes, usually forming a loose terminal corymb. Involucre hemispherical, c 3mm diameter; phyllaries

FIG. 129. **Compositae: Anthemideae.** a–b, *Hippolytia gossypina*: a, habit (× ⅔); b, capitulum (× 4). c–e, *Ajania nubigena*: c, habit (× ⅔); d, female flower (× 8), e, bisexual flower (× 8). f–h, *Artemisia campbellii*: f, habit (× ⅔); g, female flower (× 12); h, bisexual flower (× 12). Drawn by Margaret Tebbs.

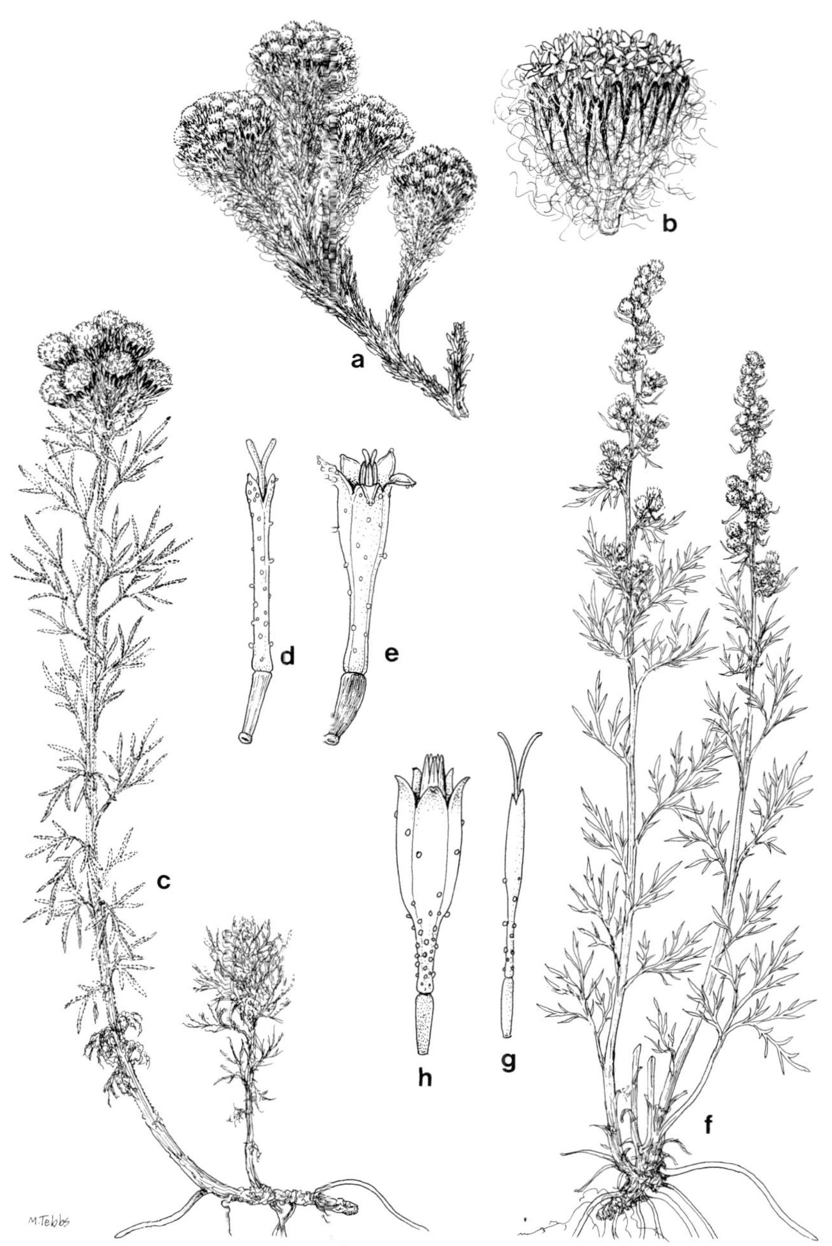

a
b
c
d
e
f
g
h
M.Tebbs

broadly ovate to oblong-elliptic, 2–3 × 1.5–2mm, obtuse, pale-margined, outer sparsely tomentose below. Corollas 2–2.5mm. Achenes c 1mm.

Bhutan: C – Ha district (Ha), Thimphu district (Tsalimaphe, Paro), Bumthang district (Bumthang). Dry banks and hillsides, 2300–3050m. November–March.

Used for incense in Bhutan.

75. ARTEMISIA L.

Aromatic annual, biennial or perennial herbs or subshrubs, glabrescent or puberulous to tomentose or villous, glandular-punctate, sometimes also glandular-stipitate. Leaves 3-fid to 3-pinnatisect or simple on inflorescence branches. Capitula (c 10–) very many, subspicate, racemose or paniculate, disciform. Involucres ovoid to hemispherical or oblate; phyllaries few-seriate, scarious-margined, glabrous to araneous. Receptacle flat or concave, naked, rarely bearing long weak hairs. Flowers white to greenish, often partly purplish; marginal female, fertile; central bisexual or male with reduced or ± absent ovaries. Corollas sparsely subsessile glandular, rarely also with some eglandular hair. Female corollas tubular, irregularly 2–3-toothed. Bisexual/male corollas tubular-campanulate or obovoid-infundibular, 5-toothed; style branches truncate. Achenes small, ellipsoid or obovoid, weakly striate, with myxogenic hairs; pappus absent.

For the purposes of comparison, the entire flowering of a stem is described as an inflorescence here, even if discrete lateral inflorescences seem present. Leaves decrease in size and amount of division from base to apex of stem and sometimes vary noticeably in form. The lowest leaves seen subtending inflorescence branches (often limited by the length of stem collected) are described, except for cases where some inflorescence branches originate below mid-stem, when mid-cauline leaves are described instead. Mature achenes (missing from most specimens) probably 1–1.8mm on species 1–12, 0.5–0.8mm on species 13–17.

1. Inner flowers bisexual: ovary at anthesis subequal to that of female flowers (usually more than 0.4mm) 2
+ Inner flowers male: ovary either rudimentary (c 0.2mm, much shorter than female ones) or absent 15

2. Receptacle covered by caducous c 1.5mm hairs .... **1. A. minor**
+ Receptacle hairless 3

3. Annual; lower inflorescence leaves 2–3-pinnatisect, primary segments with at least 5 pairs of secondary ones .... **2. A. hedinii**
+ Herbaceous perennial or subshrub; lower inflorescence leaves 3-fid or 1–2-pinnatisect, primary segments with up to 4 pairs of secondary ones (excluding teeth) 4

4. Plant not viscid; glands all sessile 5
\+ Plant viscid; stipitate glands on stems, leaves or inflorescence 12

5. Ultimate segments of lower inflorescence leaves ovate, acuminate, consistently serrate-dentate with few ovate acuminate teeth (Fig. 131a–c) 6
\+ Ultimate segments of lower inflorescence leaves usually oblong or lanceolate, entire or bearing distant irregular teeth or lobes (Fig. 130d–i, Fig. 131d–h) 8

6. Involucres 3.5–4mm diameter .... **5. A. tukuchaensis**
\+ Involucres 1.2–2.5mm diameter .... **6. A. indica**

7. Lower inflorescence leaves 7–15 × 5–9cm, at least outer 20mm of terminal segment entire 9
\+ Lower inflorescence leaves 1.5–8.5 × 1–5.5cm, teeth or lobes present within 20mm of apex of terminal segment or whole segment less than 20mm .. 10

8. Subshrub with ± exposed woody stock .... **3. A. austroyunnanensis**
\+ Perennial herb with stems solitary or few together from creeping rhizome **4. A. verlotiorum**

9. Plant 5–40cm .... **8. A. campbellii**
\+ Plant 40–90cm 11

10. Most capitula borne singly and distantly along inflorescence branches or most capitula with less bisexual than female flowers ... **7. A. moorcroftiana**
\+ Capitula crowded at end of inflorescence branches, most with more bisexual than female flowers .... **8. A. campbellii**

11. Ultimate leaf segments long acuminate; stem hispidulous throughout **9. A. bhutanica**
\+ Ultimate leaf segments shortly acuminate to obtuse, apiculate; stem araneous, tomentose or pubescent above, often glabrescent below 13

12. Leaves subglabrous above; ultimate segments of lower inflorescence leaves (remotely) dentate .... **11. A. thellungiana**
\+ Leaves pubescent above; usually ultimate segments of lower inflorescence leaves entire or bearing 1–2 teeth or lobes 14

13. Stipitate and sessile glands (and eglandular hairs) on leaves above; bisexual flowers lacking eglandular hairs .... **10. A. myriantha**
\+ Only sessile glands (and eglandular hairs) on leaves above; bisexual flowers eglandular-pubescent above .... **12. A. yadongensis**

14. Plant 2.5–40cm, rarely taller, then annual or biennial 16
\+ Plant 0.4–1.5m, perennial or suffrutescent 18

15. Involucre 3–4.5mm diameter .... **15. A. desertorum**
\+ Involucre 1.5–2.5mm diameter 16

16. Subshrub; capitula distant .... **13. A. wellbyi**
\+ Annual/biennial; capitula crowded .... **14. A. stricta**

17. Primary lateral segments of mid-cauline leaves 1–4mm wide **16. A. parviflora**
\+ Primary lateral segments of mid-cauline leaves at least 6mm .. **17. A. dubia**

1. A. minor Jacquemont. Fig. 130a–b.

Dwarf caespitose subshrub, c 10cm; stems from stout branched woody stock. Leaves bipinnatisect, 9–16mm, grey or brownish tomentose, with crowded overlapping segments; terminal segment much reduced; secondary segments 3–7, subpalmately arranged, oblanceolate, c 2.5 × 0.7mm, obtuse, entire (rarely with 1 small lobe); lower cauline leaves with 2 pairs lateral segments and broad petiole often exceeding blade, without stipuliform lobes; upper cauline ones variable, similar or with reduced secondary segments (almost teeth), often with several stipuliform lobes or subsessile with 2–4 pairs of simple segments. Capitula few in narrow racemes, ± distant, often subsessile. Involucre c 4mm diameter; phyllaries obovate, 3 × 1.3–2mm, hairy, brownish scarious at margins. Receptacle covered by caducous hairs c 1.5mm. Female flowers rather few, corollas 1.3–1.8mm. Bisexual flowers numerous; corollas widening gradually upwards, 1.8–2mm. Achenes obovoid, 1.5 × 0.5mm.

Sikkim: Yume Chhu; **Chumbi:** Phari. 4400–5100m. August–September.

2. A. hedinii Ostenfeld; *A. biennis* sensu F.B.I. p.p. non Willdenow. Fig. 130c.

Annual, somewhat succulent, 20–50cm; stem simple, erect, often purplish, sparsely glandular. Leaves oblanceolate in outline, 2–3-pinnatisect, to 13 × 5cm, sparsely glandular-hairy beneath; primary segments 5–7 pairs, elliptic in outline, to 3 × 1cm, alternating with several smaller segments; ultimate segments acute or obtuse, minutely apiculate. Inflorescence branches narrow, elongate, from most leaf axils. Involucre hemispherical, 2.5–3mm diameter; outer phyllaries mostly obovate, 2.7–3.5 × 1.7–2mm, with brown margin; innermost ones reduced, often with clear margin. Flowers pink; corollas densely glandular. Female flowers 21–28; corollas 1–1.3mm. Bisexual flowers (12–)20–28; corollas 1.3–2mm. Achenes obovoid, 1–1.3mm.

Sikkim: Phaklung, Sashethang; **Chumbi:** Gautsa–Phari, Tuna. 4265–4570m. July–September.

FIG. 130. **Compositae: Anthemideae.** a–i, lower inflorescence leaves of *Artemisia* species: a–b, *A. minor* (× 4½); c, *A. hedinii* (× ¾); d–e, *A. austroyunnanensis* (× ½); f, *A. calophylla* (× ½); g–h, *A. verlotiorum* (g × ½, h × ¾); i, *A.* sp., see *A. verlotiorum* (× ¾). Drawn by Chen Yo-Jiun.

a
b
c
d
e
f
g
h
i

3. A. austroyunnanensis Ling & Y.R. Ling. Dz: *Khempa*. Fig. 130d–e.

Subshrub (372), to 2m; stems brown pubescent at first. Lower inflorescence leaves 1-pinnatisect, 8–12 × 4–7cm, puberulous above with many glands in small deep pits, brown tomentose below; terminal segment lanceolate, 5–8cm, (briefly) acuminate, entire; lateral segments 3–5, elliptic(-oblong), to 5cm, acute or acuminate, entire; petiole 7–20mm with narrow to broad, often obcuneate wing. Capitula in moderately broad panicles. Involucre obovoid, 2.5mm in diameter; phyllaries mostly ovate to obovate, 2.2–3 × 1.1–1.8mm, sparsely arachnoid-puberulous, with few sessile glands; outer ones reduced. Female flowers (5–)8–10; corollas (1–)1.5mm. Bisexual flowers (3–)8–10, corollas 1.7–2.2mm. Achenes (immature) oblong-elliptic, c 1mm.

Bhutan: C – Thimphu district (Gunisawa*, Samakha), Punakha district (Samtengang); **Darjeeling:** Jhepi, Rangit*. Roadsides, 1200–2755m. August–October(–December).

Asterisked collections are incomplete and dubiously attributed to this species. *A. calophylla* Pampanini has been reported from Yadong district of Tibet (421). It is also suffrutescent, but differs from *A. austroyunnanensis* by the central segment of its mid-cauline leaves being scarcely longer than the upper lateral ones. A collection from Chumbi, west of Yadong, 3050m, with capitula in early bud in August may belong here (Fig. 130f).

4. A. verlotiorum Lamotte. Fig. 130g–h.

Perennial herb to 2m, sometimes with overwintering foliage, stems single or few together from creeping rhizomes, some non-flowering, erect, brown or brownish-grey pubescent at first. Lower inflorescence leaves 1-pinnatisect, 11.5–15 × 5–9cm, puberulous above with many glands in small pits, brown or brownish-grey tomentose below; terminal segment linear-lanceolate, 7.5–10cm, acuminate, entire or with 1 pair of lobes; lateral segments (1–)2 pairs, elliptic or elliptic-lanceolate, to 60mm, acuminate, with 0–2 lobes; petiole 7–12mm, obcuneately winged. Capitula in moderately broad panicles. Involucres obovoid, 1.5–2.5mm in diameter; phyllaries mostly ovate to obovate, 2–3 × 1–1.8mm, sparsely to densely araneous, with few sessile glands; outer ones reduced. Female flowers 10–13; corollas 1–1.7mm. Bisexual flowers 8–11; corollas 2–2.5mm. Achenes (immature) oblong-elliptic, c 1.1mm.

Bhutan: S – Phuntsholing district (Phuntsholing); **Darjeeling:** Darjeeling; **Darjeeling/Sikkim:** unlocalised (Hooker collection). 900–1800m.

A. austroyunnanensis, *A. calophylla* and *A. verlotiorum* form a group usually distinguished by numerous glands in deep pits on upper leaf face. Collections from Paro and Thimphu (both Thimphu district, Bhutan) with short, sometimes obtuse lower inflorescence leaves c 8 × 5cm also belong to this group (Fig. 130i). Collections with glands in shallow pits, from Singkarap (Thimphu district, Bhutan) with leaves of upper inflorescence ternate and from Darjeeling Town with all leaves apparently ternate, are doubtfully referred to the group.

5. A. tukuchaensis Kitamura. Fig. 131a.

Perennial, c 2m; stems erect, sparsely greyish araneous tomentose. Mid-cauline leaves bipinnatisect, ovate in outline, 11–20 × 6–11cm, subglabrous above, greyish tomentose below; primary segments 2–3(?–4) pairs, ovate to lanceolate in outline, often with 1–3 pairs of lobes towards base, 3.5–7cm, acute, apiculate, serrate-dentate with few, deep, ovate, acuminate teeth. Capitula rather sparse in broad panicles. Involucre 3.5–4mm diameter; phyllaries oblong-obovate, 3.5–4.5mm, at least outer ones sparsely araneous; outermost smaller, ± ovate. Female flowers c 8–9; corollas c 1.6mm. Bisexual flowers c 15–22, fertile; corollas c 2.8mm.

Bhutan: C – Thimphu district (Barshong), **N** – Upper Kuru Chu (Julu). Clearings in coniferous forests, 3050–3600m. September–October.

6. A. indica Willdenow; *A. vulgaris* sensu F.B.I. p.p. non L. Fig. 131b–c.

Perennial herb, c 0.5–1.5m; stems erect, greyish puberulous to sparsely tomentose. Lower inflorescence leaves bipinnatisect, 7–16 × 5–9cm, puberulous or glabrescent above, appressed puberulous to tomentose beneath; primary segments 2–3(–4) pairs, ovate-elliptic to lanceolate in outline, to 5 × 2cm, acuminate, with lowest occasionally stipuliform; secondary segments shallower, serrate-dentate with few, deep, ovate, acuminate teeth. Capitula in broad or narrow panicles. Involucre campanulate, 1.2–2.5mm diameter; phyllaries ovate to obovate, 1.8–3.2mm, subglabrous; outermost smaller, sparsely araneous. Female flowers c 3–8; corollas 0.7–1.3mm. Bisexual flowers c 6–12; corollas 1.7–2.2mm. Achenes ± oblong, c 1.2mm.

Bhutan: N – Upper Mo Chu district (Gasa); **Darjeeling:** Chhota Rangit; **Sikkim:** Zemu. 1800–2750m. August.

7. A. moorcroftiana DC. Dz: *Khempa*. Fig. 131d–e.

Perennial herb, 40–80cm; stems brown or purplish, thinly greyish araneous tomentose above. Lower inflorescence leaves bipinnatisect, obovate in outline, 4–8 × 3–5.5cm, with (2–)3(–4) pairs of primary segments and (1–)2–3(–4) pairs of entire or 1-lobed stipuliform segments at base, greyish araneous above, greyish araneous tomentose below; secondary segments lanceolate, entire (rarely 1-lobed), briefly acuminate. Capitula in narrow, sparse or lax panicles. Involucre ± campanulate, 2.5–4mm diameter; most phyllaries ovate to obovate, 3–4mm, all brown araneous, often with some grey hairs. Female flowers c 9–14; corollas 1.5–2.2mm. Bisexual flowers c 12–25, more numerous than female flowers in most capitula; corollas 2.2–3mm, often with purple tips. Achenes ± oblong, c 1.5mm.

Bhutan: C – Thimphu district (Cheka, Dotena), **N** – Upper Mo Chu district (Lingshi), Upper Pho Chu district (Tranza). On grassy banks or among boulders and shrubs, 2600–3950m. August–October.

Capitula are usually borne singly and distantly on elongate primary inflorescence branches. They are sometimes borne more closely on short secondary branches in the population from Tranza, which has been referred to *A. roxburghiana* Besser var. *roxburghiana* (371). Plants from Pangotang, Upper Bumthang Chu district, Bhutan and Me La, Upper Kulong Chu district, Bhutan that differ by more crowded capitula and more female than bisexual flowers in most capitula will also key out here. Those from Me La have broad leaf segments (Fig. 131f). Both populations are of uncertain identity.

8. A. campbellii Clarke. Fig. 129f–h, Fig. 131g–h.

Perennial herb with short, often exposed woody stock, 5–30(–55)cm; stems erect, thinly tomentose, purplish below, tomentum greyish below to often conspicuously reddish-brown on inflorescence. Mid-cauline leaves 2(–3)-pinnatisect, ovate to obovate in outline, 1.2–4.5 × 0.8–3cm, whitish (rather tawny) tomentose on both surfaces, with 2–3(–5) pairs of primary segments and (0–)2 pairs of stipuliform lobes near base; secondary segments oblong or lanceolate, acuminate, usually revolute at margins, rarely with 1–2 lobes. Capitula densely clustered (rarely subdistant) in short spike-like racemes to elongate interrupted panicles with lower leaves simple to 3-pinnatisect. Involucre broadly obovate or hemispherical, (3–)3.5–5.5mm diameter; phyllaries ± ovate, 2.5–4.5mm, reddish-brown araneous. Female flowers c 10–25; corollas c 1.8–2.5mm. Bisexual flowers more numerous, c 17–40, fertile; corollas c 2.5–3.2mm. Achenes obovoid, c 1.6mm.

?NE Darjeeling: locality uncertain (Dungboo collection); **Sikkim:** Bam Chhu, Chugya, Lhonak etc.; **Chumbi:** Phari. 4250–5200m. July–September.

An immature specimen from Bumthang (Bumthang district, Bhutan) at 3050m with mid-cauline leaves c 7 × 4cm may also belong here.

9. A. bhutanica Grierson. Fig. 131i.

Perennial herb at least 60cm, probably often much taller, bearing pale short-stipitate glands intermixed with longer, stiff, dark reddish, multiseptate hairs; stems erect. Lower inflorescence leaves 1–2-pinnatisect, ovate in outline, 3.5–7.5 × 2.5–5.5cm, thinly covered with multiseptate hairs above, more densely covered with mostly glandular hairs beneath; primary segments (2–)3–4(–5) pairs, upper 2 pairs pinnatifid with 1–3 pairs of simple (rarely 2-toothed) sharply acuminate segments; 2 stipuliform teeth at base; leaves of basal offsets with araneous, sometimes capitate hairs. Primary inflorescence branches spreading, racemiform, 3–8cm or fastigiate, to 20cm, simple or with short lateral branches. Involucre obovoid, 2–2.5mm diameter; phyllaries ovate to narrowly obovate, glabrous or bearing short-stipitate glands, cilia and sometimes few araneous hairs; outermost 2–2.5mm; inner 2.7–3.4mm. Female flowers c 12–17; corollas

Fig. 131. **Compositae: Anthemideae.** a–j, lower inflorescence leaves of *Artemisia* species: a, *A tukuchaensis* (× ½); b–c, *A. indica* (× ½); d–e, *A. moorcroftiana* (d × 1½, e × ¾); f, *LSH* 21060 (see *A. moorcroftiana*) (× ½); g–h, *A. campbellii* (g × ¾, h × 1½); i, *A. bhutanica* (× ¾); j, *A. myriantha* var. *myriantha* (× ½). Drawn by Chen Yo-Jiun.

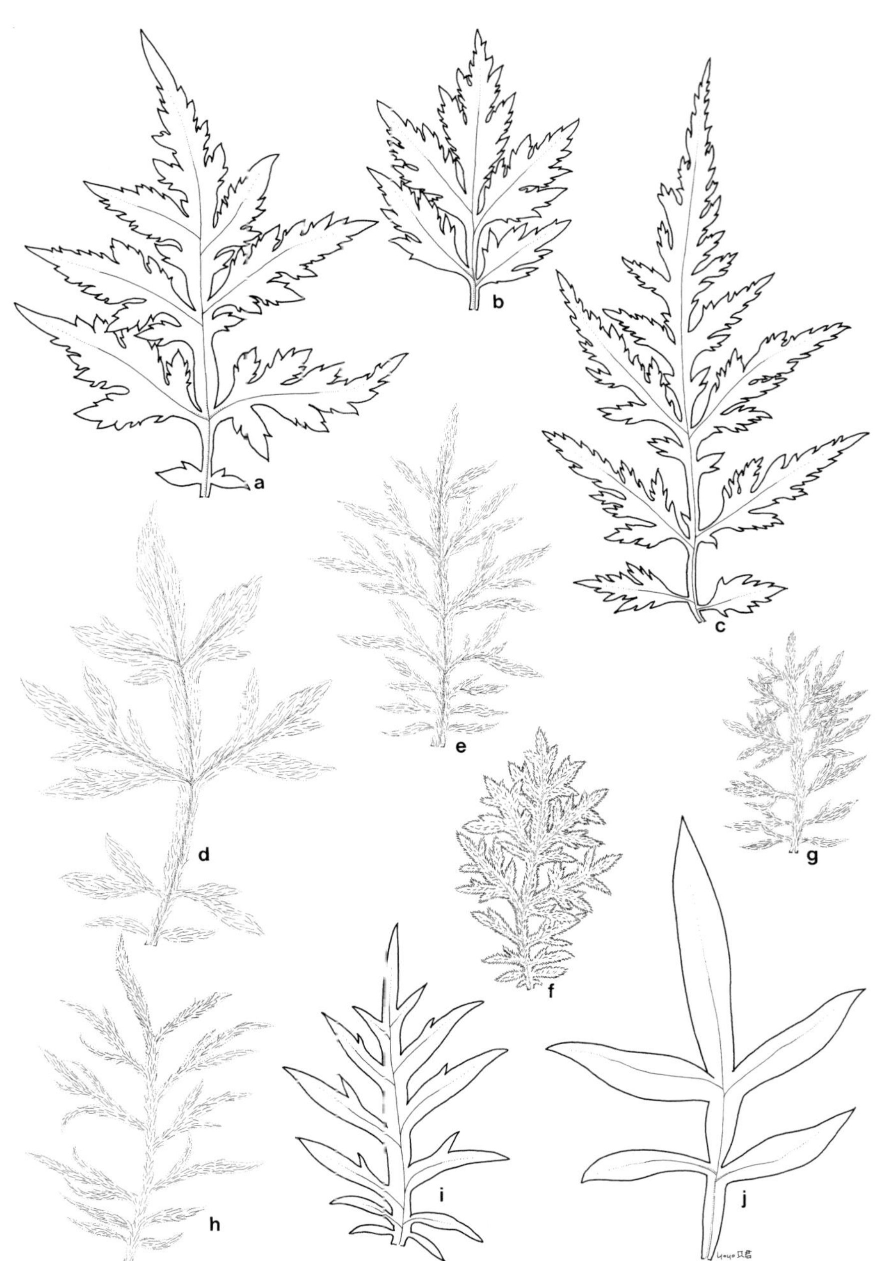
a
b
c
d
e
f
g
h
i
j

1.2–2mm, subsessile-glandular. Bisexual flowers c 8–12; corollas 1.8–2.5mm, subsessile-glandular, mostly on tube. Achenes ellipsoid-obovoid, c 1.2mm.

Bhutan: C – Thimphu (Dotena, Simtokha), **N** – Upper Mo Chu district (Laya). Open grassy slope amongst cultivation, 2750–3850m. September–October.

10. A. myriantha Besser. Nep: *Tite pali.* Fig. 131j.

Perennial herb, c 1m; stems greyish glandular- and eglandular-pubescent. Mid-cauline leaves pinnatisect, c 10 × 8.5cm, glandular- and eglandular-puberulous above, sparsely tomentose and short glandular-stipitate beneath; lateral segments 2 pairs, elliptic-lanceolate, 3–4.5cm, acute, entire. Capitula numerous but rather distant in broad panicles. Involucres c 2mm diameter; phyllaries mostly ovate, 2.3–2.8mm, araneous; outermost smaller; inner ones oblong. Female flowers c 8–11; corollas c 1.3mm. Bisexual flowers c 7–9; corollas c 2.3mm. Achenes oblong-elliptic, c 1.1mm.

Bhutan: N – Upper Mo Chu district (Gasa). 2780m. November.

The description above refers to var. **myriantha**. Var. **pleiocephala** (Pampanini) Y.R. Ling with leaves densely tomentose beneath has also been reported from Bhutan (373), without further locality details. Outside our area both varieties include plants with 2-pinnatisect mid-cauline leaves, acuminate leaf segments or very many crowded capitula.

11. A. thellungiana Pampanini; *A. vulgaris* sensu F.B.I. p.p. non L. Fig. 132a–b.

Perennial herb, c 0.8–1.5m; stems glandular puberulous at least at first. Mid-cauline leaves 1–2-pinnatisect, ± ovate(-elliptic) in outline, 7–13 × 5–8.5cm, with 3 pairs of primary segments and sometimes 1 pair of basal stipuliform lobes, subglabrous or sparsely glandular-stipitate above, sparsely (densely) tomentose below; ultimate segments subentire or bearing few acuminate teeth. Capitula rather distant in broad or narrow panicles. Involucres 2–2.5mm diameter; phyllaries ovate to obovate, 2–3.3mm, subglabrous or sparsely araneous, often purplish; outermost smaller. Female flowers c 5–10; corollas 0.7–1.4mm. Bisexual flowers c 7–16; corollas 2–2.3mm. Achenes (immature) oblong-obovate, to 1.2mm.

Bhutan: C – Thimphu/Punakha district (Dochong La), **N** – Upper Mo Chu (Laya); **Sikkim:** Lachen, Lachung. In *Picea* forest, 2750–3350m. August–September.

FIG. 132. **Compositae: Anthemideae.** a–k, lower inflorescence leaves of *Artemisia* species: a–b, *A. thellungiana* (× ½); c, *A. yadongensis* (× 1); d–e, *A. wellbyi* (d × ¾, e × 1½); f–g, *A. stricta* (f × 1½, g × 4); h–i, *A. parviflora* (h × 2½, i × 1½); j, *A. dubia* var. *dubia* (× ½); k, *A. desertorum* (× 3). Drawn by Chen Yo-Jiun; c, based on description and illustration in (370).

a
b
c
d
e
f
g
h
i
j
k

12. A. yadongensis Ling & Y.R. Ling. Fig. 132c.

Herbaceous perennial, almost suffrutescent, 50–70cm, greyish or yellowish viscid puberulous to tomentose throughout; stems purplish-brown. Lower inflorescence leaves 1(–2)-pinnatisect, ± elliptic in outline, c 5 × 3cm, puberulous and glandular-punctate above, densely arachnoid tomentose and glandular-stipitate beneath; primary segments 2–4 pairs (with 1–2 pairs of stipuliform basal lobes), lanceolate to oblanceolate and entire or ovate to obovate in outline with 1–2 broad lobes, (sub)acute, apiculate. Capitula crowded on spike-like branches in a moderately broad panicle. Involucre subglobose or broadly ovoid, 1.5–2.5mm diameter; phyllaries ovate, c 2mm, yellowish pubescent and sparsely glandular-stipitate. Corollas sparsely glandular. Female flowers 3–8. Bisexual flowers 5–10; corollas also pubescent above.

Chumbi: Yatung (370). Grasslands, 2900m. September.

No specimens seen. Description based on (370, 372, 373).

13. A. wellbyi Hemsley & Pearson. Fig. 132d–e.

Subshrub, 15–25cm; stems usually simple, erect or rather lax, brownish hairy above. Mid-cauline leaves 2-pinnatisect; 2–5 × 1.2–2.5cm, distinctly petiolate, without stipuliform basal segments; primary segments usually 3 pairs; ultimate segments linear or oblanceolate, mostly 1–7 × 0.4–1mm, hyaline-acuminate, sparsely hairy. Capitula rather few, in racemiform panicles. Involucres semi-globose, 1.5–2.5mm diameter; phyllaries ovate to obovate, 2.3–3 × 1.3–2mm, ± glabrous, with broad scarious margins. Female flowers c 15; corollas 0.9–1.6mm. Inner flowers male, c 12; corollas 1.7–2.5mm; ovaries rudimentary.

Sikkim/Tibet border: Giri, Khamba Dzong. 4700m. July.

14. A. stricta Edgeworth; *A. edgeworthii* Balakrishnan *nom. superfl.* Fig. 132f–g.

Annual or biennial, 2.5–45cm; stems erect or rather lax to ± absent, brownish hairy above. Mid-cauline leaves 1–2-pinnatisect, 1–3cm; primary segments 1–3 pairs, usually intergrading with several stipuliform basal segments; ultimate segments linear or oblanceolate, 1.5–10 × 0.2–0.5mm, hyaline acuminate, puberulous to thinly brownish villous. Capitula numerous, in dense racemiform panicles, rarely in lax panicles. Involucres ovoid, 1.5–2.5mm diameter; phyllaries mostly ovate to obovate, 1.6–2.7 × 0.8–1.5mm, villous on herbaceous part (rarely inner ones glabrous), with broad scarious margins. Female flowers c 18–40; corollas 0.5–0.7mm. Central flowers male, c 3–8; corollas 1.3–1.5mm; ovaries rudimentary. Achenes obovoid, 0.7mm.

Bhutan: **N** – Upper Mo Chu district (Laya, Gangyuel Chu); **Sikkim:** Lhonak, Phaklung; **Chumbi:** Phari–Gautsa. Grassy banks and among juniper on river bank, 3670–4570m. August–October.

The plants from Sikkim and Chumbi are short, often laxly several-branched from base and may bear simple, long-petiolate lower cauline leaves. They have been recognised as a distinct variety: *A. edgeworthii* var. *diffusa* (Pampanini)

Ling & Y.R. Ling [*A. stricta* fm. *diffusa* Pampanini]. The plants from Bhutan are tall with erect usually solitary stems and would represent the typical variety.

15. A. desertorum Sprengel. Fig. 132k.

Tufted perennial herb, 7–10cm, brown or yellowish-grey villous; stems surrounded at base by persistent petioles with dilated scarious bases, erect. Cauline leaves pinnatisect, obovate in outline, primary segments 5–7, approximate, oblong-oblanceolate, 5–9 × 0.7–1.2mm, ± acute, entire, rarely 2-lobed; petiole to 20mm. Capitula c (1–)10–12, racemose, pedunculate, distant below, crowded, subsessile above. Involucre oblate, 3–4.5mm diameter; phyllaries mostly ovate to oblong, 2–3 × 1.3–2mm, broadly scarious-margined, sparsely glandular-stipitate and sometimes pilose, outermost lanceolate. Early capitula with c 25 female flowers and c 15–30 male ones, decreasing to c 13 female:7 male in last capitulum. Female corollas c 0.7mm. Ovaries ± absent in male flowers.

Sikkim: Lhonak, Rasung. River banks, 4570–5000m. July.

16. A. parviflora D. Don; *A. japonica* auct. non Thunberg. Nep: *Tite pati*. Fig. 132h–i.

Perennial, 0.4–1m, puberulous or glabrescent; stems erect. Mid-cauline leaves pinnatisect, 1.5–4cm; primary segments 1–2 pairs, linear or linear-oblanceolate, subacute to acuminate, with 0–2 teeth or lateral lobes; 2–4 pairs of erect, linear or linear-oblanceolate, stipuliform lobes at base. Inflorescence branches many, ascending; capitula usually very numerous and crowded on spicate branchlets, rarely more distant on longer peduncles. Involucre ovoid, c 1.5mm diameter; phyllaries ovate(-elliptic), glabrous, most 1.5–2(–2.5) × 1–1.5mm, outermost smaller. Female flowers c 9–15; corollas c 0.6mm; achenes obovoid, c 0.5mm. Central flowers male, fewer, c 3–10; corollas 1.5mm; ovaries ± absent.

Bhutan: C – Thimphu district (Samana, Namselling, Thimphu), Punakha district (Ritang (357)), Bumthang district (near Bumthang); **Sikkim:** Chunthang, Lachen; **Chumbi:** Chubitang–Yatung. In sandy turf on hill-top and grassy hillsides, 2150–4350m. June–October.

17. A. dubia Besser var. **dubia**. Dz: *Khempa*. Fig. 132j.

Perennial, 1–1.5m; stems erect, minutely appressed pubescent. Lower inflorescence leaves pinnatisect, c 8 × 5cm, sparsely pubescent above, whitish araneous-pubescent beneath; lateral segments (2–)5–6, lanceolate, 2.5–4 × 0.6–0.8cm, acuminate, entire; 2–4 entire stipuliform lobes at base. Capitula numerous in dense panicles. Involucre subglobose, 1.5–2mm diameter; phyllaries ovate to oblong, c 2 × 1mm, outer ones sparsely pilose at base, outermost smaller. Female flowers c 8–11; corollas 1mm. Central flowers male, c 5–11; corollas 2mm; ovaries ± absent.

Bhutan: C – Thimphu district (Thimphu, Tsalimaphe). On open banks, 2130–2400m. August.

76. ACHILLEA L.

Perennial herb with slender elongate stolons, woolly-puberulous; stems erect. Leaves alternate, bipinnatisect, narrowly oblong-elliptic in outline; cauline ones ± sessile. Capitula small, numerous, in dense terminal corymbs, paleate. Involucre ovoid, 3-seriate; phyllaries oblong, broadly scarious. Receptacle convex, paleae oblong, hyaline, acuminate. Ray flowers usually 5, female; ligule broadly obovate, 3-toothed, pink. Disc flowers c 18, bisexual; corolla tubular-campanulate, white; style branches truncate, penicillate. Achenes narrowly obovate, weakly winged; pappus absent.

1. A. millefolium L.

Plant c 0.5m. Leaves green, tomentose beneath at first. Cauline leaves 90–120 × 20–25mm; primary segments c 20 pairs; secondary segments 3–4 pairs, acuminate, with 0–2 acuminate teeth or lobes, the ultimate segments being linear or subulate. Involucre 2mm diameter; phyllaries 2–3.5 × 0.8–1.4m, scarious margin brownish and filamentous-lacerate above. Ligule c 2.2 × 2.5mm. Disc corolla 2.2. Achenes c 2mm.

Darjeeling district: Senchal. Opening in degraded evergreen broad-leaf forest, 2350m. July.

Native of Europe and W Asia, widely introduced.

77. XANTHOPHTHALMUM Schultes Bipontinus

Erect, glabrous, annual herb. Leaves 1–2-pinnatisect, alternate. Capitula radiate, solitary at branch ends. Involucre hemispherical; phyllaries imbricate, 3–4-seriate with broad scarious margins. Receptacle convex, naked. Ray flowers female, ligules yellow. Disc flowers bisexual; corollas yellow, tubular-campanulate, laterally 2-winged, 5-toothed; style branches truncate. Achenes dimorphic; ray achenes triangular in section, angles winged; disc achenes somewhat compressed, winged on inner margin; all wings broadening upwards; pappus absent.

1. X. coronarium (L.) Trehane; *Chrysanthemum coronarium* L.

Plant 15–50(–70)cm. Leaves obovate in outline; lower leaves 5–7cm, ultimate segments oblong to linear (obovate), 1–3mm broad; upper leaves smaller, becoming 1-pinnatisect. Involucre 1.5–2cm diameter; phyllaries oblong, 4–10 × 2–4mm, with broad scarious margins. Ray flowers 15–20, ligules 15–20 × 4–5mm. Disc corollas 4–5mm. Achenes 2.5–3mm; wings of ray achenes c 1.5mm broad.

W Bengal Duars: Jalpaiguri (174).

Originally native to the Mediterranean region; cultivated and sometimes escaping.

78. LEUCANTHEMUM Miller

Erect perennial herbs. Leaves alternate, simple to weakly pinnatifid. Capitula radiate, solitary, long pedunculate. Involucre hemispherical; phyllaries imbricate, 3–4-seriate, margins scarious. Receptacle convex, naked. Ray flowers 1-seriate, female, ligules white. Disc flowers bisexual; corollas yellow, tubular-campanulate, 5-toothed, becoming spongy at base; anther bases sagittate; style branches truncate. Achenes monomorphic, oblong, ± terete, 5–10-ribbed. Pappus small, ± coroniform, irregular, usually only on ray achenes, often absent.

1. Involucre 12–20mm diameter; scarious border of phyllaries often dark purplish or dark brown towards inner margin; ligules 10–25mm **1. L. vulgare**
+ Involucre c 25mm diameter; scarious border of phyllaries pale brownish towards inner margin, never dark; ligule 35–45mm ..... **2. L. × superbum**

1. L. vulgare Lamarck; *Chrysanthemum leucanthemum* L.

Plant 15–70cm, glabrous or sparsely crisped-puberulous, with sterile rosettes of long petiolate leaves; stems simple or sparsely branched. Lower cauline leaves obovate, long petiolate, to 8 × 2.5cm, ± crenate-serrate; upper leaves oblanceolate to oblong, sessile, 2–6 × 0.5–2.5cm, usually obtuse, semi-amplexicaul, dentate or pinnatifid particularly at base. Involucre 12–20mm diameter; phyllaries lanceolate to oblong, 5–6 × 1–1.5mm, obtuse, scarious margin narrow, usually dark purplish or dark brown towards inside. Ray flowers 15–25, ligules 10–25 × 3–6mm. Disc corollas c 3mm. Achenes 2–3mm, blackish, with pale narrow ribs. Pappus sometimes present, rarely to 3mm, whitish.

Darjeeling: Senchal Lake (332a).

Originally from Europe, sometimes cultivated and escaping. Description above from European material

2. L. × superbum (J. Ingram) Kent; *L. maximum* auct. non (Ramond) DC. Eng: *Shasta daisy.*

Similar to *L. vulgare* but often coarser; leaves serrate, lower ones to 25cm, upper ones lanceolate to oblanceolate, to 10 × 2.5cm; involucre c 2.5cm diameter; phyllaries 6–10 × 3–4mm, scarious margins wide, pale brown towards inside; ray flowers c 33, ligules 35–45 × 5–6mm; disc corollas 4.5mm. Pappus present on ray achenes, 0.2mm.

Bhutan: C – Mongar district (Mongar). Cultivated in garden, 1800m. June.

Fertile garden hybrid derived from species from SW Europe.

79. COTULA L.

Annuals, often prostrate. Leaves 1–2-pinnatisect, alternate, sessile. Capitula solitary, disciform. Involucre hemispherical; phyllaries several-seriate, margins scarious. Receptacle naked, flat. Flowers sparsely subsessile-glandular. Outer flowers several-seriate, female, tangentially compressed, sometimes with thin wing, short-stalked; corolla absent or poorly developed. Disc flowers fewer, bisexual, almost sessile; corollas short, tubular-campanulate, deeply 4-toothed; style branches truncate. Achene of female flowers obovoid, dorsally compressed with pale margin or much thickened pale wing; of disc flowers scarcely winged. Pappus absent.

1. Peduncles 2–6cm in fruit, (sparsely) appressed pubescent; female flowers c 3-seriate .... **3. C. australis**
\+ Peduncles usually less than 1.5cm in fruit, glabrous or sparsely araneous; female flowers 5–6-seriate 2

2. Leaf segments acuminate to a fine point; achenes of female flowers with a pale thick margin .... **1. C. hemisphaerica**
\+ Leaf segments subacute with short apiculus; achenes of female flowers with a pale much thickened wing to 0.3mm wide .... **2. C. anthemoides**

1. C. hemisphaerica (Roxb.) Bentham & Hook.f.

Much-branched annual to 10cm, most parts subglabrous. Lower leaves 1–2-pinnatisect, to 4 × 1.5cm, ± obovate in outline, very sparsely villous beneath; primary segments 2–4(–6) pairs; ultimate segments ± narrowly oblong or linear, acuminate to a fine hyaline point. Upper leaves smaller, usually 1-pinnatisect. Capitula at anthesis 2–5mm diameter, often subsessile; peduncle c 5–10mm in fruit. Involucre 2-seriate; outer phyllaries oblong, pallid, to 2.2 × 1mm; inner ones smaller. Female flowers 5–6-seriate, without corollas; styles 0.2–0.4mm. Bisexual corollas c 0.8mm. Female achenes on stipes to 0.2mm, obovoid, c 0.8 × 0.4mm including pale thick margin.

Bhutan: S – Sarbhang district (Lam Pati); **W Bengal Duars:** Chel (175). Marshy clearing at edge of forest, fields, 410m. February–March.

One collection seen from our area. Further south plants may be larger, more villous especially at nodes or may have longer lower leaves or vestigial female corollas.

2. C. anthemoides L.

Similar to *C. hemisphaerica* but stems sometimes ± araneous; leaf segments subacute with very short hyaline point; phyllaries pale brown; disc flowers with thin wing around ovary and vestigal corolla; achenes c 1.4 × 0.8mm including pale, much thickened wing to 0.3mm wide.

Darjeeling: unlocalised (Hooker collection). 100m.

Cited from 'Sikkim' (80, 329), but only the above, dubious, very immature specimen with long peduncles seen from our area.

3. C. australis (Sprengel) Hook.f.

Similar to *C. hemisphaerica* but peduncles longer, 2–6cm in fruit, appressed pubescent; female flowers c 3-seriate; female achenes on stipes to 0.5mm, 1.2 × 0.6–0.8mm including pale thick wing c 0.15mm wide, densely minute glandular-stipitate on both faces.

Darjeeling: Ghoom, Darjeeling (419). Railway line, 2300m. July.

Native of Australia and New Zealand; possibly introduced along with sheep.

80. SOLIVA Ruiz & Pavón

Small annual herb; stems short, prostrate. Leaves alternate, 2–3-pinnatisect, spreading. Capitula sessile, solitary, axillary, subglobose, disciform. Involucre 2-seriate; phyllaries scarious. Receptacle flat, naked. Marginal flowers female, numerous; corolla absent. Central flowers fewer, male; corolla narrowly infundibular, 3-toothed; anthers 3; ovary indistinct, linear; style undivided, truncate, scarcely exserted (2- and 4-merous male flowers also recorded outside our area). Achenes obovate, dorsiventrally compressed, with wrinkled wings and persistent hardened hooked style; pappus absent.

1. S. anthemifolia (Jussieu) R. Brown

Stems up to 7cm. Leaves 4–12cm; primary segments 3–5(–7) pairs, up to 12mm; ultimate segments oblong or elliptic, 1.5–3 × 0.4–1.2mm, acuminate, sparsely brownish lanate; petiole c 3cm. Capitulum c 6mm diameter in flower, c 12mm in fruit. Phyllaries oblong to lanceolate, 3.5–4 × 1–1.6mm, acuminate, lanate. Female flowers villous; ovary c 1mm; style c 1.5mm. Male flowers c 2.5mm. Achenes obovate, incurved, 2.5 × 1.2mm, pale brown; style 2.5mm.

Bhutan: C – Punakha district (Samtengang–Tashi Choling, Wangdu Phodrang, unspecified localities (272)). Weed of winter crops, 1200–2100m (272). March–April.

SENECIONEAE

81. TUSSILAGO L.

Rhizomatous perennial herbs. Basal leaves reniform, petiolate, usually appearing after flower scapes emerge. Scapes 1-headed, bearing scale leaves. Capitulum radiate. Involucre campanulate, calyculate; phyllaries 1-seriate, equal. Receptacle flat, naked. Ray flowers marginal, several-seriate, female, narrowly ligulate; disc flowers tubular-campanulate, male; corolla 5-lobed; style undivided, rounded at apex. Pappus bristles scabrid; achenes of female flowers linear, obscurely 5–10-ribbed, pappus bristles numerous; achenes of male flowers thin, empty, pappus bristles fewer, coarser.

1. T. farfara L.

Basal leaf lamina broadly ovate, 6–15 × 6–15cm at fruiting, obtuse or subacute, base cordate, margins toothed or denticulate, ± glabrous above, white araneous beneath at least at first; petiole 7–15cm. Scapes 10–20cm in flower; scale leaves narrowly oblong, c 15 × 5mm. Involucre 7–10mm diameter, sparsely araneous at base; phyllaries linear-lanceolate, 10 × 2.5mm. Flowers yellow; tube of ray corollas 2.5–5mm, ligules 6–8 × 0.3mm, often reddish; disc corollas 7mm. Achenes oblong 4mm, pappus 13mm, white.

Sikkim: Lachen–Naram (104), Singalila Range; **Nyam Jang Chu:** Lepo. Moist slopes, 3300–3650m. April–May.

82. PETASITES Gaertner

Subdioecious rhizomatous perennial herbs. Basal leaves reniform, petiolate. Inflorescence scapose, ± paniculate, rachis bearing scale leaves; capitula disciform or discoid. Involucre campanulate, calyculate; phyllaries 1-seriate, equal. Receptacle flat, naked. Female flowers filiform; corolla slightly broadened above, with (3–)4(–5) unequal, narrow teeth in our species. Disc flowers tubular-campanulate, male; corolla ± equally 5-toothed at apex; style thickened above, abruptly narrowed to subulate branches. Achenes of female flowers linear, obscurely 5–10-ribbed; pappus bristles numerous, scabrid.

1. P. tricholobus Franchet; *P. himalaicus* Kitamura

Basal leaves scarcely developed at flowering, when lamina 5–10 × 5–10cm, acute or obtuse, base deeply cordate, margin denticulate, ± glabrous above, white araneous beneath at first, petioles 5–30cm. Scapes 10–30cm, inflorescence corymbose at first, with 6–20 capitula; basal scale leaves ovate or oblong, 2–3 × 1–1.5cm, acute, base sessile. Involucre 5–7mm diameter, sparsely araneous at base; phyllaries c 15, lanceolate, 10 × 2mm, purplish. Flowers white or purplish. Female corollas 10mm (female capitula usually with few male flowers near centre). Male corollas 8–10mm (male capitula usually without female flowers – none seen from our area). Achenes oblong, 3.5mm; pappus white, 10mm.

Bhutan: C – Thimphu district (near Thimphu), Tongsa district (Longtepang–Ritang (71)). 2250–3150m. March–April.

83. PARASENECIO W.W. Smith & J. Small

Perennial herbs. Leaves simple or palmatisect, alternate, ± palmately veined; petiole bases not sheathing, sometimes prominently auriculate. Inflorescence racemose or paniculate; capitula discoid, few-flowered. Involucres 1-seriate, calyculate; phyllaries few. Corollas tubular-campanulate, yellowish, 5-toothed. Style branches truncate, fringed with long papillae and sometimes with a central tuft of longer papillae. Achenes oblong or obovoid; pappus of capillary bristles, white or brownish.

1. Upper leaf axils and inflorescence bulbilliferous .... **2. P. quinquelobus**
\+ Bulbils absent 2

2. Leaves deeply lobed (to within 2cm of petiole) .... **1. P. palmatisectus**
\+ Leaves triangular or shallowly lobed 3

3. Leaves ± triangular, the 2 prominent basal lobes most prominent; auricles at base of petiole 1.5–2cm broad .... **3. P. chenopodifolius**
\+ Leaves ovate, shallowly 9–11-lobed; auricles at base of petiole small or absent .... **4. P. chola**

1. P. palmatisectus (Jeffrey) Chen; *Senecio palmatisectus* Jeffrey

Stems 30–60cm, sparsely glandular papillate. Leaves broadly ovate in outline, deeply (to within 10–22mm of petiole apex) (3–)5(–7)-lobed, blade 3–8(–14) × 4–9(–18)cm, base truncate or cordate, subglabrous to sparsely hirsute above, subglabrous to sparsely araneous and paler beneath; lobes usually deeply 2–7-lobulate with acute (rarely rounded), apiculate segments; petioles 1.5–10cm. Inflorescences terminal and in upper leaf axils, racemose or weakly paniculate; capitula usually rather few, widely spaced; peduncles with very few to many glandular hairs. Involucre c 3mm diameter; phyllaries (4–)5, narrowly oblong-oblanceolate, 7.5–10(–12)mm, usually sparsely glandular pubescent. Flowers 5–7; corolla tube 6–7mm; lobes recurving, Achenes oblong, 4mm, glabrous, strongly ribbed; pappus 6mm, white.

Bhutan: C – Ha district (Chelai La (364), near Ha), Thimphu district (Shodug–Barshong, Cheka–Shingkarop, Choidy Ponkay–Saka La, Bele La–Paro), **N** – Upper Bumthang Chu district (Kopub); **Chumbi:** Gowsar, Tongshong. Moist shady forests, plantations, 3000–4000m. July–September.

The application and spelling of an earlier name, published as *Senecio pelleifolius* King ex Drummond, is uncertain. Only examples of one syntype collection (of two) have so far been located. These are specimens of *P. palmatisectus* but differ from the original description in capitulum length, that given being quite exceptional for *P. palmatisectus*. The latter widely used name is accepted here with a view to the rejection of *S. pelleifolius* if the second syntype proves conspecific.

2. P. quinquelobus (DC.) Chen; *Senecio quinquelobus* (DC.) Clarke, *Cacalia pentaloba* Handel-Mazzetti, *S. bhutanicus* Balakrishnan, *S. himalaensis* Mukerjee

Similar to *P. palmatisectus* but usually bulbilliferous in axils of upper leaves and inflorescence; stem to 90cm; leaves simple, broadly ovate to triangular, blade (2.5–)5–8(–13) × (2.5–)6–10(–15)cm, acuminate, base truncate to deeply cordate, entire or shallowly (very rarely deeply) palmately 3–5-lobed, margins dentate, sometimes glabrous above; inflorescences spicate or narrowly paniculate, more congested, bulbils usually replacing most capitula; phyllaries glabrous; achenes not maturing.

Bhutan: C – Ha district (Ha La–Kyu La), **N** – Upper Mo Chu district (Laya), Upper Bumthang Chu district (Lhabja, Kopub); **Darjeeling:** Sandakphu–Phallut; **Sikkim:** Lachen, Lachung Valley, Yumthang, Yakla etc.; **Chumbi:** near Chumbi. Damp hollows amongst shrubs, 3000–4250m. August–September.

3. P. chenopodifolius (DC.) Grierson; *Senecio chenopodifolius* DC.

Stems 1–1.6m, sparsely puberulous. Leaves ± triangular, 9–15 × 10–15cm, acuminate, base truncate or cordate, margins dentate and denticulate, lateral margins also often 1–2-lobulate, sparsely puberulous on upper surface, paler and ± glabrous beneath; petioles to 10cm, with large rounded auricles 1–2.5cm broad. Inflorescence paniculate, branches elongate with many capitula on outer half. Involucres c 1.5mm diameter; phyllaries (4–)5, ovate-oblong, 2mm, sometimes exceeded by longest bract of calyculus. Flowers (2–)3–5; corolla tube 3.5–5mm; lobes c 1.2mm, erect. Achenes oblong, 3mm, glabrous; pappus c 4.5mm, brown or reddish. Occassionally phyllaries and corollas reddish above.

Bhutan: C – Thimphu district (near Barshong (363)).

No specimens seen from our area. Otherwise a W Himalayan species extending to central Nepal and Gosain Than and flowering August–September.

4. P. chola (W.W. Smith) Chen

Similar to *P. chenopodifolius* but leaves broadly ovate, 8–15 × 7–15cm, with 9–11 shallow acute lobes, base cordate, puberulous above, paler and glabrous beneath; petioles up to 10cm, often with weak oblong auricles c 0.5mm broad; inforescences denser, branches much shorter; involucre c 2mm diameter; phyllaries 5–6, oblong, 4mm; flowers (4–)5(–6); corolla tube c 6mm; pappus c 7mm, paler.

Sikkim: Chakung Chu. 3650–3960m. October.

84. LIGULARIA Cassini

Erect perennial herbs; stems leafy. Leaves alternate, simple or palmately dissected, petiolate; petioles with sheathing bases. Capitula solitary, few or numerous in racemes or corymbs, discoid or radiate. Involucres calyculate, cylindrical, obconic or campanulate; phyllaries 1-seriate, subequal. Receptacle flat, naked. Corollas yellow. Ray flowers 0–c 24, ligules ± showy. Disc flowers tubular-campanulate; anther bases shortly auriculate; style branches truncate with obtuse marginal papillae. Achenes oblong, glabrous; pappus of capillary bristles, white or reddish.

The following key includes species of *Cremanthodium* that may have more than 1 capitulum.

1. Inflorescence corymbose, branched beneath central (earliest) capitulum .. 2
+ Capitula solitary or inflorescence racemose with central axis and earliest capitulum at base or apex 6

2. Capitula discoid .... **4. L. mortonii**
+ Capitula radiate 3

3. Basal leaves broadly ovate or suborbicular, deeply cordate 4
+ Basal leaves ovate, oblong or elliptic, shallowly cordate to tapered and decurrent on petiole 5

4. Capitula 2–12; involucre c 8–20mm diameter; ray flowers c 10–24 **1. L. retusa**
+ Capitula very numerous; involucre c 2.5mm diameter; phyllaries 5–7 **4. L. kingiana**

5. Collar of stock not woolly (surrounded by dead leaf bases); leaves herbaceous; phyllaries black-margined .... **2. L. amplexicaulis**
+ Collar of stock brownish woolly; leaves thinly coriaceous; phyllaries black tipped .... **5. L. lancifera**

6. Basal leaf blades ovate-oblong to oblanceolate, cuneate to weakly cordate at base, longer than petiole 7
+ Basal leaf blades broadly ovate to triangular, hastate, cordate or sagittate, shorter than petiole .... 9.

7. Capitula 3–8; pappus 2mm .... **11. L. virgaurea**
+ Capitula 1(–3); pappus 5–8mm 8

8. Cauline leaves forming loose wide sheaths, the widest 2–6cm; phyllaries black lanate .... **Cremanthodium ellisii** (p. 1583)
+ Cauline leaves forming narrow sheaths, all c 10mm wide; phyllaries ± glabrous .... **Cremanthodium pseudo-oblongatum** (p. 1583)

9. Capitula 1–9; flowering beginning at top of raceme 10
+ Capitula (10–)15–many; flowering beginning at base of raceme 11

10. Ligule length less than 3 × width; disc corolla limb infundibular, more than (1½–)2 × length of tube, with erect teeth .... **6. L. latiligulata**
+ Ligule length more than 4 × width; disc corolla limb campanulate, less than 1½ × length of tube, with recurved teeth .... **7. L. hookeri**

11. Basal leaves narrowly ovate; blade c 7 × 3.5cm; petiole c 8cm **10. L. sagitta**
\+ Basal leaves broadly ovate; blade 7–25 × 7–22cm; petiole at least 20cm 12

12. Plants not exceeding 1m; all bracts subtending capitula linear or lowest one only broader; disc pappus reaching middle of corolla throat **8. L. atkinsonii**
\+ Plants exceeding 1m or elliptic to oblanceolate bracts subtending several lower capitula or disc pappus not exceeding base of corolla throat **9. L. fischeri**

1. L. retusa DC.; *Cremanthodium retusum* (DC.) Good, *Senecio retusus* (DC.) Hook.f.

Stems 30–60(–90)cm, blackish puberulent above. Basal leaves broadly ovate or suborbicular, blade 7–18 × 10–20cm, rounded or retuse, deeply cordate, regularly toothed, ± glabrous on both sides; petioles 15–30cm. Cauline leaves suborbicular, 3–6 × 5–10cm, with larger 5 × 6cm sheaths. Capitula 2–9 in racemose panicles, rarely solitary; involucre 1.5–2.5cm diameter; phyllaries lanceolate, 10–16 × 3–5mm, dark coloured, densely covered at base with short thick blackish hairs. Ray corollas yellow; tube c 3mm; ligule oblanceolate, 9–18 × 3–6mm, deeply toothed. Disc corollas 6–9mm. Achenes oblong 5.5mm; pappus white, 5.5–6.5mm.

Bhutan: C – Tongsa district, **N** – Upper Mangde Chu, Upper Bumthang Chu, Upper Kuru Chu and Upper Kulong Chu districts; **Sikkim:** Chakung Chhu, Kupup, Yakla etc.; **Chumbi:** Chola. Hillsides and in streams, 3950–4880m. July–August.

A variable species with differing sizes and numbers of capitula but united by the indumentum at base of involucre and by the broad cauline leaf sheath. Specimens with the indumentum extending over the whole involucre are sometimes separated as **L. cremanthodioides** Handel-Mazzetti.

2. L. amplexicaulis DC.; *Senecio yakla* Clarke, *S. ?pachycarpus* Clarke. Fig. 133c–f.

Stems 40–140cm. Leaves broadly ovate; lower ones 20–40 × 10–30cm, acute, shallowly cordate or cuneate at base, denticulate, araneous at first, petiole up to 30cm, sometimes narrowly dentate-winged; upper leaves with large sheaths. Capitula radiate, numerous in dense or lax corymbs. Involucre campanulate, c 5mm diameter; phyllaries c 10, oblong, 5–8 × 1.5mm, black-margined. Ray flowers c 7; corolla tube 3.5mm; ligule oblong, 7 × 1.2mm. Disc flowers c 15; corolla 5.5mm. Achenes oblong, 2.5mm; pappus 5mm, dirty white.

Bhutan: C – Thimphu, Tongsa, Tashigang and Sakden districts, **N** – Upper Mo Chu, Upper Pho Chu and Upper Bumthang Chu district; **Darjeeling:**

Singalila, Sandakphu, Tanglu; **Sikkim:** Kupup, Tsomogo, Phedup etc.; **Chumbi:** Phari. Marshes and damp slopes, 3350–4400m. July–September.

3. L. kingiana (W.W. Smith) Handel-Mazzetti

Similar to *L. amplexicaulis* but leaves broadly ovate or suborbicular, blades to 35 × 40cm, obtuse, deeply cordate, dentate, ± glabrous on both surfaces; petioles up to 45cm, not winged; peduncles densely brown puberulous; involucre narrowly cylindric, c 2.5mm diameter; phyllaries 5–7, oblong, 7.5–11.5 × 2–4.8mm; ray flowers (1–)2–3(–4), corolla tube 4.5–6.5mm, ligule 8–15.5mm; disc flowers 5–7, corolla 8.5–10.5mm; achenes 8–10mm; pappus 4–6mm.

Sikkim: Gnatong, Patang La, Tsomgo etc. In streams and screes, 3350–4570m. July–August.

4. L. mortonii (Clarke) Handel-Mazzetti; *Senecio mortonii* Clarke, *Cacalia mortonii* (Clarke) Koyama. Dz: *Dangbeb*; Sha: *Bong dok phu.*

Stems up to 1.5m. Radical leaves palmatisect, with 5(–7) primary segments, suborbicular in outline, 30–50(–90)cm across, sparsely pubescent on both surfaces at first, paler beneath; segments lobed or pinnatisect, coarsely serrate-dentate; petioles up to 60cm; upper cauline leaves palmate or pinnate with blade and petiole much reduced. Capitula discoid, numerous, in branched corymbs; involucres cylindrical, c 2.5mm diameter; phyllaries (4–)5, oblong-lanceolate, 9–11mm, glabrous. Flowers (3–)4–5, corollas 9mm. Achenes narrowly ellipsoid, c 7mm; pappus reddish, c 7mm.

Bhutan: C – Ha, Thimphu, Punakha, Tongsa and Sakden districts, **N** – Upper Mo Chu and Upper Kulong Chu districts; **Darjeeling:** Sandakphu (69), Tanglu; **Sikkim:** Chomnagu, Kupup, Sherabthang etc. Forest clearings, 2000–4000m. July–October.

A collection from Worpola, Sakden district, Bhutan is considered a hybrid of this species, possibly with *L. amplexicalis.* It has ± simple, deeply and irregularly toothed leaves, 6–10 irregular phyllaries, 3–6 ray flowers with irregular ± obovate ligules c 6 × 5mm, 8–18 disc flowers and 7mm achenes ?always void.

5. L. lancifera (Drummond) Grierson; *Senecio lancifer* Drummond

Similar to *L. amplexicaulis* but collar of stock densely brownish lanate; stems up to 120cm; lower leaf blades broadly ovate, rather coriaceous, 15 × 15cm, acute or obtuse, cordate at base, denticulate; involucres c 5mm diameter; phyllaries 7–8, oblong, 7 × 3mm, black tipped; ray flowers 7–8, corolla tube 3mm, ligule 10–17 × 4mm; disc flowers c 20, corollas 6–7mm.

?Chumbi.

One syntype locality possibly in Chumbi, cited in Sikkim by Drummond, but referred to Tibet by Mathur (in 344). No specimens seen from our area; description based on examples from Nepal and Tibet.

6. L. latiligulata (Good) Springate; *Cremanthodium hookeri* Clarke subsp. *hookeri* fm. *latiligulatum* Good (as subsp. *clarkei* Good fm. *latiligulatum* Good), *L. hookeri* (Clarke) Handel-Mazzetti fm. *latiligulata* (Good) Mathur

Plant 180–600mm; stems erect, glabrescent to pubescent throughout, usually surrounded by fibrous leaf remains at base. Basal leaves reniform or cordate; blade 20–100 × 25–110mm, obtuse, shallowly to deeply cordate at base, regularly and sharply dentate, glabrous to sparsely puberulous above, subglabrous to puberulous beneath; petioles 70–160(–300)mm. Lowest 1(–3) cauline leaves similar to basal leaves in size and shape but petiole usually shorter. Inflorescence erect, racemose, with hairs thickest on apex of peduncle and base of involucre, sometimes extending onto the phyllaries below; capitula 1–4(–6), opening in basipetal sequence with apical capitulum largest and earliest. Involucres (5–)7–12mm across; phyllaries oblong-elliptic, 9–15 × 3–6.5mm, acute to acuminate, inner ones with a very broad scarious margin. Corollas yellow. Ray flowers (9–)11–12(–16); corolla tube 3–5mm, ligule obdeltoid to oblong-elliptic, 8.5–15 × 4.5–8mm, length less than 3 × width, 3(–4)-toothed at apex, teeth usually very deeply and irregularly cut though sometimes obsolete. Disc corolla tube 1.8–3mm; limb narrowly infundibular, with suberect teeth, more than (1½–)2 × length of tube. Achenes oblong, 4mm; pappus 6.5–8mm, white.

Bhutan: C – Thimphu, Tongsa, Bumthang/Mongar and Tashigang districts, **N** – Upper Mo Chu and Upper Kuru Chu districts; **Darjeeling:** Phalut–Singalila; **Sikkim:** Boktak, Kupup, Yakla etc.; **Chumbi:** Chumbi. *Abies* forest to damp meadows and scree, 3050–4550m. July–September.

7. L. hookeri (Clarke) Handel-Mazzetti; *Cremanthodium hookeri* Clarke, *Senecio calthifolius* Hook.f. non (Maximowicz) Maximowicz, *Senecio sikkimensis* Franchet, *C. hookeri* Clarke subsp. *clarkei* Good fm. *angustiligulatum* Good (= *C. hookeri* subsp. *hookeri* fm. *hookeri*), *C. hookeri* subsp. *polycephalum* Good p.p., *S. flexuosus* Balakrishnan non E.D. Clarke

Similar to *L. latiligulata*, but capitula 1–9; ligule length more than 4 × width; disc corolla limb campanulate, less than 1½ × length of tube, with recurved teeth.

Bhutan: C – Ha, Thimphu and Tongsa districts, **N** – Upper Mo Chu, Upper Pho Chu, Upper Bumthang Chu, Upper Kuru Chu and Upper Kulong Chu districts; **Sikkim:** Chiya Bhanjang, Sherabthang, Yumthang etc.; **Chumbi:** Longrong. In undergrowth, *Rhododendron* scrub or forest, among boulders or tall herbs in grassland, on scree, 2450–4900m. July–September.

8. L. atkinsonii (Clarke) Liu; *L. sibirica* auct. non (L.) Cassini, *L. fischeri* auct. non (Ledebour) Turczaninov, *Senecio ligularia* Hook.f. var. *atkinsonii* (Clarke) Hook.f.

Plant 47–90cm; stems glabrescent below, sparsely crisped pubescent above, denser and sometimes mixed with white woolly hair on inflorescence. Basal leaves reniform to triangular or ovate, blade 70–150 × 70–140mm, acuminate to obtuse, prominently hastate or sagittate, regularly dentate, glabrescent or pubescent on veins beneath; petiole slender, 200–390mm. Cauline leaves 2(–3), with reduced petiole and usually reduced blade. Capitula (10–)15–50 in a narrow raceme, pendent in fruit; all bracts subtending capitula linear or the lowest one only narrowly ± elliptic. Involucre 3–4.5mm diameter; phyllaries oblong, very dark, 5.5–9.5mm, acute or acuminate, subglabrous, pubescent at apex. Ray corolla tube 3.5–5.5mm; ligule oblong, 5–11 × c 2.5mm. Disc corolla tube 3.5–5mm; limb 4.2–6.1mm including teeth 1–1.3mm. Achenes oblong, c 5.5mm (?immature); pappus white (yellowish), 5.5–8mm, reaching middle (top) of corolla throat.

Bhutan: C – Thimphu district (Barshong, Simu Dzong, Sinchu La etc.), **N** – Ha district (Ha–Chile La), Upper Bumthang Chu district (Lhabja); **Sikkim:** Kupup Tsomgo Yakla etc.; **Chumbi:** Dotag. Streamsides, meadows, marsh, turf on hill-top, 2300–4400m. July–September.

Separation from *L. fischeri* is problematic; *L. atkinsonii* has been given a narrow circumscription here.

9. L. fischeri (Ledebour) Turczaninov; *L. sibirica* auct. non (L.) Cassini, *L. racemosa* DC., *Senecio ligularia* Hook.f. var. *ligularia*.

Similar to *L. atkinsonii* but coarser plants differing by its stems to 1.8m; basal leaf blades to 25 × 22cm; capitula often more numerous, often several lower ones subtended by elliptic or lanceolate bracts; phyllaries to 13mm; ligules to 17 × 3.3mm; disc corollas to 12mm. Achenes c 7mm; pappus sometimes reddish, sometimes not exceeding base of corolla throat.

Bhutan: C – Thimphu district (Cangnana, Taba, Thimphu, Tsalimarphe), Punakha district (Nahi–Hinglai La), **N** –Upper Bumthang Chu district (Gasa); **Darjeeling:** Darjeeling; **Sikkim:** Kupup, Lachen, Samdong etc. Streamsides, meadows and marshy slopes, 2150–3650m. July–September.

A very heterogenous assemblage, provisionally grouped together here since division of our material according to length of pappus, as in (375a) etc., does not result in more consistent taxa.

10. L. sagitta (Maximowicz) Mattfeld

Stems c 50cm, araneous at first. Basal leaves narrowly ovate, subacute, sagittate; weakly denticulate, sparsely araneous beneath; blade c 7 × 3.5cm; petiole c 8cm, slender, very narrowly winged. Lower cauline leaf truncate at base. Inflorescence a simple raceme of many capitula, sparsely araneous. Involucre 3mm diameter; phyllaries lanceolate, 7mm, ± glabrous. Ray flowers c 5–7; corolla tube 6mm; ligule c 10 × 2.2mm. Disc flowers c 8–10; corolla c 13mm. Pappus whitish, c 9mm.

?Chumbi: Gyong. 3950m. September.

11. L. virgaurea (Maximowicz) Mattfield; *Senecio lagotis* W.W. Smith, *Cremanthodium plantaginifolium* Good

Stems 35–60cm, sparsely puberulous above, surrounded at base by leaf remains. Radical leaves oblanceolate, 14–25 × 4–5cm, acute or obtuse, cuneate or attenuate to petiole up to 12cm, entire, glabrous; cauline leaves several, oblong, sheathing, scarcely touching, the lowest c 3cm broad. Capitula 4–8 in a lax raceme; involucre narrowly campanulate, 5–7mm diameter; phyllaries ovate-lanceolate, 6–8 × 3–4mm, glabrous or sparsely pubescent at base. Corollas yellow. Ray flowers 5–12; corolla tube 1–2mm; ligule elliptic, 8.5–10 × 2.5mm. Disc corollas 4–5mm. Achenes ellipsoid, 2.5–5mm; pappus 2mm, white.

Bhutan: C – Thimphu district (Shodug, Barshong, Tremo La–W Kang La), **N** – Upper Mo Chu district (Lingshi); **Sikkim:** Kupup, Temu La, Tsomgo (411) etc.; **Chumbi:** Yatung. Alpine meadows, 3650–4500m. July.

Named according to the synonymy provided by Liu (375a). Our material differs considerably from the type, particularly in the short pappus, and matches the descriptions of *Cremanthodium botryocephalum* Liu and *C. spathulifolium* Liu, both based on single collections from Chumbi (no material seen). The Himalayan element may therefore need removing from *L. virgaurea* sensu Liu.

85. CREMANTHODIUM Bentham

Erect perennial herbs. Leaves most or all basal, simple, entire or dentate, to 1-sect, petiolate, petioles with sheathing base; cauline leaves similar to basal but with shorter pertiole or variously reduced down to oblong sheaths or linear bracts; stems simple mostly unbranched. Capitula 1(–3), mostly cernuous, radiate, rarely discoid. Involucre campanulate to hemispherical; phyllaries interlocked in 1 series at least at first, often separating later and sometimes aparently 2-seriate, often blackish. Receptacle ± flat, naked. Ray flowers female; corollas mostly yellow, rarely pink, ligule entire or 3-toothed. Disc flowers bisexual; corollas yellow, orange or brownish, tubular-campanulate (narrowly infundibular), 5-toothed, upper portion of corollas not completely exserted from phyllaries and pappus; style branches thick, flattened, obtuse. Achenes oblong, 5–10-ribbed; pappus of white or brownish bristles.

The following key includes species of *Ligularia* that may have 1 capitulum.

1. Basal leaves broadly ovate or suborbicular, or deeply cordate at base, rarely truncate, ± as long as broad, veins usually palmately radiating from base of lamina, on petioles longer than blades 2

\+ Basal leaves ovate, oblong or linear at least in outline, shallowly cordate, truncate or tapering into petiole, veins pinnate from a distinct midrib, petiole often shorter than blade 8

2. Ligules pink, rarely white .... **1. C. palmatum**
\+ Ligules yellow, orange or brownish 3

3. Ligule length more than 4 × width .... **Ligularia hookeri** (p. 1576)
\+ Ligule length less than 3 × width 4

4. Leaves brownish-white tomentose beneath .... **4. C. decaisnei**
\+ Leaves subglabrous or sparsely pubescent beneath 5

5. Stems leafless above base, only 1–2 small bracts present 6
\+ Stems with 1–3 leaves above base 7

6. Phyllaries large, 12–25 × 4–12mm, usually violet-black hairy at least at base, rarely glabrous .... **2. C. reniforme**
\+ Phyllaries smaller, 9–16 × 2–6mm, glabrous or minutely ciliate **3. C. thomsonii**

7. Ligules 8.5–15mm; petioles of dead leaves broken into dull blackish fibres .... **Ligularia latiligulata** (p. 1576)
\+ Ligules 17–35mm; petioles of dead leaves intact, glossy, stramineous **2. C. reniforme**

8. Leaves pinnatifid or pinnatisect .... **5. C. pinnatifidum**
\+ Leaves entire or toothed 7

9. Stems stout, 15–60cm or more, basal leaves over 2cm broad 10
\+ Stems slender, rarely reaching 15(–25)cm, basal leaves usually less than 2cm broad 13

10. Cauline leaves forming loose wide sheaths, the widest 20–60mm wide; phyllaries black lanate, rarely glabrous .... **11. C. ellisii**
\+ Cauline leaves forming narrow sheaths, all c 3–10mm wide; phyllaries ± glabrous 11

11. Stems and leaves sparsely pubescent at first, not glaucous; leaves toothed; ligules usually distinctly 3-toothed .... **12. C. oblongatum**
\+ Stems and leaves glabrous, at least thinly glaucous; leaves entire; ligules tapering to a ± untoothed point 12

12. Basal leaf blades ovate, 30–65 × 25–45mm .... **6. C. yadongense**
\+ Basal leaf blades oblong, oblong-ovate or-elliptic, 40–110 × 20–40mm **10. C. pseudo-oblongatum**

13. Leaves linear-oblanceolate; phyllaries glabrous .... **13. C. bhutanicum**
\+ Leaves narrowly ovate or oblong; phyllaries densely blackish lanate ... 14

14. Capitula discoid; basal and cauline leaves present .... **9. C. discoideum**
\+ Capitula radiate; leaves mostly cauline, basal ones reduced or absent .. 15

15. Cauline leaves dentate; basal leaves scarcely sheathing; capitula ± cernuous **7. C. humile**
\+ Cauline leaves entire or undulate, cauline leaves sheathing; capitula erect **8. C. nanum**

1. C. palmatum Bentham; *C. palmatum* var. *benthamii* Good

Plant 7–40cm. Basal leaves broadly ovate to suborbicular; blades 1–5 × 1–4.5cm, toothed or palmately divided to middle (rarely more deeply) into acute segments, truncate to cordate at base, sparsely blackish pubescent above, glabrous and purplish beneath; petiole 1–7cm; cauline leaves reniform or linear, without sheaths. Capitulum solitary; peduncle blackish puberulent above. Involucre c 6–15mm diameter; phyllaries lanceolate, 15 × 3–4mm, acute, blackish puberulent. Ray corolla tubes 3mm; ligules obcuneate, pink, rarely white, 13–25 × 5–8mm, deeply 3-toothed. Disc corollas pink, 9–10mm. Achenes oblong, 3mm; pappus white, 9.5mm.

Bhutan: C – Ha, Thimphu, Tongsa and Mongar districts, **N** – Upper Mo Chu, Upper Bumthang Chu and Upper Kulong Chu districts; **Sikkim:** Chemathang, Goecha La, Onglakthang etc.; **Chumbi:** Gyong. Open stony slopes and screes, 3000–5050m. June–October.

2. C. reniforme (DC.) Bentham

Plant 10–40cm. Basal leaves ovate to broadly reniform; blades 1–9.5 × 2–9cm, acute to ± truncate, cordate, regularly dentate, glabrous; petioles 6–23cm; cauline leaves reniform or reduced to a sheath 2–3 × 1–2cm. Capitulum solitary; peduncle brownish lanate above. Involucres 10–20mm diameter; phyllaries 12–25 × 4–12mm, usually violet-black hairy, often coarsely so, at least at base, rarely glabrous. Ray corolla tubes obconic, 1–3mm; ligules oblanceolate, yellow, 17–35 × 6–13mm, shallowly toothed. Disc corollas brownish, 7mm. Achenes narrowly obovoid, 5mm; pappus white (very rarely brown), 5.5–6mm.

Bhutan: C – Thimphu district (Pajoding), Tongsa district (Takse La, Chale La, Chendebi), Bumthang district (Dhur Chu), Bumthang/Mongar border (Rudong La); **Darjeeling:** Sandakphu, Singalila; **Sikkim:** Goecha La, Tsomgo, Zongri etc. On peaty soils in turf and among rocks, in open, in fir forest and among rhododendron scrub, 3200–4720m. June–October.

3. C. thomsonii Clarke. Fig. 117e–i.

Plant 7–40cm. Basal leaves reniform; blades 1–2.5(–4) × 1.5–3.5(–5)cm, obtuse, crenate-apiculate toothed, deeply cordate at base, glabrous or sparsely pubescent beneath; petiole 4–13cm; cauline leaves few, reniform, confined to lowest part of stem. Capitulum solitary; peduncle hairy above. Involucre c 6–15mm diameter; phyllaries ovate to lanceolate, 10–16 × 2–6mm, glabrous or minutely ciliate. Ray corolla tubes 3–4mm; ligules oblanceolate, yellow, 15–21 × 5–8mm, deeply toothed. Disc corollas c 8.5mm. Achenes oblanceolate, c 4mm; pappus brownish, c 7mm.

Bhutan: C – Thimphu, Tongsa, Bumthang and Mongar districts, **N** – Upper Mo Chu, Upper Pho Chu, Upper Mangde Chu and Upper Kulong Chu districts; **Sikkim:** Boktak, Nathang, Sherabthang etc. Open rocky hillsides, cliffs, peaty meadows, 3050–4550m. June–August.

A potential hybrid of *C. thomsonii* has been noted from Lama Pokhri, Sikkim (342). It has several characters of *C. reniforme*, including black-haired phyllaries, but is smaller in several parts than either species.

4. C. decaisnei Clarke

Plant 12–30cm. Basal leaves suborbicular or reniform; blade 1.5–3 × 2.5–4cm, crenate-apiculate toothed, deeply cordate, glabrous above, closely brownish-white tomentose beneath; petiole 2.5–15cm; cauline leaves similar, smaller. Capitulum solitary; peduncle brownish lanate above. Involucre c 6–15mm diameter; phyllaries 9–16 × 2–5mm, brownish lanate. Ray corolla tubes 3mm; ligules elliptic, yellow, 13–22 × 5–9.5mm, shortly toothed. Disc corollas c 8.5mm. Achenes oblong, 4mm; pappus white, c 6.5–7mm.

Bhutan: C – Tongsa district, **N** – Upper Mo Chu to Upper Kulong Chu districts; **Sikkim:** Yume Samdong; **Chumbi:** Chumegati, Lingshi La. Hillsides and beside streams, 3650–5150m. June–September.

5. C. pinnatifidum Bentham

Plant 8–25cm; stems and peduncles appressed blackish pubescent. Leaves pinnatifid or pinnatisect, elliptic in outline; blade 1.5–4 × 1–2cm, glabrous on both surfaces, paler beneath; segments 3–5 pairs, lanceolate, 4–12 × 1.5–3mm, acute or acuminate; petioles up to 4cm, sheaths up to 8mm broad at base; cauline leaves reduced or absent. Capitulum solitary. Involucres 7.5–15mm diameter; phyllaries lanceolate, 10 × 2–4mm, appressed blackish pubescent. Ray corolla tubes 1–1.5mm; ligules yellow, 12–20 × 4–9mm, deeply toothed. Disc corollas 5.5mm. Achenes (immature) oblanceolate, c 3mm; pappus white, 5.5mm.

Bhutan: C – Ha district (Kyu La (362)); **Sikkim:** Lachung, Sherabthang, Yakla etc. On turf covered rocks, 3950–4550m. August–September.

6. C. yadongense Liu

Plant 25–35cm, glaucous; stems glabrous below, brown-pilose above. Basal leaves ovate, 30–65 × 25–45mm, obtuse, cordate at base, denticulate or subentire, glabrous, pinnately veined; petiole 10–30mm. Cauline leaves oblong to linear, 25–50mm. Capitula solitary; involucre hemispherical, c 25mm across tips of phyllaries, brown-pilose below; phyllaries oblong or lanceolate, glabrous, blackish. Ligules to 20 × 6mm; disc corollas 7mm including tube 1mm. Achenes 4mm, glabrous; pappus as long as corolla, white.

Chumbi: without locality (421). Grassy slopes. July.

No examples seen. Details based on type description.

7. C. humile Maximowicz; *C. comptum* W.W. Smith

Stems mostly concealed, glabrous, with few partly sheathing scale leaves and elongate petioles below. Upper leaf blades, 1–4 × 0.7–2.5cm, subobtuse, attenuate (to truncate) at base, dentate, ± glabrous above, appressed white tomentose beneath; petioles 5–25mm. Capitula solitary, ± cernous; peduncle shaggily blackish pubescent; involucre c 13mm diameter; phyllaries narrow oblong-elliptic, 15–20 × 2.6–4mm, densely dark brown or blackish appressed tomentose. Ray corollas yellow, tube c 3mm, ligule elliptic, 14–21 × 5–6mm, weakly toothed. Disc corollas 7–9.5mm. Pappus white, to 10.5mm.

Bhutan: C – Thimphu district (Wasa La), **N** – Upper Mo Chu district (Lingshi), Upper Mangde Chu district (Saga La), Upper Bumthang Chu district (Marlung). Stony ground and screes, 4250–4550m. July–September.

8. C. nanum (Decaisne) W.W. Smith; *Werneria nana* (Decaisne) Hook.f.

Stems mostly concealed, greyish tomentose above. Basal and lower cauline leaves reduced to sheathing lanceolate scales or long petiolate with elliptic, entire, usually glabrous blade 1–3 × 0.6–1.2cm; upper cauline leaves elliptic, 1.5–2 × 0.5–1cm, obtuse or subacute, with sheathing base, entire or undulate, whitish tomentose on both surfaces. Capitula solitary, ± erect; involucre c 15mm diameter; phyllaries lanceolate, 15 × 3mm, blackish tomentose, intermixed with greyish hair near base. Ray flowers ± as long as phyllaries; corolla yellow, tube 3.5mm, ligule 8.5 × 3.7mm, weakly toothed. Disc corollas orange, 6.5mm, gradually widening from base. Pappus to 8mm.

Sikkim: Chakalung La, Naku La etc.; **Bhutan/Chumbi:** border E of Phari. On scree slopes, 4550–4900m. June–August.

9. C. discoideum Maximowicz; *C. cuculliferum* W.W. Smith

Stems 10–15(–25)cm, white tomentose above intermixed with blackish hairs. Basal leaf blades ovate-lanceolate, 1.5–3 × 0.8–1.5cm, subacute or obtuse, cuneate to petiole up to 5cm, revolute at margin, undulate to weakly toothed, ± fleshy, glabrous on both faces, pale beneath; cauline leaves similar, sessile, sheathing, to 4.2 × 1.8cm. Capitula solitary, cernuous, involucre 1.5–2cm diameter; phyllaries elliptic or lanceolate, 12–14 × 2.5–4mm, coarsely dark purple hairy. Ray flowers absent. Disc corollas 8mm. Achenes (immature) oblong, 3mm; pappus to 8mm, white.

Bhutan: C – Ha district (Ya La/Pya La/Gile La); **Bhutan/Chumbi:** Tsethanka; **Chumbi:** Chumegati. 4570m. July–September.

10. C. pseudo-oblongatum Good

Stems 15–45cm, glabrous. Basal leaves oblong-ovate, 3–5 × 1.5–3cm, acute, rounded at base or cuneate to petiole to 3cm, entire or obscurely denticulate, glabrous on both surfaces; cauline leaves 4–5, oblong, to 1cm broad. Capitula solitary or rarely 2; involucre 1.5–2cm diameter; phyllaries 10 × 4mm, acute. Ray corolla yellow tube c 2.5mm; ligule 15 × 4mm, apex tapering, almost untoothed. Disc corollas 6.5mm. Achenes 3–3.5mm; pappus white, 7mm.

Bhutan: C – Thimphu district (Taka La, Thimphu); **Chumbi:** Chomp Lhari, Pem La. In peaty soil, 4100–4880m. July–August.

11. C. ellisii (Hook.f.) Kitamura; *Werneria ellisii* Hook.f., *C. plantagineum* Maximowicz

Stems 10–30cm glabrous below, blackish-grey lanate above. Basal leaves oblong-elliptic, 7–20cm, obtuse or acute, attenuate below into petiole up to 10cm, denticulate to dentate, often with linear teeth, ± glabrous; cauline leaves smaller, sheathing, sessile, lower ones up to 6cm broad. Capitula solitary, involucres 1.5–2.5cm diameter; phyllaries lanceolate, 12–15 × 3–4mm, acuminate, blackish lanate, rarely glabrous. Ray corollas yellow; tube c 4mm; ligule elliptic, 12–20 × 4–5mm, shallowly toothed above. Disc corollas 7mm. Achenes oblong, 3mm, pappus white, 6.5–8mm.

Bhutan: N – Upper Pho Chu district (Kangla Karchu La); **Sikkim:** Donkiah–Samdong, Goecha La, Dzongri; **Chumbi:** Chumegati, Bhutan border E of Phari. Amongst boulders, 4870–5100m. July–September.

A group of scree plants outside our area with conspicuous purple or bronze pigment can be identified as *C. purpureifolium* Kitamura (135). Non-purple scree plants from Wasa La, Thimphu district, 4400m and Lingshi La, Upper Mo Chu/Tibet border, 4550m are very similar to *C. ellisii* but differ somewhat by the very shallow teeth on their leaves and sparser thicker phyllary hair, and should probably also be referred to a broader concept of *C. purpureifolium*.

12. C. oblongatum Clarke

Stems 10–30cm, sparsely appressed pubescent above, surrounded at base by fibrous leaf remains. Basal leaves ovate-oblong, 2–11 × 1.5–6cm, obtuse or acute, base cuneate or shallowly cordate to petiole up to 5cm, margins obscurely toothed or denticulate, puberulent, becoming ± glabrous on both surfaces; cauline leaves narrowly oblong, sheathing, up to 10mm broad. Capitula solitary, involucres 1–2cm diameter; phyllaries lanceolate, 11–18 × 2–5mm, acuminate, glabrous or sparsely puberulent. Ray corollas yellow; tube c 1.5mm; ligule oblanceolate, 10–12.5 × 2.5–4mm, shallowly toothed. Disc corollas 5.5mm. Achenes oblong, 3–3.5mm, pappus white, 5mm.

Bhutan: C – Thimphu district (Pajoding), **N** – Upper Mo Chu district (Lingshi, Laya, Pang La), Upper Kuru Chu district (Pu La); **Sikkim:** Goecha La, Lhonak, Sebu Chho; **Chumbi:** Chulong, Pem La etc. Stony hillsides, grassland and near streams, 3800–4830m. July–September.

A variable complex. *C. nepalense* Kitamura has been separated on account of the distinctly raised veins and somewhat firmer texture of its leaves, the basal ones being truncate or weakly cordate at base, the cauline ones few and well separated. No specimens showing all these characters were seen from our area. However, some specimens (none seen) from S. Tibet, including Chumbi, have been described as *C. cordatum* Liu and distinguished from *C. nepalense* by their ovate leaves with cordate base and scarcely raised nerves, brown-pilose petioles and glabrous phyllaries.

13. C. bhutanicum Ludlow

Stems 8–25cm, araneous above, base surrounded by fibrous leaf remains. Basal leaves linear-oblanceolate, 2–10 × 0.2–0.9cm, obtuse or acute, base attenuate, glabrous, margins entire, minutely reflexed, pale beneath; cauline leaves sessile, linear, 1–3cm. Involucre c 1cm diameter; phyllaries lanceolate, 8–12 × 1.5–3mm, acute, glabrous. Ray flowers about 12; corolla tube 1.5–2mm; ligule elliptic, yellow, 12.5–20 × 2.5–5mm, shallowly 3-toothed. Disc corollas 5.5–6.5mm. Achenes oblong, 2–2.5mm, pappus white, 6.5mm.

Bhutan: C – Thimphu and Tongsa districts, **N** –Upper Mo Chu, Upper Mangde Chu, Upper Bumthang Chu, Upper Kuru Chu and Upper Kulong Chu districts. Peaty turf and screes, 4100–4730m. June–August.

86. DORONICUM L.

Perennial herb; stems simple. Radical leaves petiolate, cauline alternate, ± sessile. Capitula long peduncled, radiate. Involucre hemispherical; phyllaries herbaceous, 2-seriate, subequal, acuminate. Receptacle convex, naked. Corollas yellow. Ray flowers 1-seriate, female. Disc flowers bisexual, tubular-campanulate; corolla 5-toothed above; style branches very short, broadly obovate. Achenes 10-ribbed; pappus of numerous scabrid bristles on disc flowers, absent or reduced to c 2 bristles on ray flowers.

1. D. roylei DC.

Stems erect, 45–100cm, sparsely pubescent, bearing 1–4 capitula at apex. Basal leaf blades ovate, 6–10 × 4–8cm, acute, base truncate or slightly cordate, subentire or denticulate, sparsely pubescent on upper surface, subglabrous beneath; petiole up to 20cm; cauline leaves ovate or oblanceolate, 7–13 × 3–5cm, attenuate to semi-amplexicaul base. Phyllaries lanceolate, 20 × 2mm, acuminate, densely pubescent at base. Ray corolla tubes 2–3mm; ligules 12 ×

3mm, apex entire. Disc 1.5–2cm diameter; corollas 4mm. Achenes obovoid-oblong, 2.5–3mm, ray achenes glabrous, disc pubescent on ribs; pappus reddish or sordid, 3mm.

Bhutan: C – Thimphu district (Barshong), **N** – Upper Mangde Chu district (Goktang La), Upper Kulong Chu district (Shingbe, Me La); **Sikkim:** Chamnago; **Chumbi:** Chumbi. Marshes and damp screes, 3300–4110m. June–September.

87. SYNOTIS (Clarke) Jeffrey & Chen

Erect or scrambling perennial herbs. Leaves simple, petiolate or sessile sometimes auriculate, broadly ovate to elliptic. Capitula few to numerous in axillary or terminal racemes or corymbs, usually radiate, sometimes disciform or discoid. Involucres 1-seriate, calyculate, cylindrical or campanulate; calyculus sometimes of broad, appressed segments, resembling additional series of phyllaries. Receptacle flat, sometimes fimbrillate. Phyllaries 5–12, linear-lanceolate, margins scarious. Corollas yellow. Female flowers 1–14, usually radiate, sometimes filiform or absent. Disc flowers 2–many, tubular-campanulate; corolla 5-toothed; anther bases with tail of sterile cells beneath locules; style branch apices truncate or convex with short or long marginal papillae and often with a central tuft of longer papillae. Achenes cylindrical, glabrous or pubescent; pappus of capillary bristles, present on all achenes, white or yellowish.

1. Lower part of stem not or weakly developed, with most or all leaves clustered near base 2
\+ Leaves ± equally distributed along stems 3

2. Leaves subrosulate or more scattered; petioles usually winged; female corollas filiform, 1–5mm or radiate with tube 2.5–3.5mm and ligule 2.5–4mm .... **1. S. alata**
\+ Leaves always subrosulate; petioles unwinged; female corollas tubular-infundibular, 7–8mm or very rarely radiate with 6mm tube and 2mm ligule .... **2. S. wallichii**

3. Capitula 3–12-flowered 4
\+ Capitula more than 12-flowered 7

4. Capitula usually 4-flowered 5
\+ Capitula usually 9–12-flowered 6

5. Leaves cuneate at base, petiole not more than 2cm .... **3. S. acuminata**
\+ Leaves rounded at base, petiole up to 8cm .... **4. S. tetrantha**

6. Leaves cuneate at base, ± sessile, coarsely toothed; inflorescence pubescent **5. S. triligulata**
+ Leaves rounded at base, petiole up to 1cm, finely toothed; inflorescence glabrous .... **6. S. vagans**

7. Leaves sparsely araneous to nearly glabrous beneath; stems strongly ribbed and angled .... **7. S. bhot**
+ Leaves densely white tomentose beneath; stems angular or not 8

8. Leaves up to 18 × 8cm, often auriculate at base; stems ribbed; achenes glabrous .... **8. S. cappa**
+ Leaves up to 7 × 2.5cm, never auriculate at base; stems ± smooth; achenes pubescent .... **9. S. kunthiana**

1. S. alata (DC.) Jeffrey & Chen; *Senecio alatus* DC.

Rhizomatous, (0.3–)0.5–1(–2)m; stems erect, mostly glabrescent below (villous-lanate around basal buds), cottony or short-pubescent to subglabrous above. Leaves subrosulate at base of flowering shoots to well spaced; blades 7–30 × 4–13cm, acuminate, rounded or cordate at base, coarsely dentate to remotely denticulate, very sparsely scabridulous above, subglabrous to very sparsely appressed-araneous beneath and villous along midrib; petioles winged, 4.5–20 × 0.7–4cm, often narrowly panduriform (auriculate base included). Inflorescence a panicle, usually large and open, with reduced leaves at lower nodes or linear bracts throughout, with capitula densely crowded on ultimate branchlets, sometimes densely crisped-pubescent. Capitula disciform or sometimes radiate, involucres to 2mm across; phyllaries 5, oblong, 5.5–8 × 0.8–1.5mm, obtuse to subacuminate, pubescent or sometimes glabrous. Female flowers usually 2; corolla filiform, 1–5mm or radiate with tube 2.5–3.5mm and ligule 2.5–4mm. Bisexual flowers usually 3, corollas 6.5–8mm, widening in upper ⅔. Achenes 2.5mm, glabrous or sometimes pubescent; disc pappus 3–5.5mm.

Darjeeling: Darjeeling, Khampung, Tanglu etc.; **Sikkim:** Kupup, Phusum, Zemu Chhu etc.; **Chumbi:** unlocalised (Hobson collection (353)); **Nyam Jang Chu:** Le–Pangchen (353). In woods and coniferous forests or on gravel, 1850–3950m. September–January.

Very variable. A specimen from Sonada, Darjeeling with non-rosulate leaves and short female corollas is referred here, even though it has the slender unwinged petioles of the next species. Jeffrey & Chen (353) consider a further, anomalous collection from Chumbi with rather numerous conspicuously long-caudate upper inflorescence leaves may belong here. So may three anomalous collections from NE Thimphu district, C Bhutan. *Bowes Lyon* 5093 (BM) and *Cooper* 3550 (E) have leaves puberulous beneath with long straight spreading hairs and more stiffly spreading phyllary hairs; the capitula have (2–)3 female flowers and (3–)4 bisexual ones (*Bowes Lyon* 5093) or usually 8 flower-pits on dehisced receptacles (*Cooper* 3550). *Sinclair & Long* 5545 (E) has short hairs beneath the leaves, capitula c 2.5mm across with 8 phyllaries, denser stouter phyllary hairs, some with cyanic pigments, (2–)3(–4) female flowers and (4–)5(–6) bisexual ones.

2. S. wallichii (DC.) Jeffrey & Chen; *Senecio wallichii* DC.

Similar to *S. alata* but slender, naked, ?entirely subterranean stem to 20cm sometimes present; aerial part 12–30cm; leaves always subrosulate at base of flowering stem, ovate, 3.5–15 × 2.5–15cm, acuminate, truncate or cordate at base, sinuate-dentate; petioles slender, unwinged, 5–12cm; inflorescence smaller, with ± naked flowering stem and few compact corymbs; capitula usually disciform, very rarely inconspicuously radiate; female flowers 2, corolla ± tubular but slightly dilated above, 7–8mm, 5-toothed or tube 6mm with 2mm subentire limb (Karglasa specimen only); bisexual flowers 3, corolla 8–10.5mm; achenes glabrous, disc pappus 5–7mm.

Bhutan: C – Thimphu, Punakha and Mongar districts, **N** – Upper Mo Chu and Upper Bumthang Chu districts; **Darjeeling:** Darjeeling, Senchal; **Sikkim:** Karglasa, Lachung, Phusum; **Chumbi:** Yatung (353). In forests, on rocks, 1830–3200m. August–November.

3. S. acuminata (DC.) Jeffrey & Chen; *Senecio acuminatus* DC.

Stems up to 130cm, ± glabrous. Leaves elliptic or lanceolate, 12–15 × 1.5–4cm, acuminate, cuneate or rounded at base, denticulate, ± glabrous on both surfaces; petiole 5–15mm. Capitula radiate, numerous, in corymbs; involucres c 1.5mm diameter; phyllaries 3–4, oblong, 6 × 1.5mm, acute, glabrous. Female flowers 0–1; corolla tube 2.5mm; ligule 2.5–5.5 × 0.2mm. Bisexual flowers 2–3; corollas 5mm. Achene oblong, 2.5mm; pappus 4.5mm, yellowish.

Bhutan: C – Thimphu district (Dotena, Barshong–Lingshi (or Upper Mo Chu district)), Tongsa district (Chendebi–Pele La), Bumthang district (near Bumthang); **Darjeeling:** Phalut, Tanglu; **Sikkim:** Tonglo, Yakla, Changu etc.; **Chumbi:** Yatung (353). Damp grassy swamps, 2550–4250m. June–October

4. S. tetrantha (DC.) Jeffrey & Chen; *Senecio tetranthus* DC.

Stems sparsely pubescent, subscandent; leaves ovate-lanceolate, 9–15 × 3–8cm, acuminate, rounded to weakly cordate at base, shallowly serrate or denticulate, sparsely pubescent on both surfaces but more dense beneath; petioles 1–8cm. Capitula radiate, ± cernuous, numerous, in racemes; involucres c 1.5mm diameter; phyllaries 4, oblong, 3–5 × 1–1.5mm, glabrous. Ray flowers 2; corolla tube 1.5mm; ligule 1.5–3mm. Disc flowers 2–3; corollas 5–5.5mm, broadening in upper ⅔. Achenes obovate, 1.5mm; pappus 3–5mm.

Bhutan: C – Thimphu/Punakha district (Sinchu La), **N** – Upper Bumthang Chu (Kopub); **Darjeeling:** Rechi La, Senchal, Tanglu; **Sikkim:** Tumbok, Yakla; **Chumbi:** Yatung (353). Along moist banks and under trees, 2150–3960m. September–January.

5. S. triligulata (D. Don) Jeffrey & Chen; *Senecio triligulatus* D. Don

Erect or scrambling, up to 150cm, sparsely pubescent or glabrous. Leaves ± elliptic, 10–15 × 4–6.5cm, acuminate, cuneate at base, serrate, ± glabrous on both surfaces or sparsely pubescent beneath; petiole 5mm. Capitula radiate, in axillary corymbs; involucres c 3mm diameter; phyllaries 5–6, oblong, 5 × 1.5mm, puberulous. Ray flowers 3–4; corolla tube 2mm; ligule 7.5 × 1mm. Disc flowers 6–8; corolla 6.5mm, enlarging in upper half. Achenes oblong, 1.5mm, glabrous; pappus 5–6mm, white.

Bhutan: C – Tashigang district (Nyoth–Shali (117), Sana); **Darjeeling/Sikkim:** unlocalised (Hooker and Treutler collections); **Sikkim:** Chakung. Shady woods, 1850–2000m. October–December.

6. S. vagans (DC.) Jeffrey & Chen; *Senecio vagans* DC.

Similar to *S. triligulata* but glabrous; leaves narrowly ovate, 7–12 × 2.5–4cm, rounded at base, denticulate; petioles up to 10mm; capitula in axillary corymbs; phyllaries 8, glabrous; ray flowers c 4–5, corolla tube 3mm; ligules 4 × 1.5mm; disc flowers 5–6; pappus 7mm.

Bhutan: without locality (135).

7. S. bhot (Clarke) Jeffrey & Chen; *Senecio bhot* Clarke

Stems 100–130cm, araneous, strongly angular. Leaves oblong or oblanceolate, 7–22 × 3.5–6cm, acute or shortly acuminate, attenuate at base, ± glabrous above, araneous and paler beneath. Capitula radiate, numerous, in lateral racemes or corymbs. Involucres c 6mm diameter; phyllaries c 14–16, 8 × 2.5mm, araneous, surrounded by appressed calycus of c 8 slightly shorter, more herbaceous bracts. Ray flowers 9–14; corolla tube 6mm; ligule 5 × 1.5mm. Disc flowers c 35, corollas 6.5mm. Achenes oblong, 2.5mm, glabrous; pappus 6.5mm, white.

Bhutan: S – Chukka district (near Chukka), Gaylegphug district (Rani Camp–Tama (117)); **W Bengal Duars:** Jalpaiguri district (174). Margin of forest and in open, 1220–1350m. October.

8. S. cappa (D. Don) Jeffrey & Chen; *Senecio densiflorus* DC.

Similar to *S. bhot* but stems silvery lanate; leaves oblanceolate or elliptic, 10–20(–28) × 4–8cm, shortly acuminate, tapered to petiole 3cm, often with an orbicular auricle up to 1cm broad at base, ± coarsely serrate, sparsely araneous above, densely white lanate beneath; capitula numerous in dense axillary and terminal corymbs; involucres c 4mm diameter; phyllaries lanceolate, 3.5 × 1mm, acute, white lanate outside; ray flowers 9–10, corolla tube 4–5mm, ligule 3 × 1.2mm; disc flowers 14–17, corollas 4–5mm, enlarging in upper 3/5; achenes 1mm, glabrous; pappus 3mm, white.

Bhutan: S – Samchi district (Sangura (117)), Chukka district (Chaisilaka (117)), **C** – Tashigang district (near Tashigang (117)); **Darjeeling:** Sureil, Sitong, Dumsong etc. Under shrubs, 900–1670m. October–March.

9. S. kunthiana (DC.) Jeffrey & Chen; *Senecio kunthianus* DC.

Similar to *S. bhot* and especially *S. cappa* but smaller; stems 20–45cm, sparsely pubescent or whitish araneous; leaves elliptic, 2.5–4.5 × 0.5–1.5cm, acute or acuminate, cuneate or rounded at base, serrate, upper surface glabrescent, white lanate beneath; capitula up to 10 in terminal corymbs, involucre 4mm diameter; phyllaries c 10, oblong, 6 × 1.5mm, araneous at base; ray flowers c 8, ligules 6.5 × 1.5mm; disc flowers 20, corollas 5.5mm enlarging in upper ⅔; achenes oblong, 1.5mm, glabrous; pappus 5.5mm.

Sikkim: without locality (80) 3650m.

Dubious record. May refer to juvenile shoots collected by Hooker at Lachen better treated as *Senecio kumaonensis*. Otherwise only found from Kashmir to C Nepal.

88. CISSAMPELOPSIS (DC.) Miquel

Scandent perennial herbs or subshrubs climbing by means of prehensile petioles and sometimes weakly twinning. Leaves simple, petiolate, ovate, cordate at base, palmately 3–7-veined; petioles prehensile with thickened persistent bases, exauriculate. Capitula in axillary or terminal corymbs or corymbose panicles, radiate or discoid. Involucres 1-seriate, calyculate, ± cylindrical. Receptacle flat, naked. Ray flowers 5–6, fertile, ligules yellow. Disc flowers 5–16, bisexual; corollas yellow, tubular-campanulate, 5-toothed; style branches truncate or convex with or without a central tuft of longer papillae. Achenes cylindrical, ribbed, glabrous; pappus of capillary bristles, present on all flowers, whitish.

1. Capitula discoid .... **3. C. corifolia**
\+ Capitula radiate 2

2. Disc flowers 14–16 .... **1. C. buimalia**
\+ Disc flowers 5–6 .... **2. C. sp. A**

1. C. buimalia (D. Don) Jeffrey & Chen; *Senecio buimalia* D. Don

Stems up to 5m, becoming woody. Leaves sometimes quite angular, 5–12 × 3.5–8cm, acuminate, glabrescent above, greyish or brownish arachnoid beneath, margins remotely denticulate, usually obscurely sinuate; petioles 2.5–5cm, arachnoid. Involucre 5mm diameter; phyllaries c 8, ± oblong, 7–9mm, (sub)acute, densely araneous. Ray flowers 5, ligule 6–8 × 1.5–2mm, tube c 5mm. Disc flowers 14–16; corolla 9.3–11mm, including teeth c 2mm. Style branch apices fringed with short papillae, apical tuft absent. Achenes c 4mm, glabrous; pappus 8–10mm.

Bhutan: unlocalised (329, 353); **Darjeeling/Sikkim:** unlocalised (Hooker collection). c 1700m. Flowering in November and December in C Nepal.

2. C. sp. A

Similar to *C. buimalia* but leaves not angular, thinner, sparsely brownish arachnoid beneath; petioles glabrescent; involucre 3.5mm diameter; ray flowers 5–6, disc flowers 5–6.

Darjeeling/Sikkim: unlocalised (Hooker collection). c 2300m.

Only one collection seen, lacking mature flowers and fruit.

3. C. corifolia Jeffrey & Chen; *Senecio araneosus* sensu FBI p.p. non DC. Fig. 133a–b.

Similar to *C. buimalia* but leaves subglabrous beneath, lower leaves coriaceous, 8–14 × 4.5–10.5cm, shortly acuminate, upper leaves thin, rather stiff; petioles usually glabrous; capitula discoid, numerous; peduncles with arachnoid hairs dense to absent and very few to many septate glandular hairs; involucres 3(–4)mm diameter; phyllaries 4–7.5mm, swollen at base; flowers 9–13; style branch apices with fringe and short apical tuft of papillae; achenes 2.7–3.7mm; pappus 6.5mm.

Bhutan: S – Tongsa district (Shamgong); **Darjeeling:** Sureil, Tanglu, Chunthang, Yoksam, Darjeeling etc. 1525–2600m. October–March.

89. SENECIO L.

Erect or scrambling, rarely decumbent, perennial or annual herbs; stems usually leafy. Leaves alternate, simple or lyrately, pinnately or palmately divided, radical leaves commonly absent at flowering time. Capitula few to numerous, in simple or compound corymbs, radiate or discoid, erect or cernuous. Involucres calyculate, hemispherical, campanulate or cylindrical; phyllaries 5–20, usually free, ± 1-seriate, margins scarious. Receptacle flat, naked. Ray flowers up to 20, ligules yellow, usually conspicuous, sometimes small. Disc flowers few to many, yellow, tubular-campanulate; corollas 4–5-toothed; anther bases sagittate but without tail of sterile cells beneath locules; styles branches truncate or convex with obtuse marginal papillae, without a central tuft of longer papillae. Achenes oblong, ribbed, glabrous or pubescent. Pappus of capillary bristles, uniform or sometimes with apically hooked hairs.

FIG. 133. **Compositae: Senecioneae.** a–b, *Cissampelopsis corifolia*: a, fertile portion of stem (× ⅔); b, capitulum (× 3), c–f, *Ligularia amplexicaulis*: c, leaf (× ½); d, inflorescence (× ⅔); e, ray flower (× 4); f, disc flower (× 4). Drawn by Margaret Tebbs.

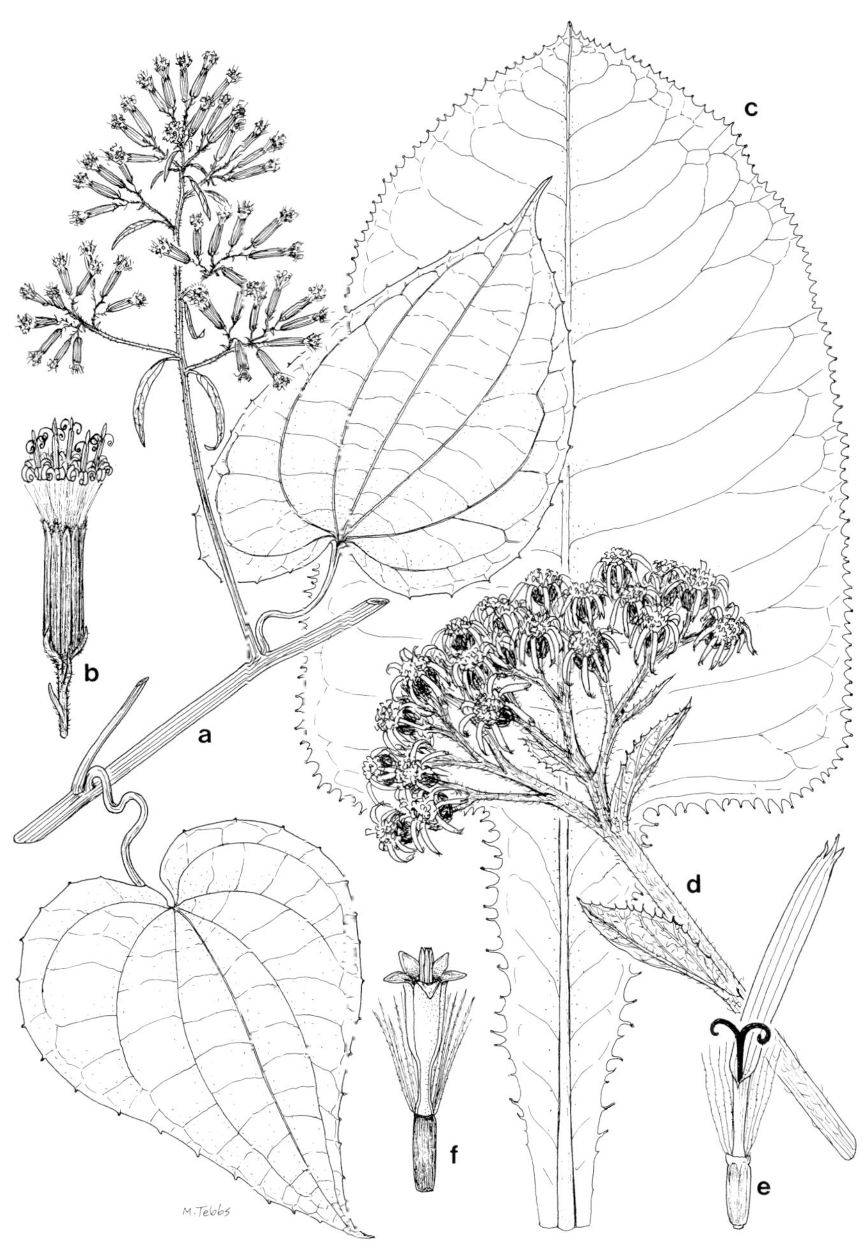
c
b
a
d
f
e
M. Tebbs

1. Leaves simple, mostly serrate, rarely with few-lobed cauline leaves smaller than unlobed persistent basal leaves; perennials 2
+ Leaves divided or lobed; annuals or perennials 5

2. Capitula discoid .... **1. S. kumaonensis**
+ Capitula radiate 3

3. Stems leafy, usually scrambling, to 5m .... **2. S. scandens**
+ Stems subscapose, to 60cm .... **3. S. nudicaulis**

4. Stems stout, mostly over 50cm; perennials 5
+ Stems slender, herbaceous, to 45(–60)cm; annuals or scree-perennials with long slender rhizomes 9

5. Involucre 3.5–10mm diameter; ligules 5–12mm 6
+ Involucre 2–3mm diameter; ligules to 3mm 7

6. Involucre 5–10mm diameter; phyllaries 5–7.5mm; ligules 7–12 × 2–4.5mm; pappus reddish, absent from ray flowers or reduced and caducous **4. S. raphanifolius**
+ Involucre 3.5–5mm diameter; phyllaries 3.5–5mm; ligules 4–7 × 1.5–2.5mm; pappus whitish or reddish, usually present and unreduced on ray flowers .... **5. S. laetus**

7. Leaves deeply lyrately divided, terminal segments much larger than laterals, triangular .... **6. S. biligulatus**
+ Leaves pinnatisect, terminal segments not or scarcely larger than laterals 8

8. Ligules not longer than involucre .... **7. S. graciliflorus**
+ Ligules c 3mm longer than involucre .... **8. S. royleanus**

9. Scree-perennial with long slender rhizomes; ligules 5–10mm **9. S. albopurpureus**
+ Annuals; ligules less than 2mm 10

10. Ligules scarcely longer than involucre; phyllaries with 2 distinct brown veins .... **10. S. ramosus**
+ Ligules 1–2mm; phyllaries indistinctly veined .... **11. S. tetrandrus**

A population of branched, annual, subglabrous to floccose plants with simple, oblanceolate to pinnatipartite leaves and 16–25 ray flowers with elliptic ligules from Chengmari in Jalpaiguri district (none seen) has been incorrectly reported as *S. vulgaris* L. (407). Its correct identity is uncertain. *S. vulgaris* would key out with *S. ramosus* and *S. tetrandrus* but has much shorter leaf segments, usually discoid capitula and longer phyllaries (5.5–7.5mm). Parker reported a solitary plant from Bhutan without locality, though with very short ray florets (383).

1. S. kumaonensis Jeffrey & Chen; *S. candolleanus* auct. non DC., *Cacalia penninervis* Koyama non *S. penninervius* DC.

Rhizomatous herbaceous perennial; stems 0.3–1m, sparsely arachnoid at first. Leaves ovate, 6–12 × 2.5–5cm, acute, rounded at base, finely dentate, araneous above, white lanate beneath; petiole 1–5cm. Capitula discoid, in racemose panicles; involucres 3–5mm diameter; phyllaries 5, lanceolate, 7 × 2mm, araneous at base. Flowers 5–6; corollas 5–6mm, widening in upper half. Achenes obovoid, 2.5mm, glabrous; pappus 5mm, whitish.

Bhutan: C – Thimphu district (Shodug, Barshong, Somana), **N** – Upper Mo Chu district (Lingshi), Upper Kulong Chu district (Me La); **Sikkim:** Goecha La, Sebozung Chhu, Sherabthang etc.; **Chumbi:** Trakarpo. Open hillsides, river banks, under forest, 3650–4570m. August–November.

2. S. scandens D. Don

Herbaceous perennial; stems scrambling, strongly flexuose, 1.5–5m (rarely free-standing, shorter, straighter), sparsely pubescent at first. Leaves ovate or triangular, 4–13 × 1.5–5cm, subentire, denticulate or pinnatifid with 1–6 pairs of lateral lobes, acuminate, truncate at base or slightly cordate or attenuate, ± glabrous on both surfaces; petiole 1–2cm. Capitula radiate, in loose divaricate corymbs; involucres c 5mm diameter; phyllaries 10, ± oblong, 6.5 × 1.5mm, ± glabrous. Ray flowers 8–10; corolla tube 3mm; ligule 5.5 × 1.2mm. Disc flowers numerous; corolla 5.5–7mm, widening in upper ⅔. Achenes oblong, 2.5–3mm, pubescent; pappus 5.5–7mm, white.

Bhutan: S – Samchi, Chukka and Deothang districts, **C** – Thimphu, Tongsa, Bumthang and Tashigang districts, **N** – Upper Mo Chu district; **Darjeeling:** Darjeeling, Kurseong (97a), Tanglu etc.; **Sikkim:** Chunthang, upper Kalej Khola, Lachung; **Chumbi:** without locality (421). Roadsides and climbing over shrubs, 1100–3800m. April–December.

There are two varieties in our area: var. **scandens** with leaves subentire or dentate without lobes at base – this is the more common, and var. **incisus** Franchet with pinnatifid leaves with 1–4 lobes at base. The latter is rarer in our area, known only from Thimphu, Tongsa and Upper Mo Chu districts.

3. S. nudicaulis D. Don

Perennial; stems often subscapose, 15–60cm, ± glabrous. Radical leaves usually present at flowering time, rosulate, oblong-obovate, 4–12 × 1–6cm, obtuse or rounded, attenuate to base, sessile, subentire or crenate-serrate, glabrous on both surfaces; cauline leaves few, smaller than radical, sometimes lobulate. Capitula few to many in terminal corymbs; involucres c 5mm diameter; phyllaries 12–13, linear-lanceolate, 5 × 2mm, glabrous. Ray flowers 12–13; corolla tube 3.5mm; ligule 7mm. Disc flowers many; corolla 3.5mm, enlarging in upper ⅔. Achenes oblong, 1.5mm, puberulent; pappus on all flowers, 3mm, white.

Bhutan: C – Punakha district (Pho Chu Valley, Ratsoo–Tsarza La, Wangdu

Phodrang–Samtengang etc.); **Darjeeling:** unlocalised (Hooker collections). Dry sandy soil and *Pinus roxburghii* forest, 300–2000m. April–June.

S. wightii (DC.) Clarke [*S. saxatilis* DC.] has been reported from Bhutan (353), but no specimens have been seen. It is similar to *S. nudicaulis*, but lacks a basal rosette of leaves, has many leaves evenly distibuted along the stems and lacks a pappus on ray flowers.

4. S. raphanifolius DC.; *S. diversifolius* DC. non Dumortier. Fig. 134a.

Rhizomatous perennial; stems 45–100(–150)cm sparsely araneous at first. Radical leaves usually absent at flowering time, with long, dentate and often winged petioles, lower and mid-cauline leaves lyrate-pinnatifid, oblanceolate in outline, sessile, 10–25 × 1.5–6.5cm, with large elliptic-oblong, lobed or dissected terminal segment and 3–5 or more pairs of oblong lateral segments decreasing in size towards base of leaf, glabrous above, araneous beneath; upper cauline leaves oblong in outline, 8–10 × 1.5–3cm, irregularly pinnatisect, sessile, auriculate at base. Capitula radiate, numerous, in terminal corymbs; involucres broadly campanulate, 4.5–8mm diameter; phyllaries 12–20, oblong, 5–7.5mm, acute, blackish above, sparsely pubescent. Ray flowers 11–20; corolla tube 2.5mm, subglabrous or hairy; ligules 7–12 × 2–4.5mm. Disc flowers numerous; corolla claw 2.5–3mm, limb c 3mm. Achenes obovoid, 2mm, glabrous; pappus 3mm, brittle, reddish, reduced and caducous on ray achenes or absent.

Bhutan: C – Thimphu district (Barshong), **N** – Upper Mangde Chu/Upper Bumthang Chu district (Ju La), Upper Kuru Chu district (Kurted), Upper Kulong Chu district (Shingbe); **Darjeeling:** Sandakphu, Singalila, Tonglu; **Sikkim:** Chola, Lachoong, Tsomgo etc. Fields, open spaces, pasture, hillsides, among shrubs, in *Abies* forest, 3050–4250m. June–November.

The Kurted specimen has larger involucres to c 12mm diameter and may be distinct.

5. S. laetus Edgeworth; *S. chrysanthemoides* DC. non Schrank. Dz: *Uma elama.*

Differs from *S. raphanifolius* by its more numerous, smaller capitula (involucres 3.3–4mm diameter) on slender peduncles; phyllaries 3.5–5mm; ligules 4–7 × 1.5–2.5mm; pappus often whitish, usually well developed on ray achenes.

Bhutan: C – Thimphu district (Dotena, Lingshi, Paro etc.), Tongsa district (Tongsa), Bumthang district (Byakar), **N** – Ha district (Kale La–Ha), Upper Bumthang Chu district (Kopub). Abandoned fields, grassy banks, 2150–3950m. May–September.

No specimens seen from Darjeeling or Sikkim. Further collections are inter-

---

FIG. 134. **Compositae: Senecioneae and Helenieae. Senecioneae.** a, *Senecio raphanifolius*: habit (× ⅔). b–c, *Gynura pseudochina*: b, habit (× ⅔); c, capitulum (× 2½). **Helenieae.** d–g, *Tagetes minuta*: d, apex of flowering shoot (× ⅔); e, ray flower (× 4); f, disc flower (× 4); g, achene (× 4). Drawn by Margaret Tebbs.

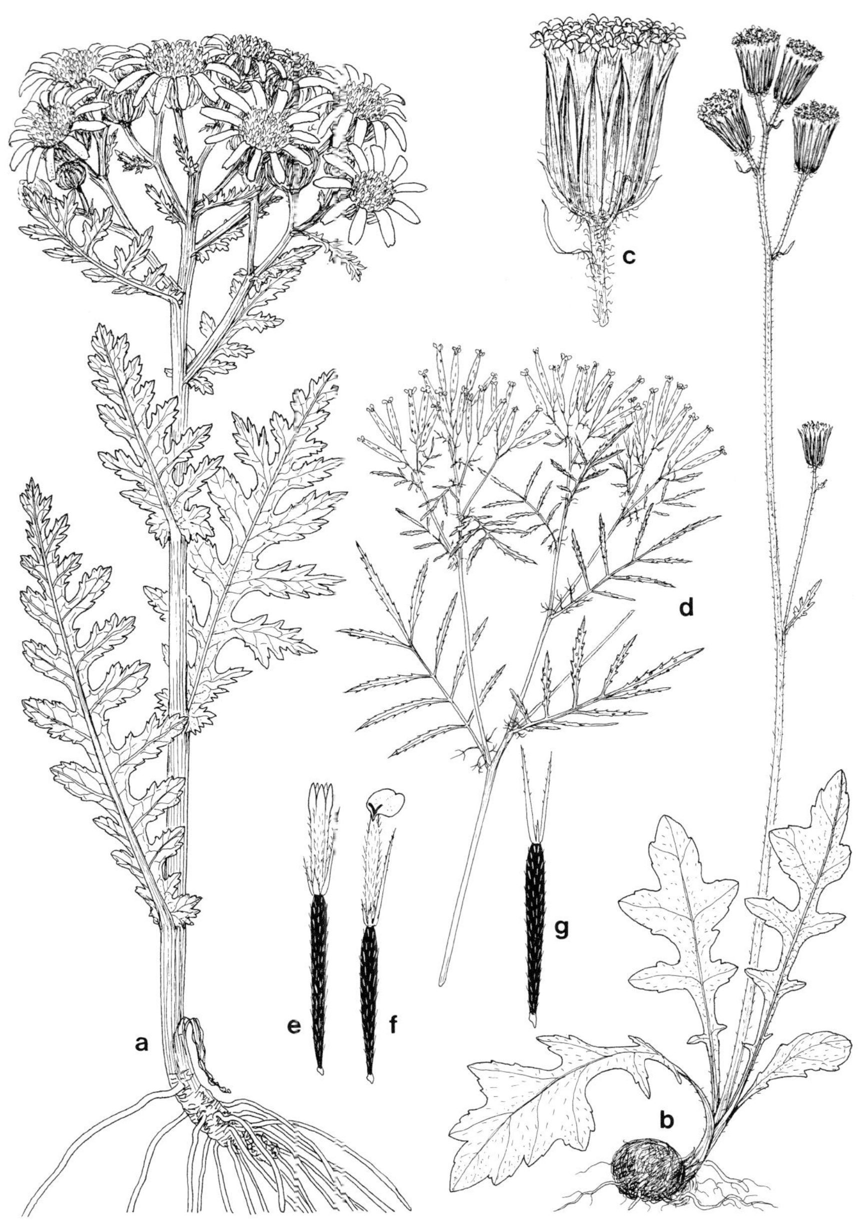
c
d
g
a
e
f
b

mediate with *S. raphanifolius*, including some from Upper Mo Chu district, Bhutan and Chumbi.

6. S. biligulatus W.W. Smith

Similar to *S. raphanifolius* but mid-cauline leaves 9–20cm, terminal segment ovate, 7–15 × 3–6.5cm, acuminate or acute, with base cordate or sharply deltoid, margins serrate, ± glabrous on both surfaces, lateral segments 2–3 pairs, abruptly smaller than terminal; capitula radiate, very numerous in dense corymbs, involucres c 1.5mm diameter; phyllaries 4, linear-lanceolate, 3 × 1mm, glabrous; ray flowers 2, corolla tube 2mm, ligule 3 × 0.2mm; disc flowers 3, corolla 3.5mm, widening in upper half; achene ellipsoid, 2mm, blackish, sparsely puberulous; pappus 3mm, white.

Bhutan: C – Thimphu, Tongsa, Bumthang, Mongar and Sakden districts, **N** – Ha district; **Sikkim:** Kapup, Tangkar La, Tsmogo etc.; **Chumbi:** Yatung. Rocky hillsides, *Pinus excelsa* forest, 2750–4570m. July–October.

7. S. graciliflorus DC.

Stems 50–100cm, sparsely pubescent when young. Leaves pinnatisect, ovate or oblong in outline; blade 6–14 × 5–12cm, acuminate, truncate at base, ± glabrous on both surfaces; lateral segments 3–4 pairs, ± lanceolate, 2–6 × 1–3cm, serrate to deeply biserrate. Capitula obscurely radiate, in terminal and axillary corymbs; involucres c 2mm diameter; phyllaries 5, linear, 6 × 0.7mm, glabrous. Ray flowers 2; corollas 3.5mm including deeply toothed ligule. Disc flowers usually 3; corollas 5mm, enlarging in upper half. Achenes oblong, 2mm, glabrous; pappus 3.5mm, white.

Sikkim: Kapup, Nathang, Lachung etc.; **Chumbi:** Chubitang (353), Yatung etc. Meadows and under trees, 2750–3950. August–November.

8. S. royleanus DC.; *S. graciliflorus* var. *hookeri* Clarke

Similar to *S. graciliflorus* but stems up to 1.5m; leaves ± whitish araneous beneath, segments coarsely serrate; capitula distinctly radiate, in dense corymbs; involucres c 2.5mm diameter; phyllaries 5–6, linear-lanceolate, 5 × 1mm; ray flowers 3–4, corolla tube 3mm, ligule 2–3 × 1–1.5mm; disc flowers 6–10; achenes 1mm; pappus 4mm.

Bhutan: C – ?Thimphu district (Tho Cu La), **N** – Upper Mo Chu district (Gasa); **Sikkim:** Lachung; **Chumbi:** Chumbi, Galing, Yatung (353). Open hillsides, 2800–3650m. August–October.

9. S. albopurpureus Kitamura; *S. bracteolatus* Hook.f. non Hooker & Arnott

Rhizomatous perennials; stems 2–15(–20)cm, araneous at first. Basal leaves usually present at flowering time, elliptic or oblanceolate in outline, serrate or palmately 3–4-lobed, 1–2.5 × 0.3–1cm, acute, attenuate to petiole up to 3cm, ± glabrous on upper surface, white-araneous beneath. Capitula solitary or up to 4, ± corymbose; involucres campanulate, 6–7mm diameter; phyllaries 10, linear-lanceolate, 8–13mm, sparsely araneous at base, often purplish at apex. Ray flowers 8–10; corolla tube 3mm, ligule 8–9 × 1.5mm. Disc flowers numerous; corolla 5.5–6.5mm, enlarging in upper half. Achenes ellipsoid, 4mm glabrous; pappus 6.5mm, white.

Bhutan: N – Upper Mangde Chu district (Saga La); **Sikkim:** Naku La, Goecha La, Prek Chhu etc. Sandy screes, 4265–5340m. August–November.

10. S. ramosus DC.

Annual; stems up to 50cm, glabrous. Leaves pinnatisect, oblong in outline, 6–7cm, with 3–4 pairs of oblong segments up to 1.5 × 0.3cm, acute, sessile, auricled at base. Capitula narrowly campanulate, 2.5mm diameter, numerous in loose corymbs; phyllaries c 12, linear-lanceolate, 3.5 × 0.7mm, with 2 distinct brown veins. Ray flowers 6–8, scarcely longer than involucre, corolla c 2mm including tube and ligule. Disc flowers numerous, corolla 2mm, 4-toothed. Achenes oblong, 1.5mm, puberulous, pappus 2.5mm, white.

Darjeeling: without locality (80). Terai.

11. S. tetrandrus DC.

Similar to *S. ramosus* but lateral segments of leaves up to 2.5cm; involucres c 3mm diameter; phyllaries 16–18, linear 5 × 0.7mm, veins indistinct; ray flowers c 12, corolla tube 2mm, ligule 1 × 0.5mm; disc corollas 3mm; achenes 2.5mm, pappus 3mm.

Darjeeling: without localities (80), 'plains at the foot of the hills'.

90. CRASSOCEPHALUM Moench

Erect annual herb. Leaves alternate, simple or lyrate-pinnatifid. Capitula in racemes or panicles, discoid. Involucre cylindrical, 1-seriate, usually with a number of filiform bracts at base; phyllaries linear. Receptacle convex, naked. Corollas narrowly tubular at base, widening slightly and 5-toothed above. Style branches recurved, filiform, with long subulate appendage of fused papillose hairs, usually with distinct whorl of hairs at junction between stigmatic part and appendage. Achenes oblong, weakly 8–10-ribbed. Pappus of fine, brittle bristles, white.

1. C. crepidioides (Bentham) S. Moore; *Gynura crepidioides* Bentham

Stems 30–50(–150)cm, brownish puberulous. Leaves elliptic or oblanceolate, 5–15 × 1–6cm, acuminate, base attenuate to petiole to 5cm or lyrately lobed with 1–2 pairs of oblong lateral segments up to 4 × 1cm, margins ± coarsely serrate, puberulous on both surfaces, denser beneath. Inflorescences terminal and from upper leaf axils; capitula few, pendulous in bud, later erect. Involucre c 5mm diameter; phyllaries linear 8–10mm, puberulous. Corollas 8–10mm, brownish-orange above; teeth (0.7–)1mm. Achenes 2.2mm, with rows of white hairs between ribs; pappus up to 12mm.

Bhutan: S – Phuntsholing district (Phuntsholing, Torsa), Chukka district (Chaisilakha, 117), **C** – Punakha district (Tinlegang (71), Bhotoka–Rinchu (71), Lobesa), Tongsa district (Shemgang); **Darjeeling:** Takvar, Happy Valley (69), Kurseong (97a) etc.; **Sikkim:** Rate Chu, Dentam–Pemayangtse (69), Saramsa (69). Roadsides, river shingle and among crops, 200–2100m. April–December.

Weed introduced from Africa. Also reported from Mongar, Tashigang and Lhuntshi administrative districts of Bhutan (383). It closely resembles *Erechtites valerianifolia* (Reichenbach) DC., a weedy species from the Americas with disciform capitula which has been recorded from Jalpaiguri district without precise locality (174).

91. EMILIA Cassini

Annual or perennial herbs, usually somewhat glaucous. Leaves alternate, often semi-amplexicaul. Capitula cylindrical or urceolate, discoid, borne on long slender peduncles in loose, often few-flowered corymbs. Involucre 1-seriate; phyllaries 8–12, coherent, outer basal bracts absent. Receptacle convex, naked. Corollas narrowly tubular, widening slightly above, teeth 5, oblong, narrow. Style branches linear or somewhat dilated towards a rounded or conic appendage. Achenes oblong, subcompressed, ± 5-angled, ribbed, glabrous or with rows of short hairs in our species; pappus of capillary bristles, silky, white.

1. Lower leaves usually lyrately pinnatifid, upper simple; achenes with 5 rows of hairs c 0.1mm .... **1. E. sonchifolia**
\+ All leaves simple; achenes glabrous .... **2. E. prenanthoidea**

1. E. sonchifolia (L.) DC.

Weak subglabrous annual or perennial, branched from base, 10–55cm. Leaves weakly dentate; basal leaves ovate, long petiolate, obtuse, sometimes absent; lower leaves lyrate, sessile, to 10 × 4cm, usually obtuse, attenuate at base, lateral segments 1–2 pairs, oblong, terminal segment ovate, to 3.5 × 2.5cm; upper leaves smaller, oblong, sessile, auriculate at base. Capitula 2.5–3.5mm diameter; phyllaries narrowly oblong-lanceolate, 8–11mm, glabrous to sparsely puberulous above. Corollas deep pink, 7–10mm, teeth 1–1.5mm. Achenes oblong, 2.5mm, with 5 longitudinal rows of short white hairs; pappus 5–7mm. Depauperate plants sometimes unbranched or subrosulate or with all leaves ovate and long petiolate.

Bhutan: S – Samchi district (Dorokha), Phuntsholing district (Phuntsholing), **C** – Punakha district (Wandu Phodrang, Mishina); **Darjeeling:** Balasun, Labha, Panchkilla etc.; **Sikkim:** Saramsa (69). Weed of cultivation, 200–1850m. All year.

2. E. prenanthoidea DC.; *E. angustifolia* DC.

Similar to *E. sonchifolia* but perennial, 30–120cm, taller plants stiffly erect above; leaves oblong-oblanceolate to narrowly triangular, sessile, 4–13 × 0.7–2cm, acute to subobtuse, sometimes auriculate at base, entire, obscurely sinuate or toothed on auricle; capitula 3–5mm diameter; phyllaries 8–11mm, sparsely puberulous to araneous above; corollas 11–12mm, teeth c 2mm; achenes to 3.5mm, glabrous; pappus 7mm.

Darjeeling: near Chenga Forest, Katambari; **Darjeeling/Sikkim:** unlocalised (Hooker collection); **W Bengal Duars:** Jalpaiguri. Marshy ground (taller forms), 75–600m. November–December.

92. GYNURA Cassini

Perennial herbs, sometimes tuberous rooted. Stems erect or scrambling. Leaves alternate, simple, serrate or pinnatifid. Capitula in 1–several ± corymbose heads in an open inflorescence, discoid. Involucre cylindrical or almost campanulate; phyllaries 10–14, 1-seriate, linear, subequal, margins scarious, usually with a number of filiform bracts at base. Receptacle convex, naked. Flowers yellow or orange, tubular-campanulate. Corollas 5-toothed above. Style branches erect, slender, tapering above into hairy subulate tips. Achenes oblong, 10-ribbed; pappus of filiform bristles, white.

1. Roots tuberous; stems ± scapose; leaves mostly basal, pinnately lobed, rarely unlobed .... **1. G. pseudochina**
\+ Roots not tuberous; stems mostly branched, leafy; leaves subentire or serrate 2

2. Leaves elliptic, subentire, petiolate, greyish pubescent; phyllaries pubescent throughout .... **2. G. nepalensis**
\+ Leaves oblanceolate, coarsely serrate, base usually sessile and auriculate, ± glabrous; phyllaries glabrous .... **3. G. bicolor**

1. G. pseudochina (L.) DC. Dz: *Nabtenkha*. Fig. 134b–c.

Tuber 2–3(–6)cm in diameter; stems (15–)30–35(–60)cm scapose, sometimes with a leaf near middle, sparsely pubescent above. Leaves pinnatisect 5–12(–35) × 2–4.5(–12)cm, acuminate or obtuse, base attenuate to petiole up to 2.5cm, sparsely pubescent on both surfaces, lobes 2–4 oblong 1–1.5 × 0.4–1cm. Capitula few, often in a single head, 5–7mm diameter; phyllaries narrowly oblong, 8–11mm, acuminate, pubescent at base. Corollas 9–12mm, throat (2.5–)3mm, teeth 1mm. Achenes oblong, 3–4mm, glabrous; pappus 7mm.

Bhutan: S – Phuntsholing/Samchi border (Torsa), **C** – Punakha district, Tongsa district (Tongsa–Tongsa Bridge), Mongar district (Lhuntsi); **Darjeeling:** Badamtam, Munsang, Pedong, Rangit Valleys; **Sikkim:** Namphak. Dry open sites, sometimes moist shade, 300–1525m. April–June.

2. G. nepalensis DC. Nep: *Tong kribi.*

Large sprawling bushy or subshrub c 1–1.5m; roots not tuberous; stems greyish-brown pubescent. Leaves ovate-elliptic, 5–10(–18) × 2–3.5(–7)cm, acute, base attenuate to petioles up to 1cm, margins distantly and shallowly serrate, pubescent on both surfaces, denser beneath. Capitula rather few, borne in loose panicles or corymbs, 6–7mm in diameter. Phyllaries narrowly oblong, 9–12mm, acuminate, pubescent. Corollas 9.5–12.5mm, throat 2–2.5mm, teeth c 1mm. Achenes 4.5mm, glabrous or with few hairs between ribs; pappus 9mm.

Bhutan: S – Samchi district, **C** – Chukka, Punakha and Tongsa district, **N** – Upper Mo Chu district; **Darjeeling:** Lebong, Tinglam, Tista etc.; **Sikkim:** Tumlang, Gangchung. On cliffs and among rocks, 250–2000m. March–June.

3. G. bicolor (Willdenow) DC.; *G. angulosa* DC. Nep: *Dhungri-phul.*

Similar to *G. nepalensis* but stems very robust, more upright, to 4.5m, subglabrous; leaves ovate to oblanceolate, 9–30 × 2–9cm, acute or acuminate, base usually sessile, auriculate (sometimes shortly petioled), sharply serrate (sometimes denticulate), subglabrous or sparsely pubescent on both surfaces; capitula more numerous, 6–9mm diameter; phyllaries 11–13mm, glabrous; corollas 11–14mm; achenes 4.5–5mm with hairs often more numerous; pappus 10mm.

Bhutan: S – Samchi (117), Phuntsholing (117), Chukka and Gaylegphug (117) districts, **C** – Thimphu and Tongsa (117) districts; **Darjeeling:** Burmake Forest, Balasun, Darjeeling etc.; **Sikkim:** Gassing–Rathang Chhu, Kalej Khola, Phodang etc. Moist forests, waste places, 1060–2440m. September–March.

Although described in a recent monograph (333a) as intergrading in all characters, two species have been recognised here by Chen (327): *G. bicolor* [*G. angulosa* var. *pedunculata* Hook.f.]: leaves with bases attenuate ± to a short petiole without auricules, undulate remotely dentate or denticulate margins and 7–9 pairs of lateral veins, capitula long-pedunculate, in a loose cyme; *G. cusimbua* (D. Don) S. Moore: leaves sessile, ± oblong in basal part or weakly panduriform, sharply often unevenly serrate, with 12–30 pairs of veins, capitula in compact cymes, with peduncles shorter than capitula. On this basis all our specimens can be referred to *G. cusimbua* except those from Mirichoma (Chukka district) and Kalej Khola.

93. STEIRODISCUS Lessing

Small subglabrous annual; stems densely corymbosely branched, slender, flexuous. Leaves pinnately cut to midrib with few linear widely spaced segments. Capitula numerous, on long slender peduncles, radiate, heterogamous. Involucre narrowly urceolate-campanulate; calyculus absent; phyllaries 1-seriate, ⅔ fused, free part triangular, acuminate; receptacle hemispherical, naked. Corollas yellow (orange). Ray corolla tube very slender, with ring of glandular papillae at apex; ligule oblanceolate, shortly toothed. Disc corolla claw slender; limb narrowly cylindric, dilated just below teeth; anther bases obtuse; style branches narrowly oblong, with conic, papillate appendage. Achenes oblanceolate, brown, glabrous, all without pappus.

1. S. tagetes (L.) Schlechter; *Gamolepis annua* Lessing

Plant 22–30cm. Leaves oblong-elliptic in outline, c 4 × 2cm; lateral segments 3–7 pairs. Phyllaries 8–14, c 5mm fused and 3mm free, brown-lined. Ray corolla tube 5mm; ligule c 8 × 2mm. Disc corollas 6mm. Achenes 3mm.

Darjeeling: Darjeeling. 2300m. April–June.

Weed introduced from S Africa, occasionally cultivated elsewhere.

HELENIEAE

94. TAGETES L.

Strongly smelling glabrous annuals. Leaves pinnate, opposite or uppermost alternate; leaflets narrowly elliptic, serrate, conspicuously gland-dotted, solitary or in corymbs, radiate. Involucres cylindric-campanulate, 1-seriate; phyllaries connate to near apex and bearing oil glands. Receptacle flat, naked. Ray flowers few to many, female. Disc flowers tubular-campanulate, bisexual; corolla 5-toothed; style branches flattened, widening slightly near apex, acute. Achenes linear, subcompressed, angular, blackish, empty and pale coloured at base. Pappus of few unequal connate scales and up to 2 elongate flattened awns.

1. Capitula in compact corymbs; involucres 1.5–2mm diameter .. **1. T. minuta**
\+ Capitula solitary; involucres 3–30mm diameter .... **2. T. erecta**

1. T. minuta L. Fig. 134d–g.

Plant 0.3–2m; stem usually simple, erect, short-branched above. Leaves 5–16cm; leaflets 4–5 pairs, 10–70 × 1.5–9.5mm, acuminate, attenuate at base, serrate, with rounded oil glands near margin at base of each tooth, others scattered near midrib. Capitula in compact corymbs. Involucre cylindrical, 9–10 × 1.5–2mm; phyllaries 3, bearing lenticular oil glands. Corollas lemon yellow. Ray flowers 2; corolla tube 3.2–5mm, hairy; ligule usually 2-lobed, 1.5–2.2 × c 2.4mm. Disc flowers 2–3; corollas 4–4.5mm. Achenes 5–6.5mm; pappus awns to 3.2mm.

Bhutan: S – Phuntsholing district (Torsa River), **C** – Thimphu district (Babesa, Paro), Punakha district (Wangdu Phodrang). River banks and cultivated ground, 200–2350m. February–October.

Native of S America now a widespread weed of warmer countries.

2. T. erecta L. Nep: *Bhaby-shaey-patri-phul* (117).

Plant 0.2–1.5m; stem simple and erect to much branched. Leaves 3–15cm, leaflets 3–9 pairs, 1–6 × 0.2–1cm, acuminate, sessile with uppermost narrowly decurrent on rachis, serrate, marginal glands situated near base of teeth. Capitula solitary at branch ends; peduncles gradually swollen, 3–12mm thick at apex. Involucres 5–30mm diameter; phyllaries 5–10, with lenticular glands at base and rounded glands above. Ray flowers c 7–10 in single forms, numerous in double forms; tube 8–10.5mm; ligule ovate-oblong, yellow, orange or brownish-red, 10–25 × 8–18mm, shallow-lobed. Disc corollas 12–19mm; lobes liguliform, hairy inside. Achenes 8mm; pappus awns c 10mm.

Bhutan: S – Phuntsholing district (Torsa), Chukka district (Chasilakha), Gaylegphug district (Betni), **C** – Thimphu district (Motithang, Paro), Punakha district (Wangdu Phodrang–Nahi), Tongsa district (Shemgang), Mongar district (Lhuntse); **Darjeeling:** Darjeeling (218); **W Bengal Duars:** Jalpaiguri. Cultivated in gardens, 1200–2350m. February–October.

Native of Mexico, now widely cultivated. Represented in our area by two cultigens, overlapping in all characters but often maintained as separate species: *T. erecta* L. and *T. patula* L. Most of our specimens can be separated by the following combination of characters:

2a. T. erecta L.

Plants coarser, to 1.5cm. Leaves 7–15cm, leaflets 4–9 pairs, lateral ones 1.5–6 × 0.3–0.8cm. Peduncles stout, leafy, to 11cm, strongly swollen above, 5–12mm diameter at apex. Involucres 9–30mm diameter; phyllaries 4–8mm across. Ray corolla tube 10–10.5mm; ligule 19–25 × 12–18mm, yellow or orange-yellow. Disc corolla tube and throat 12–13mm; lobes 5–6mm.

2b. T. patula L.

Plants less vigorous (?to 70cm). Leaves 3–8cm, leaflets 3–6 pairs, lateral ones 1–2 × 0.2–0.5cm. Peduncles more slender, with leaves obsolete, to 15cm less swollen above, 2.5–5mm diameter at apex. Involucres 7–12mm diameter; phyllaries 2–5mm across. Ray corolla tube 8–9mm; ligule 10–13 × 10–11mm, yellow, orange or deeper red. Disc corolla tube and throat 9–10mm; lobes 3–4.5mm.

95. GAILLARDIA Fougeroux

Annual, sometimes woody at base. Leaves simple, alternate, sessile, entire. Capitula radiate, paleate, solitary on long peduncles. Involucre hemispherical; phyllaries 2–3-seriate, reflexed in fruit. Receptacle convex, alveolate, with subulate stramineous scales. Ray flowers neuter; ligules 3–4-toothed, red, tipped with yellow or orange. Disc flowers bisexual; corolla tubular-campanulate, blackish-purple above, 5-toothed, teeth long acuminate, prominently ciliate; style branches hairy. Achenes obpyramidal, covered with stiff hairs on lower half. Pappus of 5 ovate-quadrate scales with long awns, strongly spreading in fruit.

1. G. pulchella Fougeroux

Stems 30–60cm, puberulous. Leaves oblanceolate, 3.5–8 × 0.5–1.5cm, ± acute, apiculate, attenuate at base, pubescent on both surfaces. Peduncles 5–15cm. Involucres 15–20mm across; phyllaries narrowly lanceolate, c 13 × 2mm, acuminate. Receptacle scales c 3mm. Ray flowers 6–10, ligules obovate-oblong, 12–20 × 12–15mm. Disc corollas 7.5mm including teeth 3.5mm. Achenes 2mm, long brownish hirsute below; pappus scales c 1.1mm at anthesis with awns to 2.5mm.

Bhutan: C – Tashigang district (Tashigang). Cultivated in garden, 1350m. June.

Native of N America.

HELIANTHEAE

96. ZINNIA L.

Annuals; stems erect, branched. Scattered glands on leaves, palea tips and corolla lobes. Leaves entire, opposite, sessile. Capitula hemispherical or campanulate, solitary, radiate; peduncles dilated and fistulose above. Involucre imbricate, several-seriate; phyllaries obovate, rounded above. Receptacle convex, paleae folded, enclosing ovary, persistent. Ray flowers 1-seriate, female; ligules ± sessile, white to red or purple, persistent on achenes. Disc flowers bisexual; corollas tubular-campanulate, 5-toothed, tube obsolete, throat dilated near base, narrowly cylindric above, teeth oblong-spathulate, densely velvety or pubescent within; style branches flattened, acuminate at apex. Ray achenes 3-angled, disc achenes laterally compressed. Pappus absent or a solitary awn.

1. Capitula hemispherical; disc corolla teeth velvety within; paleae fimbriate at apex; disc achenes awnless .... **1. Z. elegans**
\+ Capitula campanulate; disc corolla teeth pubescent within; paleae short ciliate at apex; disc achenes with a single long awn .... **2. Z. peruviana**

1. Z. elegans Jacquin

Plant to 1m; stems spreading pilose to appressed puberulous. Leaves lanceolate or ovate, 6–12 × 2.5–7cm, acute, base truncate or cordate, scabridulous. Capitula hemispherical; involucre 1–2.5cm diameter; phyllaries 4–10 × 3–6mm, most rounded and dark-margined at apex. Paleae up to 15mm, fimbriate at apex. Ray flowers 8–20; ligules narrowly obovate to oblanceolate, usually red, sometimes white to pink or purple, 20–35 × 7–12mm. Disc corollas 7–8mm. Ray achenes c 9 × 4mm; disc achenes 8.5 × 4mm, acutely angled at inner apex.

Bhutan: S – Gaylegphug district (Gaylegphug: cultivated), **C** – Thimphu district (Paro), Punakha district (Punakha). Cultivated in gardens and sometimes escaping, 300–2370m. May–October.

Native of Mexico. Double forms with all disc flowers ligulate are also cultivated in our area.

2. Z. peruviana (L.) L.; *Z. multiflora* L., *Z. pauciflora* L.

Similar to *Z. elegans* but leaves 2–4 × 0.7–1.5(–3)cm, acute or obtuse; capitula campanulate; involucre (5–)8–15mm diameter; paleae short ciliate at apex; ligules sometimes broader, 12–15 × 7–10mm; disc corollas 6mm; disc achenes 6–8mm with awn 5–6mm on inner apex.

Bhutan: locality not known (Gould collection).

Native to the Americas, now widely cultivated and naturalised.

97. ACMELLA Richard

Annual or perennial herbs. Stems 1–several from base, erect or decumbent. Leaves ovate or elliptic, opposite, petiolate. Capitula solitary or few on axillary or terminal peduncles, sometimes corymbose, radiate or discoid, paleate. Involucres 1–2-seriate; phyllaries narrowly ovate. Receptacles conical to elongate, paleae enfolding achenes. Ray flowers few or absent, female; ligules slightly longer than phyllaries, yellow or orange. Disc flowers bisexual; corollas yellow, tubular-campanulate, 4–5-lobed; style branches obtuse. Achenes obovoid, compressed, sometimes with pale ciliate border (cilia non-scabrid), disc achenes oblique; pappus absent or of 2 weak scabridulous bristles.

1. Involucre ± 1-seriate with c 6 phyllaries; capitula radiate .. **1. A. uliginosa**
\+ Involucre 2-seriate with c 10–15 phyllaries; capitula discoid 2

2. Achenes completely eciliate and epappose .... **2. A. calva**
\+ Achenes strongly ciliate; weak pappus bristles present, though difficult to distinguish from cilia .... **3. A. paniculata**

1. A. uliginosa (Schwartz) Cassini; *Spilanthes iabadicensis* A.H. Moore. Dz: *Heydonam* (272). Fig. 135d–e.

Plant erect or decumbent, 5–30cm; stems ± glabrous or sparsely pilose at nodes. Leaves 12–75 × 5–25mm, acute or subobtuse, attenuate at base, subentire or serrate, subglabrous or sparsely pubescent on both surfaces; petiole 3–15mm. Capitula 3–5mm diameter (excluding ligules). Involucre usually with 1 outer and c 5 inner phyllaries; phyllaries 1.8–2.5mm, glabrous. Ray corolla tubes c 0.9mm; ligules obovate, c 1.6 × 1.3mm, truncate, shallowly 3-dentate. Disc corollas c 1.2mm, 4-merous. Paleae c 2.5mm. Achenes with pale weakly ciliate border, 1.2–1.5 × 0.6–0.7mm excluding cilia; pappus bristles c 0.4mm or absent.

Bhutan: S – Phuntsholing district (Toribar, Torsa River), Gaylegphug district (Sham Khara), **C** – Tongsa district (Langtel, Shemgang), Mongar district

(Challi); **Darjeeling:** Badamtam, Darjeeling–Kurseong. Roadsides and cultivated ground, particulary flooded terraces, 200–2000m. February–August.

Also reported from Samchi, Chhukha, Samdrup Jongkhar, Thimphu and Punakha administrative districts (272). Discoid forms occur outside our area.

2. A. calva (DC.) Jansen; *Spilanthes acmella* L. var. *calva* (DC.) Hook.f. Nep: *Jang jurbi, Osilijat.* Fig. 135f.

Similar to *A. uliginosa* but capitula discoid, 5–8mm diameter; involucre 2-seriate with c 5 phyllaries in each whorl; phyllaries c 4mm; paleae c 3.5mm; corollas c 2mm, 5-merous; achenes without pale border or cilia, c 2 × 0.7–1mm; pappus absent.

Darjeeling: Farseng, Labha, Rongsong, Ryang Chhu, Darjeeling etc.; **Sikkim:** Gezing, Rhenok. Weedy situations, 600–2000m. February–December.

Leaves edible.

3. A. paniculata (DC.) Jansen; *Spilanthes acmella* L. var. *paniculata* (DC.) Hook.f. Fig. 135g.

Similar to *A. uliginosa* and *A. calva* but capitula discoid, 7–9mm diameter, rather many in terminal, corymbose panicles; involucres 2-seriate with c 7–8 phyllaries in each whorl; phyllaries 3–4.5mm; paleae c 3.5mm; corollas c 2mm, 4- and 5-lobed in same head; achenes with pale strongly ciliate border, c 2.4 × 1–1.2mm; pappus bristles present, to 0.8mm, often not exceeding cilia but scabridulous.

Darjeeling/Sikkim: unlocalised (Hooker collection); **W Bengal Duars:** Chel (175). November–February.

98. WEDELIA Jacquin

Perennial herbs or weak shrubs. Leaves simple, opposite, shallowly serrate, 3-veined from near base, petiolate. Capitula pedunculate, 1–4 together from upper nodes, radiate, paleate. Phyllaries few, foliaceous, (1–)2-seriate. Paleae lanceolate, enfolding disc flowers below. Corollas yellow. Ray flowers 3–10, female. Disc flowers bisexual; corollas tubular-campanulate, 5-toothed; style branches linear, acuminate. Achenes oblong to obcuneate, compressed or 4-angular, glabrous below, pilose above, smooth or tuberculate. Pappus a denticulate coronna, sometimes with 1–2 short awns, sometimes ± absent.

1. W. montana (Blume) Boerlage; *W. wallichii* Lessing, *Wollastonia montana* (Blume) DC. Fig. 135h–i.

Perennial herb, usually erect, 20–80cm; stems slender, stiffly pilose at least at nodes. Leaves ovate, 2–10 × 1–5cm, acute or acuminate, rounded and sometimes briefly attenuate at base, pilose on both surfaces; petioles to 1(–1.5)cm. Peduncles up to 6cm in fruit. Phyllaries elliptic, 4.5–7mm, pilose; 0–2 outer ones, 5 inner ones. Paleae 4.5–5.5mm, acuminate. Ray flowers (3–)5, opposite inner phyllaries; corolla tube 1.7mm, ligule oblong, 4(–5.5) × 2.5–3.2mm. Disc flowers 4–4.3mm; anthers black. Achenes obovoid, 3–3.5 × 2–2.5mm, contracted at apex, smooth; pappus very reduced.

Bhutan: S – Phuntsholing district (near Phuntsholing), unlocalised (Griffith Bhutan collection); **Darjeeling:** Kurseong, Tanglu (101), Rambi Chhu, terai etc.; **Sikkim:** Linchyam. 300–1200(–3100)m. May–December.

Two varieties have been recognised in our area: var. **wallichii** (Lessing) Koyama with short appressed hairs, and var. **pilosa** Koyama with long crisped hairs on peduncles, petioles and both leaf faces (365), though these hair-types sometimes intergrade on our material. The record of *Wollastonia biflora* (L.) DC. [*Wedelia biflora* (L.) DC.] from Rani Camp, Gaylegphug district, Bhutan, 1650m (117) may refer to this species. *Wollastonia biflora* is a coastal species differing by its obcuneate, abruptly truncate achenes and often by its more numerous longer ligules and longer, woody, straggling stems.

99. ELEUTHERANTHERA Bosc

Erect or suberect aromatic annual; stem branched. Leaves opposite, ovate to ovate-lanceolate, 3-veined, petiolate, obtuse to acute, ± attenuate at base, crenate to serrate. Capitula subsessile, 1–4 together in leafy or bracteate, short-stalked, axillary cymes, discoid, paleate. Involucre campanulate; phyllaries 5(–10), 1–2-seriate, unequal in length; paleae membranous, sheathing flowers. Flowers c 6–9, inner often male. Corollas yellow, narrowly infundibular. Achenes obovoid, obscurely to distinctly 4-angled, tuberculate especially along angles, subglabrous to puberulous; pappus a ciliate crown or cup, sometimes with 2–3 bristles.

1. E. ruderalis (Swartz) Schultes Bipontinus

Plant to 70cm; stem sparsely pubescent. Leaves 30–90 × 15–40mm, appressed white hispidulous on both faces; petiole to 20mm. Involucres c 2mm diameter at anthesis; phyllaries 6–9 × 3–5mm, hispid; palea lanceolate, 4–5mm, long pilose at apex. Corollas 2.5–4mm, teeth pilose within. Achenes 3–4mm.

Sikkim: Baluwakhani (408). Weed, 1700m.

Native to tropical America. Widely naturalised in Old World, mostly in tropics. None seen from our area, most details from Sri Lankan specimens.

100. SYNEDRELLA Gaertner

Usually annual herb, rarely persisting overwinter, appressed pilose; stems ± erect. Leaves simple, opposite. Capitula sessile or on short peduncles, 1–4 together from upper leaf axils, radiate, paleate. Involucres obovoid; phyllaries few, in 2–3 series, subequal, outer pair leafy, inner ones membranous. Receptacle convex, paleae flat. Flowers few. Corollas yellow. Ray flowers in 2 series, female;

ligules short, 2–3-toothed. Disc flowers bisexual; corolla tubular-campanulate, 4-lobed; style branches linear with long filiform appendages. Ray achenes oblanceolate, dorsally compressed, blackish with pale, narrow, thick wing with 5 upwardly pointing sharp appendages on each side; disc achenes narrowly oblanceolate, compressed, unwinged, pappus of 2 stiff awns.

1. S. nodiflora (L.) Gaertner. Nep: *Jamjobi.* Fig. 135j–k.

Plant 30–60cm. Leaves ovate, 3.5–8.5 × 1–5cm, acute, rounded or attenuate at base, margins shallowly serrate, more densely pilose beneath; petioles 0–3cm. Capitula c 4mm diameter above; phyllaries ovate-elliptic, outer pair 7–9mm, inner ones 5–6mm. Paleae ± elliptic, narrow, 5–6mm. Ray flowers c 6; corolla tube 2–3mm; ligules oblong, c 2 × 1mm. Disc corollas 3–3.5mm. Ray achenes, 3.5–4.5 × 1.5–2mm, pappus awns 1.5–2mm; disc achenes c 4mm, hispidulous above or tuberculate, pappus awns 2.5–4.5mm.

Bhutan: S – Samchi district (Samchi), Phuntsholing district (Phuntsholing, Toribar), Gaylegphug district (Gaylephug), Deothang district (Samdrup Jongkar); **Darjeeling:** Kumai, Kurseong–Balasan (97a), Ryan Chhu; **Sikkim:** Gangtok (69). Roadsides, forest, fallows, field margins, 250–600m. June–September(–March).

101. TITHONIA Jussieu

Annuals or herbaceous perennials. Leaves simple or palmately 3–5-lobed, alternate or sometimes opposite at base, subentire to serrate. Capitula mostly solitary on thick hollow peduncles, radiate, paleate. Involucres campanulate, (2–)3–4-seriate; phyllaries thickened and striate below. Receptacle convex, paleae concave, enfolding achenes. Ray flowers neuter; ligule conspicuous. Disc flowers bisexual; corolla tubular-campanulate, 5-lobed; style branches linear-lanceolate. Achenes oblong, subcompressed, weakly 4-angled; pappus 1–few awns with shorter scales between.

1. Perennial; ligules 45–70mm .... **1. T. diversifolia**
\+ Annual; ligules 20–30mm .... **2. T. rotundifolia**

1. T. diversifolia (Hemsley) A. Gray

Perennial herb c 2–3m, sometimes forming thickets; stems grey tomentose at first. Leaves simple and ovate or 3–5-lobed, usually short hispid above and araneous-puberulous below; blade 5–30 × 5–20cm, attenuate at base, sometimes with subcordate lateral lobes, subentire to crenate-serrate, segments acuminate, often broadly panduriform; petiole up to 6cm with rounded deciduous auricles. Peduncles 10–15cm, glabrous. Phyllaries (3–)4-seriate, ovate to oblong, obtuse or acute, glabrous; outer ones much reduced, ± herbaceous, 3–10 × 2–6mm; inner ones 12–20 × 4–10mm, subscarious. Disc c 30mm; paleae 10–13 × 2–4mm, acuminate. Ray flowers 10–15, ligules linear-elliptic, yellow, 45–70

× 10(–15)mm. Disc corollas yellow, 9mm. Achenes blackish, 5–8mm, brown puberulous; pappus dark brown, awns to 6mm, scales 2–3mm.

Darjeeling: Kurseong–Pankhabari, Manjitar, Rakti (97a) etc. Scrub on river bank, open slopes, 350–1200m. November–December (97a).

Native of tropical America, now a widespread weed. Only fruiting specimens seen from our area.

2. T. rotundifolia (Miller) Blake

Similar to *T. diversifolia* but annual, c 1.2m; peduncle above and involucre pubescent (in our specimens); phyllaries (2–)3-seriate, broadly lanceolate, 15–20 × 4.5–6mm, ± acuminate, outer rows not reduced and distinctly herbaceous above; ray flowers c 10, ligules oblong-obovate, orange or yellow, 20–30 × 6–15mm; disc flowers 8mm.

Bhutan: S – Samchi district (Samchi (117)), Phuntsholing district (Phuntsholing). Cultivated, 250–300m. December–February.

Native of tropical America.

102. HELIANTHUS L.

Erect, coarse annual or perennial herbs. Leaves simple, opposite or alternate, petiolate. Capitula large, solitary or corymbose, radiate, paleate; peduncles pith-filled. Involucres hemispherical or depressed, several-seriate; phyllaries herbaceous, imbricate. Receptacle flat to conic. Ray flowers neuter; ligules yellow. Disc flowers bisexual; corolla tubular-campanulate, 5-toothed, teeth yellow; style branches flattened, acuminate. Achenes obovoid, ± compressed or 4-angled. Pappus of 2 lanceolate scales, sometimes with smaller ones between them, deciduous.

1. H. annuus L. Eng: *Sunflower.*

Hispid annual, stems 1–3m, ± unbranched. Leaves ovate, 10–40 × 5–35cm, acuminate, usually cordate, serrate, only lowest ones opposite; petiole up to 15cm. Capitula solitary or few in open corymb; phyllaries 3-seriate, ovate, long acuminate, spreading much wider than disc. Disc 3–6cm diameter at flowering time, brown; paleae lanceolate, 9–13mm, 3-toothed, centre tooth elongate,

---

FIG. 135. **Compositae: Heliantheae.** a, *Acanthospermum hispidum*: disseminule (× 4). b, *Xanthium indicum*: disseminule (× 2). c, *Parthenium hysterophorus*: disseminule (× 8). d–e, *Acmella uliginosa*: achenes of two individuals (× 14). f, *Acmella calva*: achene (× 14). g, *Acmella paniculata*: achene (× 10). h–i, *Wedelia montana*: h, achene (× 8); i, capitulum (× 3). j–k, *Synedrella nodiflora* (× 8): j, ray achene; k, disc achene. l–m, *Galinsoga ciliata* (× 14): l, ray achene; m, disc achene. n–o, *Galinsoga parviflora* (× 14): n, ray achene; o, disc achene. p, *Eclipta prostrata*: achene (× 10). q, *Tridax procumbens*: achene (× 6). r, *Glossocardia bidens*: achene (× 4). s, *Bidens pilosa*: achene (× 4). t, *Bidens tripartita*: achene (× 4). u, *Cosmos sulphureus*: achene (× 4). Drawn by Margaret Tebbs.

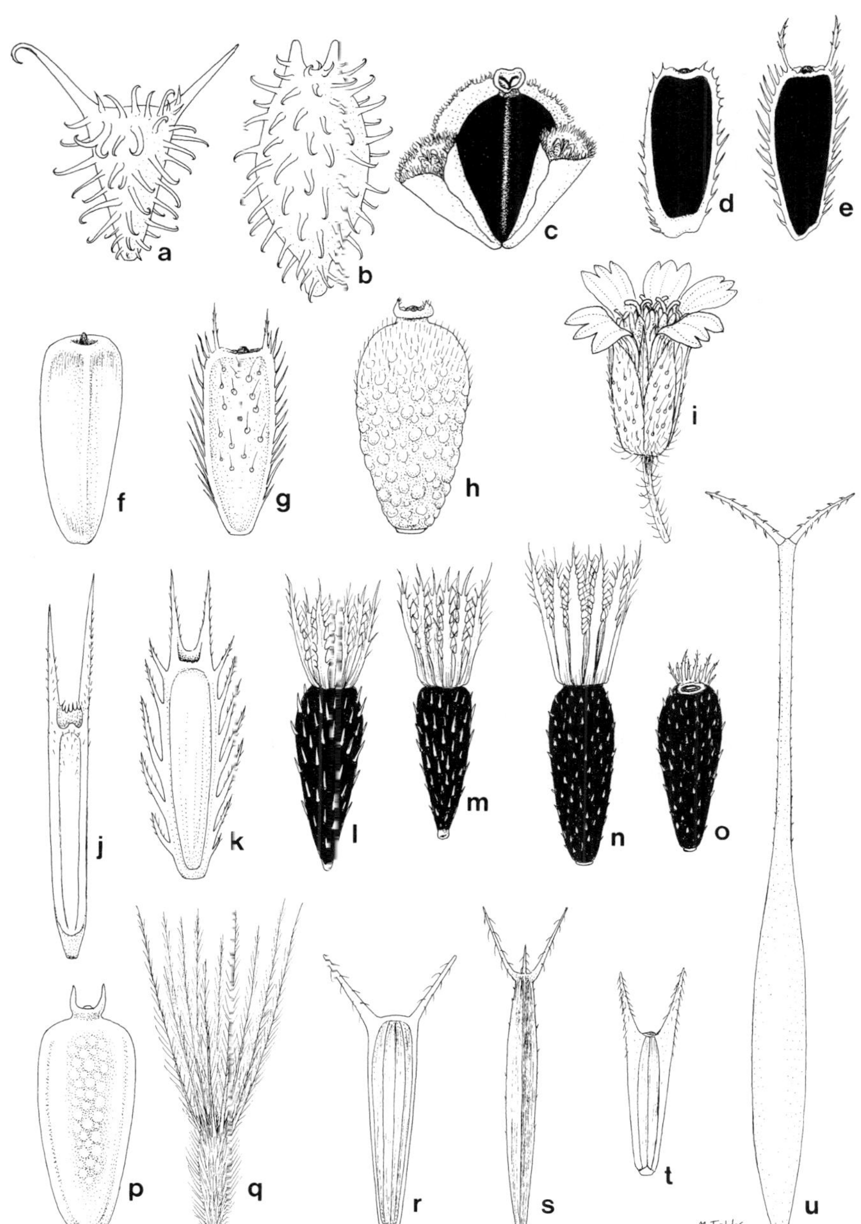
a
b
c
d
e
f
g
h
i
j
k
l
m
n
o
p
q
r
s
t
u
M.Tebbs

herbaceous. Ray flowers c 20–35, corolla tube 2–3mm, ligule 45–60 × 16–20mm. Disc corollas 6–9mm. Achenes at flowering time c 6mm, pubescent; longer pappus scales 2–4mm. Mature achenes c 12 × 7mm, ± glabrous, variously coloured.

Bhutan: C – Thimphu district (Thimphu), Tongsa district (Shemgang). Cultivated in gardens, 1980–2370m. June–September.

Only two specimens seen, other cultivated forms may differ considerably in dimensions. Genus native to N America. *H. grosseserratus* Martens has been reported to occur around Kurseong, Darjeeling (95). It is a perennial species with tuberous roots and lanceolate leaves.

103. GALINSOGA Ruiz & Pavón

Annuals. Stems erect or spreading. Leaves simple, opposite, 3-veined at base. Capitula radiate, paleate, in cymose clusters, peduncles slender. Involucre 2-seriate, hemispherical. Phyllaries herbaceous, outer ones 1–3, shorter, empty, inner ones 5, opposite a ray flower. Receptacle conical. Paleae flat, dimorphic, sometimes 2–3 outer ones connate at base with a single phyllary, enclosing the ray flower and dispersing as one unit with the achene, inner ones often trifid. Ray flowers female; corolla tube spreading-pilose; ligules obovate-quadrate, white, 3-lobed. Disc flowers bisexual; corolla weakly tubular campanulate, 5-lobed, yellow; style branches linear, flattened, acute. Achenes obconical, blackish; ray achenes flattened, pappus reduced; disc achenes angular; pappus of c 15–20 fimbriate lanceolate scales.

New World species, our plants 2 weed species now widespread throughout tropical and temperate countries. Parker (272) reports them to be among the commonest dry-land weeds in Bhutan above 1000m and records the following local names: Dz: *Jagouma*, *Jagasuju*; Nep: *Udasoy*; Sha: *Yurunpa*.

1. Hairs of peduncles eglandular, c 0.2mm; inner paleae usually trifid; ray pappus much reduced, scales bristle-like, c 0.4mm .... **1. G. parviflora**
\+ Hairs of peduncles glandular and eglandular, c 0.5mm; inner paleae usually entire; ray pappus little reduced, scales c 1mm .... **2. G. ciliata**

1. G. parviflora Cavanilles. Dz: *Jaga ima*. Fig. 135n–o, Fig. 136a.

Plant 10–60cm, stems pubescent above. Leaves ovate-lanceolate, (1.5–)2.5–6 × (1–)1.5–4cm, acute or acuminate, rounded or attenuate at base, shallowly serrate-crenate, ciliate and sparsely pilose on both surfaces, with petiole up to 1cm or upper ones subsessile. Peduncles short pubescent, eglandular (in our area). Outer phyllaries 2–3, inner ones 3.5–4mm, ovate. Outer paleae oblong-elliptic, 3.5–4mm; inner paleae oblanceolate, 2–3mm, free, trifid. Ligules c 1.5 × 1.5mm. Disc corollas c 1.4mm. Ray achenes 2mm, pappus of weak bristles c 0.4mm; disc achenes 1.5mm, pappus scales 1–1.8mm with undivided part obtuse or attenuate.

Bhutan: **S** – Samchi (117) and Chukka districts, **C** – Thimphu, Punakha, Tongsa and Tashigang (117) districts; **Darjeeling:** Darjeeling, Kurseong (97a), Rakhti (97a) etc.; **Sikkim:** Pemayangtse–Rathong Chhu (69); **Chumbi:** unlocalised (Bell collection). On waste places and cultivated ground, 900–2450m. March–December.

2. G. ciliata (Rafinesque) Blake. Fig. 135l–m.

Very similar to *G. parviflora*, differing in our area by hairs of peduncles longer (c 0.5mm), some bearing capitate glands; outer phyllaries 1–2, caducous; inner paleae usually entire; scales of ray pappus similar to those of disc, though a little shorter (c 1mm) and fewer; undivided part of disc pappus scales always attenuate.

Darjeeling: Darjeeling; **Sikkim:** Gangtok. Roadsides, 1800–2200m. July.

Probably a recent introduction to our area, only two collections from 1992 were seen. *G. ciliata* is applied to a widely introduced weedy taxon that originated in the Americas, where it is doubtfully separable from the prior described *G. quadriradiata* Ruiz & Pavón.

104. TRIDAX L.

Perennial. Stems decumbent, often rooting at nodes. Leaves simple, opposite, sometimes weakly 3-lobed. Capitula solitary on long peduncles, radiate, paleate. Involucres broadly campanulate, 2–3-seriate; outer phyllaries few, herbaceous, inner scarious. Receptacle shortly conical, paleae persistent, membranous. Ray flowers c 5, female. Disc flowers bisexual, tubular-campanulate; corolla 5-toothed; style branches slender, flattened, appendages subulate. Achenes narrowly obconic, blackish, pilose; pappus of c 18–20 unequal plumose bristles, ± alternately longer and shorter.

1. T. procumbens L. Fig. 135q.

Plant 15–30cm; stems spreading pubescent. Leaves ovate or lanceolate, 2–5 × 1–2cm, acute, cuneate at base, coarsely serrate, hirsute on both surfaces, often scabrous to touch; petioles to 1cm. Peduncles 5–18cm; involucres 5–6mm diameter; outer phyllaries ovate to obovate, c 5mm, acuminate, herbaceous, hirsute; inner phyllaries oblong, c 7.5mm. Paleae narrow oblong-lanceolate, 7.5–9mm. Ray flower corolla tubes 3–3.5mm, pilose; ligules yellowish-white, oblong, 3–4.5 × 3–3.5mm, 2–3-lobed. Disc corollas 5–7mm, partly pubescent. Achenes 2–2.5mm; longest pappus bristles subequal to tube of ray flower and whole disc flower.

Bhutan: **C** – Tashigang district (Rolong, Tashigang); **Darjeeling:** Mahaldiram, Rishap, Tindharia etc.; **Sikkim:** Gangtok. Roadsides, open rocky soil, tea gardens, 250–1750m. March–December.

Native of C America, now a widespread weed of warmer countries.

105. ACANTHOSPERMUM Schrank

Dichotomously branched annual with ± soft spreading hairs. Leaves simple, opposite, sessile. Capitula small, solitary in forks of branches, subsessile (short pedunculate in fruit), radiate, paleate. Involucre 2-seriate, broadly campanulate; outer phyllaries 5, foliaceous; inner ones each enveloping the ovary of a ray flower, forming disseminules. Receptacle small, convex, paleae membranous. Corollas yellow. Ray flowers 6–9, female; corolla tube obsolete; ligule small, 2–3-toothed. Disc flowers male; corolla tubular-campanulate, 5-toothed; style undivided, clavate. Disseminules cuneate, ± laterally compressed, echinate and minutely glandular, prickles straight or hooked with pair at apex much longer and divergent; pappus absent.

1. A. hispidum DC. Eng: *Starbur* (272). Fig. 135a.

Plant erect, 20–50cm; stems glabrescent below. Leaves obovate, 2–8(–12.5) × 0.8–3(–8)cm, obtuse or subacute, cuneate at base, shallowly serrate, hairy on both surfaces and minutely subsessile-glandular beneath. Involucre 4–5mm diameter at flowering time. Outer phyllaries elliptic-ovate, 4–5 × c 1.7mm; paleae ± obcuneate, 1.5mm. Ray flower corollas 1.3–1.7mm, ligule ± oblong. Disc flower corollas 1.5–2mm. Disseminule body 4–5mm, smaller spines c 1.2mm, apical spines 3–4mm.

Bhutan: C – Punakha district (near Punakha Dzong, Wangdi Phodrang); also reported from Mongar and Tashigang administrative districts (272). Occasional weed, mostly of dry-land crops, up to 1500m. July–September.

Of New World origin, now a widespread weed of warmer countries.

106. SIGESBECKIA L.

Annual, stems erect. Leaves simple, opposite, 3-veined above base, toothed, petiolate. Capitula in lax axillary or terminal cymes, radiate. Involucres campanulate, 2-seriate, herbaceous; outer phyllaries spreading, densely glandular-pilose; inner much smaller, concave, enfolding outer achenes, strongly glandular-pubescent. Receptacle convex, paleae concave, membranous, enfolding achenes. Corollas yellow. Ray flowers female, 1-seriate; ligule short. Disc flowers bisexual, tubular-campanulate; corolla 5-toothed; style branches short, flattened, obtuse. Achenes obovoid, 4-angular, curved, rounded at apex, dark brown or blackish, glabrous; pappus absent.

FIG. 136. **Compositae: Heliantheae.** a, *Galinsoga parviflora*: habit (× ⅔). b–d, *Sigesbeckia orientalis*: b, capitulum (× 2); c, ray flower (× 6); d, disc flower (× 6). e–g, *Bidens pilosa*: e, habit (× ⅔); f, ray flower (× 6); g, disc flower (× 6). Drawn by Margaret Tebbs.

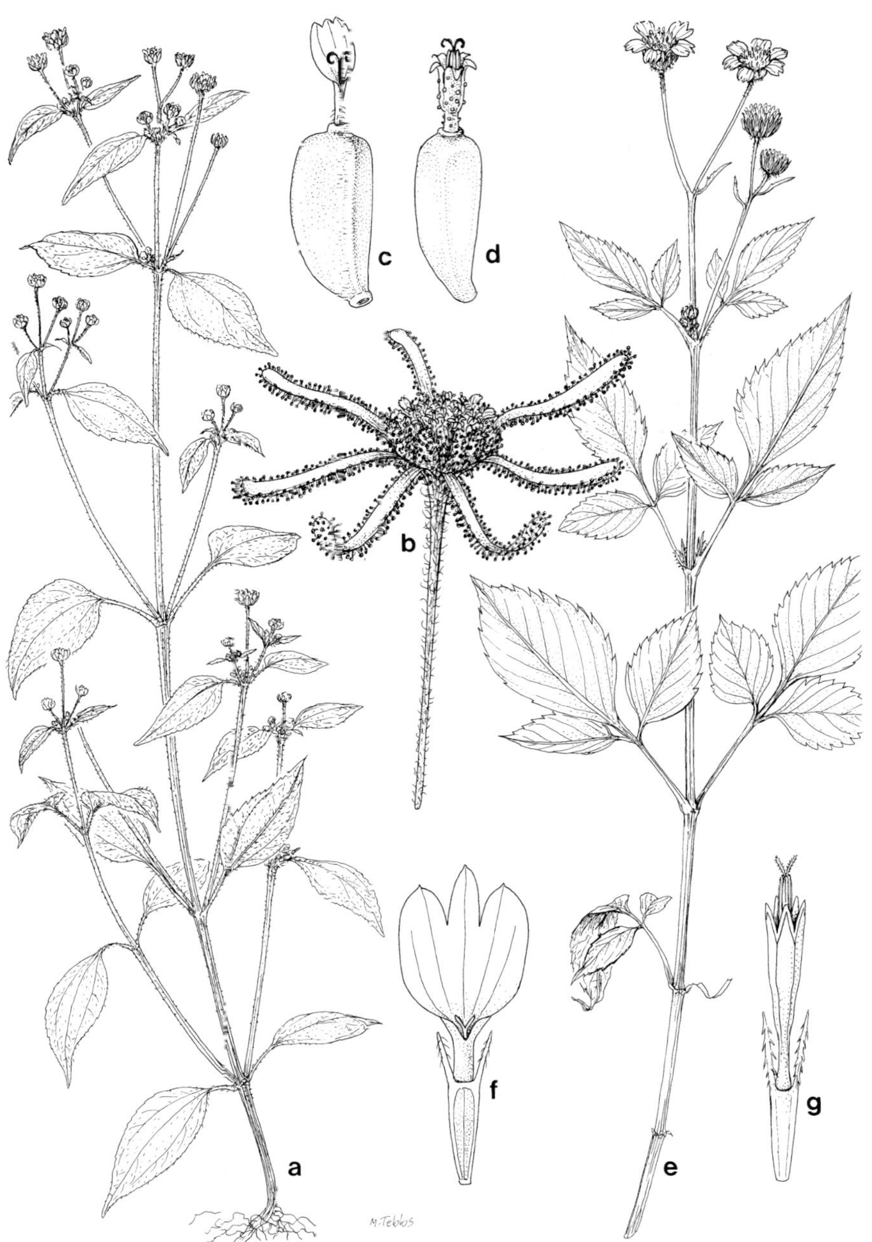

c
d
b
a
e
f
g
M. Tebbs

1. S. orientalis L. Fig. 136b–d.

Stems up to 1.5cm, purplish, spreading whitish pubescent. Leaves ovate-triangular, 6–15 × 3–10cm, acuminate, long attenuate at base, variously toothed, usually with (1–)2(–3) pairs of larger lateral teeth or lobes, pubescent on both surfaces. Capitulum, excluding outer phyllaries, 5mm diameter. Outer phyllaries oblanceolate, 8–10 × 1–1.5mm; inner phyllaries oblong, 4–5 × 1.5mm. Ray flowers 5; corolla tube 1mm; ligule 1.5 × 1mm. Disc flowers c 12; corolla c 1.75mm. Paleae 4–4.5mm, pubescent, sometimes persistent around achenes. Achenes 3–4mm.

Bhutan: S – Samchi (117), Phuntsholing (117), Chukka and Gaylegphug (117) districts, **C** – Thimphu, Punakha (71) and Tashigang (117) districts; **Darjeeling:** Darjeeling, Rakti bank, Kurseong (97a) etc.; **Sikkim:** Lachung, Rathang–Yoksam, Gangtok (69) etc.; **Chumbi:** Chumbi; **W Bengal Duars:** Jalpaiguri district (174). Wasteground near habitations, roadsides and weed of cultivation, 300–2450m. March–December.

Widely distributed in warmer countries and variously divided into infraspecific taxa or separate species. Specimens with regularly denticulate leaf margins and small lateral lobules from Darjeeling, Sikkim and Chumbi have been determined as *S. iberica* Willdenow by D.L. Schulz, a specimen from Rakti bank with shallowly crenate leaf margins without lateral lobes as typical *S. orientalis*, and specimens from Chukka and Thimphu districts with coarsely serrate-dentate leaf margins and larger lateral lobes as a new unpublished subspecies of *S. orientalis*.

107. ENYDRA Loureiro

Perennial aquatic or marsh herbs. Stems rooting at nodes. Leaves simple, opposite, subsessile. Capitula solitary, terminal but soon overtopped, sessile, semi-globose, radiate, paleate. Involucre 2-seriate; phyllaries 4, large, leafy, in opposite pairs, 2 outer larger, broadly ovate, partly enclosing capitulum, inner ± oblong. Receptacle convex, paleae enveloping flowers. Corollas white with pink tips. Ray flowers many-seriate, female; ligule small, shorter than corolla tube, deeply 3(–4)-lobed. Disc flowers fewer, bisexual or inner ones male; corolla tubular-campanulate, 5-toothed; style branches linear, acute. Achenes oblong-obovate, compressed, enclosed by palea; pappus absent.

1. E. fluctuans Loureiro; *Enhydra fluctuans* in error

Stems 10–100cm, erect or decumbent, glabrous or sparsely pubescent. Leaves oblong, 2–7 × 0.5–1cm, obtuse or acute, base abruptly narrowed, margins with few shallow teeth, glabrous or sparsely hairy on both sides, punctate beneath. Capitula 5–9mm diameter. Outer phyllaries 10–15 × 5–9mm, obtuse, ± glabrous. Paleae 3–4mm, minutely ciliate at apex, thickening and hardening after flowering. Ray corollas 2–2.5mm; disc corollas c 3.5mm. Achenes blackish, glabrous, 2–3 × 0.5mm.

Darjeeling: Siliguri; **W Bengal Duars:** Jalpaiguri district (174). Water plant. January.

Also reported from Sikkim without locality (343). Most details above from specimens collected in NE India outside our area.

108. GUIZOTIA Cassini

Erect annual. Leaves simple, opposite, lanceolate, sessile. Capitula rather few in open leafy terminal corymb, broadly campanulate, radiate, paleate; involucre 1-seriate; phyllaries 5, leafy, ovate, entire. Corollas yellow with long white hairs at base, lower ones reflexed over achene. Ray flowers few, female; ligule broad, 3-toothed. Disc flowers bisexual; corolla tubular-campanulate, 5-lobed; style branches linear with ovate appendages. Achenes oblanceolate; ray achenes 3-angled; disc achenes 4-angled; pappus absent.

1. G. abyssinica (L.f.) Cassini

Plant 0.6–2m; stems sparsely pubescent at least above. Leaves 6–11 × 1–4cm, acute, base rounded or truncate, subentire to prominently serrate, subglabrous to sparsely pubescent on both surfaces. Involucre 7–10mm diameter; phyllaries 7–10 × 4–5.5mm, obtuse, subglabrous. Paleae narrowly obovate, 7–9 × 3mm, obtuse. Ray flowers 6–8; ligules c 15 × 7mm. Disc corollas 4.5mm. Achenes black, 4mm, glossy.

Bhutan: S – Samchi district (117); **Darjeeling:** Great Rangit, terai. 300m. April, December.

Cultivated and sometimes escaping. Seeds yield an edible oil.

109. COREOPSIS L.

Erect glabrous annual herb. Leaves opposite, 1–2-pinnate. Capitula in lax cymose panicles, radiate, paleate. Involucre broadly hemispherical, 3-seriate; phyllaries ± herbaceous with scarious margins, inner whorl much larger. Receptacle convex; paleae linear, scarious, concealed at anthesis. Ray flowers 1-seriate, neuter; corolla tube obsolete; ligules conspicuous, obovate, yellow in upper half, dark crimson below. Disc flowers bisexual; corolla tubular-campanulate, reddish-purple, 4-toothed; style branches linear, slightly broadened above, ± conical. Achenes ellipsoid, dorsally compressed; pappus absent.

1. C. tinctoria Nuttall

Plants to 1m. Leaves 7–15cm, with up to 4 pairs of linear-elliptic leaflets; leaflets 2–6 × 0.2cm, acuminate, attenuate at base, lower ones divided into 3 similar leaflets. Peduncles slender, 4–10cm; involucres c 7mm diameter, outer phyllaries c 10, lanceolate or ovate, c 2mm, inner phyllaries c 8, ovate, c 5.5 × 2.5mm, brownish. Paleae c 4mm. Ray flowers c 8; corolla tube 0.5mm; ligules c 11 × 8mm, shallowly 3-lobed. Disc corollas 2.5–4.2mm; anthers strongly exserted, dark coloured. Achenes 2.4 × 0.8mm, brownish (?black when mature), glabrous.

Bhutan: S – Phuntsholing district. Cultivated in garden, 230m. October.

Native of southern USA. Elsewhere achenes may be broadly winged or papillose.

110. DAHLIA Cavanilles

Herbaceous perennial or coarse subshrub; roots tuberous; stems erect, glabrous. Leaves simple to 3-pinnate, opposite. Capitula radiate, paleate, solitary to paniculate. Involucres 2-seriate, outer phyllaries reflexed in flower, herbaceous, inner ones erect, membranous. Paleae scarious, resembling inner phyllaries. Ray flowers neuter; ligules conspicuous, white to rose purple or yellow to scarlet. Disc flowers bisexual; corolla tubular-campanulate, 5-toothed, yellow, teeth sometimes red; style branches linear, acuminate. Achenes oblanceolate or almost linear, compressed; pappus obsolete.

1. Herbaceous perennial; leaves simple to 2-pinnate .... **1. D. × pinnata**
\+ Suffrutescent with stems becoming woody and often persisting overwinter; leaves 2–3-pinnate .... **2. D. imperialis**

1. D. × pinnata Cavanilles; *D. × variabilis* (Willdenow) Desfontaines, *D. × hortensis* Guillaumin

Herbaceous perennial, c 1–2m; stems glabrous. Leaves simple to 1–2-pinnatisect (to 2-pinnate) on same plant, 12–35cm, segments broadly ovate, 5–11 × 2–5.5cm, acute, cuneate at base, coarsely serrate, glabrous above, sparsely pubescent beneath; terminal segment usually larger than lateral ones. Peduncles up to 30cm. Outer phyllaries ovate to obovate, 6–22 × 3–8mm, obtuse to acuminate; inner phyllaries ovate, 20–22 × 7–13mm, obtuse. Ray flowers several-seriate; corolla tube 3.5mm; ligule obovate, c 40–45 × 25–30mm, yellow, orange or scarlet. Disc corollas 10mm. Sometimes disorganised flowers between disc and rays. Immature achenes oblanceolate, c 11 × 4mm, blackish; pappus of 2 minute ridges.

Bhutan: C – Thimphu (Paro), Tongsa district (Shemgang). Cultivated in gardens, 1980–2370m. June–October.

Hybrids originally grown in Mexico. Plants with purple ligules are found elsewhere in India.

2. D. imperialis Ortgies

Coarse subshrub, 2–6m; stems becoming woody and often persisting for more than one season. Leaves 2–3-pinnate, 50–100cm, puberulous; ultimate leaflets ovate-oblong, acute to acuminate, tapered to subcordate-truncate at base, serrate. Capitula numerous in a terminal panicle. Outer phyllaries obovate, obtuse or acute, 6–15mm; inner phyllaries ovate (oblong-lanceolate), 15–25mm. Ligules white to rose purple, sometimes deep red at base, 35–60 × 15–38mm. Disc corollas yellow, 9–11mm, tips sometimes red. Achenes linear-oblanceolate, 13–17mm; pappus at anthesis of 2 minute ridges.

Darjeeling: Darjeeling (218), Kurseong (97a). Cultivated in gardens and sometimes escaping, 1650–1850m. September–December.

No specimens seen from our area. Details above from Mexican specimens and plant grown in Europe.

111. COSMOS Cavanilles

Erect annual herbs. Leaves opposite, pinnatisect to bipinnate. Capitula solitary on long peduncles, radiate, paleate. Involucre broadly campanulate, 2(–3)-seriate; phyllaries connate at base, outer series herbaceous, inner series scarious. Receptacle ± flat; paleae concave. Ray flowers 1-seriate, ligulate, neuter. Disc flowers bisexual; corolla tubular-campanulate, 5-toothed, yellow; style branches elongate, thick above, acutely or acuminately appendaged. Achenes linear, tetragonous or subterete, narrowed into a beak above; pappus absent or of 2–3 retrorsely barbed awns.

1. Ligules white, rose or purplish-pink.......................... **1. C. bipinnatus**
\+ Ligules yellow or orange .... **2. C. sulphureus**

1. C. bipinnatus Cavanilles

Plant 0.5–1.5m, subglabrous to stiffly puberulous. Leaves bipinnate, 6–11 × 3–6cm; segments narrowly linear, 5–30 × 1mm, acute or acuminate, attenuate at base; petioles dilated, 4–25mm. Involucres c 12mm across; phyllaries 2-seriate; outer ones c 8, with ovate base 4–6 × 3–4.5mm and linear appendage 3–17mm; inner phyllaries ovate-oblong, 8–10 × 4–4.5mm. Paleae 8.5–10mm, very narrowly elliptic below, linear above, usually yellow. Ray flowers usually 8, ligules obovate, up to 30 × 18mm, with broad shallow teeth at apex, white, rose or purplish-pink. Disc corollas c 5.5mm. Achene body 6–10mm; beak 1–6mm; pappus awns absent or present, 1–2mm, apparently both conditions in same capitulum.

Bhutan: C – Thimphu district (Simtokha, Taba, Paro); **Darjeeling:** Darjeeling (218). Grassy places, roadsides and cultivated ground, 2300–2400m. July–October.

Native of southern N America. Also reported from Bumthang and Tashigang administrative districts, mainly above 2000m (272). Fruit not seen from our area.

2. C. sulphureus Cavanilles. Fig. 135u.

Plants 20–40cm, subglabrous. Leaves (1–)2-pinnatisect with 3–4(–5) pairs of lateral segments, ovate or triangular in outline, 3–10 × 1.5–8cm, ultimate segments oblanceolate to narrowly lanceolate, apiculate; petioles 0.5–2.5cm.

Involucres 6–10m diameter, 2–3-seriate; outer phyllaries lanceolate, 5–8 × c 1.5mm, usually a row of 8 and 1 outside; inner phyllaries and paleae scarious, oblong-lanceolate, 8.5–9.5mm. Ray flowers c 8; ligule obovate, orange or yellow-orange, 18–28 × 10–17mm, prominently 3-dentate. Disc corollas c 8mm. Achenes black, body 10mm, beak 6mm, retrorse hispid; pappus bristles 2, strongly divergent, 3–4mm, minutely retrorse barbed.

Bhutan: S – Samchi district (Samchi), Phuntsholing district (Phuntsholing). Cultivated in parks and gardens, 250–500m. February–March.

Native from Mexico to Brazil. Various specimens collected in India outside our area are much larger, more hirsute or remotely spinulose-ciliate on leaves and phyllaries.

112. GLOSSOCARDIA Cassini

Perennial herb, woody at base with slender tap-root, subglabrous. Leaves mostly basal, deeply pinnatisect, Krantz venation readily apparent (as green sheaths around all veins when held against light), petiolate; cauline leaves alternate. Capitula few on long peduncles, radiate, paleate. Phyllaries 1–2-seriate, ± herbaceous; outer phyllaries 0–3; inner phyllaries c 8–10, subtending ray and disc flowers; paleae flat. Corollas yellow. Ray flowers female, few; corolla tube obsolete, ligule short, 3-toothed at apex. Disc flowers bisexual, few; corolla narrowly tubular campanulate, 4-toothed; style branches elongate, papillate. Achenes equal, oblong, compressed, 3-ribbed on each face; pappus of 2 divergent, retrorsely barbed awns.

1. G. bidens (Retzius) Veldkamp; *Glossogyne pinnatifida* DC. Fig. 135r.

Plant to 30cm; stems 1–several, ± erect, almost leafless. Basal leaves c 3–8 × 1.5–4cm; segments linear or linear-oblanceolate, sometimes with 1(–2) lobes or teeth, acuminate; petioles to 4.5cm. Phyllaries ± oblong-lanceolate, in fruit outer ones 2mm and inner ones 2.5mm, transitional to longer, narrower, subscarious paleae. Ray flowers 5–12, ligule obovate, purple-veined, c 3 × 2.5mm. Disc flowers 7–12, corollas c 3.5mm, including tube c 1.3mm. Achenes dark brown, 5–6.5mm, awns paler, to 2mm.

Bhutan: C – Tashigang district (Dangme Chu). 900m. Fruiting in August.

Only poor fruiting material seen from our area, most flower details taken from Indian specimens outside our area.

113. BIDENS L.

Erect annual herbs. Stems squarish, leafy. Leaves entire to partly tripinnatifid. Capitula 1–several, terminal or axillary, radiate or discoid, paleate. Involucres broadly campanulate, 2-seriate; phyllaries shortly connate at base, outer ones herbaceous, sometimes foliaceous; inner ones scarious; paleae similar, narrower.

Ray flowers neuter; corolla white or yellow. Disc flowers bisexual; corolla tubular-campanulate, 5-lobed, yellow; style branches linear with subulate villous appendages. Achenes obovoid or linear, 4-angled or compressed, sometimes dimorphic; pappus of 2–5 stiff retrorsely barbed awns.

1. Leaves entire to trifoliolate or pinnate 2
\+ Leaves biternate or bipinnatisect 3

2. Leaves entire to deeply divided into 3 or 5 segments; achenes narrowly obovate or obcuneate .... **1. B. tripartita**
\+ Leaves trifoliolate or pinnate; achenes linear or linear-oblong .. **2. B. pilosa**

3. All primary leaflets usually pinnatisect or lower pair bipinnatifid **3. B. bipinnata**
\+ Lower lateral primary leaflets ternate; upper leaflets simple or some 1–2-lobed .... **4. B. biternata**

1. B. tripartita L. Dz: *Yedum* (272). Fig. 135t.

Plant 20–45(–70)cm; stems subglabrous. Leaves 5–13cm, usually deeply divided into 3 (sometimes pinnately in 5) lanceolate segments or sometimes mostly entire on depauperate plants, glabrous; segments subacute, shallowly serrate, terminal one much longer, 3–5 × 0.5–2.5cm; petiole narrowly winged, 0.5–3cm. Capitula solitary or few, discoid, c 8mm across inner phyllaries. Outer phyllaries 5–9, leafy, oblanceolate or spathulate, c 7–15 × 1.5–4mm, rounded to subacute, entire or toothed, cartilaginous-denticulate, glabrous; inner phyllaries ovate to oblong, 4.5–5 × 2–3.5mm, brownish. Corollas 3mm. Achenes monomorphic, narrowly obovoid or obcuneate, 6–8 × 1.5–2.5mm; pappus awns 2, 2–3.5mm; margins of achene often weakly retrorse barbed.

Bhutan: C – Thimphu district (Papaisa, Paro, Chapcha), **N** – Upper Mo Chu district (Mo Chu). Open forest, wet sites including fields, 2150–2500m. August–October.

Also reported from Chhukha and Wangdi administrative districts, mainly above 2000m (272).

B. cernua L. has been collected at Tuna on the boundary of our area in Chumbi, at 4550m, flowering in August. It is similar to *B. tripartita*, but has simple linear-lanceolate, sessile leaves, fully developed ray flowers (corolla tube 1.5mm, ligules obovate, entire, 7 × 3.5mm) and 3–4 pappus awns.

2. B. pilosa L.; *B. pilosa* var. *pilosa* sensu F.B.I. p.p. Eng: *Blackjack* (272); Nep: *Kuro* (272). Fig. 135s, Fig. 136e–g.

Plants 20–80cm; stems glabrous or sparsely pubescent. Leaves trifoliolate, rarely pinnate with up to 7 leaflets, subglabrous; leaflets ovate or elliptic, 1–8.5 × 0.6–3.5cm, acuminate, obtuse to attenuate at base, serrate, finely ciliate; petiole 1.5–5cm, reduced on inflorescence. Capitula radiate; involucres 3.5–7mm diameter; outer phyllaries oblong-spathulate, 3–4 × 0.5–1mm, ± herbaceous; inner phyllaries 4–5 × 1–1.6mm, brownish. Ray flowers 5–6; corolla white (?or pale yellow); tube to 1.2mm; ligule ovate-oblong, truncate, variable, from 3.3 × 3mm, deeply 3-lobed, to 9 × 3.5mm, obscurely 3-dentate. Disc corollas yellow, 4mm. Achenes linear, 6–11 × c 0.8mm, blackish, glabrous below, sparsely setose above, usually dimorphic with inner achenes distinctly longer, more angular and often smoother than outer ones; pappus awns 2–3, paler, sometimes yellowish, c 2.5mm.

Bhutan: S – Phuntsholing district (Phuntsholing, Torsa River), Chukka district (Marichong), **C** – Punakha district (Lobeysa, Shenganga, Rinchu–Mischien (71) etc.), Tongsa district (Tashiling (71)); **Darjeeling:** Badamtam, Kalimpong (358), Kurseong (97a) etc.; **Sikkim:** Lachen, Temi, Yoksam. Fields, roadsides and amongst scrub, 150–2400m. April–December.

The type is discoid; our specimens have usually been treated as a distinct variety under the misapplied epithet var. *minor* (Blume) Sherff (71, 343). Reported as probably occurring in all administrative districts of Bhutan (272).

3. B. bipinnata L.; *B. pilosa* L. var. *bipinnata* (L.) Hook.f.

Plant c 1m; stems glabrous, sparsely pilose at nodes. Leaves usually bipinnate, 4–15(–20)cm, petiolate; primary leaflets usually 5, ovate or triangular in outline, pinnatifid or lowest pair bipinatifid; ultimate segments ± elliptic, subacute with apiculus to acuminate, lobulate or coarsely dentate, sparsely pilose on both surfaces. Capitula radiate, 2.5–4mm diameter at first. Outer phyllaries oblong-spathulate, ± herbaceous, 1.5–3mm, ciliate; inner phyllaries broadly oblong, 3–5.5mm, striate, broadly scarious, passing into longer, narrower paleae, 3.7–6mm. Corollas yellow. Ray flowers 3–5, ligules elliptic to obovate, 2–3 × 1.2–2mm, entire to 3-dentate. Disc corollas c 3mm. Achenes linear, 9–18mm, sparsely setose above, c 4–5 outer achenes 7–12mm, stouter and more densely setose; pappus awns (2–)3, 2–3mm.

Bhutan: C – Tashigang district (Tashigang); **Darjeeling:** unlocalised (Cowan collection); **Sikkim:** Linchyum. Among crops, 1200–2500m. June–October.

Also reported from Bhutan in Lhuntshi and Paro administrative districts (272).

4. B. biternata (Loureiro) Merrill & Sherff

Similar to *B. bipinnata* but usually basal primary leaflets ternate and upper three leaflets simple, sometimes some ultimate leaflets with 1–2 lobes.

Reported (117) from Bhutan: Samchi, Phuntsholing, Chukka, Tongsa and Tashigang districts, in open ground, 270–1500m, flowering December–April. The records need confirmation.

114. XANTHIUM L.

Monoecious annuals; stems erect, often branched. Leaves alternate, petiolate, lobed and toothed. Inflorescence terminal, leafy, paniculate, capitula unisexual.

Male capitula discoid, rounded, many-flowered, paleate, borne towards apex of inflorescence branches; phyllaries numerous, free, stellate; receptacle conical or cylindrical. Female capitula 2-flowered, sessile, borne at base of inflorescence branches, outer phyllaries free, inner mostly connate, entirely enveloping flowers except style branches, bearing spreading hooked spines, with two conic beaks at apex. Corolla of male flowers infundibular, with 5 broadly triangular lobes; filaments fused, inserted near corolla base; anthers exserted, free; style and ovary rudimentary. Style branches of female flowers emerging from inner face of beaks. Achenes 2 in each capitulum, included in hardened spiny involucre; pappus absent.

1. Fruiting involucres including beaks 15–18mm, beaks divergent at base, ± incurved, spines c 100–200, 2–4mm .... **1. X. indicum**
\+ Fruiting involucres including beaks 10–13mm, beaks usually erect, ± straight, approximate, spines c 60–90, c 1.5mm .... **2. X. sibiricum**

1. X. indicum Roxb.; *X. strumaria* sensu F.B.I. p.p. non L. Eng: *Cocklebur* (272). Fig. 135b.

Plant robust; stems to 1m, minutely white puberulous. Leaves broadly ovate, 6–12 × 6–12cm, acute, deeply cordate, ± 3-lobed, irregularly dentate, 3-veined at base, appressed stiffly puberulous, long petiolate. Male capitula 1–7, 5–7mm diameter; phyllaries lanceolate to narrowly obovate, 2–3.5mm; corollas 2–2.7mm; anthers c 1mm. Female capitula at first c 4mm, later ellipsoid, 15–18 × 7.5mm excluding spines, greyish- or greenish-brown, minutely pubescent and glandular; outer phyllaries few, lanceolate, eventually c 4mm; spines c 100–120, 2–3mm, spreading, hooked; beaks conic, c 5mm, divergent below, incurved above, finely hooked at apex.

Bhutan: S – Phuntsholing district (Torsa River), **C** – Thimphu district (Taba). Roadsides, riverside shingles, fields, 200–2450m. May, August.

Plants with fruits mid or dark brown, more sparsely pubescent and with more numerous spines are sometimes considered to be a recent introduction to the subcontinent from the West Indies: *X. occidentale* Bertoloni. This would include a specimen collected between Rongtung and Tashigang, Tashigang district, Bhutan, 1440m, with spines 3–4mm, c 200 per fruit.

2. X. sibiricum Widder

Similar to *X. indicum* but less robust; stems 30–60cm; leaves 4–8 × 3–6cm, acute, base cuneate or shallowly cordate, obscurely 3-lobed; fruit ellipsoid, 10–13 × 6.5mm excluding spines, spines slender, c 1.5mm; beaks 2(–3)mm, erect, usually appressed, usually not hooked.

Bhutan: C – Thimphu district (Paro), **N** – Upper Mo Chu district (Kencho). In pine forest, 1830–2400m.

X. indicum and *X. sibericum* are often sunk in a single Old World species: *X. strumaria* L. Further specimens are reported as *X. strumaria* from Bhutan from

Tongsa and Lhuntshi administrative districts (272) and from Jalpaiguri district (174) and require confirmation of identity.

115. PARTHENIUM L.

Erect branched short-lived annual. Leaves alternate, 1–2-pinnatifid, ovate to lanceolate in outline, lower leaves petiolate. Inflorescence terminal, open, corymbose, of many small capitula. Capitula radiate, paleate. Involucre hemispherical, phyllaries 10 in 2 series, ovate, ciliate above. Corollas white. Ray flowers 5, opposite inner phyllaries, female; corolla obsolete; ovary dorsiventrally flattened, 1 modified palea enclosing a modified disc flower attached to either ovary margin. Free paleae obcuneate, ciliate above. Disc flowers male; corolla narrowly infundibular, white, lobes 5, broadly triangular; anthers exerted; style short, capitate; ovary obsolete. Each disseminule of 1 inner phyllary, 1 ray flower and 2 adnate paleae and disc flowers; achene obovate, black, flattened; pappus of 2 small scales.

1. P. hysterophorus L. Fig. 135c.

Plant to 80cm; stems stiffly appressed white puberulous. Leaves (sparsely) white pubescent, often pilose on veins; basal leaves ovate, to 8 × 3cm, segments irregulary and coarsely rounded-dentate or -lobulate, long petiolate; cauline leaves more finely cut, short petiolate; upper leaves smaller, narrower, subsessile. Capitula 3mm. Phyllaries 2mm, scarious; outer series green above. Ray flower tube c 0.4mm; ligule c 0.7 × 1.2mm. Corolla of disc flowers 2.4mm. Achene 2.2 × 1.1mm; pappus scales 0.4mm.

Bhutan: C – Punakha district (Wangdi Phodrang), also Tongsa, Mongar and Tashigang administrative districts (272). Weed of roadsides, waste places and some perennial crops, 200–1700m. Probably all year.

Pollen and whole plant can cause intense allergic reaction in adult males (272).

116. ECLIPTA L.

Erect or prostrate annuals, white strigillose. Leaves simple, opposite, subentire, subsessile. Capitula small, terminal and in upper leaf axils, radiate, paleate. Involucres obconic-campanulate; phyllaries 2-seriate, herbaceous. Receptacles flattish; paleae linear, appressed pubescent, sometimes absent near centre. Ray flowers numerous, female, several-seriate; corolla white, tube short, ligule narrow, emarginate. Disc flowers bisexual; corolla tubular-campanulate, 4-lobed at apex, white or yellow; style branches linear, obtuse. Achenes oblong or obovoid, ± compressed, with central rib on each side, tuberculate, margins thinner, smooth; pappus of two weak scales.

1. E. prostrata (L.) L.; *E. alba* (L.) Hasskarl, *E. erecta* L. Fig. 135p.

Stems up to 50cm, often rooting at nodes. Leaves elliptic-lanceolate, 1.5–5.5 × 0.3–1cm, acute or acuminate, attenuate at base, subpetiolate, entire or obscurely serrate, strigillose on both surfaces, 3-veined at base. Peduncles 0.5–4cm. Involucres 3–5mm diameter; phyllaries ovate, short acuminate, 3.5–4.5mm. Ligule 1.5–2.5mm. Disc flowers c 10–50, 1.5–1.75mm. Achenes 2.5(–3.5)mm, sparsely and minutely pubescent at apex; pappus scales c 0.25mm, blackish.

Bhutan: S – Samchi, Phuntsholing, Chukka and Deothang districts, **C** – Punakha, Tongsa (117) and Tashigang (117) districts, **N** – Upper Mo Chu district; **Darjeeling:** Manjitar, Rayeng (69), Tista (69); **W Bengal Duars:** Apalchand (175). Roadsides and wasteground, weed of watered crops, up to 1830m. April–August.

Cosmopolitan in warmer countries.

117. MONTANOA Cervantes

Coarse shrub. Stems large, squarish in section. Leaves pinnatifid to bipinnatifid, opposite; petioles auriculate at base. Capitula radiate, paleate, in large cymes. Involucres 2-seriate, subequal, herbaceous. Receptacle convex, paleae sheathing, spine-tipped, enlarging in fruit. Ray flowers neuter; ligules showy, white, entire to 3-toothed at apex. Disc flowers bisexual; corollas tubular-campanulate, 5-toothed, yellow. Achenes obpyramidal, subterete; pappus absent.

1. M. bipinnatifida (Kunth) Koch

Stems up to 10m; herbaceous parts shortly pubescent. Leaves 12–30(–80) × 9–20(–40)cm, with 3–5 broad, often lobed segments on each side, acute or acuminate, shallowly serrate, scabrid on both surfaces; petioles 4–20cm, auricles rounded, c 1cm diameter. Involucres 15mm diameter; phyllaries oblong or ovate, 6–10 × 3–4mm, pubescent. Paleae at flowering time broadly ovate, 1–2.5mm; in fruit cultrate, 12–21mm, enveloping achene, terminating in sharp spine c 1mm. Ray flowers 10–12, ligules oblanceolate, 25–35 × 6–11mm. Disc corollas 5mm, lobes pubescent. Achenes 3 × 1.5mm, dispersed with enveloping paleae.

Darjeeling: Sepoydhura (97a); **Sikkim:** Gangtok. Roadside bank, 1800m. December.

Native of Mexico, cultivated in gardens and escaping. Only vegetative material seen from our area.

EUPATORIEAE

118. AGERATINA Spach

Herbaceous perennials, subshrubs or shrubs, partly sessile- or stipitate-glandular. Leaves opposite, simple, trullate to obovate, petiolate, subentire to coarsely toothed. Capitula in corymbs, 5–8- or 50–60-flowered, discoid. Involucres cylindric to campanulate; phyllaries imbricate, 2–3-seriate, glandular. Corollas white, tubular-campanulate. Style branches linear, exserted, obtuse. Achenes narrowly oblong, blackish, pale and empty at base, 5-angled or -ribbed; pappus of 1 series of capillary bristles, subequal to corolla, white.

1. Flowers 5–8 .... **1. A. ligustrina**
\+ Flowers 50–60 .... **2. A. adenophora**

1. A. ligustrina (DC.) King & Robinson; *Eupatorium ligustrinum* DC. Dz: *Namseeling* (272).

Erect shrubs or subshrubs, 1–5m; branchlets angular, pubescent and sessile-glandular or glabrescent. Leaves elliptic-obovate, 2–7(–8) × 1–3cm, acute or acuminate, attenuate at base, entire in the lower half, remotely serrate near apex, penninerved, glabrous above, glandular-punctate beneath; petioles 0.8–2cm. Capitula in reddish-brown glandular-pubescent corymbs, 5–8-flowered. Involucres cylindric, 2-seriate, 1.7mm diameter; phyllaries lanceolate, 2.5–5mm, sessile-glandular, outer acute, inner obtuse, scarious-margined. Corollas 5mm. Achenes 2–2.5mm, puberulous.

Darjeeling: Darjeeling (218), Lebong–Tiger Hill (138, 431). Plantations of *Cryptomeria* and *Prunus*, 2100m. August–November.

Adventive; native of C America. Description based on (431). Leaves of this species may also be lanceolate or subentire.

2. A. adenophora (Sprengel) King & Robinson; *Eupatorium adenophorum* Sprengel, *E. glandulosum* Humboldt, Bonpland & Kunth non Michaux

Herb or subshrub, 0.4–2m; stems ± erect, branched, purplish(-brown), densely glandular pubescent. Leaves trullate; blades 2–7 × 1.5–5.5cm, acuminate, obtuse and 3-veined at base, coarsely (serrate-)dentate, subglabrous above, glandular pubescent beneath; petioles slender, 0.8–3cm. Capitula campanulate, c 50–60-flowered, in dense terminal and axillary corymbs. Involucres 3-seriate, 4mm diameter; outer phyllaries few, reduced; inner ones oblanceolate, 4.3–5 × 0.7–1mm, acuminate, stipitate-glandular, 3-veined on back. Corollas 3.5–4mm. Achenes 1.5–2mm, glabrous; pappus bristles c 10–12, caducous.

Bhutan: S – Phuntsholing district (above Phuntsholing), Gaylegphug district (above Tama), **C** – Punakha district (Punakha), Tongsa district (near Dakpai); **Darjeeling:** Darjeeling, Kurseong (97a), Takdah (69) etc.; **Sikkim:** Singhik–Gangtok (71), Thinglen–Yoksam (69). Forest roadsides, cultivated sites, 750–2050m. January–June.

Native of Mexico, now a pantropical weed.

119. MIKANIA Willdenow

Scandent twining annual or short-lived perennial herb. Leaves opposite, simple, 3-veined, petiolate. Capitula small, 4-flowered, discoid, in axillary corymbs. Involucre of 4 subequal phyllaries, with a smaller additional bract at base. Receptacle naked. Corollas tubular-campanulate, 5-toothed. Anther bases obtuse. Style branches linear. Achenes oblong, 4-angled, glandular-punctate and somewhat puberulous. Pappus of 1 series of bristles, somewhat shorter than corolla.

1. M. micrantha Kunth; *M. scandens* sensu F.B.I. non (L.) Willdenow, *M. cordata* (Burman f.) Robinson var. *indica* Kitamura

Stems subglabrous. Leaf blade ovate-triangular. 5–8 × 2–6cm, acute or acuminate, cordate at base, subentire to bluntly dentate, subglabrous, sometimes glandular-punctate; petioles up to 5cm. Capitula numerous. Involucres 1.5mm diameter; phyllaries oblong, 3.5–4 × 1.2mm, acute or shortly acuminate, subglabrous. Corollas greenish-white, 3mm. Achenes c 2mm. Pappus white, 2.5mm.

Bhutan: S – Samchi district (Samchi), Phuntsholing district (Phuntsholing), Gaylegphug district (Gaylegphug (117)); **Darjeeling:** Kurseong–Ratki (97a), Sukna (400); **W Bengal Duars:** Jalpaiguri (174). Climbing over rocks, low shrubs etc. and in sal plantations. 200–500m. August–February.

Reported as common in Chukha, Mongar, Samchi and Sarbang administrative districts below 1000m (272). Introduced, native to the Americas.

120. ADENOSTEMMA Forster

Erect annual or perennial herbs. Leaves opposite or the uppermost subopposite, 3-veined, petiolate or sessile Inflorescence terminal, loose, paniculate, ± corymbose. Capitula discoid. Involucres broadly campanulate; phyllaries 2-seriate, herbaceous, subequal, connate at base. Receptacle naked. Flowers 20–40. Corolla tubular-campanulate; tube glandular-hairy; lobes pubescent. Anthers gland-tipped. Style branches clavate, exserted, prominent. Achenes obovoid 3–5 angular, verrucose or glandular tuberculate. Pappus of 3–4 smooth, clavate, gland-tipped bristles.

1. Leaves elliptic-lanceolate, 1–2cm broad, attenuate at base; capitula in flower 3–4mm diameter; pappus bristles 0.25mm .... **1. A. angustifolium**
+ Leaves ovate, obtuse to subcordate at base and narrowly decurrent on petiole, generally more than 2cm broad; capitula in flower 5–7mm diameter; pappus bristles 0.5mm .... **2. A. lavenia**

1. A. angustifolium Arnott; *A. viscosum* Forster var. *viscosum* ('var. *typica*') sensu FBI p.p., *A. lavenia* (L.) Kuntze var. *angustifolium* (Arnott) Koster

Plant 30–60cm, stems usually creeping at base, glabrous or puberulous above and at nodes. Leaves narrow elliptic-lanceolate, 4–14 × 1–2cm, acuminate, long attenuate at base, obscurely serrate, glabrous or subglabrous; lower leaves usually with a distinct petiole to 2cm. Inflorescence widely branched, ± corymbose, glandular-puberulous; capitula c 10–20. Involucre 3–4mm diameter in flower; phyllaries elliptic or oblong, 2.2–4 × 0.7–1.3mm, obtuse or subacute, sometimes briefly apiculate, sparsely hirsute at base, usually with a few cilia. Corollas 2.5–3.5mm; tube sparsely glandular-hairy; lobes white, unequally pubescent, most sparsely so. Achenes c 3mm, sparsely glandular-muricate; pappus bristles 0.25mm.

Darjeeling: Dalka Jhar, Katambari. Terai, 150m. October–November.

Reported to intergrade with the following species elsewhere in India and reduced to varietal status (382).

2. A. lavenia (L.) Kuntze; *A. viscosum* Forster var. *viscosum*, *A. viscosum* var. *elatum* (D. Don) Clarke, *A. lavenia* var. *elatum* (D. Don) Hochreutiner

Differs from *A. angustifolium* by stems to 1m, sparsely pubescent; leaves ovate, 5–22 × 1.5–12cm, acuminate, base obtuse to subcordate, narrowly decurrent on 1–5cm petiole or the uppermost sessile, margins coarsely serrate, sparsely pubescent on both surfaces; involucre 5–7mm diameter in flower; phyllaries 2.5–4.5 × 1–2mm, face and margins usually more hairy; corollas 2.5mm, lobes white or pink, all densely pubescent; achenes 3.5mm, with more prominent glandular tubercules; pappus bristles 0.5mm.

Bhutan: S – Gaylegphug district (Batase and Gaylephug (117)) Deothang district (Narfong (117)), **C** – Punakha district (Mo Chu), **N** – Upper Mo Chu district (Kancham). Roadsides, forest margins, wet forest slopes, 200–2000m. August–December.

121. AGERATUM L.

Erect branching annuals, occasionally perennating. Leaves opposite, simple, toothed, petiolate. Capitula in small clusters towards branch ends, sometimes grouped into terminal corymbs, discoid; phyllaries 2–3-seriate, subequal, oblong or linear lanceolate, 2–3-veined. Receptacle naked. Corollas tubular-campanulate, 5-toothed. Anthers acute at apex. Style branches exserted, slender, clavate. Achenes narrowly oblong, black, with 5 sparsely scabridulous black ribs; pappus usually of 5 small scales, each usually bearing a longer bristle (rarely pappus irregular).

1. Phyllaries glabrous or sparsely pubescent, eglandular (rarely acumen glandular-ciliate); corolla to 2.7mm, not exceeding pappus; style exserted

c 1mm .... **1. A. conyzoides**

\+ Phyllaries pubescent (usually densely so) and glandular; corolla 2.7–3.5mm, exceeding pappus; style exserted 2–3mm .... **2. A. houstonianum**

1. A. conyzoides L. Sha: *Rogpu-ngon.* Fig. 137a–c.

Plant rank smelling, 10–100cm; stems whitish pubescent. Leaf blades ovate, 2–8 × 1.5–6cm, obtuse or subacute, truncate or cuneate at base, crenate-serrate, sparsely pubescent on both surfaces; petioles up to 5cm. Involucre 3–5mm diameter; phyllaries oblong, 3–4 × 1mm, acute or abruptly acuminate, erose or dentate in the upper part, glabrous or with a few eglandular hairs, rarely acumen of inner phyllaries subsessile glandular-ciliate. Corollas blue, mauve or white, 1.75–2.5mm. Styles exserted c 1mm. Achenes 1.5–1.75mm; pappus of 5 bristle-tipped scales slightly exceeding corolla.

Bhutan: S – Samchi, Phuntsholing, Chukka, Sarbhang and Gaylegphug districts, **C** – Punakha district; **Darjeeling:** Balasan (97a), Kurseong (97a), Takvar etc. Roadsides, forest margins, cultivated land, 200–1900m. All year.

A common dry-land weed, probably occurring in all administrative districts of Bhutan up to 2000m (272). Now widespread in tropics and subtropics, probably native to C and S America.

2. A. houstonianum Miller; *A. conyzoides* L. var. *houstonianum* (Miller) Sahn; Nep: *Etami-paat.*

Similar to *A. conyzoides*, but leaves usually truncate to cordate; phyllaries usually gradually acuminate, sometimes to 5.5mm, entire or ciliate, not eroded, pubescent (usually densely so) and stipitate-glandular; corollas 2.7–3.5mm; styles exserted 2–3mm; pappus not exceeding corolla, sometimes irregular: 6-merous or lacking all or most bristles.

Bhutan: S – Sarbhang district (Phipsoo, Sarbhang); **Darjeeling:** Kurseong–Pankhabari; **Sikkim:** Gangtok–Temi, Tadong. Roadsides, forest clearings, 280–1230m. March–August.

Reported from Mongar, Sarbhang and Wangdi administrative districts, c 1000m (272). Although usually possible to separate from *A. conyzoides* in the field, it is considered by some botanists to be only varietally distinct, in which case the correct name is *A. conyzoides* var. *mexicanum* (Sims) DC. Mainly an escape from cultivation, native to C and S America.

122. CHROMOLAENA DC.

Rank smelling herbaceous perennial or shrub, pubescent, and glandular. Leaves opposite, ovate-triangular, toothed, 3-veined near base, petiolate. Capitula discoid, c 30-flowered, cylindrical-campanulate, numerous in ± compact corymbs. Involucres 4–5-seriate; phyllaries imbricate, stramineous, with 3 green ribs on back, inner ones narrow oblong. Corollas white or mauve. Style branches linear, exserted, obtuse. Achenes linear-oblong, blackish, with 5 pale sparsely pubescent ribs, pale and empty at base; pappus of 1 series of bristles, subequal to corolla, stramineous.

1. C. odoratum (L.) King & Robinson; *Eupatorium odoratum* L. Sha: *Nayra-ngon.*

Plant erect or scrambling, 1.5–3m; branchlets terete, crisp pubescent. Leaves 4–10 × 3–6.5cm, acuminate, obtuse to attenuate at base, coarsely serrate near the broadest part, sparsely pubescent on upper surface, pubescent and sessile glandular beneath; petioles c 1cm. Involucres c 3mm diameter; inner phyllaries c 8 × 1.2mm, acute. Corollas 5.5mm. Achenes 4mm.

Bhutan: S – Samchi district (near Samchi (117)), Phuntsholing district (Phuntsholing, Toribar); Gaylegphug district (near Gaylegphug (38)); **Darjeeling:** Kurseong (97a), Manjitar Bridge, Robatang etc. Roadsides, forest clearings and plantations, 200–1450m. August–December.

Indigenous to tropical S America, widely naturalised in India and reported to occur in most lowland districts of Bhutan (272).

123. EUPATORIUM L.

Perennial herbs or subshrubs with hoary-pubescence and fewer (sub)sessile glands. Leaves opposite, simple or trifoliolate, toothed. Capitula in terminal corymbs, discoid, 5-flowered. Involucre cylindric; phyllaries 7–10, imbricate, 2–3-seriate, at least outer ones pubescent. Receptacle naked. Corollas tubular-campanulate, 5-toothed. Style branches linear, exserted, obtuse. Achenes oblong-obconic, 5-angular, subglabrous. Pappus simple, of whitish bristles, subequal to corolla.

1. At least cauline leaves 3-foliolate; all leaves and leaflets penninerved, coarsely toothed .... **1. E. mairei**
+ All leaves simple, distinctly 3-nerved from near base (though central vein weakly penninerved), distantly serrulate 2

2. Stems simple; inflorescence compact; inner phyllaries 6–6.5mm, glabrous; corollas 4.5mm .... **2. E. nodiflorum**
+ Stems branched above; inflorescence diffuse; inner phyllaries c 3.7mm, densely hoary above; corollas 3mm .... **3. E. sp. A**

FIG. 137. **Compositae: Eupatorieae and *Cavea*. Eupatorieae.** a–c, *Ageratum conyzoides*: a, apex of flowering shoot (× ⅔); b, capitulum (× 4); c, achene (× 10). d–f, *Eupatorium mairei*: d, apex of flowering shoot (× ⅔); e, capitulum (× 6); f, achene (× 9). ***Cavea.*** g–i, *C. tanguensis*: g, habit (× ⅔); h, female flower (× 6); i, male flower (× 4). Drawn by Margaret Tebbs.

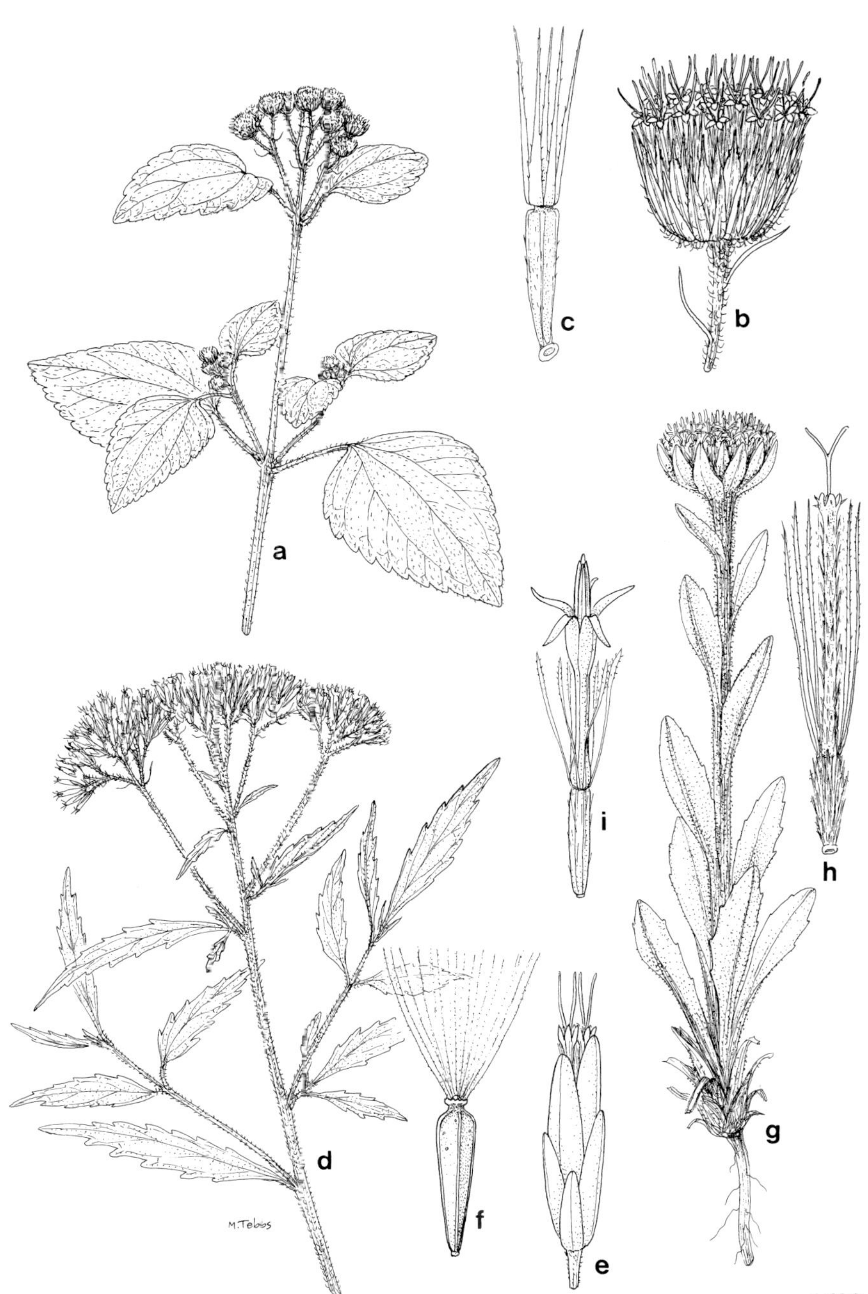

a
b
c
d
e
f
g
h
i
M.Tebbs

1. E. mairei Léveillé; *E. cannabinum* sensu F.B.I. non L., *E. heterophyllum* auct. non DC., *E. cannabinum* subsp. *asiaticum* Kitamura var. *heterophyllum* (DC.) Kitamura p.p. Dz: *Pho*, *Phomita*; Sha: *Yangrem.* Fig. 137d–f.

Stems erect, 0.3–1m, purplish, branched above, pubescent. Leaves 3-foliolate on main stems, often simple on branches; leaves and leaflets (ovate-)lanceolate, 2–9 × 0.5–2(–3)cm, acuminate, rounded or cuneate at base, pinnately veined, serrate-dentate, hoary and (sub)sessile glandular on both faces, more densely below; petioles c 2mm. Capitula in dense terminal corymbs. Involucre 2mm diameter; phyllaries sometimes pink; inner ones oblong-oblanceolate, 4–6.3 × 1.4–2mm, truncate (rarely obtuse, mucronulate), glabrous. Corollas white to pink or reddish, 4.2–5mm, sparsely subsessile glandular. Achenes 2.8-3.5mm, sparsely subsessile glandular, sometimes with few hairs above.

Bhutan: S – Chukka district, **C** – Thimphu, Punakha, Tongsa and Bumthang districts, **N** – Upper Mo Chu and Upper Kuru Chu districts; **Sikkim:** Lachen, Lachung. Grassy banks and hillsides, 1370–3050m. July–October.

Himalayan plants are considered distinct from *E. cannabinum*, which extends from Europe to Iran (355).

2. E. nodiflorum DC.

Similar to *E. mairei* but stems simple, c 0.5m; leaves all simple, distinctly 3-nerved from base, remotely serrulate; inner phyllaries acute.

Bhutan: unlocalised (Griffith collection).

3. E. sp. A; *E. longicaule* sensu Clarke non DC., *E. reevesii* sensu F.B.I. ?non DC., *E. chinense* auct. non L.

Differs from *E. mairei* by its leaves all simple, prominently 3-veined shortly above base, distantly serrulate, ovate on branches; petioles of cauline leaves to 7mm. Inflorescence very diffuse, with slender branches. Inner phyllaries elliptic, c 3.7 × 1.4mm, densely hoary on upper half. Corollas 3mm. Achenes 2.5mm, with few hairs above and very few glands; pappus barbellate.

Bhutan: S – Deothang district (Raidong). Grassy hills among pines, 850m. Fruiting in January.

Only some detached branches seen, from a doubtfully localised Griffith collection.

ASTEROIDEAE, tribe uncertain

124. ANISOPAPPUS Hooker & Arnott

Erect perennial herb. Leaves simple, alternate, toothed, petiolate. Capitula few in lax terminal corymbs, radiate, paleate. Involucre hemispherical; phyllaries c 4-seriate, lanceolate. Receptacle flat, paleate; paleae enfolding and exceeding disc flowers, acuminate. Ligules oblong, yellow, mostly 4-toothed. Disc corollas tubular-campanulate, 5-toothed. Style branches flattened, spathulate, subacute. Achenes ellipsoid or obconic, ribbed. Pappus of short broad scales and longer narrow scales.

1. A. chinensis (L.) Hooker & Arnott

Stems 30–80cm, densely pubescent at first. Leaves lanceolate, 5–10 × 1–2(–2.5)cm, mostly obtuse, truncate to subacute at base, coarsely rounded-dentate, sparsely pubescent on both surfaces with yellow sessile glands more numerous beneath; petioles slender, to 1.5cm. Phyllaries c 5 × 1mm, densely pubescent, with fewer glands. Paleae 5mm. Corolla tube of ray flowers 1.5–2mm, ligule 6–8 × c 2mm. Disc c 1cm diameter, corollas 3–3.5mm. Achenes 2.5mm, strongly ribbed; pappus with short scales c 0.5mm and long scales c 1mm.

Sikkim: without locality (320).

Description based on specimens from E India and Myanmar (Burma).

125. CAVEA W.W. Smith & J. Small

Perennial herb; stolons often present, bearing loosely appressed scale leaves; stems erect, simple, solitary or clustered. Leaves oblanceolate, mostly basal with distinct petioles, cauline ones ± sessile, alternate. Capitulum solitary, broadly campanulate, disciform with numerous marginal female flowers and central male flowers or discoid and monoecious or dioecious. Involucre several-seriate, herbaceous, outermost series largest. Receptacle convex, short fimbrillate. Female corollas tubular, shallowly 4-toothed; style branches linear, rounded at apex. Male corollas tubular-campanulate, deeply 5-toothed, teeth reflexed; style undivided, conic at apex. Achenes oblong or narrowly obovoid. Pappus of barbellate bristles, numerous on female flowers, sparse on male flowers.

1. C. tanguensis (Drummond) W.W. Smith & J. Small. Fig. 137g–i.

Plants from 5cm in flower to 25cm in fruit; stems densely glandular pubescent. Leaves 2–10 × 0.5–1.3cm, acute or obtuse, attenuate to base, subentire or denticulate, ± glabrous above, densely glandular pubescent beneath. Capitulum 1.5–2cm diameter at flowering; outer phyllaries ovate or lanceolate, 10–15 × 3–6mm, acute, glandular-pubescent; inner ones smaller. Corollas purplish, female ones c 5mm in flower, 9mm in fruit, pubescent, male ones c 6.5mm in flower, 9mm at fruiting. Achenes 4mm, densely appressed pubescent; pappus hairs purple, eventually to 11mm.

Bhutan: N – Upper Mo Chu district (Lingshi), Upper Pho Chu district (Gyophu La), Upper Mangde Chu district (Worthang), Upper Bumthang Chu district (Pangotang, Marlung); **Sikkim:** Tangu, The La, Jongsong; **Chumbi:** Chomo Lhari. Gravelly ground near streams and glaciers, 3950–5100m. May–July.

126. CENTIPEDA Loureiro

Low-growing annual herbs. Leaves alternate, ± simple, sessile. Capitula solitary, axillary, hemispherical, very small, disciform. Involucre c 2-seriate; phyllaries subequal, with scarious margins. Receptacle convex, naked. Outer flowers female, numerous; corolla shortly tubular, 2–3-toothed. Central flowers bisexual, c 20; corolla broadly tubular-campanulate, deeply 4-toothed; style branches ovate-triangular, very short, subacute. Achenes ± oblong, angular, ribbed. Pappus a minute pale corona, obscurely toothed.

1. C. minima (L.) A. Brown & Ascherson; *C. orbicularis* Loureiro.

Stems prostrate, to 15cm; stems sparsely araneous and sparsely subsessile-glandular. Leaves oblanceolate or spathulate, 4–6 × 2–2.5mm, somewhat fleshy, acute or obtuse, attenuate at base, 3–5-toothed or lobed, sparsely subsessile glandular above, denser below and puberulous or sparsely araneous.Capitula c 2mm diameter at anthesis; phyllaries oblong, c 1 × 0.4mm. Flowers yellowish-green; female corollas c 0.2mm; bisexual corollas 0.5mm. Achenes 0.75–1mm, including corona, puberulous and subsessile-glandular.

Bhutan: C – Punakha district (Punakha, Samtengang–Choojom (71), Tashigang district (Gamri Chu). Paddy field, 1200–1350m. June–August.

BIBLIOGRAPHY

Part 1: references for Volume 2 Part 3 repeated from earlier volumes.

34. Cowan, A.M. & Cowan, J.M. 1929. *The Trees of Northern Bengal.* Calcutta: Bengal Secretariat Book Depot.

38. Deb, D.B., Sen Gupta, G. & Malick, K.C. 1969. A contribution to the flora of Bhutan. *Bull. Bot. Soc. Bengal* 22(2): 169–217.

47. Gamble, J.S. 1896. *List of the Trees, Shrubs, and Large Climbers Found in the Darjeeling District, Bengal*, 2nd edn. Calcutta.

65. Griffith, W. 1847–1854. *Icones Plantarum Asiaticarum*, 4 vols. J. McClelland (ed.). Calcutta.

66. Griffith, W. 1848. *Itinerary Notes of Plants Collected in the Khasyah and Bootan Mountains, 1837–38, in Affghanisthan and Neighbouring Countries, 1839–41.* J. McClelland (ed.). Calcutta.

69. Hara, H. (ed.) 1967. *The Flora of Eastern Himalaya.* Tokyo: University of Tokyo.

71. Hara, H. (ed.) 1971. Flora of Eastern Himalaya. Second Report. *Bull. Univ. Museum, Univ. Tokyo* 2: 1–393.

73. Hara, H. & Williams, L.H.J. 1979. *An Enumeration of the Flowering Plants of Nepal*, Vol. 2. London: British Museum (Natural History).

80. Hooker, J.D. (ed.) 1875–1897. *The Flora of British India*, 7 vols. London: Reeve & Co.

97a. Mathew, K.M. 1981. *An Enumeration of the Flowering Plants in Kurseong, Darjeeling District, West Bengal, India.* Dehra Dun: Bishen Singh Mahendra Pal Singh.

101. Ohashi, H. 1975. Flora of Eastern Himalaya. Third Report. *Bull. Univ. Museum, Univ. Tokyo* 8: 1–458.

104. Rao, R.S. 1964. A botanical tour in the Sikkim state, eastern Himalayas. *Bull. Bot. Surv. India* 5: 165–205.

109. Sharma, B.D. & Ghosh, B. 1971. Contributions to the flora of Sikkim Himalayas. *Bull. Bot. Soc. Bengal* 24: 45–55.

112. Smith, W.W. 1913. The alpine and subalpine vegetation of south-east Sikkim. *Rec. Bot. Surv. India* 4: 323–431.

113. Smith, W.W. & Cave, G.H. 1911. The vegetation of the Zemu and Llonakh valleys of Sikkim. *Rec. Bot. Surv. India* 4: 141–260.

117. Subramanyam, K. (ed.) 1973. Materials for the Flora of Bhutan. *Rec. Bot. Surv. India* 20(2): 1–278.

125. Watt, G. 1889–1893. *A Dictionary of the Economic Products of India*, 6 vols. Calcutta.

135. Hara, H., Chater, A.O. & Williams, L.H.J. 1982. *Enumeration of the Flowering Plants of Nepal*, Vol. 3. London: British Museum (Natural History).

138. Tamang, K.K. & Yonzone, G.S. 1982. A brief note on the vegetation from Lebong to Tiger Hill in Darjeeling. *J. Bengal Nat. Hist. Soc.* N.S. 1(1): 93–95.
165. Nakao, S. & Nishioka, K. 1984. *Flowers of Bhutan.* Tokyo.
174. Sikdar, J.K. & Samanta, D.N. 1984. Herbaceous flora (excluding Cyperaceae, Poaceae and Pteridophytes) of Jalpaiguri District, West Bengal – a check list. *J. Econ. Taxon. Bot.* 4: 525–538.
175. Sikdar, J.K. 1984. Contributions to the flora of Baikunthapur Forest Division, Jalpaiguri District (West Bengal). *J. Econ. Taxon. Bot.* 5: 505–532.
218. Mukherjee, A. 1988. *The Flowering Plants of Darjiling.* Dehli, Lucknow: Atma Ram and Sons.
229. Sikdar, J.K. 1983. Notes on the occurrence of some plants of West Bengal. *J. Bombay Nat. Hist. Soc.* 79: 563–566.
231. Sikdar, J.K. & Rao, R.S. 1984. Further contribution to the flora of Buxa Forest Division, Jalpaiguri District (West Bengal). *J. Bombay Nat. Hist. Soc.* 81: 123–148.
272. Parker, C. 1992. *Weeds of Bhutan.* Thimphu: Royal Government of Bhutan.
276. Sikdar, J.K. 1981. Some new plant records for West Bengal from Jalpaiguri District. *J. Bombay Nat. Hist. Soc.* 78: 103–106.
298. Mandal, N.R. & Singh, P. 1993. Four new plant records for Sikkim. *J. Econ. Taxon. Bot.* 17: 555–556.

Part 2: new references for Volume 2 Part 3.

316. Anderson, T. 1867. Indian Acanthaceae. *J. Linnean Soc., Bot.* 9: 425–530.
317. Babcock, E.B. 1928. New species of *Crepis* from southern Asia. *Univ. Calif. Publ. Bot.* 14: 323–333.
318. Babcock, E.B. & Stebbins, G.L. 1937. *The Genus* Youngia. Washington: Carnegie Institution.
319. Babcock, E.B. & Stebbins, G.L. 1943. Systematic studies in the Cichorieae. *Univ. Calif. Publ. Bot.* 18: 227–240.
320. Balakrishnan, N.P. 1974, publ. 1977. Notes on some interesting plants from Jowai, Meghalaya. *Bull. Bot. Surv. India* 16: 169–174.
321. Betsche, I. 1984. Taxonomische Untersuchungen an *Kickxia* Dumortier (s.l.). Die neuen Gattungen *Pogonorrhinum* n. gen. und *Nanorrhinum* n. gen. (Phanerogamae–Scrophulariaceae). *Courier Forschungsinst. Senckenberg* 71: 125–142.
322. Bolli, R. 1994. *Revision of the Genus* Sambucus. Berlin: Cramer.
323. Bremer, K. 1994. *Asteraceae: Cladistics and Classification.* Portland: Timber Press.
324. Burbidge, N.T. 1982. A revision of *Vittadinia* A. Rich. (Compositae) together with reinstatement of *Eurybiopsis* DC. and description of a new genus *Camptacra*. *Brunonia* 5: 1–72.

325. Chatterjee, D. 1948. A review of Bignoniaceae of India and Burma. *Bull. Bot. Soc. Bengal* 2: 62–79.
326. Chen C. & Chen C.-L. 1977. On the Chinese genera *Scopolia* Jacq., *Anisodus* Link et Otto and *Atropanthe* Pascher. *Act. Phytotax. Sinica* 15(2): 58–68.
327. Chen Y.-L. 1999. Senecioneae. In Chen Y.-L. (ed.) *Flora Reipublicae Popularis Sinicae* 77(1): 1–327. Beijing: Science Press.
328. Chen Y.-L., Liang S.-Y. & Pan K.-Y. 1981. Taxa nova compositarum e flora xizanensi (tibetica). *Act. Phytotax. Sinica* 19: 85–106.
329. Clarke, C.B. 1876. *Compositae Indicae.* Calcutta: Thacker, Spink & Co.
330. Cramer, L.H. 1981. Scrophulariaceae. In: Dassanayake, M.D. (ed.) *A Revised Handbook to the Flora of Ceylon* 3: 386–449. New Dehli: Amerind Publishing.
331. Cramer, L.H. 1983. Campanulaceae. In: Dassanayake, M.D. (ed.) *A Revised Handbook to the Flora of Ceylon* 4: 160–165. New Dehli: Amerind Publishing.
332. D'Arcy, W. & Zhang Z.-Y. 1992. Notes on the Solanaceae of China and neighboring areas. *Novon* 2: 124–128.
332a. Das, D & Pramanik, B. 1970, publ. 1971. A note on *Chrysanthemum leucanthemum* Linn. (Asteraceae). *J. Bombay Nat. Hist. Soc.* 67: 613–614.
333. Das, A.P., Sengupta, G & Chanda, S. 1985. Notes on some members of Solanaceae in Darjeeling Hills, West Bengal, India. *J. Econ. Taxon. Bot.* 7: 661–663.
333a. Davies, F.G. 1979. The genus *Gynura* (Compositae) in eastern Asia and the Himalayas. *Kew Bull.* 33: 629–640.
334. Deb, D.B. 1978. Some new combinations for Indian taxa of *Lycianthes* (Solanaceae). In: Hawkes, J.G. Systematic notes on the Solanaceae, pp. 292–294. *Bot. J. Linnean Soc.* 76: 287–295.
335. Don, G. 1837/8. *A General System of Gardening and Botany*, Vol. 4. London: J.G. & F. Rivington *et al.*
336. Dutta, N.M. 1960. The genus *Veronica* Linn. of Eastern India. *J. Bombay Nat. Hist. Soc.* 57: 590–596.
337. Dutta, N.M. 1965. A revision of the genus *Torenia* Linn. of Eastern India. *Bull. Bot. Soc. Bengal* 19: 23–29.
338. Dutta, N.M. 1975. A revision of the genus *Limnophila* R. Br. of Eastern India. *Bull. Bot. Soc. Bengal* 29(1): 1–7.
339. Eldenäs, P., Källersjö, M. & Anderberg, A.A. 1999. Phylogenetic placement and circumscription of tribes Inuleae s. str. and Plucheeae (Asteraceae): evidence from sequences of chloroplast gene *ndh*F. *Molecular Phylogenetics and Evolution* 13: 50–58.
340. Elenevsky, A.G. 1977. Sistema roda *Veronica* L. [System of the genus *Veronica* L.]. *Byulletin Mosk. Obshch. Isp. Prirod., Biol.* 82(1): 149–160.
341. Firmin, I.D., Peregrine, W.T.H., Tamang, T. & Tamang, P. 1988. *Fungi*

and Plant Parasitic Bacteria, Viruses and Nematodes in Bhutan. Thimphu: Department of Agriculture.

342. Good, R. D'O. 1929. The taxonomy and geography of the Sino-Himalayan genus *Cremanthodium* Benth. *J. Linnean Soc., Bot.* 48: 259–316.
343. Hajra, P. K. *et al.* (eds) 1995. *Flora of India* 12. Calcutta: Botanical Survey of India.
344. Hajra, P. K. *et al.* (eds) 1995. *Flora of India* 13. Calcutta: Botanical Survey of India.
345. Handel-Mazzetti, H. 1928. Systematische Monographie der Gattung *Leontopodium*. *Beihefte zum Botanischen Centralblatt* 44: 1–178.
346. Hansen, H.V. 1988. A taxonomic revision of the genera *Gerbera* sect. *Isanthus*, *Leibnitzia* (in Asia), and *Uechtrichia* (Compositae, Mutisieae). *Nordic J. Bot.* 8: 61–76.
347. Haridasan, V.K. & Mukherjee, P.K. 1996. Campanulaceae. In: Hajra, P.K. & Sanjappa, M. *Fascicles of Flora of India* 22: 25–118. Calcutta: Botanical Survey of India.
348. Hong D.-Y. 1984. Taxonomy and evolution of the *Veroniceae* (*Scrophulariaceae*) with special reference to palynology. *Opera Botanica* 75.
349. Hong D.-Y. & Ma L.-M. 1991. Systematics of the genus *Cyananthus* Wall. ex Royle. *Act. Phytotax. Sinica* 29: 25–51.
350. Hong D.-Y. & Pan K.-Y. 1998. The restoration of the genus *Cyclocodon* (Campanulaceae) and its evidence from pollen and seed-coat. *Act. Phytotax. Sinica* 36: 106–110.
351. Hong D.-Y., Yang H.-B., Jin C.-L. & Holmgren, N.H. (family eds), Scrophulariaceae. In: Wu Z-Y. & Raven, P.H. (eds) *Flora of China* 18: 1–212. Beijing: Science Press, and St Louis: Missouri Botanical Garden Press.
352. Jain, S.K. 1991. *Dictionary of Indian Folk Medicine and Ethnobotany.* New Dehli: Deep Publications.
353. Jeffrey, C. & Chen Y.-L. 1984. Taxonomic studies on the tribe *Senecioneae* (Compositae) of Eastern Asia. *Kew Bull.* 39: 205–446.
354. Judd, W.S., Campbell, C.S., Kellogg, E.A. & Stevens, P.F. 1999. *Plant Systematics: A phylogenetic approach.* Sunderland, Massachusetts: Sinauer Associates.
355. Kawahara, T., Yahara, T. & Watanabe, K. 1989. Distribution of sexual and agamospermous populations of *Eupatorium* (Compositae) in Asia. *Plant Species Biol.* 4: 37–46.
356. Kirschner, J. & Štěpánek, J. 1996. Interpretation of some older *Taraxacum* names from Asia. *Edinb. J. Bot.* 53: 215–221.
357. Kitamura, S. 1968. Compositae of Southeast Asia and Himalayas, 1. *Act. Phytotax. Geobot.* 23: 1–19.

358. Kitamura, S. 1958. Compositae of Southeast Asia and Himalayas, 2. *Act. Phytotax. Geobot.* 23: 65–81.
359. Kitamura, S. 1958. Compositae of Southeast Asia and Himalayas, 3. *Act. Phytotax. Geobot.* 23: 129–152.
360. Kitamura, S. 1959. Compositae of Southeast Asia and Himalayas, 4. *Act. Phytotax. Geobot.* 24: 1–27.
361. Kitamura, S. 1990. Compositae Asiaticae 4. *Act. Phytotax. Geobot.* 41: 169–187.
362. Koyama, H. 1968. Taxonomic studies on the tribe Senecionieae of Eastern Asia. 2. Enumeration of the species of Eastern Asia. *Mem. Fac. Sci., Kyoto Univ., Ser. Biol.* 2: 19–60.
363. Koyama, H. 1968. Taxonomic studies on the tribe Senecionieae of Eastern Asia. 2. Enumeration of the species of Eastern Asia. *Mem. Fac. Sci., Kyoto Univ., Ser. Biol.* 2: 137–183.
364. Koyama, H. 1978. Notes on some Chinese species of *Cacalia* 2. *Act. Phytotax. Geobot.* 29: 17[illegible]–178.
365. Koyama, H. 1985. Taxonomic studies in the Compositae of Thailand 6. *Act. Phytotax. Geobot.* 36: 167–172.
366. Li, H.-L. 1947. Relationship and taxonomy of the genus *Brandisia*. *J. Arnold Arb.* 28: 127–136.
367. Li, H.-L. 1948. A revision of the genus *Pedicularis* in China. Part I. *Proc. Acad. Nat. Sci. Philadelphia* 100: 205–378.
368. Li, H.-L. 1949. A revision of the genus *Pedicularis* in China. Part II. *Proc. Acad. Nat. Sci. Philadelphia* 101: 1–214.
369. Ling Y. 1965. Notulae de nonnulis generibus tribus Inulearum familiae Compositarum florae sinicae. *Act. Phytotax. Sinica* 10: 167–181.
370. Ling Y.-R. 1980. Taxa nova generum *Artemisiae* et *Seriphidii* xizangensis. *Act. Phytotax. Sinica* 18: 504–513.
371. Ling Y.-R. 1991. An enumeration of *Artemisia* L. and *Seriphidium* (Bess.) Poljak. (Compositae) in Himalayas and the south Asian subcontinent. *Bull. Bot. Surv. India* 33: 296–308.
372. Ling Y.-R. 1992. *Artemisia* L. In: Ling Y. & Ling Y.-R. (eds) *Flora Reipublicae Popularis Sinicae* 76(2): 1–253. Beijing: Science Press.
373. Ling Y.-R. 1996. Cladistics and biogeography of *Artemisia* sect. *Viscidipubes* and sect. *Albibractea* (Compositae). *Act. Phytotax. Sinica* 34: 610–620.
374. Lipschitz, S. 1958. Revisio critica specierum sectionis *Taraxifoliae* Lipsch. generis *Saussurea* DC. *Novit. Syst. Plant. Vasc.* 9: 194–229.
375. Lipschitz, S. 1979. *Genus* Saussurea *DC. (Asteraceae)*. Leningrad: Nauka.
375a. Liu S.-W. 1989. Senecioninae. In: Ling Y. & Liu S.-W. (eds) *Flora Reipublicae Popularis Sinicae* 77(2): 1–184. Beijing: Science Press.
376. Maheshwari, J K. 1961. The genus *Wightia* Wall. in India with a discussion on its systematic position. *Bull. Bot. Surv. India* 3: 31–35.

377. Maximowicz, C.J. 1888. Diagnoses des plantes nouvelles asiatiques. VII. *Bull. Acad. Imp. Sci. St-Pétersbourg* 32: 477–629.
378. Mill, R.R. 2001. Notes relating to the Flora of Bhutan: XLIII. Scrophulariaceae (*Pedicularis*). *Edinb. J. Bot.* 58: 57–98.
379. Molau, U. 1981. The genus *Calceolaria* in NW South America VIII. The section *Calceolaria* and appendices to parts I–VIII. *Nordic J. Bot.* 1: 595–615.
380. Morris, K.E. & Lammers, T.G. 1997. Circumscription of *Codonopsis* and the allied genera *Campanumoea* and *Leptocodon* (Campanulaceae: Campanuloideae). I. Palynological data. *Bot. Bull. Acad. Sinica* 38: 277–284.
381. Nee, M. 1991. *Datura* or *Brugmansia*: two genera or one? *Solanaceae Newsl.* 3(2): 27–35.
382. Panigrahi, G. 1975, publ. 1976. The genus *Adenostemma* (Compositae) in the Indian Region. *Kew Bull.* 30: 647–655.
383. Parker, C. 1991. *A First Manual of Bhutan Weeds.* (Privately published.)
384. Pennell, F.W. 1943. The Scrophulariaceae of the Western Himalayas. *Acad. Nat. Sci. Philadelphia Monographs No. 5.*
385. Philcox, D. 1968. Revision of the Malesian species of *Lindernia* All. (Scrophulariaceae). *Kew Bull.* 22: 1–72.
386. Philcox, D. 1970. A taxonomic revision of the genus *Limnophila* R.Br. (Scrophulariaceae). *Kew Bull.* 24: 101–170.
387. Pilger, R. 1937. Plantaginaceae. In: Engler, A., *Das Pflanzenreich* 4. 269 (102 Heft). Leipzig: Engelmann.
388. Prain, D. 1889. Some additional species of *Pedicularis*. *J. Asiatic Soc. Bengal* 58 (2, no. 3): 255–278.
389. Prain, D. 1890. Noviciae Indicae. I. The species of *Pedicularis* of the Indian Empire. *Ann. Bot. Gard., Calcutta* 3: 1–196 [= *Noviciae Indicae*, pp. 1–26].
390. Prain, D. 1896. Noviciae Indicae. XI. Two additional species of *Lagotis*. *J. Asiatic Soc. Bengal* 65(2, no. 2): 57–66 [= *Noviciae Indicae*, pp. 162–171].
391. Prain, D. 1896. Noviciae Indicae. XIV. Some additional Solanaceae. *J. Asiatic Soc. Bengal* 65 (2, no. 3): 541–543 [= *Noviciae Indicae*, pp. 177–179].
392. Prain, D. 1903. Noviciae Indicae. XX. Some additional Scrophularineae. *J. Asiatic Soc. Bengal* 72 (2, no. 2): 11–38 [= *Noviciae Indicae*, pp. 393–405].
393. Prakash, V. 1999. *Indian Valerianaceae*. Jodhpur: Scientific Publishers (India).
394. Prijanto, B. 1969. The Asiatic species of *Lindenbergia* Lehm. (Scrophulariaceae). *Reinwardtia* 7: 543–560.
395. Randeria, A.J. 1960. The Composite genus *Blumea*, a taxonomic revision. *Blumea* 10: 176–317.

396. Rao, A.N. 1966. A note on *Ellisiophyllum pinnatum* (Benth.) Makino. *Bull. Bot. Surv. India* 8: 94–96.

397. Rao, R.R., Chowdhery, H.J., Hajra, P.K. *et al.* 1988. *Florae Indicae Enumeratio – Asteraceae.* Calcutta: Botanical Survey of India.

398. Rehder, A. 1903 Synopsis of the genus *Lonicera. Ann. Rep. Missouri Bot. Gard.* 14: 27–232.

399. Reveal, J.L., Judd, W.S. & Olmstead, R. 1999. (1405) Proposal to conserve the name Antirrhinaceae against Plantaginaceae. *Taxon* 48: 182.

399a. Robinson, H. 1999. Revisions in palaeotropical Vernonieae (Asteraceae). *Proc. Biol. Soc. Washington* 112: 220–247.

400. Rudra, A.B. 1958. Distribution of *Mikania scandens* Willd. (syn. *Eupatorium scandens* L.) in India. *Indian Forester* 84: 648.

401. Sandina, I.B. 1980. [A critical analysis of the genus *Scopolia* (Solanaceae).] *Botanicheskii Zhurnal* 65: 485–496.

402. Sanjappa, M. & Raju, D.C.S. 1989. *Przewalskia* Maxim. (Solanaceae) – a new generic record for India. *Bull. Bot. Surv. India* 31: 175–177.

403. Sastri, B.N. (ed.) 1952. *The Wealth of India. Raw Materials*, Vol. 3. New Dehli: Council of Industrial and Scientific Research.

404. Shih, C. 1987. Cynareae. In: Ling, Y. & Shih, C. (eds) *Flora Reipublicae Popularis Sinicae* 78(1): 20–209. Beijing: Science Press.

405. Shih, C. 1997. Lactuceae. In: Ling Y. & Shih C. (eds) *Flora Reipublicae Popularis Sinicae* 80(1): 1–331. Beijing: Science Press.

406. Shrestha, K.K. 1997. Taxonomic revision of the Sino-Himalayan genus *Cyananthus* (Campanulaceae). *Act. Phytotax. Sinica* 35: 396–433.

407. Sikdar, J.K. & Maiti, G.G. 1979. Two new records of Compositae for West Bengal. *Bull. Bot. Surv. India* 21: 218–220.

408. Singh, P. & Dash, S.S. 1998. Two new generic records for the flora of Sikkim. *J. Econ. Taxon. Bot.* 22: 741–744.

409. Singh, S. & Singh, P. 1998. *Lonicera litangensis* Batalin (Caprifoliaceae) – first report from India. *J. Econ. Taxon. Bot.* 22: 694–696.

410. Sink, K.C. 1984. *Petunia. Monogr. Thoer. Appl. Genet.* 9: 3–9.

411. Smith, W.W. 1911. Plantarum novarum in herbario horti regii calcuttensis cognitarum decas. *J. Asiatic Soc. Bengal* N.S. 7: 69–75.

412. Srivastrava, R.C. 1997. A taxononomic study of the genus *Saussurea* DC. (Asteraceae) in Sikkim. In: Gupta, P.K. (ed.) *Plants of the Indian Subcontinent* 6: 11–29. Dehra Dun: Bishen Singh Mahendra Pal Singh.

413. Sutton, D. 1988. *Revision of the Tribe Antirrhineae.* London: British Museum (Natural History), and Oxford: Oxford University Press.

414. Symon, D. E. & Haegi, A. R. 1991. *Brugmansia*: to be or not to be. *Solanaceae Newsl.* 3(2): 25–26.

415. Taylor, P.J. 1989. *Utricularia.* London: HMSO.

416. Tsoong, P.C. 1954. Some new species of Scrophulariaceae from eastern Asia. *Kew Bull.* 1954: 443–450.

417. Tsoong, P.C. 1963. *Pedicularis*. In: Tsoong, P.C. (ed.) *Flora Reipublicae Popularis Sinicae* 68. Beijing: Science Press.
418. Ungricht, S., Knapp, S. & Press, J.R. 1998. A revision of the genus *Mandragora* (Solanaceae). *Bull. Nat. Hist. Museum (London), Botany* 28: 17–40.
419. Vaid, K.M. & Naithani, H.B. 1970. *Cotula australis* (Sieb. ex Spreng.) Hook. f. – a new record for the north-western and eastern Himalayas. *Indian Forester* 96: 426–428.
420. Winkler, C. 1895. *Carpesii* L. Generis species adhuc notas. *Act. Hort. Petropolitani* 14: 2–22.
421. Wu, C.-Y. 1985. *Flora Xizangica* 4. Beijing: Science Press.
422. Yamazaki, T. 1955. Notes on *Lindernia*, *Vandellia*, *Torenia* and their allied genera in eastern Asia. *J. Japan. Bot.* 30: 171–180.
423. Yamazaki, T. 1962. Notes on Scrophulariaceae of Asia. *J. Japan. Bot.* 37: 263–273.
424. Yamazaki, T. 1965. A revision of the genera *Limnophila* and *Torenia* from Indochina. *J. Fac. Sci., Univ. Tokyo* 13: 575–625.
425. Yamazaki, T. 1971. New and noteworthy plants of Scrophulariaceae in Himalaya (3). *J. Japan. Bot.* 46: 242–250.
426. Yamazaki, T. 1988. A revision of the genus *Pedicularis* in Nepal. In: Ohba, H. & Malla, S.B. (eds) *The Himalayan Plants* 1: 91–161. Tokyo: The University Museum, The University of Tokyo.
427. Yamazaki, T. 1990. Scrophulariaceae. In: Smitinand, T. & Larsen, K. (eds) *Flora of Thailand* 5(2). Bangkok: Forest Herbarium, Royal Forest Department.
428. Yang H.-B. 1995. New series of the genus *Pedicularis*. *Act. Phytotax. Sinica* 33(3): 244–250.
429. Yang H.-B., Holmgren, N.H. & Mill, R.R. 1998. *Pedicularis*. In: Wu Z.-Y. & Raven, P.H. (eds) *Flora of China* 18: 97–209. Beijing: Science Press, and St Louis: Missouri Botanical Garden Press.
430. Yeo, P.F. 1978. A taxonomic revision of *Euphrasia* in Europe. *Bot. J. Linnean Soc.* 77: 223–334.
431. Yonzone, G.S., Babu, C.R. & Das, D. 1970. On the occurrence of *E. ligustrinum* in India. *Indian Forester* 96: 351.
432. Yu G.-T., Li X.-C., Wang Y.-F. *et al.* 1996. Spermicidal saponins from *Oreosolen wattii*. *Act. Bot. Yunnanica* 18(2): 229–233.
433. Zhang Z.-Y., Lu A.-M. & D'Arcy 1994. Solanaceae. In: Wu Z.-Y. & Raven, P.H. (eds) *Flora of China* 17: 300–332. Beijing: Science Press, and St Louis: Missouri Botanical Garden Press.
434. Zheng W.-L. & Luo D.-Q. 1998. A new species of *Lobelia* (Campanulaceae) from Xizang (Tibet), China. *Act. Phytotax. Sinica* 36: 549–551.

Taxonomic papers relevant to this volume have been published in the 'Notes relating to the Flora of Bhutan' series in the *Edinburgh Journal of Botany* as follows:

XXII A new species of *Torenia* (Scrophulariaceae) by R.R. Mill 50: 211–215 (1993).

XXIX Acanthaceae, with special reference to *Strobilanthes* by J.R.I. Wood 51: 175–273 (1994).

XLI Compositae (Asteraceae) by A.J.C. Grierson & L.S. Springate 57: 397–412 (2000).

XLII Scrophulariaceae, excluding *Pedicularis* by R.R. Mill 57: 413–428 (2000).

XLIII Scrophulariaceae (*Pedicularis*) by R.R. Mill 58: 57–98.

INDEX OF BOTANICAL NAMES

Page numbers in bold refer to an illustration.

INDEX OF COMMON NAMES

INDEX OF FAMILIES

All parts published except Volume 3 Part 3 (to be published mid 2001).